Lecture Notes in Mathematics 1979

Editors:
J.-M. Morel, Cachan
F. Takens, Groningen
B. Teissier, Paris

T0242452

Catherine Donati-Martin · Michel Émery ·
Alain Rouault · Christophe Stricker (Eds.)

Séminaire de Probabilités XLII

 Springer

Editors
Catherine Donati-Martin
Laboratoire de Probabilités et
    Modèles Aléatoires
Université Paris VI
Boîte Courrier 188, 4 place Jussieu
75252 Paris Cedex 05
France
catherine.donati@upmc.fr

Alain Rouault
Laboratoire de Mathématiques
Université de Versailles
45 av. des États-Unis
78035 Versailles Cedex
France
Alain.Rouault@math.uvsq.fr

Michel Émery
IRMA
Université de Strasbourg
7 rue René Descartes
67084 Strasbourg Cedex
France
emery@math.u-strasbg.fr

Christophe Stricker
Laboratoire de Mathématiques
Université de Franche Comté
16 route de Gray
25030 Besançon Cedex
France
christophe.stricker@univ-fcomte.fr

ISBN: 978-3-642-01762-9   e-ISBN: 978-3-642-01763-6
DOI: 10.1007/978-3-642-01763-6

Lecture Notes in Mathematics ISSN print edition: 0075-8434
                ISSN electronic edition: 1617-9692

Library of Congress Control Number: 2009286035

Mathematics Subject Classification (2000): 60Gxx, 60Hxx, 60Kxx, 60J80, 81S25, 11M41, 39B72, 47D07, 93-02

Cover drawing by Anthony Phan

Cover design: SPi Publisher Services

Printed on acid-free paper

springer.com

Photo by George Bergman.

Marc Yor, one of the most prominent members of the French probabilistic school, is turning 60. For the last 33 years, he contributed to the Séminaire by his own articles and his counselling of other authors; his methods and style permeate throughout the volumes of this series. He was a tireless member of the Rédaction during a quarter of a century, from Volume XIV to XXXIX, careful to maintain the highest quality, from broad mathematical ideas to minutest details. We wish we were able to keep up with the high standards he has set! Since Volume XL, he is no longer an official rédacteur, but keeps helping us with the editorial work and the refereeing.

Marc, nous te souhaitons un joyeux anniversaire, et nous sommes heureux de te dédier ce volume.

Catherine Donati-Martin, Michel Émery,
Alain Rouault, Christophe Stricker

Preface

Nine volumes ago, in Séminaire de Probabilités XXXIII, a series of advanced courses was started; nine such courses have appeared since. Two of them are due to Antoine Lejay, including his *Introduction to rough paths* in volume XXXVII. This unrepentant recidivist now strikes again, with *Yet another introduction to rough paths*, which sheds a more algebraic light on the same matter.

The various contributions which constitute the rest of the volume exemplify the rôle the Séminaire intends to play on the probabilistic stage: junior authors go side by side with older contributors, with a predominance from French or francophile ones; short notes mix with real research articles; and the themes are well in the traditional spirit of the Séminaire, ranging over the broad spectrum of interest of the readership of the Séminaire.

Catherine Donati-Martin, Michel Émery,
Alain Rouault, Christophe Stricker

Contents

List of Contributors

Stéphane Attal
Université Lyon 1
Institut Camille Jordan
43 bld du 11 novembre 1918
69622 Villeurbanne Cedex, France
attal@math.univ-lyon1.fr

Philippe Biane
CNRS, Laboratoire d'Informatique
Institut Gaspard Monge
Université Paris-Est
5 bd Descartes
Champs-sur-Marne
77454 Marne-la-Vallée Cedex 2
France
Philippe.Biane@univ-mlv.fr

Pierre Debs
Institut Élie Cartan Nancy
B.P. 239, 54506 Vandœuvre-lès-
Nancy Cedex, France
Pierre.Debs@iecn.u-nancy.fr

Nizar Demni
Fakultät für Mathematik
Universität Bielefeld
Postfach 100131
Bielefeld, Germany
demni@math.uni-bielefeld.de

Ameur Dhahri
Ceremade, UMR CNRS 7534
Université Paris Dauphine
Place de Lattre de Tassigny
75775 Paris Cedex 16, France
dhahri@ceremade.dauphine.fr

Michel Émery
IRMA, Université de Strasbourg
et C.N.R.S.
7 rue René Descartes
67 084 Strasbourg Cedex
France
emery@math.u-strasbg.fr

M. Erraoui
Université Cadi Ayyad
Faculté des Sciences Semlalia
Département de Mathématiques
B.P. 2390, Marrakech, Maroc
erraoui@ucam.ac.ma

E.H. Essaky
Université Cadi Ayyad
Faculté Poly-disciplinaire
Département de Mathématiques
et d'Informatique, B.P 4162
Safi, Maroc
essaky@ucam.ac.ma

P.J. Fitzsimmons
Department of Mathematics
0112; University of California
San Diego, 9500 Gilman Drive
La Jolla, CA 92093–0112
USA
pfitzsim@ucsd.edu

R.K. Getoor

Nadine Guillotin-Plantard
Université Lyon 1
Institut Camille, Jordan
43 bld du 11 novembre 1918
69622 Villeurbanne Cedex, France
nadine.guillotin@univ-lyon1.fr

Robert Hardy
Department of Mathematical
Sciences, University of Bath
Claverton Down, Bath BA2 7AY
UK
S.C.Harris@bath.ac.uk

Simon C. Harris
Department of Mathematical
Sciences, University of Bath
Claverton Down, Bath BA2 7AY
UK
S.C.Harris@bath.ac.uk

Nathalie Krell
Laboratoire de Probabilités
et Modèles Aléatoires
Université Paris 6
175 rue du Chevaleret
75013 Paris, France
nathalie.krell@upmc.fr

Antoine Lejay
Équipe-Projet TOSCA
Institut Élie Cartan
(Nancy-Université, CNRS, INRIA)
Campus scientifique, BP 239
54506 Vandœuvre-lès-Nancy Cedex
France
Antoine.Lejay@iecn.u-nancy.fr

Laurent Miclo
Laboratoire d'Analyse
Topologie, Probabilités, UMR 6632
CNRS, 39, rue F. Joliot-Curie
13453 Marseille Cedex 13
France
miclo@latp.univ-mrs.fr

Miklós Rásonyi
Computer and Automation Institute
of the Hungarian Academy of
Sciences
rasonyi@sztaki.hu

Walter Schachermayer
Vienna University of Technology
Wiedner Hauptstrasse 8–10
1040 Vienna, Austria
wschach@fam.tuwien.ac.at

Uwe Schmock
Vienna University of Technology
Wiedner Hauptstrasse 8–10
1040 Vienna, Austria
schmock@fam.tuwien.ac.at

Josef Teichmann
Vienna University of Technology
Wiedner Hauptstrasse 8–10
1040 Vienna, Austria
jteichma@fam.tuwien.ac.at

Fangjun Xu
Department of Mathematics
University of Connecticut
196 Auditorium Road
Unit 3009, Storrs
CT 06269-3009, USA
fangjun@math.uconn.edu

Kouji Yano
Department of Mathematics
Graduate School of Science
Kobe University, Kobe, Japan
kyano@math.kobe-u.ac.jp

Yuko Yano
Research Institute for
Mathematical Sciences
Kyoto University, Kyoto
Japan

Marc Yor
Laboratoire de Probabilités
et Modèles Aléatoires

Université Paris VI
Paris, France
and
Institut Universitaire de France
and
Research Institute for
Mathematical Sciences
Kyoto University, Kyoto
Japan

Yet Another Introduction to Rough Paths

Antoine Lejay

Équipe-Projet TOSCA, Institut Élie Cartan (Nancy-Université, CNRS, INRIA)
Campus scientifique, BP 239, 54506 Vandœuvre-lès-Nancy Cedex, France
e-mail: *Antoine.Lejay@iecn.u-nancy.fr*

Summary. This specialized course provides another point of view on the theory of rough paths, starting with simple considerations on ordinary integrals, and stressing the importance of the Green-Riemann formula, as in the work of D. Feyel and A. de La Pradelle. This point of view allows us to gently introduce the required algebraic structures and provides alternative ways to understand why the construction of T. Lyons *et al.* is a natural generalization of the notion of integral of differential forms, in the sense that it shares the same properties as integrals along smooth paths, when we use the "right notion" of a path.

Key words: Rough paths; integral of differential forms along irregular paths; controlled differential equations; Lie algebra; Lie group; Chen series; sub-Riemannian geometry

1 Introduction

The theory of rough paths [42, 44, 52, 55] is now an active field of research, especially among the probabilistic community. Although this theory is motivated by stochastic analysis, it takes its roots in analysis and control theory, and is also connected to differential geometry and algebra.

Given a path x of finite p-variation with $p \geqslant 2$ on $[0, T]$ with values in $\mathbb{R}^d$ or an α-Hölder continuous path with $\alpha \leqslant 1/2$, this theory allows us to define the integral $\int_x f$ of a differential form f along x, which is $\int_x f = \int_0^T f(x_s)\, \mathrm{d}x_s$. Using a fixed point theorem, it is then possible to solve differential equations driven by x of type

$$y_t = y_0 + \int_0^t g(y_s)\, \mathrm{d}x_s.$$

The case $1 \leqslant p < 2$ (or $\alpha > 1/2$) is covered by the Young integrals introduced by L.C. Young in [73]. Some of the most common stochastic processes, including Brownian motion, have trajectories that are of finite p-variation with

C. Donati-Martin et al. (eds.), *Séminaire de Probabilités XLII*,
Lecture Notes in Mathematics 1979, DOI 10.1007/978-3-642-01763-6_1,
© Springer-Verlag Berlin Heidelberg 2009

$p > 2$. So, being able to define almost surely an integral along such irregular paths is of great practical interest, both theoretically and numerically. Yet we know this is not possible in general, and integrals of Itô and Stratonovich type are defined only as limits in probability of Riemann sums.

Introduced in the 50's by K.-T. Chen (see for example [11]), the notion of iterated integrals provides an algebraic tool to deal with a geometrical object which is a smooth path, and allows us to manipulate controlled differential equations using formal computations (see for example [23, 39]).

The main feature of the rough paths theory is then to assert that, if it is possible to consider not only a path x but a path $\mathbf{x}$ which encodes the iterated integrals (that cannot be canonically defined if x is of finite p-variation with $p \geqslant 2$), then one may properly define the integral $z_t = \int_0^t f(x_s)\,\mathrm{d}\mathbf{x}_s$ and solve the differential equation $y_t = y_0 + \int_0^t g(y_s)\,\mathrm{d}\mathbf{x}_s$ provided that f and g are smooth enough. In addition, the maps $\mathbf{x} \mapsto z$ and $\mathbf{x} \mapsto y$ are continuous, with respect to the topology induced by the p-variation distance. The dimension of the path $\mathbf{x}$, or equivalently the number of "iterated integrals" to be considered, depends on the regularity of x. For $p \in [2, 3)$ (or $\alpha \in (1/3, 1/2]$), one only has to consider the iterated integrals of x along itself. This can be justified by the first order Taylor expansion of $\int_s^t f(x_r)\,\mathrm{d}x_r$:

$$\sum_{i=1}^d \int_s^t f_i(x_r)\,\mathrm{d}x_r^i \approx \sum_{i=1}^d f_i(x_s)(x_t^i - x_s^i) + \sum_{i,j=1}^d \frac{\partial f_i}{\partial x_j}(x_s) \int_s^t (x_r^j - x_s^j)\,\mathrm{d}x_r^i.$$

If x is α-Hölder continuous with $\alpha \in (1/3, 1/2]$ and one has succeeded in constructing $K_{s,t}^{i,j}(x) = \int_s^t (x_r^j - x_s^j)\,\mathrm{d}x_r^j$, then one can expect that $|K_{s,t}^{i,j}(x)| \leqslant C|t - s|^{2\alpha}$. So, to approximate $\int_0^T f(x_r)\,\mathrm{d}x_r$, we will use the sum

$$\sum_{k=0}^{n-1} \sum_{i=1}^d f_i(x_{kT/n})(x_{(k+1)T/n}^i - x_{kT/n}^i)$$

$$+ \sum_{k=0}^{n-1} \sum_{i,j=1}^d \frac{\partial f_i}{\partial x_j}(x_{kT/n}) K_{kT/n,(k+1)T/n}^{i,j}(x)$$

and show its convergence as $n \to \infty$. Hence, the integral will be defined not along a path x, but along $\mathbf{x}_{s,t}$ given by

$$\mathbf{x}_{s,t} = (1, x_t^i - x_s^i, \ldots, x_t^d - x_s^d, K_{s,t}^{1,1}(x), \ldots, K_{s,t}^{d,d}(x)),$$

where the first component 1 is here for algebraic reasons. The element $\mathbf{x}$ can be seen as an element of the truncated tensor space $\mathrm{T}(\mathbb{R}) = \mathbb{R} \oplus \mathbb{R}^d \oplus (\mathbb{R}^d \otimes \mathbb{R}^d)$. By similarity with what happens for the power series constructed from the iterated integrals—sometimes called the *signature* of the path —, one has that for all $0 \leqslant s \leqslant r \leqslant t \leqslant T$,

$$\mathbf{x}_{s,t} = \mathbf{x}_{s,r} \otimes \mathbf{x}_{r,t},$$

where $\otimes$ is the tensor product on $T(\mathbb{R})$ (where the tensor products of more than 2 terms are killed). In addition, it is possible to consider the formal logarithm of $\mathbf{x}$, and following also the properties of the Chen series, we look for paths $\mathbf{x}$ such that $\log(\mathbf{x}_{s,t})$ belongs to $A(\mathbb{R}^d) = \mathbb{R}^d \oplus [\mathbb{R}^d, \mathbb{R}^d]$, where $[\mathbb{R}^d, \mathbb{R}^d]$ is the space generated by all Lie brackets between two elements of $\mathbb{R}^d$. This algebraic property allows us to give proper definitions of rough paths and geometric rough paths from an algebraic point of view. The articles [44, 55] and the books [42, 52] use this point of view.

As first noted by N. Victoir, since $(T_1(\mathbb{R}^d), \otimes)$, the subset of $T(\mathbb{R}^d)$ whose elements have a first term equal to 1, is a Lie group, one may describe $\mathbf{x}_{s,t}$ by $\mathbf{x}_{s,t} = (-\mathbf{x}_{0,s})^{-1} \otimes \mathbf{x}_{0,t}$, and then, instead of considering the family $(\mathbf{x}_{s,t})_{0 \leqslant s < t \leqslant T}$, one may work with the path $\mathbf{x}_t = \mathbf{x}_{0,t}$, which lives in the non-commutative space $(T_1(\mathbb{R}^d), \otimes)$. This provides some simplifications on the statement of some theorems, but also opens the door to look for more connections with differential geometry.

Shortly after the article [55] was published, other authors provided alternative constructions of the differential equations and integrals, still by using some of the ideas provided by the theory of rough paths. One of these works— from D. Feyel and A. de la Pradelle [21]—uses a point of view from differential geometry and stresses the importance of the Gauss/Green-Riemann/Stokes formula to understand the need to "enhance" the path with more information to get a rigorous definition. Another approach, by M. Gubinelli, rather relies on algebraic considerations [36].

The idea of this article is then to justify the construction of the algebraic structures (tensor space, Lie groups) needed in the theory of rough paths from basic considerations on integrals of differential forms. To simplify, we consider that the dimension d of the state space is $d = 2$ (for $d = 1$, there is no real problem since (i) Any differential form is the differential of a function; (ii) The controlled differential equation $y_t = g(y_t) \, dx_t$ is solved under reasonable assumptions on g by $y_t = \Phi(x_t)$ with $\Phi'(z) = g(\Phi(z))$ if x is smooth, so a density argument may be used. On that topic, see for example the work of Doss and Sussmann [19]). By considering all pairs of components, it is easy to pass from $d = 2$ to $d > 2$. In addition, we restrict ourselves to α-Hölder continuous paths, which is not a stringent assumption at all, since a time change allows us to transform any path with p-finite variation into a path which is $(1/p)$-Hölder continuous.

Given a differential form, we wish to construct a map $x \mapsto \int_x f$ which is continuous on the space C^α of α-Hölder continuous paths. If $\alpha > 1/2$, the existence of $\int_x f$ is provided by the theory of Young integrals. We also get that $x \mapsto \int_x f$ is continuous on C^α equipped with the α-Hölder norm. Yet we construct some sequence $(x^n)_{n \in \mathbb{N}}$ of functions in C^α that converges to x in C^β with $\beta < 1/2$, and such that $\int_0^T f(x_s^n) \, dx_s^n$ does not converges to $\int_x f$, but to $\int_0^T f(x_s) \, dx_s + \int_0^T [f, f](x_s) \, d\varphi_s$ where $[f, f]$ is the Lie bracket of f and φ is an arbitrary function. This counter-example makes use of the

Green-Riemann functions, and one can see that, if one considers not a path x, but a path (x, φ) with values in $\mathbb{R}^3$, then one can extend the notion of the integral to C^α with $\alpha \in (1/3, 1/2]$. In some sense, the third component records the area enclosed between that path and its chord between times s and t. We can then provide an algebraic setting for describing such paths, still with a non-commutative operation. Then, we construct paths with values in $A(\mathbb{R}^2)$, a space of dimension 3, where the first two coordinates correspond to an "ordinary" path in the Euclidean vector space $\mathbb{R}^2$. The non-commutativity comes from the fact that the area enclosed between $x \cdot y$ (the concatenation of two paths x and y) and its chord is different from the area enclosed between $y \cdot x$ and its chord. The degree of freedom we gain comes from the fact that small loops allow us to move in the third direction while staying roughly at the same position in $\mathbb{R}^2$. Any α-Hölder continuous path with values in $A(\mathbb{R}^2)$ (with the right distance) with $\alpha > 1/3$ may be approximated by smooth paths lifted in $A(\mathbb{R}^2)$ using their area. In addition, the convergence of paths with values in $A(\mathbb{R}^2)$ in the α-Hölder topology implies that the corresponding integrals form a Cauchy sequence in C^β for any $\beta < \alpha$. It is then possible to extend the notion of Young integrals to α-Hölder continuous functions with values in $A(\mathbb{R}^2)$, and also to get the continuity result we need.

The basic idea to approximate some α-Hölder continuous path $\mathbf{x}$ taking its values in $A(\mathbb{R}^2)$ with $\alpha > 1/3$ consists in lifting paths x^n that take the same values as $\mathbf{x}$ on the points of a partition of $[0, T]$ and that link two successive times by a loop and a straight line. The loop is a way to "encode the area". One may then be tempted to look for real geodesics. For this, we will interpret the space $A(\mathbb{R}^2)$ as the subspace of the tangent space at any point of the tensor space $T(\mathbb{R}^2)$, and we will look for simple curves linking two points in $T(\mathbb{R}^2)$. There are several possibilities. One consists in using tools from sub-Riemannian geometry [29, 32]. Another one consists in studying paths with values in a sub-manifold $G(\mathbb{R}^2)$ of $T(\mathbb{R}^2)$, which is also a subgroup of $(T(\mathbb{R}^2), \otimes)$, and which is the Lie group whose Lie algebra may be identified with $A(\mathbb{R}^2)$. We give another way to define the integral by extending the differential form f to a differential form on $G(\mathbb{R}^2)$ and construct curves that connect two points of $G(\mathbb{R}^2)$. Hence, instead of considering paths with values in $A(\mathbb{R}^2)$, we will consider paths with values in $G(\mathbb{R}^2)$, and the difference between two points in $A(\mathbb{R}^2)$ then corresponds to a direction.

With this, we may redefine the integral as the limit of some Riemann sums—which is the original definition given by T. Lyons—, but where the addition has been replaced by some tensor product. Moreover, it becomes then possible to extend the notion of integrals to paths living in the bigger space $T_1(\mathbb{R}^2)$.

Consequently, using the concept of path living in a non-commutative space, the rough path theory provides a way to define an integral $\int f(x_s) \, dx_s$ that shares the same properties as ordinary integrals:

(a) It is a limit of expressions similar to Riemann sums.

(b) It is a limit of integrals along approximations of the path obtained by sampling the path at finitely many points and connecting successive sample points by "simple" curves.

In addition, this map $x \mapsto \int f(x_s) \, \mathrm{d}x_s$ is continuous from $C^\alpha([0, T]; T_1(\mathbb{R}^2))$ to $C^\alpha([0, T]; T_1(\mathbb{R}^2))$ and may be used to solve differential equations driven by x, still with a continuity property.

The theory of rough paths turns out to be the natural extension of integrals on the space of α-Hölder continuous paths with $\alpha \in (1/3, 1/2]$, in the same way Young integrals are the natural notion of integral against α-Hölder continuous paths with $\alpha \in (1/2, 1]$.

Outline

In Section 2, we introduce our notations and recall some elementary facts about integrals of differential forms along smooth paths as well as about Hölder continuous paths. In section 3, we quickly present results about Young integrals, and thus show the properties of integrals along α-Hölder continuous paths with $\alpha > 1/2$. In Section 4, we assume that one can integrate differential forms along α-Hölder continuous path with $\alpha \in (1/3, 1/2]$, and we show how to transform this integral into a continuous one with respect to the path. In Section 5, we consider paths taking their values in $A(\mathbb{R}^2)$, and show how to define the integral $\int_x f$ as limits of ordinary integrals. In Section 6, we continue our analysis of the space $A(\mathbb{R}^2)$ and introduce the tensor space $T(\mathbb{R}^2)$. In Section 7, we give another definition of the integral of f along x, using an expression of Riemann sum type. This construction corresponds to the original one of T. Lyons [42, 52, 55]. In Section 8, we give some related results: case of the d-dimensional space, Chen series, other constructions for paths with quadratic variation, link with stochastic integrals. In Section 9, we solve differential equations. We end this article with appendix on the Heisenberg group and we recall a technical result about almost rough paths, on which the original construction of $\int_x f$ is based.

Acknowledgement

The author wishes to thank Laure Coutin, Lluís Quer y Sardanyons and Jérémie Unterberger whose remarks helped to improve this article.

2 Notations

2.1 Differential Forms

Let $f_1, \ldots, f_d$ be some functions from $\mathbb{R}^d$ to $\mathbb{R}^m$. Consider the differential form

$$f(x) = f_1(x^1) \, \mathrm{d}x^1 + \cdots + f_d(x^d) \, \mathrm{d}x^d$$

on $\mathbb{R}^d$.

Definition 1. *For $\gamma > 0$, f is said to be γ-Lipschitz if the f_i's for $i = 1, \ldots, d$ are of class $C^{\lfloor\gamma\rfloor}(\mathbb{R}^d; \mathbb{R}^m)$ with bounded derivative up to order $\lfloor\gamma\rfloor$, and the $f_i^{\lfloor\gamma\rfloor}$'s are $(\gamma - \lfloor\gamma\rfloor)$-Hölder continuous with a $(\gamma - \lfloor\gamma\rfloor)$-Hölder constant $H_\gamma^i(f)$. The class of γ-Lipschitz differential forms is denoted by $\mathrm{Lip}(\gamma; \mathbb{R}^d \to \mathbb{R}^m)$.*

For $f \in \mathrm{Lip}(\gamma; \mathbb{R}^d \to \mathbb{R}^m)$, define

$$\|f\|_{\mathrm{Lip}} = \max_{i=1,\ldots,d} \max\{\|f_i^{(0)}\|_\infty, \ldots, \|f_i^{(\lfloor\gamma\rfloor)}\|_\infty, H_\gamma^i(f)\},$$

which is a norm on $\mathrm{Lip}(\gamma; \mathbb{R}^d \to \mathbb{R}^m)$.

Remark 1. If $\gamma = 1$, this definition is slightly different from the notion of Lipschitz functions, since this definition implies that f is of class $C^1(\mathbb{R}^d; \mathbb{R}^m)$, while the definition that $|f(x) - f(y)|/|x - y|$ is bounded as $x \to y$ for all $y \in \mathbb{R}^d$ only means that f is almost everywhere differentiable. Anyway, in our context, the case $\gamma \in \mathbb{N}$ is never considered.

Given a path $x \in C^1([0, T]; \mathbb{R}^d)$ and a continuous differential form f, define the *integral of f along x* by

$$\int_x f = \int_0^T f(x_s) \left.\frac{\mathrm{d}x}{\mathrm{d}t}\right|_{t=s} \mathrm{d}s = \sum_{i=1}^d \int_0^T f_i(x_s) \left.\frac{\mathrm{d}x^i}{\mathrm{d}t}\right|_{t=s} \mathrm{d}s.$$

Recall a few facts on such integrals, that will be heavily used:

(i)   If $\varphi : \mathbb{R}_+ \to \mathbb{R}_+$ is strictly increasing and continuous, then $\int_{x \circ \varphi} f = \int_x f$. In other words, the integral of f along x does not depend on the parametrization of x.

(ii) If $\varphi : [0, T] \to [0, T]$ is $\varphi(t) = T - t$, then $\int_{x \circ \varphi} f = -\int_x f$. In other words, reversing time changes the sign of $\int_x f$.

(iii) If $x, y \in C_p^1([0, T]; \mathbb{R}^d)$ (the class of functions from $[0, T]$ to $\mathbb{R}^d$ which are piecewise C^1) and $x \cdot y$ is the concatenation of x and y, then $\int_{x \cdot y} f = \int_x f + \int_y f$. This is the *Chasles relation*.

(iv) If $x \in C_p^1([0, T]; \mathbb{R}^2)$ is a closed loop in $\mathbb{R}^2$, that is, $x_T = x_0$, then

$$\int_x f = \iint_{\mathrm{Surface}(x)} [f, f](x^1, x^2) \, \mathrm{d}x^1 \, \mathrm{d}x^2, \tag{1}$$

where $\mathrm{Surface}(x)$ is the oriented surface surrounded by x and

$$[f, f] = \frac{\partial f_1}{\partial x_2} - \frac{\partial f_2}{\partial x_1}.$$

This is the *Green-Riemann/Stokes/Gauss formula*.

2.2 Paths of Finite p-Variation

Fix $T > 0$. Let x be a continuous path from $[0, T]$ to $\mathbb{R}^d$ and $\Pi = \{t_i\}_{i=0,\ldots,k}$ be a partition of $[0, T]$ with k elements. For $p \geqslant 1$, define

$$\mathfrak{P}(x; \Pi, p) = \sum_{i=0}^{k-1} |x_{t_{i+1}} - x_{t_i}|^p.$$

The p-variation of x on $[s, t] \subset [0, T]$ is defined by

$$\mathrm{Var}_{p,[s,t]}(x) = \sup_{\Pi \text{ partition of } [0,T]} \mathfrak{P}(x_{|[s,t]}; \Pi \cap [s, t], p)^{1/p}.$$

Definition 2. *A function* $x : [0, T] \to \mathbb{R}^d$ *is said to be of* finite p-variation *if* $\mathrm{Var}_{p,[0,T]}(x)$ *is finite.*

If x is of finite p-variation, then we easily get

$$\mathrm{Var}_{q,[0,T]}(x) \leqslant 2^{(q-p)/q} \|x\|_\infty^{(q-p)/q} (\mathrm{Var}_{p,[0,T]}(x))^{p/q} \qquad (2)$$

and then x is of finite q-variation for all $q > p$. Note that $\mathrm{Var}_{p,[0,T]}(x)$ defines a semi-norm on the space of functions of finite p-variation, but not a norm, since $\mathrm{Var}_{p,[0,T]}(x) = 0$ only implies that x is constant. In addition, on the space of functions x with $x_0 = 0$ and $\mathrm{Var}_{p,[0,T]}(x) < +\infty$, $\mathrm{Var}_{p,[0,T]}$ defines a norm which is however *not* equivalent to the uniform norm $\| \cdot \|_\infty$, and counter-examples are easily constructed.

Following a recent remark due to P. Friz [26], we may work with a more precise norm than the norm constructed from p-variation. Indeed, to simplify our approach, we work only with Hölder continuous paths and the Hölder norm.

If x is a path of finite p-variation and

$$\varphi(t) = \inf \left\{ s > 0 \,\middle|\, \mathrm{Var}_{p,[0,s]}(x)^p > t \right\},$$

then φ is increasing and $x \circ \varphi$ is $(1/p)$-Hölder continuous. As the integral of a differential form keeps the same value under a continuous, increasing time change, there is no difficulty in considering the $(1/p)$-Hölder norm, which is simpler to use than the p-variation norm (for some results on the relationship between p-variation and $(1/p)$-Hölder continuity, see for example [9]). Yet for convergence problems, this is not the most general framework, and dealing with the p-variation norm allows us to obtain more complete results (for example, in [45, 46], we only prove convergence in p-variation although the path is α-Hölder continuous, and this is due to a singularity at 0 of some term).

Denote by $H_\alpha(x)$ the Hölder continuity modulus of a path $x : [0, T] \to \mathbb{R}^d$ which is α-Hölder continuous, that is

$$H_\alpha(x) = \sup_{0 \leqslant s < t \leqslant T} \frac{|x_t - x_s|}{|t - s|^\alpha}.$$

Of course, an α-Hölder continuous path is also β-Hölder continuous for any $\beta \leqslant \alpha$. In addition, the equivalent of (2) is

$$\text{for } \beta \leqslant \alpha, \ H_\beta(x) \leqslant 2^{1-\beta/\alpha} \|x\|_\infty^{1-\beta/\alpha} H_\alpha(x)^{\beta/\alpha}. \tag{3}$$

If $H_\alpha(x) = 0$ then x is constant, and H_α defines only a semi-norm.

Notation 1. If $x : [0,T] \to \mathbb{R}^d$ α-Hölder continuous, then we set

$$\|x\|_\alpha = |x_0| + H_\alpha(x)$$

and call $C^\alpha([0,T];\mathbb{R}^d)$ the subset of functions x in $C([0,T];\mathbb{R}^d)$ such that $\|x\|_\alpha$ is finite.

Equipped with $\|\cdot\|_\alpha$, this space $C^\alpha([0,T],\mathbb{R}^d)$ is a Banach space. In addition, we get the following Lemma which is a consequence of the Ascoli Theorem and (3).

Lemma 1. *Let $(x^n)_{n\in\mathbb{N}}$ be such that $x^n \in C^\alpha([0,T];\mathbb{R}^d)$ and $(\|x^n\|_\alpha)_{n\in\mathbb{N}}$ is bounded. Then there exist x in $C^\alpha([0,T];\mathbb{R}^d)$ and a subsequence of $(x^n)_{n\in\mathbb{N}}$ that converges to x with respect to $\|\cdot\|_\beta$ for each $\beta < \alpha$.*

Remark 2. It is important to note that here, we used the $\|\cdot\|_\beta$ norm for the space $C^\alpha([0,T];\mathbb{R}^d)$ with $\beta < \alpha$. When equipped with this norm, $(C^\alpha([0,T];\mathbb{R}^d), \|\cdot\|_\beta)$ becomes a separable space, while $(C^\alpha([0,T];\mathbb{R}^d), \|\cdot\|_\alpha)$ is not separable: See [61] for example.

The next corollary follows easily.

Corollary 1. *Let Π be a partition of $[0,T]$ and x^Π be the linear approximation of $x \in C^\alpha([0,T];\mathbb{R}^d)$ along Π. Then $\|x^\Pi\|_\alpha \leqslant 3^{1-\alpha}\|x\|_\alpha$.*

If $(\Pi^n)_{n\in\mathbb{N}}$ is a sequence of partitions of $[0,T]$ whose meshes converge to 0, then $(x^{\Pi^n})_{n\in\mathbb{N}}$ converges to x in $(C^\alpha([0,T];\mathbb{R}^d), \|\cdot\|_\beta)$ for all $\beta < \alpha$.

Proof. Let $\Pi = \{t_i\}_{i=1,\ldots,J}$. For $0 \leqslant s < t \leqslant T$, let $s' = \min \Pi \cap [s,T]$ and $t' = \max \Pi \cap [0,t]$. If $s \notin \Pi$ (resp. $t \notin \Pi$), denote by $s'' = \max \Pi \cap [0,s]$ (resp. $t'' = \min \Pi \cap [t,T]$). As $s',t' \in \Pi$, if $s,t \notin \Pi$,

$$\begin{aligned}
|x_t^\Pi - x_s^\Pi| &\leqslant |x_t^\Pi - x_{t'}^\Pi| + |x_{t'}^\Pi - x_{s'}^\Pi| + |x_{s'}^\Pi - x_s^\Pi| \\
&\leqslant \frac{t-t'}{t''-t'}|x_{t''} - x_{t'}| + |x_{t'} - x_{s'}| + \frac{s-s'}{s''-s'}|x_{s'} - x_s| \\
&\leqslant \|x\|_\alpha(t-t')^\alpha + \|x\|_\alpha(s'-s)^\alpha + \|x\|_\alpha(t'-s')^\alpha \\
&\leqslant 3^{1-\alpha}\|x\|_\alpha(t-s)^\alpha,
\end{aligned}$$

the last inequality coming from the Jensen inequality applied by $x \mapsto x^{1/\alpha}$. The case where s or t belongs to Π is treated similarly. This proves that $\|x^\Pi\|_\alpha \leqslant 3^{1-\alpha}\|x\|_\alpha$.

The second part of this corollary is an immediate consequence of Lemma 1.

Remark 3. One may wonder whether it is possible to approximate a function $x \in C^\alpha([0,T]; \mathbb{R}^2)$ by piecewise linear functions that converge with respect to $\| \cdot \|_\alpha$, and not with respect to $\| \cdot \|_\beta$ for $\beta < \alpha$. As shown in [61] (see also [18, § 4.3]), this is possible only if x belongs to the class of functions such that

$$\lim_{\delta \to 0} \sup_{\substack{0 \leqslant s < t \leqslant T \\ |t-s| \leqslant \delta}} \frac{|x_t - x_s|}{(t-s)^\alpha} = 0,$$

or, in other words, if $|x(t+h)-x(t)| = o(h^\alpha)$. Of course, this class of functions is strictly included in $C^\alpha([0,T]; \mathbb{R}^d)$: the function $f(x) = \sum_{k=0}^{+\infty} c^{-k\alpha} \sin(c^k x)$ for c large enough yields a counter-example, as is easily proved using the results from [12].

3 Integrals Along α-Hölder Continuous Paths, $\alpha \in (1/2, 1]$

For the sake of simplicity, consider $d = 2$. The construction of $\mathfrak{I}$ on $C^\alpha([0,T]; \mathbb{R}^2)$ for $\alpha > 1/2$ is first deduced from the Young integral.

3.1 Defining the Integrals

We recall here the construction of the integral of a β-Hölder continuous path driven by a α-Hölder continuous path, provided that $\alpha + \beta > 1$. This theorem is due to L.C. Young [73] (see also [18] for example).

Theorem 1. *Let $\alpha, \beta \in (0,1]$ with $\alpha + \beta > 1$. Then*

$$(x, y) \mapsto \left(t \mapsto \int_0^t y_s \, dx_s \right)$$

is bilinear and continuous from $C^\alpha([0,T]; \mathbb{R}) \times C^\beta([0,T]; \mathbb{R})$ to $C^\alpha([0,T]; \mathbb{R})$.

Proof (Sketch of the proof). Fix $n \in \mathbb{N}^*$, and set, for $t_k^n = Tk/2^n$,

$$J^n = \sum_{k=0}^{2^n-1} y_{t_k^n} (x_{t_{k+1}^n} - x_{t_k^n}).$$

Then

$$|J^{n+1} - J^n| = \left| \sum_{k=0}^{2^n-1} (y_{t_{2k+1}^{n+1}} - y_{t_{2k}^{n+1}})(x_{t_{2k+2}^{n+1}} - x_{t_{2k+1}^{n+1}}) \right|$$

$$\leqslant \sum_{k=0}^{2^n-1} H_\beta(y) H_\alpha(x) T^{\alpha+\beta} 2^{-(n+1)(\alpha+\beta)}$$

$$\leqslant 2^{-n(\alpha+\beta-1)} H_\beta(y) H_\alpha(x).$$

As $\alpha + \beta - 1 > 0$, we deduce that the series $\sum_{n \geqslant 0}(J_{n+1} - J_n)$ converges and thus that, if $J \overset{\text{def}}{=} J_0 + \sum_{n \geqslant 0}(J_{n+1} - J_n)$, then

$$|J - y_0(x_T - x_0)| \leqslant \zeta(\alpha + \beta - 1)T^{\alpha+\beta}H_\beta(y)H_\alpha(x), \tag{4}$$

where $\zeta(\theta) = \sum_{n \geqslant 0} 1/n^\theta$. Of course, we define $\int_0^T y_s \, dx_s$ as J. From the last inequality in which t is substituted to T and s to 0, this also proves that $t \mapsto \int_0^t y_s \, dx_s$ is α-Hölder continuous.

The other properties of the integral are easily, although tediously, deduced from this construction.

Remark 4. Indeed, using the argument of Lemma 2.2.1, p. 244 [55], there is no need to consider dyadic partitions, but we keep them for simplicity. Note that however, especially when dealing with stochastic processes, some results in the rough paths theory do depend on the choice of a dyadic partition (see for example [13]).

One may then define for $0 \leqslant s \leqslant t \leqslant T$,

$$\mathfrak{I}(x; s, t) = \int_{x_{|[s,t]}} f = \int_s^t f_1(x_r) \, dx_r^1 + \int_s^t f_2(x_r) \, dx_r^2 \tag{5}$$

as Young integrals with $y_t = f(x_t)$. Yet a global regularity condition is imposed on (x, y) with implies in particular that $\alpha > 1/2$ and the minimal assumptions on the regularity of f also depends on α.

Notation 2. For a path x defined on the time interval $[S, T]$, we will use $\mathfrak{I}(x; s, t)$ to denote the integral $\int_{x_{|[s,t]}} f$ when $S \leqslant s < t \leqslant T$, and $\mathfrak{I}(x)$ to denote the function $t \in [S, T] \mapsto \mathfrak{I}(x; S, t)$.

The following corollaries follow from the construction of the Young integrals and (4): see in particular [44, 56].

Corollary 2. *Fix $\alpha \in (1/2, 1]$ and $f \in \text{Lip}(\gamma; \mathbb{R}^2 \to \mathbb{R}^m)$ with $\gamma > 1/\alpha - 1$. Then $\mathfrak{I}$ defined in (5) is well defined as a Young integral on $C^\alpha([0, T]; \mathbb{R}^2)$ and is a locally Lipschitz map from $(C^\alpha([0, T]; \mathbb{R}^2), \|\cdot\|_\alpha)$ to $(C^\alpha([0, T]; \mathbb{R}^m), \|\cdot\|_\alpha)$.*

Corollary 3. *Fix $\alpha \in (1/3, 1/2]$ and let $f \in \text{Lip}(\gamma; \mathbb{R}^2 \to \mathbb{R}^m)$ with $\gamma > 1/\alpha - 1$. Then*

$$C^{2\alpha}([0, T]; \mathbb{R}) \times C^\alpha([0, T]; \mathbb{R}^2) \to C^{2\alpha}([0, T]; \mathbb{R}^m)$$

$$(\varphi, x) \mapsto \left(t \mapsto \int_0^t [f, f](x_s) \, d\varphi_s\right)$$

is well defined as a Young integral and is a locally Lipschitz map from $(C^{2\alpha}([0, T]; \mathbb{R}^2), \|\cdot\|_{2\alpha}) \times (C^\alpha([0, T]; \mathbb{R}^2), \|\cdot\|_\alpha)$ to $(C^{2\alpha}([0, T]; \mathbb{R}^2), \|\cdot\|_{2\alpha})$.

3.2 A Problem of Continuity

We have to take great care of the meaning of the continuity result in Corollary 2: the norm $\| \cdot \|_\alpha$ is *not* equivalent to the uniform norm. Convergence in C^α implies uniform convergence but the converse is not true.

The following counter-example is the cornerstone to understand how $\mathfrak{I}$ will be defined so as to deal with irregular paths.

Let $(x^n)_{n\in\mathbb{N}}$ and x be continuous paths such that x^n converges to x in $C^\alpha([0,T];\mathbb{R}^2)$ with $\alpha \in (1/2, 1]$.

Let φ be a function in $C^\beta([0,T];\mathbb{R})$ with $\beta \in (2/3, 1]$. Assume also that f belongs to $\mathrm{Lip}(\gamma; \mathbb{R}^d \to \mathbb{R})$ where

$$(\gamma + 1)\beta > 2,$$

which implies that $2 > \gamma > 1$. Let $\Pi^n = \{t^n_k\}_{k=0,\dots,2^n-1}$ be the dyadic partition of $[0,T]$ at level n, that is, $t^n_k = Tk/2^n$. For each $n = 1, 2, \dots$, denote by $\Phi^n = \{y^n_k\}_{k=0,\dots,2^n-1}$ a set of functions piecewise of class C^1 such that for a fixed $\kappa > 1$,

$$y^n_k : [t^n_k, t^n_{k+1}] \to \mathbb{R}^2 \text{ with } y^n_k(t^n_k) = y^n_k(t^n_{k+1}) = x^n(t^n_k), \tag{6a}$$

$$\sup_{n=1,2,\dots,\ k=0,\dots,2^n} \|y^n_k\|_{\beta/2} < +\infty, \tag{6b}$$

$$\text{uniformly in } n, k, \ |\operatorname{Area}(y^n_k) - (\varphi(t^n_{k+1}) - \varphi(t^n_k))| \leqslant CT^\kappa 2^{-n\kappa}, \tag{6c}$$

where $\operatorname{Area}(y^n_k)$ is the algebraic area of the loop y^n_k defined by

$$\operatorname{Area}(y^n_k) = \frac{1}{2} \int_{t^n_k}^{t^n_{k+1}} (y^{1,n}_k(s) - y^{1,n}_k(t^n_k))\, dy^{2,n}_k(s)$$

$$- \frac{1}{2} \int_{t^n_k}^{t^n_{k+1}} (y^{2,n}_k(s) - y^{2,n}_k(t^n_k))\, dy^{1,n}_k(s).$$

For such a sequence, we say that φ encodes asymptotically the areas of $(\Phi^n)_{n\in\mathbb{N}}$.

Denote by $x^n \bowtie \Phi^n$ the path from $[0, 2T]$ to $\mathbb{R}^2$ defined by

$$x^n \bowtie \Phi^n = y^n_0 \cdot x^n_{|[t^n_0, t^n_1]} \cdot y^n_1 \cdot x^n_{|[t^n_1, t^n_2]} \cdots y^n_{2^n-1} \cdot x^n_{|[t^n_{2^n-1}, t^n_{2^n}]},$$

where $x \cdot y$ is the concatenation between two path x and y (see Figure 1). This path $x^n \bowtie \Phi^n$ is defined on the time interval $[0, 2T]$.

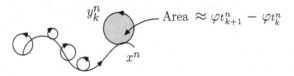

Fig. 1. The path $x^n \bowtie \Phi^n$.

Then, by the Chasles property of the integral,

$$\Im(x^n \bowtie \Phi^n; 0, 2T) = \Im(x^n; 0, T) + \sum_{k=0}^{2^n-1} \int_{t_k^n}^{t_{k+1}^n} f(y_k^n(s)) \, dy_k^n(s).$$

By the Green-Riemann formula (1),

$$\int_{t_k^n}^{t_{k+1}^n} f(y_k^n(s)) \, dy_k^n(s) = \iint_{\text{Surface}(y_k^n)} [f, f](x^1, x^2) \, dx^1 \, dx^2.$$

The idea is now the following:

$$\iint_{\text{Surface}(y_k^n)} [f, f](x^1, x^2) \, dx^1 \, dx^2 \approx [f, f](x_{t_k^n}) \operatorname{Area}(y_k^n)$$

$$\approx [f, f](x_{t_k^n})(\varphi(t_{k+1}^n) - \varphi(t_k^n)).$$

To be more precise, using our hypotheses on f and Φ^n, with $\Delta_n t = T 2^{-n}$,

$$\left| \iint_{\text{Surface}(y_k^n)} [f, f](x^1, x^2) \, dx^1 \, dx^2 - [f, f](x_{t_k^n})(\varphi(t_{k+1}^n) - \varphi(t_k^n)) \right|$$

$$\leqslant 2\|\nabla f\|_{\gamma-1} \|y_k^n\|_{\beta/2}^{\gamma-1} (C\Delta_n t^\kappa + \|\varphi\|_\beta \Delta_n t^\beta) \Delta_n t^{(\gamma-1)\beta/2} + 2C\|\nabla f\|_\infty \Delta_n t^\kappa$$

$$\leqslant 2\|\nabla f\|_{\gamma-1} \|y_k^n\|_{\beta/2}^{\gamma-1} (C\Delta_n t^{\kappa-\beta} + \|\varphi\|_\beta) \Delta_n t^{(\gamma+1)\beta/2} + 2C\|\nabla f\|_\infty \Delta_n t^\kappa. \quad (7)$$

There are now 2^n such terms to sum. By hypothesis, $(\gamma + 1)\beta/2 > 1$ and $\kappa > 1$, so the sum of the right-hand side of (7) vanishes as $n \to \infty$. In addition, necessarily $\beta + \gamma\alpha > 1$, so $\int [f, f](x_s) \, d\varphi_s$ can be considered as a Young integral. Thus, we easily get that

$$\sum_{k=0}^{2^n-1} \int_{t_k^n}^{t_{k+1}^n} f(y_k^n(s)) \, dy_k^n(s) \xrightarrow[n \to \infty]{} \int_0^T [f, f](x_s) \, d\varphi_s.$$

In other words,

$$\Im(x^n \bowtie \Phi^n; 0, 2T) \xrightarrow[n \to \infty]{} \Im(x; s, t) + \int_0^T [f, f](x_r) \, d\varphi_r.$$

It is important to note that here, $(x^n \bowtie \Phi^n)_{n \in \mathbb{N}}$ is in general not bounded in $C^\alpha([0, 2T]; \mathbb{R}^2)$, but it is bounded in $C^{\beta/2}([0, 2T]; \mathbb{R}^2)$. Remark that for $t \in [0, 2T]$, if $t/2 \in [t_k^{n+1}, t_{k+1}^{n+1}]$ and k is odd, then $x^n \bowtie \Phi^n(t) = x^n(t/2)$. If k is even, then $x^n \bowtie \Phi^n(t) = y_k^n(t/2)$. Thus,

$$|x^n \bowtie \varPhi^n(t) - x^n \bowtie \varPhi^n(s)|$$

$$\leqslant \begin{cases}
|x^n(t/2) - x^n(s/2)| \\
\quad \text{if } s/2 \in [t_{2k+1}^{n+1}, t_{2k+2}^{n+1}], \ t/2 \in [t_{2\ell+1}^{n+1}, t_{2\ell+2}^{n+1}], \\
|y_\ell^n(t/2) - y_\ell^n(t_\ell^n)| + |x^n(t_k^n) - x^n(s/2)| \\
\quad \text{if } s/2 \in [t_{2k+1}^{n+1}, t_{2k+2}^{n+1}], \ t/2 \in [t_{2\ell}^{n+1}, t_{2\ell+1}^{n+1}], \\
|y_\ell^n(t/2) - y_\ell^n(t_\ell^n)| + |y_\ell^n(t_\ell^n) - y_k^n(t_k^n)| + |y_k^n(t_k^n) - y_k^n(s/2)| \\
\quad \text{if } s/2 \in [t_{2k}^{n+1}, t_{2k+1}^{n+1}], \ t/2 \in [t_{2\ell}^{n+1}, t_{2\ell+1}^{n+1}], \ k \neq \ell, \\
|y_\ell^n(t/2) - y_\ell^n(s/2)| \\
\quad \text{if } s/2 \in [t_{2\ell}^{n+1}, t_{2\ell+1}^{n+1}], \ t/2 \in [t_{2\ell}^{n+1}, t_{2\ell+1}^{n+1}], \\
|x^n(t/2) - x^n(t_k^n)| + |y_k^n(t_k^n) - y_k^n(s/2)| \\
\quad \text{if } s/2 \in [t_{2k}^{n+1}, t_{2k+1}^{n+1}], \ t/2 \in [t_{2\ell+1}^{n+1}, t_{2\ell+2}^{n+1}].
\end{cases}$$

Using the convexity inequality, one gets that for some constant C that depends only on α and β,

$$|x^n \bowtie \varPhi^n(t) - x^n \bowtie \varPhi^n(s)|$$
$$\leqslant C \max\{\|x\|_\alpha, \sup_{k=0,\dots,2^n-1} \|y_k^n\|_{\beta/2}\} \max\{(t-s)^{\beta/2}, (t-s)^\alpha\}.$$

Since $\beta/2 \leqslant \alpha$, it follows that $(x^n \bowtie \varPhi^n)_{n \in \mathbb{N}}$ is bounded in $C^{\beta/2}([0, 2T]; \mathbb{R}^2)$ assuming of course that the y_k^n have a $\beta/2$-Hölder norm different from zero. As we required that φ is β-Hölder continuous, and if we choose for y_k^k some circles with area $\varphi(t_{k+1}^n) - \varphi(t_k^n)$, then their radius are $\sqrt{|\varphi(t_{k+1}^n) - \varphi(t_k^n)|/\pi}$ and this is why we look for y_k^n's that are $\beta/2$-Hölder continuous.

This also means that when one considers a sequence $(x^n)_{n \in \mathbb{N}}$ of elements in $C^\alpha([0, T]; \mathbb{R}^2)$ and a path x of $C^\alpha([0, T]; \mathbb{R}^2)$ with $\alpha > 1/2$, one has to consider the fact that $(x^n)_{n \in \mathbb{N}}$ may converge to x with respect to some β-Hölder norm with $\beta \leqslant 1/2$. In addition, this counter-example ruins all hopes to extend $\mathfrak{I}$ naturally to $C^\alpha([0, T]; \mathbb{R}^2)$ for $\alpha < 1/2$, since one may construct at least two bounded sequences $(x^n)_{n \in \mathbb{N}}$ and $(z^n)_{n \in \mathbb{N}}$ in $C^\alpha([0, T]; \mathbb{R}^2)$ with $\alpha < 1/2$ converging uniformly to x—hence that converge to x in $C^\beta([0, T]; \mathbb{R}^2)$ for any $\beta < \alpha$—such that $\mathfrak{I}(x^n; 0, T) \xrightarrow[n \to \infty]{} \mathfrak{I}(x; 0, T)$ and $\mathfrak{I}(z^n; 0, T) \xrightarrow[n \to \infty]{}$ $\mathfrak{I}(x; 0, T) + \int_0^T [f, f](x_s) \, d\varphi_s$, which is different from $\mathfrak{I}(x; 0, T)$ unless $[f, f] = 0$ or φ is constant.

3.3 A Practical Counter-example in the Stochastic Setting

In [43, 48], we give a stochastic example of such a phenomenon coming from homogenization theory. Consider some coefficients σ from $\mathbb{R}^d$ to the space of

$d \times d$-matrices and $b : \mathbb{R}^d \to \mathbb{R}^d$ smooth enough which are 1-periodic. Consider the SDE

$$X_t^\varepsilon = \int_0^t \sigma(X_s^\varepsilon/\varepsilon)\, dB_s + \frac{1}{\varepsilon} \int_0^t b(X_s^\varepsilon/\varepsilon)\, ds$$

for some Brownian motion B. It is well known from homogenization theory (see [7] for example) that X^ε converges as $\varepsilon \to 0$ to $\overline{\sigma}W$ for some Brownian motion W and a $d \times d$-matrix $\overline{\sigma}$ which is constant, provided that the drift b satisfies some averaging property. One of the applications of this theory is to provide a tool to replace (for modelling or numerical computations) a PDE of type $\partial_t u^\varepsilon(t,x) + L^\varepsilon u^\varepsilon(t,x) = 0$, $u^\varepsilon(T,x) = g(x)$ with $L^\varepsilon = \sum_{i,j=1}^d \frac{1}{2}a_{i,j}(\cdot/\varepsilon)\partial^2_{x_i x_j} + \sum_{i=1}^d \frac{1}{\varepsilon}b_i(\cdot/\varepsilon)\partial_{x_i}$ and $a = \sigma\sigma^{\mathrm{t}}$ by the simpler PDE $\partial_t u(t,x) + \overline{L}u(t,x) = 0$ with $\overline{L} = \sum_{i,j=1}^d \frac{1}{2}\overline{a}_{i,j}\partial^2_{x_i x_j}$ and $\overline{a} = \overline{\sigma}\,\overline{\sigma}^{\mathrm{t}}$. From the probabilistic point of view, this means that X^ε behaves—thanks to a functional Central Limit Theorem and the ergodic behavior of its projection on the torus $\mathbb{R}^d/[0,1]^d$—like a non-standard Brownian motion. However, one has to take care when using X^ε as the driver of some SDE, since

$$i,j = 1,\ldots,d, \quad \mathfrak{A}^{i,j}(X^\varepsilon;0,t) \xrightarrow[\varepsilon\to0]{} \mathfrak{A}^{i,j}(\overline{\sigma}W;0,t) + t\overline{c}_{i,j}$$

uniformly and in p-variation for $p > 2$, where $(\overline{c}_{i,j})_{i,j=1,\ldots,d}$ is a $d \times d$-antisymmetric matrix that can be computed from a and b, and $\mathfrak{A}^{i,j}$ is the Lévy area of (Y^i, Y^j), i.e.,

$$\mathfrak{A}^{i,j}(Y;0,t) = \frac{1}{2}\int_0^t (Y_s^i - Y_0^i) \circ dY_s^j - \frac{1}{2}\int_0^t (Y_s^j - Y_0^j) \circ dY_s^i$$

for a d-dimensional semi-martingale Y. If $b = 0$, then $\overline{c} = 0$, so this effect comes from the presence of the drift.

From the Wong-Zakai theorem (see for example [40]), the Stratonovich integral appears as the natural extension of $\mathfrak{I}$ on the subset $\mathrm{SM}([0,T];\mathbb{R}^2)$ of $C^\alpha([0,T];\mathbb{R}^2)$ with $\alpha < 1/2$ that contains trajectories of semi-martingales. Note however that for $Y \in \mathrm{SM}([0,T];\mathbb{R}^2)$ and $(f_1, f_2) = \frac{1}{2}(-x_j, x_i)$,

$$\mathfrak{I}(Y;0,t) = \mathfrak{A}^{1,2}(Y;0,t)$$

for $t \in [0,T]$, if $\mathfrak{I}$ is defined on $\mathrm{SM}([0,T];\mathbb{R}^2)$ as the Stratonovich integral $\mathfrak{I}(Y;0,t) = \int_0^t f(Y_s) \circ dY_s$. Since both X^ε and $\overline{\sigma}W$ belong to $\mathrm{SM}([0,T];\mathbb{R}^2)$, the previous example shows that $\mathfrak{I}(X^\varepsilon;0,t)$ does not converge in general to $\mathfrak{I}(B;0,t)$. This proves that $\mathfrak{I}$ cannot be continuous on $\mathrm{SM}([0,T];\mathbb{R}^2) \subset C^\alpha([0,T];\mathbb{R}^2)$.

Counter-examples to the Wong-Zakai theorem (see [40, 59]) also rely on the construction of approximations of the Brownian trajectories by "perturbating" the piecewise linear approximation that gives rise, in the limit, to a non-vanishing supplementary area and then, for the SDE, to a drift term. The theory of rough paths gives a better understanding of this phenomenon [48].

This problem of convergence may arise in a natural setting and then be of practical interest.

4 Integrals along α-Hölder Continuous Paths, $\alpha \in (1/3, 1/2]$: Heuristic Considerations

We present in this section a construction of the integral which is not the best possible one, but which allows to understand the main ideas and problems.

The counter-example of Section 3.2 has yielded a few ideas: (1) We may use the Green-Riemann formula to deal with close loops. (2) For some $\alpha > 1/2$, we may add to our paths small loops whose radii are of order $2^{-n\alpha/2}$ and thus whose area are of order $2^{-n\alpha}$. (3) As many loops are added, the sum of the areas does not vanish and gives rise to an extra term.

Our construction will now take these facts into account.

4.1 Construction of the Integral along a Subset of $C^\alpha([0, T]; \mathbb{R}^2)$

As we wish our definition of the integral to be continuous, a naive construction is the following: Fix $K > 0$, $\alpha \in (1/3, 1/2]$ and $f \in \text{Lip}(\gamma; \mathbb{R}^2 \to \mathbb{R})$ with $\gamma > 1/\alpha - 1$ (and then $\gamma > 1$). Denote by Π^n the dyadic partition of $[0, T]$ at level n, and by $L^\alpha([0, T]; \mathbb{R}^2)$ the set of functions $x \in C^\alpha([0, T]; \mathbb{R}^2)$ for which the linear approximations $(x^{\Pi^n})_{n \in \mathbb{N}}$ satisfy

$$\mathfrak{I}(x) \stackrel{\text{def}}{=} \lim_{n \in \mathbb{N}} \mathfrak{I}(x^{\Pi^n}) \text{ exists in } C^\alpha([0, T]; \mathbb{R})$$

and $|\mathfrak{I}(x_{|[s,t]}) - \mathfrak{I}(x^{\Pi^n}_{|[s,t]})| \leqslant K \|x - x^{\Pi^n}\|_\alpha |t - s|^\alpha$, $0 \leqslant s < t \leqslant T$.

If K is large enough, it follows from Corollary 2 that $L^\alpha([0, T]; \mathbb{R}^2)$ contains subsets of $C^\beta([0, T]; \mathbb{R}^2)$ for all $\beta > 1/2$ (this depends on f and on the choice of K, since from Corollary 2, $x \mapsto \mathfrak{I}(x)$ is locally Lipschitz) and it is also known (but for this, we need a more complete theory) that it contains paths that are not β-Hölder continuous for $\beta > 1/2$, such as Brownian trajectories (see for example [13, 65]). Any element x of $L^\alpha([0, T]; \mathbb{R}^2)$ may be identified with the sequence $(x^{\Pi^n})_{n \in \mathbb{N}}$.

Now, consider $\varphi \in C^{2\alpha}([0, T]; \mathbb{R}^2)$ and $(\Phi^n)_{n \in \mathbb{N}}$ a sequence of loops at each level n whose areas are asymptotically encoded by φ. Then, as previously,

$$\mathfrak{I}(x^{\Pi^n} \bowtie \Phi^n) \xrightarrow[n \to \infty]{C^\alpha} \mathfrak{I}(x, \varphi) \stackrel{\text{def}}{=} \mathfrak{I}(x) + \int [f, f](x_s) \, \mathrm{d}\varphi_s.$$

For $(x, \varphi) \in L^{\bowtie, \alpha}([0, T]; \mathbb{R}^3) \stackrel{\text{def}}{=} L^\alpha([0, T]; \mathbb{R}^2) \times C^{2\alpha}([0, T]; \mathbb{R})$, we may then define

$$\mathfrak{I}(x, \varphi) = \lim_{n \to \infty} \mathfrak{I}(x^{\Pi^n} \bowtie \Phi^n)$$

where φ encodes asymptotically the areas of $(\varPhi^n)_{n\in\mathbb{N}}$. The space $L^{\bowtie,\alpha}([0,T];\mathbb{R}^3)$ is naturally a Banach space when equipped with the norm $\|(x,\varphi)\|_{\bowtie,\alpha} = \|x\|_\alpha + \|\varphi\|_{2\alpha}$.

The interesting point with this definition of the map $(x,\varphi)\mapsto \Im(x,\varphi)$ is that its continuity follows naturally from its very construction.

Proposition 1. *For all $\beta < \alpha$ with $\alpha \in (1/3,1/2]$, the map $\Im$ is continuous from $(L^{\bowtie,\alpha}([0,T];\mathbb{R}^3), \|\cdot\|_{\bowtie,\alpha})$ to $(C^\alpha([0,T];\mathbb{R}), \|\cdot\|_\beta)$*

Proof. Let $(x^n,\varphi^n)_{n\in\mathbb{N}}$ be a sequence of paths converging to (x,φ) in the space $L^{\bowtie,\alpha}([0,T];\mathbb{R}^3)$.

By definition, $\Im(x^n,\varphi^n;s,t) = \Im(x^n;s,t) + \int_s^t [f,f](x_r^n)\,\mathrm{d}\varphi_r^n$. From Corollary 3, we know that $\int_0^\cdot [f,f](x^n)\,\mathrm{d}\varphi^n$ converges to $\int_0^\cdot [f,f](x)\,\mathrm{d}\varphi$ in the space $C^{2\alpha}([0,T];\mathbb{R})$.

From the very definition of $L^\alpha([0,T];\mathbb{R}^2)$,

$$\|\Im(x^{n,\varPi^m}) - \Im(x^n)\|_\alpha \leqslant K\|x^{n,\varPi^m} - x^n\|_\alpha.$$

But it is easily shown with Corollary 1 that for all $\beta < \alpha$ and some constant K_2, $\|x^{n,\varPi^m} - x^n\|_\beta \leqslant K_2\|x^n\|_\alpha/2^{m(\beta-\alpha)}$ and thus $(\Im(x^{n,\varPi^m}))_{m\in\mathbb{N}}$ converges to $\Im(x^n)$ in $C^\beta([0,T];\mathbb{R})$ at a rate which is uniform in n since $(\|x^n\|_\alpha)_{n\in\mathbb{N}}$ is bounded.

It follows that for all $\beta < \alpha$, $\Im(x^{n,\varPi^m})$ converges uniformly in n to $\Im(x^n)$ in $C^\beta([0,T];\mathbb{R})$ as $m \to \infty$.

For $s < t$ fixed, there exist some integers i_m and j_m such that $t_{i_m-1}^m \leqslant s < t_{i_m}^m$ and $t_{j_m}^m < t \leqslant t_{j_m+1}^m$. To simplify the notations, set $t_{i_m-1}^m = s$ and $t_{j_m+1}^m = t$. For $k = i_m - 1,\dots,j_m + 1$, denote by $z_k^{n,m}$ the following path (see Figure 2)

$$z_k^{n,m} = x^{\varPi^m}_{|[t_k^m,t_{k+1}^m]} \cdot \overline{x^{\varPi^m}_{t_{k+1}^m} x^{n,\varPi^m}_{t_{k+1}^m}} \cdot x^{n,\varPi^m}_{|[t_{k+1}^m,t_k^m]} \cdot \overline{x^{n,\varPi^m}_{t_k^m} x^{\varPi^m}_{t_k^m}}.$$

Hence, with the previous convention on $t_{i_m-1}^m$ and $t_{j_m}^m$,

$$\Im(x^{\varPi^m};s,t) - \Im(x^{n,\varPi^m};s,t)$$

$$= \sum_{k=i_m-1}^{j_m} \int_{z_k^{n,m}} f + \int_{\overline{x_s^{\varPi^m} x_s^{n,\varPi^m}}} f + \int_{\overline{x_t^{n,\varPi^m} x_t^{\varPi^m}}} f. \tag{8}$$

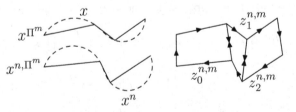

Fig. 2. The paths $z_k^{n,m}$.

Note that

$$\left| \int_{z_k^{n,m}} f \right| = \left| \iint_{\mathrm{Surface}(z_k^{n,m})} [f,f](x^1, x^2)\, \mathrm{d}x^1\, \mathrm{d}x^2 \right|$$

$$\leqslant \frac{1}{2} \|f\|_{\mathrm{Lip}} |x_{t_{k+1}^m} - x_{t_k^m}| \times |x_{t_k^m} - x_{t_k^m}|$$

$$\leqslant \frac{(t_{k+1}^m - t_k^m)^{\alpha}}{2} \|x\|_{\alpha} \|f\|_{\mathrm{Lip}} \|x^n - x\|_{\infty}.$$

Using the convexity inequality with $x \mapsto x^{1/\alpha}$, since there are at most 2^m terms in the series in the right-hand-side of (8), we get

$$\sum_{k=i_m-1}^{j_m} \left| \int_{z_k^{n,m}} f \right| \leqslant 2^{m(1-\alpha)} \left(\sum_{k=i_m-1}^{j_m} \left(\left| \int_{z_k^{n,m}} f \right| \right)^{1/\alpha} \right)^{\alpha}$$

$$\leqslant \frac{2^m}{2} \|f\|_{\mathrm{Lip}} \|x\|_{\alpha} \|x^n - x\|_{\infty} (t-s)^{\alpha}.$$

On the other hand, setting $\Delta_r^n = x_r^{n,\Pi^m} - x_r^n$ for $r \in \{s,t\}$,

$$\left| \int_{x_s^{\Pi^m} x_s^{n,\Pi^m}} f + \int_{x_t^{n,\Pi^m} x_t^{\Pi^m}} f \right| = \left| \int_{x_s^{\Pi^m} x_s^{n,\Pi^m}} f - \int_{x_t^{\Pi^m} x_t^{n,\Pi^m}} f \right|$$

$$\leqslant \left| \int_0^1 (f(x_s^{\Pi^m} + r\Delta_s^n) - f(x_t^{\Pi^m} + r\Delta_t^n)) \Delta_s^n\, \mathrm{d}r \right|$$

$$+ \left| \int_0^1 f(x_t^{\Pi^m} + r\Delta_t^n)(\Delta_t^n - \Delta_s^n)\, \mathrm{d}r \right|$$

$$\leqslant \|f\|_{\mathrm{Lip}} |\Delta_s^n| (\|x^n\|_{\alpha} + \|x^{n,\Pi^m}\|_{\alpha})(t-s)^{\alpha} + \|f\|_{\mathrm{Lip}} |\Delta_t^n - \Delta_s^n|.$$

But, for any $\delta \in [0,1)$,

$$|\Delta_t^n - \Delta_s^n| \leqslant |x_t^{\Pi^m} - x_s^{\Pi^m} - x^{n,\Pi^m} + x_s^{n,\Pi^m}|$$

$$\leqslant (|x_t^{\Pi^m} - x_s^{\Pi^m}|^{\delta} + |x_t^{n,\Pi^m} - x_s^{n,\Pi^m}|^{\delta}) 2 \|x^{\Pi^m} - x^{n,\Pi^m}\|_{\infty}^{1-\delta}$$

$$\leqslant (t-s)^{\alpha\delta} 2 \max\{\|x^{\Pi^m}\|_{\alpha}^{\delta}, \|x^{n,\Pi^m}\|_{\alpha}^{\delta}\} 2 \|x^{\Pi^m} - x^{n,\Pi^m}\|_{\infty}^{1-\delta}.$$

This proves convergence of $\Im(x^{n,\Pi^m})$ to $\Im(x^{\Pi^m})$ in $C^{\beta}([0,T]; \mathbb{R})$ as $n \to \infty$ for any m and any $\beta < \alpha$.

It is now possible to complete the following diagram

$$
\begin{array}{ccc}
\Im(x^{n,\Pi^m}) & \xrightarrow[n\to\infty]{\|\cdot\|_{\beta}} & \Im(x^{\Pi^m}) \\
\|\cdot\|_{\beta} \downarrow {\substack{m\to\infty \\ \text{unif. in } n}} & & \|\cdot\|_{\beta} \downarrow m\to\infty \\
\Im(x^n) & & \Im(x)
\end{array}
$$

to obtain that $\Im(x^n, \varphi^n)$ converges in $C^{\beta}([0,T]; \mathbb{R})$ to $\Im(x, \varphi)$.

Fig. 3. The paths x, x^{Π^n}, $x^{\Pi^{n+1}}$ and the areas defined by $\Phi^{n,n+1}$ (in gray).

Moreover, the following stability result is easily proved.

Lemma 2. *If ψ (resp. φ) is given in $C^{2\alpha}([0,T];\mathbb{R})$ and if it asymptotically encodes the areas of $(\Psi^n)_{n\in\mathbb{N}}$ (resp. $(\Phi^n)_{n\in\mathbb{N}}$), then*

$$\lim_{n\to+\infty} \mathfrak{I}(x^{\Pi^n} \bowtie \Phi^n \bowtie \Psi^n) = \mathfrak{I}(x, \varphi + \psi).$$

The function φ can be arbitrarily chosen, so we have gained a degree of freedom. In other words, to get a proper definition of $\mathfrak{I}$ that respects continuity, we have to consider not a path with values in $\mathbb{R}^2$ but a path with values in $\mathbb{R}^3$. Indeed, this construction is far from optimal, i.e., the set $L^{\bowtie,\alpha}([0,T];\mathbb{R}^3)$ is not the biggest one that can be considered. Yet it gives a proper understanding of the problem.

4.2 Is this Construction Natural?

Of course, the real question is to consider whether or not is it natural to extend $\mathfrak{I}$ on (at least) a subset of $C^\alpha([0,T];\mathbb{R}^2)$ with $\alpha \in (1/3, 1/2]$ by considering paths valued not in $\mathbb{R}^2$ but in $\mathbb{R}^3$.

Consider a path $x \in C^\alpha([0,T];\mathbb{R}^2)$. The piecewise linear path x^{Π^n} is an approximation of x, and for each $m \geqslant n$, we may define

$$\widehat{x}^{\Pi^m} \overset{\text{def}}{=} (x^{\Pi^m}_{|[t^n_0, t^n_1]} \cdot x^{\Pi^n}_{|[t^n_1, t^n_0]}) \cdot x^{\Pi^n}_{|[t^n_0, t^n_1]} \cdots (x^{\Pi^m}_{|[t^n_{2^n-1}, t^n_{2^n}]} \cdot x^{\Pi^n}_{|[t^n_{2^n}, t^n_{2^n-1}]}) \cdot x^{\Pi^n}_{|[t^n_{2^n-1}, t^n_{2^n}]}$$

on the time interval $[0, 3T]$. As we go back and forth on the segments composing x^{Π^n}, we get that $\mathfrak{I}(\widehat{x}^{\Pi^m}; 0, 3T) = \mathfrak{I}(x^{\Pi^m}; 0, T)$. We then define $y^{n,m}_k = x^{\Pi^m}_{|[t^n_k, t^n_{k+1}]} \cdot x^{\Pi^n}_{|[t^n_{k+1}, t^n_k]}$, that satisfies (6a)–(6b) and $\Phi^{n,m} = \{y^{n,m}_k\}_{k=0,\ldots,2^n-1}$. Since $\widehat{x}^{\Pi^m} = x^n \bowtie \Phi^{n,m}$,

$$\mathfrak{I}(x^{\Pi^m}; 0, T) = \mathfrak{I}(\widehat{x}^{\Pi^m}; 0, 3T) = \mathfrak{I}(x^{\Pi^n} \bowtie \Phi^{n,m}; 0, 3T).$$

If we now set for example $m = n^2$, then a priori nothing ensures, unless $x \in L^\alpha([0,T];\mathbb{R}^2)$, that the areas of $(\Phi^{n,n^2})_{n\in\mathbb{N}}$ are asymptotically encoded by the function $\varphi \equiv 0$, nor that there exists a function $\varphi \in C^{2\alpha}([0,T];\mathbb{R})$ that encodes the areas of $(\Phi^{n,n^2})_{n\in\mathbb{N}}$. In the last two cases, how then is the limit of $\mathfrak{I}(x^{\Pi^{n^2}})$ to be considered, since it may differ from the limit of $\mathfrak{I}(x^{\Pi^n} \bowtie \Phi^{n,n^2})$? Indeed,

$$\mathfrak{I}(x^{\Pi^n} \bowtie \Phi^{n,n^2}; 0, T) = \mathfrak{I}(x^{\Pi^n}; 0, T) + \sum_{k=0}^{2^n-1} \mathfrak{I}(y^{n,n^2}_k; t^n_k, t^n_{k+1}).$$

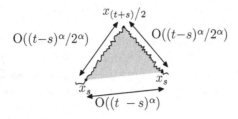

Fig. 4. The area of some α-Hölder continuous path between times s and t is of order $(t - s)^{2\alpha}$.

Yet with the Green-Riemann formula,

$$\Im(y_k^{n,n^2}; t_k^n, t_{k+1}^n) \approx [f, f](x_{t_k^n}) \operatorname{Area}(y_k^{n,n^2}).$$

As already seen, the function $\mathfrak{A}$ on $C^\beta([0,T]; \mathbb{R}^2)$, $\beta > 1/2$, defined by

$$\mathfrak{A}(x; s, t) = \frac{1}{2} \int_s^t (x_r^1 - x_s^1) \, \mathrm{d}x_r^2 - \frac{1}{2} \int_s^t (x_r^2 - x_s^2) \, \mathrm{d}x_r^1 \tag{9}$$

is *not* continuous with respect to the uniform norm: One only has to take $f(x) = \frac{1}{2}x^1 \, \mathrm{d}x^2 - \frac{1}{2}x^2 \, \mathrm{d}x^1$ and to use the previous counter-examples. As $\operatorname{Area}(y_k^{n,n^2}) = \mathfrak{A}(x^{n^2}; t_k^n, t_{k+1}^n)$ and although y_k^{n,n^2} converges uniformly to 0, it may happen that $\operatorname{Area}(y_k^{n,n^2}; t_k^n, t_{k+1}^n)$ is of order $2^{-2\alpha n}$ (this is possible since the distance between $x_{t_k^n}$ and $x_{t_{k+1}^n}$ is roughly of order $2^{-\alpha n}$ if x is α-Hölder continuous, see Figure 4). In this case, $\sum_{k=0}^{2^n-1} \Im(y_k^{n,n^2})$ may have a limit different from 0, or no limit at all.

In other words, the area contained between a path x and its chord for all couple of times (s, t) is "hidden" in x and has to be determined in an arbitrary manner[1].

For some $(x, \varphi) \in L^{\bowtie,\alpha}([0,T]; \mathbb{R}^3)$, which is identified with a sequence converging uniformly to x, the element φ means in some sense that some area has been chosen and then that our integral is properly determined. Once this choice of φ has been performed, Lemma 2 says how to construct different integrals by choosing other areas.

4.3 Justifications for a New Setting

The previous construction does not answer our main question: "How to construct an integral for paths in $C^\alpha([0,T]; \mathbb{R}^2)$ for $\alpha \in (1/3, 1]$?". Yet it yields

---

[1] Consider the case of Brownian trajectories, where the Lévy area is a natural choice, but not the only one, and was the first example of a stochastic integral [47]. In addition, it is then defined as a limit in probability.

the fact that one cannot define a map $x \mapsto \mathfrak{J}$ which extends the map $x \mapsto \int_x f$ on $C^\alpha([0,T];\mathbb{R}^2)$ with $\alpha > 1/2$ unless some extra information is added. Here, this information corresponds to the choice of a function φ, so that we consider indeed a subset of $C^\alpha([0,T];\mathbb{R}^2) \times C^{2\alpha}([0,T];\mathbb{R})$ (for $\alpha \leqslant 1/2$) such that, when equipped with the norm $\|(x,\varphi)\| = \|x\|_\alpha + \|\varphi\|_{2\alpha}$, the map $\mathfrak{J}$ is continuous.

We have also seen in Section 4.2 above that for considering an integral along a path in $C^\alpha([0,T];\mathbb{R}^2)$ with $\alpha \in (1/3, 1/2]$, it is natural to consider the area contained between the path and its chord in view of defining some integral, although there is no way to define it canonically in general.

The drawback of our construction is that we assumed convergence of the integrals along piecewise linear approximations of x.

The idea is now to construct directly a path in $\mathbb{R}^3$ so that it may be identified with a limit of converging sequence of piecewise smooth paths in $\mathbb{R}^2$ whose integrals also converge. This allows us to to get rid of the loops themselves, since the only information we need is the asymptotic limit of the area, while keeping enough information to construct the integral. Besides, this proves that the choice of a converging subsequence does not depend on the choice of the differential form which is integrated.

5 Integrals along α-Hölder Continuous Paths, $\alpha \in (1/3, 1/2]$: Construction by Approximations

It is time to turn to the full picture, now that the importance of knowing the area has been shown.

5.1 Motivations

The main idea in the previous approach was to replace an irregular path $(x, \varphi) \in L^{\bowtie, \alpha}([0,T];\mathbb{R}^3)$ with a simpler path $x^n \in C^1_p([0,T];\mathbb{R}^2)$ which "approximates" x in the following sense: $x^n_{t^n_k} = x_{t^n_k}$ for the dyadic points $\{t^n_k\}_{k=0,\dots,2^n}$ of $[0,T]$, and on $[t^n_k, t^n_{k+1}]$, x^n is composed of a loop y^n_k : $[t^n_k, t^n_k + T2^{-n-1}] \to \mathbb{R}^2$ and then a segment joining $x_{t^n_k}$ and $x_{t^n_{k+1}}$.

Once this family $(x^n)_{n \in \mathbb{N}}$ has been constructed, one may study the convergence of the ordinary integrals $\mathfrak{J}(x^n)$, where the integrals of f on the loops have been transformed with the Green-Riemann formula into double integrals approximately given by the areas of the loops times the Lie brackets of f at the starting points of the loop.

If x^n is defined on $[0,T]$ with loops on $[t^n_k, t^n_{k+1} + T2^{-n-1}]$ and straight lines on $[t^n_k + T2^{-n-1}, t^n_{k+1}]$, a simple approximation of $\mathfrak{J}(x)$ is then given by

$$J^n = \sum_{k=0}^{2^n-1} \left(\int_{t^n_k + T2^{-n-1}}^{t^n_{k+1}} f(x^n_s) \, dx^n_s + [f,f](x_{t^n_k})\mathfrak{A}(x^n; t^n_k, t^n_k + T2^{-n-1}) \right), \quad (10)$$

where $\mathfrak{A}(x; s, t)$ has been defined by (9). Now, following the heuristic reasoning of Section 4.2, we replace the assumption

(H1) The path (x, φ) belongs to $L^{\bowtie, \alpha}([0, T]; \mathbb{R}^3)$.

by the assumption

(H2) There exists some function $\mathfrak{A}(x; s, t)$ which is the limit of $\mathfrak{A}(x^n; s, t)$ for all $0 \leqslant s \leqslant t \leqslant T$.

Note that the assumption (H1) implies (H2) if f is the differential form $f(x) = \frac{1}{2}(x^1 \, dx^2 - x^2 \, dx^1)$. In (H2), there is no more reference to f, while a priori the set $L^{\bowtie, \alpha}([0, T]; \mathbb{R}^3)$ depends on f.

The assumption (H2) means that $\mathfrak{A}(x^{n^2}; t_k^n, t_{k+1}^n)$ (which is equal to $\mathfrak{A}(x^{n^2}; t_k^n, t_{k+1}^n + T2^{-n-1}))$ is equivalent to $\mathfrak{A}(x; t_k^n, t_{k+1}^n)$ as $n \to \infty$. Hence, one may replace (10) by

$$J^n = \sum_{k=0}^{2^n-1} \left(\int_{t_k^n + T2^{-n-1}}^{t_{k+1}^n} f(x_s^{\varPi^n}) \, dx_s^{\varPi^n} + [f, f](x_{t_k^n}) \mathfrak{A}(x; t_k^n, t_{k+1}^n) \right). \qquad (11)$$

This form has the following advantage over the previous one: Under (H2), one can study, as was done in proof of the Young integrals, the convergence of J^n by studying $J^{n+1} - J^n$ in order to prove that $\sum_{n \geqslant 0}(J^{n+1} - J^n)$ converges and to define the integral of x as the limit of this series plus J^0. This method is central in the theory of rough paths.

Still using some approximation, we change (11) into

$$J^n = \sum_{k=0}^{2^n-1} \int_{t_k^n + T2^{-n-1}}^{t_{k+1}^n} f(x_s^{\varPi^n})(x_{t_{k+1}^n} - x_{t_k^n}) \frac{ds}{\varDelta_n t}$$

$$+ \sum_{k=0}^{2^n-1} \int_{t_k^n}^{t_{k+1}^n} [f, f](x_s^{\varPi^N}) \mathfrak{A}(x; t_k^n, t_{k+1}^n) \frac{ds}{\varDelta_n t},$$

with $\varDelta_n t = T2^{-n}$. We use this expression to motivate our introduction of some algebraic structures. Our wish is then to interpret $\mathfrak{A}(x; s, t)$ as some "vector", in the same way as $x_t - x_s$ can be seen, from a geometrical point of view, as the vector that links the two points x_s and x_t, and $\mathbb{R}^2$ as some affine space. As will appear below, $\mathfrak{A}(x; s, t)$ is in general different from $\mathfrak{A}(x; 0, t) - \mathfrak{A}(x; 0, s)$. Hence, the Euclidean structure is not adapted.

We will now construct some space $A(\mathbb{R}^2)$ of dimension 3, that will play the role both of an affine and a vector space, and the kind of vectors we will consider will be $(x_t^1 - x_s^1, x_t^2 - x_s^2, \mathfrak{A}(x; s, t))$. Nevertheless, they will be constructed from the paths $(x_t^1, x_t^2, \mathfrak{A}(x; 0, t))_{t \geqslant 0}$ living in $A(\mathbb{R}^2)$ seen as some affine space.

Firstly, we define this space $A(\mathbb{R}^2)$, then we study the approximation of paths living in this space, and finally we define an integral as a limit of ordinary integrals using the previously constructed approximations.

5.2 What Happens to the Area?

For a continuous path $x \in C^\alpha([0,T]; \mathbb{R}^2)$ with $\alpha > 1/2$, let $y_t = \mathfrak{A}(x; 0, t)$ be the area enclosed between the curve $x_{|[0,t]}$ and its chord $\overline{x_0 x_t}$, where $\mathfrak{A}$ has been defined by (9). This path y is well defined by (9) and belongs to $C^\alpha([0,T]; \mathbb{R})$.

As we have seen that $x \mapsto \mathfrak{A}(x; 0, \cdot)$ is not continuous in general on $C^\alpha([0,T]; \mathbb{R}^2)$ for $\alpha \leqslant 1/2$, we are nonetheless willing to define the equivalent of a process y for an irregular path. This can be achieved using an algebraic setting. Remark first that if $x \in C^\alpha([0,T]; \mathbb{R}^2)$ with $\alpha \in (1/2, 1]$,

$$\mathfrak{A}(x; s, t) = \mathfrak{A}(x; s, u) + \mathfrak{A}(x; u, t) + \frac{1}{2}(x_u - x_s) \wedge (x_t - x_u) \qquad (12)$$

for all $0 \leqslant s < u < t \leqslant T$ (See Figure 5). Here, $\wedge$ is the vector product between two vectors: $a \wedge b = a^1 b^2 - a^2 b^1$.

5.3 Linking Points

We first consider, for a piecewise smooth path x, the path $(x^1, x^2, \mathfrak{A}(x))$ living in a three dimensional space. If u belongs to $\mathbb{R}$, then we set

$$\mathfrak{C}(x, u; t) = (x_t^1, x_t^2, u + \mathfrak{A}(x; 0, t)) \qquad (13)$$

for $t \in [0,T]$. In the following, we may think that x represents a 2-dimensional control trajectory of the position of a particle moving in $\mathbb{R}^3$.

Given two points $a = (a^1, a^2, a^3)$ and $b = (b^1, b^2, b^3)$, we wish to construct a piecewise smooth path x from $[0,1]$ to $\mathbb{R}^2$ such that the continuous path $(x_t, a^3 + \mathfrak{A}(x; 0, t))$ from $[0,1]$ to $\mathbb{R}^3$ goes from a to b.

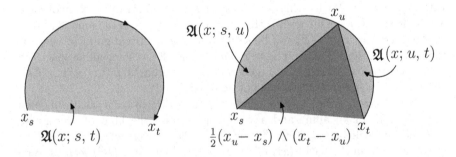

Fig. 5. A geometrical illustration of (12).

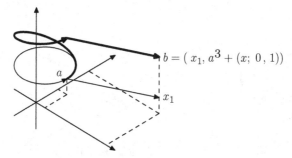

Fig. 6. A simple path (x, y) from a to b controlled by a path x in $\mathbb{R}^2$.

Such a path is easily constructed. We give here a simple example, that serves as a prototype for our approach. Our choice, drawn in Figure 6, is

$$x_t = \begin{bmatrix} a^1 \\ a^2 \end{bmatrix} + \frac{\sqrt{|b^3 - a^3|}}{\sqrt{\pi}} \begin{bmatrix} \cos(4\pi t) - 1, \\ \mathrm{sgn}(b^3 - a^3)\sin(4\pi t) \end{bmatrix} \text{ if } t \in \left[0, \frac{1}{2}\right],$$

$$\text{and } x_t = \begin{bmatrix} a^1 \\ a^2 \end{bmatrix} + (2t - 1) \begin{bmatrix} b^1 - a^1 \\ b^2 - a^2 \end{bmatrix} \text{ if } t \in \left[\frac{1}{2}, 1\right].$$

Given two points a and b in $\mathbb{R}^3$, consider two paths x and y in $C^1_p([0, T]; \mathbb{R}^2)$ such that $x_0 = y_0 = 0$ and $\mathfrak{C}(x, 0; T) = a$, $\mathfrak{C}(y, 0; T) = b$.

The concatenation $x \cdot y$ of x and y gives rise to a path that goes from 0 to $\pi(a + b)$ through $\pi(a)$, where π is the projection $\pi(a^1, a^2, a^3) = (a^1, a^2)$. What can then be said on $\mathfrak{C}(x \cdot y; 0, 2T)$? Due to (12), we get that $\mathfrak{C}(x \cdot y)$ is a path that goes from 0 to the point denoted by $a \boxplus b$ and defined by

$$a \boxplus b = \left(a^1 + b^1, a^2 + b^2, a^3 + b^3 + \frac{1}{2} \begin{bmatrix} a^1 \\ a^2 \end{bmatrix} \wedge \begin{bmatrix} b^1 \\ b^2 \end{bmatrix}\right).$$

With this notation, $\boxplus$ clearly defines an operation on $\mathbb{R}^3$, which is different from the usual addition (geometrically equivalent to some translation) in this space $\mathbb{R}^3$. In addition, $\mathfrak{C}(x \cdot y, 0)$ passes through the point a.

As illustrated in Figures 6a–6d, this gives rise to a different path as the one obtained by concatenation of $\mathfrak{C}(x, 0)$ and $\mathfrak{C}(\pi(a) + y; a^3)$, which ends at $a + b$.

5.4 The Space $\mathbb{R}^3$ as a Non-Commutative Group

We have now equipped $\mathbb{R}^3$ with an operation $\boxplus$, which is easily proved to be associative. When equipped with this operation $\boxplus$, we denote $\mathbb{R}^3$ by $A(\mathbb{R}^2)$. We also set

$$[a, b] = a \boxplus b - b \boxplus a = \left(0, 0, \frac{1}{2} \begin{bmatrix} a^1 \\ a^2 \end{bmatrix} \wedge \begin{bmatrix} b^1 \\ b^2 \end{bmatrix}\right).$$

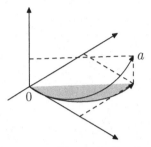

Fig. 6a. The path $\mathfrak{C}(x,0)$.

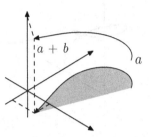

Fig. 6b. The path $\mathfrak{C}(\pi(a)+y,a^3)$.

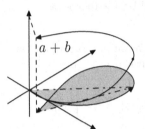

Fig. 6c. The path $\mathfrak{C}(x,0) \cdot \mathfrak{C}(\pi(a)+y,a^3)$.

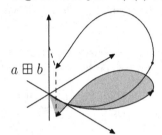

Fig. 6d. The path $\mathfrak{C}(x \cdot y,0)$.

This bracket $[\cdot, \cdot]$ is of course linked to the fact that $(A(\mathbb{R}^2), \boxplus)$ is a non-commutative group, called the *Heisenberg group* (see Section 6.3).

Lemma 3. *The space* $(A(\mathbb{R}^2), \boxplus)$ *is a non-commutative group with* 0 *as the neutral element. The inverse of any element* $a = (a^1, a^2, a^3)$ *is* $-a = (-a^1, -a^2, -a^3)$.

Proof. That the inverse of a is $-a$ is easily verified since

$$(-a^1, -a^2, -a^3) \boxplus (a^1, a^2, a^3) = -\frac{1}{2}[a, a] = 0.$$

The non-commutativity of $\boxplus$ in general follows from $b \boxplus a = a \boxplus b \boxplus [b, a]$.

The non-commutativity of $\boxplus$ is illustrated in Figures 6e–6f. Of course, if $a, b \in \mathbb{R}^3$ are of type $a = (a^1, a^2, 0)$ and $b = (b^1, b^2, 0)$, then $a \boxplus b = b \boxplus a$: the non-commutativity concerns only the third component. If $x : [0, 1] \to \mathbb{R}^2$ goes from a to b and $y : [0, 1] \to \mathbb{R}^2$ goes from b to c, then $x \cdot y$ goes from a to c and $(y - b + a) \cdot (b - a + x)$ goes from also from a to c. Yet the area enclosed between these two paths and its chord is not the same.

It is now easy to remark that $A(\mathbb{R}^2)$ is both a Lie algebra and a Lie group. For some introduction on these notions, see [17, 37, 67, 69, 71] among many other books.

Lemma 4. *The space* $(A(\mathbb{R}^2), [\cdot, \cdot])$ *is a Lie algebra.*

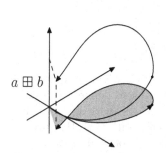

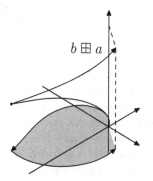

Fig. 6e. The path $\mathfrak{C}(x \cdot y, 0)$.

Fig. 6f. The path $\mathfrak{C}(y \cdot x, 0)$.

Proof. Clearly, $(a, b) \mapsto [a, b]$ is bilinear, $[a, b] = -[b, a]$ and the Jacobi identity is easily satisfied:

$$[a, [b, c]] + [b, [c, a]] + [c, [a, b]] = 0, \ \forall a, b, c \in A(\mathbb{R}^2).$$

This proves the Lemma.

As for $\mathbb{R}^3$, $A(\mathbb{R}^2)$ may be equipped with the multiplication by a scalar, which is $(\lambda, x) = \lambda \cdot x \overset{\text{def}}{=} (\lambda x^1, \lambda x^2, \lambda x^3)$ if $x = (x^1, x^2, x^3) \in A(\mathbb{R}^2)$ and $\lambda \in \mathbb{R}$. But unlike $\mathbb{R}^3$, this operation is not distributive, since

$$\lambda \cdot (x \boxplus y) = (\lambda x) \boxplus (\lambda y) + \lambda(1 - \lambda)[x, y].$$

Thus, $(A(\mathbb{R}^2), \boxplus, \cdot)$, where $\cdot$ denotes the multiplication by a scalar, is not a module.

Another natural external law equips naturally $A(\mathbb{R}^2)$, namely, the *dilation*. Given $\lambda \in \mathbb{R}$, set

$$\delta_\lambda x = (\lambda x^1, \lambda x^2, \lambda^2 x^3) \text{ for } x = (x^1, x^2, x^3) \in A(\mathbb{R}^2). \tag{14}$$

Note that

$$\delta_\lambda(x \boxplus y) = (\delta_\lambda x) \boxplus \delta_\lambda y \text{ and } \delta_\lambda \delta_\mu x = \delta_{\lambda\mu} x$$

for $\lambda, \mu \in \mathbb{R}$ and $x \in A(\mathbb{R}^2)$. However, we do not have that $\delta_{\lambda+\mu} x = \delta_\lambda x \boxplus \delta_\mu x$. Hence, $(A(\mathbb{R}^2), \boxplus, \delta)$ is not a module.

This space $A(\mathbb{R}^2)$ is equipped with a norm defined by

$$|a|_\star = \max\{|a^1|, |a^2|, |a^3|\} \tag{15}$$

and a *homogeneous norm* defined by

$$|a| = \max\left\{|a^1|, |a^2|, \sqrt{\frac{1}{2}|a^3|}\right\}, \tag{16}$$

which means that $|a| = 0$ if and only if $a = 0$,

$$|\delta_\lambda x| = |\lambda| \cdot |x| \text{ for } \lambda \in \mathbb{R} \text{ and } x \in A(\mathbb{R}^2),$$

and $|-x| = |x|$ for all $x \in A(\mathbb{R}^2)$ (see also Section A).

Remark that this choice ensures that $|a \boxplus b| \leqslant (3/2) (|a| + |b|)$. We will see below in Sections 5.9 and A that this homogeneous norm is equivalent to another homogeneous norm $\| \cdot \|_{CC}$ which allows us to define a distance between two points a and b in $A(\mathbb{R}^2)$ by $\|(-a) \boxplus b\|_{CC}$ (with $\| \cdot \|_{CC}$, the triangular inequality is satisfied, which is not the case with $| \cdot |$). Because of the square root in the definition of $| \cdot |$, this distance is not equivalent to the one generated by $| \cdot |_\star$. Yet it generates the same topology.

Remark 5. Because $| \cdot |$ does not satisfy the triangle inequality, $d : (a, b) \mapsto |(-a) \boxplus b|$ does not define a distance. However, this may be called a *near-metric* because $d(a, b) \leqslant C(d(a, c) + d(c, b))$ for some constant $C > 0$ and all $a, c, b \in A(\mathbb{R}^2)$.

From this, we easily deduce that $A(\mathbb{R}^2)$ is also a Lie group. We recall that a *Lie group* $(G, \times)$ is a group with a differentiable manifold structure (and in particular a norm) such that $(x, y) \mapsto x \times y$ and $x \mapsto x^{-1}$ are continuous (see for example [67, 71, 69, 37] and many other books).

Lemma 5. *The space* $(A(\mathbb{R}^2), \boxplus)$ *is a Lie group.*

Proof. The continuity of $(x, y) \mapsto x \boxplus y$ and $x \mapsto -x$ is easily proved.

5.5 Enhanced Paths and their Approximations

Of course, we have constructed the space $A(\mathbb{R}^2)$ with the idea of considering paths living in $A(\mathbb{R}^2)$, the third component giving all the information we need.

Basically, a continuous path with values in $A(\mathbb{R}^2)$ is a continuous path with values in the Euclidean space $\mathbb{R}^3$ (recall that the norm $| \cdot |_\star$ we put on $A(\mathbb{R}^2)$ is equivalent to the Euclidean norm). However, we will use the group operation $\boxplus$ of $A(\mathbb{R}^2)$ in replacement as the translation by a vector in $\mathbb{R}^3$, and thus the paths we consider will be seen differently from the usual paths.

Recall that $(\mathbb{R}^2, +)$ is in some sense contained in $(A(\mathbb{R}^2), \boxplus)$, and then plays a special role.

Definition 3. *Given a continuous path x with values in $\mathbb{R}^2$, a continuous path* **x** *with values in* $A(\mathbb{R}^2)$ *with* $x = (\mathbf{x}^1, \mathbf{x}^2)$ *may then be called an* enhanced path, *or a path lying above x. Given a path $x : [0, T] \to \mathbb{R}^2$, a path* $\mathbf{x} : [0, T] \to A(\mathbb{R}^2)$ *with lies above x is called a* lift *of x.*

Let x and y be two smooth paths lifted as $\mathbf{x} = \mathfrak{C}(x, 0)$ and $\mathbf{y} = \mathfrak{C}(y, 0)$, where $\mathfrak{C}$ has been defined by (13). We have seen that the usual concatenation $\mathbf{x} \cdot \mathbf{y}$ of $\mathbf{x}$ and $\mathbf{y}$ seen as paths with values in $\mathbb{R}^3$ is different from the path

$\mathfrak{C}(x \cdot y, 0)$. We introduce then a new kind of concatenation of two paths $\mathbf{x}$: $[0, T] \to A(\mathbb{R}^2)$ and $\mathbf{y} : [0, S] \to A(\mathbb{R}^2)$. This concatenation is defined by

$$(\mathbf{x} \,\square\, \mathbf{y})_t = \begin{cases} \mathbf{x}_t & \text{if } t \in [0, T], \\ \mathbf{x}_T \boxplus ((-\mathbf{y}_0) \boxplus \mathbf{y}_{t-T}) & \text{if } t \in [T, S + T] \end{cases}$$

and gives rise to a continuous path from $[0, T + S]$ to $A(\mathbb{R}^2)$ when $\mathbf{x}$ and $\mathbf{y}$ are continuous. In addition, $\mathbf{x} \,\square\, \mathbf{y}$ lies above $x \cdot y$ if $\mathbf{x}$ (resp. $\mathbf{y}$) lies above x (resp. y). Yet we have to be warned of an important fact: this concatenation is different from the usual concatenation in $\mathbb{R}^3$.

If $x : [0, T] \to \mathbb{R}^2$ and $y : [0, S] \to \mathbb{R}^2$ are two piecewise smooth paths, then this concatenation satisfies

$$\mathfrak{C}(x \cdot y, 0) = \mathfrak{C}(x, 0) \,\square\, \mathfrak{C}(y, 0).$$

For two points a and b in $A(\mathbb{R}^2)$, let $\psi_{a,b} \in C_p^1([0, 1]; \mathbb{R}^3)$ be a smooth path joining a and b lying above $\zeta_{a,b} : [0, 1] \to \mathbb{R}^2$ (for example, we can use the one of Section 5.3). By definition of $\zeta_{a,b}$ and $\psi_{a,b}$, $\psi_{a,b}(t) = \mathfrak{C}(\zeta_{a,b}, a^3; t)$. Moreover, for a, b, c in $A(\mathbb{R}^2)$,

$$\psi_{a,b} \,\square\, \psi_{b,c} = \mathfrak{C}(\zeta_{a,b} \cdot \zeta_{b,c}, a^3).$$

Thus, $\psi_{a,b} \,\square\, \psi_{b,c}$ is a path that goes from a to c through b.

Let $\mathbf{x}$ be a continuous path from $[0, T]$ living in $A(\mathbb{R}^2)$. It is then natural to look for an approximation of $\mathbf{x}$ given by the sequence of paths

$$\mathbf{x}^n = \psi_{\mathbf{x}_{t_0^n}, \mathbf{x}_{t_1^n}} \,\square\, \psi_{\mathbf{x}_{t_1^n}, \mathbf{x}_{t_2^n}} \,\square\, \cdots \,\square\, \psi_{\mathbf{x}_{t_{n-1}^n}, \mathbf{x}_{t_n^n}}.$$

The path $\mathbf{x}^n$ satisfies $\mathbf{x}^n(t) = \mathbf{x}(t)$ for the dyadic times t at level n. In addition,

$$\mathbf{x}^n = \mathfrak{C}(\zeta^n, \mathbf{x}_0^3) \text{ with } \zeta^n = \zeta_{\mathbf{x}_{t_0^n}, \mathbf{x}_{t_1^n}} \cdot \zeta_{\mathbf{x}_{t_1^n}, \mathbf{x}_{t_2^n}} \cdots \cdots \zeta_{\mathbf{x}_{t_{n-1}^n}, \mathbf{x}_{t_n^n}},$$

and it is easily proved that ζ^n converges uniformly to x, the path above which $\mathbf{x}$ lives (See Figure 7).

Now, there are two natural questions: (1) Provided $\mathbf{x}$ is regular enough, does $\mathbf{x}^n$ converge to $\mathbf{x}$, in which sense? (2) Is it possible to construct $\mathfrak{I}(\mathbf{x})$ as the limit of the $\mathfrak{I}(\zeta^n)$'s, which are then ordinary integrals?

5.6 Hölder Continuous Enhanced Paths

We have defined the space $A(\mathbb{R}^2)$ as the space $\mathbb{R}^3$ with a special non-commutative group structure, which is different from the translation.

Let $x \in C^\alpha([0, T]; \mathbb{R}^2)$ with $\alpha > 1/2$ and $x_0 = 0$. Set $\mathbf{x} = (x^1, x^2, \mathfrak{A}(x))$. With (12),

$$(-\mathbf{x}_s) \boxplus \mathbf{x}_t = (x_t^1 - x_s^1, x_t^2 - x_s^2, \mathfrak{A}(x; s, t)),$$

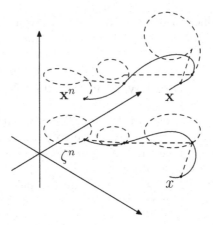

Fig. 7. Approximation of a path $\mathbf{x}$ in $A(\mathbb{R}^2)$.

which means that $(-\mathbf{x}_s) \boxplus \mathbf{x}_t$ can be constructed from the path x restricted to $[s, t]$. The same is true even if $x_0 \neq 0$.

For a path $\mathbf{x}$ from $[0, T]$ to $A(\mathbb{R}^2)$, $\mathbf{x}_{s,t} \overset{\text{def}}{=} (-\mathbf{x}_s) \boxplus \mathbf{x}_t$ may then be interpreted as an "increment" of $\mathbf{x}$, and indeed we get the following trivial identity $\mathbf{x}_t = \mathbf{x}_s \boxplus \mathbf{x}_{s,t}$ for all $0 \leqslant s \leqslant t \leqslant T$, which is the equivalent to $x_t = x_s + (x_t - x_s)$ in $\mathbb{R}^2$. Note that in general $\mathbf{x}_{s,t}^3$ is different from $\mathbf{x}_t^3 - \mathbf{x}_t^3$, although $\mathbf{x}_{s,t}^i = \mathbf{x}_t^i - \mathbf{x}_s^i$ for $i = 1, 2$.

Similarly, we may write the value of $\mathbf{x}_t$ at time t as a function of the values of $\mathbf{x}$ at times $s \leqslant r \leqslant t$:

$$\mathbf{x}_t = \mathbf{x}_s \boxplus \mathbf{x}_{s,r} \boxplus \mathbf{x}_{r,t} \tag{17}$$

for all $0 \leqslant s \leqslant r \leqslant t \leqslant T$. When one sees $\mathbf{x}$ as a geometric object, (17) yields

$$\mathbf{x}_{|[s,t]} = \mathbf{x}_{|[s,r]} \boxdot \mathbf{x}_{|[r,t]}, \tag{18}$$

for all $0 \leqslant s \leqslant r \leqslant t \leqslant T$.

From now on, to take into account the fact that we work in $A(\mathbb{R}^2)$, we have to think of paths from $[0, T]$ to $A(\mathbb{R}^2)$ as continuous paths $\mathbf{x}$ satisfying (18), although this relation is satisfied by any continuous path from $[0, T]$ to $\mathbb{R}^3$ (which also means that there are infinitely many paths lying above a continuous path from $[0, T]$ to $\mathbb{R}^2$). But we will see below that if $\mathbf{x}$ lies above a smooth path x and is also quite regular (in a sense to be defined), then (18) and the regularity condition will impose some "constraint" on the path $\mathbf{x}$.

Lemma 6 ([55, Lemma 2.2.3, p. 250]). *Let $\mathbf{x}$ and $\mathbf{y}$ be two continuous paths from $[0, T]$ to $A(\mathbb{R}^2)$ such that $(\mathbf{x}^1, \mathbf{x}^2) = (\mathbf{y}^1, \mathbf{y}^2)$. Then there exists a continuous path $\varphi : [0, T] \to \mathbb{R}$ such that $\mathbf{y} = (\mathbf{x}^1, \mathbf{x}^2, \mathbf{x}^3 + \varphi)$, which means that*

$$((-\mathbf{y}_s) \boxplus \mathbf{y}_t)^3 = ((-\mathbf{x}_s) \boxplus \mathbf{x}_t)^3 + \varphi_t - \varphi_s \tag{19}$$

for all $0 \leqslant s \leqslant t \leqslant T$.

Proof. It is sufficient to put $\varphi_t = ((-\mathbf{y}_0) \boxplus \mathbf{y}_t)^3 - ((-\mathbf{x}_0) \boxplus \mathbf{x}_t)^3$, which clearly satisfies (19).

Notation 3. Denote by $C^\alpha([0,T]; A(\mathbb{R}^2))$ the set of continuous paths $\mathbf{x}$: $[0,T] \to A(\mathbb{R}^2)$ and such that

$$\|\mathbf{x}\|_\alpha = |\mathbf{x}_0| + \sup_{0 \leqslant s < t \leqslant T} \frac{|(-\mathbf{x}_s) \boxplus \mathbf{x}_t|}{|t-s|^\alpha}$$

is finite. If $x = (x^1, x^2)$ is a path in $C^\alpha([0,T]; \mathbb{R}^2)$ and $\mathbf{x} = (x^1, x^2, y)$ a path in $C^\alpha([0,T]; A(\mathbb{R}^2))$, then we say that $\mathbf{x}$ *lies above* x.

Lemma 7. *Let* $x \in C^\alpha([0,T]; \mathbb{R}^2)$ *with* $\alpha > 1/2$. *Then* $\mathbf{x} = (x, \mathfrak{A}(x; 0, \cdot))$ *belongs to* $C^\alpha([0,T]; A(\mathbb{R}^2))$. *In addition the map* $x \mapsto \mathbf{x}$ *is Lipschitz continuous from* $(C^\alpha([0,T]; \mathbb{R}^2), \|\cdot\|_\alpha)$ *to* $(C^\alpha([0,T]; A(\mathbb{R}^2)), \|\cdot\|_\alpha)$.

Proof. By construction, $\mathbf{x}$ is a path with value in $A(\mathbb{R}^2)$. Note that $(-\mathbf{x}_s) \boxplus \mathbf{x}_t = (x_t^1 - x_s^1, x_t^2 - x_s^2, \mathfrak{A}(x; s, t))$. From the construction of the Young integral (more specifically, from a variation of (4)),

$$|\mathfrak{A}(x; s, t)| \leqslant \zeta(2\alpha - 1)(t-s)^{2\alpha} \|x\|_\alpha^2 \tag{20}$$

and then the result is proved.

Note that in the previous proof, (20) does not mean that $t \mapsto \mathfrak{A}(x; 0, t)$ is 2α-Hölder continuous (in which case $2\alpha > 1$!). Indeed, $t \mapsto \mathfrak{A}(x; 0, t)$ is only α-Hölder continuous, since x is α-Hölder continuous.

On the other hand, any path in $C^\alpha([0,T]; A(\mathbb{R}^2))$ with $\alpha > 1/2$ can be expressed as a path $x \in C^\alpha([0,T]; \mathbb{R}^2)$ lifted using its area $\mathfrak{A}(x)$.

Lemma 8. *Let* $\mathbf{x} \in C^\alpha([0,T]; A(\mathbb{R}^2))$ *with* $\alpha > 1/2$. *Then* $\mathbf{x} = \mathfrak{C}(x, \mathbf{x}_0^3) = (x, \mathbf{x}_0^3 + \mathfrak{A}(x))$ *with* $x = (\mathbf{x}^1, \mathbf{x}^2)$.

Remark 6. If for some $\alpha > 1/2$, $(\mathbf{x}^n)_{n \in \mathbb{N}}$ belongs to $C^\alpha([0,T]; A(\mathbb{R}^2))$ is composed of paths of type $\mathbf{x}^n = (x^n, \mathfrak{A}(x^n))$ with $x^n \in C^\alpha([0,T]; \mathbb{R}^2)$ and $\mathbf{x}^n$ converges in $C^\alpha([0,T]; A(\mathbb{R}^2))$ to some $\mathbf{x}$, then $\mathbf{x} \in C^\alpha([0,T]; A(\mathbb{R}))$ is necessarily of type $\mathbf{x} = (x, \mathfrak{A}(x))$ with $x \in C^\alpha([0,T]; \mathbb{R}^2)$. In Proposition 2 below, we will see how to construct a family of paths x^n in $C^1([0,T]; \mathbb{R}^2)$ for which $\mathbf{x}^n = (x, \mathfrak{A}(x))$ converges to $\mathbf{x} \in C^\alpha([0,T]; A(\mathbb{R}^2))$ with $\alpha > 1/3$. Thus, if one considers a path with values in $A(\mathbb{R}^2)$ which is not of type $(x, \mathfrak{A}(x))$ but which is piecewise smooth, one has to interpret it as a path in $C^{1/2}([0,T]; A(\mathbb{R}^2))$ in order to identify it with a family of converging paths.

Proof. From Lemma 7, $\mathbf{y} = \mathfrak{C}(x, \mathbf{x}_0^3)$ belongs to $C^\alpha([0,T]; A(\mathbb{R}^2))$, and from Lemma 6, there exists a function $\varphi : [0,T] \to \mathbb{R}$ such that $((-\mathbf{x}_s) \boxplus \mathbf{x}_t)^3 = ((-\mathbf{y}_s) \boxplus \mathbf{y}_t)^3 + \varphi_t - \varphi_s$ for all $0 \leqslant s \leqslant t \leqslant T$. Hence, $\sqrt{|\varphi_t - \varphi_s|} \leqslant \|\mathbf{x}\|_\alpha |t-s|^\alpha$ and then $|\varphi_t - \varphi_s| \leqslant \|\mathbf{x}\|_\alpha^2 |t-s|^{2\alpha}$. As $\alpha > 1/2$, necessarily φ is constant.

As we saw earlier, one can add a path with values in $\mathbb{R}$ to the third component of a path with values in $A(\mathbb{R}^2)$ to get a new path with values in $A(\mathbb{R}^2)$. Although a path with values in $\mathbb{R}^2$ which is regular enough can be naturally lifted as a path with values in $\mathbb{R}^3$, we gain one degree of freedom: there are infinitely many paths that lie above a path in $\mathbb{R}^2$. The next lemma, whose proof is immediate, specifies the kind of paths we have to use to stay in $C^\alpha([0,T]; A(\mathbb{R}^2))$.

Lemma 9. *For* $\alpha \leqslant 1/2$, *let* $\mathbf{x} \in C^\alpha([0,T]; A(\mathbb{R}^2))$ *and* $\varphi \in C^{2\alpha}([0,T]; \mathbb{R})$. *Then* $\mathbf{y} = (\mathbf{x}^1, \mathbf{x}^2, \mathbf{x}^3 + \varphi)$ *belongs to* $C^\alpha([0,T]; A(\mathbb{R}^2))$.

Any path in $C^\alpha([0,T]; A(\mathbb{R}^2))$ can be seen as a limit of paths naturally constructed above a path of finite variation. Before proving this, we state a lemma on relative compactness, which is just an adaptation of Lemma 2.

Lemma 10. *Let* $(\mathbf{x}^n)_{n \in \mathbb{N}}$ *be such that* $\mathbf{x}^n \in C^\alpha([0,T]; A(\mathbb{R}^2))$ *and is bounded. Then there exist* $\mathbf{x}$ *in* $C^\alpha([0,T]; A(\mathbb{R}^2))$ *and a subsequence of* $(\mathbf{x}^n)_{n \in \mathbb{N}}$ *that converges to* $\mathbf{x}$ *in* $(C^\alpha([0,T]; A(\mathbb{R}^2)), \|\cdot\|_\beta)$ *for each* $\beta < \alpha$.

We shall now prove the main result of this section: any path $\mathbf{x}$ in $C^\alpha([0,T]; A(\mathbb{R}^2))$ with $\alpha \in (1/3, 1/2)$ may be identified as the limit of $\mathfrak{C}(x^n, \mathbf{x}_0^3)$, where x^n are paths in $C_p^\infty([0,T]; \mathbb{R}^2)$. Paths taking their values in $A(\mathbb{R}^2)$ are then objects that are easier to deal with than sequences of paths with loops as we did previously.

Let $\mathbf{x} \in C^\alpha([0,T]; A(\mathbb{R}^2))$ with $\alpha \in (1/3, 1/2)$ lying above x. Denote by x^{Π^n} the linear interpolation of x along the dyadic partition $\Pi^n = \{t_k^n\}_{k=0,\ldots,2^n}$ at level n, with $t_k^n = Tk/2^n$. Also define

$$\theta_k^n = ((-\mathbf{x}_{t_{k+1}^n}) \boxplus \mathbf{x}_{t_k^n})^3. \tag{21a}$$

Set $\Phi^n = \{y_k^n\}_{k=0,\ldots,2^n-1}$ with $y_k^n : [t_k^n, t_{k+1}^n] \to \mathbb{R}^2$ and

$$y_k^n(t) = \sqrt{\frac{|\theta_k^n|}{\pi}} \begin{bmatrix} \cos\left(2\pi \frac{t-t_k^n}{t_{k+1}^n - t_k^n}\right) - 1 \\ \operatorname{sgn}(\theta_k^n) \sin\left(2\pi \frac{t-t_k^n}{t_{k+1}^n - t_k^n}\right) \end{bmatrix}. \tag{21b}$$

Finally, set

$$x_t^n = x^{\Pi^n} \bowtie \Phi^n(t/2) \text{ for } t \in [0,T] \text{ and } \mathbf{x}^n = (x^n, \mathbf{x}_0^3 + \mathfrak{A}(x^n; 0, \cdot)). \tag{21c}$$

This corresponds to joining the points of $\{\mathbf{x}_{t_k^n}\}_{k=0,\ldots,2^n}$ by the simple paths constructed in Section 5.3 (see Figure 6).

Proposition 2. *With the previous notations (21a)-(21c),* $(\mathbf{x}^n)_{n \in \mathbb{N}}$ *is uniformly bounded in* $C^\alpha([0,T]; A(\mathbb{R}^2))$ *and converges to* $\mathbf{x}$ *with respect to* $\|\cdot\|_\beta$ *for all* $\beta < \alpha$.

Remark 7. We have considered a path $\mathbf{x}$ in $\mathrm{C}^\alpha([0,T]; \mathrm{A}(\mathbb{R}^2))$ above a path $x \in \mathrm{C}^\alpha([0,T]; \mathbb{R}^2)$, but we have not shown how to construct such a path, except when $\alpha > 1/2$. For that, we may either use the results in [54], that assert it is always possible to do so, or study particular cases. For example, many trajectories of stochastic processes have been dealt with (Brownian motion [65], semi-martingales [13], fractional Brownian motion [14, 15, 62], Wiener process [50], Gaussian processes [30, 31], free Brownian motion [70], ... The book [28] contains many such constructions). In general, these results are obtained in connection with an approximation of Wong-Zakai type.

Choosing a path $\mathbf{x}$ above x corresponds to a determination of the limit of $\mathfrak{A}(x^n; s, t)$ where x^n converges to x, and is then a slightly weaker hypothesis than (H2).

Proof (Proof of Proposition 2). Note first that $\mathbf{x}^n_{t^n_k} = \mathbf{x}_{t^n_k}$. For $t \in [0,T)$, let $\underline{M}(t,n)$ be the largest integer such that $t^n_{\underline{M}(t,n)} \leqslant t$. Then, for $0 \leqslant t < T$,

$$
|\mathbf{x}^n_t - \mathbf{x}_t| \leqslant |\mathbf{x}^n_t - \mathbf{x}^n_{t^n_{\underline{M}(t,n)}}| + |\mathbf{x}_t - \mathbf{x}_{t^n_{\underline{M}(t,n)}}|
$$

$$
\leqslant \max\{\sqrt{|\theta^n_k|/\pi}, |\mathbf{x}_{t^n_{\underline{M}(t,n)+1}} - \mathbf{x}_{t^n_{\underline{M}(t,n)}}|\} + \|\mathbf{x}\|_\alpha (t - t^n_{\underline{M}(t,n)})^\alpha
$$

$$
\leqslant 2\|\mathbf{x}\|_\alpha T^\alpha 2^{-\alpha n}.
$$

This proves that $\mathbf{x}^n$ converges uniformly to $\mathbf{x}$.

Convergence in $\mathrm{C}^\beta([0,T]; \mathrm{A}(\mathbb{R}^2))$ follows from the uniform boundedness of the α-Hölder norm of $\mathbf{x}^n$ and Lemma 10.

So, it remains to estimate the α-Hölder norm of $\mathbf{x}^n$ in $\mathrm{A}(\mathbb{R}^2)$. For $0 \leqslant s < t \leqslant T$, let $\overline{M}(s,n)$ be the smallest integer such that $s \leqslant t^n_{\overline{M}(s,n)}$. Then, unless s, t belongs to the same dyadic interval $[t^n_k, t^n_{k+1}]$ for some $k = 0, \ldots, 2^n - 1$,

$$
\mathbf{x}^n_{s,t} = \mathbf{x}^n_{s,t^n_{\overline{M}(s,n)}} \boxplus \mathbf{x}^n_{t^n_{\overline{M}(s,n)}, t^n_{\underline{M}(t,n)}} \boxplus \mathbf{x}^n_{t^n_{\underline{M}(t,n)}, t}.
$$

for all $0 \leqslant s < t \leqslant T$. In addition, $\mathbf{x}^n_{t^n_{\overline{M}(s,n)}, t^n_{\underline{M}(t,n)}} = \mathbf{x}_{t^n_{\overline{M}(s,n)}, t^n_{\underline{M}(t,n)}}$ for any integer n. Since $|\cdot|$ is a homogeneous norm on $\mathrm{A}(\mathbb{R}^2)$, it follows that for some universal constant C_0,

$$
|\mathbf{x}^n_{s,t}| \leqslant C_0 |\mathbf{x}^n_{s,t^n_{\overline{M}(s,n)}}| + C_0 |\mathbf{x}_{t^n_{\overline{M}(s,n)}, t^n_{\underline{M}(t,n)}}| + C_0 |\mathbf{x}^n_{t^n_{\underline{M}(t,n)}, t}|
$$

$$
\leqslant C_0 |\mathbf{x}^n_{s,t^n_{\overline{M}(s,n)}}| + C_0 \|\mathbf{x}\|_\alpha (t^n_{\underline{M}(t,n)} - t^n_{\overline{M}(s,n)})^\alpha + C_0 |\mathbf{x}^n_{t^n_{\underline{M}(t,n)}, t}|.
$$

Assume that we have proved that for some constant K,

$$
|\mathbf{x}^n_{s,t}| \leqslant K(t - s)^\alpha \text{ for all } t^n_k \leqslant s \leqslant t \leqslant t^n_{k+1}, \ k = 0, \ldots, 2^n - 1, \tag{22}
$$

then boundedness of $(\|\mathbf{x}^n\|_\alpha)_{n \in \mathbb{N}}$ follows easily as in the proof of Corollary 1 by applying (22) to s, t in the same dyadic interval, and to $|\mathbf{x}^n_{s,t^n_{\overline{M}(s,n)}}|$ as well as to $|\mathbf{x}^n_{t^n_{\underline{M}(t,n)}, t}|$.

We now turn to the proof of (22). First, consider that for some $k \in \{0, \ldots, 2^n - 1\}$, either $s, t \in [t_k^n, t_k^n - T2^{-n-1}]$ or $s, t \in [t_k^n + T2^{-n-1}, t_k^n]$. In the latter case,

$$
\mathbf{x}_{s,t}^n \overset{\text{def}}{=} (-\mathbf{x}_s^n) \boxplus \mathbf{x}_t^n = \begin{bmatrix} T^{-1}2^{n+1}(t-s)(x_{t_{k+1}^n}^1 - x_{t_k^n}^2) \\ T^{-1}2^{n+1}(t-s)(x_{t_{k+1}^n}^1 - x_{t_k^n}^2) \\ 0 \end{bmatrix}
$$

and then $|\mathbf{x}_{s,t}^n| \leqslant \|\mathbf{x}\|_\alpha |t-s|^\alpha$. In the former case, setting $\Delta_n t = T2^{-n}$,

$$
\mathbf{x}_{s,t}^n = \begin{bmatrix} \sqrt{\frac{|\theta_k^n|}{\pi}} \left(\cos\left(\frac{\pi}{\Delta_{n+1}t}(t-t_k^n)\right) - \cos\left(\frac{\pi}{\Delta_{n+1}t}(s-t_k^n)\right) \right) \\ \operatorname{sgn}(\theta_k^n)\sqrt{\frac{|\theta_k^n|}{\pi}} \left(\sin\left(\frac{\pi}{\Delta_{n+1}t}(t-t_k^n)\right) - \sin\left(\frac{\pi}{\Delta_{n+1}t}(s-t_k^n)\right) \right) \\ \theta_k^n \frac{t-s}{\Delta_{n+1}t} \end{bmatrix}.
$$

Thus, for some universal constant C_1,

$$
|\mathbf{x}_{s,t}^n| \leqslant C_1 2^{n+1}\sqrt{|\theta_k^n|}\frac{t-s}{T} \leqslant 2C_1 2^{n(1-\alpha)}\|\mathbf{x}\|_\alpha\frac{t-s}{T} \leqslant C_2\|\mathbf{x}\|_\alpha(t-s)^\alpha,
$$

where C_2 depends only on C_1 and T.

Now, if $t_k^n \leqslant s \leqslant t_k^n + T2^{-n-1} \leqslant t \leqslant t_{k+1}^n$, we get by combining the previous estimates that

$$
\begin{aligned}
|\mathbf{x}_{s,t}^n| &\leqslant C_0 C_2\|\mathbf{x}\|_\alpha((t-T2^{-n-1})^\alpha + (T2^{-n-1}-s)^\alpha) \\
&\leqslant 2^{\alpha-1}C_0 C_2\|\mathbf{x}\|_\alpha(t-s)^\alpha.
\end{aligned}
$$

We have then proved (22) with a constant which is in addition proportional to $\|\mathbf{x}\|_\alpha$.

Let us come back to the Remark 6 following Lemma 8. For $\alpha \in (1/3, 1/2]$, consider $\mathbf{x}_t = (0, 0, \varphi_t)$ where $\varphi \in C^{2\alpha}([0,T]; \mathbb{R})$, then one can find $x^n \in C_p^1([0,T]; \mathbb{R})$ such that x^n converges uniformly to 0, $\mathbf{x}^n = (x^n, \mathfrak{A}(x^n; 0, \cdot))$ is uniformly bounded in $C^\alpha([0,T]; A(\mathbb{R}^2))$ and converges in $C^\beta([0,T]; A(\mathbb{R}^2))$ to $\mathbf{x}$ for any $\beta < \alpha$. For this, one may simply consider (see Figure 8)

$$
z_t^n = \frac{1}{n\sqrt{\pi}}(\cos(2\pi t n^2) - 1, \sin(2\pi t n^2)),
$$

and then set $x_t^n = z_{\varphi_t}^n$.

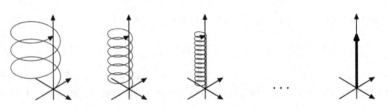

Fig. 8. Moving freely in the third direction.

Thus, moving freely in the "third direction" is equivalent to accumulating areas of small loops. Using the language of differential geometry, which we develop below, this new degree of freedom comes from the lack of commutativity of $(A(\mathbb{R}^2), \boxplus)$: a small loop of radius $\sqrt{\varepsilon}$ around the origin in the plane $\mathbb{R}^2$ is equivalent in some sense to a small displacement of length ε in the third direction. To rephrase Remark 6, even if $\varphi \in C^1([0,T];\mathbb{R})$, one has to see $\mathbf{x}$ as a path in $C^{1/2}([0,T];A(\mathbb{R}^2))$ that may be approximated by paths in $C_p^1([0,T];A(\mathbb{R}^2))$ (here, Lipschitz continuous paths with values in $A(\mathbb{R}^2)$) which converge to $\mathbf{x}$ only in $\|\cdot\|_\beta$ for any $\beta < 1/2$. Hence, we recover the problem underlined in Section 3.2.

5.7 Construction of the Integral

If $\mathbf{x} \in C^\alpha([0,T];A(\mathbb{R}^2))$ with $\alpha > 1/2$, then from Lemma 8, $\mathbf{x} = (x, x_0^3 + \mathfrak{A}(x))$ with $x = (x^1, x^2)$. For a differential form $f \in \mathrm{Lip}(\gamma;\mathbb{R}^2 \to \mathbb{R})$ with $\gamma > 1/\alpha - 1$, we set $\mathfrak{I}(\mathbf{x}) \overset{\text{def}}{=} \mathfrak{I}(x) = \int_{x_{|[0,\cdot]}} f$ which is well defined as a Young integral.

The next proposition will be refined later.

Proposition 3. *Let $\mathbf{x} \in C^\alpha([0,T];A(\mathbb{R}^2))$ with $\alpha \in (1/3, 1/2]$ and f be a differential form in $\mathrm{Lip}(\gamma;\mathbb{R}^2 \to \mathbb{R})$ with $\gamma > 1/\alpha - 1$. Let $(\mathbf{x}^n)_{n \in \mathbb{N}}$ be constructed by (21a)–(21c). Then $(\mathfrak{I}(x^n))_{n \in \mathbb{N}}$ has a unique limit in $(C^\alpha([0,T];\mathbb{R}), \|\cdot\|_\beta)$ for all $\beta < \alpha$, which we denote by $\mathfrak{I}(\mathbf{x})$ (of course, the limit does not depend on β). Both the α-Hölder continuity modulus of $\mathfrak{I}(\mathbf{x})$ and the rate of convergence with respect to $\|\cdot\|_\beta$ depend only on T, α, γ, β, $\|\mathbf{x}\|_\alpha$ and $\|f\|_{\mathrm{Lip}}$.*

Other properties of this map $x \mapsto \mathfrak{I}(\mathbf{x})$ will be proved below. Indeed, this map is obviously an extension of the one we have constructed beforehand on $L^{\bowtie,\alpha}([0,T];\mathbb{R}^3)$, with a more convenient way to encode the loops.

Proof. Fix a dyadic level n. Remark first that for $k \in \{0, \dots, 2^n - 1\}$, $t_k^n \leqslant s < t \leqslant t_{k+1}^n$,

$$\mathfrak{I}(x^n; s, t)$$

$$= \begin{cases} \iint_{\mathrm{Part}^n(s,t)} [f,f](z^1, z^2)\, dz^1\, dz^2 + \int_{x_s^n x_t^n} f \\ \qquad \text{if } t_k^n \leqslant s \leqslant t \leqslant t_k^n + T2^{-n-1}, \\[2mm] \iint_{\mathrm{Part}^n(s,t_k^n + T2^{-n-1})} [f,f](z^1, z^2)\, dz^1\, dz^2 + \int_{x_s^n x_{t_k^n + T2^{-n-1}}^n} f \\ \qquad + \int_{t_k^n}^{t_k^n + 2(t - t_k^n - 2^{-n+1}T)} f(x_r^{\Pi^n})\, dx_r^{\Pi^n} \text{ if } t_k^n \leqslant s \leqslant t_k^n + T2^{-n-1} \leqslant t \leqslant t_{k+1}^n, \\[2mm] \int_{t_k^n + 2(s - t_k^n - 2^{-n+1}T)}^{t_k^n + 2(t - t_k^n - 2^{-n+1}T)} f(x_r^{\Pi^n})\, dx_r^{\Pi^n} \text{ if } t_k^n + T2^{-n-1} \leqslant s < t \leqslant t_{k+1}^n, \end{cases}$$

where $\mathrm{Part}^n(s,t)$ stands for the portion of the disk enclosed between the loop $x^n_{|[t^n_k,t^n_k+T2^{-n-1}]}$ and the segment $\overline{x^n_s x^n_t}$. Of course, the integral of f over $\mathrm{Part}^n(t^n_k, t^n_k + T2^{-n-1})$ is the integral of $[f,f]$ over the surface of the loop $x^n_{|[t^n_k,t^n_k+T2^{-n-1}]}$.

If $t^n_k \leqslant s < t \leqslant t^n_k + T2^{-n-1}$, then the algebraic area of $\mathrm{Part}^n(s,t)$ is $\theta^n_k(t-s)2^{n+1}/T$. In addition, the maximal distance between two points in $\mathrm{Part}^n(s,t)$ is smaller than $\sqrt{|\theta^n_k|}\sqrt{2(t-s)2^{n+1}/T}$. As $[f,f]$ is $(\gamma-1)$-Hölder continuous, we deduce that for $r \in [s,t]$, there exists a constant C that depends only on T such that

$$\left| \iint_{\mathrm{Part}^n(s,t)} [f,f](z^1,z^2)\,\mathrm{d}z^1\,\mathrm{d}z^2 - [f,f](x_s)\theta^n_k \frac{t-s}{T2^{-n-1}} \right|$$
$$\leqslant C\|f\|_{\mathrm{Lip}}\|\mathbf{x}\|^{1+\gamma}_\alpha (t-s)^{\alpha(1+\gamma)} \quad (23)$$

since $|\theta^n_k| \leqslant \|\mathbf{x}\|^2_\alpha 2^{-2n\alpha}$. We also deduce that for some constant C' that depends only on T, $\|\mathbf{x}\|_\alpha$ and $\|f\|_{\mathrm{Lip}}$,

$$\left| \iint_{\mathrm{Part}^n(s,t)} [f,f](z^1,z^2)\,\mathrm{d}z^1\,\mathrm{d}z^2 \right| \leqslant C'(t-s)^{2\alpha}. \quad (24)$$

In addition, since from Proposition 2, x^n is α-Hölder continuous with some constant that depends only on $\|\mathbf{x}\|_\alpha$, there exists some constant C'' such that

$$\left| \int_{\overline{x^n_s x^n_t}} f \right| \leqslant \|f\|_\infty C''(t-s)^\alpha. \quad (25)$$

If $t^n_k + T2^{-n-1} \leqslant s < t \leqslant t^n_{k+1}$, then

$$\left| \int_{t^n_k+2(s-t^n_k-2^{n+1}T)}^{t^n_k+2(t-t^n_k-2^{n+1}T)} f(x^{\Pi^n}_r)\,\mathrm{d}x^{\Pi^n}_r \right|$$
$$\leqslant \|f\|_{\mathrm{Lip}}\|\mathbf{x}\|_\alpha (T2^n)^{1-\alpha}(t-s) \leqslant \|f\|_{\mathrm{Lip}}\|\mathbf{x}\|_\alpha(t-s)^\alpha. \quad (26)$$

It follows from (24), (25) and (26) that for some constant C_1 that depends only on $\|f\|_{\mathrm{Lip}}$ and $\|\mathbf{x}\|_\alpha$,

$$|\mathfrak{I}(x^n;s,t)| \leqslant C_1(t-s)^\alpha \quad (27)$$

for all $t^n_k \leqslant s \leqslant t \leqslant t^n_{k+1}$, $k=0,\ldots,2^n-1$.

Yet this is not sufficient to bound $|\mathfrak{I}(x^n;s,t)|$ by $C(t-s)^\alpha$ for all $0 \leqslant s < t \leqslant T$. We then use another computation.

First remark that $t^{n+1}_{2k} = t^n_k$, $t^{n+1}_{2k+2} = t^n_{k+1}$ and that

$$\mathfrak{I}(x^{\Pi^{n+1}};t^{n+1}_{2k},t^{n+1}_{2k+1}) + \mathfrak{I}(x^{\Pi^{n+1}};t^{n+1}_{2k+1},t^{n+1}_{2k+2}) - \mathfrak{I}(x^{\Pi^n};t^{n+1}_{2k},t^{n+1}_{2k+2})$$
$$= \iint_{T^n_k} [f,f](z)\,\mathrm{d}z,$$

where $T_k^n = \text{Triangle}\,(x_{t_{2k}^n}, x_{t_{2k+1}^n}, x_{t_{2k+2}^n})$ with area

$$\text{Area}(T_k^n) = -\frac{1}{2}(x_{t_{2k+1}^{n+1}} - x_{t_{2k}^{n+1}}) \wedge (x_{t_{2k+2}^{n+1}} - x_{t_{2k+1}^{n+1}}).$$

In addition,

$$\Im(x^n; t_k^n, t_k^n + T2^{-n-1}) = \iint_{\text{Part}^n(t_k^n, t_k^n + T2^{-n-1})} [f, f](z^1, z^2)\, \mathrm{d}z^1\, \mathrm{d}z^2$$
$$= [f, f](x_{t_k^n})\theta_k^n + \zeta_k^n,$$

where, from (23), $|\zeta_k^n| \leqslant C_2 2^{-n\alpha(1+\gamma)}$ for some constant C_2 that depends only on $\|\mathbf{x}\|_\alpha$, $\|f\|_{\text{Lip}}$ and T.

Recall that from (12),

$$\theta_{2k}^{n+1} + \theta_{2k+1}^{n+1} + \frac{1}{2}(x_{t_{2k+1}^{n+1}} - x_{t_{2k}^{n+1}}) \wedge (x_{t_{2k+2}^{n+1}} - x_{t_{2k+1}^{n+1}}) = \theta_k^n.$$

Hence, we easily get

$$\Im(x^{n+1}; t_{2k}^{n+1}, t_{2k+1}^{n+1}) + \Im(x^{n+1}; t_{2k+1}^{n+1}, t_{2k+2}^{n+1}) - \Im(x^n; t_{2k}^{n+1}, t_{2k+2}^{n+1})$$
$$= \zeta_{2k}^{n+1} + \zeta_{2k+1}^{n+1} - \zeta_k^n + ([f, f](x_{t_{2k+1}^{n+1}}) - [f, f](x_{t_{2k}^{n+1}}))\theta_{2k+1}^{n+1} + \xi_k^n,$$

where

$$\xi_k^n = \iint_{T_k^n} [f, f](z^1, z^2)\, \mathrm{d}z^1\, \mathrm{d}z^2 - [f, f](x_{t_{2k+1}^n})\,\text{Area}(T_k^n).$$

As in (23),

$$|\xi_k^n| \leqslant \|f\|_{\text{Lip}} \|\mathbf{x}\|_\alpha^{1+\gamma} \Delta_n t^{\alpha(\gamma+1)},$$

where $\Delta_n t = T2^{-n}$. Thus, for some constant C_3 that depends only on $\|f\|_{\text{Lip}}$ and $\|\mathbf{x}\|_\alpha$,

$$|\Im(x^{n+1}; t_{2k}^{n+1}, t_{2k+1}^{n+1}) + \Im(x^{n+1}; t_{2k+1}^{n+1}, t_{2k+2}^{n+1}) - \Im(x^n; t_{2k}^{n+1}, t_{2k+2}^{n+1})|$$
$$\leqslant C_3 2^{-n\alpha(\gamma+1)}. \qquad (28)$$

For $m \leqslant n$ and $k \in \{0, \ldots, 2^m - 1\}$,

$$\Im(x^n; t_k^m, t_{k+1}^m) - \Im(x^m; t_k^m, t_{k+1}^m) = \sum_{\ell=m}^{n-1} (\Im(x^{\ell+1}; t_k^m, t_{k+1}^m) - \Im(x^\ell; t_k^m, t_{k+1}^m)).$$

As there are exactly $2^{\ell-m}$ dyadics intervals of the form $[t_i^\ell, t_{i+1}^\ell]$ contained in $[t_k^m, t_{k+1}^m]$ for all $\ell \geqslant m$, we deduce from the Chasles relation and (28) that

$$|\Im(x^n; t_k^m, t_{k+1}^m) - \Im(x^m; t_k^m, t_{k+1}^m)| \leqslant C_3 \sum_{\ell=m}^{n-1} \frac{2^{\ell-m}}{2^{\ell\alpha(\gamma+1)}} \leqslant \frac{C_4}{2^{m\alpha(\gamma+1)}}, \qquad (29)$$

where C_4 depends on C_3 and on the choice of α and γ (note that our choice of α and γ ensures that the involved series converges as $n \to \infty$).

We now choose for $m(0)$ the smallest integer such that there exists some $k \in \{0, \ldots, 2^{m(0)} - 1\}$ for which $[t_k^{m(0)}, t_{k+1}^{m(0)}] \subset [t_{\overline{M}(s,n)}^n, t_{\underline{M}(t,n)}^n]$, where $\overline{M}(s,n)$ (resp. $\underline{M}(t,n)$) is the smallest (resp. the largest) integer such that $s \leqslant t_{\overline{M}(s,n)}^n$ (resp. $t \geqslant t_{\underline{M}(t,n)}^n$).

From the Chasles relation,

$$\mathfrak{I}(x^n; t_{\overline{M}(s,n)}^n, t_{\underline{M}(t,n)}^n)$$
$$= \mathfrak{I}(x^n; t_{\overline{M}(s,n)}^n, t_k^{m(0)}) + \mathfrak{I}(x^n; t_k^{m(0)}, t_{k+1}^{m(0)}) + \mathfrak{I}(x^n; t_{k+1}^{m(0)}, t_{\underline{M}(t,n)}^n).$$

By combining (27) and (29), we get $|\mathfrak{I}(x^n; t_k^{m(0)}, t_{k+1}^{m(0)})| \leqslant C_5 2^{-m(0)\alpha}$ for some constant C_5 that depends only on T, α, γ, $\|f\|_{\mathrm{Lip}}$ and $\|\mathbf{x}\|_\alpha$.

We may now find some integers $m(1)$ and $k(1)$ such that $[t_{k(1)}^{m(1)}, t_{k(1)+1}^{m(1)}]$ is the biggest interval of this type contained in $[t_{\overline{M}(s,n)}^n, t_k^{m(0)}]$, in order to estimate $\mathfrak{I}(x^n; t_{\overline{M}(s,n)}^n, t_k^{m(0)})$. Similarly, we can find some integers $m'(1)$ and $k'(1)$ such that $[t_{k'(1)}^{m'(1)}, t_{k'(1)+1}^{m'(1)}]$ is the biggest interval of this type contained in $[t_{k+1}^{m(0)}, t_{\underline{M}(t,n)}^n]$, in order to estimate $\mathfrak{I}(x^n; t_{k+1}^{m(0)}, t_{\underline{M}(t,n)}^n)$. Note that necessarily, $m(1)$ and $m'(1)$ are strictly greater than $m(0)$.

Hence, proceeding recursively, we obtain with (23) and (29) that

$$|\mathfrak{I}(x^n; t_{\overline{M}(s,n)}^n, t_{\underline{M}(t,n)}^n)| \leqslant \frac{C_5}{2^{m(0)\alpha}} + \sum_{j \in J} \frac{C_5}{2^{m(j)\alpha}} + \sum_{j \in J'} \frac{C_5}{2^{m'(j)\alpha}},$$

where $(m(j))_{j \in J}$ and $(m'(j))_{j \in J'}$ are two finite increasing families of integers, that are bounded by n and greater than $m(0)$. This kind of computation is the core of the proof of the Kolmogorov Lemma (see for example Corollary of Theorem 4.5 in [40]) and is also an important tool in the theory of rough paths. It also is close to the one used in [21].

For some constant C_6, we then obtain that

$$|\mathfrak{I}(x^n; t_{\overline{M}(s,n)}^n, t_{\underline{M}(t,n)}^n)| \leqslant \frac{C_6}{2^{m(0)\alpha}}.$$

Note that $T2^{-m(0)} \leqslant t_{\underline{M}(t,n)}^n - t_{\overline{M}(s,n)}^n < T2^{-m(0)+1}$. With (27) and the Chasles relation, we then obtain that

$$|\mathfrak{I}(x^n; s, t)| \leqslant C_1(t_{\overline{M}(s,n)}^n - s)^\alpha + \frac{C_6}{2^{m\alpha}} + C_1(t - t_{\underline{M}(t,n)}^n)^\alpha$$
$$\leqslant \max\{C_1, C_6/T^\alpha\}(t - s)^\alpha. \quad (30)$$

Since $\mathfrak{I}(x^n; 0) = 0$, this proves that $\mathfrak{I}(x^n; s, t)$ is uniformly bounded in $(C^\alpha([0, T]; \mathbb{R}), \|\cdot\|_\alpha)$. It follows that there exists a convergent subsequence in $(C^\alpha([0, T]; \mathbb{R}), \|\cdot\|_\beta)$, whose limit is denoted by $\mathfrak{I}(x)$, which is also a α-Hölder continuous function.

We may however give more information on the limit. With (29) and (30), for some constant C_7 and any integers $0 \leqslant m \leqslant n$ and any $0 \leqslant s \leqslant t \leqslant T$ with $t - s > T2^{-m}$,

$$|\mathfrak{I}(x^n; s, t) - \mathfrak{I}(x^m, s, t)| \leqslant C_7(t^m_{\overline{M}(s,m)} - s)^\alpha + C_7(t - t^m_{\underline{M}(t,m)})^\alpha$$
$$+ \frac{C_4(\underline{M}(t, m) - \overline{M}(s, m))}{2^{m\alpha(\gamma+1)}}.$$

As $\underline{M}(t, m) - \overline{M}(s, m) \leqslant 2^m$ and $\varepsilon = \alpha(\gamma + 1) - 1 > 0$, it follows that

$$|\mathfrak{I}(x^n; s, t) - \mathfrak{I}(x^m, s, t)|$$
$$\leqslant C_7(t^m_{\overline{M}(s,m)} - s)^\alpha + C_7(t - t^m_{\underline{M}(t,m)})^\alpha + \frac{C_4}{2^{m\varepsilon}}. \quad (31)$$

Set

$$R_m(s, t; \alpha, \varepsilon) = \max \left\{ C_7(t^m_{\overline{M}(s,m)} - s)^\alpha, C_7(t - t^m_{\underline{M}(t,m)})^\alpha, \frac{C_4}{2^{m\varepsilon}} \right\}.$$

As $R_m(s, t; \alpha, \varepsilon)$ converges to 0 when $m \to \infty$, the sequence $(\mathfrak{I}(x^n; s, t))_{n \in \mathbb{N}}$ is a Cauchy sequence for any $0 \leqslant s \leqslant t \leqslant T$, an so has a unique limit. Necessarily, this limit is $\mathfrak{I}(\mathbf{x}; s, t)$. Besides, we get from (31) that for some constant C_8 and any $\beta < \min\{\alpha, \varepsilon\}$,

$$|\mathfrak{I}(x; s, t) - \mathfrak{I}(x^m, s, t)| \leqslant C_8(t - s)^\beta R_m(s; t, \alpha - \beta, \varepsilon - \beta),$$

when m is large enough so that $T2^{-m} < t - s$. If $T2^{-m} > t - s$, then there is at most one point t^m_k such that $s \leqslant t^m_k \leqslant t$ and then for some constant C_9,

$$|\mathfrak{I}(x; s, t) - \mathfrak{I}(x^m, s, t)| \leqslant |\mathfrak{I}(x; s, t)| + |\mathfrak{I}(x^m, s, t)|$$
$$\leqslant C_9(t - s)^\alpha \leqslant \frac{C_9 T^{\alpha \wedge \varepsilon - \beta}}{2^{-m(\alpha \wedge \varepsilon - \beta)}}(t - s)^\beta.$$

We get that the whole sequence $(\mathfrak{I}(x^n))_{n \in \mathbb{N}}$ converges to $\mathfrak{I}(x)$ in the space $(C^\alpha([0, T]; \mathbb{R}), \| \cdot \|_\beta)$ for any $\beta < \alpha \wedge \varepsilon$. Since $(\mathfrak{I}(x^n))_{n \in \mathbb{N}}$ is bounded in $C^\alpha([0, T]; \mathbb{R})$ and since $C^\varepsilon([0, T]; \mathbb{R})$ is contained in $C^\alpha([0, T]; \mathbb{R})$ for $\varepsilon < \alpha$, $(\mathfrak{I}(x^n))_{n \in \mathbb{N}}$ converges to $\mathfrak{I}(x)$ in the space $(C^\alpha([0, T]; \mathbb{R}), \| \cdot \|_\beta)$ for any $\beta < \alpha$.

The proposition is thus proved.

Corollary 4. *Let $(\mathbf{x}^n)_{n \in \mathbb{N}}$ be a sequence of paths converging to $\mathbf{x}$ in the space $(C^\alpha([0, T]; \mathrm{A}(\mathbb{R}^2)), \| \cdot \|_\alpha)$. Then for all $\beta < \alpha$, $\mathfrak{I}(\mathbf{x}^n; 0, \cdot)$ converges to $\mathfrak{I}(\mathbf{x}; 0, \cdot)$ in $(C^\alpha([0, T]; \mathbb{R}), \| \cdot \|_\beta)$.*

Proof. The proof follows the same line as the proof of Proposition 1.

To simplify the notation, denote $\mathbf{x}$ by $\mathbf{x}^\infty$.

Since $\mathbf{x}^n$ is convergent in $C^\alpha([0, T]; \mathrm{A}(\mathbb{R}^2))$, the sequence $(\|\mathbf{x}^n\|_\alpha)_{n \in \mathbb{N}}$ is bounded and then, from Proposition 3, $(\mathfrak{I}(\mathbf{x}^n))_{n \in \mathbb{N}}$ is bounded in the space $(C^\alpha([0, T]; \mathbb{R}), \| \cdot \|_\alpha)$.

For $n \in \mathbb{N} \cup \{\infty\}$, let $(\mathbf{x}^{n,m})_{m \in \mathbb{N}}$ be the sequence of paths converging to $\mathbf{x}^n$ given by Proposition 2. We have seen in Proposition 3 for for any $\beta < \alpha$, there exists some constant K^n that depends on $\|\mathbf{x}^n\|_\alpha$ such that $\|\mathfrak{I}(\mathbf{x}^{n,m}) - \mathfrak{I}(\mathbf{x}^n)\|_\beta \leqslant K^n 2^{m(\beta-\alpha)}$. In addition, the sequence $(K^n)_{n \in \mathbb{N}}$ is bounded if $(\|\mathbf{x}^n\|_\alpha)_{n \in \mathbb{N}}$ is bounded. As $\mathfrak{I}(\mathbf{x}^{n,m})$ is a Young integral, it follows from Corollary 2 that $\mathfrak{I}(\mathbf{x}^{n,m})$ converges to $\mathfrak{I}(\mathbf{x}^{\infty,m})$ in $(C^\alpha([0,T];\mathbb{R}), \|\cdot\|_\beta)$. Hence, this is sufficient to prove that $\mathfrak{I}(\mathbf{x}^n)$ converges to $\mathfrak{I}(\mathbf{x})$ in $(C^\alpha([0,T];\mathbb{R}), \|\cdot\|_\beta)$, as in the proof of Proposition 1.

Remark 8. Consider the following equivalence relation $\sim$ between two sequences $(x^n)_{n \in \mathbb{N}}$ and $(y^n)_{n \in \mathbb{N}}$ of paths converging in $(C^\alpha([0,T];\mathbb{R}^2), \|\cdot\|_\beta)$ with $\alpha > 1/2$ and $\beta \in (1/3,1]$: $(x^n)_{n \in \mathbb{N}} \sim (y^n)_{n \in \mathbb{N}}$ if $\mathbf{x} \stackrel{\text{def}}{=} \lim_{n \in \mathbb{N}} \mathfrak{C}(x^n, 0) = \lim_{n \in \mathbb{N}} \mathfrak{C}(y^n, 0)$ in $(C^\gamma([0,T];\mathrm{A}(\mathbb{R}^2), \|\cdot\|_\beta)$ for some $\gamma > \beta$. This implies that $\mathfrak{I}(x^n; s, t)$ and $\mathfrak{I}(y^n; s, t)$ converge to the same limit $\mathfrak{I}(\mathbf{x}; s, t)$. Hence, it is possible to identify $C^\gamma([0,T];\mathrm{A}(\mathbb{R}^2), \|\cdot\|_\gamma)$ with the quotient space $(C^\alpha([0,T];\mathbb{R}^2), \|\cdot\|_\beta)^{\mathbb{N}}/\sim$, and two elements in the same equivalence class give rise to the same integral.

Here, we have used dyadics partitions[2], so one may ask whether $\mathfrak{I}(\mathbf{x}; s, t)$ is equal to $\mathfrak{I}(\mathbf{x}_{|[s,t]})$. As this is true for ordinary integrals, we easily get the following result.

Lemma 11. *Let* $\mathbf{x}$ *in* $C^\alpha([0,T];\mathrm{A}(\mathbb{R}^2))$. *Then, for all* $0 \leqslant s \leqslant t \leqslant T$, $\mathfrak{I}(\mathbf{x}; s, t) = \mathfrak{I}(\mathbf{x}_{|[s,t]})$.

From this lemma, we deduce that if $\mathbf{x} \in C^\alpha([0,T];\mathrm{A}(\mathbb{R}^2))$ and $\mathbf{y} \in C^\alpha([0,S];\mathrm{A}(\mathbb{R}^2))$, then

$$\mathfrak{I}(\mathbf{x} \square \mathbf{y}; 0, t) = \begin{cases} \mathfrak{I}(\mathbf{x}; 0, t) & \text{if } t \in [0,T], \\ \mathfrak{I}(\mathbf{x}; 0, T) + \mathfrak{I}(\mathbf{y}; 0, t - T) & \text{if } t \in [S,T]. \end{cases}$$

Proof. This lemma means that the integral constructed using the dyadics on $[0,T]$ but restricted to $[s,t]$ corresponds to the integral constructed using the dyadics on $[s,t]$. One knows that such a relation holds for ordinary integrals, since the integral does not depend on the choice of the family of partitions on which approximations of the integrals are defined.

Let $(\mathbf{x}^n)_{n \in \mathbb{N}}$ be the approximation of $\mathbf{x}$ given by Proposition 2. Then $\mathfrak{I}(\mathbf{x}^n)$ is an ordinary integral. Hence $\mathfrak{I}(\mathbf{x}^n; s, t) = \mathfrak{I}(\mathbf{x}^n_{|[s,t]}; 0, t - s)$ (the last integral means that T is replaced by $t - s$ and thus that we consider the dyadic partitions of $[0, t - s]$. The result follows from passing to the limit.

Let us end this section with an important remark. Consider $\mathbf{x}$ in the space $C^\alpha([0,T];\mathrm{A}(\mathbb{R}^2))$ with $\alpha \in (1/3, 1/2)$ and φ in $C^{2\alpha}([0,T];\mathbb{R})$. We saw in Lemma 9 that $\mathbf{y} = (\mathbf{x}^1, \mathbf{x}^2, \mathbf{x}^3 + \varphi)$ also belongs to $C^\alpha([0,T];\mathrm{A}(\mathbb{R}^2))$.

---

[2] We will give below another construction of $\mathfrak{I}$ for which a family of partitions different from the dyadics ones can be used.

Hence, we set $y_t^n = x^{\Pi^n} \bowtie \varPhi^n \bowtie \varPsi^n(t/3)$ for $t \in [0, 3T]$ where $\varPsi^n = \{z_k^n\}_{k=0,\dots,2^n-1}$ with $z_k^n : [t_k^n, t_{k+1}^n] \to \mathbb{R}^2$ defined by

$$z_k^n(t) = \frac{\varphi_{t_{k+1}^n} - \varphi_{t_k^n}}{\sqrt{\pi}} \begin{bmatrix} \cos\left(2\pi \frac{t-t_k^n}{t_{k+1}^n - t_k^n}\right) - 1 \\ \sin\left(2\pi \frac{t-t_k^n}{t_{k+1}^n - t_k^n}\right) \end{bmatrix},$$

so that φ asymptotically encodes the area of $(\varPsi^n)_{n\in\mathbb{N}}$.

Similarly as in Section 3.2, it is then easily shown that

$$\mathfrak{I}(y^n; 0, t) \xrightarrow[n\to\infty]{} \mathfrak{I}(\mathbf{y}; 0, t) = \mathfrak{I}(\mathbf{x}; 0, t) + \int_0^t [f, f](x_s) \, d\varphi_s.$$

Hence, adding a path φ to the third component of $\mathbf{x}$ amounts to adding a term $\int_0^t [f, f](x_s) \, d\varphi_s$ to $\mathfrak{I}(\mathbf{x})$.

5.8 A Sub-Riemannian Point of View

Our definition of $\mathfrak{I}$ consists in approximating a path $\mathbf{x} \in C^\alpha([0, T]; A(\mathbb{R}^2))$ by a family of paths $(\mathbf{x}^n)_{n\in\mathbb{N}}$ in $C^1([0, T]; A(\mathbb{R}^2))$ such that $\mathfrak{I}(\mathbf{x}^n)$ converges with respect to the β-Hölder norm in $C^\alpha([0, T]; \mathbb{R})$ as $n \to \infty$ for all $\beta < \alpha$. The integral $\mathfrak{I}(\mathbf{x})$ is then defined as the limit if $\mathfrak{I}(\mathbf{x}^n)$. In addition, necessarily, it follows from Lemma 8 that $\mathbf{x}^n = (x^{1,n}, x^{2,n}, x_0^3 + \mathfrak{A}(x^n))$, where x^n is a family of functions in $C_p^1([0, T]; \mathbb{R}^2)$.

The paths $\mathbf{x}^n$ were constructed by replacing $\mathbf{x}_{|[t_k^n, t_{k+1}^n]}$ with some paths obtained by combining loops and segments. Of course, other choices are possible, and a natural one consists in using geodesics.

Let a be a point in $A(\mathbb{R}^2)$. How to find a path $\mathbf{x} : [0, 1] \to A(\mathbb{R}^2)$ with $\mathbf{x}_0 = 0$, $\mathbf{x}_1 = a$ and whose length (or whose energy) is minimal? Of course, one can use the segment $\mathbf{y} = (ta^1, ta^2, ta^3)_{t\in[0,1]}$ that goes from 0 to a, which is the natural geodesic in $\mathbb{R}^3$. But $\mathfrak{A}(\mathbf{y}^1, \mathbf{y}^2; t) = 0$ and thus $\mathbf{y}$ is not of type $(y, \mathfrak{A}(y))$ and does not belong to $C^1([0, T]; A(\mathbb{R}^2))$. We will use this point of view in Section 7.2, and this will help us to bridge our construction with another one of Riemann sum type. So, we may reformulate our question by imposing the condition that $\mathbf{y}$ is of type $\mathbf{y} = (y, \mathfrak{A}(y))$, which means that $y_t^3 = \mathfrak{A}(\mathbf{y}^1, \mathbf{y}^2; 0, t)$ for $t \in [0, 1]$. This kind of problem is related to sub-Riemannian geometry: see [4, 5, 35, 60] for example.

The notion of length we use is then the length of the path $(\mathbf{y}^1, \mathbf{y}^2)$:

$$\text{Length}(\mathbf{y}) = \int_0^1 \sqrt{(\dot{\mathbf{y}}_s^1)^2 + (\dot{\mathbf{y}}_s^2)^2} \, ds.$$

Such a path—which will be characterized from the differentiable point of view in the next section—, is called *horizontal*. It is then possible to introduce a distance between two points of $A(\mathbb{R}^2)$ by

$$d(a, b) = \inf_{\substack{\mathbf{y}:[0,1]\to A(\mathbb{R}^2) \text{ horizontal} \\ \mathbf{y}_0 = a, \ \mathbf{y}_1 = b}} \text{Length}(\mathbf{y}),$$

which is called the *Carnot-Carathéodory* distance. We may then define $\|x\|_{\text{CC}} = d(0,x)$, which becomes a homogeneous sub-additive norm on $A(\mathbb{R}^2)$ (see Section A) i.e., $\|x\|_{\text{CC}} = 0$ if and only $x = 0$ and for all $x, y \in A(\mathbb{R}^2)$ and $\lambda \in \mathbb{R}$, $\|\delta_\lambda x\|_{\text{CC}} = |\lambda| \cdot \|x\|_{\text{CC}}$, $\|x^{-1}\|_{\text{CC}} = \|x\|_{\text{CC}}$ and $\|x \boxplus y\|_{\text{CC}} \leqslant \|x\|_{\text{CC}} + \|y\|_{\text{CC}}$, which is the *sub-additive property*.

For any $a \in A(\mathbb{R}^2)$, we succeeded in Section 5.3 in constructing a path that goes from 0 to a, so that $\|a\|_{\text{CC}}$ is finite. Of course, $d(a,b) = \|a^{-1} \boxplus b\|_{\text{CC}}$ for all $a, b \in A(\mathbb{R}^2)$. If $a^3 = 0$, then the shortest horizontal path from 0 to a is the segment going from 0 to a. If $a = (0, 0, a^3)$ with $a^3 \neq 0$, this problem is equivalent to the isoperimetric problem, whose solution is known to be the circle.

In the general case, this problem is called the *Dido problem*, and the solutions are known to be arcs of circle (see for example [60, 66]), but they are less practical to use than our construction with circles and loops (see below in the proof of Proposition 4).

These solutions are not real geodesics in $A(\mathbb{R}^2)$, but they are called *sub-Riemannian geodesics*. The sub-Riemannian geodesic that links a to b is then denoted by $\psi_{a,b}$ and belongs to $C^1([0, T]; A(\mathbb{R}^2))$.

If we define the energy of a path by $\text{Energy}(\mathbf{y}) = \frac{1}{2} \int_0^1 ((\dot{\mathbf{y}}_s^1)^2 + (\dot{\mathbf{y}}_s^2)^2) \, ds$, then $\psi_{a,b}$ is also energy minimizing among all paths with constant speed $\text{Length}(\psi_{a,b})$.

To a path $\mathbf{x}$ in $C^\alpha([0, T]; A(\mathbb{R}^2))$, we associate

$$\mathbf{x}_t^n = \psi_{\mathbf{x}_{t_k^n}, \mathbf{x}_{t_{k+1}^n}} \left(\frac{t - t_k^n}{t_{k+1}^n - t_k^n} \right) \text{ for } t \in [t_k^n, t_{k+1}^n], \tag{32}$$

for $n = 0, 1, 2, \ldots$.

Proposition 4. *The sequence of paths $(\mathbf{x}^n)_{n \in \mathbb{N}}$ constructed by (32) is a family of paths in $C^1([0, T]; A(\mathbb{R}^2))$ which converges to $\mathbf{x}$ in $C^\alpha([0, T]; A(\mathbb{R}^2))$ with respect to $\|\cdot\|_\beta$ for any $\beta < \alpha$.*

Proof. The proof is similar to the one of Corollary 1 or of Proposition 2.

Obviously, $(\mathbf{x}^n)_{n \in \mathbb{N}}$ converges uniformly to $\mathbf{x}$. Remark that $\mathbf{x}_{s,t}^n = \mathbf{x}_{s,t_k^n}^n \boxplus \mathbf{x}_{t_k^n, t_{k+1}^n}^n \boxplus \mathbf{x}_{t_{k+1}^n, t}^n$ and that $\mathbf{x}_{t_k^n, t_{k+1}^n}^n = \mathbf{x}_{t_k^n, t_{k+1}^n}$. Using the same argument as in Corollary 1, the α-Hölder norm of $\mathbf{x}^n$ is then deduced from estimates on $\mathbf{x}_{s, t_k^n}^n$ and $\mathbf{x}_{t_{k+1}^n, t}^n$ for $t \in [t_k^n, t_{k+1}^n]$ for $k = 0, \ldots, 2^n - 1$.

After a translation, we would like to establish an estimate of type $|\psi_{0,x}(t)| \leqslant Ct|x|$ for $t \in [0,1]$ for some constant C. If this holds, then for $t \in [t_k^n, t_{k+1}^n]$,

$$|\psi_{\mathbf{x}_{t_k^n}, \mathbf{x}_{t_{k+1}^n}}(t/\Delta_n t)| \leqslant C \frac{t}{\Delta_n t} |\mathbf{x}_{t_k^n, t_{k+1}^n}| \leqslant C \frac{t \Delta_n t^\alpha}{\Delta_n t} \|\mathbf{x}\|_\alpha \leqslant Ct^\alpha \|\mathbf{x}\|_\alpha.$$

We now give two proofs: one is done "by hand", and the second one uses the properties of the Carnot-Carathéodory distance.

○ If $x^3 = 0$, then $\psi_{0,x}(t)$ is a segment and for $t \in [0,1]$,

$$|\psi_{0,x}(t)| \leqslant |x|t,$$

which gives the desired result.

Now, if $x^3 \neq 0$, observe first that for some constants $a \neq 0$ and $r, \varphi \in [0, 2\pi)$,

$$\begin{cases} \psi^1_{0,x}(t) = a(\cos(rt + \varphi) - \cos(\varphi)), \\ \psi^2_{0,x}(t) = a(\sin(rt + \varphi) - \sin(\varphi)), \\ \psi^3_{0,x}(t) = a^2 rt \end{cases}$$

since the minimizers lie above arcs of circles. Hence, $a^2 r = x^3$ and

$$(x^1)^2 + (x^2)^2 = \psi^1_{0,x}(1)^2 + \psi^2_{0,x}(1)^2 = 2a^2(1 - \cos(r)).$$

It is easily seen that one may find a and r so as to satisfy $\psi_{0,x}(1) = x$.

If $r \in [\pi/2, 3\pi/2]$, then $1 \leqslant 1 - \cos(r) \leqslant 2$, $a^2 \leqslant \max\{|x^1|^2, |x^2|^2\}$ and

$$\max\{|\psi^1_{0,x}(t)|, |\psi^2_{0,x}(t)|\} \leqslant \sqrt{2}\pi t \max\{|x^1|, |x^2|\},$$

and $|\psi^3_{0,t}(t)| \leqslant 4\pi^{-1}t \max\{|x^1|, |x^2|\}^2$. This is sufficient to conclude.

In the other case, since cos and sin are Lipschitz continuous and $|a^2 r| \leqslant |x^2|$, we get

$$\psi^1_{0,x}(t)^2 + \psi^2_{0,x}(t)^2 = 2a^2(1 - \cos(rt)) \leqslant 2|x^3|t \leqslant 2|x|^2 t.$$

Hence, $|\psi_{0,x}(t)| \leqslant \sqrt{2}|x|t$.

It follows that $(\mathbf{x}^n)_{n\in\mathbb{N}}$ is bounded in $C^\alpha([0,T]; A(\mathbb{R}^2))$ and this is sufficient to conclude.

○ (Alternative proof). As the Carnot-Carathéodory norm is equivalent to any homogeneous norm (see Proposition 10 in Section A), it follows that for some universal constants C and C',

$$\forall t \in [0,1], \ |\psi_{0,x}(t)| \leqslant C\|\psi_{0,x}(t)\|_{CC} = Ct\|x\|_{CC} \leqslant CC't|x|, \tag{33}$$

since $\psi_{0,x}(t)$ is a sub-Riemannian geodesic and then $\|\psi_{0,x}(t)\|_{CC} = td(0,x)$. The inequalities (33) yields the result.

The point of view of the sub-Riemannian geometry, which is natural in the context of Heisenberg groups, has been used by P. Friz and N. Victoir in [29] and [32].

5.9 A Sub-Riemannian Point of View: Differentiable Paths in $A(\mathbb{R}^2)$

We have introduced the set of paths $C^\alpha([0,T]; A(\mathbb{R}^2))$ for $\alpha \in [1/2, 1/3)$, but we have that the value of α does not really refer to the regularity of the path $\mathbf{x}$

in such a set, but to the norms to be used to approximate $\mathbf{x}$ by a family of paths x^n that are naturally lifted as $\mathbf{x}^n = (x^n, \mathfrak{A}(x^n))$. It is then possible to consider paths $\mathbf{x} \in C^\alpha([0, T]; A(\mathbb{R}^2))$ with $\alpha < 1/2$ that are differentiable: for example, if $\mathbf{x}$ in $C^1([0, T]; A(\mathbb{R}^2))$ and φ in $C^1([0, T]; \mathbb{R})$, then $\mathbf{y}_t = (\mathbf{x}_t^1, \mathbf{x}_t^2, \mathbf{x}_t^3 + \varphi_t)$ is almost everywhere differentiable, in the sense that

$$i = 1, 2, 3, \qquad \lim_{\varepsilon \to 0} \frac{\mathbf{y}_{t+\varepsilon}^i - \mathbf{y}_t^i}{\varepsilon} = \alpha^i(t) \tag{34}$$

exists for almost every t. Another natural way of thinking the derivative of $\mathbf{y}$ consists in setting

$$i = 1, 2, 3, \qquad \lim_{\varepsilon \to 0} \frac{1}{\varepsilon}(-\mathbf{y}_t) \boxplus \mathbf{y}_{t+\varepsilon}^i = \beta^i(t) \tag{35}$$

when this limit exists. If $t \in [0, T]$ is such that (34) holds, then $\beta^i(t)$ exists and

$$\beta(t) = \alpha(t) - \frac{1}{2}[\mathbf{y}_t, \alpha(t)].$$

Conversely, if (35) holds, then (34) also holds and

$$\alpha(t) = \beta(t) + \frac{1}{2}[\mathbf{y}_t, \beta(t)].$$

Of course, $(\alpha^1(t), \alpha^2(t)) = (\beta^1(t), \beta^2(t))$ for all t at which $\mathbf{y}_t$ is differentiable. If the path $\mathbf{y}$ is of type $(y, \mathfrak{A}(y))$, then

$$\alpha^1(t) = \frac{dy_t^1}{dt}, \quad \alpha^2(t) = \frac{dy_t^2}{dt} \text{ and } \alpha^3(t) = \frac{1}{2} y_t \wedge \frac{dy_t}{dt} = \frac{1}{2} y_t \wedge \begin{bmatrix} \alpha^1(t) \\ \alpha^2(t) \end{bmatrix}.$$

To each point a of $A(\mathbb{R}^2)$, we associate the 2-dimensional vector space

$$\Theta(a) = \left\{ (v^1, v^2, v^3) \in \mathbb{R}^3 \,\middle|\, v^3 = \frac{1}{2} \begin{bmatrix} a^1 \\ a^2 \end{bmatrix} \wedge \begin{bmatrix} v^1 \\ v^2 \end{bmatrix} \right\}$$

as well as the space $\Xi(a)$ orthogonal to $\Theta(a)$ with respect to the usual scalar product in $\mathbb{R}^3$. The one-dimensional space $\Xi(a)$ is generated by the vector $(-a^2/2, a^1/2, 1)^{\mathrm{T}}$. It is easily seen that $a \mapsto (a, \Xi(a))$ and $a \mapsto (a, \Theta(a))$ form two sub-bundles of the tangent bundle of $A(\mathbb{R}^2)$.

We then obtain the next result.

Lemma 12. *A differentiable curve $\mathbf{y}$ is the natural lift $(y, \mathfrak{A}(y))$ of a differentiable curve y if and only if $\dot{\mathbf{y}}_t$ belongs to $\Theta(\mathbf{y}_t)$ for each $t \in [0, T]$.*

For a differentiable path $\mathbf{y} : [0, T] \to A(\mathbb{R}^2)$, let $\beta(t)$ be given by (35). The condition that $\dot{\mathbf{y}}_t$ belongs to $\Theta(\mathbf{y}_t)$ is equivalent to $\beta(t) = (\dot{y}_t^1, \dot{y}_t^2, 0)$. More generally, if $\pi_{\Xi(a)}$ is the projection from $\mathbb{R}^3$ identified with the tangent space of $A(\mathbb{R}^2)$ at a onto $\Xi(a)$, then for $t \in [0, T]$,

$$\beta(t) = (\dot{\mathbf{y}}_t^1, \dot{\mathbf{y}}_t^2, \pi_{\Xi(\mathbf{y}_t)}(\dot{\mathbf{y}}_t)).$$

Thus, a differentiable path $\mathbf{y}$ from $[0, T]$ to $(A(\mathbb{R}^2), \boxplus)$ is necessarily of type $(y, \mathfrak{A}(y) + \varphi)$ where $y = (\mathbf{y}^1, \mathbf{y}^2)$ and φ is differentiable, and $\beta(t) = (\dot{y}_t^1, \dot{y}_t^2, \dot{\varphi}_t)$ for $t \in [0, T]$.

We will see in Section 6.12 how to interpret this condition.

6 Geometric and Algebraic Structures

6.1 Motivations

Up to now, we have introduced a space $A(\mathbb{R}^2)$ and considered paths in $C^\alpha([0, T]; A(\mathbb{R}^2))$. For a path $\mathbf{x} \in C^\alpha([0, T]; A(\mathbb{R}^2))$, we have seen how to construct a sequence $(\mathbf{x}^n)_{n \in \mathbb{N}}$ of paths converging to $C^\beta([0, T]; A(\mathbb{R}^2))$ with $\beta < \alpha$ such that $x^n = (\mathbf{x}^{1,n}, \mathbf{x}^{2,n})$ is piecewise smooth and $\mathbf{x}^{3,n} = \mathbf{x}_0^3 + \mathfrak{A}(x^n)$. As $\mathbf{x}^n$ lies above a piecewise smooth path x^n, $\mathfrak{I}(\mathbf{x}^n)$ is well defined as a Young integral, and we have shown in Proposition 3 that the sequence $(\mathfrak{I}(\mathbf{x}^n))_{n \in \mathbb{N}}$ converges and its limit defines $\mathfrak{I}(\mathbf{x})$.

On the other hand, we may rewrite

$$\mathfrak{I}(\mathbf{x}^n; 0, T) = \sum_{k=0}^{2^n-1} \mathfrak{I}(\mathbf{x}_{|[t_k^n, t_{k+1}^n]}^n) \quad \text{and} \quad \mathfrak{I}(\mathbf{x}; 0, T) = \sum_{k=0}^{2^n-1} \mathfrak{I}(\mathbf{x}_{|[t_k^n, t_{k+1}^n]}).$$

The path $\mathbf{x}_{|[t_k^n, t_{k+1}^n]}^n$ was constructed in Section 5.3 from the values of $\mathbf{x}_{t_{k+1}^n}$ and $\mathbf{x}_{t_k^n}$. Hence, $\int_{t_k^n}^{t_{k+1}^n} f(x_s^n) \, dx_s^n$ is an approximation of $\mathfrak{I}(\mathbf{x}_{|[t_k^n, t_{k+1}^n]})$, and $\mathfrak{I}(\mathbf{x}^n)$ is constructed from the values of $\{\mathbf{x}_{t_k^n}\}_{k=0,\dots,2^n-1}$ only.

We have proposed two constructions of integrals that rely on path approximation. We are now looking for a Riemann sum like expression, which consists in finding approximations of $\mathfrak{I}(\mathbf{x}; 0, T)$ and summing them over the dyadic partitions of $[0, T]$.

First note that if $\mathbf{x}$ belongs to $C^\alpha([0, T]; A(\mathbb{R}^2))$ with $\alpha > 1/2$ and x^{Π^n} is the piecewise linear approximation of x along the dyadic partition Π^n, then

$$|\mathfrak{I}(x^{\Pi^n}; t_k^n, t_{k+1}^n) - \mathfrak{I}(\mathbf{x}; t_k^n, t_{k+1}^n)| \leqslant \|f\|_{\text{Lip}} |\mathfrak{A}(x; t_k^n, t_{k+1}^n)| \leqslant \frac{T^{2\alpha} \|f\|_{\text{Lip}} \|x\|_\alpha^2}{2^{2n\alpha}}$$

and thus, since $\alpha > 1/2$,

$$\mathfrak{I}(\mathbf{x}; 0, T) = \lim_{n \to \infty} \sum_{k=0}^{2^n-1} \int_{t_k^n}^{t_{k+1}^n} f(x_s^{\Pi^n}) \frac{x_{t_{k+1}^n} - x_{t_k^n}}{t_{k+1}^n - t_k^n} \, ds$$

$$= \lim_{n \to \infty} \sum_{k=0}^{2^n-1} \int_{t_k^n}^{t_{k+1}^n} f(x_s^{\Pi^n}) \frac{dx_s^{\Pi^n}}{ds} \, ds \qquad (36)$$

where x is the path above which $\mathbf{x}$ lies.

The first idea is then to find a formulation similar to (36), by looking for another way of drawing a piecewise differentiable path $\mathbf{y}^n$ lying above a path $y^n : [0, T] \to \mathbb{R}^2$ with $y^n(t_k^n) = x_{t_k^n}$ for $k = 0, \dots, 2^n$ and for which the expression $\xi_k^n = \int_{t_k^n}^{t_{k+1}^n} f(y_s^n) \frac{dy_s^n}{ds} \, ds$ provides a good approximation of $\mathfrak{I}(\mathbf{x}; t_k^n, t_{k+1}^n)$, in the sense that for some $\theta > 1$ and $C > 0$,

$$\left| \xi_k^n - \mathfrak{I}(\mathbf{x}; t_k^n, t_{k+1}^n) \right| \leqslant \frac{C}{2^{n\theta}}.$$

The space in which $\mathbf{y}$ lives has to be specified, but it is natural to assume that $\frac{dy_k^n(s)}{ds}$ belongs to $A(\mathbb{R}^2)$, and one then has to accordingly extend the definition of f into a differential form on $A(\mathbb{R}^2)$.

The second idea is then to get an expression of type $\sum_{k=0}^{2^n-1} f(x_{t_k^n}) \Delta_k^n \mathbf{x}$ where $\Delta_k^n \mathbf{x}$ depends only on $\mathbf{x}_{t_{k+1}^n}$ and $\mathbf{x}_{t_k^n}$. As we deal with second-order calculus, things are not that simple: think of the difference between Stratonovich and Itô integrals for the Brownian motion.

6.2 Another Formulation for the Integral

We rewrite $\mathfrak{I}(\mathbf{x}^n; t_k^n, t_{k+1}^n)$ as

$$\mathfrak{I}(\mathbf{x}^n; t_k^n, t_{k+1}^n) = \int_{t_k^n}^{t_{k+1}^n} f(x_s^{\Pi^n}) \, dx_s^{\Pi^n} + \iint_{\mathrm{Surface}(y_k^n)} [f, f](z) \, dz$$

where y_k^n has been defined by (21b). Setting $\mathbf{x}_{s,t} = (-\mathbf{x}_s) \boxplus \mathbf{x}_t$ and $\Delta_n t = T2^{-n}$, we have already seen that

$$\left| \iint_{\mathrm{Surface}(y_k^n)} [f, f](z) \, dz - \mathbf{x}_{t_k^n, t_{k+1}^n}^3 [f, f](x_{t_k^n}) \right| \leqslant \Delta_n t^{\alpha(1+\gamma)} \|f\|_{\mathrm{Lip}} \|\mathbf{x}\|_\alpha^{1+\gamma}.$$

On the other hand,

$$\left| \mathbf{x}_{t_k^n, t_{k+1}^n}^3 [f, f](x_{t_k^n}) - \mathbf{x}_{t_k^n, t_{k+1}^n}^3 \int_{t_k^n}^{t_{k+1}^n} [f, f](x_s^{\Pi^n}) \frac{ds}{\Delta_n t} \right| \leqslant \Delta_n t^{\alpha(1+\gamma)} \|f\|_{\mathrm{Lip}} \|\mathbf{x}\|_\alpha^{1+\gamma}.$$

Hence, this means that one can replace $\mathfrak{I}(\mathbf{x}^n; t_k^n, t_{k+1}^n)$ by

$$\xi_k^n = \mathbf{x}_{t_k^n, t_{k+1}^n}^1 \int_{t_k^n}^{t_{k+1}^n} f_1(x_s^{\Pi^n}) \frac{ds}{\Delta_n t}$$

$$+ \mathbf{x}_{t_k^n, t_{k+1}^n}^2 \int_{t_k^n}^{t_{k+1}^n} f_2(x_s^{\Pi^n}) \frac{ds}{\Delta_n t} + \mathbf{x}_{t_k^n, t_{k+1}^n}^3 \int_{t_k^n}^{t_{k+1}^n} [f, f](x_s^{\Pi^n}) \frac{ds}{\Delta_n t},$$

in the sense that $\mathfrak{I}(\mathbf{x}; 0, T) = \lim_{n \to \infty} \sum_{k=0}^{2^n-1} \xi_k^n$.

Call $\{e_1, e_2, [e_1, e_2]\}$ the canonical basis of $A(\mathbb{R}^2)$, and $\{e^1, e^2, [e^1, e^2]\}$ its dual basis. For $z = (z^1, z^2, z^3) \in A(\mathbb{R}^2)$, define the differential form

$$\mathfrak{E}_{A(\mathbb{R}^2)}(f)(z) = f_1(z^1, z^2)e^1 + f_2(z^1, z^2)e^2 + [f, f](z^1, z^2)[e^1, e^2]. \qquad (37)$$

With $\mathbf{x}^{\Pi^n} = (x^{\Pi^n}, 0)$, the term ξ_k^n may be put in a more synthetic form

$$\xi_k^n = \int_{t_k^n}^{t_{k+1}^n} \mathfrak{E}_{A(\mathbb{R}^2)}(f)(\mathbf{x}_s^{\Pi^n}) \mathbf{x}_{t_k^n, t_{k+1}^n} \frac{ds}{\Delta_n t}.$$

Remark 9. We have to note the following point: using the same technique as in Corollary 1, one can show that for $\mathbf{x} \in C^\alpha([0, T]; A(\mathbb{R}^2))$, the path $\mathbf{x}^n$ defined by

$$\mathbf{x}_t^n = \mathbf{x}_{t_k^n} \boxplus \delta_{(t-t_k^n)/(t_{k+1}^n - t_k^n)}((-\mathbf{x}_{t_k^n}) \boxplus \mathbf{x}_{t_{k+1}^n}) \text{ for } t \in [t_k^n, t_{k+1}^n]$$

converges to $\mathbf{x}$ in $(C^\alpha([0, T]; A(\mathbb{R}^2)), \| \cdot \|_\beta)$ for any $\beta < \alpha$ when the mesh of the partition $\{t_k^n\}_{k=0,\dots,n}$ converges to 0. Here, δ. is the dilation operator introduced in (14). We have then that $\mathfrak{I}(\mathbf{x}^n)$ converges to $\mathfrak{I}(\mathbf{x})$ in $(C^\alpha([0, T]; A(\mathbb{R}^2)), \| \cdot \|_\beta)$ for any $\beta < \alpha$ if $\alpha \in (1/3, 1]$.

Here, we consider the piecewise linear approximation

$$\widehat{\mathbf{x}}_t^n = \mathbf{x}_{t_k^n} \boxplus \frac{t - t_k^n}{t_{k+1}^n - t_k^n}((-\mathbf{x}_{t_k^n}) \boxplus \mathbf{x}_{t_{k+1}^n}) \text{ for } t \in [t_k^n, t_{k+1}^n]$$

which is a piecewise smooth path with values in $A(\mathbb{R}^2)$. If $\alpha > 1/2$, we may show that $(\widehat{\mathbf{x}}^n)_{n \in \mathbb{N}}$ is bounded in $C^\beta([0, T]; A(\mathbb{R}^2))$ with $\beta = 2\alpha - 1$. We do not know whether or not $\widehat{\mathbf{x}}^n$ is bounded in $C^\beta([0, T]; A(\mathbb{R}^2))$ when $\alpha < 1/2$ for $\beta < \alpha$. However, we may define $\mathfrak{I}(\mathbf{x})$ using $(\widehat{\mathbf{x}}^n)_{n \in \mathbb{N}}$ by changing the definition of the integral.

The important point is the following: as we primarily want to focus on the increments of the paths, we leave the world of sub-Riemannian geometry, where paths in $A(\mathbb{R}^2)$ are basically seen as 2-dimensional paths with a constraint on their areas. We are now willing to deal with paths that are seen directly as paths with values in $A(\mathbb{R}^2)$ (or other spaces that will be introduced later).

We are now looking for a curve $\mathbf{y}^n(t)$ on $[0, T]$ which is piecewise differentiable and such that

$$\frac{d\mathbf{y}^n(t)}{dt} = \frac{1}{\Delta_n t} \mathbf{x}_{t_k^n, t_{k+1}^n}, \ t \in (t_k^n, t_{k+1}^n). \qquad (38)$$

Of course, from (38), such a path lies above x^{Π^n}. The problem is now to find the space in which $\mathbf{y}^n$ lives.

Recall the results from Section 5.4: The space $(A(\mathbb{R}^2))$ is a non-commutative group when equipped with $\boxplus$, and it is also a *Lie algebra* when equipped with the brackets $[\cdot, \cdot]$.

We have already denoted the basis of $A(\mathbb{R}^2)$ by $\{e_1, e_2, [e_1, e_2]\}$. The choice of $[e_1, e_2]$ to denote the third component naturally follows from the bilinearity of $[\cdot, \cdot]$.

The Lie algebra structure is particularly important here, since one knows that $A(\mathbb{R}^2)$ may be identified with the tangent space at any point of a Lie group. We will now construct such a Lie group.

6.3 Matrix Groups

We give here a very brief presentation of matrix groups. This part can also serve as a presentation of Lie groups, for which matrix groups are a prototype with the advantage of having an explicit coordinate system. For a more detailed insight, there are many books (see specifically [3, 68] or some books on Lie groups as [17]).

Consider a matrix group M, that is, a subset of $d \times d$-matrices such that for $p, q \in$ M, $p \times q$ also belongs to M and p^{-1} belongs to M, and which is closed. This matrix group can be equipped with the topology induced by the set $M_d(\mathbb{R})$ of $d \times d$-matrices.

A general result is that a matrix group forms a *smooth manifold* [68, Theorem 7.17, p. 106], which means that around each point p of M, there exists an open set $U(p)$ in $\mathbb{R}^m$ (for some fixed m) and an open neighbourhood V_p of p in $M_d(\mathbb{R})$ (see as $\mathbb{R}^{d^2}$) such that there exists a map Φ_p which is a homeomorphism from U_p to $V_p \cap$ M. In addition, we require that for two points p and q of M, $V_p \cap V_q \neq \emptyset$, $\Phi_p \circ \Phi_q^{-1}$ and $\Phi_q \circ \Phi_p^{-1}$ are smooth on their domain of definition. In other word, one can describe locally M using a smooth one-to-one map from an open set of $\mathbb{R}^m$ (indeed, the dimension m does not depend on the points around which the neighbourhood is considered) to M.

Example 1. Basic examples of Lie group are given by the sets of invertible matrices, of orthogonal matrices, ...

Example 2. A particular example for us is the *Heisenberg group* H, which is the set of matrices

$$\mathrm{H} = \left\{ \begin{bmatrix} 1 & a & c \\ 0 & 1 & b \\ 0 & 0 & 1 \end{bmatrix} \;\middle|\; a, b, c \in \mathbb{R} \right\}.$$

which is easily seen to be stable under matrix multiplication.

The Heisenberg group has been widely studied, and appears in sub-Riemannian geometry, quantum physics, ... (see for example [4, 25, 60]).

For a given point p in M, we can consider a smooth path γ from $(-\varepsilon, \varepsilon)$ to M $\subset M_d(\mathbb{R})$ for some $\varepsilon > 0$ and with $\gamma(0) = p$. As $\gamma(t) = [\gamma_{i,j}(t)]_{i,j=1,\dots,d}$, we may consider its derivative $\gamma'(t) = [\gamma'_{i,j}(t)]_{i,j=1,\dots,d}$.

As γ moves only on M, $\gamma'(t)$ can only belong to a subspace of $M_d(\mathbb{R})$ at each time. Denote by T_pM the subset of $M_d(\mathbb{R})$ given by all the derivatives of the possible curves γ as above. This is the *tangent space*, which is obviously a vector space.

Example 3. For the Heisenberg group, it is easily computed that the tangent space T_pM at each point $p \in H$ is

$$T_pH = \left\{ \begin{bmatrix} 0 & a & c \\ 0 & 0 & b \\ 0 & 0 & 0 \end{bmatrix} \middle| a, b, c \in \mathbb{R} \right\}$$

Consider now a map φ from a matrix group M to a matrix group M'. Let p a point of M and set $p' = \varphi(p')$. Given two neighbourhood V_p and $V'_{p'}$ of p and p' in M' and the associated maps Φ_p and $\Phi'_{p'}$ defined on open subset of $\mathbb{R}^m$ and $\mathbb{R}^{m'}$, we assume that $(\Phi'_{p'})^{-1} \circ \varphi \circ \Phi_p$ is smooth. We may then define the *differential* $d_p\varphi$ of φ at p as the linear map from T_pM to $T_{\varphi(p)}M'$ given by

$$d_p\varphi(v) = \left. \frac{d\varphi \circ \gamma'}{dt} \right|_{t=0}$$

where $\gamma : (-\varepsilon, \varepsilon) \to M$ is any smooth path such that $\gamma(0) = p$ and $\gamma'(0) = v$ for $v \in T_pM$.

Remark 10. The advantage with matrix groups is that $M_d(\mathbb{R})$ gives a global systems of coordinates for M and for each tangent space. However, as usual in differential geometry, even if we may identify T_pM with T_qM, they are really different spaces.

Two particular smooth maps are the following: for a given p in M, set

$$R_p(q) = q \times p \text{ and } L_p(q) = p \times q$$

for all $q \in M$.

The differentials of $R_p : T_qM \to T_{q \times p}M$ and $L_p : T_qM \to T_{p \times q}M$ are easily computed:

$$d_q R_p(v) = v \times p \text{ and } d_q R_p(v) = p \times v \text{ for any } q \in M, \ v \in T_qM.$$

In particular, this implies that the left or right multiplication of an element of T_qM by an element of M gives an element in some tangent space of M.

Using for p the inverse q^{-1} of $q \in M$, we deduce that the tangent space T_qM at any q is in bijection with the tangent space $T_{Id}M$ at the identity matrix Id (which necessarily belongs to M). Hence, the dimension of T_qM does not depend on q, and the dimension of $T_{Id}M$ is then called the *dimension of the matrix group* M.

Denote by TM the set $\cup_{p \in M} T_pM$, and call it the *tangent bundle of* M. This set has itself a manifold structure. A smooth *vector field* is an application that associates with any point p of M a tangent vector X_p in T_pM and such that the dependence is smooth (the precise definition uses local coordinates, as above). An *integral curve* along X is a smooth path $\gamma : [0, T] \to M$ such that $\gamma'(t) = X_{\gamma(t)}$.

Given two matrix groups M and M' with a smooth map φ between them and two vectors fields X and X' on M and M', we say that X and X' are *related* if $X'_\varphi(p)$ is equal to $d_p\varphi(X_p)$ at any point p of M. In particular, this means that if γ is an integral curve of X, then $\varphi \circ \gamma$ is an integral curve of X'.

A *left-invariant vector field* is a vector field X such that $d_q L_p(X_q) = X_{L_p q}$. For a matrix group, this means that $p \times X_q = X_{p \times q}$. Using $q = \mathrm{Id}$, the value of a left-invariant vector field X may be deduced from the value of X at Id, that is, from a vector in $T_{\mathrm{Id}}M$.

Let γ be the integral curve of a left-invariant vector field X, with $\gamma(0) = p$ (and then $\gamma'(0) = X_p = p \times X_{\mathrm{Id}}$). We obtain that

$$\gamma'(t) = X_{\gamma(t)} = \gamma(t) \times X_{\mathrm{Id}} = \gamma(t) \times p^{-1} \times X_p.$$

When $p = \mathrm{Id}$ and $X_{\mathrm{Id}} = v$, we deduce that $\gamma'(t) = \gamma(t) \times v$ which we know how to solve:

$$\gamma(t) = \exp(tv) \text{ for } t \geqslant 0,$$

where exp is the matrix exponential:

$$\exp(v) = \mathrm{Id} + \sum_{k \geqslant 1} \frac{1}{k!} v^k.$$

As $\exp(-v)$ is the inverse of $\exp(v)$, one can extend γ to $\mathbb{R}$. In addition, we also easily obtain that $\gamma(t + s) = \gamma(t) \times \gamma(s)$, so that $\gamma : \mathbb{R} \to$ M is a group homomorphism.

Proposition 5 (See for example [17, Proposition 1.3.4, p. 19]). *There exist some open neighbourhood U of 0 in $T_{\mathrm{Id}}M$ and some neighbourhood V of Id in M such that the application exp is a C^1 diffeomorphism between U and V.*

Example 4. For the Heisenberg group H, we have that $P^3 = 0$ for $P \in T_{\mathrm{Id}}H$ (which means that H is a step 2 nilpotent group) and then

$$\exp(P) = \mathrm{Id} + P + \frac{1}{2}P^2.$$

In addition, for $Q \in H$, $P = \mathrm{Id} - Q \in T_{\mathrm{Id}}H$ and one can define

$$\log(\mathrm{Id} + P) = P - \frac{1}{2}P^2.$$

Here, both $\exp : T_{\mathrm{Id}}H \to H$ and $\log : H \to T_{\mathrm{Id}}H$ are one-to-one map that are inverse to each other, and exp is a global C^1 diffeomorphism.

More generally, the inverse of the exponential is also denoted by log, and as it maps a neighbourhood of V_{Id} of M containing V_{Id} to the vector space $T_{\mathrm{Id}}M$, this gives a local system of coordinates $\Psi_{\mathrm{Id}} : V_{\mathrm{Id}} \to \mathbb{R}^m$ (where m is the dimension of the matrix group) by $\Psi_{\mathrm{Id}} = i \circ \log$, where $i : T_{\mathrm{Id}}M \mapsto \mathbb{R}^m$ is the map which naturally identifies $T_{\mathrm{Id}}M$ with $\mathbb{R}^m$. This function $\Phi : V \to \mathbb{R}^m$ is called the *normal chart* or the *logarithmic chart*.

We then deduce a local system of coordinates in a neighbourhood V of a point p of M by $\Phi_p : V_p \to \mathbb{R}^m$ with $\Phi_p(x) = i(\log(p^{-1} \otimes x))$ for $x \in V_p$.

Another map from M to M of interest is the *adjoint* defined by

$$\mathrm{Ad}(p)(q) = p \times q \times p^{-1} \text{ for } p, q \in \mathrm{M}.$$

Of course, the interest of this map comes from the fact that in general, M is not an Abelian group and then that $p \times q \neq q \times p$. It can be turned into a map from $T_{\mathrm{Id}}M$ to $T_{\mathrm{Id}}M$, still denoted by $\mathrm{Ad}(p)$, by setting $\mathrm{Ad}(p)(q) = p \times q \times p^{-1}$ for $q \in T_{\mathrm{Id}}M$. This new map $\mathrm{Ad}(p)$ is simply the differential at Id of $\mathrm{Ad}(p)$.

Given some smooth path $\gamma : (-\varepsilon, \varepsilon) \to \mathrm{M}$ with $\gamma(0) = \mathrm{Id}$ and $\gamma'(0) = p \in T_{\mathrm{Id}}M$,

$$\mathrm{ad}(p)(q) \stackrel{\mathrm{def}}{=} \left. \frac{\mathrm{d}\,\mathrm{Ad}(\gamma(t))(q)}{\mathrm{d}t} \right|_{t=0} = p \times q - q \times p.$$

For two matrices $p, q \in \mathrm{M}_d(\mathbb{R})$, denote by $[p, q]$ their bracket—called their *Lie bracket*—$[p, q] = p \times q - q \times p$. Hence, $\mathrm{ad}(p)(q) = [p, v]$, and we see that from the definition of ad, $[p, q]$ belongs to $T_{\mathrm{Id}}M$ when $p, q \in T_{\mathrm{Id}}M$.

The space $(T_{\mathrm{Id}}M, [\cdot, \cdot])$ then has a Lie algebra structure.

The Lie brackets are useful for the following property: let p and q in $T_{\mathrm{Id}}M$, and let t be small enough. Then

$$\exp(tp) \times \exp(tq)$$

$$= \exp\left(tp + tq + \frac{t^2}{2}[p, q] + \frac{t^3}{12}[p, [p, q]] + \frac{t^3}{12}[q, [q, p]] + \cdots \right). \quad (39)$$

This is the Dynkin formula (also called the Baker-Campbell-Hausdorff formula), for which the complete (infinite) expansion may be given in terms of Lie brackets (See for example [17, § 1.7, p. 29]).

If we identify an element p of the tangent space $T_{\mathrm{Id}}M$ with the flow $t \mapsto \exp(tp)$ is generates, a geometric interpretation of the Lie bracket follows from (39), as for ε small enough,

$$\exp(\varepsilon p) \times \exp(\varepsilon q) \times \exp(-\varepsilon p) \times \exp(-\varepsilon q) = \exp(\varepsilon^2[p, q] + \mathrm{o}(\varepsilon^2)) ,$$

which means that if we follow the flow $t \mapsto \exp(tp)$ in direction of p up to a time ε, then the flow $t \mapsto \exp(q)$ in the direction of q before coming back in the direction of $-p$ and then of $-q$, always up to a time ε, we arrive close to a point given by the value of the flow $t \mapsto \exp(t[p, q])$ at time ε^2.

Example 5. For the Heisenberg group H, we easily obtain that the product of two matrices P and Q in $T_{\mathrm{Id}}H$ is of type

$$PQ = \begin{bmatrix} 0 & 0 & c \\ 0 & 0 & 0 \\ 0 & 0 & 0 \end{bmatrix} \text{ for some } c \in \mathbb{R}$$

and then that the product of the matrices P, Q and R in $T_{\mathrm{Id}}H$ is equal to 0. Then formula (39) becomes an exact formula

$$\exp(P) \times \exp(Q) = \exp\left(P + Q + \frac{1}{2}[P, Q]\right)$$

and is true whatever the norms of P and Q.

We now consider an element $x = (a, b, c) \in \mathrm{A}(\mathbb{R}^2)$, and

$$\Phi(x) = \begin{bmatrix} 0 & a & c \\ 0 & 0 & b \\ 0 & 0 & 0 \end{bmatrix}. \tag{40}$$

Clearly, Φ is a one-to-one map between $\mathrm{A}(\mathbb{R}^d)$ and $T_{\mathrm{Id}}H$. In addition, it is easily obtained that

$$\Phi([x, y]) = [\Phi(x), \Phi(y)] \text{ for all } x, y \in \mathrm{A}(\mathbb{R}^2),$$

or in other words, that Φ is a Lie algebra isomorphism between $(\mathrm{A}(\mathbb{R}^d), [\cdot, \cdot])$ and $(T_{\mathrm{Id}}H, [\cdot, \cdot])$. With the exponential application exp, we may then identify a path $\mathbf{x}$ in $\mathrm{A}(\mathbb{R}^2)$ with a path $\mathbf{y} = \exp(\mathbf{x})$ living in the Heisenberg group. The path $\mathbf{x}$ is valued in the vector space $\mathrm{A}(\mathbb{R}^d)$ and $\mathbf{x}_t$ gives the "direction" to follow so as to reach $\mathbf{y}_t$ via integral curves of left-invariant vector fields.

6.4 Lie Groups

We have already seen that $(\mathrm{A}(\mathbb{R}^2), \boxplus)$ is a Lie group, that is, a group $(G, \times)$ such that $(x, y) \mapsto x \times y$ and $x \mapsto x^{-1}$ are continuous. Denote by 1 the neutral element of G.

Here, we consider groups $(G, \times)$ that are finite-dimensional manifolds of class C^2 and such that $(x, y) \mapsto x \times y$ and $x \mapsto x^{-1}$ are also of class C^2. Any matrix group is a Lie group.

We recall here some general results about G, which are merely a copy of the previous statements on matrix groups. For $x \in G$, denote by $T_x(G)$ the tangent space at x. A *vector field* X is a differentiable application $X : x \in G \mapsto X_x \in T_x G$.

A *left-invariant* vector field X is a vector field such that $X_{L_x(y)} = \mathrm{d}_y L_x X_y$ for all $x, y \in G$, where $L_x(y) = x \times y$. It is easily shown that for such a vector field,

$$X_x = \mathrm{d}_1 L_x X_1, \ \forall x \in G,$$

where 1 is the neutral element of the Lie group G. In other words, a left-invariant vector field is fully characterized by the tangent vector X_1 in the tangent space $T_1(G)$ at the identity of G.

An *integral curve of* X is a differentiable curve $\gamma : \mathbb{R}_+ \to G$ such that

$$\frac{\mathrm{d}\gamma(t)}{\mathrm{d}t} = X_{\gamma(t)}.$$

A *one-parameter subgroup of* G is a differentiable curve $\gamma : \mathbb{R} \to G$ such that $\gamma(t+s) = \gamma(t) \times \gamma(s)$ for all $s, t \in \mathbb{R}$ (note that $\gamma(-t) = \gamma(t)^{-1}$ for all $t \in \mathbb{R}$). This implies in particular that $\gamma(0) = 1$. If γ is an integral curve of a left-invariant vector field X, then γ is deduced from the tangent vector $X_1 \in T_1 G$ at the identity 1 of G. This vector X_1 is then called the *generator* of γ. Given a vector v in $T_1 G$, it is usual to denote by $(\exp(tv))_{t \in \mathbb{R}}$ the one-parameter subgroup of G generated by v.

One may define a map Ad on G such that $\mathrm{Ad}(x) : y \mapsto x \times y \times x^{-1}$. Its differential $\mathrm{Ad}'(x) \overset{\text{def}}{=} \mathrm{d}_1 \mathrm{Ad}(x)$ at 1 maps $T_1 G$ to $T_1 G$, which is linear. Hence, $x \mapsto \mathrm{Ad}'(x)$ can be seen as a map from G to $\mathrm{L}(T_1 G, T_1 G)$, the vector space of linear maps from $T_1 G$ to $T_1 G$, and its differential $\mathrm{ad}(x) \overset{\text{def}}{=} \mathrm{d}_1 \mathrm{Ad}'$ is a linear map from $T_1 G$ to $\mathrm{L}(T_1 G, T_1 G)$. Thus, for $(x, y) \in T_1 G^2 \mapsto \mathrm{ad}(x)(y)$ is a bilinear map with values in $T_1 G$, which is anti-symmetric: $\mathrm{ad}(y)(x) = -\mathrm{ad}(x)(y)$. We then define by $[x, y] \overset{\text{def}}{=} \mathrm{ad}(x)(y)$ the Lie bracket of x and y, and $(T_1 G, [\cdot, \cdot])$ is a Lie algebra. This space is called the *Lie algebra* of G.

For a matrix group, this Lie bracket correspond to the Lie bracket of matrices.

6.5 Tensor Algebra

We have introduced matrix groups, and we have seen that $(\mathrm{A}(\mathbb{R}^d), [\cdot, \cdot])$ is isomorphic to the Lie algebra $T_{\mathrm{Id}} H$ of the Heisenberg group. We will now construct a bigger space, that will contain also the Heisenberg group.

Consider now the following *tensor algebra* $\mathrm{T}(\mathbb{R}^2) = \mathbb{R} \oplus \mathbb{R}^2 \oplus (\mathbb{R}^2 \otimes \mathbb{R}^2)$ where $\mathbb{R}^2 \otimes \mathbb{R}^2$ is the tensor product of $\mathbb{R}^2$ (on this notion see for example [20]). If $\{e_1, e_2\}$ is the canonical basis of $\mathbb{R}^2$, then $\mathbb{R}^2 \otimes \mathbb{R}^2$ is the vector space of dimension 4 with basis $\{e_1 \otimes e_1, e_1 \otimes e_2, e_2 \otimes e_1, e_2 \otimes e_2\}$. For $x, y \in \mathbb{R}^2$,

$$x \otimes y = (x^1 e_1 + x^2 e_2) \otimes (y^1 e_1 + y^2 e_2) = \sum_{i,j=1,2} x^i y^j e_i \otimes e_j,$$

$$\lambda(x \otimes y) = (\lambda x) \otimes y = x \otimes (\lambda y), \quad \forall \lambda \in \mathbb{R}.$$

Any element $x \in \mathrm{T}(\mathbb{R}^2)$ may be decomposed as $x = (x^0, x^1, x^2)$ where $x^0 \in \mathbb{R}$, $x^1 \in \mathbb{R}^2$ and $x^2 \in \mathbb{R}^2 \otimes \mathbb{R}^2$. This space $\mathrm{T}(\mathbb{R}^2)$ is equipped with the term-wise addition $+$, and the multiplication $\otimes$ defined by the tensor product between two elements of $\mathbb{R}^2$ and

$$x \otimes y = xy \text{ if } x \in \mathbb{R}, \ y \in \mathrm{T}(\mathbb{R}^2),$$

$$x \otimes y \otimes z = 0 \text{ if } x, y, z \in \mathbb{R}^2.$$

The element $e_0 = 1 = (1, 0, 0)$ is the neutral element of $T(\mathbb{R}^2)$ for $\otimes$, while $0 = (0, 0, 0)$ is the neutral element of $+$. The space $(T(\mathbb{R}^2), +, \otimes)$ is an associative algebra, which is obtained by quotienting the tensor algebra $\mathbb{R} \oplus \mathbb{R}^2 \oplus \mathbb{R}^2 \otimes \mathbb{R}^2 \oplus \cdots$ by the ideal formed by all the elements which belongs to $(\mathbb{R}^2)^{\otimes 3} \oplus (\mathbb{R}^2)^{\otimes 4} \oplus \cdots$.

Remark 11. Consider the space $\mathbb{R}\langle X_1, X_2\rangle$ of polynomials with two non-commutative variables X_1 and X_2, as well as the equivalence relation $\sim$ on $\mathbb{R}\langle X_1, X_2\rangle$ defined by $P \sim Q$ if $P - Q$ is a sum of terms of total degree at least 3. Then there exists an isomorphism Φ between the associative algebras $(T(\mathbb{R}^2), +, \otimes)$ and $(\mathbb{R}\langle X_1, X_2\rangle/\sim, +, \times)$ such that $\Phi(e_i) = X_i$ for $i = 1, 2$. In other words, the elements of $T(\mathbb{R}^2)$ are manipulated as polynomials where only the terms of total degree $\leqslant 2$ are kept.

For $\xi \in \{0, 1\}$, denote by $T_\xi(\mathbb{R}^2)$ the subset of $T(\mathbb{R}^2)$ defined by

$$T_\xi(\mathbb{R}^2) = \left\{ (\xi, x^1, x^2) \,\middle|\, x^1 \in \mathbb{R}^2, x^2 \in \mathbb{R}^2 \otimes \mathbb{R}^2 \right\}.$$

Lemma 13. *The space $(T_1(\mathbb{R}^2), \otimes)$ is a non-commutative group.*

Proof. Clearly, if $x, y \in T_1(\mathbb{R}^2)$, then $x \otimes y \in T_1(\mathbb{R}^2)$. That $(T_1(\mathbb{R}^2), \otimes)$ is non-commutative follows from the very definition of $\otimes$. To show it is a group, it remains to compute the inverse of each element. If $x = (1, x^1, x^2)$, then $x^{-1} = (1, -x^1, -x^2 + x^1 \otimes x^1)$ is the inverse of x.

For $x, y \in T(\mathbb{R}^2)$, define the bracket of x and y by

$$[x, y] = x \otimes y - y \otimes x.$$

If $x = (x^0, x^1, x^2)$ and $y = (y^0, y^1, y^2)$ belong to $T(\mathbb{R}^2)$, then

$$[x, y] = [x^1, y^1] = (x^1 \wedge y^1)[e_1, e_2].$$

Note also that $[x, y] = -[y, x]$.

A natural sub-vector space of $(T_0(\mathbb{R}^2), +) \subset (T(\mathbb{R}^2), +)$ is then

$$g(\mathbb{R}^2) = \left\{ x \in T_0(\mathbb{R}^2) \,\middle|\, x = x^1 + x^a[e_1, e_2], \ x^1 \in \mathbb{R}^2, \ x^a \in \mathbb{R} \right\}.$$

Although $g(\mathbb{R}^2)$ is not stable under $\otimes$, it is stable under $[\cdot, \cdot]$: if $x = (x^1, x^a)$ and $y = (y^1, y^a)$ are in $g(\mathbb{R}^2)$, then

$$[x, y] = x^1 \wedge y^1[e_1, e_2] \in g(\mathbb{R}^2).$$

This space $g(\mathbb{R}^2)$ is of dimension 3. For $x = x^1 + x^a[e_1, e_2]$ and $y = y^1 + x^a[e_1, e_2]$, set

$$x \boxplus y = x^1 + y^1 + (x^a + y^a)[e_1, e_2] + \frac{1}{2}[x^1, y^1]$$

$$= x^1 + y^1 + \left(x^a + y^a + \frac{1}{2}x^1 \wedge y^1\right)[e_1, e_2].$$

Finally, define $i_{g(\mathbb{R}^2),A(\mathbb{R}^2)}$ by

$$i_{g(\mathbb{R}^2),A(\mathbb{R}^2)}(x) = (x^{1,1}, x^{1,2}, x^a) \text{ if } x = x^{1,1}e_1 + x^{1,2}e_2 + x^a[e_1, e_2].$$

It is clear that $i_{g(\mathbb{R}^2),A(\mathbb{R}^2)}$ is one-to-one from $g(\mathbb{R}^2)$ to $A(\mathbb{R}^2)$, and an additive group homomorphism from $(g(\mathbb{R}^2), \boxplus)$ to $(A(\mathbb{R}^2), \boxplus)$. In addition, $[i_{g(\mathbb{R}^2),A(\mathbb{R}^2)}(x), i_{g(\mathbb{R}^2),A(\mathbb{R}^2)}(y)] = i_{g(\mathbb{R}^2),A(\mathbb{R}^2)}[x, y]$ for all $x, y \in g(\mathbb{R}^2)$, which means that $i_{g(\mathbb{R}^2),A(\mathbb{R}^2)}$ is also a Lie homomorphism. Hence, we identify the spaces $g(\mathbb{R}^2)$ and $A(\mathbb{R}^2)$. Lemmas 4 and 5 are then rewritten in the following way.

Lemma 14. *The space* $(g(\mathbb{R}^2), [\cdot, \cdot])$ *is a Lie algebra, and* $(g(\mathbb{R}^2), \boxplus)$ *is a Lie group with* 0 *as neutral element.*

On $T_0(\mathbb{R}^2)$, define

$$\exp(x) = 1 + x^1 + x^2 + \frac{1}{2}x^1 \otimes x^1 \text{ for } x = (0, x^1, x^2). \tag{41}$$

This map exp is given by the first terms of the formal expansion of the exponential, since we are working in a truncated tensor algebra.

Similarly, define on $T_1(\mathbb{R}^2)$

$$\log(x) = x^1 + x^2 - \frac{1}{2}x^1 \otimes x^1 \text{ for } x = (1, x^1, x^2) \in T_1(\mathbb{R}^2).$$

It is easily seen that $\exp \circ \log$ and $\log \circ \exp$ are equal to the identity respectively on $T_1(\mathbb{R}^2)$ and on $T_0(\mathbb{R}^2)$.

If $x, y \in T_0(\mathbb{R}^2)$,

$$\exp(x) \otimes \exp(y) = 1 + x^1 + y^1 + x^2 + y^2 + \frac{1}{2}x^1 \otimes x^1 + \frac{1}{2}y^1 \otimes y^1 + x^1 \otimes y^1$$

and then

$$\log(\exp(x) \otimes \exp(y)) = x \boxplus y \tag{42}$$

with

$$x \boxplus y = x^1 + y^1 + x^2 + y^2 + \frac{1}{2}x^1 \otimes y^1 - \frac{1}{2}y^1 \otimes x^1 = x + y + \frac{1}{2}[x, y].$$

This is the truncated version of the *Baker-Campbell-Hausdorff-Dynkin formula* (see for example [37, 63]).

Lemma 15. *If* $G(\mathbb{R}^2) = \exp(g(\mathbb{R}^2))$, *then* $G(\mathbb{R}^2)$ *is a subgroup of* $(T_1(\mathbb{R}^2), \otimes)$ *and* exp *is a group isomorphism from* $(g(\mathbb{R}^2), \boxplus)$ *to* $(G(\mathbb{R}^2), \otimes)$.

Note that $\exp(-x)$ is the inverse of $\exp(x)$ in $G(\mathbb{R}^2)$, for all $x \in g(\mathbb{R}^2)$.

For a sub-vector space V of $T(\mathbb{R}^2)$, π_V denotes the projection onto V. If $V = \text{Vect}(e)$ for some $e \in T(\mathbb{R}^2)$, then denote $\pi_{\text{Vect}(e)}$ simply by π_e. For $x \in T(\mathbb{R}^2)$, set

$$\mathfrak{s}(x) = \sum_{i,j=1,2} \frac{1}{2} (\pi_{e_i \otimes e_j}(x) + \pi_{e_j \otimes e_i}(x)) e_i \otimes e_j,$$

$$\mathfrak{a}(x) = \frac{1}{2} (\pi_{e_1 \otimes e_2}(x) - \pi_{e_2 \otimes e_1}(x))[e_1, e_2].$$

If x belongs to $\mathbb{R}^2 \otimes \mathbb{R}^2$, then

$$x = \mathfrak{s}(x) + \mathfrak{a}(x), \tag{43}$$

and $\mathfrak{s}(x)$ (resp. $\mathfrak{a}(x)$) corresponds to the symmetric (resp. anti-symmetric) part of x. Finally, note that for $x \in T(\mathbb{R}^2)$,

$$\mathfrak{s}(x \otimes x) = \pi_{\mathbb{R}^2 \otimes \mathbb{R}^2}(x \otimes x). \tag{44}$$

For $z = \exp(x) \in G(\mathbb{R}^2)$, we have

$$\mathfrak{s}(z) = \frac{1}{2}\mathfrak{s}(x \otimes x) = \frac{1}{2} x \otimes x \tag{45}$$

and

$$\mathfrak{a}(z) = \pi_{[e_1,e_2]}(x)[e_1, e_2].$$

Hence, for $x \in g(\mathbb{R}^2)$, one may rewrite

$$\exp(x) = 1 + \pi_{\mathbb{R}^2}(x) + \frac{1}{2} x \otimes x + \mathfrak{a}(x) \text{ and } x = \pi_{\mathbb{R}^2}(x) + \mathfrak{a}(x). \tag{46}$$

In particular, for $z \in G(\mathbb{R}^2)$, $\mathfrak{a}(\log(z)) = \mathfrak{a}(z)$.

6.6 The Tensor Space as a Lie Group

It is possible to find a norm $|\cdot|$ on $\mathbb{R}^2 \otimes \mathbb{R}^2$ such that $|x \otimes y| \leqslant |x| \cdot |y|$ for all $x, y \in \mathbb{R}^2$ (there are indeed several possibilities [64]).

For $x = (1, x^1, x^2) \in T_1(\mathbb{R}^2)$ or for $x = (0, x^1, x^2) \in T_0(\mathbb{R}^2)$, set

$$\|x\|_\star = \max\{|x^1|, |x^2|\}$$

and

$$\|x\| = \max\left\{|x^1|, \sqrt{\frac{1}{2}|x^2|}\right\}.$$

Then $\|\cdot\|$ is a *homogeneous gauge* for the dilation operator δ_t defined by $\delta_t x = (1, tx^1, t^2 x^2)$, $t \in \mathbb{R}$, since $\|\delta_t x\| = |t| \cdot \|x\|$ (see Section A). Besides, $\|x \otimes y\| \leqslant (3/2)(\|x\| + \|y\|)$ for all $x, y \in T_1(\mathbb{R}^2)$. We have introduced in Section 5.4 a dilation operator, also denoted by δ, in a similar way. Note that for $x \in A(\mathbb{R}^2)$ and $t \in \mathbb{R}$, $\exp(\delta_t x) = \delta_t \exp(x)$.

The next lemma is easily proved.

Lemma 16. *With the norm* $\|\cdot\|_*$, *the spaces* $(\mathrm{T}_1(\mathbb{R}^2), \otimes)$ *and* $(\mathrm{G}(\mathbb{R}^2), \otimes)$ *are Lie groups, and* $\mathrm{G}(\mathbb{R}^2)$ *is a closed subgroup of* $\mathrm{T}_1(\mathbb{R}^2)$.

For $x \in \mathfrak{g}(\mathbb{R}^2)$, $t \in \mathbb{R} \mapsto \gamma_x(t) \overset{\text{def}}{=} \exp(tx) \in \mathrm{G}(\mathbb{R}^2)$ is a one-parameter subgroup of $(\mathrm{G}(\mathbb{R}^2), \otimes)$. The point x is the tangent vector to $\gamma_x(t)$ for $t = 0$:

$$\left. \frac{\mathrm{d}\gamma_x}{\mathrm{d}t} \right|_{t=0} = x.$$

Hence, $\mathfrak{g}(\mathbb{R}^2)$ may be identified with the tangent space of $\mathrm{G}(\mathbb{R}^2)$ at point 1, and in fact at any point $y \in \mathrm{G}(\mathbb{R}^2)$.

The bracket allows us to characterize the lack of commutativity of $\mathrm{G}(\mathbb{R}^2)$, as follows from the next result, which is classical in the theory of Lie groups (see Figure 9): For $x, y \in \mathfrak{g}(\mathbb{R}^2)$ and for $t \geqslant 0$, set

$$\theta_{x,y}(t) = \gamma_x(\sqrt{t}) \otimes \gamma_y(\sqrt{t}) \otimes (\gamma_x(-\sqrt{t})) \otimes \gamma_y(-\sqrt{t}).$$

Then $\theta_{x,y}(0) = 1$ and

$$\left. \frac{\mathrm{d}\theta_{x,y}}{\mathrm{d}t} \right|_{t=0} = [x, y].$$

In our case, it follows from the truncated version the Baker-Campbell-Hausdorff-Dynkin formula (42) that $\theta_{x,y}(t) = \exp(t[x, y])$ for all $t \geqslant 0$.

To any Lie group corresponds a Lie algebra, which is identified with the tangent space at the neutral element, and then at any point. Of course, $\mathfrak{g}(\mathbb{R}^2) \cong A(\mathbb{R}^2)$ has been constructed to be the tangent space of $\mathrm{G}(\mathbb{R}^2)$ at any point.

Lemma 17. *The tangent space of* $\mathrm{G}(\mathbb{R}^2)$ *at any point may be identified with* $A(\mathbb{R}^2)$, *and the tangent space of* $\mathrm{T}_1(\mathbb{R}^2)$ *at any point may be identified with* $\mathrm{T}_0(\mathbb{R}^2)$.

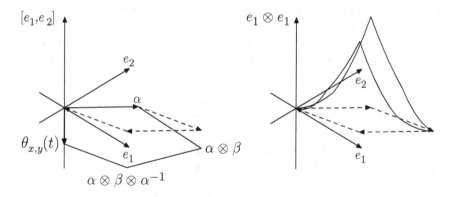

Fig. 9. Illustration of the non-commutativity with $\alpha = \gamma_x(\sqrt{t})$ and $\beta = \gamma_y(\sqrt{t})$.

Remark 12. We have seen that $(A(\mathbb{R}^d), [\cdot, \cdot])$ is isomorphic to the Lie algebra $(T_{\mathrm{Id}}H, [\cdot, \cdot])$ of the Heisenberg group.

Consider the map $\Psi : T(\mathbb{R}^2)$ to H defined by

$$\Psi(x) = \begin{bmatrix} 1 & x_1 & x_{1,2} \\ 0 & 1 & x_2 \\ 0 & 0 & 1 \end{bmatrix} \text{ for } x = x_0 e_0 + \sum_{i=1}^{2} x_i e_i + \sum_{i,j=1}^{2} x_{i,j} e_i \otimes e_j.$$

Then note that

$$\Psi(x \otimes y) = \Psi(x) \times \Psi(y) \text{ for } x, y \in T(\mathbb{R}^2),$$

so that Ψ is a group homomorphism from $(T_1(\mathbb{R}^2), \otimes)$ or $(G(\mathbb{R}^2), \otimes)$ to $(H, \times)$. As Ψ is linear, we easily get that $\Psi(\exp(x)) = \exp(\Phi(x))$, where Φ is the Lie algebra isomorphism given by (40). We then deduce that Ψ is indeed an isomorphism between $(G(\mathbb{R}^2), \otimes)$ and the Heisenberg group $(H, \times)$. The Heisenberg group is then a representation of the group $(G(\mathbb{R}^2), \otimes)$.

This section ends with a very useful lemma, whose proof is straightforward. The notion of Lipschitz functions on spaces with homogeneous gauges is similar to the notion of Lipschitz functions (See Definition 9 in Section A).

Lemma 18. *The application* $\exp$ *is Lipschitz continuous from* $(A(\mathbb{R}^2), |\cdot|)$ *to* $(G(\mathbb{R}^2), \|\cdot\|)$, *and* $\log$ *is Lipschitz continuous from* $(G(\mathbb{R}^2), \|\cdot\|)$ *to the space* $(A(\mathbb{R}^2), |\cdot|)$.

The application $\exp$ *is locally Lipschitz continuous from* $(A(\mathbb{R}^2), |\cdot|_\star)$ *to* $(G(\mathbb{R}^2), \|\cdot\|_\star)$, *and* $\log$ *is locally Lipschitz continuous from* $(G(\mathbb{R}^2), \|\cdot\|_\star)$ *to* $(A(\mathbb{R}^2), |\cdot|_\star)$.

6.7 The Riemannian Structure on $T_1(\mathbb{R}^2)$ Induced by Euclidean Coordinates

A natural system of coordinates—which we call the Euclidean chart—follows from the identification of $T_1(\mathbb{R}^2)$ with the vector space $\mathbb{R}^2 \oplus (\mathbb{R}^2 \otimes \mathbb{R}^2)$. If $\gamma(t) = 1 + \sum_{i=1,2} \gamma_i(t)e_i + \sum_{i,j=1,2} \gamma_{i,j}(t)e_i \otimes e_j$ is a smooth path with from $(-\varepsilon, \varepsilon)$ to $T_1(\mathbb{R}^2)$ with $\gamma(0) = x \in T_1(\mathbb{R}^2)$, then the derivative $\gamma'(0)$ of γ at time 0 may be simply expressed as

$$\gamma'(0) = \sum_{i=1,2} \gamma_i'(0)e_i(x) + \sum_{i,j=1,2} \gamma_{i,j}'(0)e_{i,j}(x),$$

where $e_i(x) \in T_x T_1(\mathbb{R}^2)$ is the tangent vector at 0 of the path $\varphi_i(t) = x + te_i$ and $e_{i,j}(x) \in T_x T_1(\mathbb{R}^2)$ is the tangent vector at 0 of the path $\varphi_{i,j}(t) = x + te_i \otimes e_j$.

Introduce the natural *attach* map A_x from $T_0(\mathbb{R}^2)$ to $T_x T_1(\mathbb{R}^2)$ which is linear and satisfies $A_x(e_i) = e_i(x)$ and $A_x(e_i \otimes e_j) = e_{i,j}(x)$ for $i, j = 1, 2$.

With this map, the derivative of γ at $t = 0$ is easily computed by

$$\gamma'(0) = A_x \left(\lim_{t \to 0} \frac{1}{t} (\gamma(t) - \gamma(0)) \right). \tag{47}$$

Hence, it is possible to endow $T_1(\mathbb{R}^2)$ with a Riemannian structure $\langle \cdot, \cdot \rangle$ by setting for $x \in T_1(\mathbb{R}^2)$,

$$\langle e_i(x), e_j(x) \rangle_x = \delta_{i,j}, \ \langle e_i(x), e_{j,k}(x) \rangle_x = 0, \ \langle e_{i,j}(x), e_{k,\ell}(x) \rangle_x = \delta_{i,k} \delta_{j,\ell}$$

for $i, j, k, \ell = 1, 2$, where $\delta_{i,j} = 1$ if $i = j$ and $\delta_{i,j} = 0$ otherwise. We then define $\langle \cdot, \cdot \rangle_x$ as a bilinear form on $T_x T_1(\mathbb{R}^2)$.

6.8 The Left-Invariant Riemannian Structure on $T_1(\mathbb{R}^2)$

We have defined the logarithm map log as a map from $T_1(\mathbb{R}^2)$ to the vector space $T_0(\mathbb{R}^2) \cong \mathbb{R}^2 \oplus (\mathbb{R}^2 \otimes \mathbb{R}^2)$. Given a point $x \in T_1(\mathbb{R}^2)$, another system of coordinates Φ_x from $T_1(\mathbb{R}^2)$ to $\mathbb{R}^2 \oplus (\mathbb{R}^2 \otimes \mathbb{R}^2)$ around x is given by

$$\Phi_x(y) = i_{T_0(\mathbb{R}^2) \to \mathbb{R}^2 \oplus (\mathbb{R}^2 \otimes \mathbb{R}^2)} \left(\log(x^{-1} \otimes y) \right),$$

where $i_{T_0(\mathbb{R}^2) \to \mathbb{R}^2 \oplus (\mathbb{R}^2 \otimes \mathbb{R}^2)}$ is the natural identification of $T_0(\mathbb{R}^2)$ with $\mathbb{R}^2 \oplus (\mathbb{R}^2 \otimes \mathbb{R}^2)$ for which we use the basis $\{e_i, e_j \otimes e_k\}_{i,j,k=1,2}$. For $y \in T_1(\mathbb{R}^2)$, we then set

$$\Phi_x(y) = \sum_{i=1,2} \Phi_x^i(y) e_i + \sum_{i,j=1,2} \Phi_x^{i,j}(y) e_i \otimes e_j.$$

This system of coordinates is called the *normal chart* or the *logarithmic chart*.

Let $\gamma : (-\varepsilon, \varepsilon) \to T_1(\mathbb{R}^2)$ be a smooth map with $\gamma(0) = x$. The derivative $\gamma'(0)$ of γ at 0 in this system of coordinate is then given by

$$\gamma'(0) = \sum_{i=1,2} (\Phi_x^i \circ \gamma)'(0) \widetilde{e}_i(x) + \sum_{i,j=1,2} (\Phi_x^{i,j} \circ \gamma)'(0) \widetilde{e}_{i,j}(x),$$

where $\widetilde{e}_i(x)$ (resp. $\widetilde{e}_{i,j}(x)$) is the tangent vector in $T_x T_1(\mathbb{R}^2)$ which is the derivative at 0 of the path ψ_x^i (resp. $\psi_x^{i,j}$) such that $(\Phi_x \circ \psi_x^i)'(0) = e_i$ (resp. $(\Phi_x \circ \psi_x^{i,j})'(0) = e_i \otimes e_j$). These paths are easily computed: $\psi_x^i(t) = x \otimes \exp(te_i)$ for $i = 1, 2$ and $\psi_x^{i,j}(t) = x \otimes \exp(te_i \otimes e_j)$ for $i, j = 1, 2$.

If we write $\gamma(t) = x \otimes \exp(\lambda(t))$ for $\lambda : (-\varepsilon, \varepsilon) \to T_0(\mathbb{R}^2)$ with $\lambda(0) = 0$ and

$$\lambda(t) = \sum_{i=1,2} \lambda_i(t) e_i + \sum_{i,j=1,2} \lambda_{i,j}(t) e_i \otimes e_j,$$

then

$$\gamma'(0) = \sum_{i=1,2} \lambda_i'(0) \widetilde{e}_i(x) + \sum_{i,j=1,2} \lambda_{i,j}'(0) \widetilde{e}_{i,j}(x).$$

In the Euclidean structure, it follows from (47) that if $x = 1 + \sum_{i=1,2} x_i e_i + \sum_{i,j=1,2} x_{i,j} e_i \otimes e_j$, then

$$\widetilde{e}_i(x) = A_x(x \otimes e_i) = e_i(x) + \sum_{j=1,2} x_j e_{j,i}(x)$$

$$\text{and } \widetilde{e}_{i,j}(x) = A_x(x \otimes (e_i \otimes e_j)) = e_{i,j}(x). \tag{48}$$

Let D_x (for *detach*) be the linear map from $T_x T_1(\mathbb{R}^2)$ which is the inverse of A_x, that is, which transforms $e_i(x)$ (resp. $e_{i,j}(x)$) into e_i (resp. $e_i \otimes e_j$).

For $x \in T_1(\mathbb{R}^2)$, let $L_x(y) = x \otimes y$ be the left multiplication on $T_1(\mathbb{R}^2)$. Its differential at point y maps $T_y T_1(\mathbb{R}^2)$ to $T_{x \otimes y} T_1(\mathbb{R}^2)$ and is defined by

$$d_y L_x(v) = A_{x \otimes y}(x \otimes D_y(v)).$$

A left-invariant vector field X on $T_1(\mathbb{R}^2)$ satisfies $X_x = d_1 L_x(X_1)$ and then $X_x = A_x(x \otimes D_1(X_1))$. From (48),

$$\widetilde{e}_i(x) = d_1 L_x(e_i(1)) \text{ and } \widetilde{e}_{i,j}(x) = d_1 L_x(e_{i,j}(1)).$$

In other words, the vector field $\widetilde{e}_i$ (resp. $\widetilde{e}_{i,j}$)—it is easily verified that they vary smoothly—is then the left-invariant vector field generated by $e_i(1)$ (resp. $e_{i,j}(1)$) in the Lie group $(T_1(\mathbb{R}^2), \otimes)$.

We may then define another bilinear form $\langle\!\langle \cdot, \cdot \rangle\!\rangle_x$ at any point x of $T_1(\mathbb{R}^2)$ by

$$\langle\!\langle \widetilde{e}_i(x), \widetilde{e}_j(x) \rangle\!\rangle_x = \delta_{i,j}, \ \langle\!\langle \widetilde{e}_i(x), \widetilde{e}_{j,k}(x) \rangle\!\rangle_x = 0, \ \langle\!\langle \widetilde{e}_{i,j}(x), \widetilde{e}_{k,\ell}(x) \rangle\!\rangle_x = \delta_{i,k} \delta_{j,\ell}$$

for $i, j, k, \ell = 1, 2$. These bilinear forms induce another Riemannian structure $\langle\!\langle \cdot, \cdot \rangle\!\rangle$ on $T_1(\mathbb{R}^2)$.

Note that for $v, w \in T_1 T_1(\mathbb{R}^2)$ and $x \in T_1(\mathbb{R}^2)$,

$$\langle\!\langle d_1 L_x(v), d_1 L_x(w) \rangle\!\rangle_x = \langle\!\langle v, w \rangle\!\rangle_1,$$

which means that $\langle\!\langle \cdot, \cdot \rangle\!\rangle$ is a *left-invariant metric*. For a left-invariant vector field X, the norm $\langle\!\langle X_x, X_x \rangle\!\rangle_x$ is constant.

Introduce the linear maps $\widetilde{A}_x : T_0(\mathbb{R}^2) \to T_x T_1(\mathbb{R}^2)$ and $\widetilde{D}_x : T_x T_1(\mathbb{R}^2) \to T_0(\mathbb{R}^2)$ such that $\widetilde{A}_x(e_i) = \widetilde{e}_i(x)$, $\widetilde{A}_x(e_i \otimes e_j) = \widetilde{e}_{i,j}(x)$ and $\widetilde{D}_x$ is the inverse of $\widetilde{A}_x$.

If $(\cdot|\cdot)$ is the natural scalar product on $T_0(\mathbb{R}^2)$ for which $\{e_i, e_j \otimes e_k\}_{i,j,k=1,2}$ is orthonormal, then for $x \in T_1(\mathbb{R}^2)$ and $v, w \in T_x T_1(\mathbb{R}^2)$,

$$\langle v, w \rangle_x = (D_x(v)|D_x(w)) \text{ and } \langle\!\langle v, w \rangle\!\rangle_x = (\widetilde{D}_x(v)|\widetilde{D}_x(w)) \tag{49}$$

To conclude this section, remark that it is very easy to express a vector $v \in T_x T_1(\mathbb{R}^2)$ in the basis $\{\widetilde{e}_i(x), \widetilde{e}_{j,k}(x)\}_{i,j,k=1,2}$ when we know its decomposition in $\{e_i(x), e_{j,k}(x)\}_{i,j,k=1,2}$: If $v = \sum_{i=1,2} v^i e_i(x) + \sum_{i,j=1,2} v^{i,j} e_{i,j}(x)$, then

$$v = \widetilde{A}_x(x^{-1} \otimes D_x(v)), \tag{50}$$

so that with (49),

$$\langle\!\langle v, w \rangle\!\rangle_x = (x^{-1} \otimes D_x(v) | x^{-1} \otimes D_x(w)).$$

Moreover, if γ is a smooth path from $(-\varepsilon, \varepsilon)$ to $T_1(\mathbb{R}^2)$, then we get from (50) a simple expression for the derivative γ' of γ at time $t \in (-\varepsilon, \varepsilon)$ in the basis $\{\widetilde{e}_i(x), \widetilde{e}_{j,k}(x)\}_{i,j,k=1,2}$ by

$$\gamma'(t) = \lim_{h \to 0} \widetilde{A}_{\gamma(t)}\left(\frac{1}{h}(\gamma(t)^{-1} \otimes \gamma(t+h) - 1)\right). \tag{51}$$

6.9 The Exponential Map Revisited

Consider an integral curve γ along a left-invariant vector field X with $\gamma(0) = 1$. If for $t \geqslant 0$, the path $\gamma(t)$ is written

$$\gamma(t) = 1 + \sum_{i=1,2} \gamma_i(t)e_i + \sum_{i,j=1,2} \gamma_{i,j}(t)e_i \otimes e_j,$$

then

$$\gamma'(t) = X_{\gamma(t)} = d_1 L_{\gamma(t)}(X_1) = A_{\gamma(t)}(\gamma(t) \otimes D_1(X_1))$$

and, if $X_1 = \sum_{i=1,2} v_i e_i + \sum_{i,j=1,2} v_{i,j} e_i \otimes e_j$,

$$\gamma_i'(t) = v_i e_i, \quad \gamma_{i,j}' v_{i,j} + \gamma_i(t)v_j$$

for $i, j = 1, 2$. It follows that

$$\gamma_i(t) = tv_i \text{ and } \gamma_{i,j}(t) = tv_{i,j} + \frac{t^2}{2}v_i v_j$$

which means that $\gamma(t) = \exp(tX_1)$ where exp has been defined by (41). Note that $\exp(tX_1) \otimes \exp(sX_1) = \exp((t+s)X_1)$, since $(tX_1) \boxplus (sX_1) = (t+s)X_1$. Hence, the one-parameter subgroup of $T_1(\mathbb{R}^d)$ generated by v is given by $t \in \mathbb{R} \mapsto \exp(tX_1)$.

In the sytem of left-invariant coordinates, we get

$$\gamma'(t) = \widetilde{A}_{\gamma(t)}(\gamma(t)^{-1} \otimes D_{\gamma(t)}(\gamma'(t))) = \widetilde{A}_{\gamma(t)}(\gamma(t)^{-1} \otimes \gamma(t) \otimes D_1(X_1))$$
$$= \widetilde{A}_{\gamma(t)}(D_1(X_1)),$$

which means that $\gamma'(t)$ is constant in the system of left-invariant coordinates.

It follows that for any $y \in T_1(\mathbb{R}^2)$, it is always possible to construct an integral curve γ along a left-invariant vector field that connects x to y and which is given by $(x \otimes \exp(tv))_{t \in [0,1]}$ with $v = \log(x^{-1} \otimes y)$.

6.10 Some Particular Curves for the Left-Invariant Riemannian Metric

For two points x and y in $T_1(\mathbb{R}^2)$ and a smooth path γ from $[0,1]$ to $T_1(\mathbb{R}^2)$ with $\gamma(0) = x$ and $\gamma(1) = y$, define the energy $\mathrm{Energy}(\gamma)$ of the path γ as

$$\mathrm{Energy}(\gamma) \stackrel{\text{def}}{=} \frac{1}{2} \int_0^1 \langle\!\langle \gamma'(s), \gamma'(s) \rangle\!\rangle_{\gamma(s)} \, ds.$$

For $t \in [0,1]$, set $\varphi(t) = \log(a^{-1} \otimes \gamma(t))$ so that $\gamma(t) = a \otimes \exp(\varphi(t))$ and then $\varphi(0) = 0$. The path φ belongs to $T_0(\mathbb{R}^2)$. With (51), we get

$$\gamma'(t) = \widetilde{A}_{\gamma(t)} \left(\lim_{h \to 0} \frac{1}{h} (\exp((-\varphi(t)) \boxplus \varphi(t+h)) - 1) \right)$$

$$= \widetilde{A}_{\gamma(t)} \left(\varphi'(t) + \frac{1}{2}[\varphi'(t), \varphi(t)] \right),$$

where $\varphi(t) = \sum_{i=1,2} \varphi_i(t) e_i + \sum_{i,j=1,2} \varphi_{i,j}(t) e_i \otimes e_j$ and $\varphi'(t) = \sum_{i=1,2} \varphi_i'(t) e_i + \sum_{i,j=1,2} \varphi_{i,j}'(t) e_i \otimes e_j$.

Thus, the energy of γ is given by

$$\mathrm{Energy}(\gamma) = \frac{1}{2} \int_0^1 \left\| \varphi'(s) + \frac{1}{2}[\varphi'(s), \varphi(s)] \right\|_{\mathrm{Euc}}^2 ds.$$

where $\| \cdot \|_{\mathrm{Euc}}$ is the Euclidean norm of $T_0(\mathbb{R}^2)$ identified with $\mathbb{R}^6$.

We now consider the particular path γ such that $\varphi(0) = 0$, $\varphi(1) = \log(a^{-1} \otimes b)$ and $\varphi'(t) + \frac{1}{2}[\varphi'(t), \varphi(t)]$ is constant over $[0,1]$. This means that $\varphi(t) = tv$ for some $v \in T_0(\mathbb{R}^2)$. This comes from the fact that that the projection of φ on $\mathbb{R}^2$ is then constant, since $[\phi'(t), \phi(t)]$ lives in $\mathbb{R}^2 \otimes \mathbb{R}^2$ and then $[\varphi'(t), \varphi(t)] = 0$ for $t \in [0,1]$. With the condition on $\varphi(1)$, $\varphi(t) = t \log(a^{-1} \otimes b)$ and $\gamma(t) = a \otimes \exp(t \log(a^{-1} \otimes b))$.

Let also $\psi : [0,1] \to T_0(\mathbb{R}^d)$ be a differentiable path with $\psi(0) = \psi(1) = 0$. Set for $\varepsilon > 0$,

$$\Gamma_\varepsilon(t) = a \otimes \exp(\varphi(t) \boxplus (\varepsilon\psi(t)))$$

so that

$$\Gamma_\varepsilon'(t) = \widetilde{A}_{\Gamma_\varepsilon(t)} \Big(\varphi'(t) + \varepsilon\psi'(t) + \varepsilon[\varphi'(t), \psi(t)]$$

$$+ \frac{1}{2}[\varphi'(t), \varphi(t)] + \frac{\varepsilon^2}{2}[\psi'(t), \psi(t)] \Big).$$

Thus, if $\varphi(t) = tv$ for some $v \in T_0(\mathbb{R}^2)$, we get

$$\mathrm{Energy}(\Gamma_\varepsilon(t)) = \frac{1}{2}\|v\|_{\mathrm{Euc}}^2 + \varepsilon \int_0^1 (v|\psi'(t)) \, dt + \frac{\varepsilon}{2} \int_0^1 (v|[v, \psi(t)]) \, dt$$

$$+ \frac{\varepsilon^2}{2} \int_0^1 \left\| \psi'(t) + [v, \psi(t)] + \frac{\varepsilon}{2}[\psi'(t), \psi(t)] \right\|_{\mathrm{Euc}}^2 dt$$

$$+ \frac{\varepsilon^2}{4} \int_0^1 (v|[\psi'(t), \psi(t)]) \, dt.$$

Since $\psi(0) = \psi(1) = 0$, $\int_0^1 (v|\psi'(t)) \, dt = 0$. But the term $\frac{\varepsilon}{2} \int_0^1 (v|[v, \psi(t)]) \, dt$ may be different from 0, as well as $\frac{\varepsilon^2}{4} \int_0^1 (v|[\psi'(t), \psi(t)]) \, dt$. Hence, we see that γ is not necessarily a path with minimal energy.

Remark 13. At first sight, this seems to contradicts the result that $t \mapsto \exp(tv)$ is a path with a constant derivative in the left-invariant system of coordinates seen in Section 6.9 above. Indeed, the geodesics ξ associated to the left-invariant Riemannian structure are those for which $\nabla_{\xi'(t)} \xi'(t) = 0$ where ∇ is the Levi-Civita connection associated to $\langle\!\langle \cdot, \cdot \rangle\!\rangle$. Since there exist some elements x, y and z such that

$$\langle\!\langle [z, x], y \rangle\!\rangle \neq \langle\!\langle x, [z, y] \rangle\!\rangle$$

(consider for example $x = e_1$, $x = e_2$ and $y = e_1 \otimes e_2$), this connection differs from the Cartan-Schouten (0) connection ∇^{CS} which is such that all paths of type $\gamma(t) = \exp(tv)$ are geodesics in the sense that $\nabla^{CS}_{\gamma'(t)}(\gamma'(t)) = 0$. On this topic, see for example [58].

However, if v belongs to $\mathrm{Vect}\{e_1, e_2\}$, then $(v|[v, \psi'(t)]) = 0$ and so

$$\mathrm{Energy}(\varGamma_\varepsilon(t)) \geqslant \mathrm{Energy}(\gamma) = \frac{1}{2} \| \log(a^{-1} \otimes b) \|_{\mathrm{Euc}}^2, \ \forall \varepsilon > 0$$

and thus γ is a *geodesic*, that is, a curve with minimal energy. As usual, it can also be shown that it is a path with minimal length, and the length

$$\mathrm{Length}(\gamma) \overset{\mathrm{def}}{=} \int_0^1 \sqrt{\langle\!\langle \gamma'(s), \gamma'(s) \rangle\!\rangle_{\gamma(s)}} \, ds$$

is then equal to $\| \log(a^{-1} \otimes b) \|_{\mathrm{Euc}}$. Another simple situation is when $v \in \mathrm{Vect}\{e_i \otimes e_i\}_{i,j=1,2}$; in this case $[v, w] = 0$ for all $w \in T_0(\mathbb{R}^2)$ and we also obtain that γ is a geodesic.

We also deduce that the length of the geodesic between a and b for $\langle\!\langle \cdot, \cdot \rangle\!\rangle$ is smaller than $\| \log(a^{-1} \otimes b) \|$.

Remark also that if a and b belong to $G(\mathbb{R}^2)$, then $\gamma(t)$ belongs to $G(\mathbb{R}^2)$ for $t \in [0, 1]$.

Of course, if we see $T_1(\mathbb{R}^2)$ with its Euclidean structure $\langle \cdot, \cdot \rangle$ then the geodesics are simply $\varphi(t) = a + t(b - a)$. In this case, $\varphi(t)$ does not belong to $G(\mathbb{R}^2)$ in general when a and b are in $G(\mathbb{R}^2)$.

6.11 A Transverse Decomposition of the Tensor Space

We have introduced a subgroup $G(\mathbb{R}^2)$ of $T_1(\mathbb{R}^2)$. Is this subgroup strict or not?

The tangent space of $T_1(\mathbb{R}^2)$ at any point may be identified with the vector space $(T_0(\mathbb{R}^2), +)$, which has dimension 6. We have also seen that the tangent space of $G(\mathbb{R}^2)$ at any point may be identified with $A(\mathbb{R}^2)$, and thus has dimension 3. Then, of course, $G(\mathbb{R}^2) \neq T_1(\mathbb{R}^2)$. Indeed, we may be more precise on the decomposition of $T_1(\mathbb{R}^2)$.

Denote by $S(\mathbb{R}^2)$ the subset of $T_0(\mathbb{R}^2)$ defined by

$$S(\mathbb{R}^2) = \left\{ x = (0,0,x^2) \in T_0(\mathbb{R}^2) \middle| \begin{array}{c} x^2 = \lambda e_1 \otimes e_1 + \mu e_2 \otimes e_2 \\ +\nu(e_1 \otimes e_2 + e_2 \otimes e_1), \\ \lambda, \mu, \nu \in \mathbb{R} \end{array} \right\}.$$

In other words, an element of $S(\mathbb{R}^2)$ belongs to $\mathbb{R}^2 \otimes \mathbb{R}^2$ and is symmetric. Of course, $S(\mathbb{R}^2)$ is linear, stable under $\otimes$ and $+$ (indeed, if $x, y \in S(\mathbb{R}^2)$, then $x \otimes y = x + y$), and is a vector space of dimension 3.

For an element e of the basis of $T(\mathbb{R}^2)$, call π_e the projection from $T(\mathbb{R}^2)$ to $T(\mathbb{R}^2)$, such that $x = \pi_1(x) + \sum_{i=1,2} \pi_{e_i}(x)e_i + \sum_{i,j=1,2} \pi_{e_i \otimes e_j}(x)e_i \otimes e_j$.

The next result follows easily from the construction of the projection operator $\widehat{\Upsilon}_s : T_0(\mathbb{R}^2) \to S(\mathbb{R}^2)$ and $\widehat{\Upsilon}_a : T_0(\mathbb{R}^2) \to A(\mathbb{R}^2)$ defined by

$$\widehat{\Upsilon}_s(x) = \mathfrak{s}(x) \text{ and } \widehat{\Upsilon}_a(x) = \pi_{\mathbb{R}^2}(x) + \mathfrak{a}(x).$$

Proposition 6. *The space $T_0(\mathbb{R}^2)$ is the direct sum of $A(\mathbb{R}^2)$ and $S(\mathbb{R}^2)$.*

This decomposition holds at the level of the tangent spaces at any point of $T_1(\mathbb{R}^2)$.

Proposition 7. *Any element x of $T_1(\mathbb{R}^2)$ may be written as a sum $x = y + z$ for some $y \in G(\mathbb{R}^2)$ and $z \in S(\mathbb{R}^2)$.*

Proof. For $x \in T(\mathbb{R}^2)$, set

$$\Upsilon_s(x) = \mathfrak{s}(x) - \frac{1}{2}x \otimes x \quad \text{and} \quad \Upsilon_a(x) = 1 + \pi_{\mathbb{R}^2}(x) + \mathfrak{a}(x) + \frac{1}{2}x \otimes x.$$

With (43), $\Upsilon_a(x) + \Upsilon_s(x) = x$ for all $x \in T(\mathbb{R}^2)$. Also, thanks to (44) and (46), $\Upsilon_s(T_1(\mathbb{R}^2)) \subset S(\mathbb{R}^2)$ and $\Upsilon_a(T_1(\mathbb{R}^2)) \subset G(\mathbb{R}^2)$.

We have to note that with the previous decomposition, $G(\mathbb{R}^2)$ is not a linear subspace of $T_1(\mathbb{R}^2)$, and Υ_a and Υ_s are not linear projections, since they involve quadratic terms. This is why we do not write $T_1(\mathbb{R}^2)$ as the direct sum of $G(\mathbb{R}^2)$ and $S(\mathbb{R}^2)$. However, as the tangent space of $S(\mathbb{R}^2)$ is $S(\mathbb{R}^2)$ itself, if $G(\mathbb{R}^2)$ and $\exp(S(\mathbb{R}^2)) = \{1 + x | x \in S(\mathbb{R}^2)\}$ are sub-manifolds of $T_1(\mathbb{R}^2)$, we get that $G(\mathbb{R}^2)$ and $\exp(S(\mathbb{R}^2))$ provides a *transverse decomposition* of $T_1(\mathbb{R}^2)$, in the sense that their tangent spaces at any point x provides an orthogonal decomposition (with respect to $\langle\!\langle \cdot, \cdot \rangle\!\rangle_x$) of the tangent space of $T_1(\mathbb{R}^2)$ at x.

Define a homogeneous norm $\| \cdot \|_{G(\mathbb{R}^2) \times S(\mathbb{R}^2)}$ by

$$\|x\|_{G(\mathbb{R}^2) \times S(\mathbb{R}^2)} = \max \left\{ \|\Upsilon_a(x)\|, \sqrt{\frac{1}{2}\|\Upsilon_s(x)\|} \right\}. \tag{52}$$

It is easily shown that this homogeneous norm is equivalent to the homogeneous gauge $\| \cdot \|$ on $T^1(\mathbb{R}^2)$.

6.12 Back to the Sub-Riemannian Point of View

We now come back to the result of Section 5.9, in order to bring some precision on the sub-Riemannian geometric framework. We have already seen that $(A(\mathbb{R}^2), \boxplus)$ is a Lie group (here, we no longer consider the space $T(\mathbb{R}^2)$). In addition, it is a vector space and then a smooth manifold with a natural system of coordinates given by the decomposition of $a \in A(\mathbb{R}^2)$ on the basis $\{e_1, e_2, e_3\}$, where e_3 corresponds to $[e_1, e_2]$.

If $\varphi_i(t; a) = a + te_i$ for $i = 1, 2, 3$ and $t \in \mathbb{R}$ and $a \in A(\mathbb{R}^2)$, denote by $e_i(a)$ the derivative $\varphi_i'(0; a)$ at time 0 of $\varphi_i(\cdot, a)$.

As in Sections 6.7, define for $a \in A(\mathbb{R}^2)$ two linear maps A_a and D_a by $A_a(e_i) = e_i(a)$ and $D_a = A_a^{-1}$.

We now proceed as in Section 6.8. The left multiplication is $L_a(y) = a \boxplus y$, and its differential $d_b L_a : T_b A(\mathbb{R}^2) \to T_{a \boxplus b} A(\mathbb{R}^2)$ at any point b is given by

$$d_b L_a(v) = A_{a \boxplus b}\left(D_b(v) + \frac{1}{2}[a, D_b(v)]\right),$$

Here $[a, v] = (a^1 v^2 - a^2 v^1)e_3$ for $a = a^1 e_1 + a^2 e_2 + a^3 e_3$.

Thus, any left-invariant vector field $(V_a)_{a \in A(\mathbb{R}^2)}$ satisfies $V_a = d_0 L_a(V_0)$. The left-invariant vector fields $\widetilde{e}_1$, $\widetilde{e}_2$ and $\widetilde{e}_3$ associated to e_1, e_2 and e_3 are given by

$$\widetilde{e}_1(a) = e_1(a) - \frac{1}{2}a^2 e_3(a), \ \widetilde{e}_2(a) = e_2(a) + \frac{1}{2}a^1 e_3(a) \text{ and } \widetilde{e}_3(a) = e_3(a)$$

for $a = a^1 e_1 + a^2 e_2 + a^3 e_3$. The space $\Theta(a)$ introduced in Section 5.9 is then the vector space generated by $\widetilde{e}_1(a)$ and $\widetilde{e}_2(a)$.

Let $\widetilde{A}_a$ be the linear map from $A(\mathbb{R}^2)$ to $T_a A(\mathbb{R}^2)$ defined by $\widetilde{A}_a(e_i) = \widetilde{e}_i(a)$. Then a vector v in $T_a A(\mathbb{R}^2)$ is easily expressed in the left-invariant basis $\{\widetilde{e}_1(a), \widetilde{e}_2(a), \widetilde{e}_3(a)\}$ by

$$v = \widetilde{A}_a((-a) \boxplus D_a(v)).$$

Similarly, if $\gamma : (-\varepsilon, \varepsilon) \to A(\mathbb{R}^2)$ is a smooth path, then it is easily checked that

$$\gamma'(t) = \widetilde{A}_a\left(\lim_{h \to 0} \frac{1}{\varepsilon}(-\gamma(t)) \boxplus \gamma(t + h)\right)$$

$$= \widetilde{A}_{\gamma(t)}\left(D_{\gamma(t)}(\gamma'(t)) + \frac{1}{2}[D_{\gamma(t)}(\gamma'(t)), \gamma(t)]\right).$$

For a differentiable path $\mathbf{y}_t$ in $A(\mathbb{R}^2)$ we have introduced in (34) and (35) some paths α and β that correspond indeed to the coordinates of the derivative of $\mathbf{y}$ in the bases $\{e_1, e_2, e_3\}$ and $\{\widetilde{e}_1, \widetilde{e}_2, \widetilde{e}_3\}$, in the sense that

$$\frac{d\mathbf{y}_t}{dt} = \sum_{i=1}^{3} \alpha^i(t) e_i(\mathbf{y}_t) = \sum_{i=1}^{3} \beta^i(t) \widetilde{e}_i(\mathbf{y}_t).$$

7 Rough Paths and their Integrals

7.1 What are Rough Paths?

If $x \in G(\mathbb{R}^2)$, then it is easily seen that for some universal constants c and c', $c\|x\| \leqslant |\log(x)| \leqslant c'\|x\|$, where $|\cdot|$ is the homogeneous norm defined on $A(\mathbb{R}^2)$ by (16).

Definition 4. *A rough path is a continuous path $\mathbf{x}$ with values in $T_1(\mathbb{R}^2)$.*

Denote by $C^\alpha([0,T], T_1(\mathbb{R}^2))$ the set of rough paths $\mathbf{x} : [0,T] \to T_1(\mathbb{R}^2)$ such that

$$\|\mathbf{x}\|_\alpha \overset{\text{def}}{=} \sup_{0 \leqslant s < t \leqslant T} \frac{\|\mathbf{x}_s^{-1} \otimes \mathbf{x}_t\|}{|t - s|^\alpha}$$

is finite.

A particular class of paths is the notion of geometric rough paths. The next definition follows from [29], which corrects some result of [55].

Definition 5. *A geometric rough path is a continuous path with values in $G(\mathbb{R}^2)$.*

A smooth rough path *is an element of the set*

$$C^\infty([0,T]; G(\mathbb{R}^2)) = \left\{ \exp(\mathbf{x}) \,\middle|\, \begin{array}{l} \mathbf{x} = x + \mathfrak{A}(x)[e_1, e_2] \\ \text{with } x \in C_p^\infty([0,T]; \mathbb{R}^2) \end{array} \right\}.$$

A weak geometric p-rough path with Hölder control *is a path with values in $G(\mathbb{R}^2)$ which is $(1/p)$-Hölder continuous.*

A geometric p-rough path *is the closure of the set of smooth rough paths with respect to the $\|\cdot\|_{1/p}$-norm.*

Remark 14. As discussed in Section 5.9, a rough path which is smooth is not necessarily a smooth rough path.

The space of weak geometric $1/\alpha$-rough paths with Hölder control is denoted by $C^\alpha([0,T]; G(\mathbb{R}^2))$, while the space of $1/\alpha$-rough paths with Hölder control is denoted by $C^{0,\alpha}([0,T]; G(\mathbb{R}^2))$. This latter space is strictly included in $C^\alpha([0,T]; G(\mathbb{R}^2))$. In addition $C^{0,\alpha}([0,T]; G(\mathbb{R}^2))$ is a Polish space, while $C^\alpha([0,T]; G(\mathbb{R}^2))$ is not a Polish space (this space is not separable: See [29]). The difference between weak geometric p-rough paths and geometric p-rough paths comes from an extension of the properties of Hölder continuous paths given in Remarks 2 and 3. For practical applications, the difference between weak geometric rough paths and geometric rough paths is not that important, if one accepts to weaken the Hölder norm (this is in general cost free).

Of course, there is a one-to-one correspondence between the paths with values in $G(\mathbb{R}^2)$ and the ones with values in $A(\mathbb{R}^2)$: Since exp and log are Lipschitz continuous (see Lemma 18), we easily get the following lemma, which is illustrated by Figure 10.

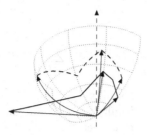

Fig. 10. From the tangent space $A(\mathbb{R}^2)$ at point 1 (perpendicular to the vertical axis) to the manifold $G(\mathbb{R}^2)$: the paths **x** (dashed) and $\log(\mathbf{x})$ (plain).

Lemma 19. *A path* **x** *belongs to* $C^\alpha([0,T]; G(\mathbb{R}^2))$ *if and only if* $\log(\mathbf{x})$ *belongs to* $C^\alpha([0,T]; A(\mathbb{R}^2))$.

A path $\mathbf{y} = (a(t), b(t), c(t))_{t \in [0,T]}$ with value in $A(\mathbb{R}^2)$ is then transformed into a path $\mathbf{x}_t = \exp(\mathbf{y}_t)$ with value in $G(\mathbb{R}^2)$ by the relation

$$\mathbf{x}_t = a(t)e_1 + b(t)e_2 + \frac{1}{2}a(t)^2 e_1 \otimes e_1 + \frac{1}{2}b(t)^2 e_2 \otimes e_2$$
$$+ (a(t)b(t) + c(t))e_1 \otimes e_2 + (a(t)b(t) - c(t))e_2 \otimes e_1.$$

Similarly, a path **x** with values in $G(\mathbb{R}^2)$ is transformed into a path **y** with values in $A(\mathbb{R}^2)$ by setting $\mathbf{y}_t = \log(\mathbf{x}_t)$.

In addition, note that $\mathbf{x}_{s,t} \stackrel{\text{def}}{=} \mathbf{x}_s^{-1} \otimes \mathbf{x}_t = \exp((-\mathbf{y}_s) \boxplus \mathbf{y}_t)$ and then

$$\mathfrak{s}(\mathbf{x}_{s,t}) = \frac{1}{2}(x_t - x_s) \otimes (x_t - x_s)$$

where $x_t = a(t)e_1 + b(t)e_2$ is the path above which **x** lies.

Assume now that **y** belongs $C^\alpha([0,T]; A(\mathbb{R}^2))$ with $\alpha > 1/2$. We have seen in Lemma 7 that necessarily, $c(t) = \mathfrak{A}(x; 0, t)$. Hence, from (46),

$$\mathbf{x}_t = 1 + x_t + \mathfrak{A}(x; 0, t)[e_1, e_2] + \frac{1}{2}(x_t - x_0) \otimes (x_t - x_0). \tag{53}$$

As for $0 \leqslant s \leqslant t \leqslant T$,

$$\mathfrak{A}(x; s, t) = \frac{1}{2}\int_s^t (x_r^1 - x_s^1)\, dx_r^2 - \frac{1}{2}\int_s^t (x_r^2 - x_s^2)\, dx_r^1$$

$$\text{and } \frac{1}{2}(x_t^i - x_s^i)^2 = \int_s^t (x_r^i - x_s^i)\, dx_r^i \text{ for } i = 1, 2,$$

we may rewrite (53) as

$$\mathbf{x}_t = 1 + x_t + \sum_{i,j=1,2} \left(\int_0^t (x_r^i - x_0^i)\, dx_r^j \right) e_i \otimes e_j. \tag{54}$$

Note also that

$$\mathbf{x}_{s,t} \overset{\text{def}}{=} (-\mathbf{x}_s) \otimes \mathbf{x}_t = 1 + x_t - x_s + \sum_{i,j=1,2} \left(\int_s^t (x_r^i - x_s^i) \, dx_r^j \right) e_i \otimes e_j.$$

This means that the terms of $\mathbf{x}_t$ in $\mathbb{R}^2 \otimes \mathbb{R}^2$ are the *iterated integrals* of x. When $\alpha < 1/2$, the difficulty comes from the fact that these iterated integrals are not canonically constructed. As the iterated integrals have some nice algebraic properties (see Section 8.2), we replace them by an object—a rough path—which shares the same algebraic properties, whose existence is not discussed in this article.

Let us end this Section with a result on paths that are not geometric. If $\mathbf{x}$ belongs to $C^\alpha([0,T]; T_1(\mathbb{R}^2))$ (with $\alpha \in (0,1]$) and $\mathbf{x}_t - 1 \in S(\mathbb{R}^2)$ for all t, then

$$\|\mathbf{x}_s^{-1} \otimes \mathbf{x}_t\| = \sqrt{\frac{1}{2}|\mathbf{x}_t^2 - \mathbf{x}_s^2|} \leqslant \|\mathbf{x}\|_\alpha |t - s|^\alpha$$

with $\mathbf{x}_t = (1, \mathbf{x}_t^1, \mathbf{x}_t^2)$. This implies that $\mathbf{x}_t$ can be identified with a path in $C^{2\alpha}([0,T]; \mathbb{R}^3)$ (note that if $\alpha > 1/2$, then $\mathbf{x}$ is constant).

7.2 Joining Two Points by Staying in $G(\mathbb{R}^2)$

We have seen that the integral of a differential form f along a path $x : [0,T] \to \mathbb{R}^2$ may be written as the limit of the following scheme: consider the family of dyadic partitions $\{t_k^n\}_{k=0,\dots,2^n}$ of $[0,T]$, and construct approximations x^n of x such that $x_{t_k^n} = x_{t_k^n}^n$ for $k = 0, \dots, 2^n$, and two successive points $x_{t_k^n}^n$ and $x_{t_{k+1}^n}^n$ are linked by a path that depends only on these two points. Then the integral $\mathfrak{I}(x)$ of f along x is defined as the limit of the integrals of f along x^n.

When x is a α-Hölder continuous path with values in $\mathbb{R}^2$ with $\alpha > 1/2$, then the "natural" family of approximations is given by piecewise linear approximations. If $\alpha \in (1/3, 1/2]$, we have seen that we need to replace x by a path $\mathbf{x}$ with values in $A(\mathbb{R}^2)$ that projects onto x, and to construct x^n by joining two successive points $x_{t_k^n}^n$ and $x_{t_{k+1}^n}^n$ of x^n with some sub-Riemannian geodesic computed from $\mathbf{x}_{t_k^n}^n$ and $\mathbf{x}_{t_{k+1}^n}^n$. Such a path x^n is automatically lifted to a path $(x^n, \mathfrak{A}(x^n))$ in $C^\alpha([0,T]; A(\mathbb{R}^2))$, and the integral $\mathfrak{I}(\mathbf{x})$ is defined as the limit of the $\mathfrak{I}(x^n)$.

Computations in Sections 6.1 and 7.5 have shown that it may be advisable to work with piecewise linear approximations of paths of $C^\alpha([0,t]; A(\mathbb{R}^2))$. For this, we have extended the differential form f to a differential form $\mathfrak{E}_{A(\mathbb{R}^2)}(f)$ on $A(\mathbb{R}^2)$. We have subsequently introduced some tensor space $T(\mathbb{R}^2)$, as well as a Lie groups $G(\mathbb{R}^2)$ and $T_1(\mathbb{R}^2)$ whose Lie algebras are $A(\mathbb{R}^2)$ and $T_0(\mathbb{R}^2)$. We have also introduced in Section 6.8 an operator $\widetilde{D}_x : T_x T_1(\mathbb{R}^2) \mapsto T_0(\mathbb{R}^2)$ such that $\widetilde{D}_x(T_x G(\mathbb{R}^2)) \subset A(\mathbb{R}^2)$.

For a piecewise smooth path $\mathbf{x} : [0,T] \to G(\mathbb{R}^2)$ with values that project onto $x : [0,T] \to \mathbb{R}^2$, it is then natural to define

$$\mathfrak{L}(\mathbf{x};0,t) = \int_0^t \mathfrak{E}_{A(\mathbb{R}^2)}(f)(x_s)\tilde{D}_{\mathbf{x}_s}\left(\frac{d\mathbf{x}_s}{ds}\right)ds \qquad (55)$$

for $t \in [0,T]$, where $\mathfrak{E}_{A(\mathbb{R}^2)}(f)$ has been defined in (37).

Remark 15. Note that here, we use the operator $\tilde{D}_x$ to transfer all problems to $T_0(\mathbb{R}^2)$ identified with the tangent space $T_1 T_1(\mathbb{R}^2)$ at point 1. If one wants to avoid this formulation, as we have seen it in Sections 6.7 and 6.8, $\mathfrak{E}_{A(\mathbb{R}^2)}(f)$ can be defined as the differential form

$$\mathfrak{E}_{A(\mathbb{R}^2)}(f)(x) = f_1(x)\tilde{e}^1(x) + f_2(x)\tilde{e}^2(x) + [f,f](x)\tilde{e}^3(x), \qquad (56)$$

where $\tilde{e}^i(x)$ is the dual element of $\tilde{e}_i(x)$ in $T_x T_1(\mathbb{R}^2)$ for $i = 1,2,3$. Formula (55) may then be rewritten

$$\mathfrak{L}(\mathbf{x};0,t) = \int_0^t \mathfrak{E}_{A(\mathbb{R}^2)}(f)(x_s)\frac{d\mathbf{x}_s}{ds}ds.$$

Now, given a path $\mathbf{x} \in C^\alpha([0,T];G(\mathbb{R}^2))$, define the equivalent of the piecewise linear approximation $\mathbf{x}^n$ by using the curves constructed in Section 6.10 (see Figures 10a–10b for an illustration. Note that unlike sub-Riemannian geodesics, $\mathbf{x}^n$ is not necessarily a smooth rough path, but it is a rough path which is smooth): set $\varphi_{a,b}(t) = a \otimes \exp(t\log(a^{-1} \otimes b))$ for $t \in [0,1]$, and

$$\mathbf{x}_t^n = \varphi_{\mathbf{x}_{t_k^n},\mathbf{x}_{t_{k+1}^n}}\left(\frac{t - t_k^n}{t_{k+1}^n - t_k^n}\right) \text{ for } t \in [t_k^n, t_{k+1}^n], \qquad (57)$$

for $n \in \mathbb{N}^*$ and $t_k^n = Tk/2^n$, $k = 0, \ldots, 2^n$.

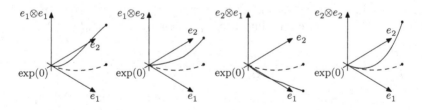

Fig. 10a. A sub-Riemannian geodesic in $G(\mathbb{R}^2)$ as constructed from Section 5.9.

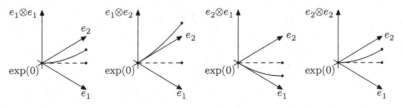

Fig. 10b. The path $\varphi_{x,y}$ with $x = \exp(0)$ and $y = \exp((1,1,1))$.

Proposition 8. *For* $\mathbf{x} \in C^\alpha([0,T]; G(\mathbb{R}^2))$ *with* $\alpha > 1/3$, *let* $\mathbf{x}^n$ *be the path defined above by* (57). *Then*

$$\Im(\log(\mathbf{x}); 0, t) = \lim_{n\to\infty} \mathfrak{L}(\mathbf{x}^n; 0, t)$$

uniformly in $t \in [0, T]$.

Proof. This follows from the computations of Sections 6.1 and 7.5, and from the definition of $\widetilde{D}_x$, since we have seen in Section 6.10 that $\widetilde{D}_{\varphi_{a,b}(t)}(\varphi'_{a,b}(t)) = \log(a^{-1} \otimes b)$ for $t \in [0, 1]$.

As there is an identification between $\log(\mathbf{x})$ and $\mathbf{x}$, one can set for $\mathbf{x} \in C^\alpha([0,T]; G(\mathbb{R}^2))$, $\Im(\mathbf{x}) = \Im(\log(\mathbf{x}))$.

7.3 A Riemann Sum Like Definition

We are now willing to give another definition of the integral in the spirit of Riemann sums, to get rid of the integrals between the successive times t_k^n and t_{k+1}^n for $k = 0, \ldots, 2^n - 1$. For this, we use the Taylor expansion of f: For $x, y \in \mathbb{R}^2$ and $i = 1, 2$,

$$f_i(y^1, y^2) = f_i(x^1, x^2) + \sum_{j=1,2} \frac{\partial f_i}{\partial x_j}(x^1, x^2) z^j + \kappa_1^i(z)$$

with $|\kappa_1^i(z)| \leq \|f\|_{\text{Lip}} |z|^{1+\gamma}$ and $z = y - x$. In addition,

$$[f, f](y^1, y^2) = [f, f](x^1, x^2) + \kappa_2(z) \text{ with } |\kappa_2(z)| \leq \|f\|_{\text{Lip}} |z|^{\gamma-1}.$$

Set $\mathbf{x} \in C^\alpha([0,T]; G(\mathbb{R}^2))$ with $\alpha > 1/3$ and $\mathbf{x}^n$ constructed as in Proposition 8. In addition, define x and x^n by $x = \pi_{\mathbb{R}^2}(\mathbf{x})$ and $x^n = \pi_{\mathbb{R}^d}(\mathbf{x})$. Remark that $x^n = x^{\Pi^n}$, the piecewise linear interpolation of x. For $\Delta_n t = T2^{-n}$,

$$\left| \int_{t_k^n}^{t_{k+1}^n} [f, f](x_s^n) \mathfrak{a}(\mathbf{x}_{t_k^n, t_{k+1}^n}) \frac{ds}{\Delta_n t} - [f, f](x_{t_k^n}) \mathfrak{a}(\mathbf{x}_{t_k^n, t_{k+1}^n}) \right|$$

$$\leq \Delta_n t^{\alpha(1+\gamma)} \|\mathbf{x}\|_\alpha^{\alpha(1+\gamma)} \|f\|_{\text{Lip}}.$$

In addition, with the Taylor formula,

$$\left| \int_{t_k^n}^{t_{k+1}^n} f_i(x_s^n)(x_{t_{k+1}^n} - x_{t_k^n}) \frac{ds}{\Delta_n t} - f_i(x_{t_k^n})(x_{t_{k+1}^n}^i - x_{t_k^n}^i) \right.$$

$$\left. - \sum_{i,j=1,2} \frac{1}{2} \frac{\partial f_i}{\partial x_j}(x_{t_k^n}) \pi_{e_j}(x_{t_{k+1}^n} - x_{t_k^n}) \pi_{e_i}(x_{t_{k+1}^n} - x_{t_k^n}) \right|$$

$$\leq \Delta_n t^{\alpha(1+\gamma)} \|f\|_{\text{Lip}} \|\mathbf{x}\|_\alpha^{1+\gamma}. \quad (58)$$

If $e^i(x)$ is the dual element of $e_i(x)$, denote by $f(x)$ the linear operator $f = f_1(x)e^1(x) + f_2(x)e^2(x)$. If $e^i(x) \otimes e^j(x)$ is the dual element of $e_i(x) \otimes e_j(x)$ for $i, j = 1, 2$, denote by ∇f the linear operator

$$\nabla f(x) = \sum_{i,j=1,2} \frac{\partial f_i}{\partial x_j}(x) e^j(x) \otimes e^i(x)$$

so that with (45),

$$\frac{1}{2} \sum_{i,j=1,2} \frac{\partial f_i}{\partial x_j}(x_{t_k^n}) \pi_{e_j}(x_{t_{k+1}^n} - x_{t_k^n}) \pi_{e_i}(x_{t_{k+1}^n} - x_{t_k^n}) = \nabla f(x_{t_k^n}) \mathfrak{s}(\mathbf{x}_{t_k^n, t_{k+1}^n}).$$

Hence, with (43), we deduce that

$$\int_{t_k^n}^{t_{k+1}^n} \mathfrak{E}_{A(\mathbb{R}^2)}(f)(\mathbf{x}_s^n) \widetilde{D}_{\mathbf{x}_s^n} \left(\frac{d\mathbf{x}_s^n}{ds} \right) ds$$

$$= f(x_{t_k^n}) \pi_{\mathbb{R}^2}(\mathbf{x}_{t_k^n, t_{k+1}^n}) + \nabla f(x_{t_k^n}) \pi_{\mathbb{R}^2 \otimes \mathbb{R}^2}(\mathbf{x}_{t_k^n, t_{k+1}^n}) + \theta_k^n \qquad (59)$$

with $|\theta_k^n| \leqslant \|f\|_{\mathrm{Lip}} \|\mathbf{x}\|_\alpha^{1+\gamma} \Delta_n t^{\alpha(1+\gamma)}$. As $\alpha(\gamma+1) > 1$, $\lim_{n\to\infty} \sum_{k=0}^{2^n-1} |\theta_k^n| = 0$. We then define a differential form $\mathfrak{E}_{T_1(\mathbb{R}^2)}(f)$ on $T_1(\mathbb{R}^2)$ by

$$\mathfrak{E}_{T_1(\mathbb{R}^2)}(f)(x) = \sum_{i=1,2} f_i(\pi_{\mathbb{R}^2}(x)) e^i(x) + \sum_{i,j=1,2} \frac{\partial f_i}{\partial x_j}(\pi_{\mathbb{R}^2}(x)) e^i(x) \otimes e^j(x),$$

With (59) and the property of the θ_k^n's, we get that, after having identified $\mathfrak{I}(\mathbf{x}; 0, T)$ with $\mathfrak{I}(\log(\mathbf{x}); 0, T)$ for $\mathbf{x} \in C^\alpha([0, T]; G(\mathbb{R}^2))$,

$$\mathfrak{I}(\mathbf{x}; 0, T) = \lim_{n\to\infty} \sum_{k=0}^{2^n-1} \mathfrak{E}_{T_1(\mathbb{R}^2)}(f)(x_{t_k^n}) \mathbf{x}_{t_k^n, t_{k+1}^n}, \qquad (60)$$

which is a Riemann sum like expression.

This means also that $\mathfrak{E}_{T_1(\mathbb{R}^2)}(f)(x_{t_k^n}) \mathbf{x}_{t_k^n, t_{k+1}^n}$ is a "good" approximation of $\mathfrak{I}(\mathbf{x}; t_k^n, t_{k+1}^n)$.

7.4 Another Construction of the Integral

Assume that the functions f_1, f_2 take their values in the space $\mathbb{R}^m$ with $m > 1$. For the sake of simplicity, assume that $m = 2$. The integral $\mathfrak{I}(\mathbf{x}) = (\mathfrak{I}^1(\mathbf{x}), \mathfrak{I}^2(\mathbf{x}))$ then becomes a path in $\mathbb{R}^2$, and we are interested in constructing its iterated integrals.

If x belongs to $C^\alpha([0, T]; \mathbb{R}^2)$ with $\alpha > 1/2$, then $\mathfrak{I}(\mathbf{x})$ also corresponds to a Young integral and belongs to $C^\alpha([0, T]; \mathbb{R}^2)$. Hence, we use the natural lift in (54), which means that we only need to define $t \mapsto \mathfrak{A}(\mathfrak{I}(\mathbf{x}; 0, t); 0, t)$, or equivalently, $\int_s^t \mathfrak{I}^i(\mathbf{x}; s, r) \, d\mathfrak{I}^j(\mathbf{x}; s, r)$ for $i, j = 1, 2$.

Remark that, if $x_r = x_s + (r-s)(t-s)^{-1}(x_t - x_s)$,

$$\int_s^t \left(\int_s^r f_i^k(x_u)\,dx^i \right) f_j^\ell(x_r)\,dx_r^j = f_i^j(x_s) f_j^\ell(x_s) \int_s^t (x_r^i - x_s^i)\,dx_r^j$$

$$+ \int_s^t \left(\int_s^r (f_i^k(x_r) - f_i^k(x_s))\,dx_r^i \right) f_j^\ell(x_r)\,dx_r^j$$

$$+ \int_s^t f_i^k(x_s)(x_r^i - x_s^i)(f_j^\ell(x_r) - f_j^\ell(x_s))\,dx_r^j.$$

This suggests to approximate $\int_{t_k^n}^{t_{k+1}^n} \mathfrak{I}^i(\mathbf{x}; t_k^n, s)\,d\mathfrak{I}^j(\mathbf{x}; t_k^n, s)$ by the quantity

$$\mathbf{y}_{t_k^n, t_{k+1}^n}^{k,\ell} = \sum_{i,j=1,2} f_i^k(x_{t_k^n}) f_k^\ell(x_{t_k^n}) \mathbf{x}_{t_k^n, t_{k+1}^n}^{2,i,j}, \quad k, \ell = 1, 2.$$

With (60), we also set

$$\mathbf{y}_{t_k^n, t_{k+1}^n}^i = \mathfrak{E}_{T_1(\mathbb{R}^2)}(f^i)(x_{t_k^n}) \mathbf{x}_{t_k^n, t_{k+1}^n}, \quad i = 1, 2.$$

Let $\{e_1, e_2\}$ be the canonical basis of $\mathbb{R}^2$ and $\{\check{e}^1, \check{e}^2\}$ be its dual basis, which we distinguish from $\{e_1, e_2\}$ to refer to the space in which f takes its values. Then we introduce the differential form $\mathfrak{E}_{T_1(\mathbb{R}^2), T_1(\mathbb{R}^2)}(f)$ with value in $T_1(\mathbb{R}^2)$ defined by, for $z \in T_1(\mathbb{R}^2)$,

$$\mathfrak{E}_{T_1(\mathbb{R}^2), T_1(\mathbb{R}^2)}(f)(z) = 1 + \mathfrak{E}_{T_1(\mathbb{R}^2)}(f^1)(z)\check{e}^1 + \mathfrak{E}_{T_1(\mathbb{R}^2)}(f^2)(z)\check{e}^2$$

$$+ \sum_{i,j=1,2} f^i(\pi_{\mathbb{R}^2}(z)) f^j(\pi_{\mathbb{R}^2}(z))\check{e}^1 \otimes \check{e}^2,$$

or, more concisely,

$$\mathfrak{E}_{T_1(\mathbb{R}^2)}(f)(z) = \mathfrak{E}_{T_1(\mathbb{R}^2)}(f)(z) + f(\pi_{\mathbb{R}^2}(z)) \otimes f(\pi_{\mathbb{R}^2}(z))$$

with $f = f^1 \check{e}^1 + f^2 \check{e}^2$. Hence, in order to approximate $\mathfrak{I}(\mathbf{x}; s, t)$ and its iterated integral, we may then set

$$\mathbf{y}_{s,t} = \mathfrak{F}(f, \mathbf{x}; s, t) \overset{\text{def}}{=} \mathfrak{E}_{T_1(\mathbb{R}^2), T_1(\mathbb{R}^2)}(f)(x_s)\mathbf{x}_{s,t} \qquad (61)$$

and set, for $t \in (t_{\underline{M}(t,n)-1}^n, t_{\underline{M}(t,n)}^n]$ and $s \in [t_{\overline{M}(s,n)}^n, t_{\overline{M}(s,n)}^n)$,

$$\mathfrak{I}^n(\mathbf{x}; s, t)$$

$$\overset{\text{def}}{=} \mathfrak{F}(\mathbf{x}; s, t_{\overline{M}(s,n)}^n) \otimes \left(\overset{\underline{M}(t,n)-1}{\underset{k=\overline{M}(s,n)}{\bigotimes}} \mathfrak{F}(\mathbf{x}; t_k^n, t_{k+1}^n) \right) \otimes \mathfrak{F}(\mathbf{x}; t_{\underline{M}(t,n)}^n, t). \qquad (62)$$

Finally, set

$$\mathfrak{I}(\mathbf{x}; s, t) = \lim_{n \to \infty} \mathfrak{I}^n(\mathbf{x}; s, t) \tag{63}$$

when this limit exists.

In the definition of $\mathfrak{F}(\mathbf{x})$, we have assumed that $\mathbf{x}$ is a path with values in $G(\mathbb{R}^2)$. In fact, this definition may be extended to paths with values in $T_1(\mathbb{R}^2)$. In addition, note that if $\mathbf{x}$ takes its values in $G(\mathbb{R}^2)$, then $\mathfrak{F}(\mathbf{x}; s, t) \in G(\mathbb{R}^2)$. The analysis of $\mathfrak{F}(\mathbf{x}; s, t)$ for $\mathbf{x}$ in $S(\mathbb{R}^2)$ is performed in Section 7.5.

We will see below that the integral defined by (62)-(63) satisfies the relation

$$\mathfrak{I}(\mathbf{x}; s, t) = \mathfrak{I}(\mathbf{x}; s, r) \otimes \mathfrak{I}(\mathbf{x}; r, t), \quad \forall 0 \leqslant s \leqslant r \leqslant t \leqslant T, \tag{64}$$

which means that $t \in [0, T] \mapsto \mathfrak{I}(\mathbf{x}; 0, t)$ is a path with values in $T_1(\mathbb{R}^2)$ and $\mathfrak{I}(\mathbf{x}; s, t)$ represents its increments.

But $\mathfrak{I}^n(\mathbf{x})$ does not satisfy (64), unless s, r, t belong to $\{t_k^n\}_{k=0,\dots,2^n-1}$.

The next results are borrowed from [52, Section 3.2, p. 40] or from [55, Section 3.1, p. 273].

Definition 6. *A function* $\mathbf{y}_{s,t}$ *from* $\Delta_+ = \{(s, t) \in [0, T]^2 \mid 0 \leqslant s \leqslant t \leqslant T\}$ *to* $T_1(\mathbb{R}^2)$ *is an almost rough path if there exist some constants* $C > 0$ *and* $\theta > 1$ *such that*

$$\|\mathbf{y}_{s,t} - \mathbf{y}_{s,r} \otimes \mathbf{y}_{r,t}\|_\star \leqslant C|t - s|^\theta, \quad \forall 0 \leqslant s \leqslant r \leqslant t \leqslant T.$$

where $\| \cdot \|_\star$ *is the norm defined by* $\|x\|_\star = \max\{|x^1|, |x^2|\}$.

An almost rough path is the "basic brick" for constructing a rough path. We give a proof of the next theorem in Section C in the appendix.

Theorem 2. *Let* $\mathbf{y} : \Delta_+ \to T_1(\mathbb{R}^2)$ *be an almost rough path such that* $\|\mathbf{y}_{s,t}\| \leqslant C|t - s|^\alpha$ *for* $\alpha \in (1/3, 1]$ *and* $C > 0$. *Set*

$$\mathbf{y}_{s,t}^n \stackrel{def}{=} \mathbf{y}_{s,\overline{M}(s,n)} \otimes \left(\bigotimes_{k=\underline{M}(s,n)}^{M(t,n)-1} \mathbf{y}_{t_k^n, t_{k+1}^n} \right) \otimes \mathbf{y}_{\underline{M}(t,n),t}, \quad \forall (s, t) \in \Delta_+.$$

Then there exist a unique path $\mathbf{z}$ *in* $C^\alpha([0, T]; T_1(\mathbb{R}^2))$ *and a sequence* $(K_n)_{n \in \mathbb{N}}$ *decreasing to* 0 *such that*

$$\|\mathbf{z}_{s,t} - \mathbf{y}_{s,t}^n\|_\star \leqslant K_n |t - s|^\theta.$$

If $\mathbf{y}$ *is an almost rough path in* $G(\mathbb{R}^2)$, *then* $\mathbf{z}$ *is a weak geometric rough path with* α-*Hölder control.*

In addition, if $\mathbf{y}$ *and* $\mathbf{y}'$ *are both almost rough paths with*

$$|\pi_{\mathbb{R}^2}(\mathbf{y}_{s,t} - \mathbf{y}'_{s,t})| \leqslant \varepsilon|t - s|^\alpha, \quad |\pi_{\mathbb{R}^2 \otimes \mathbb{R}^d}(\mathbf{y}_{s,t} - \mathbf{y}'_{s,t})| \leqslant \varepsilon|t - s|^{2\alpha}$$

for all $(s, t) \in \Delta_+$, *then the corresponding rough paths* $\mathbf{z}$ *and* $\mathbf{z}'$ *satisfy*

$$|\pi_{\mathbb{R}^d}(\mathbf{z}_{s,t} - \mathbf{z}'_{s,t})| \leqslant K(\varepsilon)|t - s|^\alpha, \quad |\pi_{\mathbb{R}^d \otimes \mathbb{R}^d}(\mathbf{z}_{s,t} - \mathbf{z}'_{s,t})| \leqslant K(\varepsilon)|t - s|^{2\alpha}$$

for some function $K(\varepsilon)$ *decreasing to* 0 *as* $\varepsilon \to 0$ *and depending on* T, α *and* θ *only.*

The existence of $\mathfrak{I}(\mathbf{x})$ in (63) as a (weak geometric) rough path when $\mathbf{x}$ is a (weak geometric) rough path is then justified by the next proposition and the application of Theorem 2.

Roughly speaking, the proof follows the same line as the for the Young integral: the reader is referred to [55, Section 3.2.2, p. 289], [52, Section 5.2, p. 117], [44, Section 3] or [53].

Proposition 9. *For* $\mathbf{x} \in C^\alpha([0,T]; T_1(\mathbb{R}^2))$ *with* $\alpha \in (1/3, 1]$, *the function* $(s,t) \in \Delta_+ \mapsto \mathfrak{F}(\mathbf{x}; s, t)$ *is an almost rough path. In addition, if* $\mathbf{x} \in C^\alpha([0,T]; G(\mathbb{R}^2))$, *then* $\mathfrak{F}(\mathbf{x}; s, t)$ *belongs to* $G(\mathbb{R}^2)$. *Hence,* $\mathfrak{I}(\mathbf{x})$ *given by* (63) *exists and belongs to* $C^\alpha([0,T]; T_1(\mathbb{R}^2))$ *(resp.* $C^\alpha([0,T]; G(\mathbb{R}^2))$*) if* $\mathbf{x} \in C^\alpha([0,T]; T_1(\mathbb{R}^2))$ *(resp.* $C^\alpha([0,T]; G(\mathbb{R}^2))$*).*

We have already seen that the integral $\mathfrak{I}(\mathbf{x})$ lies above the integral we constructed in Section 6 using some approximation of $\mathbf{x}$. With Theorem 2, we not only have continuity of $\mathbf{x} \mapsto \mathfrak{I}(\mathbf{x})$, but we also get that it is locally Lipschitz under a stronger assumption on f and we are not bound to use the $\|\cdot\|_\beta$ norm with $\beta < \alpha$ while working with α-Hölder paths. In addition, we may consider any family of partitions whose meshes decrease to zero (see Remark 4 or the proof of Theorem 5 in Appendix C).

We introduce a new norm $\|\cdot\|_{\ast,\alpha}$ on $C^\alpha([0,T]; T_1(\mathbb{R}^d))$, which is not equivalent to $\|\cdot\|_\alpha$ but which generates the same topology: for $\mathbf{x} \in C^\alpha([0,T]; T_1(\mathbb{R}^d))$,

$$\|\mathbf{x}\|_{\ast,\alpha} = \sup_{0 \leqslant s < t \leqslant T} \max \left\{ \frac{|\pi_{\mathbb{R}^d}(\mathbf{x}_{s,t})|}{(t-s)^\alpha}, \frac{|\pi_{\mathbb{R}^d \otimes \mathbb{R}^d}(\mathbf{x}_{s,t})|}{(t-s)^{2\alpha}} \right\}.$$

The next theorem summarizes Proposition 9 and some continuity results.

Theorem 3. *If* $f \in \mathrm{Lip}(\gamma; \mathbb{R}^2 \to \mathbb{R}^2)$ *with* $\alpha(\gamma + 1) > 1$ *and* $\alpha > 1/3$, *then the limit of* $(\mathfrak{I}^n(\mathbf{x}; s, t))_{n \in \mathbb{N}}$ *in* (63) *exists and is unique. Besides,* $\mathfrak{I}$ *maps continuously* $(C^\alpha([0,T]; T_1(\mathbb{R}^2)), \|\cdot\|_{\ast,\alpha})$ *to* $(C^\alpha([0,T]; T_1(\mathbb{R}^2)), \|\cdot\|_{\ast,\alpha})$. *If* f *is of class* $C^2(\mathbb{R}^2; \mathbb{R}^2)$ *with a* κ-*Hölder continuous second-order derivative where* $\alpha(\kappa + 2) > 1$, *then* $\mathfrak{I}$ *is locally Lipschitz continuous.*

In addition, if $\mathbf{x}$ *is a smooth rough path, then*

$$\mathfrak{I}(\mathbf{x}; 0, t) = \exp\left(\int_0^t f(x_s)\, \mathrm{d}x_s + \mathfrak{A}(\mathfrak{I}(\mathbf{x}; 0, t); 0, t)[\check{e}_1, \check{e}_2] \right)$$

and $\mathfrak{I}(\mathbf{x})$ *is also a smooth rough path.*

Hence, for $\mathbf{x} \in C^{0,\alpha}([0,T]; G(\mathbb{R}^2))$, there exists a sequence of paths $\mathbf{x}^n \in C_p^\infty([0,T]; G(\mathbb{R}^2))$ convergent to $\mathbf{x}$ in $\|\cdot\|_\alpha$, then $\mathfrak{I}(\mathbf{x}) = \lim_{n \to \infty} \mathfrak{I}(\mathbf{x}^n)$, $\mathfrak{I}(\mathbf{x}^n)$ is a smooth rough path and $\mathfrak{I}(\mathbf{x})$ belongs to $C^{0,\alpha}([0,T]; G(\mathbb{R}^2))$.

Now, if $\mathbf{x}$ is only a weak geometric $1/\alpha$-rough path with Hölder control, then we have seen that $\mathbf{x}$ may be approximated by some smooth rough paths $\mathbf{x}^n$ in the β-Hölder norm $\|\cdot\|_\beta$ with $\beta < \alpha$. Hence, $\mathfrak{I}(\mathbf{x}^n)$ converges to $\mathfrak{I}(\mathbf{x})$ in $\|\cdot\|_\beta$ with $\beta < \alpha$. Anyway, $\mathfrak{I}(\mathbf{x})$ belongs to $C^\alpha([0,T]; G(\mathbb{R}^2))$.

We then deduce the following stability result.

Corollary 5. *If $\mathfrak{J}$ is defined by Theorem 3, then $\mathfrak{J}$ maps $C^{\alpha}([0,T]; G(\mathbb{R}^2))$ into $C^{\alpha}([0,T]; G(\mathbb{R}^2))$ and $C^{0,\alpha}([0,T]; G(\mathbb{R}^2))$ into $C^{0,\alpha}([0,T]; G(\mathbb{R}^2))$.*

We end this section with a lemma similar to Lemma 11.

Lemma 20. *For any $\mathbf{x} \in C^{\alpha}([0,T]; T_1(\mathbb{R}^2))$, $\mathfrak{J}(\mathbf{x}; s,t) = \mathfrak{J}(\mathbf{x}_{|[s,t]})$ for all $0 \leqslant s < t \leqslant T$.*

Proof. If $\mathbf{x} \in C^{\alpha}([0,T]; G(\mathbb{R}^2))$, the proof of this Lemma is similar to the one of Lemma 11. If $\mathbf{x} \in C^{\alpha}([0,T]; T_1(\mathbb{R}^2))$, the results at the end of Section 7.5 allow us to conclude in the same way.

7.5 Integral along a Path Living in the Tensor Space

Now, consider $\mathbf{x} \in C^{\alpha}([0,T]; T_1(\mathbb{R}^2))$ with $\alpha \in (1/3, 1/2)$. What can be said about $\mathfrak{J}(\mathbf{x})$? From Proposition 7, one may decompose $\mathbf{x}_t$ as the sum $\mathbf{x}_t = \mathbf{y}_t + \mathbf{z}_t$ with $\mathbf{y} = \Upsilon_{\mathrm{a}}(\mathbf{x})$ and $\mathbf{z} = \Upsilon_{\mathrm{s}}(\mathbf{x})$. In addition, $(\mathbf{y}, \mathbf{z})$ belongs to $C^{\alpha}([0,T]; G(\mathbb{R}^2) \times S(\mathbb{R}^2))$, where the homogeneous norm on $G(\mathbb{R}^2) \times S(\mathbb{R}^2)$ has been defined by (52). In particular, this implies that $\pi_{e_i \otimes e_j}(\mathbf{z})$ belongs to $C^{2\alpha}([0,T]; \mathbb{R})$, i.e., each of its components is 2α-Hölder continuous. In addition, for $(s,t) \in \Delta_+$, $\mathfrak{E}_{T_1(\mathbb{R}^2), T_1(\mathbb{R}^2)}(f)(\mathbf{x}_s)(\mathbf{y}_t - \mathbf{y}_s)$ belongs to $G(\mathbb{R}^2)$, while

$$\mathfrak{E}_{T_1(\mathbb{R}^2), T_1(\mathbb{R}^2)}(f)(\mathbf{x}_s)(\mathbf{z}_t - \mathbf{z}_s) = \sum_{k,i,j=1,2} \frac{\partial f_i^k}{\partial x_j}(x_s)\pi_{e_i \otimes e_j}(\mathbf{z}_t - \mathbf{z}_s)\check{e}_k$$

$$+ \sum_{k,\ell,i,j=1,2} f_i^k(x_s)f_j^{\ell}(x_s)\pi_{e_i \otimes e_j}(\mathbf{z}_t - \mathbf{z}_s)\check{e}_k \otimes \check{e}_\ell.$$

Since $\pi_{e_i \otimes e_j}(\mathbf{z}_t - \mathbf{z}_s) = \pi_{e_j \otimes e_i}(\mathbf{z}_t - \mathbf{z}_s)$, we get that $\mathfrak{E}_{T_1(\mathbb{R}^2), T_1(\mathbb{R}^2)}(f)(\mathbf{x}_s)$ $(\mathbf{z}_t - \mathbf{z}_s)$ belongs to $\mathbb{R}^2 \oplus S(\mathbb{R}^2)$. Besides, for $t \in [0,T]$,

$$\lim_{n \to \infty} \sum_{k=0}^{M(t,n)} \pi_{\mathbb{R}^2}\left(\mathfrak{E}_{T_1(\mathbb{R}^2), T_1(\mathbb{R}^2)}(f)(\mathbf{x}_{t_k^n})(\mathbf{z}_{t_{k+1}^n} - \mathbf{z}_{t_k^n})\right)$$

$$= \sum_{k=1,2} \check{e}^k \int_0^t \frac{\partial f_1^k}{\partial x_1}(x_s)\, d\pi_{e_1 \otimes e_1}(z_s) + \check{e}^k \int_0^t \frac{\partial f_2^k}{\partial x_2}(x_s)\, d\pi_{e_2 \otimes e_2}(z_s)$$

$$+ \check{e}^k \int_0^t \left(\frac{\partial f_1^k}{\partial x_2}(x_s) + \frac{\partial f_2^k}{\partial x_1}(x_s)\right) d\pi_{e_1 \otimes e_2}(z_s)$$

which we can more concisely write $\int_0^t \nabla f(x_s)\, dz_s$.

In addition, if $\{\alpha_k\}_{k=0,\ldots,m}$ and $\{\beta_k\}_{k=0,\ldots,m}$ belongs to $T_0(\mathbb{R}^2)$, then

$$\bigotimes_{k=0}^{m}(1+\alpha_k+\beta_k) = \bigotimes_{k=1}^{m}(1+\alpha_k) + \sum_{k=0,\ldots,m} \alpha_k^1 \otimes \left(\sum_{\ell=k+1,\ldots,m}\beta_\ell^1\right)$$

$$+ \sum_{k=0,\ldots,m}\beta_k^1 \otimes \left(\sum_{\ell=k+1,\ldots,m}\alpha_\ell^1\right) + \sum_{k=0,\ldots,m}\beta_k^1 \otimes \left(\sum_{\ell=k+1,\ldots,m}\beta_\ell^1\right)$$

with $\alpha_k^1 = \pi_{\mathbb{R}^2}(\alpha_k)$ and $\beta_k^1 = \pi_{\mathbb{R}^2}(\beta_k)$. Remark that

$$\sum_{k=0,\ldots,m}\beta_k^1 \otimes \left(\sum_{\ell=k+1,\ldots,m}\alpha_\ell^1\right) = \sum_{\ell=1}^{m}\sum_{k=0}^{\ell}\alpha_k \otimes \beta_\ell,$$

and

$$\sum_{k=0,\ldots,m}\beta_k^1 \otimes \left(\sum_{\ell=k+1,\ldots,m}\beta_\ell^1\right) = \frac{1}{2}\left(\sum_{k=0}^{m}\beta_k\right) \otimes \left(\sum_{k=0}^{m}\beta_k\right).$$

We set

$$1+\alpha_k = \mathfrak{F}(f,\mathbf{y};t_k^n,t_{k+1}^n) \text{ and } \beta_k = \mathfrak{E}_{T_1(\mathbb{R}^d)}(f)(\mathbf{x}_{t_k^n})(\mathbf{z}_{t_{k+1}^n}-\mathbf{z}_{t_k^n}).$$

Furthermore, $\sum_{k=0,\ldots,\underline{M}(t,n)}\alpha_k^1 \longrightarrow \pi_{\mathbb{R}^2}(\mathfrak{I}(\mathbf{x};0,t))$, while $\sum_{k=0,\ldots,\underline{M}(t,n)}\beta_k^1$ converges to $\int_0^t \nabla f(x_s)\,\mathrm{d}\mathbf{z}_s$. Remark also that if $\beta_k^2 = \pi_{\mathbb{R}^2 \otimes \mathbb{R}^2}(\beta_k)$, then

$$\sum_{k=0}^{\underline{M}(t,n)}\beta_k^2 \xrightarrow[n\to\infty]{} \int_0^t f(x_s) \otimes f(x_s)\,\mathrm{d}\mathbf{z}_s.$$

By combining all these facts and using techniques similar to those in [52, Section 3.3.3, p. 56] or in [53], since the components of $\int_0^t \nabla f(x_s)\,\mathrm{d}\mathbf{z}_s$ are 2α-Hölder continuous, we can get

$$\bigotimes_{k=0}^{\underline{M}(t,n)-1} \mathfrak{F}(f,\mathbf{x};t_k^n,t_{k+1}^n) \xrightarrow[n\to\infty]{} \int_0^t f(x_s)\,\mathrm{d}\mathbf{y}_s + \mathfrak{K}(\mathbf{y},\mathbf{z};0,t)$$

with

$$\mathfrak{K}(\mathbf{y},\mathbf{z};0,t) = \int_0^t \nabla f(y_s)\,\mathrm{d}\mathbf{z}_s + \int_0^t f(y_s) \otimes f(y_s)\,\mathrm{d}\mathbf{z}_s$$

$$+ \sum_{k,\ell=1,2}\check{e}_k \otimes \check{e}_\ell \int_0^t \nabla f^k(y_s)\left(\int_0^s f^\ell(y_s)\,\mathrm{d}\mathbf{y}_s\right)\mathrm{d}\mathbf{z}_s$$

$$+ \sum_{k,\ell=1,2}\check{e}_k \otimes \check{e}_\ell \int_0^t f^k(y_s)\left(\int_0^s \nabla f^\ell(y_s)\,\mathrm{d}\mathbf{z}_s\right)\mathrm{d}\mathbf{y}_s$$

$$+ \frac{1}{2}\left(\int_0^t \nabla f(y_s)\,\mathrm{d}\mathbf{z}_s\right) \otimes \left(\int_0^t \nabla f(y_s)\,\mathrm{d}\mathbf{z}_s\right). \quad (65)$$

In the previous expression, we have to remember that $\mathbf{x}$ and $\mathbf{y}$ live above the same path $x = y$.

Thus, if for each $n \in \mathbb{N}$, $\mathbf{z}^n$ belongs to $C_p^{\infty}([0, T]; S(\mathbb{R}^2))$ and converges to $\mathbf{z}$, while $\mathbf{y}^n \in C_p^{\infty}([0, T]; G(\mathbb{R}^2))$ converges to $\mathbf{y}$, one gets that $\mathbf{x}^n = \mathbf{y}^n + \mathbf{z}^n$ converges to $C^{\alpha}([0, T]; T_1(\mathbb{R}^2))$ and

$$\mathfrak{I}(\mathbf{x}) = \lim_{n \to \infty} (\mathfrak{I}(\mathbf{y}^n) + \mathfrak{K}(\mathbf{y}^n, \mathbf{z}^n)),$$

where the limit is in $C^{\beta}([0, T]; T_1(\mathbb{R}^2))$ for all $\beta < \alpha$. Of course, both $\mathfrak{K}(\mathbf{y}^n, \mathbf{z}^n)$ and $\mathfrak{I}(\mathbf{y}^n)$ correspond to integrals of differential forms along piecewise smooth paths, and hence to ordinary integrals.

Yet the following fact has to be noted: If $\mathbf{x} \in C^{\alpha}([0, T]; T_1(\mathbb{R}^2))$ but $\mathbf{x} \notin C^{\alpha}([0, T]; G(\mathbb{R}^2))$, then it is not possible to find a family $(\mathbf{x}^n)_{n \in \mathbb{N}}$ of smooth rough paths such that $\mathfrak{I}(\mathbf{x}^n)$ converges to $\mathfrak{I}(\mathbf{x})$. This means that $\mathfrak{I}(\mathbf{x})$ cannot be approximated by the ordinary integrals $\mathfrak{I}(\mathbf{x}^n)$. This motivates our definition of geometric rough paths. However, using the decomposition of $T_1(\mathbb{R}^2)$ as $G(\mathbb{R}^2) \times S(\mathbb{R}^2)$, it is then possible to interpret any α^{-1}-rough path as a geometric $(1/\alpha, 2/\alpha)$-rough path in the sense defined in [53].

7.6 On Geometric Rough Paths Lying Above the Same Path

We have seen in Lemma 6 that if $\mathbf{x}$ and $\mathbf{y}$ are two paths in $C^{\alpha}([0, T]; A(\mathbb{R}^2))$ with $\alpha \in (1/3, 1/2)$ and lying above the same path taking its values in $\mathbb{R}^2$ (i.e., $\pi_{\mathbb{R}^2}(\mathbf{x}) = \pi_{\mathbb{R}^2}(\mathbf{y})$), then there exists a path $\varphi \in C^{2\alpha}([0, T]; \mathbb{R})$ such that $\mathbf{x} = \mathbf{y} + \varphi[e_1, e_2]$. In addition, $(-\mathbf{x}_s) \boxplus \mathbf{x}_t = (-\mathbf{y}_s) \boxplus \mathbf{y}_t + (\varphi_t - \varphi_s)[e_1, e_2]$. Now, if we lift $\mathbf{x}$ and $\mathbf{y}$ as paths in $C^{\alpha}([0, T]; G(\mathbb{R}^2))$ by $\widehat{\mathbf{x}}_t = \exp(\mathbf{x}_t)$ and $\widehat{\mathbf{y}}_t = \exp(\mathbf{y}_t)$, we deduce that there exists $\psi \in C^{\alpha}([0, T]; T_0(\mathbb{R}^2))$ such that $\widehat{\mathbf{x}}_t = \widehat{\mathbf{y}}_t + \psi_t$ and in addition, $\widehat{\mathbf{x}}_s^{-1} \otimes \widehat{\mathbf{x}}_t = \widehat{\mathbf{y}}_s^{-1} \otimes \widehat{\mathbf{y}}_t + \psi_t - \psi_s$. This path is given by

$$\psi_t = \varphi_t e_1 \otimes e_2 - \varphi_t e_2 \otimes e_1 = \varphi_t[e_1, e_2].$$

Each component of ψ is 2α-Hölder continuous. Using the map $\mathfrak{K}$ previously defined by (65), we get

$$\mathfrak{I}(\mathbf{x}) = \mathfrak{I}(\mathbf{y}) + \mathfrak{K}(\mathbf{y}, \psi).$$

Finally, using the fact that ψ is anti-symmetric, setting $[f, f] = \check{e}_1[f^1, f^1] + \check{e}_2[f^2, f^2]$, we get

$$\mathfrak{K}(\mathbf{y}, \psi; s, t) = \int_s^t [f, f](y_s) \, d\varphi_s$$

$$+ \sum_{k, \ell = 1, 2} \check{e}_k \otimes \check{e}_\ell \int_0^t [f^k, f^k](y_s) \left(\int_0^s f^\ell(y_s) \, dy_s \right) d\varphi_s$$

$$+ \sum_{k,\ell=1,2} \check{e}_k \otimes \check{e}_\ell \int_0^t f^\ell(y_s) \left(\int_0^s [f^k, f^k](y_s) \, d\varphi_s \right) dy_s$$

$$+ \frac{1}{2} \left(\int_0^t [f, f](y_s) \, d\varphi_s \right) \otimes \left(\int_0^t [f, f](y_s) \, d\varphi_s \right)$$

for all $0 \leqslant s \leqslant t \leqslant T$.

If $[f, f] = 0$, we deduce that $\mathfrak{K}(\mathbf{y}, \psi) = 0$ and then that $\mathfrak{I}(\mathbf{y}) = \mathfrak{I}(\mathbf{x})$. In other words, any rough path lying above the same path x gives rise to the same integral.

With the results in [54], which assert that it is always possible to lift a path $x \in C^\alpha([0, T]; \mathbb{R}^2)$ to a path $\mathbf{x} \in C^\alpha([0, T]; G(\mathbb{R}^2))$ when $\alpha \in (1/3, 1/2]$, this means that if $[f, f] = 0$, one may define $\mathfrak{I}$ only on $C^\alpha([0, T]; \mathbb{R}^2)$ for $\alpha \in (1/3, 1/2]$; but the continuity of $\mathfrak{I}$ remains an open question.

8 Variations in the Construction of the Integral

8.1 Case of a Path Living in a d-Dimensional Space

The case of a space of dimension d is not harder to treat than the case when $d = 2$; one only has to consider the area between the components grouped in pairs.

The tensor space $T(\mathbb{R}^d)$ then becomes the space $T(\mathbb{R}^d) = \mathbb{R} \oplus \mathbb{R}^d \oplus (\mathbb{R}^d \otimes \mathbb{R}^d)$ whose basis is, if $\{e_1, \ldots, e_d\}$ is a basis of $\mathbb{R}^d$,

$$1, e_1, \ldots, e_d, e_1 \otimes e_1, e_1 \otimes e_2; \ldots, e_d \otimes e_d,$$

hence $T(\mathbb{R}^d)$ is a space of dimension $1 + d + d^2$.

The space $A(\mathbb{R}^2)$ is a space of dimension $d + d(d-1)/2$, with basis

$$\{e_i | i = 1, \ldots, d\} \cup \{[e_i, e_j] | i \neq j, \ i, j = 1, \ldots, d\},$$

where $[e_i, e_j] = e_i \otimes e_j - e_j \otimes e_i$. The space $A(\mathbb{R}^2)$ is then $\mathbb{R}^d \oplus [\mathbb{R}^d, \mathbb{R}^d]$, where $[\mathbb{R}^d, \mathbb{R}^d] = \{[x, y] | x, y \in \mathbb{R}^d\}$.

The applications exp and log are defined as previously:

$$\exp(x) = 1 + x + \frac{1}{2} x \otimes x \quad \text{for } x \in A(\mathbb{R}^d)$$

$$\text{and} \quad \log(1 + x) = x - \frac{1}{2} x \otimes x \quad \text{for } x \in T(\mathbb{R}^d), \ \pi_1(x) = 0.$$

The space $G(\mathbb{R}^d) = \exp(A(\mathbb{R}^d))$ is a subgroup of $(T_1(\mathbb{R}^d), \otimes)$, where $T_1(\mathbb{R}^d) = \{x \in T(\mathbb{R}^d) | \pi_1(x) = 1\}$, and $(A(\mathbb{R}^d), [\cdot, \cdot])$ is the Lie algebra of $(G(\mathbb{R}^d), \otimes)$. It may also be identified with its tangent space at any point.

A smooth path x in $\mathbb{R}^d$ is then lifted into a path $\widehat{\mathbf{x}}$ in $A(\mathbb{R}^d)$ by

$$\widehat{\mathbf{x}}_t = x_t + \sum_{i,j=1,\ldots,d,\ i<j} \mathfrak{A}((x^i, x^j); 0, t)[e_i, e_j],$$

where (x^i, x^j) is the two dimensional path composed of the i-th and j-th component of x. Remark that $\mathfrak{A}((x^i, x^j)) = -\mathfrak{A}((x^j, x^i))$ and $\mathfrak{A}((x^i, x^i)) = 0$.

The path $\widehat{\mathbf{x}}$ is then lifted into a path $\mathbf{x}$ in $\mathbb{R}^d$ by $\mathbf{x} = \exp(\widehat{\mathbf{x}})$, and thus

$$\mathbf{x}_t = 1 + x_t + \sum_{i,j=1}^{d} \int_0^t (x_s^j - x_0^j)\, dx_s^i e_j \otimes e_i.$$

The symmetric part $\mathfrak{s}(\mathbf{x})$ of $\pi_{\mathbb{R}^d \otimes \mathbb{R}^d}(\mathbf{x})$ is

$$\mathfrak{s}(\mathbf{x}_t) = \frac{1}{2}(x_t - x_0) \otimes (x_t - x_0)$$

while the anti-symmetric part $\mathfrak{a}(\mathbf{x})$ of $\pi_{\mathbb{R}^d \otimes \mathbb{R}^d}(\mathbf{x})$ is

$$\mathfrak{a}(\mathbf{x}_t) = \sum_{i,j=1}^{d} \mathfrak{A}((x^i, x^j); 0, t) e_i \otimes e_j = \sum_{\substack{i=1,\ldots,d \\ i<j}} \mathfrak{A}((x^i, x^j); 0, t)[e_i, e_j].$$

Hence, all previous notions and results easily extend to this case.

Finally, note that the theory of rough paths also applies to the infinite dimensional case (see [50] for example).

8.2 Using Iterated Integrals

We saw in Sections 7.1 and 8.1 that a path $x \in C^\alpha([0, T]; \mathbb{R}^d)$ with $\alpha > 1/2$ may be naturally lifted as a path $\mathbf{x}$ in $G(\mathbb{R}^d)$ with

$$\mathbf{x}_t = 1 + x_t + \sum_{i,j=1}^{d} \left(\int_0^t (x_r^i - x_0^i)\, dx_r^j \right) e_i \otimes e_j.$$

The term $K^{i,j}(x; 0, t) = \pi_{e_i \otimes e_j}(\mathbf{x}_t)$ is called an *iterated integral* of x.

Fix $d \geqslant 1$ and consider the tensor space $T^\infty(\mathbb{R}^d)$ defined by

$$T^\infty(\mathbb{R}^d) = \mathbb{R} \oplus \mathbb{R}^d \oplus (\mathbb{R}^d \otimes \mathbb{R}^d) \oplus (\mathbb{R}^d \otimes \mathbb{R}^d \otimes \mathbb{R}^d) \oplus \cdots,$$

and, for a smooth path $x : [0, T] \to \mathbb{R}^d$, the iterated integrals

$$K^{i_1,\ldots,i_\ell}(x; 0, t) = \int_0^t \int_0^{t_1} \cdots \int_0^{t_{\ell-1}} dx_{i_{t_\ell}}^{i_1} \cdots dx_{t_1}^{i_\ell}$$

for each integer ℓ and each $(i_1, \ldots, i_\ell) \in \{1, \ldots, d\}^\ell$. It was noted first by K.T. Chen in the 50's [10, 11] that the formal power series

$$\Psi(x;0,t) = \sum_{\ell \geq 0} \sum_{(i_1,\ldots,i_\ell) \in \{1,\ldots,d\}^\ell} K^{i_1,\ldots,i_\ell}(x;0,t) e_{i_1} \otimes \cdots \otimes e_{i_\ell}$$

in $T^\infty(\mathbb{R}^d)$ provides an algebraic way to encode the geometric object which is the path x. Therefore, $\Psi(x;0,t)$ is sometimes called the *signature* of the path.

With the tensor product $\otimes$, $T^\infty(\mathbb{R}^d)$ remains a group, and thus if x : $[0,T] \to \mathbb{R}^d$ and $y : [0,S] \to \mathbb{R}^d$ are two smooth paths, $\Psi(x \cdot y; 0, T + S) = \Psi(x; 0, T) \otimes \Psi(y; 0, S)$. In addition, if $\overline{x}$ is the path $\overline{x}_t = x_{T-t}$, then $\Psi(\overline{x}; T) = \Psi(x; T)^{-1}$. The signature characterizes x in the sense that there is a one-to-one equivalence[3] between the algebraic object $\Psi(x)$ and the geometric object x in $C_p^\infty([0,T]; \mathbb{R}^d)$ (see also [38] for some extension).

Let $[x,y]$ be the Lie bracket $[x,y] = x \otimes y - y \otimes x$. Denote by $A^\infty(\mathbb{R}^d)$ the subset of $T^\infty(\mathbb{R}^d)$ defined by

$$A^\infty(\mathbb{R}^d) = \mathbb{R}^d \oplus [\mathbb{R}^d, \mathbb{R}^d] \oplus [\mathbb{R}^d, [\mathbb{R}^d, \mathbb{R}^d]] \oplus \cdots.$$

This subset is stable under Lie brackets $[\cdot, \cdot]$. The tensor space $T^\infty(\mathbb{R}^d)$ is the universal Lie algebra of $A^\infty(\mathbb{R}^d)$ (see [63] for example). One may then define two maps $\exp : A^\infty(\mathbb{R}^d) \to T_1^\infty(\mathbb{R}^d)$ and $\log : T_1^\infty(\mathbb{R}^d)$, where $T_1^\infty(\mathbb{R}^d)$ is the subset of $T^\infty(\mathbb{R}^d)$ such that $\pi_\mathbb{R}(x) = 1$, which are given by

$$\exp(x) = 1 + x + \frac{1}{2}x \otimes x + \frac{1}{6}x \otimes x \otimes x + \cdots,$$

$$\log(1+x) = x - \frac{1}{2}x \otimes x + \frac{1}{3}x \otimes x \otimes x - \cdots.$$

In particular, if $G^\infty(\mathbb{R}^d) = \exp(A^\infty(\mathbb{R}^d))$, then $(G^\infty(\mathbb{R}^d), \otimes)$ is a closed subgroup of $(T_1^\infty(\mathbb{R}^d), \otimes)$. In addition, $\exp$ is one-to-one from $A^\infty(\mathbb{R}^d)$ to $G^\infty(\mathbb{R}^d)$, and $\log$ is its inverse.

One of the striking results of K.T. Chen, which uses some properties of the iterated integrals, is that $\Psi(x;0,t)$ belongs to $G^\infty(\mathbb{R}^d)$, or equivalently, with the Baker-Campbell-Hausdorff-Dynkin formula, that $\log(\Psi(x;0,t))$ belongs to $A^\infty(\mathbb{R}^d)$.

This approach proved to be very useful, since it allows to consider equations driven by smooth paths or differential equations in an algebraic setting, and allows formal computations. Numerous topics in control theory use this point of view (see for example [23, 39, 41]). It was also used in the stochastic context to deal with flows of stochastic differential equations (see for example [2, 8, 24, 72]... or the book [4]).

For some integer k, we may truncate Ψ by considering that $e_{i_1} \otimes \cdots \otimes e_{i_\ell} = 0$ for all $\ell > k$. For such a truncated power series $\Psi_k(x)$ we still get the relationship $\Psi_k(x \cdot y; 0, T + S) = \Psi_k(x; 0, T) \otimes \Psi_k(y; 0, S)$. In particular, we deduce that

---

[3] In fact, for this equivalence to be exactly one-to-one, one has to eliminate the paths such that, one some time interval, x goes from a point a to a point b and then back to a by reversing the path.

$$\Psi_k(x_{|[s,t]}; s, t) = \Psi_k(x_{|[s,r]}; s, r) \otimes \Psi_k(x_{|[r,t]}; r, t)$$

for all $0 \leqslant s \leqslant r \leqslant t \leqslant T$. With $k = 2$, we get exactly that our natural lift of $\mathbf{x}_t = \Psi_2(x; 0, t)$ satisfies the relationship $\mathbf{x}_{s,t} = \mathbf{x}_{s,r} \otimes \mathbf{x}_{r,t}$.

Thus, given a path $\mathbf{x}$ in $T_1(\mathbb{R}^2)$, one can think of $\pi_{e_i \otimes e_j}(\mathbf{x}_t)$ as the iterated integrals of x^j against x^i. Of course, one knows that for irregular paths, there is no canonical way to define them (think of Brownian motion trajectories). Anyway, for weak geometric rough paths, these iterated integrals are approximated by iterated integrals of some smooth paths.

We may now present another heuristic argument to derive the expression of $\mathfrak{F}(f, \mathbf{x}; s, t)$ and then (62). This argument is the historical one (see [42, 44, 52, 55]). Consider a smooth path $x : [0, T] \to \mathbb{R}^d$ and a smooth function $f = (f_1, \ldots, f_d)$. Then using a Taylor expansion, one gets

$$\sum_{i=1}^{d} \int_s^t f_i(x_r) \, \mathrm{d}x_r^i = \sum_{i=1}^{d} f_i(x_s)(x_t^i - x_s^i)$$

$$+ \sum_{\ell \geqslant 1} \sum_{(i_1, \ldots, i_\ell) \in \{1, \ldots, d\}^\ell} \frac{\partial^\ell f_i}{\partial x_{i_1} \cdots \partial x_{i_\ell}}(x_0) K^{i_\ell, \ldots, i_1, i}(x; s, t)$$

$$= \mathfrak{E}_{T^\infty(\mathbb{R}^d)}(f)(x_s) \Psi(x; s, t)$$

with, for $z \in \mathbb{R}^d$,

$$\mathfrak{E}_{T^\infty(\mathbb{R}^d)}(f)(z) = \sum_{\ell \geqslant 0} \sum_{(i_1, \ldots, i_\ell) \in \{1, \ldots, d\}^\ell} \frac{\partial^\ell f_i}{\partial x_{i_1} \cdots \partial x_{i_\ell}}(z) e^{i_1} \otimes \cdots \otimes e^{i_\ell}.$$

In the usual case, we keep only the first term $\sum_{i=1}^{d} f_i(x_s)(x_t^i - x_s^i)$ as an approximation of $\sum_{i=1}^{d} \int_s^t f_i(x_r) \, \mathrm{d}x_r^i$, and we use it as the term in a Riemann sum. Keeping higher order terms has no influence, since $K^{i_1, \ldots, i_\ell}(x; s, t) \leqslant (1/\ell!) \|x'\|_\infty^\ell (t - s)^\ell$.

The idea is then to keep enough terms, if x is α-Hölder continuous and we get an object $\mathbf{x}^{(k)}$ having the same algebraic properties as $\Psi_k(x; s, t)$ for some integer k, to get a Riemann sum that converges. In [52, 55], T. Lyons and co-authors proved that the number of terms must be $k = \lfloor 1/\alpha \rfloor$. In particular, from $\mathbf{x}^{(k)}$, it is possible to reconstruct an object living in $T^\infty(\mathbb{R}^d)$, equal to $\Psi(x)$ when x is smooth, and possessing the same algebraic properties as $\Psi(x)$.

For $k = 2$ and using the path $\mathbf{x}$ as the object $\mathbf{x}^{(2)}$, we get the expression (62).

8.3 Paths with Quadratic Variation

For Brownian motion or a semi-martingale, one knows how to construct several integrals—the major ones are the Itô and the Stratonovich integrals—whose difference depends on the fact that their trajectories have finite quadratic variation.

With the theory of rough paths, we can indeed construct a pathwise equivalent theory of the Itô integral. For this, we need the path to have a quadratic variation.

Definition 7. *Given* $\alpha \in (1/3, 1/2]$, *a path* $x \in C^\alpha([0, T]; \mathbb{R}^2)$ *has a quadratic variation if there exists a process* $\mathfrak{Q}(x) \in C^\alpha([0, T]; S(\mathbb{R}^2))$ *such that* $\xi_0 = 0$ *and, writing* $z^{\otimes 2}$ *instead of* $z \otimes z$ *for* $z \in \mathbb{R}^2$,

$$\mathfrak{Q}_n(x; t) = \frac{t - t^n_{\underline{M}(t,n)}}{t^n_{\underline{M}(t,n)+1} - t^n_{\underline{M}(t,n)}} (x_{t^n_{\underline{M}(t,n)+1}} - x_{t^n_{\underline{M}(t,n)}})^{\otimes 2}$$

$$+ \sum_{k=0}^{\underline{M}(t,n)-1} (x_{t^n_{k+1}} - x_{t^n_k})^{\otimes 2}$$

and $\mathfrak{Q}(x) = \lim_{n \to \infty} \mathfrak{Q}_n(x)$ *where the limit holds in* $C^\alpha([0, T]; S(\mathbb{R}^2))$.

Remark 16. Note that with the norm we use, this means that the components of $\mathfrak{Q}(x)$ are 2α-Hölder continuous.

Remark 17. If $x \in C^\alpha([0, T]; \mathbb{R}^2)$ with $\alpha > 1/2$, then it is easily seen that necessarily, $\mathfrak{Q}(x; t) = 0$ for $t \in [0, T]$.

The trajectories of the Brownian motion and of Hölder continuous martingales present this feature (see [13, 65]).

Thus, a natural expression for the equivalent of the Itô integral consists in considering the path $\mathbf{x}^n$ defined in (57), and in setting

$$\mathfrak{D}(\mathbf{x}^n; 0, t) = \sum_{k=0 \text{ s.t. } t^n_k \leq t} \mathfrak{E}_{A(\mathbb{R}^2)}(f)(\mathbf{x}^n_{t^n_k}) \frac{d\mathbf{x}^n(t^n_k)}{dt} \Delta_n t$$

where $\mathfrak{E}_{A(\mathbb{R}^2)}(f)(\mathbf{x}^n_{t^n_k})$ has been defined by (55). This construction differs from (55), since

$$\sum_{k=0 \text{ s.t. } t^n_k \leq t} \mathfrak{E}_{A(\mathbb{R}^2)}(f)(\mathbf{x}_{t^n_k}) \frac{d\mathbf{x}^n(t^n_k)}{dt} \Delta_n t$$

$$= \sum_{k=0 \text{ s.t. } t^n_i \leq t} \int_{t^n_i}^{t^n_{i+1}} \mathfrak{E}_{A(\mathbb{R}^2)}(f)(\mathbf{x}_{t_k}) \log(\mathbf{x}_{t^n_k, t^n_{k+1}}) \, ds.$$

Comparing with (59) leads to

$$\mathfrak{D}(\mathbf{x}^n; 0, t) = \sum_{k=0 \text{ s.t. } t^n_i \leq t} \mathfrak{F}(f, \mathbf{x}, t^n_k, t^n_{k+1}) - \nabla f(x_{t^n_k}) \mathfrak{s}(\mathbf{x}_{t^n_k, t^n_{k+1}}).$$

If x has a quadratic variation $\mathfrak{Q}(x)$, then the components of $\mathfrak{Q}(x)$ are 2α-Hölder continuous. In addition, the components of ∇f belongs to the space $\text{Lip}(\gamma - 1; \mathbb{R}^2 \to \mathbb{R}^2)$. Hence, since $\mathfrak{Q}_n(x)$ converges to $\mathfrak{Q}(x)$ and

$$\left| \nabla f(x_{t_k^n}) \mathfrak{s}(\mathbf{x}_{t_k^n, t_{k+1}^n}) - \int_{t_k^n}^{t_{k+1}^n} \nabla f(x_s) \, \mathrm{d}\, \mathfrak{Q}_n(x; s) \right|$$

$$\leqslant \Delta_n t^{\alpha(1+\gamma)} \|f\|_{\mathrm{Lip}} \|\mathbf{x}\|_\alpha^{\alpha(1+\gamma)},$$

we easily get convergence of the last term to the Young integral defined by $\frac{1}{2} \int_0^T \nabla f(x_r) \, \mathrm{d}\, \mathfrak{Q}(x; r)$.

Thus, the limit of $\mathfrak{D}(\mathbf{x}; 0, t)$ is $\mathfrak{J}(\mathbf{x}; 0, t) - \int_0^t \frac{1}{2} \nabla f(x_s) \, \mathrm{d}\, \mathfrak{Q}(x_s)$ for $t \in [0, T]$. The integral $\mathfrak{D}(\mathbf{x})$ thus constructed is the same at the first level as if we had used the $(1/\alpha, 2/\alpha)$-Hölder continuous rough path $(\mathbf{x}, -\frac{1}{2} \mathfrak{Q}(x))$ (see [53]).

8.4 Link with Stochastic Integrals

Itô and Stratonovich integrals are defined as limits in probability of Riemann sums. On the other hand, the rough path theory gives a pathwise definition of the integral, but the price to pay is to add a supplementary information. Is there some link between both integrals?

Let B be a d-dimensional Brownian motion (a semi-martingale may just as well be used). A natural way to construct a rough path $\mathbf{B}$ lying above B is to set

$$\pi_{e_i \otimes e_j}(\mathbf{B}_t) = \int_0^t (B_r^i - B_0^i) \circ \mathrm{d}B_r^j.$$

for $i, j = 1, \ldots, d$. For the construction of $\mathbf{B}$ as a rough path, see for example [13, 44, 52, 65]. The process $\log(\mathbf{B})$ is called the *Brownian motion on the Heisenberg group*, and has been widely studied (See references in Section B).

Continuity of the rough path integral and the Wong-Zakai theorem allow us to identify the integral $\mathfrak{J}(\mathbf{B}; 0, T)$ with the Stratonovich integral given by

$$\int f(B_s) \circ \mathrm{d}B_s = \lim_{n \to \infty} \sum_{k=0}^{2^n - 1} \frac{1}{2} (f(B_{t_{k+1}^n}) + f(B_{t_k^n}))(B_{t_{k+1}^n} - B_{t_k^n})$$

where the limit is a limit in probability. We will see here that there is another relationship between both integrals without invoking this continuity result, and that the construction of the Stratonovich and Itô integrals (although under stronger condition on the function f than the one required by the "classical" theory) can be deduced from the rough paths theory.

The theory of rough paths also gives a better intuitive understanding of the counter-examples to the Wong-Zakai theorem (see [40, 59] for SDEs and [48] in the context of rough paths).

The projection on $\mathbb{R}^d$ of $\mathfrak{J}(\mathbf{B}; 0, T)$ is given by

$$\pi_{\mathbb{R}^d}(\mathfrak{J}(\mathbf{B}; 0, T))$$

$$= \lim_{n \to \infty} \sum_{k=0}^{2^n - 1} \left(f(B_{t_k^n})(B_{t_{k+1}^n} - B_{t_k^n}) + \nabla f(B_{t_k^n}) \pi_{\mathbb{R}^d \otimes \mathbb{R}^d} (\mathbf{B}_{t_k^n, t_{k+1}^n}) \right) \quad (66)$$

which we rewrite using $\mathfrak{a}$ and $\mathfrak{s}$ as

$$\pi_{\mathbb{R}^d}(\mathfrak{I}(\mathbf{B};0,T)) = \lim_{n\to\infty} \sum_{k=0}^{2^n-1} \Big(f(B_{t_k^n})(B_{t_{k+1}^n} - B_{t_k^n}) + \nabla f(B_{t_k^n})\mathfrak{a}(\mathbf{B}_{t_k^n,t_{k+1}^n})$$
$$+ \nabla f(B_{t_k^n})\mathfrak{s}(\mathbf{B}_{t_k^n,t_{k+1}^n}) \Big).$$

But we have seen that

$$f(B_{t_k^n})(B_{t_{k+1}^n} - B_{t_k^n}) + \nabla f(B_{t_k^n})\mathfrak{s}(\mathbf{B}_{t_k^n,t_{k+1}^n}) \approx \int_{t_k^n}^{t_{k+1}^n} f(B_s^{\Pi^n})\,\mathrm{d}B_s^{\Pi^n}$$

with B^{Π^n} the piecewise linear approximation of B along the dyadic partition Π^n, and $\approx$ meaning that the difference between the two terms is less than $C2^{-n\theta}$ with $\theta > 1$.

On the other hand, using $f(x) - f(y) = \int_0^1 \nabla f(x + \tau(y-x))(y-x)\,\mathrm{d}\tau$ and the change of variable $\tau' = 2^n\tau$, we get that for $k = 0,\dots,2^n$,

$$\sum_{i=1}^{d}(f_i(B_{t_{k+1}^n}) - f_i(B_{t_k^n}))(B_{t_{k+1}^n}^i - B_{t_k^n}^i)$$

$$= \int_{t_k^n}^{t_{k+1}^n} \sum_{i,j=1}^{d} \frac{\partial f_i}{\partial x_j}(B_s^{\Pi^n})(B_{t_{k+1}^n}^j - B_{t_k^n}^j)(B_{t_{k+1}^n}^i - B_{t_k^n}^i)2^n\,\mathrm{d}s$$

$$\approx \nabla f(B_{t_k^n})(B_{t_{k+1}^n} - B_{t_k^n}) \otimes (B_{t_{k+1}^n} - B_{t_k^n}).$$

With (45), $\mathfrak{s}(\mathbf{B}_{s,t}) = \frac{1}{2}(B_t - B_s) \otimes (B_t - B_s)$. This implies that

$$\sum_{i=1}^{d} \frac{1}{2}(f_i(B_{t_{k+1}^n}) - f_i(B_{t_k^n}))(B_{t_{k+1}^n}^i - B_{t_k^n}^i) \approx \nabla f(B_{t_k^n})\mathfrak{s}(\mathbf{B}_{t_k^n,t_{k+1}^n}).$$

Now, remark that if $M_k = \mathfrak{a}(\mathbf{B}_{t_k^n,t_{k+1}^n})$, then $(\sum_{\ell=0}^{k} M_\ell)_{k=0,\dots,2^n}$ forms a martingale with respect to $(\mathcal{F}_k)_{k=0,\dots,2^n}$, where $(\mathcal{F}_t)_{t\geqslant 0}$ is the filtration of the Brownian motion. In addition,

$$\mathbb{E}[(M_k)^2] \leqslant \frac{6T^2}{2^{2n}}.$$

Hence,

$$\mathbb{E}\left[\left(\sum_{k=0}^{2^n-1} \nabla f(B_{t_k^n})\mathfrak{a}(\mathbf{B}_{t_k^n,t_{k+1}^n})\right)^2\right] \leqslant \frac{6T^2}{2^n}\|\nabla f\|_\infty$$

and the latter term converges to 0 in probability. Convergence in probability of the Stratonovich integral follows from the last convergence and the almost sure convergence of the rough path approximation given in (66).

Regarding the Itô integrals, we lift the Brownian motion B as a Brownian motion $\mathbf{B}'$ with $\pi_{\mathbb{R}^2}(\mathbf{B}') = B$ and

$$\pi_{e_i \otimes e_j}(\mathbf{B}'_t) = \int_0^t (B^i_r - B^i_0)\, \mathrm{d}B^j_r = \int_0^t (B^i_r - B^i_0) \circ \mathrm{d}B^j_r - \frac{1}{2}\delta_{i,j}t.$$

However, note that the anti-symmetric part $\mathfrak{a}(\mathbf{B}')$ is equal to the anti-symmetric part of $\mathfrak{a}(\mathbf{B})$. Indeed, due to the Wong-Zakai theorem [40], $\mathbf{B}$ is a geometric rough path, while $\mathbf{B}'$ is not a geometric rough path. From the previous computations, we easily get

$$\pi_{\mathbb{R}^d}(\mathfrak{I}(\mathbf{B}'; 0, T)) = \mathfrak{I}_{\mathbb{R}^d}(\mathfrak{I}(\mathbf{B}; 0, T)) - \frac{1}{2}\sum_{i=1}^d \int_0^T \frac{\partial f_i}{\partial x_i}(B_s)\, \mathrm{d}s = \int_0^T f(B_s)\, \mathrm{d}B_s$$

and thus $\mathbf{B}'$ gives rise to the Itô integral. The effect of the bracket terms $t \mapsto \langle B^i, B^j \rangle_t = \delta_{i,j}t$ on $\mathfrak{I}(\mathbf{B}')$ with respect to $\mathfrak{I}(\mathbf{B})$ in studied in Section 7.5.

9 Solving a Differential Equations

The theory of rough paths may be applied to solve differential equations, since one can transform integrals into differential equations using a fixed point principle. Indeed, as noted in Section 8.2, most ideas from the rough path theory come from developments around iterated integrals as a way to deal formally with ordinary differential equations. Thus, the algebraic structures we used were introduced in the context of differential equations, not integrals (see for example [10, 57, 66]... and also [2, 8, 24, 72]... on Stratonovich stochastic differential equations).

We wish now to consider the following differential equation

$$y_t = y_0 + \int_0^t g(y_s)\, \mathrm{d}x_s, \tag{67}$$

where x is an irregular path. We assume that x lives in $\mathbb{R}^d$, and y lives in $\mathbb{R}^m$. Denote by $\{e_1, \ldots, e_d\}$ (resp. $\{\bar{e}_1, \ldots, \bar{e}_m\}$) the canonical basis of $\mathbb{R}^d$ (resp. $\mathbb{R}^m$). If one wishes to interpret this integral as a rough path, one first has to transform the vector field

$$g(z) = \sum_{\substack{i=1,\ldots,d \\ k=1,\ldots,m}} \bar{e}_k g_i^k(z) \frac{\partial}{\partial x_i}$$

into a differential form h which is integrated along a path (x, y) living in $\mathbb{R}^d \oplus \mathbb{R}^m$. For this, the natural extension is

$$h(z, z') = \sum_{\substack{i=1,\ldots,d \\ k=1,\ldots,m}} \bar{e}_k g_i^k(z')e^i + \sum_{i=1}^d e_i \cdot e^i, \quad z \in \mathbb{R}^d, \; z' \in \mathbb{R}^m.$$

Hence, if x is smooth and (67) has a smooth solution y,

$$(x_t, y_t) = (x_0, y_0) + \int_0^t h(x_s, y_s)\, \mathrm{d}(x_s, y_s) = (x_0, y_0) + \int_{(x,y)_{|[0,t]}} h.$$

In order to deal with an irregular path x, the last integral will be defined as a rough path, which means that we shall consider a rough path $\mathbf{z}$ living above (x, y), in the tensor space $\mathrm{T}_1(\mathbb{R}^d \oplus \mathbb{R}^m)$. We also have to extend the differential form h. For $(z, z') \in \mathbb{R}^d \oplus \mathbb{R}^m$, define by $\mathfrak{E}_{\mathrm{T}_1(\mathbb{R}^d \oplus \mathbb{R}^m)}(h)(z, z')$ the linear form on $\mathrm{T}_0(\mathbb{R}^d \oplus \mathbb{R}^m)$ by

$$\mathfrak{E}_{\mathrm{T}_1(\mathbb{R}^d \oplus \mathbb{R}^m)}(h)(z, z') = h(z, z') + \sum_{\substack{i=1,\ldots,d \\ k,\ell=1,\ldots,m}} \overline{e}_k \frac{\partial g_i^k}{\partial x_\ell}(z')\overline{e}^\ell \otimes e^i$$

$$+ \sum_{\substack{k,\ell=1,\ldots,m \\ i,j=1,\ldots,d}} \overline{e}_k \otimes \overline{e}_\ell g_i^k(z')g_j^\ell(z')e^i \otimes e^j + \sum_{i,j=1,\ldots,d} e_i \otimes e_j \cdot e^i \otimes e^j$$

$$+ \sum_{\substack{k=1,\ldots,m \\ i,j=1,\ldots,d}} \overline{e}_k \otimes e_j g_i^k(z)e^i \otimes e^j + \sum_{\substack{k=1,\ldots,m \\ i,j=1,\ldots,d}} e_i \otimes \overline{e}_k g_j^k(z)e^i \otimes e^j.$$

Then, use Remark 15 to transform this linear form into a differential form on $\mathrm{T}_1(\mathbb{R}^d \oplus \mathbb{R}^m)$. The idea is now to apply a Picard iteration scheme. Define by $\mathfrak{J}$ the integral with respect to the differential form h. If $\mathbf{z}^0$ is a rough path in $C^\alpha([0,T]; \mathrm{T}_1(\mathbb{R}^d \oplus \mathbb{R}^m))$ lying above (x, y^0) for some path $y^0 \in C^\alpha([0,T]; \mathbb{R}^m))$ and $\pi_{\mathrm{T}_1(\mathbb{R}^d)}(\mathbf{z}^0) = \mathbf{x}$, then set recursively $\mathbf{z}^{k+1} = \mathfrak{J}(\mathbf{z}^k)$. The problem is to study the convergence of $(\mathbf{z}^k)_{k \in \mathbb{N}}$.

Definition 8. *A solution of* (67) *is a rough path* $\mathbf{z}$ *living in* $\mathrm{T}_1(\mathbb{R}^d \oplus \mathbb{R}^m)$ *with* $\mathbf{z}_0 = (x_0, y_0, 0)$ *and such that* $\mathfrak{J}(\mathbf{z}; s, t) = \mathbf{z}_{s,t}$ *for all* $0 \leqslant s \leqslant t \leqslant T$ *and* $\pi_{\mathrm{T}_1(\mathbb{R}^d)}(\mathbf{z}) = \mathbf{x}$.

Let us start our study by the following observation: from the choice of h, $\pi_{\mathrm{T}_1(\mathbb{R}^d)}(\mathbf{z}^k)$ is equal to $\mathbf{x}$, whatever k. In addition, to compute $\mathbf{z}^{k+1}$, we need $\mathbf{x} = \pi_{\mathrm{T}_1(\mathbb{R}^d)}(\mathbf{z}^k)$, $\pi_{\mathbb{R}^m}(\mathbf{z}^k)$ and $\pi_{\mathbb{R}^m \otimes \mathbb{R}^d}(\mathbf{z}^k)$. If $\mathbf{z}^k$ lies above (x, y^k), the last term corresponds to the iterated integrals of y^k against x.

For proofs, the reader is referred to [55, Section 4.1, p. 296], [52, Chapter 6, p. 148] and to [53].

Theorem 4. *Let* $\mathbf{x}$ *be a rough path in* $C^\alpha([0,T]; \mathrm{T}_1(\mathbb{R}^d))$. *Let* $g_1, \ldots, g_d$ *be vector fields on* $\mathbb{R}^m$ *with derivatives bounded and* κ-*Hölder continuous where* $\alpha(2 + \kappa) > 1$. *Then there exists at least one solution to* (67) *in* $C^\alpha([0,T]; \mathrm{T}_1(\mathbb{R}^d \oplus \mathbb{R}^m))$.

If $g_1, \ldots, g_d$ *are vector fields on* $\mathbb{R}^m$ *that are twice differentiable, and if, for* $i, j, k = 1, \ldots, d$, $\partial_{x_j} g_i$ *is bounded and* $\partial^2_{x_k, x_j} g_i$ *is bounded and* κ-*Hölder continuous where* $\alpha(2+\kappa) > 1$, *then the solution of* (67) *is unique and* $\mathbf{x} \mapsto \mathbf{z}$ *is continuous from* $(C^\alpha([0,T]; \mathrm{T}_1(\mathbb{R}^d)), \|\cdot\|_\alpha)$ *to* $(C^\alpha([0,T]; \mathrm{T}_1(\mathbb{R}^d \oplus \mathbb{R}^m)), \|\cdot\|_\alpha)$.

Remark 18. The map $\mathbf{x} \mapsto \mathbf{z}$ is called the *Itô map*. Its differentiability is studied in [51, 52], in [49] (for $\alpha > 1/2$) and in [32].

Here again, because of the continuity of $\mathbf{x} \mapsto \mathbf{z}$, we get that if $\mathbf{x}$ belongs to $C^\alpha([0,T]; G(\mathbb{R}^d))$, then $\mathbf{z} \in C^\alpha([0,T]; G(\mathbb{R}^d \oplus \mathbb{R}^m))$ and if $\mathbf{x}$ belongs to $C^{0,\alpha}([0,T]; G(\mathbb{R}^d))$, then $\mathbf{z} \in C^{0,\alpha}([0,T]; G(\mathbb{R}^d \oplus \mathbb{R}^m))$.

Finally, the solution of (67) may also be interpreted using an Euler scheme, as in [32], following [16]. In addition, A.M. Davie proved in [16] that there exists a unique solution if g_i are of class C^2, and that the solution may not be unique if g_i only has Hölder continuous derivatives.

A Carnot Groups and Homogeneous Gauges and Norms

Let $(G, \times)$ be a Lie group, and $(g, [\cdot, \cdot])$ be its Lie algebra. G is a *Carnot group of step k* [4, 60] if for some positive integer k, $g = V_1 \oplus V_2 \oplus \cdots \oplus V_k$—this decomposition being called a *stratification*—with

$$[V_1, V_i] = V_{i+1} \text{ for } i = 1, \ldots, k-1 \text{ and } [V_1, V_k] = \{0\},$$

where $[V_i, V_j] = \{[x,y] \mid x \in V_i, y \in V_j\}$. A Carnot group is naturally equipped with a *dilation* operator $\delta_\lambda(x) = (\lambda^{\alpha_1} x^1, \cdots, \lambda^{\alpha_k} x^k)$ with $x^i \in \exp(V^i)$ and some positive real numbers $\alpha_1, \ldots, \alpha_k$, where exp is the map from g to G. This dilation operator must verify $\delta_\lambda(x \times y) = \delta_\lambda(x) \times \delta_\lambda(y)$. If the dimension of V_1 is finite, the real number $N = \alpha_1 \dim(V_1) + \cdots + \alpha_k \dim(V_k)$ is called the *homogeneous dimension*.

On G equipped with a dilation operator δ, a *homogeneous gauge* is a continuous function which maps x into a non-negative real number $\|x\|$ such that $\|x\| = 0$ if and only if x is the neutral element of G, and for all $\lambda \in \mathbb{R}$, $\|\delta_\lambda(x)\| = |\lambda| \cdot \|x\|$.

A homogeneous gauge is a *homogeneous norm* if $\|x^{-1}\| = \|x\|$ for all $x \in G$. In addition, this homogeneous norm is said to be *sub-additive* if $\|x \times y\| \leqslant \|x\| + \|y\|$ for all $x, y \in G$.

If V_1 is of finite dimension, then a homogeneous norm always exists [27]. For this, equip the Lie algebra g with the Euclidean norm $|\cdot|$ and denote by exp the canonical diffeomorphism from g to G. For $x \in g$, let $r(x)$ be the smallest positive real such that $|\delta_{r(x)} x| = 1$, which exists, since $|\delta_r x|$ is increasing from $[0, +\infty)$ to $[0, +\infty)$. Then, for $y \in G$, $\|y\| = 1/r(\exp^{-1} y)$ defines a symmetric homogeneous norm.

Two homogeneous gauges $\|\cdot\|$ and $\|\cdot\|'$ are said to be equivalent if for some constants C and C', $C\|x\|' \leqslant \|x\| \leqslant C'\|x\|'$ for all $x \in G$.

Proposition 10 ([34]). *If the dimension of V_1 is finite, then all homogeneous gauges are equivalent. In addition, for a homogeneous gauge $\|\cdot\|$, there exist some constants C and C' such that $\|x^{-1}\| \leqslant C\|x\|$ and $|x \times y| \leqslant C'(|x| + |y|)$, for all $x, y \in G$.*

Proof. If $\exp^{-1}(x)$ is decomposed as $y_1, \ldots, y_k$ with $y_i \in V_i$ for $x \in G$, then set $|x|' = \sum_{i=1}^{k} |y_i|^{1/i}$, where $|\cdot|$ denotes the Euclidean norm on each of the finite-dimensional vector spaces V_i. It is easily verified that $|x|$ is a homogeneous gauge. Let $\|\cdot\|$ be another homogeneous gauge. Set $\varphi(x) = \|x\|/|x|'$. Then φ and $1/\varphi$ are continuous on $G \setminus \{1\}$, where 1 is the neutral element of G. As $\{x \in G \mid |x|' = 1\}$ is compact, we easily get that φ and $1/\varphi$ are bounded, and then that for some constants C and C', $C \leqslant \|x\| \leqslant C'$ when $|x|' = 1$. This implies that $\|\cdot\|$ and $|\cdot|'$ are equivalent by using the dilation $\delta_{1/|x|'}$ for a general x.

The other results are proved in a similar way by using $\varphi(x) = \|x^{-1}\|/\|x\|$ and $\varphi(x, y) = \|x \times y\|/(\|x\| + \|y\|)$.

It follows that any homogeneous gauge can be transformed in an equivalent homogeneous norm by setting $\|x\|' = \|x\| + \|x^{-1}\|$.

The notion a *Lipschitz* function is then extended to homogeneous gauges.

Definition 9. *If $(G, \times)$ and $(G', \times)$ are two nilpotent Carnot groups with homogeneous gauges $\|\cdot\|$ and $\|\cdot\|'$, then $f : G \to G'$ is said to be* Lipschitz *if for some constant C,*

$$\|f(x)^{-1} \times f(y)\|' \leqslant C\|x^{-1} \times y\|$$

for all $x, y \in G$.

The group $(A(\mathbb{R}^2), \boxplus)$ (and thus $(G(\mathbb{R}^2), \otimes)$) is obviously a Carnot group of step 2 with $V_1 = \mathbb{R}^2$ and $V_2 = [\mathbb{R}^2, \mathbb{R}^2]$, and $\delta_\lambda(x) = (\lambda x^1, \lambda^2 x^2)$. Its homogeneous dimension is 4. Homogeneous norms and gauges are easily constructed. It is sufficient to consider $\|x\| = |x^1| + \sqrt{|x^2|}$, $\|x\| = \max\{|x^1|, \sqrt{|x^2|}\}$ either on $A(\mathbb{R}^2)$ or $G(\mathbb{R}^2)$. Of course, if $\|\cdot\|$ is a homogeneous gauge on $A(\mathbb{R}^2)$, then $\|\cdot\|'$ defined by $\|x\|' = \|\log(x)\|$ is a homogeneous gauge on $G(\mathbb{R}^2)$.

B Brownian Motion on the Heisenberg Group

We have seen in Section 8.4 that Brownian motion is naturally lifted as a rough path and then that the integrals correspond to the usual Itô or Stratonovich integrals.

The tangent space of $A(\mathbb{R}^2)$ may be identified with $A(\mathbb{R}^2)$, and we denote by ∂_x, ∂_y and ∂_z the basis of $T_x A(\mathbb{R}^2)$ at a point x which is deduced from the canonical coordinates e_1, e_2 and $[e_1, e_2]$.

Let V^1, V^2 and V^3 be the left invariant vector fields that go through 0 and that coincide respectively with ∂_x, ∂_y and ∂_z at this point.

For example, for $a \in A(\mathbb{R}^2)$, for all $x \in A(\mathbb{R}^2)$ and all smooth functions f on $A(\mathbb{R}^2)$, $V^i f(a \boxplus x) = V^i f \circ L_a(x)$ where $L_a(x) = a \boxplus x$ for $i = 1, 2, 3$. We have seen in Section 6.12 that the V^i's are decomposed in the basis $\{\partial_x, \partial_y, \partial_z\}$ as

$$V^1 = \partial_x - \frac{1}{2}y\partial_z, \ V^2 = \partial_y + \frac{1}{2}x\partial_y \ \text{and} \ V^3 = \partial_z.$$

Remark that $[V^1, V^2] = V^3$ and $[V^i, V^j] = 0$ in all other cases. The tangent space at any point of $A(\mathbb{R}^2)$ is then equipped with a scalar product $\langle\!\langle \cdot, \cdot \rangle\!\rangle$ such that $\langle\!\langle V^i, V^j \rangle\!\rangle = \delta_{i,j}$ for $i, j = 1, 2, 3$, i.e., for which $\{V^1, V^2, V^3\}$ forms an orthonormal basis. With this scalar product, $A(\mathbb{R}^2)$ becomes a Riemannian manifold.

Let $B = (B^1, B^2)$ be a two dimensional Brownian motion, and $B^n = (B^{n,1}, B^{n,2})$ for $n = 1, 2, \ldots$ be a family of piecewise linear approximations of B along a family of deterministic partitions whose meshes decrease to 0.

We then consider $\mathbf{X}$ the solution of the Stratonovich SDE

$$\mathbf{X}_t = \int_0^t V^1(\mathbf{X}_s) \circ dB_s^1 + \int_0^t V^2(\mathbf{X}_s) \circ dB_s^2$$

as well as the solutions $\mathbf{X}^n$ of the ordinary differential equations

$$\mathbf{X}_t^n = \int_0^t V^1(\mathbf{X}_s^n) \circ dB_s^{1,n} + \int_0^t V^2(\mathbf{X}_s^n) \circ dB_s^{2,n}.$$

Using the decomposition of the V^i on the coordinates $\{\partial_x, \partial_y, \partial_z\}$, we get

$$\mathbf{X}_t = B_t^1 e_1 + B_t^2 e_2 + \mathfrak{A}(B^1, B^2; 0, t)[e_1, e_2]$$

where

$$\mathfrak{A}(B^1, B^2; 0, t) = \frac{1}{2}\int_0^t B_s^1 \circ dB_s^2 - \frac{1}{2}\int_0^t B_s^2 \circ dB_s^1$$

is the Lévy area of (B^1, B^2). As already mentioned in Section 8.4, the process $\mathbf{X}$ is the *Brownian motion on the Heisenberg group*. Similarly, we get

$$\mathbf{X}_t^n = B_t^{1,n} e_1 + B_t^{2,n} e_2 + \mathfrak{A}(B^{1,n}, B^{2,n}; 0, t)[e_1, e_2]$$

and it is known from the Wong-Zakai theorem [40] that $\mathbf{X}^n$ converges in probability to $\mathbf{X}$ (with a dyadic partition, we get an almost sure convergence in the α-Hölder norm for any $\alpha < 1/2$ [13, 65]). Note that the piecewise smooth curves $\mathbf{X}^n$ are horizontal curves, so that in this case, the natural approximation of $\mathbf{X} \in C^\alpha([0, T]; A(\mathbb{R}^2))$ is provided by the piecewise linear approximations of (B^1, B^2) naturally lifted as paths in $A(\mathbb{R}^2)$. Many processes share this property : see for example [13, 15, 45].

This is a special case of a Brownian motion in a Lie group. Its short time behavior and its density have been already widely studied: see for example [1, 2, 4, 6, 33], ... From the Hörmander theorem, as $\{V^1, V^2, [V^1, V^2]\}$ spans the tangent space at any point, one knows that for any $t > 0$, $\mathbf{X}_t$ has a density on the three dimensional space $A(\mathbb{R}^2)$, although it is constructed from a two dimensional Brownian motion. The infinitesimal generator of $\mathbf{X}$ is

$$\mathcal{L} = \frac{1}{2}(V^1)^2 + \frac{1}{2}(V^2)^2$$
$$= \frac{1}{2}\partial_x^2 + \frac{1}{2}\partial_y^2 + \frac{1}{2}x\partial_{zy}^2 - \frac{1}{2}y\partial_{zx}^2 + \frac{1}{8}(x^2 + y^2)\partial_z^2.$$

This is an hypo-elliptic generator.

C From Almost Rough Paths to Rough Paths

C.1 Theorems and Proofs

In this Section we prove Theorem 2 on almost rough paths, which we rewrite in a more general setting than with Hölder continuous norms.

We set $\Delta_+ = \{(s,t) \in [0,T]^2 | 0 \leqslant s \leqslant t \leqslant T\}$. A *control* is a function $\omega : \Delta_+ \to \mathbb{R}_+$ such that ω is continuous, ω is super-additive, i.e.,

$$\forall 0 \leqslant s < t < u \leqslant T, \ \omega(s,t) + \omega(t,u) \leqslant \omega(s,u)$$

and $\omega(t,t) = 0$ for all $t \in [0,T]$. If ω is super-additive and $\theta \geqslant 1$, then ω^θ is also super-additive.

Recall that for $x = (\xi, x^1, x^2)$ in $T_\xi(\mathbb{R}^d)$ with $\xi = 0$ or $\xi = 1$, we have defined $\|x\| = \max\{|x^1|, \sqrt{\frac{1}{2}|x^2|}\}$. We also set $\|x\|_\star = \max\{|x^1|, |x^2|\}$. These two norms are not equivalent, but they define the same topology.

For a continuous path $\mathbf{x}$ with values in $T_1(\mathbb{R}^2)$, introduce the norms

$$\|\mathbf{x}\|_{p,\omega} = \sup_{0 \leqslant s < t \leqslant T} \frac{\|\mathbf{x}_{s,t}\|}{\omega(s,t)^{1/p}}$$

and

$$\|\mathbf{x}\|_{\star,p,\omega} = \sup_{0 \leqslant s < t \leqslant T} \max\left\{\frac{|\mathbf{x}_{s,t}^1|}{\omega(s,t)^{1/p}}, \frac{|\mathbf{x}_{s,t}^2|}{\omega(s,t)^{2/p}}\right\}$$

with $\mathbf{x}^1 = \pi_{\mathbb{R}^d}(\mathbf{x})$ and $\mathbf{x}^2 = \pi_{\mathbb{R}^d \otimes \mathbb{R}^d}(\mathbf{x})$. Note that $\|\mathbf{x}\|_{\star,p,\omega}$ is finite if and only if $\|\mathbf{x}\|_{p,\omega}$ is finite. Hence, we denote by $C^{p,\omega}([0,T]; T_1(\mathbb{R}^d))$ the space of continuous paths with values in $T_1(\mathbb{R}^d)$ for which $\|\mathbf{x}\|_{p,\omega}$ (or equivalently $\|\mathbf{x}\|_{\star,p,\omega}$ is finite. We rewrite the first part of Theorem 2 with a control ω.

Remark 19. The case of α-Hölder continuous paths corresponds to $\omega(s,t) = t - s$ and $p = 1/\alpha$. All the results we gave about existence of the integral, solving a differential equation, ... may be written using a control $\omega(s,t)$ instead of $\omega(s,t) = t - s$ and the appropriate norms $\|\cdot\|_{p,\omega}$ and $\|\cdot\|_{\star,p,\omega}$. Similarly, we are not bound to use dyadic partitions, although some results may be related to dyadic partitions (see for example [13] for an application to semi-martingales), and it is in general computationally simpler.

Theorem 5. *Let* $(\mathbf{x}_{s,t})_{(s,t)\in\Delta_+}$ *be a family of elements of* $\mathrm{T}_1(\mathbb{R}^d)$ *such that for some* $\theta > 1$, $K > 0$,

$$\|\mathbf{x}\|_{p,\omega} < +\infty \text{ and } \|\mathbf{x}_{s,t} - \mathbf{x}_{s,r} \otimes \mathbf{x}_{r,t}\|_\star \leqslant K\omega(s,t)^\theta \qquad (68)$$

for all $(s,t) \in \Delta_+$. *We call such a family an* almost rough path *controlled by* ω.

Then there exists a rough path $\mathbf{y}$ *in* $\mathrm{C}^{p,\omega}([0,T];\mathrm{T}_1(\mathbb{R}^d))$ *such that*

$$\|\mathbf{y}_{s,t} - \mathbf{x}_{s,t}\|_\star \leqslant C\omega(s,t)^\theta \qquad (69)$$

for some constant C *that depends only on* K, θ, p, $\omega(0,T)$ *and* $\|\mathbf{x}\|_{\star,p,\omega}$. *In addition,* $\mathbf{y}$ *is unique up to the value of* $\mathbf{y}_0$.

In addition, if $\mathbf{x}_{s,t}$ *belongs to* $\mathrm{G}(\mathbb{R}^d)$ *for any* $0 \leqslant s \leqslant t \leqslant T$, *then* $\mathbf{y}$ *is a weak geometric rough path with p-variation controlled by* ω.

We give two proofs of this theorem. The first proof concerns the general case, and is taken from [55]. The other proof is a simpler proof in the case $\omega(s,t) = t - s$, which is adapted from [22]. For integrals, where $\mathbf{x}_{s,t} = f(z_s)\mathbf{z}_{s,t}$ for some rough path $\mathbf{z}$ of finite p-variation, one can find some increasing, continuous function $\varphi : [0,T] \to \mathbb{R}_+$ such that $\mathbf{z} \circ \varphi$ is Hölder continuous (See [9] and [13] for an example of application in the context of rough paths), so that in many cases, one can consider that $\omega(s,t) = t - s$ (as the integral of a differential form along a path in insensitive to change of time).

Proof. Remark first that if $\alpha^{(n)} = \bigotimes_{i=1}^n (1 + \alpha_i)$ with $\alpha_i \in \mathrm{T}_0(\mathbb{R}^d)$, then

$$\alpha^{(n)} = 1 + \sum_{i=1}^n \alpha_i + \sum_{i=1}^n \sum_{j=i+1}^n \alpha_i \otimes \alpha_j.$$

Hence, if $\overline{\alpha}^{(n)} = \bigotimes_{i=1}^n (1 + \overline{\alpha}_i)$ with $\overline{\alpha}_k = \alpha_k + \zeta$ for some $k \in \{1,\ldots,n\}$ and $\overline{\alpha}_i = \alpha_i$, $i \neq k$, then

$$\overline{\alpha}^{(n)} = \alpha^{(n)} + \zeta + \zeta \otimes \sum_{j=k+1}^n \alpha_i + \sum_{i=1}^{k-1} \alpha_i \otimes \zeta. \qquad (70)$$

For a partition $\pi = \{t_k\}_{k=1}^{n+1}$ of $[s,t]$ with $t_1 = s$ and $t_{n+1} = t$, set

$$\mathbf{x}_{s,t}^{(\pi)} = \bigotimes_{k=1}^n \mathbf{x}_{t_k,t_{k+1}}. \qquad (71)$$

Put $\mathbf{x}_{s,t}^{1,(\pi)} = \pi_{\mathbb{R}^d}(\mathbf{x}_{s,t}^{(\pi)})$ and $\mathbf{x}_{s,t}^1 = \pi_{\mathbb{R}^d}(\mathbf{x}_{s,t})$. Let t_i be some point of π (except s, t), and set $\widehat{\pi} = \pi \setminus \{t_i\}$. Then

$$\mathbf{x}_{s,t}^{1,(\pi)} - \mathbf{x}_{s,t}^{1,(\widehat{\pi})} = \mathbf{x}_{t_{i_1},t_i}^1 + \mathbf{x}_{t_i,t_{i+1}}^1 - \mathbf{x}_{t_{i-1},t_{i+1}}^1.$$

For a partition $\pi = \{t_i\}_{i=1,\ldots,n+1}$ with $n+1$ points in $[s,t]$ and $t_1 = s$, $t_{n+1} = t$, pick a point t_i such that $\omega(t_{i-1}, t_{i+1}) \leqslant 2\omega(s,t)/n$. This is possible if $n > 3$ thanks to Lemma 2.2.1 from [55, p. 244]. Then

$$|\mathbf{x}_{s,t}^{1,(\pi)} - \mathbf{x}_{s,t}^{1,(\widehat{\pi})}| \leqslant K\frac{2^\theta}{n^\theta}\omega(s,t)^\theta.$$

If π has 3 elements $\{t_1, t_2, t_3\}$ with $t_1 = s$ and $t_3 = s$, then $|\mathbf{x}_{s,t}^{1,(\pi)} - \mathbf{x}_{s,t}| \leqslant K\omega(s,t)^\theta$. Thus, by summing from $k = 1, \ldots, n$ by choosing carefully which element of the partition is suppressed, we get that

$$|\mathbf{x}_{s,t}^{1,(\pi)} - \mathbf{x}_{s,t}^1| \leqslant 2^\theta \zeta(\theta) K\omega(s,t)^\theta \tag{72}$$

with $\zeta(\theta) = \sum_{n \geqslant 1} 1/n^\theta$. This is true for any partition π, whatever its size.

Consider now a sequence of partitions π^n of $[0,T]$ whose meshes decrease to 0. We set $\pi^n[s,t] = (\pi^n \cap [s,t]) \cup \{s,t\}$. Then for any $(s,t) \in \Delta_+$, $(\mathbf{x}_{s,t}^{1,(\pi^n[s,t])})_{n \in \mathbb{N}}$ has a convergent subsequence.

One can extract a subsequence $(n_k)_{k \in \mathbb{N}}$ such that $(\mathbf{x}_{s,t}^{1,(\pi^{n_k}[s,t])})_{k \in \mathbb{N}}$ converges for any $(s,t) \in \Delta_+$, $s,t \in \mathbb{Q}$. Denote by $y_{s,t}$ one of the possible limits for $(s,t) \in \Delta_+$, $s,t \in \mathbb{Q}$. With $K_1 = K2^\theta\zeta(\theta)$ and with (72), we get that

$$|y_{s,t} - \mathbf{x}_{s,t}^1| \leqslant K_1\omega(s,t)^\theta. \tag{73}$$

As ω is continuous and $\mathbf{x}_{s,t}^1$ converges to 0 as $|t-s| \to 0$, we may extend y by continuity on Δ_+.

In addition, for $0 \leqslant s < r < t \leqslant T$ and $r \in \pi^n$, then

$$\mathbf{x}_{s,t}^{1,(\pi^n[s,t])} = \mathbf{x}_{s,r}^{1,(\pi^n[s,r])} + \mathbf{x}_{r,t}^{1,(\pi^n[r,t])}.$$

Choosing the partitions π^n such that $\pi^n \subset \pi^{n+1}$ and $\pi^n \subset \mathbb{Q}$ for each $\mathbb{Q}$, we get that, by passing to the limit for $r \in \pi^{n_{k_0}}$ for some k_0 and $s,t \in \mathbb{Q}$, we get $y_{s,t} = y_{s,r} + y_{r,t}$. Using the continuity of y, this is true for any $0 \leqslant s < r < t \leqslant T$. Define $y_t = y_{0,t}$ and remark that $y_{s,t} = y_t - y_s$.

Now, consider another function z on $[0,T]$ with values in $\mathbb{R}^d$ and satisfying $|z_t - z_s - \mathbf{x}_{s,t}^1| \leqslant 2^\theta\zeta(\theta)K\omega(s,t)^\theta$ for all $(s,t) \in \Delta_+$.

Since

$$|(y_t - y_s) - (z_t - z_s)| \leqslant |(y_r - y_s) - \mathbf{x}_{r,s}^1| + |(z_r - z_s) - \mathbf{x}_{r,s}^1|$$

for $r \in [s,t]$, $(s,t) \in \Delta_+$,

$$|(y_t - y_s) - (z_t - z_s)| \leqslant 2K_1\omega(s,t)^\theta.$$

Thus, $\widehat{y}_t = y_t - z_t$ is controlled by ω^θ with $\theta > 1$ and is necessarily constant. Otherwise,

$$|\widehat{y}_t - \widehat{y}_0| \leqslant \sum_{k=0,\ldots,2^n-1,\ t_k^n \leqslant t} |\widehat{y}_{t_{k+1}^n} - \widehat{y}_{t_k^n}| \leqslant 2K_1\omega(s,t) \sup_{k=0,\ldots,2^n-1} \omega(t_k^n, t_{k+1}^n)^{\theta-1}$$

and this converges to 0.

We now have to construct the second level of the rough path. For this purpose, set $\mathbf{z}_{s,t} = 1 + y_t - y_s + \pi_{\mathbb{R}^d \otimes \mathbb{R}^d}(\mathbf{x}_{s,t})$, and, for a partition π with s and t as endpoints, define $\mathbf{z}_{s,t}^{(\pi)}$ as $\mathbf{x}_{s,t}^{(\pi)}$ in (71) with $\mathbf{z}$ instead of $\mathbf{x}$. Note that $\mathbf{z}$ is also an almost rough path, since

$$\mathbf{z}_{s,t} - \mathbf{z}_{s,r} \otimes \mathbf{z}_{r,t} = \mathbf{x}_{s,t}^2 - \mathbf{x}_{s,r}^2 - \mathbf{x}_{r,t}^2 - \mathbf{x}_{s,r}^1 \otimes \mathbf{x}_{r,t}^1$$
$$- (\mathbf{z}_{s,r}^1 - \mathbf{x}_{s,t}^1) \otimes \mathbf{z}_{r,t}^1 + \mathbf{z}_{s,r}^1 \otimes (\mathbf{z}_{r,t}^1 - \mathbf{x}_{r,t}^1)$$

and therefore with (73),

$$\|\mathbf{z}_{s,t} - \mathbf{z}_{s,r} \otimes \mathbf{z}_{r,t}\|_\star \leqslant K_2 \omega(s,t)^\theta$$

where $K_2 = \left\{ K + 2K_1(K_1 + \|\mathbf{x}\|_{\star,p,\omega})\omega(0,T)^{1/p} \right\} \omega(s,t)^\theta$.

For $0 \leqslant s < t \leqslant T$ and $\pi = \{t_i\}_{i=1,\ldots,n+1}$ a partition of $[s,t]$ with $n+1$ points and $t_1 = s$, $t_{n+1} = t$, then for some $i \in \{2, \ldots, n\}$ and $\widehat{\pi} = \pi \setminus \{t_i\}$,

$$\|\mathbf{z}_{s,t}^{(\pi)} - \mathbf{z}_{s,t}^{(\widehat{\pi})}\|_\star \leqslant K_2 \omega(t_{i-1}, t_{i+1})^\theta.$$

One may choose t_i such that $\omega(t_{i-1}, t_{i+1}) \leqslant 2\omega(s,t)/n$. Hence, as previously,

$$\|\mathbf{z}_{s,t}^{(\pi)} - \mathbf{z}_{s,t}\|_\star \leqslant 2^\theta \zeta(\theta) K_2 \omega(s,t)^\theta. \tag{74}$$

Then, the same arguments apply and one can show that for all $(s,t) \in \Delta_+$, there exists $\mathbf{y}_{s,t} \in T_1(\mathbb{R}^d)$ such that $\pi_{\mathbb{R}^d}(\mathbf{y}_{s,t}) = y_t - y_s$, where y was the function previously defined at the first level, for all $0 \leqslant s \leqslant r \leqslant t \leqslant T$, $\mathbf{y}_{s,t} = \mathbf{y}_{s,r} \otimes \mathbf{y}_{r,t}$ and $\|\mathbf{y}_{s,t} - \mathbf{x}_{s,t}\|_\star \leqslant K_3 \omega(s,t)^\theta$ with $K_3 = K_2 2^\theta \zeta(\theta)$. In particular, $\mathbf{y}$ is continuous on Δ_+ and $t \mapsto \mathbf{y}_{0,t}$ is a rough path in $C^{p,\omega}([0,T]; T_1(\mathbb{R}^d))$ lying above y.

Let $\widehat{\mathbf{y}}$ be another rough path in $C^{p,\omega}([0,T]; T_1(\mathbb{R}^d))$ lying above y and such that $\|\widehat{\mathbf{y}}_{s,t} - \mathbf{x}_{s,t}\|_\star \leqslant K_3 \omega(s,t)^\theta$. Hence,

$$\|\mathbf{y}_{s,t} - \widehat{\mathbf{y}}_{s,t}\|_\star \leqslant |\mathbf{y}_{s,r}^2 - \mathbf{z}_{s,r}^2| + |\mathbf{y}_{r,t}^2 - \mathbf{z}_{r,t}^2| + |\widehat{\mathbf{y}}_{s,r}^2 - \mathbf{z}_{s,r}^2| + |\widehat{\mathbf{y}}_{r,t}^2 - \mathbf{z}_{r,t}^2|$$

for all $0 \leqslant s \leqslant r \leqslant t \leqslant T$. As previously, it follows that $\mathbf{y}_{s,t} = \widehat{\mathbf{y}}_{s,t}$ for all $(s,t) \in \Delta_+$.

This proves that $\mathbf{y}$ is unique up to to an additive constant.

The question is now to know whether or not $\mathbf{y}$ is also the limit of $(\mathbf{x}^{(\pi^n)})_{n \in \mathbb{N}}$ for a family of partitions $(\pi^n)_{n \in \mathbb{N}}$ whose meshes decrease to 0.

With the notation from the beginning of the proof, if $\{\alpha_i\}_{i=1,\ldots,n}$ is a family of elements in $T_0(\mathbb{R}^d)$ and $\{\eta_i\}_{i=1,\ldots,n}$ belongs to $\mathbb{R}^d$, then

$$\bigotimes_{i=1}^n (1 + \alpha_i + \eta_i) = \bigotimes_{i=1}^n (1 + \alpha_i) + \sum_{i=1}^n \eta_i + \sum_{i=1}^{n-1} \eta_i \otimes \sum_{j=i+1}^n \alpha_j$$
$$+ \sum_{i=1}^{n-1} \alpha_i \otimes \sum_{j=i+1}^n \eta_j + \sum_{i=1}^{n-1} \eta_i \otimes \sum_{j=i+1}^n \eta_j.$$

Now, set $\alpha_i = \mathbf{x}_{t_i,t_{i+1}}$ and $\eta_i = \mathbf{y}^1_{t_i,t_{i+1}}$ for some partition $\pi = \{t_i\}_{i=1,\dots,n+1}$ of $[s,t]$. Then for some constant C_1,

$$\left| \sum_{i=1}^n \eta_i \right| \leq \sum_{i=1}^n C_1 \omega(t_i, t_{i+1})^\theta \leq C\omega(0,T) \sup_{i=1,\dots,n} \omega(t_i, t_{i+1})^{\theta-1}.$$

This last term converges to 0. Finally, remark that

$$\sum_{i=1}^{n-1} \alpha_i \otimes \sum_{j=i+1}^n \eta_j = \sum_{j=2}^n \left(\sum_{i=1}^{j-1} \alpha_i \right) \otimes \eta_j.$$

But from (72), for $k \in \{2,\dots,n\}$,

$$\left| \sum_{i=1}^k \alpha_i \right| = |\mathbf{x}^{1,(\pi \cap [s,t_k])}_{s,t_k}| \leq K_1 \omega(s,t)^\theta + \|\mathbf{x}\|_{\star,p,\omega} \omega(s,t)^{1/p}.$$

It follows that for some constant C_2 depending only on $\|\mathbf{x}\|_{\star,p,\omega}$, K_1, $\omega(0,T)$, θ and p,

$$\left| \sum_{j=2}^n \left(\sum_{i=1}^{j-1} \alpha_i \right) \otimes \eta_j \right| \leq C_2 \omega(s,t) \sup_{i=2,\dots,n} \omega(t_i, t_{i+1})^{\theta-1}.$$

Similarly,

$$\left| \sum_{i=1}^{n-1} \eta_i \otimes \sum_{j=i+1}^n \alpha_j \right| \leq C_2 \omega(s,t) \sup_{i=1,\dots,n-1} \omega(t_i, t_{i+1})^{\theta-1}$$

and

$$\left| \sum_{i=1}^{n-1} \eta_i \otimes \sum_{j=i+1}^n \eta_j \right| \leq K_1 \omega(s,t)^2 \sup_{i=1,\dots,n} \omega(t_i, t_{i+1})^{2\theta-2}.$$

It follows that for some constant C_3 depending on C_2, K_1, θ and $\omega(0,T)$,

$$\|\mathbf{x}^{(\pi)}_{s,t} - \mathbf{z}^{(\pi)}_{s,t}\|_\star \leq C_3 \omega(s,t) \sup_{i=1,\dots,n} \omega(t_i, t_{i+1})^{\theta-1}.$$

This proves that if $(\pi^n)_{n\in\mathbb{N}}$ is a family of partitions whose meshes converge to 0 as $n \to \infty$, then $\mathbf{x}^{(\pi^n)}_{s,t}$ converges to $\mathbf{y}_{s,t}$. In addition, combined with (73) and (74), this gives (69).

The last assertion of this theorem follows from the fact that $\mathbf{x}^{(\pi)}$ belongs to $G(\mathbb{R}^d)$ if $\mathbf{x}_{s,t}$ belongs to $G(\mathbb{R}^d)$, which is a closed subgroup of $T_1(\mathbb{R}^d)$.

Proof (Proof of Theorem 5: alternative proof when $\omega(s,t) = K_1(t-s)$). Define a distance on $T_1(\mathbb{R}^d)$ by $d(x,y) = \|x - y\|_\star$. Note that

$$d(x \otimes z, y \otimes z) \leqslant d(x, y)(1 + \|z\|_\star) \tag{75}$$
$$\text{and } d(z \otimes x, z \otimes z) \leqslant d(x, y)(1 + \|z\|_\star) \tag{76}$$

for all $x, y, z \in T_1(\mathbb{R}^d)$.

For $0 \leqslant s \leqslant t \leqslant T$, set $r = (t + s)/2$, $\mathbf{x}_{s,t}^0 = \mathbf{x}_{s,t}$ and recursively,

$$\mathbf{x}_{s,t}^{n+1} = \mathbf{x}_{s,r}^n \otimes \mathbf{x}_{r,t}^n.$$

By the triangular inequality,

$$d(\mathbf{x}_{s,t}^{n+2}, \mathbf{x}_{s,t}^{n+1}) \leqslant d(\mathbf{x}_{s,r}^{n+1} \otimes \mathbf{x}_{r,t}^{n+1}, \mathbf{x}_{s,r}^{n+1} \otimes \mathbf{x}_{r,t}^n) + d(\mathbf{x}_{s,r}^{n+1} \otimes \mathbf{x}_{r,t}^n, \mathbf{x}_{s,r}^n \otimes \mathbf{x}_{r,t}^n).$$

With (75) and (76),

$$d(\mathbf{x}_{s,t}^{n+2}, \mathbf{x}_{s,t}^{n+1}) \leqslant d(\mathbf{x}_{r,t}^{n+1}, \mathbf{x}_{r,t}^n)(1 + \|\mathbf{x}_{s,r}^{n+1}\|_\star)$$
$$+ d(\mathbf{x}_{s,r}^{n+1}, \mathbf{x}_{s,r}^n)(1 + \|\mathbf{x}_{r,t}^n\|_\star). \tag{77}$$

Set

$$V_n(\tau) = \sup_{0 \leqslant s \leqslant t \leqslant s + \tau} d(\mathbf{x}_{s,t}^{n+1}, \mathbf{x}_{s,t}^n) \text{ and } h_n(\tau) = \sup_{0 \leqslant s \leqslant t \leqslant s + \tau} \|\mathbf{x}_{s,t}^n\|_\star.$$

From (77),

$$V_{n+1}(\tau) \leqslant (2 + h_n(\tau/2) + h_{n+1}(\tau/2))V_n(\tau/2)$$

Choose $2 < \kappa < 2^\theta$. As $V_0(\tau) = K(t - s)^\theta$, the quantity

$$\overline{V}(\tau) = \sum_{k=0}^{+\infty} \kappa^n V_0(\tau/2^n)$$

is finite. Remark that

$$h_{n+1}(\tau) \leqslant h_n(\tau) + V_n(\tau) \leqslant h_0(\tau) + \overline{V}(\tau).$$

Fix τ_0 such that $1 + h_0(\tau_0)\overline{V}(\tau_0) < \kappa/2$. This is possible since $h_0(\tau)$ and $\overline{V}(\tau)$ converge to 0 as τ decreases to 0. Assume that

$$V_n(\tau) \leqslant \kappa^n V_0(\tau/2^n) \text{ for } \tau \leqslant \tau_0. \tag{78}$$

For $\tau \leqslant \tau_0$, $2 + h_n(\tau/2) + h_{n+1}(\tau/2) \leqslant \kappa$ and then $V_{n+1}(\tau) \leqslant \kappa^{n+1} V_0(\tau/2^{n+1})$. Then, (78) is true for any n and $\sum_{n \geqslant 0} V_n(\tau) \leqslant \overline{V}(\tau)$ converges. This means that $(\mathbf{x}_{s,t}^n)_{n \in \mathbb{N}}$ is a Cauchy sequence for all $(s, t) \in \Delta_+$ such that $t - s \leqslant \tau$.

Denote by $\mathbf{y}_{s,t}$ the limit of $(\mathbf{x}_{s,t}^n)_{n \in \mathbb{N}}$, which is continuous in s and t. This limit satisfies $\mathbf{y}_{s,t} = \mathbf{y}_{s,r} \otimes \mathbf{y}_{r,t}$ with $r = (t + s)/2$. In addition, $d(\mathbf{y}_{s,t}, \mathbf{x}_{s,t}) \leqslant C(t - s)^\theta$ for some constant C. We extend $\mathbf{y}_{s,t}$ to $(s, t) \in \Delta_+$ by setting $\mathbf{y}_{s,t} = \mathbf{y}_{t_0^m, t_1^m} \otimes \cdots \otimes \mathbf{y}_{t_{2^m-1}^m, t_{2^m}^m}$ for the partition $t_i^m = s + i(t - s)2^{-m}$, $i = 0, \ldots, 2^m$ when m is large enough so that $(t - s) \leqslant \tau_0 2^m$. We easily get that $\mathbf{y}$ is *mid-point additive*, that $\mathbf{y}$ does not depend on m, is such that $\mathbf{y}_{s,t} = \mathbf{y}_{s,r} \otimes \mathbf{y}_{r,t}$ for $r = (t+s)/2$ and satisfies $d(\mathbf{y}_{s,t}, \mathbf{x}_{s,t}) \leqslant C'(t-s)^\theta$ for $(s,t) \in \Delta_+$ with possibly another constant C'.

Let us now prove that $\mathbf{y}$ is unique. Let $\mathbf{z}$ be another function from Δ_+ to $\mathrm{T}_1(\mathbb{R}^d)$ which satisfies

$$\mathbf{z}_{s,t} = \mathbf{z}_{s,r} \otimes \mathbf{z}_{r,t} \text{ for } r = (t+s)/2 \text{ and } d(\mathbf{z}_{s,t}, \mathbf{x}_{s,t}) \leqslant C'(t-s)^\theta \qquad (79)$$

for some $C' > 0$ and any $(s,t) \in \Delta_+$. For $(s,t) \in \Delta_+$ and $r = (t+s)/2$,

$$\begin{aligned}
d(\mathbf{y}_{s,t}, \mathbf{z}_{s,t}) &\leqslant d(\mathbf{y}_{s,t}, \mathbf{y}_{s,r} \otimes \mathbf{z}_{r,t}) + d(\mathbf{y}_{s,r} \otimes \mathbf{z}_{r,t}, \mathbf{z}_{s,t}) \\
&\leqslant d(\mathbf{y}_{r,t}, \mathbf{z}_{r,t})(1 + \|\mathbf{y}_{s,r}\|_\star) + d(\mathbf{y}_{s,r}, \mathbf{z}_{s,r})(1 + \|\mathbf{z}_{r,t}\|_\star) \\
&\leqslant \kappa(\tau/2)W(\tau/2),
\end{aligned}$$

where $W(\tau) = \sup_{s \leqslant t \leqslant s+\tau} d(\mathbf{y}_{s,t}, \mathbf{z}_{s,t})$ and

$$\kappa(\tau) = 2 + \sup_{0 \leqslant t-s \leqslant \tau} \|\mathbf{y}_{s,t}\|_\star + \sup_{0 \leqslant t-s \leqslant \tau} \|\mathbf{z}_{s,t}\|.$$

Thus, $W(\tau) \leqslant \kappa W(\tau/2)$. Now, note that

$$W(\tau) \leqslant \sup_{s \leqslant t \leqslant s+\tau} \left(d(\mathbf{y}_{s,t}, \mathbf{x}_{s,t}) + d(\mathbf{z}_{s,t}, \mathbf{x}_{s,t}) \right) \leqslant 2C\tau^\theta.$$

Then, if $\tau < \tau_0$ with $\kappa(\tau_0) < 2^\theta$,

$$W(\tau) \leqslant \kappa(\tau_0)^n W(\tau/2^n) \leqslant \frac{C\kappa(\tau_0)^n \tau^\theta}{2^{(n+1)(\theta-1)}} \xrightarrow[n\to\infty]{} 0,$$

which means that $W(\tau) = 0$ for $\tau \in [0, \tau_0]$. Using the fact that both $\mathbf{y}$ and $\mathbf{z}$ are mid-point additive, we get that $\mathbf{y}_{s,t} = \mathbf{z}_{s,t}$ for all $(s,t) \in \Delta_+$ and that $\mathbf{y}$ is unique.

Now, fix $(s,t) \in \Delta_+$ and $n \in \mathbb{N}$. Set

$$\mathbf{z}_{s,t} = \mathbf{y}_{t_0^n, t_1^n} \otimes \cdots \otimes \mathbf{y}_{t_{n-1}^n, t_n^n}$$

for $t_i^n = s + (t-s)i/n$. Note that for $r = (t+s)/2$,

$$\mathbf{z}_{s,t} = \mathbf{z}_{s,r} \otimes \mathbf{z}_{r,t} \text{ for } s = \frac{t+s}{2}, \ (s,t) \in \Delta_+.$$

It follows that

$$d(\mathbf{z}_{s,t}, \mathbf{x}_{s,t}) \leqslant d\left(\bigotimes_{i=0}^{n-1} \mathbf{y}_{t_i^n, t_{i+1}^n}, \mathbf{y}_{t_0^n, t_1^n} \otimes \mathbf{x}_{t_1^n, t_2^n} \right)$$

$$+ d(\mathbf{y}_{t_0^n, t_1^n} \otimes \mathbf{x}_{t_1^n, t_2^n}, \mathbf{x}_{t_0^n, t_1^n} \otimes \mathbf{x}_{t_1^n, t_2^n}) + d(\mathbf{x}_{t_0^n, t_1^n} \otimes \mathbf{x}_{t_1^n, t_2^n}, \mathbf{x}_{t_0^n, t_1^n})$$

$$\leqslant d\left(\bigotimes_{i=1}^{n-1} \mathbf{y}_{t_i^n, t_{i+1}^n}, \mathbf{x}_{t_1^n, t_2^n} \right)(1 + \|\mathbf{y}_{t_0^n, t_1^n}\|_\star)$$

$$+ d(\mathbf{y}_{t_0^n, t_1^n}, \mathbf{x}_{t_0^n, t_1^n})(1 + \|\mathbf{x}_{t_1^n, t_2^n}\|_\star) + K|t-s|^\theta$$

$$\leqslant C_1 d\left(\bigotimes_{i=1}^{n-1} \mathbf{y}_{t_i^n, t_{i+1}^n}, \mathbf{x}_{t_1^n, t_2^n} \right) + K|t-s|^\theta + C_2 \frac{|t-s|^\theta}{n^\theta}$$

for some constants C_1 and C_2 that depend only on T, K and K_1. Applying recursively the same computation leads to

$$d(\mathbf{z}_{s,t}, \mathbf{x}_{s,t}) \leqslant C_3 |t - s|^\theta$$

for some constant C_3 that depends on K, T, K_1 and n. We have previously proved that any function $\mathbf{z} : \Delta_+ \to T_1(\mathbb{R}^d)$ which satisfies (79) is equal to $\mathbf{y}$, so $\mathbf{y}_{s,t} = \bigotimes_{i=1}^{n-1} \mathbf{y}_{t_i^n, t_{i+1}^n}$. Then, $\mathbf{y}_{s,t} = \mathbf{y}_{s,s+p(t-s)} \otimes \mathbf{y}_{s+p(t-s),t}$ for all $p \in \mathbb{Q}$. From the continuity of $(s,t) \in \Delta_+ \mapsto \mathbf{y}_{s,t}$, we deduce that $\mathbf{y}_{s,r} \otimes \mathbf{y}_{r,t} = \mathbf{y}_{s,t}$ for any $r \in [s,t]$, $(s,t) \in \Delta_+$.

Theorem 6. *Let $\mathbf{x}$ and $\mathbf{y}$ be two almost rough paths, both satisfying (68) with the same constants K and θ.*

(i) *Assume that there exists an $\varepsilon > 0$ such that*

$$\|\mathbf{x} - \mathbf{y}\|_{\star,p,\omega} \leqslant \varepsilon.$$

Then there exists some function $\varepsilon \mapsto K(\varepsilon)$ that depends only on K, θ, p, $\omega(0,T)$, $\|\mathbf{x}\|_{\star,p,\omega}$ and $\|\mathbf{y}\|_{\star,p,\omega}$ such that the two rough paths $\tilde{\mathbf{x}}$ and $\tilde{\mathbf{y}}$ associated to $\mathbf{x}$ and $\mathbf{y}$ by Theorem 5 satisfy

$$\|\tilde{\mathbf{x}} - \tilde{\mathbf{y}}\|_{\star,p,\omega} \leqslant K(\varepsilon)$$

with $K(\varepsilon) \to 0$ as $\varepsilon \to 0$.

(ii) *If in addition for all $(s,t) \in \Delta_+$,*

$$\|\mathbf{x}_{s,t} - \mathbf{x}_{s,r} \otimes \mathbf{x}_{r,t} - (\mathbf{y}_{s,t} - \mathbf{y}_{s,r} \otimes \mathbf{y}_{r,t})\|_\star \leqslant \varepsilon \omega(s,t)^\theta,$$

then $K(\varepsilon) = K'\varepsilon$ for some constant K' depending only on K, θ, p, $\omega(0,T)$, $\|\mathbf{x}\|_{\star,p,\omega}$ and $\|\mathbf{y}\|_{\star,p,\omega}$.

Proof. We first prove statement (ii) of this theorem. We use the same notations as previously. For a partition $\pi = \{t_i\}_{i=1,\dots,n}$ of $[s,t]$ with $t_1 = s$, $t_{n+1} = t$, consider $\mathbf{x}_{s,t}^{(\pi)}$ and $\mathbf{y}_{s,t}^{(\pi)}$ as above.

Pick a point t_i in π such that $\omega(t_{i-1}, t_{i+1}) \leqslant 2\omega(s,t)/n$. For

$$\xi = \mathbf{x}_{t_{i-1},t_i} \otimes \mathbf{x}_{t_i,t_{i+1}} - \mathbf{x}_{t_{i-1},t_{i+1}} - \mathbf{y}_{t_{i-1},t_i} \otimes \mathbf{y}_{t_i,t_{i+1}} + \mathbf{y}_{t_{i-1},t_{i+1}},$$

we get that, with (70),

$$\|\mathbf{x}_{s,t}^{(\pi)} - \mathbf{x}_{s,t}^{(\hat{\pi})} - (\mathbf{y}_{s,t}^{(\pi)} - \mathbf{y}_{s,t}^{(\hat{\pi})})\|_\star \leqslant \|\xi\|_\star \left(1 + \sum_{j=1,\dots,n \; j \neq i} |\mathbf{x}_{t_j,t_{j+1}}^1 - \mathbf{y}_{t_j,t_{j+1}}^1|\right)$$

where $\mathbf{x}_{s,t}^1$ (resp. $\mathbf{y}_{s,t}^1$) is the projection of $\mathbf{x}_{s,t}$ (resp. $\mathbf{y}_{s,t}$) on $\mathbb{R}^d$.

With (72), we get that, for some constant C that depends only on K, θ, p, $\omega(0,T)$, $\|\mathbf{x}\|_{p,\omega}$ and $\|\mathbf{y}\|_{p,\omega}$,

$$\sum_{j=1,\ldots,n\ j\neq i} |x^1_{t_j,t_{j+1}} - y^1_{t_j,t_{j+1}}| \leqslant |x^{1,(\pi)}_{s,t}| + |y^{1,(\pi)}_{s,t}|$$

$$\leqslant (C\omega(s,t)^{\theta-1/p} + \|y\|_{\star,p,\omega} + \|x\|_{\star,p,\omega})\omega(s,t)^{1/p}.$$

Thus, for some constant K,

$$\|x^{(\pi)}_{s,t} - x^{(\widehat{\pi})}_{s,t} - (y^{(\pi)}_{s,t} - y^{(\widehat{\pi})}_{s,t})\|_\star \leqslant \varepsilon \frac{K}{n^\theta}\omega(s,t)^\theta.$$

It follows that by carefully removing all points of π one after the other,

$$\|x^{(\pi)}_{s,t} - x_{s,t} - (y^{(\pi)}_{s,t} - y_{s,t})\|_\star \leqslant \varepsilon\zeta(\theta)K\omega(s,t)^\theta.$$

As we have seen that $x^{(\pi)}$ and $y^{(\pi)}$ converges to $\widetilde{x}_{s,t}$ and $\widetilde{y}_{s,t}$, we deduce that

$$\|\widetilde{x}_{s,t} - x_{s,t} - (\widetilde{y}_{s,t} - y_{s,t})\|_\star \leqslant \varepsilon\zeta(\theta)K\omega(s,t)^\theta.$$

The result is then easily deduced.

Now, to prove the statement (i), we just have to remark that for some $1/\theta < \eta < 1$,

$$\|x_{s,t} - x_{s,r} \otimes x_{r,t} - y_{s,t} + y_{s,r} \otimes y_{r,t}\|_\star$$
$$\leqslant 2^{\eta-1}(\|x_{s,t} - x_{s,r} \otimes x_{r,t}\|^\eta_\star + \|y_{s,t} - y_{s,r} \otimes y_{r,t}\|^\eta_\star)$$
$$\times(\|x_{s,t} - y_{s,t}\|_\star + \|x_{s,r} \otimes x_{r,t} - y_{s,r} \otimes y_{r,t}\|_\star)^{1-\eta}$$
$$\leqslant C\omega(s,t)^{\eta\theta+(1-\eta)/p}\varepsilon^{1-\eta}$$

for some constant C that depends only on η, θ, $\omega(0,T)$, and then to apply the result of (ii) by replacing ε by $\varepsilon^{1-\eta}$ and θ by $\eta\theta$.

C.2 An Algebraic Interpretation

We now give an algebraic interpretation of this construction, which is strongly inspired from the one given by M. Gubinelli in [36].

Consider the sets

$$\Delta_1 = [0,T], \qquad \Delta_2 = \{(s,t)|0 \leqslant s \leqslant t \leqslant T\}$$
$$\text{and} \qquad \Delta_3 = \{(s,r,t)|0 \leqslant s \leqslant r \leqslant t \leqslant T\},$$

and call $\mathcal{C}_i$ the set of functions from Δ_i to $T_1(\mathbb{R}^d)$ for $i = 1, 2, 3$.

Introduce the operator from $\mathcal{C}_1 \cup \mathcal{C}_2$ to $\mathcal{C}_2 \cup \mathcal{C}_3$ defined by

$$\delta(x)_{s,t} = x^{-1}_s \otimes x_t, \quad (s,t) \in \Delta_2, \quad x \in \mathcal{C}_1,$$
$$\delta(x)_{s,r,t} = x_{s,t} - x_{s,r} \otimes x_{r,t}, \quad (s,r,t) \in \Delta_3, \quad x \in \mathcal{C}_2,$$

hence δ maps $\mathcal{C}_i$ to $\mathcal{C}_{i+1}$, $i = 1, 2$. Note that if $x \in \mathcal{C}_1$, then $\delta(\delta(x)) = 0$, so the range $\text{Range}(\delta_{|\mathcal{C}_1})$ of $\delta_{|\mathcal{C}_1}$ is contained in the kernel $\text{Ker}(\delta_{|\mathcal{C}_2})$ of $\delta_{|\mathcal{C}_2}$. Indeed, we get a better result.

Lemma 21. *The range of $\delta_{|\mathcal{C}_1}$ is equal to the kernel of $\delta_{|\mathcal{C}_2}$, and δ is injective from $\mathcal{C}_1(x)$ into $\mathcal{C}_2$ where $\mathcal{C}_1(x)$ is the set of paths $\mathbf{x}$ in $\mathcal{C}_1$ with $\mathbf{x}_0 = x$ for $x \in T_1(\mathbb{R}^2)$. In particular, when restricted to $\mathrm{Range}(\delta_{\mathcal{C}_1(x)})$, δ is invertible.*

Proof. We have already seen the inclusion of $\mathrm{Range}(\delta_{|\mathcal{C}_1})$ in $\mathrm{Ker}(\delta_{|\mathcal{C}_2})$.

Now, let $\mathbf{x} \in \mathcal{C}_2$ belong to the kernel $\mathrm{Ker}(\delta_{|\mathcal{C}_2})$, and set $\mathbf{y}_t = \mathbf{x}_{0,t}$. As $\delta(\mathbf{x})_{0,s,t} = \mathbf{y}_t - \mathbf{y}_s \otimes \mathbf{x}_{s,t} = 0$, we get $\mathbf{x}_{s,t} = \mathbf{y}_s^{-1} \otimes \mathbf{y}_t$ and thus $\mathbf{x} = \delta(\mathbf{y})$. This proves the result.

If two paths $\mathbf{x}$ and $\mathbf{y}$ are distinct in $\mathcal{C}_1(x)$, then $\delta(\mathbf{x})_{0,t} = x^{-1} \otimes \mathbf{x}_t$ is different from $\delta(\mathbf{y})_{0,t} = x^{-1} \otimes \mathbf{y}_t$ and δ is injective from $\mathcal{C}_1(x)$ into $\mathcal{C}_2$.

Given a rough path $\mathbf{x}$, which then belongs to $\mathcal{C}_1$ and a differential form f, the integral $\mathfrak{I}(\mathbf{x}) = \int f(x) \, \mathrm{d}x$ is also a path in $\mathcal{C}_1(0)$. The idea is then to consider an approximation of $\mathfrak{I}(\mathbf{x}; s, t)$ for $t - s$ small, and to project it on the range of $\delta_{|\mathcal{C}_1(x)}$. Of course, the approximation of $\mathfrak{I}(\mathbf{x}; s, t)$ has to be close enough to the range of $\delta_{|\mathcal{C}_1(x)}$.

For $p \geqslant 1$ and $\theta > 1$, define the distance $d_{\theta,\omega}$ on $\mathcal{C}_2$ by

$$D_{\star,\theta,\omega}(\mathbf{x},\mathbf{y}) = \sup_{(s,t) \in \Delta^+} \frac{\|\mathbf{x}_{s,t} - \mathbf{y}_{s,t}\|_\star}{\omega(s,t)^\theta}.$$

To simplify the notation, extend $\delta_{|\mathcal{C}_2}$ as a function defined on $\Delta^2 \times [0,T]$ by setting $\delta(\mathbf{x})_{s,r,t} = 1$ if $r \notin [s,t]$. For a fixed $r \in [0,T]$, $\delta_{\cdot,r,\cdot}(\mathbf{x})$ is then a function in $\mathcal{C}_2$.

Theorem 5 is the rewritten the following way.

Theorem 7. *For $K, K' > 0$ and $\theta > 1$, denote by $B(K, K', \theta, p, \omega)$ the subsets of functions $\mathbf{x} \in \mathcal{C}_2$ for which*

$$\|\mathbf{x}\|_{\star,p,\omega} \leqslant K \quad and \quad \sup_{r \in [0,T]} D_{\star,p,\omega}(\delta(\mathbf{x})_{\cdot,r,\cdot}, 0) \leqslant K'.$$

Then to any $\mathbf{x}$ in $B(K, K', \theta, p, \omega)$ is associated a unique element $\widehat{\mathbf{x}}$ of $\mathrm{Ker}(\delta_{|\mathcal{C}_2})$. In addition, for some constants C_1 and C_2 that depend only on K, K', θ, p and $\omega(0,T)$,

$$\|\widehat{\mathbf{x}}\|_{\star,p,\omega} \leqslant C_1 \quad and \quad D_{\star,p,\omega}(\mathbf{x}, \widehat{\mathbf{x}}) \leqslant C_2.$$

Moreover, if one defines a distance $\Theta_{\star,p,\theta,\omega}$ on $\cup_{K,K'>0} B(K, K', \theta, p, \omega)$ by

$$\Theta_{\star,p,\theta,\omega}(\mathbf{x},\mathbf{y}) = \max\{\|\mathbf{x} - \mathbf{y}\|_{\star,p,\omega}, d_{\star,\theta,\omega}(\mathbf{x},\mathbf{y})\},$$

then this map $\Pi : \mathbf{x} \mapsto \widehat{\mathbf{x}}$ is locally Lipschitz with respect to $\Theta_{\star,p,\theta,\omega}$.

From the definition, an almost rough path $\mathbf{x}$ of p-variation controlled by ω belongs to $\cup_{K,K'>0} B(K, K', \theta, p, \omega)$. It is then "projected" on an element $\Pi(\mathbf{x})$ in $\mathcal{C}_2$ in the kernel of $\delta_{\mathcal{C}_2}$, which is also equal to the image of $\delta_{\mathcal{C}_1(1)}$. The inverse image of $\Pi(\mathbf{x})$ then gives a rough path in $C^{p,\omega}([0,T]; T_1(\mathbb{R}^d))$.

Given an element f in $\mathrm{Lip}(\gamma; \mathbb{R}^d \to \mathbb{R}^m)$ with $\gamma > p - 1$, the map $\mathfrak{F}(f, \mathbf{x})$ defined by (61) defines an element of $\mathcal{C}_2$. The integral $\mathfrak{I}$ may then be defined as the composition of the maps

$$\mathfrak{I} = \delta_{|\mathcal{C}_1(1)}^{-1} \circ \varPi \circ \mathfrak{F}(f, \cdot),$$

which corresponds to the construction given in Section 7.4.

References

1. R. AZENCOTT. *Géodésiques et diffusions en temps petit*, vol. 84 of *Astérisque*. Société Mathématique de France, Paris, 1981.
2. G. BEN AROUS. Flots et séries de Taylor stochastiques. *Probab. Theory Related Fields*, **81**:1, 29–77, 1989.
3. A. BAKER. *Matrix groups: An introduction to Lie group theory*. Springer Undergraduate Mathematics Series. Springer-Verlag London Ltd., London, 2002.
4. F. BAUDOIN. *An introduction to the geometry of stochastic flows*. Imperial College Press, London, 2004.
5. D. BURAGO, Y. BURAGO and S. IVANOV. *A course in metric geometry*, vol. 33 of *Graduate Studies in Mathematics*. American Mathematical Society, Providence, RI, 2001.
6. J.-M. BISMUT. *Large deviations and the Malliavin calculus*, vol. 45 of *Progress in Mathematics*. Birkhäuser Boston Inc., Boston, MA, 1984.
7. A. BENSOUSSAN, J.-L. LIONS and G. PAPANICOLAOU. *Asymptotic Analysis for Periodic Structures*. North-Holland, 1978.
8. F. CASTELL. Asymptotic expansion of stochastic flows. *Probab. Theory Related Fields*, **96**:2, 225–239, 1993.
9. V.V. CHISTYAKOV and O.E. GALKIN. On maps of bounded p-variations with $p > 1$. *Positivity*, **2**:1, 19–45, 1998.
10. K.-T. CHEN. Integration of paths, geometric invariants and a generalized Baker-Hausdorff formula. *Ann. of Math. (2)*, **65**, 163–178, 1957.
11. K.-T. CHEN. Integration of Paths–A Faithful Representation, of Paths by Non-commutative Formal Power Series. *Trans. Amer. Math. Soc.*, **89**:2, 395–407, 1958.
12. Z. CIESIELSKI. On the isomorphisms of the spaces H_α and m. *Bull. Acad. Polon. Sci. Sér. Sci. Math. Astronom. Phys.*, **8**, 217–222, 1960.
13. L. COUTIN and A. LEJAY. Semi-martingales and rough paths theory. *Electron. J. Probab.*, **10**:23, 761–785, 2005.
14. L. COUTIN. An introduction to (stochastic) calculus with respect to fractional Brownian motion. In *Séminaire de Probabilités XL*, vol. 1899 of *Lecture Notes in Math.*, pp. 3–65. Springer, 2007.
15. L. COUTIN and Z. QIAN. Stochastic analysis, rough path analysis and fractional Brownian motions. *Probab. Theory Related Fields*, **122**:1, 108–140, 2002. <DOI: 10.1007/s004400100158>.
16. A.M. DAVIE. Differential Equations Driven by Rough Signals: an Approach via Discrete Approximation. *Appl. Math. Res. Express. AMRX*, **2**, Art. ID abm009, 40, 2007.

17. J. J. DUISTERMAAT and J. A. C. KOLK. *Lie groups*. Universitext. Springer-Verlag, Berlin, 2000.
18. R.M. DUDLEY and R. NORVAIŠA. *An introduction to p-variation and Young integrals—with emphasis on sample functions of stochastic processes*, 1998. Lecture given at the Centre for Mathematical Physics and Stochastics, Department of Mathematical Sciences, University of Aarhus. Available on the web site <www.maphysto.dk>.
19. H. DOSS. Liens entre équations différentielles stochastiques et ordinaires. *Ann. Inst. H. Poincaré Sect. B (N.S.)*, **13**:2, 99–125, 1977.
20. C. T. J. DODSON and T. POSTON. *Tensor geometry: The geometric viewpoint and its uses*, vol. 130 of *Graduate Texts in Mathematics*. Springer-Verlag, Berlin, 2nd edition, 1991.
21. D. FEYEL and A. de LA PRADELLE. Curvilinear integrals along enriched paths. *Electron. J. Probab.*, **11**:35, 860–892, 2006.
22. D. FEYEL, A. de LA PRADELLE and G. MOKOBODZKI. A non-commutative sewing lemma. *Electron. Commun. Probab.*, **13**, 24–34, 2008. <ARXIV: 0706.0202>.
23. M. FLIESS. Fonctionnelles causales non linéaires et indéterminées non commutatives. *Bull. Soc. Math. France*, **109**:1, 3–40, 1981.
24. M. FLIESS and D. NORMAND-CYROT. Algèbres de Lie nilpotentes, formule de Baker-Campbell-Hausdorff et intégrales itérées de K. T. Chen. In *Séminaire de Probabilités, XVI*, vol. 920, pp. 257–267. Springer, Berlin, 1982.
25. G. B. FOLLAND. *Harmonic analysis in phase space*, vol. 122 of *Annals of Mathematics Studies*. Princeton University Press, Princeton, NJ, 1989.
26. P. FRIZ. Continuity of the Itô-Map for Hölder rough paths with applications to the support theorem in Hölder norm. In *Probability and Partial Differential Equations in Modern Applied Mathematics*, vol. 140 of *IMA Volumes in Mathematics and its Applications*, pp. 117–135. Springer, 2005.
27. G. B. FOLLAND and E. M. STEIN. *Hardy spaces on homogeneous groups*, vol. 28 of *Mathematical Notes*. Princeton University Press, 1982.
28. P. FRIZ and N. VICTOIR. *Multidimensional Stochastic Processes as Rough Paths. Theory and Applications*. Cambridge University Press, 2009.
29. P. FRIZ and N. VICTOIR. A Note on the Notion of Geometric Rough Paths. *Probab. Theory Related Fields*, **136**:3, 395–416, 2006. <DOI: 10.1007/s00440-005-0487-7>, <ARXIV: math.PR/0403115>.
30. P. FRIZ and N. VICTOIR. Differential Equations Driven by Gaussian Signals I. <ARXIV: 0707.0313>, Cambridge University (preprint), 2007.
31. P. FRIZ and N. VICTOIR. Differential Equations Driven by Gaussian Signals II. <ARXIV: 0711.0668>, Cambridge University (preprint), 2007.
32. P. FRIZ and N. VICTOIR. Euler Estimates of Rough Differential Equations. *J. Differential Equations*, **244**:2, 388–412, 2008. <ARXIV: math.CA/0602345>.
33. B. GAVEAU. Principe de moindre action, propagation de la chaleur et estimées sous elliptiques sur certains groupes nilpotents. *Acta Math.*, **139**:1-2, 95–153, 1977.
34. R. GOODMAN. Filtrations and asymptotic automorphisms on nilpotent Lie groups. *J. Differential Geometry*, **12**:2, 183–196, 1977.
35. M. GROMOV. Carnot-Carathéodory spaces seen from within. In *Sub-Riemannian geometry*, vol. 144 of *Progr. Math.*, pp. 79–323. Birkhäuser, 1996.
36. M. GUBINELLI. Controlling rough paths. *J. Funct. Anal.*, **216**:1, 86–140, 2004.

37. B. C. HALL. *Lie groups, Lie algebras, and representations*, vol. 222 of *Graduate Texts in Mathematics*. Springer-Verlag, New York, 2003.

38. B.M. HAMBLY and T.J. LYONS. Uniqueness for the signature of a path of bounded variation and continuous analogues for the free group. Oxford University (preprint), 2006.

39. A. ISIDORI. *Nonlinear control systems*. Springer-Verlag, Berlin, 3rd edition, 1995.

40. N. IKEDA and S. WATANABE. *Stochastic differential equations and diffusion processes*, vol. 24 of *North-Holland Mathematical Library*. North-Holland Publishing Co., Second edition, 1989.

41. M. KAWSKI. Nonlinear control and combinatorics of words. In *Geometry of feedback and optimal control*, vol. 207 of *Monogr. Textbooks Pure Appl. Math.*, pp. 305–346. Dekker, New York, 1998.

42. T. LYONS, M. CARUANA and T. LÉVY. Differential Equations Driven by Rough Paths. In *École d'été de probabilités de Saint-Flour XXXIV—2004*, edited by J. PICARD, vol. 1908 of *Lecture Notes in Math.*, Berlin, 2007. Springer.

43. A. LEJAY. On the convergence of stochastic integrals driven by processes converging on account of a homogenization property. *Electron. J. Probab.*, **7**:18, 1–18, 2002.

44. A. LEJAY. An introduction to rough paths. In *Séminaire de probabilités, XXXVII*, vol. 1832 of *Lecture Notes in Mathematics*, pp. 1–59. Springer-Verlag, 2003.

45. A. LEJAY. Stochastic Differential Equations driven by processes generated by divergence form operators I: a Wong-Zakai theorem. *ESAIM Probab. Stat.*, **10**, 356–379, 2006. <DOI: 10.1051/ps:2006015>.

46. A. LEJAY. Stochastic Differential Equations driven by processes generated by divergence form operators II: convergence results. *ESAIM Probab. Stat.*, **12**, 387–411, 2008. <DOI: 10.1051/ps:2007040>.

47. P. LÉVY. *Processus stochastiques et mouvement brownien*. Gauthier-Villars & Cie, Paris, 2e édition, 1965.

48. A. LEJAY and T. LYONS. On the importance of the Lévy area for systems controlled by converging stochastic processes. Application to homogenization. In *New Trends in Potential Theory, Conference Proceedings, Bucharest, September 2002 and 2003*, edited by D. BAKRY, L. BEZNEA, Gh. BUCUR and M. RÖCKNER, pp. 63–84. The Theta Foundation, 2006.

49. X.D. LI and T.J. LYONS. Smoothness of Itô maps and diffusion processes on path spaces. I. *Ann. Sci. École Norm. Sup.*, **39**:4, 649–677, 2006.

50. M. LEDOUX, T. LYONS and Z. QIAN. Lévy area of Wiener processes in Banach spaces. *Ann. Probab.*, **30**:2, 546–578, 2002.

51. T. LYONS and Z. QIAN. Flow of diffeomorphisms induced by a geometric multiplicative functional. *Proba. Theory Related Fields*, **112**:1, 91–119, 1998.

52. T. LYONS and Z. QIAN. *System Control and Rough Paths*. Oxford Mathematical Monographs. Oxford University Press, 2002.

53. A. LEJAY and N. VICTOIR. On (p, q)-rough paths. *J. Differential Equations*, **225**:1, 103–133, 2006. <DOI: 10.1016/j.jde.2006.01.018>.

54. T. LYONS and N. VICTOIR. An Extension Theorem to Rough Paths. *Ann. Inst. H. Poincaré Anal. Non Linéaire*, **24**:5, 835–847, 2007. <DOI: 10.1016/j.anihpc.2006.07.004>.

55. T.J. LYONS. Differential equations driven by rough signals. *Rev. Mat. Iberoamericana*, **14**:2, 215–310, 1998.

56. T.J. LYONS and T. ZHANG. Decomposition of Dirichlet Processes and its Application. *Ann. Probab.*, **22**:1, 494–524, 1994.

57. W. MAGNUS. On the exponential solution of differential equations for a linear operator. *Comm. Pure Appl. Math.*, **7**, 649–673, 1954.

58. R. MAHONY and J.H. MANTON. The Geometry of the Newton Method on Non-Compact Lie Groups *J. Global Optim.*, **23**, 309–327, 2002.

59. E. J. MCSHANE. Stochastic differential equations and models of random processes. In *Proceedings of the Sixth Berkeley Symposium on Mathematical Statistics and Probability (Univ. California, Berkeley, Calif., 1970/1971), Vol. III: Probability theory*, pp. 263–294. Univ. California Press, 1972.

60. R. MONTGOMERY. *A tour of subriemannian geometries, their geodesics and applications*, vol. 91 of *Mathematical Surveys and Monographs*. American Mathematical Society, 2002.

61. J. MUSIELAK and Z. SEMADENI. Some classes of Banach spaces depending on a parameter. *Studia Math.*, **20**, 271–284, 1961.

62. A. MILLET and M. SANZ-SOLÉ. Approximation of rough paths of fractional Brownian motion. In *Seminar on Stochastic Analysis, Random Fields and Applications V*, pp. 275–303, Progr. Probab., Birkhäuser, 2008.

63. C. REUTENAUER. *Free Lie algebras*, vol. 7 of *London Mathematical Society Monographs. New Series*. Oxford University Press, 1993.

64. R.A. RYAN. *Introduction to Tensor Products of Banach Spaces*. Springer-Verlag, 2002.

65. E.-M. SIPILÄINEN. *A pathwise view of solutions of stochastic differential equations*. PhD thesis, University of Edinburgh, 1993.

66. R. S. STRICHARTZ. The Campbell-Baker-Hausdorff-Dynkin formula and solutions of differential equations. *J. Funct. Anal.*, **72**:2, 320–345, 1987.

67. D. H. SATTINGER and O. L. WEAVER. *Lie groups and algebras with applications to physics, geometry, and mechanics*, vol. 61 of *Applied Mathematical Sciences*. Springer-Verlag, New York, 1993. Corrected reprint of the 1986 original.

68. K. TAPP. *Matrix groups for undergraduates*, vol. 29 of *Student Mathematical Library*. American Mathematical Society, 2005.

69. V. S. VARADARAJAN. *Lie groups, Lie algebras, and their representations*, vol. 102 of *Graduate Texts in Mathematics*. Springer-Verlag, New York, 1984. Reprint of the 1974 edition.

70. N. VICTOIR. Lévy area for the free Brownian motion: existence and non-existence. *J. Funct. Anal.*, **208**:1, 107–121, 2004.

71. F.W. WARNER. *Foundations of differentiable manifolds and Lie groups*, vol. 94 of *Graduate Texts in Mathematics*. Springer-Verlag, New York, 1983. Corrected reprint of the 1971 edition.

72. Y. YAMATO. Stochastic differential equations and nilpotent Lie algebras. *Z. Wahrsch. Verw. Gebiete*, **47**:2, 213–229, 1979.

73. L.C. YOUNG. An inequality of the Hölder type, connected with Stieltjes integration. *Acta Math.*, **67**, 251–282, 1936.

Monotonicity of the Extremal Functions for One-dimensional Inequalities of Logarithmic Sobolev Type

Laurent Miclo

Laboratoire d'Analyse, Topologie, Probabilités, UMR 6632, CNRS
39, rue F. Joliot-Curie, 13453 Marseille cedex 13, France
e-mail: miclo@latp.univ-mrs.fr

Summary. In various one-dimensional functional inequalities, the optimal constants can be found by considering only monotone functions. We study the discrete and continuous settings (and their relationships); we are interested in Poincaré or logarithmic Sobolev inequalities, and several variants obtained by modifying entropy and energy terms.

Keywords: Poincaré inequality, (modified) logarithmic Sobolev inequality, monotonicity of extremal functions, linear diffusions, birth and death process

MSC2000: 46E35, 46E39, 49R50, 26A48, 26D10, 60E15

1 Introduction and Result

On the Borel σ-field of $\mathbb{R}$, let μ be a probability and ν a positive measure. We are interested in the logarithmic Sobolev constant $C(\mu, \nu)$ defined (with the usual conventions $1/\infty = 0$, $1/0 = \infty$ and, most important, $0 \cdot \infty = 0$) by

$$C(\mu, \nu) := \sup_{f \in \mathcal{C}} \frac{\mathrm{Ent}(f^2, \mu)}{\nu[(f')^2]} \quad \in \bar{\mathbb{R}}_+ \tag{1}$$

where $\mathcal{C}$ is the set of all absolutely continuous functions f on $\mathbb{R}$; f' denotes the weak derivative of f. Recall that in general the entropy of a positive, measurable function f with respect to a probability μ is defined as

$$\mathrm{Ent}(f, \mu) := \begin{cases} \mu[f \ln(f)] - \mu[f] \ln(\mu[f]) & \text{if } f \ln(f) \text{ is } \mu\text{-integrable} \\ +\infty & \text{else} \end{cases}$$

and that this quantity belongs to $\bar{\mathbb{R}}_+$, as an immediate consequence of Jensen's inequality with the convex map $\mathbb{R}_+ \ni x \mapsto x \ln(x) \in \mathbb{R}$.

C. Donati-Martin et al. (eds.), *Séminaire de Probabilités XLII*,
Lecture Notes in Mathematics 1979, DOI 10.1007/978-3-642-01763-6_2,
© Springer-Verlag Berlin Heidelberg 2009

One of our aims is to show that the above definition of $C(\mu, \nu)$ is not modified when restricted to monotone functions:

Theorem 1. *Calling $\mathcal{D}$ the cone in $\mathcal{C}$ consisting of all functions f such that $f' \geqslant 0$ a.e., one has*

$$C(\mu, \nu) := \sup_{f \in \mathcal{D}} \frac{\text{Ent}(f^2, \mu)}{\nu[(f')^2]} \quad \in \bar{\mathbb{R}}_+.$$

This can be illustrated by the most famous case of the logarithmic Sobolev inequality (due to Gross [10]), where $\mu = \nu$ is a Gaussian (non degenerate) distribution; then the maximising functions are exactly the exponentials $\mathbb{R} \ni x \mapsto \exp(ax + b)$ with $a \in \mathbb{R}^*$ and $b \in \mathbb{R}$ (see Carlen's article [4]).

We shall also be interested in the following discrete version of the preceding result. For a given $N \in \mathbb{N}^*$, consider the discrete segment $E := \{0, 1, ..., N\}$ as a linear non-oriented graph; call $A := \{\{l, l+1\} : 0 \leqslant l < N\}$ the set of its edges. Denote by $\mathcal{C}$ the set of functions defined on E. If $f \in \mathcal{C}$, its discrete derivative f' is defined on A by

$$\forall \, 0 \leqslant l < N, \qquad f'(\{l, l+1\}) := f(l+1) - f(l)$$

Let also be given a probability μ on E and a measure ν on A. These notations enable us to reinterpret (1) in this new setting, and, as above, our main concern will be to prove:

Theorem 2. *In this discrete framework, one has*

$$C(\mu, \nu) = \sup_{f \in \mathcal{D}} \frac{\text{Ent}(f^2, \mu)}{\nu[(f')^2]} \quad \in \bar{\mathbb{R}}_+$$

where $\mathcal{D}$ is the cone in $\mathcal{C}$ consisting of those functions with positive derivative.

In fact, using interlinks between the continuous and discrete contexts, one can pass from one result to the other. So we shall start with the discrete situation, which is more immediate and better illustrates our itinerary; then similar properties in the continuous framework will derive from the discrete one. The discrete proof can also be directly translated, but precautions must be taken; more on this later.

These monotonicity properties will also be extended to some modified logarithmic Sobolev inequalities (discrete, as in Wu [18] or coutinuous in the sense of Gentil, Guillin and Miclo [9]).

More precisely, in the discrete case, one would like to replace the energy term $\nu[(f')^2]$ by the quantity $\mathcal{E}_\nu(f^2, \ln(f^2))$ defined for $f \in \mathcal{C}$ by

$$\sum_{\{l,l+1\} \in A} \nu(\{l, l+1\})[f^2(l+1) - f^2(l)][\ln(f^2(l+1)) - \ln(f^2(l))];$$

observe that this quantity is quadratically homogeneous. This will be done in

Theorem 3. *Consider the case that $E = \mathbb{Z}$, with the previous notations extended to this setting. One has*

$$\sup_{f \in \mathcal{C}} \frac{\mathrm{Ent}(f^2, \mu)}{\mathcal{E}_\nu(f^2, \ln(f^2))} = \sup_{f \in \mathcal{D}} \frac{\mathrm{Ent}(f^2, \mu)}{\mathcal{E}_\nu(f^2, \ln(f^2))}$$

In the continuous framework, let $H : \mathbb{R}_+ \to \mathbb{R}_+$ be a convex function such that $H(0) = 0$ and $H'(0) = 1$. We now wish to replace the energy term with the following quadratically homogeneous quantity:

$$\forall f \in \mathcal{C}, \qquad \mathcal{E}_{H,\nu}(f) := \int H\left(\left(\frac{f'}{f} \right)^2 \right) f^2 \, d\nu$$

where by convention the integrand equals $(f')^2$ on the set where f vanishes. As before, one then has

Theorem 4. *If μ is a probability on $\mathbb{R}$ and ν a measure on $\mathbb{R}$, one has*

$$\sup_{f \in \mathcal{C}} \frac{\mathrm{Ent}(f^2, \mu)}{\mathcal{E}_{H,\nu}(f)} = \sup_{f \in \mathcal{D}} \frac{\mathrm{Ent}(f^2, \mu)}{\mathcal{E}_{H,\nu}(f)}.$$

Similar results will be obtained when it is the entropy which is modified; for a precise statement, see sub-section 5.3.

But our main motivation comes from the modified logarithmic Sobolev inequalities in Theorems 3 and 4, because we hope that the monotonicity properties we have established eventually allow to apply Hardy inequalities. Indeed, the link between Hardy and modified logarithmic Sobolev inequalities is still poorly understood, whereas that between Hardy and Poincaré, or classical logarithmic Sobolev, inequalities is clear (see for instance Bobkov and Götze's article [3]).

Besides, let us mention that similar results for the Poincaré constant have already been obtained, in the discrete case by Chen (in the proof of Theorem 3.2 in [7]) and in the continuous case by Chen and Wang (Proposition 6.4 in [6], see also the end of the proof of Theorem 1.1 in Chen [8]), for diffusions which are regular enough. Their method partially rests on the equation satisfied by a maximising function (which then is an eigenvector associated to the spectral gap). But it does not clearly adapt to logarithmic Sobolev inequalities, nor even, in the case of the Poincaré constant, to the irregular situations considered above (see for instance the continuity hypothesis needed in the second part of Theorem 1.3 of Chen [8]); therefore we prefer another approach. In particular, we do not a priori deal with the problem of existence of a maximising function (which is crucial in the approach by Chen and Wang [6, 8]). Furthermore, it may be preferable to attack this existence question a posteriori, when discussion is restricted to increasing functions; for rather regular situations, see also the last remark in Section 4.

Still in the case of the Poincaré constant, observe that the equation giving the maximising functions (if they exist) is not easily exploited, for it already involves the Poincaré constant which is unknown in general. Moreover, if in this equation the constant is replaced by the inverse of an eigenvalue other than 0 and the spectral gap, the functions which satisfy this new equation are the corresponding eigenvectors, which are not monotone (under irreducibility hypotheses; see for instance [12]). Therefore we prefer to base our approach on Dirichlet forms rather than on the equation possibly satisfied by the maximising functions.

Let us add that, at least in the case of the Poincaré constant, some monotonicity properties can also be obtained when the underlying graph is a tree. See [12] for a description of the eigenspace associated to the spectral gap (in the discrete case).

The outline of the article is as follows: the next section deals with monotonicity properties for the spectral gap; they have to be considered first to treat the case when no extremal function exists in the above logarithmic Sobolev inequalities. The situations when it exists will then be studied in Section 3, still in the discrete setting. Then Section 4 will extend discussion to the continuous setting, by two different ways. The last section will be devoted to extensions with modified entropy or energy.

Last, I wish to thank the referee whose sugestions led to a better presentation.

2 Poincaré Inequality

In the discrete setting presented in the introduction, we consider the inverse of the spectral gap (also called Poincaré constant) associated to μ and ν, defined by

$$A(\mu, \nu) := \sup_{f \in C} \frac{\mathrm{Var}(f, \mu)}{\nu[(f')^2]} \quad \in \bar{\mathbb{R}}_+, \tag{2}$$

where we recall that the variance of a measurable function f with respect to a probability μ is defined by

$$\mathrm{Var}(f, \mu) = \int \big(f(y) - f(x)\big)^2 \mu(dx)\, \mu(dy) \quad \in \bar{\mathbb{R}}_+.$$

The interest for us of $A(\mu, \nu)$ comes from Theorem 2.2.3 in Saloff-Coste's course [17], where a result due to Rothaus [13, 14, 15] is adapted to the continuous case (in a more general framework than our one-dimensional one). It says that either $C(\mu, \nu) = 2A(\mu, \nu)$, or there exists a function $f \in C$ such that $C(\mu, \nu) = \mathrm{Ent}(f^2, \mu)/\nu[(f')^2]$. This alternative is shown by considering a maximising sequence in (1). So, keeping in mind the aim presented in the introduction, it is useful and instructive to start with its analogue for the spectral gap:

Proposition 1. *Definition (2) is not changed when $\mathcal{C}$ is replaced with $\mathcal{D}$, that is, when only monotone functions are considered.*

But first observe that the supremum featuring in (2) is always achieved. To establish that, two situations will be distinguished.

a) The non-degenerate case where $\nu(a) > 0$ for every $a \in A$. As the expressions $\mathrm{Var}(f, \mu)$ and $\nu[(f')^2]$ are invariant when a constant is added to f and as they are quadratically homogeneous, in (2) one may consider only functions f such that $f(0) = 0$ and $\nu[(f')^2] = 1$. Let now $(f_n)_{n \in \mathbb{N}}$ be a maximising sequence for (2) which satisfies those two conditions. The hypothesis on ν clearly ensures boundedness in $\mathbb{R}^{1+N}$ of the sequence $(f_n)_{n \in \mathbb{N}}$. Hence a convergent subsequence can be extracted, with limit a function f. This limit also verfies $\nu[(f')^2] = 1$, wherefrom one easily deduces that $A(\mu, \nu) = \mathrm{Var}(f, \mu)/\nu[(f')^2]$, showing existence of an extremal function for (2).

b) If $\nu(\{i, i+1\}) = 0$ for some $\{i, i+1\} \in A$, two sub-cases can be considered:

b1) If $\mu(\{0, ..., i\}) > 0$ and $\mu(\{i+1, ..., N\}) > 0$, putting $f = \mathbb{1}_{\{i+1,...,N\}}$, one has $\mathrm{Var}(f, \mu) > 0$ and $\nu[(f')^2] = 0$, hence $C(\mu, \nu) = +\infty$ and f is extremal.

b2) Else, one among $\mu(\{0, ..., i\})$ and $\mu(\{i+1, ..., N\})$ vanishes, and the problem can be restricted to the segment $\{0, ..., i\}$ or $\{i+1, ..., N\}$, whichever has mass 1. By iteration, one is then back to one of the preceding cases.

Note that in the above case (b1), Proposition 1 is established; so we can henceforth assume that $\nu > 0$ on A. This observation is also valid for the logarithmic Sobolev constant, and it almost makes it possible to assume the irreducibility hypothesis of Theorem 2.2.3 of Saloff-Coste [17], except that μ was a priori not supposed to be strictly positive on E. Yet, one always can revert to this situation: call $0 \leqslant x_0 < x_2 < \cdots < x_n \leqslant N$ the elements of E with strictly positive μ-weight. Given some real numbers $y_0, y_1, ..., y_n$, consider the affine sub-space of $\mathcal{C}$ consisting of those functions f such that $f(x_i) = y_i$ for each $0 \leqslant i \leqslant n$, and try minimising $\nu[(f')^2]$ therein. For fixed $0 \leqslant i < n$, this leads to look for the functions g on $\{x_i, x_i + 1, ..., x_{i+1}\}$ which minimise $\sum_{x_i \leqslant x < x_{i+1}} \nu(\{x, x+1\})(g'(\{x, x+1\}))^2$ under the constraints $g(x_i) = y_i$ and $g(x_{i+1}) = y_{i+1}$. By a simple application of the equality case in the Cauchy-Schwarz inequality, this optimisation problem admits the following unique solution:

$\forall\, x_i \leqslant x \leqslant x_{i+1},$

$$g(x) = y_i + \left(\sum_{x_i \leqslant y < x_{i+1}} \frac{1}{\nu(\{y, y+1\})} \right)^{-1} \sum_{x_i \leqslant y < x} \frac{y_{i+1} - y_i}{\nu(\{y, y+1\})}. \tag{3}$$

So, setting

$$\forall\, 0 \leqslant i \leqslant n, \qquad \tilde{\mu}(i) := \mu(x_i)$$

$$\forall\, 0 \leqslant i < n, \qquad \tilde{\nu}(\{i, i+1\}) := \left(\sum_{x_i \leqslant y < x_{i+1}} \frac{1}{\nu(\{y, y+1\})} \right)^{-1},$$

one would be reduced to a situation where the underlying probability is everywhere strictly positive; moreover, using (3), one easily switches back and forth
between extremal functions for both problems. This would also fully justify
the reminder before Proposition 1.

On the other hand, we shall also discard the trivial case when μ is a Dirac
mass; this ensures that $A(\mu, \nu) > 0$.

We can now be a little more precise on the maximising functions in (2):

Lemma 1. *Let f be a function realizing the maximum in (2). Assuming that
$\nu > 0$ on A and that μ is not a Dirac mass, every maximising function has
the form $af + b\mathbb{1}$ where $a \in \mathbb{R}^*$, $b \in \mathbb{R}$ and $\mathbb{1}$ denotes the constant function
with value 1.*

Proof. Clearly, if f is maximising and if $a \in \mathbb{R}^*$ and $b \in \mathbb{R}$, $af + b\mathbb{1}$ is also
maximising in (2).

Conversely, let g be maximising in (2); by subtracting a constant, we may
suppose that $\mu[g] = 0$. By variational calculus around g (i.e., by considering
$g + \epsilon h$, with $\epsilon \in \mathbb{R}$ and any $h \in \mathcal{C}$, and taking a first order expansion when
$\epsilon \to 0$ of the ratio $\mathrm{Var}(g + \epsilon h, \mu)/\nu[(g' + \epsilon h')^2])$, one easily sees that for each
$i \in E$, g satisfies

$$A(\mu, \nu)\left[\nu(\{i, i+1\})(g(i) - g(i+1)) + \nu(\{i-1, i\})(g(i) - g(i-1))\right] = \mu(i)g(i)$$

with the conventions $\nu(\{-1, 0\}) = 0 = \nu(\{N, N+1\})$.

Now, since $A(\mu, \nu) > 0$ and $\nu > 0$ on A, starting from $g(0)$ these equations
inductively determine $g(1)$, $g(2)$, up to $g(N)$. Note that $g(0) \neq 0$, else we
would end up with $g \equiv 0$, contradicting $A(\mu, \nu) > 0$. So there is at most one
minimising function g for (2) which satisfies $\mu[g] = 0$ and $g(0) = 1$. This is
exactly what the lemma asserts. □

Given a maximising f for (2), our strategy to show its monotonicity will
be as follows: supposing on the contrary f not to be monotone, we shall
decompose f as $\widetilde{f} + \widehat{f}$, with $\widetilde{f}$ (and hence also $\widehat{f}$) not belonging to the linear
span $\mathrm{Vect}(\mathbb{1}, f)$, and with

$$\mathrm{Var}(f, \mu) = \mathrm{Var}(\widetilde{f}, \mu) + \mathrm{Var}(\widehat{f}, \mu)$$
$$\nu[(f')^2] \geqslant \nu[(\widetilde{f}')^2] + \nu[(\widehat{f}')^2].$$

Clearly, these two relations imply that $\widetilde{f}$ and $\widehat{f}$ also are maximising for (2), a
contradiction since $\widetilde{f}$ and $\widehat{f}$ do not have the form required by Lemma 1.

So let f be maximising for (2) but not monotone.

A point $i \in E$ will be called a local maximum of f if for each $j \in E$ verifying
$f(j) > f(i)$, the segment $[\![i, j]\!]$ (the sub-segment of E with endpoints i and j)
contains an element k such that $f(k) < f(i)$. By definition, a local minimum
of f will be a local maximum of $-f$.

We shall now construct $\widetilde{f}$ by splitting f at a particular level. Replacing
f by $-f$ if necessary, we may choose a local maximum i in $[\![1, N-1]\!]$ such

that f has a local minimum in $[\![0, i]\!]$ and another one in $[\![i, N]\!]$. Among such local maxima i, choose one which minimises $f(i)$, and call it i_0. Denote by i_1 (respectively i_{-1}) the closest local minimum on the right (respectively on the left) of i_0. By possibly reversing the order of $[\![0, N]\!]$, one can suppose that $f(i_{-1}) \leqslant f(i_1)$. Also, set $i_2 := \max\{y \geqslant i_1 : \forall\, i_1 \leqslant x \leqslant y,\ f(x) = f(i_1)\}$.

For $s \in [f(i_1), f(i_0)]$, let $S_s := [\![a_s, b_s]\!]$ be the discrete segment whose ends are defined by

$$a_s := \min\{x \in [\![i_{-1}, i_0]\!] : f(x) \geqslant s\}$$
$$b_s := \min\{x \in [\![i_2, N]\!] : f(x) \geqslant s\} - 1$$

(with the convention that $b_s = N$ if the latter set is empty).

By those choices, particularly by minimality of i_0, one easily verifies that for any $s \in [f(i_1), f(i_0)]$, f is increasing (this is always understood in the wide sense) on $[\![a_s, i_0]\!]$, decreasing on $[\![i_0, i_2]\!]$ and increasing on $[\![i_2, b_s + 1]\!]$ (the reader is urged to draw a picture).

Still for $s \in [f(i_1), f(i_0)]$, set for $x \in E$

$$\widetilde{f}_s(x) = f(x)\mathbb{1}_{S_s^c}(x) + s\mathbb{1}_{S_s}(x)$$
$$\widehat{f}_s(x) = (f(x) - s)\mathbb{1}_{S_s}(x).$$

One has indeed $f_s = \widetilde{f}_s + \widehat{f}_s$, and the claimed decomposition will be obtained owing to the following two lemmas.

Lemma 2. *For any $s \in\,]f(i_1), f(i_0)[$, one has*

$$\nu[(f')^2] \geqslant \nu[(\widetilde{f}'_s)^2] + \nu[(\widehat{f}'_s)^2].$$

Proof. An immediate calculation first gives

$$\nu[(f')^2] = \nu[(\widetilde{f}'_s + \widehat{f}'_s)^2] = \nu[(\widetilde{f}'_s)^2] + \nu[(\widehat{f}'_s)^2] + 2\nu[\widetilde{f}'_s\widehat{f}'_s]$$

and then

$$\nu[\widetilde{f}'_s\widehat{f}'_s] = \nu(\{a_s - 1, a_s\})\big(s - f(a_s - 1)\big)\big(f(a_s) - s\big)$$
$$+ \nu(\{b_s, b_s + 1\})\big(f(b_s + 1) - s\big)\big(s - f(b_s)\big)$$

(still with the convention that $\nu(\{N, N + 1\}) = 0$). Now, from the fact that $s \in\,]f(i_1), f(i_0)[$, it appears that $f(i_{-1}) \leqslant f(a_s - 1) < s \leqslant f(a_s) \leqslant f(i_0)$ and $f(i_2) \leqslant f(b_s) < s \leqslant f(b_s + 1)$, which allows to notice that $\nu[\widetilde{f}'_s\widehat{f}'_s] \geqslant 0$, wherefrom the claimed inequality derives. $\square$

Lemma 3. *There exists $s_0 \in\,]f(i_1), f(i_0)[$ such that*

$$\mathrm{Var}(f, \mu) = \mathrm{Var}(\widetilde{f}_s, \mu) + \mathrm{Var}(\widehat{f}_s, \mu).$$

Proof. The difference between the left and right hand sides is but twice the covariance of $\widetilde{f}_s$ and $\widehat{f}_s$ under μ, which equals

$$\mu\left[\left(\widetilde{f}_s - \mu[\widetilde{f}_s]\right)\left(\widehat{f}_s - \mu[\widehat{f}_s]\right)\right] = \mu\left[\left(\widetilde{f}_s - \mu[\widetilde{f}_s]\right)\widehat{f}_s\right]$$
$$= \left(s - \mu[\widetilde{f}_s]\right)\mu\left[(f - s)\mathbb{1}_{S_s}\right]. \qquad (4)$$

Hence, it suffices to find an $s \in \,]f(i_1), f(i_0)[$ such that $\mu[(f - s)\mathbb{1}_{S_s}] = 0$. Put $i_3 := b_{f(i_0)} + 1$; from the increasingness of f on $[\![i_{-1}, i_0]\!]$ and on $[\![i_2, i_3]\!]$, one is easily convinced that the map $\Psi : [f(i_1), f(i_0)] \ni s \mapsto \mu[(f - s)\mathbb{1}_{S_s}]$ is continuous. Now, the pattern of f on $[\![i_{-1}, i_3]\!]$ implies that $\Psi(f(i_1)) > 0$ and $\Psi(f(i_0)) < 0$, so there exists $s_0 \in \,]f(i_1), f(i_0)[$ such that $\Psi(s_0) = 0$. $\qquad\square$

Notate $\widetilde{f} = \widetilde{f}_{s_0}$ and $\widehat{f} = \widehat{f}_{s_0}$, where s_0 is chosen as in the preceding lemma. To finalize the proof of Proposition 1, it remains to see that $\widehat{f}$ is not in $\mathrm{Vect}(f, \mathbb{1})$. To this end, notice that i_1 is no longer a local minimum for $\widetilde{f}$ (this function may go down from i_1 to i_{-1}, and yet $\widetilde{f}(i_1) = s_0 > f(i_1) \geqslant f(i_{-1}) = \widetilde{f}(i_{-1})$), and consequently $\widehat{f}$ cannot be written as $af + b\mathbb{1}$ with $a > 0$ and $b \in \mathbb{R}$. On the other hand, the inequalities $\widehat{f}(i_{-1}) < \widehat{f}(i_0)$ and $f(i_{-1}) < f(i_0)$ also show that $\widehat{f}$ cannot be written as $af + b\mathbb{1}$ with $a \leqslant 0$ and $b \in \mathbb{R}$. Therefore the claimed result follows.

3 Splitting up the Entropy

Our aim here is to establish (2) in the discrete setting. According to the results from the preceding section, it suffices to consider the case when there exists a (non constant) maximising f for (1). For else, a maximising family for the logarithmic Sobolev inequality is $(1 + f/(n + 1))_{n\in\mathbb{N}}$, where f is a maximising function for the corresponding Poincaré inequality (and hence f is monotone). Globally, the scheme of our proof will be similar to that of the previous section, most of whose notation will be kept in use.

First of all, observe that one may from now on suppose that $f \geqslant 0$, by possibly replacing f with $|f|$, since one has $\nu[(|f|')^2] \leqslant \nu[(f')^2]$. Assume now the hypothesis (to be refuted) that f is not monotone. Two possibilities arise: either f has a local maximum i in $[\![1, N-1]\!]$ such that there is a local minimum in $[\![0, i]\!]$ and one in $[\![i, N]\!]$, or the same holds for $-f$. We shall consider the first case only; the second one is very similar and left to the reader (one has to work with the negatively valued function $-f$).

As in section 2, i_{-1}, i_0, i_1, i_2 and i_3 are defined, then, for $s \in [f(i_1), f(i_0)]$, S_s, $\widetilde{f}_s$ and $\widehat{f}_s$. Our main task will consist in "splitting up" the entropy:

Lemma 4. *There exists* $s_1 \in \,]f(i_1), f(i_0)[$ *such that*

$$\mathrm{Ent}(f^2, \mu) = \mathrm{Ent}(\widetilde{f}_{s_1}^2, \mu) + \mathrm{Ent}\left((s + \widehat{f}_{s_1})^2, \mu\right).$$

Proof. First remark that for all $s \in [f(i_1), f(i_0)]$ and for all function $F :$ $\mathbb{R}_+ \to \mathbb{R}$, one has

$$\mu[F(f)] = \mu[F(\widetilde{f_s})] + \mu[F(s + \widehat{f_s})] - F(s). \tag{5}$$

Indeed, by definition, one can perform the following expansion:

$$\begin{aligned} \mu[F(f)] &= \mu[\mathbb{1}_{S_s^c} F(\widetilde{f_s})] + \mu[\mathbb{1}_{S_s} F(s + \widehat{f_s})] \\ &= \mu[F(\widetilde{f_s})] - \mu[\mathbb{1}_{S_s} F(s)] + \mu[F(s + \widehat{f_s})] - \mu[\mathbb{1}_{S_s^c} F(s)] \\ &= \mu[F(\widetilde{f_s})] + \mu[F(s + \widehat{f_s})] - F(s). \end{aligned}$$

In particular, applying this to $F : \mathbb{R}_+ \ni u \mapsto u^2 \ln(u^2)$, it appears that

$$\mathrm{Ent}(f^2, \mu) - \mathrm{Ent}(\widetilde{f}_{s_1}^2, \mu) - \mathrm{Ent}\big((s + \widehat{f}_{s_1})^2, \mu\big) = \varphi(y_s') + \varphi(x_s') - \varphi(y) - \varphi(x_s)$$

with φ the convex map given by $\varphi : \mathbb{R}_+ \ni u \mapsto u \ln(u)$ and

$$\begin{aligned} y_s' &= \mu[\widetilde{f_s^2}] & y &= \mu[f^2] \\ x_s' &= \mu[(s + \widehat{f_s})^2] & x_s &= s^2. \end{aligned}$$

Resorting again to (5), but with $F(s) = s^2$, it appears that $x_s + y = x_s' + y_s'$, which means that both segments $[x_s, y]$ and $[x_s', y_s']$ have the same midpoint. So, by convexity of φ, the inequality $\varphi(x_s) + \varphi(y) \geqslant \varphi(x_s') + \varphi(y_s')$ is equivalent to $|y - x_s| \geqslant |y_s' - x_s'|$. Or also, if some $s_1 \in \,]f(i_1), f(i_0)[$ happens to be such that $|y - x_s| = |y_s' - x_s'|$, then the equality in Lemma 3 holds (without even using the convexity of φ). Now one computes (still owing to (5) with $F(s) = s^2$) that

$$\begin{aligned} y_s' - x_s' &= \mu[\widetilde{f_s^2}] - \mu[(s + \widehat{f_s})^2] = \mu[f^2] + s^2 - 2\mu[(s + \widehat{f_s})^2] \\ &= \mu[f^2] - s^2 - 2\mu[\widehat{f_s^2}] - 4s\mu[\widehat{f_s}] = y - x_s - 2\mu[\widehat{f_s}(\widehat{f_s} + 2s)]. \end{aligned}$$

Hence it suffices to find an $s \in \,]f(i_1), f(i_0)[$ such that $\mu[\widehat{f_s}(\widehat{f_s} + 2s)] = 0$. But $\widehat{f_s} + 2s$ is a positive function, whereas $\widehat{f_s}$ is positive for $s = f(i_1)$ and negative for $s = f(i_0)$. The claim follows by continuity of the application $[f(i_1), f(i_0)] \ni s \mapsto \mu[\widehat{f_s}(\widehat{f_s} + 2s)]$, which is easily seen not to vanish at the endpoints. $\qquad\square$

Besides, according to Lemma 2, one has for all $s \in \,]f(i_1), f(i_0)[$

$$\nu[(f')^2] \geqslant \nu[(\widetilde{f_s'})^2] + \nu[(\widehat{f_s'})^2] = \nu[(\widetilde{f_s'})^2] + \nu[((s + \widehat{f_s})')^2].$$

Using the notation and proof of that Lemma again, one can even say a little more: equality can hold only if for all edges $a \in A$ one has $\widetilde{f_s'}(a)\widehat{f_s'}(a) = 0$, which in particular entails that $f(a_s) = s$. So, for $s \in \,]f(i_1), f(i_0)[$, the discrete segment S_s contains at least three different points, a_s, i_0 and i_1.

Now, what we saw just before implies that $\widetilde{f}_{s_1}$ and $s_1 + \widehat{f}_{s_1}$ also are maximising functions for (1), and that necessarily

$$\nu\big[(f')^2\big] = \nu\big[(\widetilde{f}'_{s_1})^2\big] + \nu\big[\big((s_1 + \widehat{f}_{s_1})'\big)^2\big],$$

for else, one would have

$$\frac{\mathrm{Ent}(f^2, \mu)}{\nu\big[(f')^2\big]} < \frac{\mathrm{Ent}(\widetilde{f}_{s_1}^2, \mu) + \mathrm{Ent}\big((s + \widehat{f}_{s_1})^2, \mu\big)}{\nu\big[(\widetilde{f}'_{s_1})^2\big] + \nu\big[\big((s_1 + \widehat{f}_{s_1})'\big)^2\big]}$$

$$\leqslant \max\left(\frac{\mathrm{Ent}(\widetilde{f}_{s_1}^2, \mu)}{\nu\big[(\widetilde{f}'_{s_1})^2\big]}, \frac{\mathrm{Ent}\big((s + \widehat{f}_{s_1})^2, \mu\big)}{\nu\big[\big((s_1 + \widehat{f}_{s_1})'\big)^2\big]}\right)$$

(the first inequality uses that by construction $\mathrm{Ent}(f^2, \mu) > 0$). Therefore there exist three successive points in S_{s_1} where $\widetilde{f}_{s_1}$ assumes the same value (namely, s_1) and we shall now verify that this is not possible, more precisely that this would imply constancy of $\widetilde{f}_{s_1}$, which does not hold (for $\widehat{f}_{s_1}(i_{-1}) < \widetilde{f}_{s_1}(i_0)$). Indeed, by variational calculus around a maximising function f, one sees that f must verify for all $i \in E$ (with the usual conventions)

$$C(\mu, \nu)\big[\nu(\{i, i+1\})(f(i) - f(i+1)) + \nu(\{i-1, i\})(f(i) - f(i-1))\big]$$
$$= \mu(i)f(i)\ln\left(\frac{f^2(i)}{\mu[f^2]}\right).$$

Recall that discussion has been reduced to the situation that μ, ν and $C(\mu, \nu)$ are strictly positive (see before Lemma 1); so if f takes the same value v at three successive points $y - 1$, y and $y + 1$, with $0 < y < N$, then the preceding equation taken at $i = y$ forces $v\ln(v^2/\mu[f^2]) = 0$, that is to say, $v = 0$ or $v = \sqrt{\mu[f^2]}$. Applying then the equation at $i = y + 1$ instead, one obtains $f(y+2) = f(y+1)$, at least if $y \leqslant N - 2$. Similarly, for $i = y - 1$, $f(y-2) = v$ if $y \geqslant 2$. So equality $f(i) = v$ propagates everywhere and f is constanty equal to v.

These arguments terminate the proof of (2) by replacing the recourse to Lemma 1. For even though the knowledge of $\mu[f^2]$ and of $f(0)$ determines a maximising function f for (1) owing to the linear structure of the graph E (still for fixed μ and ν verifying $C(\mu, \nu) > 0$ and $\nu > 0$ on A, as we were allowed to suppose in the preceding section), here this no longer implies Lemma 1 because the term $\mu(i)f(i)\ln(f^2(i)/\mu[f^2])$ above is not affine in $f(i)$. Besides, this lemma never holds in the context of logarithmic Sobolev inequalities. Indeed, let again f be a positive function which maximises (1). Perturbating f by a constant function and performing a variational computation, one obtains $\mu[f\ln(f/\mu[f^2])] = 0$. Set $F(t) = \mu[(f + t)\ln((f + t)/\mu[(f + t)^2])]$ for all $t \geqslant 0$. Differentiating twice this expression on $\mathbb{R}^*_+$, one obtains

$$F''(t) = 2\int \frac{1}{f + t}\,d\mu - 2\frac{\mu[f + t]}{\mu[(f + t)^2]}\left(2 - \frac{\mu[f + t]^2}{\mu[(f + t)^2]}\right).$$

Using Jensen's inequality $\mu[1/(f+t)] \geqslant 1/\mu[f+t]$ and the fact that the map $[0,1] \ni x \mapsto x(2-x)$ is bounded by 1, it appears that F'' is strictly positive on $\mathbb{R}_+^*$ if f is not μ-a.s. constant (consider the case when Jensen's inequality is an equality). So, there may exist at most two $t \geqslant 0$ such that $F(t) = 0$.

Remark 1. The inequality $\mu[\widehat{f}_{f(i_1)}(\widehat{f}_{f(i_1)} + 2f(i_1))] > 0$ does not allow to deduce that $\mathrm{Ent}(f^2, \mu) < \mathrm{Ent}(\widehat{f}_{f(i_1)}^2, \mu) + \mathrm{Ent}((f(i_1) + \widehat{f}_{f(i_1)})^2, \mu)$; this is true only under additional conditions concerning the signs of $y'_{f(i_1)} - x'_{f(i_1)}$ and $y - x_{f(i_1)}$ (a similar observation holds at $s = f(i_0)$). The possibility for $y'_s - x'_s$ and $y - x_s$ to change sign when s ranges over $[f(i_1), f(i_0)]$ (the worst case is when such changes precisely occur where $\mu[\widehat{f}_s(\widehat{f}_s + 2s)]$ vanishes) is as much a nuisance as the the factor $s - \mu[\widetilde{f}_s]$ which appeared in (4). Therefore we are a priori not sure of the existence of some $s \in [f(i_1), f(i_0)]$ making one of the functions $\widetilde{f}_s$ and $s + \widehat{f}_s$ "strictly more maximising" than f. On the opposite, in the spectral gap case, this conclusion was nonetheless reachable, by using the extra fact that the map $[f(i_1), f(i_0)] \ni s \mapsto s - \mu[\widetilde{f}_s]$ is increasing (more precisely, a further analysis easily shows that $[f(i_1), f(i_0)] \ni s \mapsto s - \mu[\widetilde{f}_s]$ is increasing).

4 Continuous Situation

So we come back to the framework first considered in the introduction. We shall only deal with the case of the logarithmic Sobolev constant; the Poincaré constant can be treated in a very similar way. As already explained, the continuous situation will be reduced to the discrete one, thus giving the proof a slight probabilistic touch. We shall also consider the other possibility, to adapt the previous proofs, which leads to further analysing the (almost) minimising functions. But whichever way is chosen, the beginning of the proof appears to need some regularization as its first step.

For $M > 0$, let $\mathcal{C}_{[-M,M]}$ (respectively $\mathcal{D}_{[-M,M]}$) be the sub-set of $\mathcal{C}$ (respectively of $\mathcal{D}$) consisting of the absolutely continuous functions with weak derivative a.e. null on $]-\infty, -M] \cup [M, +\infty[$. Also, put

$$C_{[-M,M]}(\mu, \nu) := \sup_{f \in \mathcal{C}_{[-M,M]}} \frac{\mathrm{Ent}(f^2, \mu)}{\nu[(f')^2]}$$

$$D_{[-M,M]}(\mu, \nu) := \sup_{f \in \mathcal{D}_{[-M,M]}} \frac{\mathrm{Ent}(f^2, \mu)}{\nu[(f')^2]}.$$

One is easily convinced that these two quantities increase with $M > 0$ and that they respectively converge for large M to $C(\mu, \nu)$ and

$$D(\mu, \nu) := \sup_{f \in \mathcal{D}} \frac{\mathrm{Ent}(f^2, \mu)}{\nu[(f')^2]} \quad \in \bar{\mathbb{R}}_+.$$

Call $\nu_{[-M,M]}$ the restriction of ν to $[-M, M]$ (it vanishes outside this interval) and $\mu_{[-M,M]}$ the probability obtained by accumulating on the endpoints $-M$ and M the mass outside $[-M, M]$; i.e., $\mu_{[-M,M]}$ is defined by

$$\mu_{[-M,M]}(B) := \mu(B \cap]-M, M[) + \mu(]-\infty, M])\delta_{-M}(B) + \mu([M, +\infty[)\delta_M(B)$$

for B any Borel set in $\mathbb{R}$. The interest of these measures is that $C_{[-M,M]}(\mu, \nu) = C(\mu_{[-M,M]}, \nu_{[-M,M]})$ and $D_{[-M,M]}(\mu, \nu) = D(\mu_{[-M,M]}, \nu_{[-M,M]})$, so the convergences seen above allow restriction to the case that μ and ν are supported in the compact $[-M, M]$, where $M > 0$ is fixed from now on. We shall also content ourselves with only considering functions defined on $[-M, M]$.

Denote by λ the restriction of the Lebesgue measure to $[-M, M]$ and, by abuse of language, still call ν the Radon-Nikodym derivative of ν with respect to λ (which exists without any restriction on ν, provided the value $+\infty$ is allowed; see for instance [11]). As weak derivatives are only a.e. defined, it is well known that $C(\mu, \nu)$ (or $D(\mu, \nu)$) is not modified when ν is replaced with the measure having ν as density with respect to λ, which we henceforth assume. One can also without loss suppose the function ν to be minorated by an a.e. strictly positive constant. Indeed, this derives from the fact that for any $f \in \mathcal{C}$, one has

$$\lim_{\eta \to 0+} \frac{\mathrm{Ent}(f^2, \mu)}{\int (f')^2 (\eta \wedge \nu) d\lambda} = \frac{\mathrm{Ent}(f^2, \mu)}{\nu[(f')^2]}$$

and that this convergence is monotone. So, by exchanging suprema, equality is preserved in the limit. Hence $\eta > 0$ wil be fixed in the sequel, so that $\nu \geqslant \eta$ everywhere on $[-M, M]$, i.e., a suitable version of ν is chosen; but beware, ν may still assume the value $+\infty$ (remark that obtaining the corresponding majorization of ν would be more delicate).

The next procedure consists in modifying μ and is a little less immediate; a general preparation is needed:

Lemma 5. *On some measurable space, let μ be a probability and f and g two bounded, measurable functions. Suppose that $\|g - f\|_\infty \leqslant \epsilon \leqslant 1$ (uniform norm) and that the oscillation of f (i.e., $\mathrm{osc}(f) := \sup f - \inf f$) is majorized by a, where ϵ and a are positive real numbers. Then there exists a number $b(a) \geqslant 0$, depending only upon a, such that*

$$\left| \mathrm{Ent}(g^2, \mu) - \mathrm{Ent}(f^2, \mu) \right| \leqslant b(a) \epsilon.$$

Proof. Note that $|f|$ and $|g|$ fulfill the same hypotheses as f and g; so no generality is loss by further supposing f and g to be positive.

Two situations are then distinguished, according to $\mu[f]$ being "large" or "small". We shall start with the case when $\mu[f] \leqslant 2 + 2a$. This ensures that f is majorized by $2 + 3a$ and g by $3 + 3a$. Now, on the interval $[0, 3 + 3a]$, the derivative of the map $t \mapsto t^2 \ln(t^2)$ is bounded by a finite quantity $b_1(a)$; this entails that

$$\left|\mu[g^2 \ln(g^2)] - \mu[f^2 \ln(f^2)]\right| \leqslant \mu\left[\left|g^2 \ln(g^2) - f^2 \ln(f^2)\right|\right]$$
$$\leqslant b_1(a)\, \mu[|g - f|] \leqslant b_1(a)\, \epsilon.$$

Similarly, the norm inequality $\left|\sqrt{\mu[g^2]} - \sqrt{\mu[f^2]}\right| \leqslant \sqrt{\mu[(g - f)^2]}$ in $\mathbb{L}^2(\mu)$ yields

$$\left|\mu[g^2] \ln(\mu[g^2]) - \mu[f^2] \ln(\mu[f^2])\right| \leqslant b_1(a)\, \epsilon,$$

wherefrom finally the claimed inequality with $b(a) = 2b_1(a)$.

Consider now the case when $\mu[f] > 2 + 2a$. It seems more convenient to deal with the map $\mathbb{R}_+ \ni t \mapsto t \ln(t)$. Performing an expansion with first-order remainder, centred at $\mu[f^2]$, one finds a $\theta \in [0, 1]$ such that $\mu[g^2] \ln(\mu[g^2])$ equals

$$\mu[f^2] \ln\left(\mu[f^2]\right) + \left(1 + \ln\left[\mu[f^2] + \theta(\mu[g^2] - \mu[f^2])\right]\right) \left(\mu[g^2] - \mu[f^2]\right).$$

The same operation performed pointwise yields another measurable function $\widetilde{\theta}$ with values in $[0, 1]$ such that one has everywhere

$$g^2 \ln(g^2) = f^2 \ln(f^2) + \left(1 + \ln\left(f^2 + \widetilde{\theta}(g^2 - f^2)\right)\right) (g^2 - f^2).$$

Integrating this against μ and taking into account the preceding equality, it appears that

$$\mathrm{Ent}(g^2, \mu) - \mathrm{Ent}(f^2, \mu)$$
$$= \mu\left[\left(\ln(f^2 + \widetilde{\theta}(g^2 - f^2)) - \ln\left[\mu[f^2] + \theta(\mu[g^2] - \mu[f^2])\right]\right)(g^2 - f^2)\right]. \quad (6)$$

However, observe that

$$f^2 + \widetilde{\theta}(g^2 - f^2) \geqslant f^2 \wedge g^2 \geqslant \left(\mu[f] - \mathrm{osc}(f) - 1\right)^2$$
$$\geqslant \left(\mu[f] - a - 1\right)^2 \geqslant \frac{\mu[f]^2}{4}$$

and similarly

$$\mu[f^2] + \theta\left(\mu[g^2] - \mu[f^2]\right) \geqslant \frac{\mu[f]^2}{4}.$$

So one obtains the pointwise inequality

$$\left|\ln\left(f^2 + \widetilde{\theta}(g^2 - f^2)\right) - \ln\left(\mu[f^2] + \theta(\mu[g^2] - \mu[f^2])\right)\right|$$
$$\leqslant 4\,\mu[f]^{-2}\,\left|f^2 + \widetilde{\theta}(g^2 - f^2) - \mu[f^2] - \theta(\mu[g^2] - \mu[f^2])\right|.$$

Let us look at the last absolute value. It can be majorized by

$$\left(f + \sqrt{\mu[f^2]}\right)\left|f - \sqrt{\mu[f^2]}\right| + (f + g)\,|f - g| +$$
$$\left(\sqrt{\mu[g^2]} + \sqrt{\mu[f^2]}\right)\left|\sqrt{\mu[g^2]} - \sqrt{\mu[f^2]}\right|$$
$$\leqslant 2\left(\mu[f] + a\right)a + (2\mu[f] + 2a + 1)\,\epsilon + (2\mu[f] + 2a + 1)\,\epsilon$$
$$\leqslant (2\mu[f] + 2a + 1)(a + 2).$$

On the other hand, one has as above

$$\left|g^2 - f^2\right| \leqslant (2\mu[f] + 2a + 1)\,\epsilon,$$

wherefrom, coming back to (6), it appears that

$$\left|\text{Ent}(g^2, \mu) - \text{Ent}(f^2, \mu)\right| \leqslant 4\,\frac{(a + 2)\left(2\mu[f] + 2a + 1\right)^2}{\mu[f]^2}\,\epsilon$$

and in that case the lemma holds with $b(a) = b_2(a)$, where

$$b_2(a) := \sup_{t \geqslant 2 + 2a} 4\,\frac{(a + 2)(2t + 2a + 1)^2}{t^2} < +\infty. \qquad \square$$

This technical result will be used to measure how certain modifications of μ influence $C(\mu, \nu)$. More precisely, for fixed $n \in \mathbb{N}^*$, for any $0 \leqslant i \leqslant n$ put $x_{n,i} := -M + i2M/n$ and introduce the probability

$$\mu_n := \sum_{0 \leqslant i \leqslant n} \mu\left([x_{n,i}, x_{n,i+1})\right)\delta_{x_{n,i}}$$

with the convention that $x_{n,n+1} = +\infty$.

Lemma 6. *With the notation of Lemma 5, for all $n \in \mathbb{N}^*$ one has*

$$\left|C(\mu_n, \nu) - C(\mu, \nu)\right| \leqslant b\left(\sqrt{2M}\right)\sqrt{\frac{2M}{n}}.$$

Proof. Calling $\mathcal{C}(\nu)$ the set of absolutely continuous functions f such that $\nu[(f')^2] = 1$, one has

$$C(\mu, \nu) = \sup_{f \in \mathcal{C}(\nu)}\,\text{Ent}(f^2, \mu)$$

and one also has a similar formula for $C(\mu_n, \nu)$. Thus, to obtain the claimed bound, it suffices to see that for all $f \in \mathcal{C}(\nu)$, one has

$$\left|\text{Ent}(f^2, \mu_n) - \text{Ent}(f^2, \mu)\right| \leqslant b\left(\sqrt{2M}\right)\sqrt{\frac{2M}{n}}.$$

To that end, rewrite $\text{Ent}(f^2, \mu_n)$ as $\text{Ent}(f_n^2, \mu)$, where f_n is the function which equals $f(x_{n,i})$ on $[x_{n,i}, x_{n,i+1}[$ for all $0 \leqslant i \leqslant n$. To apply Lemma 5, it remains to evaluate $\text{osc}(f)$ and $\|f_n - f\|_\infty$. These estimates, and consequently also the claimed result, easily follow from the following application of the Cauchy-Schwarz inequality:

$$\forall \, x, y \in [-M, M],$$

$$|f(y) - f(x)| = \left| \int_{[x,y]} f' \, d\lambda \right| \leqslant \sqrt{\int_{[x,y]} (f')^2 \, d\nu} \sqrt{\int_{[x,y]} \frac{1}{\nu} d\lambda}$$

$$\leqslant \eta^{-1/2} \sqrt{|y - x|},$$

where the last estimate holds for any function belonging to $\mathcal{C}(\nu)$. □

Evidently, the above proof also shows that

$$|D(\mu_n, \nu) - D(\mu, \nu)| \leqslant b(\sqrt{2M}) \sqrt{\frac{2M}{n}} \, ;$$

so, to get convinced of the equality $C(\mu, \nu) = D(\mu, \nu)$, it suffices to see that $C(\mu_n, \nu) = D(\mu_n, \nu)$ for all $n \in \mathbb{N}^*$. But this problem reduces to the discrete context. Indeed, as before Lemma 1, the values of $f(x_{n,i})$ being fixed, one has to minimise the quantity $\int_{x_{n,i}}^{x_{n,i+1}} (f')^2 \nu \, d\lambda$ for each given $0 \leqslant i < n$. This optimisation problem is simply solved; the minimal value is

$$\left(\int_{x_{n,i}}^{x_{n,i+1}} \frac{1}{\nu} d\lambda \right)^{-1} \left(f(x_{n,i+1}) - f(x_{n,i}) \right)^2$$

and is achieved by a function which is monotone on the segment $[x_{n,i}, x_{n,i+1}]$. Hence we are back to the discrete problem on $n+1$ points with the probability $\tilde{\mu}_n$ and the measure $\tilde{\nu}_n$ respectively defined by

$$\forall \, 0 \leqslant i \leqslant n, \qquad \tilde{\mu}_n(i) := \mu_n(x_{n,i})$$

$$\forall \, 0 \leqslant i < n, \qquad \tilde{\nu}_n(\{i, i+1\}) := \left(\int_{x_{n,i}}^{x_{n,i+1}} \frac{1}{\nu} d\lambda \right)^{-1}.$$

Sections 2 and 3 now allow to conclude.

From a possibly more analytically-minded point of view, remark that Lemmas 5 and 6 could also allow to regularize μ, which could be supposed to admit a $\mathcal{C}^\infty$ density with respect to λ.

Let us now mention another possible approach, directly inspired from the method of sections 2 and 3. A priori two problems arise in this perspective: on the one hand, whether a minimising function exists (even in the case of the Poincaré inequality), and on the other hand, when it exists, whether the set of its global minima and maxima can have infinitely many connected components (this means, the function oscillates infinitely often; this is inconvenient for us, see the considerations before Lemma 2). These problems can be bypassed as follows. We put ourselves back in the framework preceding Lemma 5.

First, the notion of local minimum or maximum introduced in section 2 will be extended to the continuous case, with discrete segments replaced by continuous ones. For $f \in \mathcal{C}$, $\mathcal{M}(f)$ will denote the set of local minima and maxima of f. For $p \in \mathbb{N}^*$, call $\mathcal{C}_p$ the set of functions $f \in \mathcal{C}$ such that $\mathcal{M}(f)$ has at most p connected components. So one verifies that $\mathcal{C}_1$ (respectively $\mathcal{C}_2$) is the set of constant (respectively monotone) functions. Set also $\mathcal{C}_\infty := \cup_{p \in \mathbb{N}^*} \mathcal{C}_p$, for which one has the following preliminary result:

Lemma 7. *One has*

$$C(\mu, \nu) = \sup_{f \in \mathcal{C}_\infty} \frac{\mathrm{Ent}(f^2, \mu)}{\nu[(f')^2]}.$$

Proof. Let $\mathcal{F}$ denote the set of all measurable functions $g : [-M, M] \to \mathbb{R}$ belonging to $\mathbb{L}^1([-M, M], \lambda)$ and for which one can find $n \in \mathbb{N}^*$ and $-M = x_0 < x_1 < \cdots < x_n = M$ such that for all $0 \leqslant i < n$, g has a constant sign on $]x_i, x_{i+1}[$ (0 is considered as having at the same time a positive and negative sign). So $\mathcal{C}_\infty$ is nothing but the set of antiderivatives of elements of $\mathcal{F}$.

It then suffices to verify that $\{g \in \mathcal{F} : \nu[g^2] \leqslant 1\}$ is dense in the $\mathbb{L}^2(\nu)$ sense in the unit ball of this space. Indeed, let $f \in \mathcal{C}$ with $\nu[(f')^2] = 1$. According to the preceding property, there exists a sequence $(g_n)_{n \in \mathbb{N}}$ of elements of $\mathcal{F}$ converging to f'. Put for all $n \in \mathbb{N}$

$$\forall\, x \in [-M, M], \qquad G_n(x) = f(-M) + \int_{-M}^{x} g_n(y)\, dy.$$

Due to the minorization $\nu \geqslant \eta$, it is clear that the G_n converge uniformly to f for large n. And since $\mathrm{osc}(f) < +\infty$, Lemma 5 applies and shows that

$$\lim_{n \to \infty} \mathrm{Ent}(G_n^2, \mu) = \mathrm{Ent}(f^2, \mu),$$

wherefrom follows the equality in the lemma.

To show the claimed density, take $g \in \mathbb{L}^2(\nu)$ with $\nu[g^2] = 1$; for $n \in \mathbb{N}$, put

$$g_n := g \mathbb{1}_{\{\nu \leqslant n, |g| \leqslant n\}}.$$

By dominated convergence, the sequence $(g_n)_{n \in \mathbb{N}}$ converges in $\mathbb{L}^2(\nu)$ to g. Now, for fixed $n \in \mathbb{N}$, the measure $(\nu \wedge n) d\lambda$ is regular (in the sense of inner and outer approximation of Borel sets; see for instance Rudin's book [16]), so one can find a sequence $(\tilde{g}_{n,m})_{m \in \mathbb{N}}$ in $\mathcal{F}$ such that

$$\lim_{m \to \infty} \int (\tilde{g}_{n,m} - g_n)^2 (\nu \wedge n)\, d\lambda = 0.$$

So, setting for all $m \in \mathbb{N}$, $\widehat{g}_{n,m} := \widetilde{g}_{n,m}\mathbb{1}_{\{\nu \leqslant n, |g| \leqslant n\}}$, which still belongs to $\mathcal{F}$, one also has

$$\lim_{m \to \infty} \int (\widehat{g}_{n,m} - g_n)^2 \, d\nu = 0$$

and the claimed density is established.   □

The lemma entails that

$$C(\mu, \nu) = \lim_{p \to \infty} \sup_{f \in \mathcal{C}_p} \frac{\mathrm{Ent}(f^2, \mu)}{\nu[(f')^2]}.$$

However, for $p \geqslant 3$ and $f \in \mathcal{C}_p \setminus \mathcal{C}_2$, the considerations from the preceding section applied to f yield $\widetilde{f} \in \mathcal{C}_{p-1}$ and $\widehat{f} \in \mathcal{C}_4$ such that

$$\nu[(f')^2] = \nu[(\widetilde{f}')^2] + \nu[(\widehat{f}')^2]$$
$$\mathrm{Ent}(f^2, \mu) = \mathrm{Ent}(\widetilde{f}^2, \mu) + \mathrm{Ent}(\widehat{f}^2, \mu).$$

Let us make this more precise. For $g \in \mathcal{C}$, a connected component of $\mathcal{M}(g)$ will be called internal if it contains neither $-M$ nor M. The union of the internal connected components of $\mathcal{M}(g)$ will be denoted by $\widetilde{\mathcal{M}}(g)$. One then introduces a set $\mathcal{C}_3 \subset \widehat{\mathcal{C}}_4 \subset \mathcal{C}_4$ by imposing that $\widehat{\mathcal{C}}_4 \cap (\mathcal{C}_4 \setminus \mathcal{C}_3)$ consists of the functions $g \in \mathcal{C}_4 \setminus \mathcal{C}_3$ such that $\min_{\widetilde{\mathcal{M}}(g)} g \leqslant g(-M), g(M) \leqslant \max_{\widetilde{\mathcal{M}}(g)} g$. The interest of this set $\widehat{\mathcal{C}}_4$ will be twofold for us: on the one hand, in the above construction, one has $\widehat{f} \in \widehat{\mathcal{C}}_4$, and on the other hand, if $g \in \widehat{\mathcal{C}}_4 \setminus \mathcal{C}_2$ then $\widetilde{g}$ obtained from the preceding procedure is monotone.

However, the sole fact that $\widehat{f} \in \mathcal{C}_4$ already showed that for $p \geqslant 5$, one has

$$\sup_{f \in \mathcal{C}_p} \frac{\mathrm{Ent}(f^2, \mu)}{\nu[(f')^2]} = \sup_{f \in \mathcal{C}_{p-1}} \frac{\mathrm{Ent}(f^2, \mu)}{\nu[(f')^2]},$$

and by induction, one ends up with the fact that this quantity is nothing but $\sup_{f \in \mathcal{C}_4} \mathrm{Ent}(f^2, \mu)/\nu[(f')^2]$. More precisely, the preceding observations even imply that

$$C(\mu, \nu) = \sup_{f \in \widehat{\mathcal{C}}_4} \frac{\mathrm{Ent}(f^2, \mu)}{\nu[(f')^2]}.$$

So let $(f_n)_{n \in \mathbb{N}}$ be a sequence of elements from $\widehat{\mathcal{C}}_4$ satisfying $\nu[(f_n')^2] = 1$ for all $n \in \mathbb{N}$ and $C(\mu, \nu) = \lim_{n \to \infty} \mathrm{Ent}(f_n^2, \mu)$. Two situations can be distinguished: either one can extract from $(f_n)_{n \in \mathbb{N}}$ a subsequence (still denoted $(f_n)_{n \in \mathbb{N}}$) such that $(f_n(0))_{n \in \mathbb{N}}$ converges in $\mathbb{R}$, or one has $\liminf_{n \to \infty} |f_n(0)| = +\infty$. The latter case corresponds to the equality $C(\mu, \nu) = A(\mu, \nu)/2$, whose treatment amounts to that of the Poincaré constant, left to the reader. Thus, from now on, we assume to be in the first situation described above. By weak compactness of the unit ball of $\mathbb{L}^2(\nu)$, one can extract a subsequence of $(f_n)_{n \in \mathbb{N}}$, such that $(f'_n)_{n\in\mathbb{N}}$ is weakly convergent in $\mathbb{L}^2(\nu)$. Together with the convergence of $(f_n(0))_{n\in\mathbb{N}}$, this weak convergence implies that the sequence $(f_n)_{n\in\mathbb{N}}$ converges pointwise on $[-M, M]$ to a function f which has a weak derivative f' satisfying $\nu[(f')^2] \leqslant 1$ (because the norm is weakly lower semi-continuous). However, the uniform continuity of the f_n for $n \in \mathbb{N}$ (due to the majorization by $\eta^{-1/2}$ of their Hölder coefficient of order $1/2$) ensures, via Ascoli's theorem, that the convergence of the f_n towards f is in fact uniform on the compact $[-M, M]$. In particular, one obtains

$$\text{Ent}(f^2, \mu) = \lim_{n\to\infty} \text{Ent}(f_n^2, \mu) = C(\mu, \nu).$$

Discarding the trivial situation that $C(\mu, \nu) = 0$ (which corresponds to the cases when μ is a Dirac mass or $\nu = +\infty$ a.s. on the convex hull of the support of μ), one then obtains

$$\frac{\text{Ent}(f^2, \mu)}{\nu[(f')^2]} \geqslant C(\mu, \nu),$$

with strict inequality if $0 \leqslant \nu[(f')^2] < 1$, wherefrom necessarily $\nu[(f')^2] = 1$. So f is a maximising function for (1), which, moreover, belongs to $\widehat{\mathcal{C}}_4$, whereof one is easily convinced: at the cost of extracting a subsequence, one can require that the number (between 0 and 2) of internal connected components is the same for each f_n and that there exists a point in each of these components which converges in $[-M, M]$ for large n, and this allows to see a posteriori that $f \in \widehat{\mathcal{C}}_4$). If f is not already monotone, the procedure of the preceding section can be applied again to construct $\widetilde{f}$ and $\widehat{f}$. As f is maximising, so must be these two functions too; now, owing to f belonging to $\widehat{\mathcal{C}}_4$, $\widetilde{f}$ is necessarily monotone. So these arguments allow to conclude that $C(\mu, \nu) = D(\mu, \nu)$.

Remark 2. The latter proof rests partially on the existence of a maximising function for (1), but, contrary to the approach by Chen and Wang [6, 8] (in the case of the Poincaré constant), we have not tried to exploit the equation it fulfills.

More generally, call $S(\mu)$ the convex hull of the support of μ and $[s_-, s_+]$ its closure in the compactified real line $\mathbb{R} \sqcup \{-\infty, +\infty\}$. Still denoting by ν the density of ν with respect to λ, assume that

$$\int_{S(\mu)} \frac{1}{\nu} \, d\lambda < +\infty.$$

One can then show that if $C(\mu, \nu) > A(\mu, \nu)/2$, a maximising function for (1) exists (but these two conditions are not sufficient as can be seen by taking for μ and ν the standard Gaussian distribution). Indeed, fix $o \in S(\mu)$ and define

$$\forall\, x \in S(\mu), \qquad F(x) := \int_o^x \frac{1}{\nu(y)} \, dy.$$

By the preceding condition, F is continuously extendable to $[s_-, s_+]$. Consider then an absolutely continuous function f whose weak derivative satisfies $\int (f')^2 \, d\nu \leqslant 1$. Applying as above a Cauchy-Schwarz, inequality, one gets that

$$\forall \, x, y \in S(\mu), \qquad |f(y) - f(x)| \leqslant \sqrt{|F(y) - F(x)|},$$

and consequently, by Cauchy's criterion, f too is continuously extendable to $[s_-, s_+]$. One can then repeat the preceding arguments on this compact (taking into account that $\nu^{-1}\mathbb{1}_I \in \mathbb{L}^2(S(\mu), \nu)$ for each segment $I \subset [s_-, s_+]$, this alowing to obtain pointwise convergence from the weak compactness of the unit ball of $\mathbb{L}^2(S(\mu), \nu)$), and see that except when $C(\mu, \nu) = A(\mu, \nu)/2$, there exists a maximising function f for (1) (and since it is known that dealing with monotone functions is sufficient, Ascoli's theorem can even be replaced with one of Dini's ones). Performing a variational calculation around this function, one realizes that it satisfies two conditions:

$$\int_{S(\mu)} f \ln\left(\frac{f^2}{\mu[f^2]}\right) d\mu = 0$$

and for a.a. $x \in S(\mu)$,

$$C(\mu, \nu)\nu(x)f'(x) = \int_{[s_-, x]} f \ln\left(\frac{f^2}{\mu[f^2]}\right) d\mu. \tag{7}$$

Obviously, if moreover the function ν is assumed to be absolutely continuous and μ absolutely continuous with respect to λ, a further differentiation yields a second-order equation (non linear in the zeroth order term) satisfied by f.

Last, if in addition $[s_-, s_+] \subset \mathbb{R}$, $\nu(s_-) > 0$ and $\nu(s_+) > 0$, equation (7) allows to recover a Neumann condition for f, namely $f'(s_-) = f'(s_+) = 0$.

5 Extensions

We present here a few generalisations of the preceding results, corresponding to modifications of the quantities featuring in (1).

5.1 Modification of the Energy in the Discrete Case

We shall show here Theorem 3, whose context is now assumed, and we put

$$E(\mu, \nu) := \sup_{f \in \mathcal{C}} \frac{\mathrm{Ent}(f^2, \mu)}{\mathcal{E}_\nu(f^2, \ln(f^2))}.$$

Considering $\mathbb{Z}$ brings no further difficulty, since, as in section 4, one can without loss consider only the finite situation where $E = \{0, ..., N\}$ with $N \in \mathbb{N}^*$, at the cost of accumulating mass on the endpoints and translating the obtained segment. However, we take this opportunity to point out the most famous infinite example where the preceding constant is finite, namely the Poisson laws on $\mathbb{N}$: fix $\alpha > 0$ and take

$$\forall\, l \in \mathbb{N}, \qquad \mu(\{l\}) := \frac{\alpha^l}{l!}\exp(-\alpha)$$

$$\nu(\{l, l+1\}) := \mu(\{l\}).$$

It is then known (see for instance section 1.6 of the book [1] by Ané, Blachère, Chafaï, Fougères, Gentil, Malrieu, Roberto and Scheffer) that $E(\mu, \nu)$ equals α.

To get convinced of Theorem 3, on has to inspect again the three-step proof in sections 2 and 3.

• As in the case of the logarithmic Sobolev inequality, one is brought back, up to a multiplicative constant, to the problem of estimating the Poincaré constant when there exists a minimising sequence $(f_n)_{n \in \mathbb{N}}$ verifying

$$\forall\, n \in \mathbb{N}, \qquad \mathcal{E}_\nu(f_n^2, \ln(f_n^2)) = 1$$

$$\lim_{n \to \infty} |f_n(0)| = +\infty.$$

Indeed, it is well known (see for instance Lemma 2.6.6 in the book by Ané and al. [1]) that

$$\forall\, f \in \mathcal{C}, \qquad \mathcal{E}_\nu(f^2, \ln(f^2)) \geqslant 4\nu[(f')^2];$$

so the first condition above ensures that the oscillations of the f_n are bounded in $n \in \mathbb{N}$ (the situation should have been beforehand reduced to the case when $\nu > 0$). This observation allows to perform finite order expansions showing the following equivalent for large n:

$$\frac{\mathrm{Ent}(f_n^2, \mu)}{\mathcal{E}_\nu\big(f_n^2, \ln(f_n^2)\big)} \sim \frac{\mathrm{Var}(f_n, \mu)}{8\nu[(f_n')^2]},$$

wherefrom one easily deduces

$$\sup_{f \in \mathcal{C}} \frac{\mathrm{Ent}(f^2, \mu)}{\mathcal{E}_\nu\big(f^2, \ln(f^2)\big)} = \frac{A(\mu, \nu)}{8} = \sup_{f \in \mathcal{D}} \frac{\mathrm{Ent}(f^2, \mu)}{\mathcal{E}_\nu\big(f^2, \ln(f^2)\big)}.$$

Thus it suffices to consider the situations where there exists a minimising sequence $(f_n)_{n \in \mathbb{N}}$ such that

$$\forall\, n \in \mathbb{N}, \qquad \mathcal{E}_\nu\big(f_n^2, \ln(f_n^2)\big) = 1$$

$$\limsup_{n \to \infty} |f_n(0)| < \infty,$$

in which cases one can extract a subsequence that converges toward a maximiser for the supremum we are interested in.

• Calling f this maximiser, one is easily convinced that it cannot vanish, at least in the relevant situations where $E(\mu, \nu) > 0$. Performing then a variational computation around f shows it to verify for each $i \in E$ the following equation:

$$\mu(i)\, f(i)\, \ln\!\left(\frac{f^2(i)}{\mu[f^2]}\right)$$
$$= E(\mu, \nu)\Big[f(i) \big[\nu(\{i, i+1\}) \big(\ln(f^2(i)) - \ln(f^2(i+1))\big) \big.$$
$$\big. + \nu(\{i-1, i\}) \big(\ln(f^2(i)) - \ln(f^2(i-1))\big) \big]$$
$$+ \frac{\nu(\{i, i+1\})\big(f^2(i) - f^2(i+1)\big) + \nu(\{i-1, i\})\big(f^2(i) - f^2(i-1)\big)}{f(i)} \Big]$$

(as usual, $\nu(\{-1, 0\}) = 0 = \nu(\{N, N+1\})$, hence the terms $f(-1)$ and $f(N+1)$ never show up). If μ does not vanish, the form of this equation enables to apply the arguments of the end of section 3, taking advantage of the fact that a maximising function for $E(\mu, \nu)$ cannot take the same value at three consecutive points, unless it is constant (which won't do either). Remark also that contrary to sections 2 and 3, this equation does not allow to recursively compute f from the values of $f(0)$ and $\mu[f^2]$, for the right-hand side is not injective as a function of $f(i+1)$ (for $0 \leqslant i < N$), but only as a function of $f^2(i+1)$. But this could be forseeen, since the signs of the functions really play no role in the quantities considered here. There remain the cases when μ vanishes at some (interior) points; they cannot be discarded as before Lemma 1. The simplest is to bypass the argument of the consecutive three points with same value, by adapting the second proof of the preceding section (by classifying the functions according to the maximal number of segments included in their set of local extrema); this is immediate enough.

• The last point to be verified, which is also the most important, is the possibility of modifying Lemma 2; namely, with the notations therein, is it true that for all $s \in\,]f(i_1), f(i_0)[$,

$$\mathcal{E}_\nu\big(f^2, \ln(f^2)\big) \geqslant \mathcal{E}_\nu\big((\widetilde{f}_s')^2, \ln((\widetilde{f}_s')^2)\big) + \mathcal{E}_\nu\big((\widehat{f}_s')^2, \ln((\widehat{f}_s')^2)\big) \qquad (8)$$

for any function f with a constant sign (the situation should have been reduced to that case). This question amounts to asking if for all $0 \leqslant x \leqslant y \leqslant z$, one has

$$\varphi_{x,z}(y) \leqslant (z - x)\big(\ln(z) - \ln(x)\big), \qquad (9)$$

where $\varphi_{x,z}$ is the function defined by

$$\forall\, y \in [x, z], \qquad \varphi_{x,z}(y) := (y - x)\big(\ln(y) - \ln(x)\big) + (z - y)\big(\ln(z) - \ln(y)\big).$$

Now, differentiating this function twice shows it to be strictly convex, and (9) then derives from the fact that $\varphi_{x,z}(x) = \varphi_{x,z}(z) = (z - x)\big(\ln(z) - \ln(x)\big)$. One also derives therefrom that equality in (8) can hold only if $\widetilde{f}_s'(a)\widehat{f}_s'(a) = 0$ for every edge $a \in A$.

The other arguments of section 3 are valid without modification, since they only involve entropy. Theorem 3 follows.

5.2 Modification of the Energy in the Continuous Case

Our aim here is to prove Theorem 4. Recall that $H : \mathbb{R}_+ \to \mathbb{R}_+$ is a convex function such that $H(0) = 0$ and $H'(0) = 1$ (besides these two equalities, we shall only use the bound $x \leqslant H(x)$, valid for all $x \geqslant 0$). In particular, it appears that

$$\forall f \in \mathcal{C}, \qquad \mathcal{E}_{H,\nu}(f) \geqslant \nu\big[(f')^2\big]. \tag{10}$$

For μ a probability and ν a measure on $\mathbb{R}$, put

$$F(\mu,\nu) := \sup_{f \in \mathcal{C}} \frac{\mathrm{Ent}(f^2,\mu)}{\mathcal{E}_{H,\nu}(f)} \in \bar{\mathbb{R}}_+.$$

In view of the second proof in the preceding section, the only non immediate point in the proof of Theorem 4 concerns the cases that can be reduced to that of the Poincaré constant. Indeed, after having supposed without loss that μ is supported in $[-M, M]$ and that $\nu \geqslant \eta$, with $M, \eta > 0$, we have to see that if $(f_n)_{n \in \mathbb{N}}$ is a maximising sequence for $F(\mu, \nu)$ such that

$$\forall\, n \in \mathbb{N}, \qquad \mathcal{E}_{H,\nu}(f) = 1$$
$$\lim_{n \to \infty} |f_n(0)| = +\infty,$$

then $F(\mu, \nu) = A(\mu, \nu)/2$. But, again, such a sequence will satisfy $\nu[(f')^2] \leqslant 1$ for all $n \in \mathbb{N}$, and the oscillations of the f_n will be bounded, allowing to obtain for large n the equivalent

$$\mathrm{Ent}(f_n^2, \mu) \sim \frac{\mathrm{Var}(f_n, \mu)}{2}.$$

By extracting a subsequence (first, by relative compactness of the f_n, then, by Ascoli's theorem), one may suppose that the f_n converge uniformly to $f \in \mathcal{C}$, with $\nu[(f')^2] \leqslant 1$, wherefrom

$$F(\mu,\nu) = \lim_{n \to \infty} \mathrm{Ent}(f_n^2, \mu) = \lim_{n \to \infty} \frac{\mathrm{Var}(f_n, \mu)}{2}$$
$$= \frac{\mathrm{Var}(f, \mu)}{2} \leqslant \frac{\mathrm{Var}(f, \mu)}{2\nu[(f')^2]} \leqslant \frac{A(\mu, \nu)}{2}.$$

However, the reverse inequality always holds. Indeed, note first that one may content oneself in only dealing, for the supremum defining $A(\mu, \nu)$, with functions having a weak derivative essentially bounded in the sense of the Lebesgue measure on $[-M, M]$. This is because only functions such that $\nu[(f')^2] < +\infty$ need to be considered, and such functions can be approximated in the traditional way. Let $f \in \mathcal{C}$ with $f \geqslant 0$ and f' bounded. For $n \in \mathbb{N}$, consider $f_n := n + f$. The oscillation of f being finite, for large n one has $\mathrm{Ent}(f_n^2, \mu) \sim \mathrm{Var}(f_n, \mu)/2 = \mathrm{Var}(f, \mu)/2$. On the other hand, since $H'(0) = 1$, one has by dominated convergence

$$\lim_{n \to \infty} \mathcal{E}_{H,\nu}(f_n) = \lim_{n \to \infty} \int H\left(\frac{(f')^2}{(n+f)^2}\right)(n+f)^2 \, d\nu = \int (f')^2 \, d\nu.$$

It ensues therefrom that

$$\frac{\mathrm{Var}(f, \mu)}{2\nu[(f')^2]} \leqslant F(\mu, \nu),$$

then the claimed inequality, by taking the supremum over such functions f.

Similar results hold when $\mathcal{C}$ is replaced with $\mathcal{D}$. It therefore suffices to deal with sequences $(f_n)_{n \in \mathbb{N}}$ maximising for $F(\mu, \nu)$, satisfying $\mathcal{E}_{H,\nu}(f_n) = 1$ for all $n \in \mathbb{N}$, and such that $\lim_{n \to \infty} f_n(0)$ exists in $\mathbb{R}$. But in this situation, the arguments in the second proof in section 4 easily adapt (after one has noted that for each function $f \in \mathcal{C}$ which splits as $\widetilde{f} + \widehat{f}$, with $\widetilde{f}, \widehat{f} \in \mathcal{C}$ and $\widetilde{f}'\widehat{f}' = 0$ a.s., one trivially has $\mathcal{E}_{H,\nu}(f) = \mathcal{E}_{H,\nu}(\widetilde{f}) + \mathcal{E}_{H,\nu}(\widehat{f})$).

Remark 3. One may wonder if there is a link between the discrete modified logarithmic Sobolev inequalities, and the continuous ones as above. As an attempt to shed light on such a link, consider again the approximation procedure used in the first proof of section 4. Thus we work with a probability μ of the form $\sum_{0 \leqslant n \leqslant N} \mu(n)\delta_n$. The constant $F(\mu, \nu)$ can then be rewritten

$$\sup_{f \in \mathcal{C}} \frac{\mathrm{Ent}(f^2, \mu)}{\mathcal{E}_J(f)} \tag{11}$$

with for each $f \in \mathcal{C}$ in the discrete context

$$\mathcal{E}_J(f) := \sum_{0 \leqslant n < N} J_{n,n+1}\big(f(n), f(n+1)\big)$$

et where the maps $(J_{n,n+1})_{0 \leqslant n < N}$ are defined on $\mathbb{R}^2$ by

$$\forall \, x, y \in \mathbb{R}, \quad J_{n,n+1}(x, y) := \inf_{\substack{g \in \mathcal{C}([n,n+1]) : \\ g(n)=x,\, g(n+1)=y}} \int_n^{n+1} H\left(\left(\frac{g'}{g}\right)^2\right) g^2 \, \nu \, d\lambda.$$

Obviously, the supremum (11) is not changed by restricting it to monotone functions, since this "discrete" problem can be interpreted in the continuous context where this property has just been verified. But one could certainly also show it directly; note in particular that for any $0 \leqslant n < N$ and all real numbers réels $x \leqslant y \leqslant z$, one has indeed $J_{n,n+1}(x,z) \geqslant J_{n,n+1}(x,y) + J_{n,n+1}(y,z)$ (it suffices to split any function going from x to z as the sum of two functions, the first one being its restriction going from x to y and remaining there).

This leads to ponder on the possibility of rewriting $\mathcal{E}_\nu$ as an $\mathcal{E}_J$, for a suitable choice of the continuous measure μ (the discrete one being given), and of the function H.

5.3 Modification of the Entropy

We now aim to change the entropy term in (1); this leads to logarithmic Sobolev inequalities modified in another sense (see for instance Chafaï [5]). This will give the opportunity to test the limits of the arguments in section 3. We shall content ourselves by treating the discrete case with the usual energy given by the quadratic form $\mathcal{C} \ni f \mapsto \nu[(f')^2]$, although one may think that similar considerations should allow to extend the following to the continuous situation or to energies modified as above. Let $\varphi : \mathbb{R}_+ \to \mathbb{R}$ be a convex function, of class $\mathcal{C}^3$ on $]0, +\infty[$. The corresponding modified entropy is the functional which to any map $f \in \mathcal{C}$, $f \geqslant 0$ associates the quantity (positive by Jensen's inequality)

$$E_\varphi[f] = \mu[\varphi(f)] - \varphi(\mu[f]).$$

Unfortunately the expression $E_\varphi(f^2)$ is no longer quadratically homogeneous in f (unless it is proportional to the usual entropy in f^2). To remedy this flaw, we shall need two additional hypotheses. Call ψ the map defined by

$$\forall\, x > 0, \qquad \psi(x) := x\varphi'(x) - \varphi(x).$$

One says that ψ is asymptotically concave if for some $R > 0$ the function ψ remains below its tangents at points larger than R:

$$\forall\, y \geqslant R, \forall\, x > 0, \qquad \psi(x) \leqslant \psi(y) + \psi'(y)(y - x).$$

This notably implies that ψ is concave on $[R, +\infty[$ (which is not sufficient, but becomes sufficient if moreover $\lim_{x\to+\infty} \psi(x) - x\psi'(x) = +\infty$). We shall first suppose ψ to be asymptotically concave. The second additional hypothesis states the existence of a constant $\eta > 0$ such that for any $0 < x < \eta$, one has $\varphi''(x) + x\varphi'''(x) \geqslant 0$ (if φ is $\mathcal{C}^3$ on $\mathbb{R}_+$, this is ensured by $\varphi''(0) > 0$; more generally, if one does not even want to suppose φ to be of class $\mathcal{C}^3$ on $\mathbb{R}_+^*$, it can be seen that it suffices to suppose that the map $x \mapsto x\varphi''(x)$ is increasing on some interval $]0, \eta[$). An example of a function φ satisfying these conditions is $\mathbb{R}_+ \ni x \mapsto x\ln(\ln(e + x))$.

Remark that

$$\forall\, x > 0, \qquad \psi'(x) = x\varphi''(x) \geqslant 0$$

and that this quantity decreases for $x \geqslant R$; hence it admits a limit $L \geqslant 0$ at $+\infty$. So $\varphi''(x) \leqslant (1 + L)/x$ for x large, which shows that up to a constant factor, $\varphi(x)$ is dominated by $x \ln(x)$. Somehow, the usual entropy is an upper bound for the modified entropies to be considered here.

For μ a probability on $E = \{0, ..., N\}$ and ν a measure on the corresponding set A of edges, we are interested in the quantity

$$G(\mu, \nu) := \sup_{f \in \mathcal{C}} \frac{\mathrm{E}_\varphi(f^2)}{\nu[(f')^2]}$$

and our aim here is to prove

Proposition 2. *One has as usual*

$$G(\mu, \nu) = \sup_{f \in \mathcal{D}} \frac{\mathrm{E}_\varphi(f^2)}{\nu[(f')^2]}.$$

The main annoyance comes from the inhomogeneity of E_φ, which a priori forbids to only consider maximising sequences for $G(\mu, \nu)$ with energy bounded above and below by a strictly positive constant. To remedy to that, observe that nothing here hinders us from supposing μ and ν to be strictly positive on E. This property ensures the existence of a constant $b_1 > 0$ such that

$$\forall\, g \in \mathcal{C}, \qquad \nu[(g')^2] = 1 \quad \Rightarrow \quad \mu[g^2] \geqslant b_1.$$

Fix a function g satisfying $\nu[(g')^2] = 1$ and consider the function

$$F : \mathbb{R}_+^* \ni t \mapsto \mathrm{E}_\varphi[tg^2]/t. \tag{12}$$

A computation gives its derivative as

$$\forall\, t > 0, \qquad F'(t) = t^{-2} \left(\mu[\psi(tg^2)] - \psi(t\mu[g^2]) \right).$$

So by our hypothesis that ψ is asymptotically concave, F is decreasing on $[R/b_1, +\infty[$. This shows that

$$G(\mu, \nu) = \sup_{f \in \mathcal{C}\, :\, \nu[(f')^2] \leqslant R/b_1} \frac{\mathrm{E}_\varphi(f^2)}{\nu[(f')^2]},$$

which enables us to only consider maximising sequences $(f_n)_{n \in \mathbb{N}}$ satisfying $\nu[(f_n')^2] \leqslant R/b_1$ for all $n \in \mathbb{N}$. One can also suppose that these functions f_n are positive. Write $f_n = \sqrt{t_n} g_n$, with $t_n > 0$ (discarding the trivial cases that $t_n = 0$) and $g_n \in \mathcal{C}$ satisfying $\nu[(g_n')^2] = 1$. Extracting a sub-sequence reduces to the situation when the sequences $(t_n)_{n \in \mathbb{N}}$ and $(f_n(0))_{n \in \mathbb{N}}$ are respectively convergent in $[0, R/b_1]$ and $\bar{\mathbb{R}}_+$. Several cases will be distinguished:

• If $\lim_{n \to \infty} t_n = 0$, we shall verify that we may without loss suppose that $\lim_{n \to \infty} f_n(0) > 0$. Indeed, our second hypothesis on ψ ensures that for $g \in \mathcal{C}$, $g \geqslant 0$, the function F defined in (12) is increasing on $]0, \eta/\max g^2]$. This is obtained via a second-order expansion with remainder: for fixed $t > 0$, there exists a function $\theta_t : E \to \,]0, t \max g^2[$ such that

$$\psi(tg^2) = \psi(t\mu[g^2]) + \psi'(t\mu[g^2])t(g^2 - \mu[g^2]) + \frac{\psi''(\theta_t)}{2}t^2(g^2 - \mu[g^2])^2.$$

When this inequality is integrated with respect to μ, it appears that $F'(t)$ is positive as soon as $t \max g^2 \leqslant \eta$. On the other hand, there exists a constant $b_2 > 0$ such that if g satisfies $\nu[(g')^2] = 1$, then $\mathrm{osc}(g) \leqslant b_2$ and hence, if moreover g is positive, $\max g^2 \leqslant (g(0) + b_2)^2$. Consequently, if one constructs a new sequence $(\widetilde{t}_n)_{n\in\mathbb{N}}$ by setting

$$\forall\, n \in \mathbb{N}, \qquad \widetilde{t}_n := \begin{cases} t_n & \text{si } t_n(g_n(0) + b_2)^2 > \eta \\ \eta/(g_n(0) + b_2)^2 & \text{else,} \end{cases}$$

the sequence $(\widetilde{f}_n)_{n\in\mathbb{N}}$ defined by $\widetilde{f}_n := \widetilde{t}_n g_n$ for $n \in \mathbb{N}$ remains maximising for $G(\mu, \nu)$. We consider from now on this sequence, still called $(f_n)_{n\in\mathbb{N}}$. Then one has

$$\forall\, n \in \mathbb{N}, \qquad t_n(g_n(0) + b_2)^2 \geqslant \eta,$$

that is to say $f_n^2(0) + 2b_2\sqrt{t_n}f_n(0) + b_2^2 t_n \geqslant \eta$, which prevents the convergence $\lim_{n\to\infty} f_n(0) = 0$.

One can now perform a second-order expansion with remainder for $\mathrm{E}_\varphi(f_n^2)$; there exists a new function θ_n valued in $[f_n(0) - \sqrt{t_n}b_2, f_n(0) + \sqrt{t_n}b_2]$ and such that

$$\mathrm{E}_\varphi(f_n^2) = \mu\left[\varphi''(\theta_n)(f_n^2 - \mu[f_n^2])^2\right]/2.$$

First consider the case that $l := \lim_{n\to\infty} f_n(0)$ is finite. Since $l > 0$, one has uniformly on E

$$\lim_{n\to\infty} \varphi''(\theta_n)(f_n + \sqrt{\mu[f_n^2]})^2/2 = 2l^2\varphi''(l^2).$$

If $l^2\varphi''(l^2) > 0$, one draws therefrom the equivalent for large n

$$\mathrm{E}_\varphi(f_n^2) \sim 2l^2\varphi''(l^2)\mu[(f_n - \sqrt{\mu[f_n^2]})^2] \leqslant 2l^2\varphi''(l^2)\,\mathrm{Var}(f_n, \mu),$$

wherefrom

$$\lim_{n\to\infty} \frac{\mathrm{E}_\varphi(f_n^2)}{\nu[(f_n')^2]} \leqslant 2l^2\varphi''(l^2)\limsup_{n\to\infty} \frac{\mathrm{Var}(f_n, \mu)}{\nu[(f_n')^2]} \leqslant 2l^2\varphi''(l^2)A(\mu, \nu).$$

Similarly, one gets

$$\lim_{n\to\infty} \frac{\mathrm{E}_\varphi(f_n^2)}{\nu[(f_n')^2]} = 0$$

when $l^2\varphi''(l^2) = 0$. So it appears that one always has

$$G(\mu,\nu) \leqslant \sup_{l>0} 2l^2\varphi''(l^2)A(\mu,\nu) = \sup_{l>\eta} 2l^2\varphi''(l^2)A(\mu,\nu),$$

where the latter equality comes from the map $x \mapsto x\varphi''(x)$ being increasing on $]0,\eta]$. Conversely the inequality $G(\mu,\nu) \geqslant \sup_{l>\eta} 2l^2\varphi''(l^2)A(\mu,\nu)$ is satisfied under all circumstances: for all l larger than some given η, in the supremum defining $G(\mu,\nu)$, it suffices to consider functions of the form $l + \epsilon f$, with $f \in \mathcal{C}$ and $\epsilon > 0$ which is made to tend to 0. The above argument also holds if $\lim_{n\to\infty} f_n(0) = +\infty$, by existence and finiteness of $L = \lim_{x\to+\infty} x\varphi''(x)$. Thus, in all cases, the convergence entails the equality $G(\mu,\nu) = \sup_{l>\eta^2} 2l\varphi''(l)A(\mu,\nu)$. Then, one also has

$$\sup_{f\in\mathcal{D}} \frac{E_\varphi(f^2)}{\nu[(f')^2]} = \left(\sup_{l>\eta^2} 2l\varphi''(l)\right)\sup_{f\in\mathcal{D}} \frac{\mathrm{Var}(f,\mu)}{\nu[(f')^2]} = \left(\sup_{l>\eta^2} 2l\varphi''(l)\right)A(\mu,\nu),$$

the claimed identity (2) follows.

• If $\lim_{n\to\infty} t_n \in]0, R/b_1]$, one is back in a more classical framework, and, as already happened several times, two sub-cases will be considered.

- If $\lim_{n\to\infty} f_n(0) = +\infty$, the boundedness in $n \in \mathbb{N}$ of the oscillations of the f_n and the convergence $\lim_{t\to+\infty} x\varphi''(x) = L$ allow again to perform a second-order expansion with remainder, yielding for large n the equivalent

$$E_\varphi(f_n^2) \sim \frac{L}{2}\mathrm{Var}(f_n,\mu)$$

if $L > 0$. On the other hand, if $L = 0$, it appears that

$$E_\varphi(f_n^2) \ll \mathrm{Var}(f_n,\mu).$$

Since $A(\mu,\nu) < +\infty$, the latter possibility implies that one is in the trivial situation that $G(\mu,\nu) = 0$. If $L > 0$, one also obtains $G(\mu,\nu) = LA(\mu,\nu)/2$. So one is reduced to the case of the Poincaré inequality.

- If $\lim_{n\to\infty} f_n(0)$ exists in $\mathbb{R}$, one easily shows existence of some minimising function. But the proof of Lemma 4 immediately adapts to this situation, in view of the form of the modified entropy E_φ. Then the quickest way to conclude that (2) holds is to adapt the second proof of section 4.

References

1. Cécile Ané, Sébastien Blachère, Djalil Chafaï, Pierre Fougères, Ivan Gentil, Florent Malrieu, Cyril Roberto and Grégory Scheffer. *Sur les inégalités de Sobolev logarithmiques*, volume 10 of *Panoramas et Synthèses*. Société Mathématique de France, Paris, 2000. With a preface by Dominique Bakry and Michel Ledoux.

2. F. Barthe and C. Roberto. Sobolev inequalities for probability measures on the real line. *Studia Math.*, 159(3):481–497, 2003. Dedicated to Professor Aleksander Pełczyński on occasion of his 70th birthday (Polish).

3. S. G. Bobkov and F. Götze. Exponential integrability and transportation cost related to logarithmic Sobolev inequalities. *J. Funct. Anal.*, 163(1):1–28, 1999.

4. Eric A. Carlen. Superadditivity of Fisher's information and logarithmic Sobolev inequalities. *J. Funct. Anal.*, 101(1):194–211, 1991.

5. Djalil Chafaï. Entropies, convexity, and functional inequalities. *J. Math. Kyoto Univ.*, 44(2):325–363, 2004.

6. Mu-Fa Chen and Feng-Yu Wang. Estimation of spectral gap for elliptic operators. *Trans. Amer. Math. Soc.*, 349(3):1239–1267, 1997.

7. Mufa Chen. Analytic proof of dual variational formula for the first eigenvalue in dimension one. *Sci. China Ser. A*, 42(8):805–815, 1999.

8. Mufa Chen. Variational formulas and approximation theorems for the first eigenvalue in dimension one. *Sci. China Ser. A*, 44(4):409–418, 2001.

9. Ivan Gentil, Arnaud Guillin and Laurent Miclo. Modified logarithmic Sobolev inequalities and transportation inequalities. *Probab. Theory Related Fields* 133 (2005), no. 3, 409–436.

10. Leonard Gross. Logarithmic Sobolev inequalities. *Amer. J. Math.*, 97(4):1061–1083, 1975.

11. Laurent Miclo. Quand est-ce que des bornes de Hardy permettent de calculer une constante de Poincaré exacte sur la droite? *Annales de la Faculté des Sciences de Toulouse*, Sér. 6, no. 17(1):121–192, 2008.

12. Laurent Miclo. On eigenfunctions of Markov processes on trees. *Probab. Theory Related Fields*, 142(3-4):561–594, 2008.

13. O. S. Rothaus. Logarithmic Sobolev inequalities and the spectrum of Sturm-Liouville operators. *J. Funct. Anal.*, 39(1):42–56, 1980.

14. O. S. Rothaus. Diffusion on compact Riemannian manifolds and logarithmic Sobolev inequalities. *J. Funct. Anal.*, 42(1):102–109, 1981.

15. O. S. Rothaus. Logarithmic Sobolev inequalities and the spectrum of Schrödinger operators. *J. Funct. Anal.*, 42(1):110–120, 1981.

16. Walter Rudin. *Real and complex analysis*. McGraw-Hill Book Co., New York, third edition, 1987.

17. Laurent Saloff-Coste. Lectures on finite Markov chains. In *Lectures on probability theory and statistics (Saint-Flour, 1996)*, volume 1665 of *Lecture Notes in Math.*, pages 301–413. Springer, Berlin, 1997.

18. Liming Wu. A new modified logarithmic Sobolev inequality for Poisson point processes and several applications. *Probab. Theory Related Fields*, 118(3):427–438, 2000.

Non-monotone Convergence in the Quadratic Wasserstein Distance

Walter Schachermayer[1], Uwe Schmock[2], and Josef Teichmann[3]

[1] Vienna University of Technology
   Wiedner Hauptstrasse 8–10, 1040 Vienna, Austria
   email: wschach@fam.tuwien.ac.at
[2] Vienna University of Technology
   Wiedner Hauptstrasse 8–10, 1040 Vienna, Austria
   email: schmock@fam.tuwien.ac.at
[3] Vienna University of Technology
   Wiedner Hauptstrasse 8–10, 1040 Vienna, Austria
   email: jteichma@fam.tuwien.ac.at

Summary. We give an easy counterexample to Problem 7.20 from C. Villani's book on mass transport: in general, the quadratic Wasserstein distance between n-fold normalized convolutions of two given measures fails to decrease monotonically.

We use the terminology and notation from [5]. For Borel measures μ, ν on $\mathbb{R}^d$ we define the quadratic Wasserstein distance

$$\mathcal{T}(\mu, \nu) := \inf_{(X,Y)} \mathbb{E}\big[\|X - Y\|^2\big]$$

where $\|\cdot\|$ is the Euclidean distance on $\mathbb{R}^d$ and the pairs (X, Y) run through all random vectors defined on some common probability space $(\Omega, \mathcal{F}, \mathbb{P})$, such that X has distribution μ and Y has distribution ν. By a slight abuse of notation, we define $\mathcal{T}(U, V) := \mathcal{T}(\mu, \nu)$ for two random vectors U, V, such that U has distribution μ and V has distribution ν. The following theorem (see [5, Proposition 7.17]) is due to Tanaka [4].

Theorem 1. *For $a, b \in \mathbb{R}$ and square integrable random vectors X, Y, X', Y' such that X is independent of Y, and X' is independent of Y', and $\mathbb{E}[X] = \mathbb{E}[X']$ or $\mathbb{E}[Y] = \mathbb{E}[Y']$, we have*

$$\mathcal{T}(aX + bY, aX' + bY') \leq a^2 \mathcal{T}(X, X') + b^2 \mathcal{T}(Y, Y').$$

For a sequence of i.i.d. random vectors $(X_i)_{i \in \mathbb{N}}$ we define the normalized partial sums

$$S_m := \frac{1}{\sqrt{m}} \sum_{i=1}^{m} X_i, \qquad m \in \mathbb{N}.$$

C. Donati-Martin et al. (eds.), *Séminaire de Probabilités XLII*,
Lecture Notes in Mathematics 1979, DOI 10.1007/978-3-642-01763-6_3,
© Springer-Verlag Berlin Heidelberg 2009

If μ denotes the law of X_1, we write $\mu^{(m)}$ for the law of S_m. Clearly $\mu^{(m)}$ equals, up to the scaling factor $\sqrt{m}$, the m-fold convolution $\mu * \mu * \cdots * \mu$ of μ.

We shall always deal with measures μ, ν with vanishing barycenter. Given two measures μ and ν on $\mathbb{R}^d$ with finite second moments, we let $(X_i)_{i \in \mathbb{N}}$ and $(X'_i)_{i \in \mathbb{N}}$ be i.i.d. sequences with law μ and ν, respectively, and denote by S_m and S'_m the corresponding normalized partial sums. From Theorem 1 we obtain

$$T\big(\mu^{(2m)}, \nu^{(2m)}\big) \leq T\big(\mu^{(m)}, \nu^{(m)}\big), \qquad m \in \mathbb{N},$$

from which one may quickly deduce a proof of the central limit theorem (compare [5, Ch. 7.4] and the references given there).

However, we can *not* deduce from Theorem 1 that the inequality

$$T\big(\mu^{(m+1)}, \nu^{(m+1)}\big) \leq T\big(\mu^{(m)}, \nu^{(m)}\big) \tag{1}$$

holds true for all $m \in \mathbb{N}$. Specializing to the case $m = 2$, an estimate, which we can obtain from Tanaka's theorem, is

$$T\big(\mu^{(3)}, \nu^{(3)}\big) \leq \frac{1}{3}\big[2T\big(\mu^{(2)}, \nu^{(2)}\big) + T(\mu, \nu)\big] \leq T(\mu, \nu).$$

This contains some valid information, but does not imply (1). It was posed as Problem 7.20 of [5], whether inequality (1) holds true for all probability measures μ, ν on $\mathbb{R}^d$ and all $m \in \mathbb{N}$.

The subsequent easy example shows that the answer is no, even for $d = 1$ and symmetric measures. We can choose $\mu = \mu_n$ and $\nu = \nu_n$ for sufficiently large $n \geq 2$, as the following proposition (see also Remark 1) shows.

Proposition 1. *Denote by μ_n the distribution of $\sum_{i=1}^{2n-1} Z_i$, and by ν_n the distribution of $\sum_{i=1}^{2n} Z_i$ with $(Z_i)_{i \in \mathbb{N}}$ i.i.d. and $\mathbb{P}(Z_1 = 1) = \mathbb{P}(Z_1 = -1) = \frac{1}{2}$. Then*

$$\lim_{n \to \infty} \sqrt{n}\, T(\mu_n * \mu_n, \nu_n * \nu_n) = \frac{2}{\sqrt{2\pi}}, \tag{2}$$

while $T(\mu_n * \mu_n * \mu_n, \nu_n * \nu_n * \nu_n) \geq 1$ for all $n \in \mathbb{N}$.

Remark 1. If one only wants to find a counterexample to Problem 7.20 of [5], one does not really need the full strength of Proposition 1, i.e., the estimate that $T(\mu_n * \mu_n, \nu_n * \nu_n) = \mathcal{O}(1/\sqrt{n})$. In fact, it is sufficient to consider the case $n = 2$ in order to contradict the monotonicity of inequality (1). Indeed, a direct calculation reveals that

$$T(\mu_2 * \mu_2, \nu_2 * \nu_2) = 0.625 < \frac{2}{3} \leq \left(\frac{\sqrt{2}}{\sqrt{3}}\right)^2 T(\mu_2 * \mu_2 * \mu_2, \nu_2 * \nu_2 * \nu_2).$$

Proof (of Proposition 1). We start with the final assertion, which is easy to show. The 3-fold convolutions of the measures μ_n and ν_n, respectively, are supported on odd and even numbers, respectively. Hence they have disjoint supports with distance 1 and so the quadratic transportation costs are bounded from below by 1.

For the proof of (2), fix $n \in \mathbb{N}$, define $\sigma_n = \mu_n * \mu_n$ and $\tau_n = \nu_n * \nu_n$, and note that σ_n and τ_n are supported by the even numbers. For $k = -(2n - 1)$, $\ldots, (2n - 1)$ we denote by $p_{n,k}$ the probability of the point $2k$ under σ_n, i.e.

$$p_{n,k} = \binom{4n - 2}{k + 2n - 1} \frac{1}{2^{4n-2}}.$$

We define $p_{n,k} = 0$ for $|k| \geq 2n$. We have $\tau_n = \sigma_n * \rho$, where ρ is the distribution giving probability $\frac{1}{4}$, $\frac{1}{2}$, $\frac{1}{4}$ to -2, 0, 2, respectively. We deduce that for $0 \leq k \leq 2n - 2$,

$$
\begin{aligned}
\tau_n(2k + 2) &= \frac{1}{4}p_{n,k} + \frac{1}{4}p_{n,k+2} + \frac{1}{2}p_{n,k+1} \\
&= \frac{1}{4}(p_{n,k} - p_{n,k+1}) + \frac{1}{4}(p_{n,k+2} - p_{n,k+1}) + \sigma_n(2k + 2) \qquad (3) \\
&= \frac{1}{4}p_{n,k}\left(1 - \frac{p_{n,k+1}}{p_{n,k}}\right) + \frac{1}{4}p_{n,k+1}\left(\frac{p_{n,k+2}}{p_{n,k+1}} - 1\right) + \sigma_n(2k + 2).
\end{aligned}
$$

Notice that $p_{n,k} \geq p_{n,k+1}$ for $0 \leq k \leq 2n - 1$. The term in the first parentheses is therefore non-negative. It can easily be calculated and estimated via

$$0 \leq 1 - \frac{p_{n,k+1}}{p_{n,k}} = 1 - \frac{\binom{4n-2}{k+2n}}{\binom{4n-2}{k+2n-1}} = 1 - \frac{2n - k - 1}{k + 2n} = \frac{2k + 1}{2n + k} \leq \frac{2k + 1}{2n},$$

for $0 \leq k \leq 2n - 1$.

Following [5] we know that the quadratic Wasserstein distance $\mathcal{T}$ can be given by a cyclically monotone transport plan $\pi = \pi_n$. We define the transport plan π via an intuitive transport map T. It is sufficient to define T for $0 \leq k \leq 2n - 1$, since it acts symmetrically on the negative side. T moves mass $\frac{1}{4}p_{n,k}\frac{2k+1}{2n+k}$ from the point $2k$ to $2k + 2$ for $k \geq 1$. At $k = 0$ the transport T moves $\frac{1}{8n}p_{n,0}$ to every side, which is possible, since there is enough mass concentrated at 0.

By equation (3) we see that the transport T moves σ_n to τ_n, since, for $1 \leq k \leq 2n - 2$, the first terms corresponds to the mass, which arrives from the left and is added to σ_n, and the second term to the mass, which is transported away: summing up one obtains τ_n. For $k = 2n - 1$, mass only arrives from the left. At $k = 0$ mass is only transported away. By the symmetry of the problem around 0 and by the quadratic nature of the cost function (the distance of the transport is 2, hence cost 2^2), we finally have

$$\mathcal{T}(\sigma_n, \tau_n) \leq 2 \sum_{k=0}^{2n-1} \frac{2^2}{4}p_{n,k}\frac{2k + 1}{2n + k} \leq \sum_{k=0}^{2n-1} p_{n,k}\frac{2k + 1}{n}.$$

By the central limit theorem and uniform integrability of the function $x \mapsto x_+ := \max(0, x)$ with respect to the binomial approximations, we obtain

$$\lim_{n \to \infty} \frac{1}{2\sqrt{n}} \sum_{k=0}^{2n-1} (2k) p_{n,k} = \int_0^\infty \frac{x}{\sqrt{2\pi}} e^{-x^2/2} \, dx.$$

Hence

$$\limsup_{n \to \infty} \sqrt{n} \, T(\sigma_n, \tau_n) \leq \frac{2}{\sqrt{2\pi}} \approx 0.79788 \,.$$

In order to obtain equality we start from the local monotonicity of the respective transport maps on non-positive and non-negative numbers. It easily follows that the given transport plan is cyclically monotone and hence optimal (see [5, Ch. 2]). The subsequent equality allows also to consider estimates from below. Rewriting (3) yields

$$\tau_n(2k+2) = \frac{1}{4} p_{n,k+1} \left(\frac{p_{n,k}}{p_{n,k+1}} - 1 \right) + \frac{1}{4} p_{n,k+2} \left(1 - \frac{p_{n,k+1}}{p_{n,k+2}} \right) + \sigma_n(2k+2)$$

for $0 \leq k \leq 2n - 3$, and

$$\tau_n(2k+2) = \frac{1}{4} p_{n,k+1} \left(\frac{p_{n,k}}{p_{n,k+1}} - 1 \right) + \sigma_n(2k+2)$$

for $k = 2n - 2$. Furthermore,

$$\frac{p_{n,k}}{p_{n,k+1}} - 1 = \frac{\binom{4n-2}{k+2n-1}}{\binom{4n-2}{k+2n}} - 1 = \frac{k+2n}{2n-k-1} - 1 = \frac{2k+1}{2n-k-1} \geq \frac{2k+1}{2n}$$

for $0 \leq k \leq 2n - 2$. This yields by a reasoning similar to the above that

$$T(\sigma_n, \tau_n) \geq \sum_{k=0}^{2n-2} p_{n,k+1} \frac{2k+1}{n},$$

hence

$$\liminf_{n \to \infty} \sqrt{n} \, T(\sigma_n, \tau_n) \geq \frac{2}{\sqrt{2\pi}}.$$

Remark 2. Let $p \geq 2$ be an integer. By slight modifications of the proof of Proposition 1 we can construct sequences of measures $(\mu_n)_{n \in \mathbb{N}}$ and $(\nu_n)_{n \in \mathbb{N}}$, such that the quadratic Wasserstein distances of k-fold convolutions are bounded from below by 1 for all k which are not multiples of p, while

$$\lim_{n \to \infty} T(\mu_n^{(p)}, \nu_n^{(p)}) = 0.$$

Remark 3. Assume the notations of [5]. In the previous considerations we can replace the quadratic cost function by any other lower semi-continuous cost function $c : \mathbb{R}^2 \to [0, +\infty]$, which is bounded on parallels to the diagonal and vanishes on the diagonal. For example, if we choose $c(x,y) = |x - y|^r$ for $0 < r < \infty$, then we obtain the same asymptotics as in Proposition 1 (with a different constant).

Remark 4. We have used in the above proof that τ_n is obtained from σ_n by convolving with the measure ρ. In fact, this theme goes back (at least) as far as L. Bachelier's famous thesis from 1900 on option pricing [2, p. 45]. Strictly speaking, L. Bachelier deals with the measure assigning mass $\frac{1}{2}$ to -1, 1 and considers consecutive convolutions, instead of the above ρ. Hence convolutions with ρ correspond to Bachelier's result after two time steps. Bachelier makes the crucial observation that this convolution leads to a *radiation* of probabilities: Each stock price x radiates during a time unit to its neighboring price a quantity of probability proportional to the difference of their probabilities. This was essentially the argument which allowed us to prove (1). Let us mention that Bachelier uses this argument to derive the fundamental relation between Brownian motion (which he was the first to define and analyse in his thesis) and the heat equation (compare e.g. [3] for more on this topic).

Remark 5. Having established the above counterexample, it becomes clear how to modify Problem 7.20 from [5] to give it a chance to hold true. This possible modification was also pointed out to us by C. Villani.

Problem 1. Let μ be a probability measure on $\mathbb{R}^d$ with finite second moment and vanishing barycenter, and γ the Gaussian measure with same first and second moments. Does $(\mathcal{T}(\mu^{(n)}, \gamma))_{n \geq 1}$ decrease monotonically to zero?

When entropy is considered instead of the quadratic Wasserstein distance, the corresponding question on monotonicity was answered affirmatively in the recent paper [1].

One may also formulate a variant of Problem 7.20 as given in (1) by replacing the measure ν through a log-concave probability distribution. This would again generalize Problem 1.

Acknowledgement. Financial support from the Austrian Science Fund under grant P 15889 and Y 328, from the Vienna Science and Technology Fund under grant MA 13, from the European Union under grant HPRN-CT-2002-00281 is gratefully acknowledged. Furthermore, this work was financially supported by the Christian Doppler Research Association (CDG) via PRisMa Lab (www.prismalab.at). The authors gratefully acknowledge a fruitful collaboration and continued support by Bank Austria and the Austrian Federal Financing Agency through CDG.

References

1. S. Artstein, K. M. Ball, F. Barthe and A. Naor, *Solution of Shannon's Problem on the Monotonicity of Entropy*, Journal of the AMS **17**(4), 2004, pp. 975–982.
2. L. Bachelier, *Théorie de la Spéculation*, Annales scientifiques de l'École Normale Supérieure Série 3, **17**, 1900, pp. 21–86. Also available from the site http://www.numdam.org/

3. W. Schachermayer, *Introduction to the Mathematics of Financial Markets*, LNM 1816 - Lectures on Probability Theory and Statistics, Saint-Flour summer school 2000 (Pierre Bernard, editor), Springer-Verlag, Heidelberg, 2003, pp. 111–177.

4. H. Tanaka, *An inequality for a functional of probability distributions and its applications to Kac's one-dimensional model of a Maxwell gas*, Zeitschrift für Wahrscheinlichkeitstheorie und verwandte Gebiete **27**, 1973, pp. 47–52.

5. C. Villani, *Topics in Optimal Transportation*, Graduate Studies in Mathematics **58**, American Mathematical Society, Providence Rhode Island, 2003.

On the Equation $\mu = S_t\mu * \mu_t$

Fangjun Xu[*]

Department of Mathematics, University of Connecticut
196 Auditorium Road, Unit 3009, Storrs, CT 06269-3009, USA
e-mail: fangjun@math.uconn.edu

Summary. We discuss solutions of equation $\mu = S_t\mu * \mu_t$ and study their structure. The relationship with Ornstein-Uhlenbeck processes will also be considered.

Keywords: C_0-semigroup; Infinitely divisible; Mehler semigroup; Ornstein-Uhlenbeck processes

1 Introduction

Let E be a real separable Banach space and E^* its dual. We define $P(E)$ and $ID(E)$ to be the sets of all probability measures and infinitely divisible probability measures on E, respectively. Let $OS(E)$ be the set of all probability measures on E which satisfy the following equation (1). For $\mu \in P(E)$, the Fourier transform of μ is

$$\widehat{\mu}(\lambda) = \int_E e^{i<\lambda,x>}d\mu(x) \quad \text{for all } \lambda \in E^*.$$

Define $H(\mu) = \{\lambda \in E^* : \widehat{\mu}(\lambda) = 0\}$. Obviously, $H(\mu)$ is closed but may be empty. Let $(S_t, t \geq 0)$ be a C_0-semigroup of linear operators acting on E with infinitesimal generator J. Using the notation $S_t\mu$ for the induced probability measure $\mu \circ S_t^{-1}$, we say that $\mu \in OS(E)$, if for each $t \geq 0$, there exists $\mu_t \in P(E)$ such that

$$\mu = S_t\mu * \mu_t. \tag{1}$$

As far as we know, there are several papers which studied solutions of the above equation. In these papers, under some given assumptions, solutions of equation (1) are called *operator-selfdecomposable distributions* and can be expressed as limit distributions. Moreover, integral expressions of operator-selfdecomposable solutions were found in some of these papers. For the case of

---

[*] Work carried out at Nankai University

C. Donati-Martin et al. (eds.), *Séminaire de Probabilités XLII*,
Lecture Notes in Mathematics 1979, DOI 10.1007/978-3-642-01763-6_4,
© Springer-Verlag Berlin Heidelberg 2009

infinite dimensional Banach space, see [BRS96], [Cho87], [JV83] and [Urb78]. For the case of finite dimensional Banach space, see [JM93] and [SY84]. A most recent article on operator-selfdecomposable distribution and its relationship with associated Ornstein-Uhlenbeck processes is [App07]. Throughout this article, if not mentioned otherwise, the topology we consider is the weak convergence topology.

Proposition 1.1 *If $\mu \in OS(E)$ and $H(\mu) = \emptyset$, then*

$$\mu_{t+s} = S_t \mu_s * \mu_t \quad \text{for all } t, s \geq 0. \tag{2}$$

Proof. We use similar arguments as in [JV83]. From equation (1), we have

$$\mu = S_{t+s}\mu * \mu_{t+s} = S_t(S_s\mu * \mu_s) * \mu_t = S_{t+s}\mu * S_t\mu_s * \mu_t.$$

Thus, we obtain

$$S_{t+s}\mu * \mu_{t+s} = S_{t+s}\mu * S_t\mu_s * \mu_t. \tag{3}$$

By the assumption, it can be easily concluded that

$$H(S_{t+s}\mu) = \emptyset.$$

Taking Fourier transforms on both sides of equation (3), we can easily obtain

$$\widehat{\mu}_{t+s}(\lambda) = \widehat{S_t\mu_s}(\lambda)\widehat{\mu}_t(\lambda)$$

for all $\lambda \in E^*$, i.e.,

$$\mu_{t+s} = S_t\mu_s * \mu_t.$$

$\square$

Remark 1.1 *In [JV83], $\lim_{t\to\infty} e^{-t} = 0$ implies $H(\mu) = \emptyset$. However, the assumption $H(\mu) = \emptyset$ here cannot be dropped. In fact, Theorem 5.1.1 in [Luk60] shows that the cancellation law cannot be applied in the convolution semigroup $P(E)$.*

The semigroup $(\mu_t, t \geq 0)$ satisfying equation (2) is called the *Mehler semigroup* (or (S_t)-*skew convolution semigroup*). For a recent study of the Mehler semigroup, we recommend [BRS96], [FR00], [SS01], [Jur04] and [Jur07].

In this paper we mainly consider solutions of the above equation (1), the structure of its solutions and its relationship with associated Ornstein-Uhlenbeck processes.

2 The Structure of Solutions

In this section, we first show that $OS(E)$ is a closed sub semigroup of $P(E)$. Then we will consider the existence of limits of $S_t\mu$ and μ_t as t tends to infinity and some related problems.

Proposition 2.1 $OS(E)$ *is a closed subsemigroup of* $P(E)$.

Proof. For any μ and $\nu \in OS(E)$, it is obvious that $\mu * \nu \in OS(E)$. Suppose $\{\mu_k, k \in \mathbb{N}\} \subset OS(E)$ and $\mu_k \Rightarrow \mu$ (weak convergence as $k \to \infty$). Then, for any k in $\mathbb{N}$ and all $t \geq 0$, there exist $\mu_{k,t}$ in $P(E)$ such that

$$\mu_k = S_t\mu_k * \mu_{k,t}.$$

Since

$$\int_E f(x)dS_t\mu_k(x) = \int_E f(S_tx)d\mu_k(x)$$

for all $f \in C_b(E)$, we have

$$\int_E f(x)dS_t\mu_k(x) \xrightarrow{k \to \infty} \int_E f(S_tx)d\mu(x) = \int_E f(x)dS_t\mu(x)$$

for all $f \in C_b(E)$. This means $S_t\mu_k \Rightarrow S_t\mu$ as $k \to \infty$. Theorem 2.1 in Chap.III of [Par67] shows that $\{\mu_{k,t}, k \in \mathbb{N}\}$ is conditionally compact. Let μ_t be a cluster point of $\{\mu_{k,t}, k \in \mathbb{N}\}$. Then it is obvious that μ satisfies equation (1) and thus $\mu \in OS(E)$. Consequently, $OS(E)$ is a closed subsemigroup of $P(E)$. $\square$

Remark 2.1 δ_0 *is the identity in this closed subsemigroup. Thus,* $OS(E)$ *is a closed monic subsemigroup of* $P(E)$.

Proposition 2.2 *If* $\lim_{t\to\infty} S_tx = 0$ *for each* $x \in E$, *then* μ *and* μ_t *are infinitely divisible. Moreover, the n-th convolution root of* μ *also belongs to* $OS(E)$.

Proof. Using the same arguments as in Lemma 4.4 of [Urb78], we know that μ is infinitely divisible. So $H(\mu) = \emptyset$. Using Proposition 1.1, we have

$$\mu_{t+s} = S_t\mu_s * \mu_t \quad \text{for all } t, s \geq 0.$$

Then Proposition 1 in [SS01] implies that μ_t is infinitely divisible. Assume $\mu = \mu_n^{*n}$ and $\mu_t = \mu_{t,n}^{*n}$. From equation (1), we have

$$\mu_n^{*n} = S_t\mu_n^{*n} * \mu_{t,n}^{*n} = (S_t\mu_n * \mu_{t,n})^{*n}.$$

This means

$$(\widehat{\mu_n}(\lambda))^n = (\widehat{S_t\mu_n * \mu_{t,n}}(\lambda))^n$$

for all $\lambda \in E^*$. Using the continuity of Fourier transform and the fact that $H(\mu) = \emptyset$, we obtain

$$\widehat{\mu_n}(\lambda) = \widehat{S_t\mu_n * \mu_{t,n}}(\lambda)$$

for all $\lambda \in E^*$. This implies

$$\mu_n = S_t\mu_n * \mu_{t,n}.$$

Hence $\mu_n \in OS(E)$. $\square$

Remark 2.2 *The above proposition is a generalization of Lemma 4.4 in [Urb78].*

We say that μ_t is *shift convergent* as t tends to infinity, if there exists a family $\{x_t, t \in \mathbb{R}^+\} \subset E$ such that $\mu_t * \delta_{x_t}$ is convergent as t tends to infinity; we further say that ν is *dominated* by μ if ν is a factor of μ. For notational convenience, put $OS_0(E) = \{\mu \in OS(E) : H(\mu) = \emptyset\}$.

Proposition 2.3 *Suppose $\mu \in OS_0(E)$, then $S_t\mu$ and μ_t are shift convergent as t tends to infinity. Moreover, the shift limit of μ_t is infinitely divisible.*

Proof. Equation (1) shows that μ_t is dominated by μ; meanwhile, equation (2) shows that, for any $0 \leq s_1 < s_2$, μ_{s_1} is dominated by μ_{s_2}. Hence, by Theorem 5.3 in Chap.III of [Par67], μ_t is shift convergent. Moreover, by Theorem 2.1 in Chap.III of [Par67], $S_t\mu$ is shift convergent as well.

Proposition 1.1 and Proposition 1 in [SS01] show that $\mu_t (t \geq 0)$ is infinitely divisible. Thus, the shift limit of μ_t is also infinitely divisible.   □

From Proposition 2.3, we can denote the shift convergent limits of $S_t\mu$ and μ_t by $S_\infty\mu$ and μ_∞, respectively. From the definition of shift convergence and equation (1), we know that there exists a set $\{x_t, t \in \mathbb{R}^+\} \subset E$ such that

$$\lim_{t \to \infty} S_t\mu * \delta_{-x_t} = S_\infty\mu \tag{4}$$

and

$$\lim_{t \to \infty} \mu_t * \delta_{x_t} = \mu_\infty. \tag{5}$$

Lemma 2.1 *For $\mu \in OS_0(E)$ and $\{x_t, t \in \mathbb{R}^+\}$ mentioned above, we have*

$$\lim_{h \to \infty} x_{t+h} - S_t x_h \text{ exists for all } t \geq 0$$

and then can put $x_t' := \lim_{h \to \infty} x_{t+h} - S_t x_h$.

Proof. For any fixed $t \geq 0$, we have

$$
\begin{aligned}
S_t S_\infty\mu &= S_t \lim_{h \to \infty} S_h\mu * \delta_{-x_h} \\
&= \lim_{h \to \infty} S_t(S_h\mu * \delta_{-x_h}) \\
&= \lim_{h \to \infty} S_{t+h}\mu * \delta_{-S_t x_h} \\
&= \lim_{h \to \infty} S_{t+h}\mu * \delta_{-x_{t+h}} * \delta_{x_{t+h} - S_t x_h}. \tag{6}
\end{aligned}
$$

By Proposition 2.3, (4) and Corollary 2.2.4 in [Hey04], we obtain the existence of $\lim_{h \to \infty} x_{t+h} - S_t x_h$.   □

Theorem 2.1 *For $\mu \in OS_0(E)$, we have*
(a) $S_t S_\infty \mu = S_\infty \mu * \delta_{x_t'}$ for all $t \geq 0$;
(b) μ_∞ satisfies

$$\mu_\infty = S_t(\mu_\infty) * \mu_t * \delta_{x_t'} \quad \text{for all } t \geq 0;$$

(c) $\lim_{t\to\infty} S_t(\mu_\infty) * \delta_{x_t' - x_t} = \delta_0$;
(d) $\lim_{t\to\infty} S_t(\mu_s * \delta_{x_s}) * \delta_{x_{t+s} - S_t x_s - x_t} = \delta_0$ for all $s \geq 0$.

Proof. By (6), (a) holds. From equation (2), we have

$$\mu_{t+s} * \delta_{x_{t+s}} = S_t(\mu_s * \delta_{x_s}) * (\mu_t * \delta_{x_t}) * \delta_{x_{t+s} - S_t x_s - x_t} \tag{7}$$

for all $t, s \geq 0$. Letting s tend to infinity in both sides of equation (7), we have

$$\mu_\infty = S_t(\mu_\infty) * (\mu_t * \delta_{x_t}) * \delta_{x_t' - x_t}. \tag{8}$$

This establishes (b). (c) follows from letting t tend to infinity in the right side of equation (8) and employing Corollary 2.2.4 in [Hey04] while (d) follows from letting t tend to infinity in both sides of equation (7) and using Corollary 2.2.4 in [Hey04]. $\qquad\square$

For each $\mu \in P(E)$, we define the *adjoint* μ^- of μ and the *symmetrization* $\widetilde{\mu}$ of μ by

$$\mu^-(B) := \mu(-B)$$

for all $B \in \mathscr{B}(E)$ and $\widetilde{\mu} := \mu * \mu^-$, respectively. It is obvious that $\mu \in OS(E)$ implies μ^- and $\widetilde{\mu} \in OS(E)$. Moreover, we refer to μ as *symmetric* if $\mu^- = \mu$ and have the following proposition.

Proposition 2.4 *For $\mu \in OS_0(E)$, if μ is symmetric, then $S_t\mu$ and μ_t are convergent as t tends to infinity. Moreover, the limits of $S_t\mu$ and μ_t are also symmetric.*

Proof. By Proposition 2.3 and Theorem 2.2.20 in [Hey04], we obtain that μ_t is convergent as t tends to infinity. Then we employ Corollary 2.2.4 in [Hey04] to see that $S_t\mu$ is convergent as t tends to infinity. Moreover, $\lim_{t\to\infty} S_t\mu = \lim_{t\to\infty} S_t\mu^-$ and $\lim_{t\to\infty} \widehat{\mu}_t(\lambda) = \lim_{t\to\infty} \dfrac{\widehat{\mu}(\lambda)}{\widehat{S_t\mu}(\lambda)} = \lim_{t\to\infty} \dfrac{\widehat{\mu^-}(\lambda)}{\widehat{S_t\mu^-}(\lambda)} = \lim_{t\to\infty} \widehat{\mu_t^-}(\lambda) = \lim_{t\to\infty} \widehat{\widetilde{\mu}_t}(\lambda) = \lim_{t\to\infty} \widehat{\widetilde{\mu}_t}(\lambda)$ for all $\lambda \in E^*$. Therefore, the limits of $S_t\mu$ and μ_t are both symmetric. $\qquad\square$

From the above Proposition 2.4, the limits of $S_t\mu$ and μ_t can also be denoted by $S_\infty\mu$ and μ_∞, respectively. Moreover, $S_\infty\mu$ and μ_∞ are symmetric.

Theorem 2.2 *For $\mu \in OS_0(E)$, if μ is symmetric, then we have*
(a) $S_\infty\mu$ is the invariant measure of $(S_t, t \geq 0)$;
(b) μ_∞ satisfies

$$\mu_\infty = S_t\mu_\infty * \mu_t \quad \text{for all } t \geq 0;$$

(c) $\lim_{t\to\infty} S_t(\mu_\infty) = \delta_0$;
(d) $\lim_{t\to\infty} S_t(\mu_s) = \delta_0$ for all $s \geq 0$.

Proof. Employing Proposition 2.4 and using similar arguments as in Theorem 2.1.   □

Theorem 2.3 *For* $\mu \in OS_0(E)$, μ *can be expressed as the convolution of* $S_\infty \mu$ *and* μ_∞ :

$$\mu = S_\infty \mu * \mu_\infty,$$

where $S_\infty \mu = \lim\limits_{t \to \infty} S_t \mu * \delta_{-x_t}$ *and* $\mu_\infty = \lim\limits_{t \to \infty} \mu_t * \delta_{x_t}$. *Moreover, if* μ *is symmetric, then* $S_\infty \mu = \lim\limits_{t \to \infty} S_t \mu$ *and* $\mu_\infty = \lim\limits_{t \to \infty} \mu_t$.

Proof. By Proposition 2.3 and Proposition 2.4.   □

Corollary 2.1 *For* $\mu \in OS_0(E)$, *we have*

$$\mu \in ID(E) \text{ if and only if } S_\infty \mu \in ID(E).$$

Proof. By Proposition 2.3 and Theorem 2.3.   □

In many cases, such as in the papers mentioned in the introduction, $S_\infty \mu$ is degenerate and μ appears as the limit distribution for an infinite triangular array. Therefore, μ is infinitely divisible.

Proposition 2.5 *Let* A *be a linear operator on* E. *Suppose there exists* $\mu \in P(E)$ *such that* $\mu \neq \delta_0$ *and* μ *satisfies*

$$\mu = A\mu. \tag{9}$$

Then, we have $\|A\| \geq 1$.

Proof. From equation (9), we have $\mu = A^n \mu$. Moreover, if $\|A\| < 1$, then $\|A\|^n \to 0$, which yields $\mu = \delta_0$.   □

In the last part of this section, we only need to consider the nondegenerate symmetric μ of $OS_0(E)$. Since δ_0 is the trivial solution of equation (1) and we can consider the symmetrization of μ when μ is not symmetric.

Case one: $S_\infty \mu = \delta_0$, $\mu_\infty = \mu$.

By proposition 2.3, we see that μ is infinitely divisible.

Example 2.1

1. $(S_t, t \geq 0)$ is stable, i.e., $\lim\limits_{t \to \infty} S_t x = 0$ for each $x \in E$.
2. $(S_t, t \geq 0)$ is exponentially stable, i.e., $\lim\limits_{t \to \infty} \|S_t\| = 0$.

Remark 2.3 *In the above example, "stable" and "exponentially stable" imply* $H(\mu) = \emptyset$.

Case two: $S_\infty\mu = \mu$, $\mu_\infty = \delta_0$.

From equation (2) and $\mu_\infty = \delta_0$, it can be easily verified that $\mu_t = \delta_0$ for all $t \geq 0$. Therefore, in this case, we have

$$\mu = S_t\mu \quad \text{for all } t \geq 0.$$

Moreover, by Proposition 2.5, we have

$$\|S_t\| \geq 1 \quad \text{for all } t \geq 0.$$

Example 2.2 $(S_t = I, t \geq 0)$.

Case three: $S_\infty\mu \neq \delta_0$, $\mu_\infty \neq \delta_0$.

By Theorem 2.2, we have

$$S_\infty\mu = S_t S_\infty\mu \quad \text{for all } t \geq 0.$$

Thus, Proposition 2.5 and the above equation imply

$$\|S_t\| \geq 1 \quad \text{for all } t \geq 0.$$

3 Relationship with Ornstein-Uhlenbeck Process

In this section, we assume that E is a Hilbert Space and mainly consider the relationship between solutions of equation (1) and Ornstein-Uhlenbeck processes.

Here we introduce the infinite dimensional Langevin equation:

$$dY(t) = JY(t)dt + dX(t), \ Y(0) = Y_0 \text{ a.s.}, \tag{10}$$

where $X = (X(t), \ t \geq 0)$ is an E-valued Lévy process (see [App07]).
The Ornstein-Uhlenbeck process

$$Y(t) = S_t Y(0) + \int_0^t S_{t-s} dX(s) \tag{11}$$

is the unique weak solution to equation (10)-see [Cho87]. Obviously, $Y = (Y(t), t \geq 0)$ is a Markov process. It induces a generalized Mehler semigroup $(\mathcal{T}_t, t \geq 0)$ on $C_b(E)$:

$$(\mathcal{T}_t f)(x) = E(f(Y(t))|Y_0 = x) = \int_E f(S_t x + y)\mu_t(dy) \tag{12}$$

(cf. [App06]). Linear operators defined in (12) form a semigroup if and only if μ_t satisfies equation (2), i.e.,

$$\mu_{t+s} = S_t\mu_s * \mu_t \quad \text{for all } t, s \geq 0.$$

The above Mehler semigroup is not (in general) continuous for the norm topology-see p.111 of [DZ02]. We need to introduce a mixed topology τ_m and recommend [GK01] and [App07] for the definition of this topology. Theorem 4.1 in [App07] shows that $(\mathcal{T}_t, t \geq 0)$ is strongly continuous on $(C_b(E), \tau_m)$. Thus, we can define the infinitesimal generator $\mathcal{A}$, which is densely defined and closed with respect to τ_m.

Theorem 3.1 *Suppose that* $H(\mu) = \emptyset$, *then the following three conditions are equivalent:*
(i) $\mu \in OS(E)$;
(ii) μ *is an invariant measure for* $(\mathcal{T}_t, t \geq 0)$;
(iii) $\int_E \mathcal{A}f(x)\mu(dx) = 0$, $f \in D(\mathcal{A})$.

Proof. $(ii) \Rightarrow (i)$: Since μ is an invariant measure, we have

$$\int_E (\mathcal{T}_t f)(x)\mu(dx) = \int_E \int_E f(S_t x + y)\mu_t(dy)\mu(dx) = \int_E f(x)\mu(dx)$$

for all $f \in C_b(E)$. So $\mu = S_t \mu * \mu_t$.
$(i) \Rightarrow (ii)$: Use Proposition 1.1 and similar arguments as above to show that μ is an invariant measure;
$(ii) \Leftrightarrow (iii)$: See [IW89](p. 292). $\square$

Remark 3.1 *The equivalence of* (i) *and* (ii) *is due to D. Applebaum (private communication).*

Acknowledgement. This work resulted from communication with Professor David Applebaum. I am very grateful to him for reading early versions and giving many useful comments. I am also grateful to Professor Zbigniew J. Jurek for some good comments.

References

[App06]   Applebaum, D.: Martingale-valued measures, Ornstein-Uhlenbeck processes with jumps and operator self-decomposability in Hilbert space. *In Memoriam Paul-André Meyer, Séminaire de Probabilités* **39**, ed. M.Emery and M.Yor, Lecture Notes in Math Vol., **1874**, 173–198 Springer-Verlag (2006)

[App07]   Applebaum, D.: On the infinitesimal generators of Ornstein-Uhlenbeck processes with jumps in Hilbert Space. Potential Analysis, Vol., **26**, 79–100 (2007)

[BRS96]   Bogachev, V.I., Röckner, M., Schmuland, B.: Generalized Mehler semigroups and applications. Probab. Theory Relat Fields, **105**, 193–225 (1996)

[Cho87]   Chojnowska-Michalik, A.: On processes of Ornstern-Uhlenbeck type in Hilbert space. Stochastics, Vol., **21**, 251–286 (1987)

[DZ02]   Da Prato, G., Zabczyk, J.: Second Order Partial Differential Equation in Hilbert Space. Cambridge University Press (2002)

[FR00]   Fuhrman, M., Röckner, M.: Generalized Mehler semigroups: the non-Gaussian case. Potential Anal., **12**, 1–47 (2000)

[GK01]   Goldys, B., Kocan, M.: Diffusion Semigroups in Spaces of Continuous Functions with Mixed Topology. Journal of Differential Equations, **173**, 17–39 (2001)

[Hey04]   Heyer, H.: Structural aspects in the theory of probability: a primer in probabilities on algebraic-topological structures. World Scientific (2004)

[IW89]   Ikeda, N., Watanabe, S.: Stochastic Differential Equations and Diffusion Processes (Second Edition). Amsterdam: North-Holland Pub.Co. (1989)

[Jur82]   Jurek, Z.J.: An integral representation of operator-selfdecomposable random variables. Bull. Acad. Pol. Sci., **30**, 385–393 (1982)

[JV83]   Jurek, Z.J., Vervaat, W.: An integral representation for selfdcomposable Banach space valued random variables. Z.Wahrscheinlichkeitstheorie verw. Gebiete, **62**, 247–262 (1983)

[JM93]   Jurek, Z.J., Mason, J.D.: Operator-Limit Distributions in Probability Theory. John Wiley and Sons.Inc (1993)

[Jur04]   Jurek, Z.J.: Measure valued cocycles from my papers in 1982 and 1983 and Mehler semigroups. www.math.uni.wroc.pl/ zjjurek (2004)

[Jur07]   Jurek, Z.J.: Remarks on relations between Urbanik and Mehler semigroups. www.math.uni.wroc.pl/ zjjurek (2007)

[Luk60]   Lukacs, E.: Characteristic functions. Griffin's Statistical Monographs and Courses, **5**. Charles Griffin, London (1960)

[Par67]   Parthasarathy, K.R.: Probability measures on metric spaces. Academic Press (1967)

[SS01]   Schmuland, B., Sun, W.: On the equation $\mu_{s+t} = \mu_s * T_s\mu_t$. Stat. Prob. Lett., **52**, 183–188 (2001)

[SY84]   Sato, K-I., Yamazato, M.: Operator-Selfdecomposable distributions as limit distributions of processes of Ornstern-Uhlenbeck type. Stochastic Processes and Their Applications, **17**, 73–100 (1984)

[Urb78]   Urbanik, K.: Lévy's probability measures on Banach spaces. Studia Math, **63**, 283–308 (1978)

Shabat Polynomials and Harmonic Measure

Philippe Biane[1]

CNRS, Laboratoire d'Informatique Institut Gaspard Monge Université Paris-Est
5 bd Descartes, Champs-sur-Marne, 77454 Marne-la-Vallée cedex 2
Philippe.Biane@univ-mlv.fr

1 Introduction

This note is inspired by [BZ], which describes *the true shape of a tree*. Each planar tree (remember that a planar tree is a tree in which, for each vertex, the adjacent edges are cyclically ordered) has a distinguished embedding in the complex plane (up to similitude).

Theorem 1. *For each finite planar tree Γ there exists a complex polynomial P having at most 0 and 1 as critical values, and such that the inverse image of $[0,1]$ by this polynomial is the union of the edges of a tree which is isomorphic to Γ whereas its vertices are the inverse images of 0 and 1. This polynomial is unique up to a change of variable $z \mapsto az+b$ or the substitution $P \mapsto 1-P$.*

Recall that the critical values of a polynomial are the numbers $P(w)$ where w are the zeros of P'. The polynomial of the theorem is called the Shabat polynomial of the tree, and we shall call Shabat embedding the corresponding embedding of the tree. The proof of the theorem uses Grothendieck's theory of *Dessins d'enfants*. Some of these trees are depicted in [BZ] or [LZ], page 89. In this short note I will give a potential theoretic characterization of the shape of these trees, which explains some aspects which can be observed on the pictures mentioned above, for example the respective sizes of the different branches of the trees, as well a their curvature. Meanwhile, I shall sketch a proof of this theorem, which uses some results on Löwner's equation.

We start in section 2 by recalling a well known correspondance between planar trees and noncrossing partitions, then in section 3 we study the conformal mapping of the exterior of a tree and give the sought for interpretation of the Shabat polynomial.

I would like to thank Alexander Zvonkin for his remarks and comments on the first version of this paper.

C. Donati-Martin et al. (eds.), *Séminaire de Probabilités XLII*,
Lecture Notes in Mathematics 1979, DOI 10.1007/978-3-642-01763-6_5,
© Springer-Verlag Berlin Heidelberg 2009

2 Planar Trees and Noncrossing Partitions

Let us consider, in the complex plane, the $2n^{th}$ roots of unity and the arcs of the unit circle joining them. Let π be a partition of these arcs into n pairs, which is noncrossing. This means that if one draws the n segments joining the middles of the arcs which are in the same pair of π, these segments do not cross. The quotient of the unit circle by the equivalence relation identifying the two arcs of each pair is a planar tree. Each planar tree can be so obtained, and the corresponding partition is unique, up to some rotation of the circle. Here is an example, for $n = 4$, of a noncrossing partition and its associated planar tree.

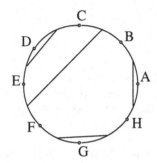

Fig. 1. A noncrossing partition

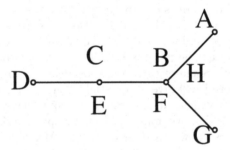

Fig. 2. The planar tree

Let us now assume that we identify each pair of arcs according to the natural length. This means that if we identify $[\frac{2k\pi}{2n}, \frac{(2k+1)\pi}{2n}]$ with $[\frac{(2l-1)\pi}{2n}, \frac{2l\pi}{2n}]$ (for parity reasons, these are the only possible identifications) we have to match the points $\frac{(2k+\theta)\pi}{2n}$ and $\frac{(2l-\theta)\pi}{2n}$, for $\theta \in [0, 1]$.

3 Conformal Mapping and Harmonic Measure

Proposition 1. *Let π be a noncrossing pair partition of the unit circle. Then there exists a unique conformal mapping from the outside of the unit disk to $\mathbf{C}$, with a Laurent expansion*

$$z + \dots \tag{1}$$

which extends continuously to the boundary of the circle and such that the equivalence relation on the unit circle induced by this map is the noncrossing pair partition π. The image of the unit circle by this map is an embedding of the tree associated with π.

The equivalence relation on the circle in the proposition is the one for which two points are in the same class if they have the same image by the continuous extension.

    Sketch of proof. The conformal mapping of the proposition can be constructed in the following way. Choose a leaf of the tree (a vertex with only one adjacent edge), it corresponds to some $2n^{th}$ root of unity whose adjacent arcs are in the same part of the partition π. We can assume, without loss of generality, that this root of unity is 1, then the maps $\phi_\theta, \theta \in [0, \frac{\pi}{n}]$ given by

$$\phi_\theta(z) = \left(z^2 + 1 + 2\sin^2(\theta/2)z + (z+1)\sqrt{z^2 + 1 - 2z\cos\theta} \right) / (2z)$$

glue the two intervals according to their natural length. These maps define a conformal mapping from the exterior of the disk to a domain which is the complement of the disk centered at 0, with radius $\cos^2(\theta/2)$, and of the segment $[\cos^2(\theta/2), (1+\sin(\theta/2))^2]$. For $\theta = \pi/n$ we have glued the two intervals.

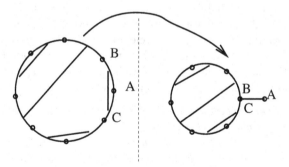

Fig. 3. The domain

    Let us map the partition π to the new circle C_1. This gives a noncrossing partition of $n - 1$ pairs, corresponding to the tree obtained by erasing the initial leaf from the original tree. These arcs are no longer identified through their arc-length, since the transformation $\phi_{\pi/n}$ does not preserve distances, but one can still find maps analogous to the ϕ_θ which glue two arcs coming from a leaf of the new tree. These maps can be constructed using Löwner's equation (see e.g. part 3 of [MR]), but the details of this construction are beyond the scope of this small note. One gets a conformal mapping from the exterior of C_1 to the exterior of a smaller circle and an analytic arc coming out of this circle. Iterating the process, after n steps we obtain ψ.   □

Since it is normalized at infinity by (1), this conformal mapping is unique up to a translation, furthermore it maps the outside of the unit circle to the outside of a planar tree. We have associated, in a canonical way, to each planar tree a conformal mapping ψ, and an embedding of the planar tree in the complex plane. This embedding can be characterized as in the following theorem. Before stating this result, I need to recall some facts about harmonic measures. Let K be a non polar compact set in the complex plane, there exists a unique probability measure μ_K, on K, which minimizes the logarithmic energy

$$E(\mu) = \int_{K \times K} \log(|z_1 - z_2|^{-1}) d\mu(z_1) d\mu(z_2).$$

(K is non polar if there exists a probability measure μ such that $E(\mu) < +\infty$). The measure μ_K is called the harmonic measure of K. If the complement of K is connected, and if φ is a conformal map from the exterior of the disc to the exterior of K, this conformal mapping extends by continuity to the unit circle, and the image of the uniform probability measure on the circle by φ is μ_K, which can be also described as the law of the hitting position of K by a Brownian starting from infinity, see e.g. [D] for more on these relations between classical potential theory and Brownian motion. Consider a planar tree embedded in **C**, such that each edge is given by a C^1 arc, then each edge has two sides, and each of these sides carries a part of the harmonic measure, corresponding to the probability that Brownian motion hits the edge on this side. Using the conformal mapping given by proposition 1, the n vertices of the tree come from $2n$ points on the circle, and each edge comes from 2 of the $2n$ arcs joining consecutive points. Each of these arcs corresponds to a side of the edge and the harmonic measure on each side is the image of the uniform measure on the corresponding arc. Therefore we conclude.

Theorem 2. *Given a planar tree with n edges, the exists an embedding of this tree in the complex plane such that*
i) the harmonic measure of each edge is $\frac{1}{n}$
ii) the harmonic measures on both sides of each edge coincide.
   This embedding is unique up to a similarity transform.

The first proprerty explains why on the pictures of the trees, the vertices tend to accumulate around the leaves, this is a well known electrostatic effect, which concentrates the harmonic measure towards extremities. The second property helps us understand why edges tend to become incurvate in order to let Brownian motion approach both sides with the same probability.

It remains to check that the embedding is that of Shabat. Consider the map $z \mapsto z^n$, which wraps the unit circle n times over itself, sending the $2n^{th}$ roots of unity to -1 and 1, and the arcs between these roots to the two half-circles between -1 and 1. Then $z \mapsto \frac{1}{4}(z + \frac{1}{z} + 2)$ maps conformally the exterior of the unit disc to the exterior of the segment $[0,1]$, and extends continuously to the circle, identifying points of this circle with the same abscissa. Let now η be the inverse map of ψ, which is defined on the exterior of the tree. The preceding considerations imply that $P(z) = \frac{1}{4}(\eta(z)^n + \eta(z)^{-n} + 2)$ maps conformally the exterior of the tree to the exterior of $[0,1]$. Furthermore, from the construction of the tree, one can extend P continuously to the whole of $\mathbf{C}$. Since P is analytic outside the tree, and continuous everywhere, Morera's theorem (cf [R]) implies that P is entire. Since $P(z) = z^n + O(z^{n-1})$ at infinity it is a polynomial of degree n, and the embedded tree is $P^{-1}([0,1])$. It is easy to check that the only critical values of this polynomial are 0 and 1.

References

[BZ]   Bétréma, J., Zvonkin, A.: *La vraie forme d'un arbre.* TAPSOFT '93: theory and practice of software development (Orsay, 1993), 599–612, Lecture Notes in Comput. Sci., 668, Springer, Berlin, 1993. 05C05.

[D]   Doob, J.L.: Classical potential theory and its probabilistic counterpart. Grundlehren der Mathematischen Wissenschaften, 262. Springer-Verlag, New York, 1984.

[LZ]   Lando, S., Zvonkin, A.: Graphs on surfaces and their applications, Encyclopedia of Mathematical Sciences, Low dimensional topology, II. Springer-Verlag, berlin, Heidelberg, 2004.

[MR]   Marshall, D.E.; Rohde, S.: The Löwner differential equation and slit mappings. J. Amer. Math. Soc. 18 (2005), no. 4, 763–778.

[R]   Rudin, W.: Real and complex analysis. Third edition. McGraw-Hill Book Co., New York, 1987.

Radial Dunkl Processes Associated with Dihedral Systems

Nizar Demni

Fakultät für Mathematik, Universität Bielefeld, Postfach 100131, Bielefeld, Germany.
demni@math.uni-bielefeld.de

Summary. We are interested in radial Dunkl processes associated with dihedral systems. We write down the semi-group density and as a by-product the generalized Bessel function and the W-invariant generalized Hermite polynomials. Then, a skew product decomposition, involving only independent Bessel processes, is given and the tail distribution of the first hitting time of boundary of the Weyl chamber is computed.

1 A Quick Reminder

We refer the reader to [11] and [16] for facts on root systems and to [5], [20] for facts on radial Dunkl processes. Let R be a reduced root system in a finite Euclidean space $(V, \langle, \rangle)$ with positive system R_+ and simple system S. Let W be its reflection group and C be its positive Weyl chamber. The radial Dunkl process X associated with R is a pathwise continuous Markov process valued in $\overline{C}$ whose generator acts on $C^2(\overline{C})$-functions as

$$\mathscr{L}_k u(x) = \frac{1}{2}\Delta u(x) + \sum_{\alpha \in R_+} k(\alpha)\frac{\langle \nabla u(x), \alpha \rangle}{\langle x, \alpha \rangle}$$

with $\langle \nabla u(x), \alpha \rangle = 0$ whenever $\langle x, \alpha \rangle = 0$, where Δ, ∇ respectively denote the Euclidean Laplacian and gradient and k is a positive multiplicity function, that is, a $\mathbb{R}_+$-valued W-invariant function. The semi-group density of X with respect to the Lebesgue measure in V is given by

$$p_t^k(x,y) = \frac{1}{c_k t^{\gamma+m/2}} e^{-(|x|^2+|y|^2)/2t} D_k^W\left(\frac{x}{\sqrt{t}}, \frac{y}{\sqrt{t}}\right) \omega_k^2(y), \quad x, y \in \overline{C} \quad (1)$$

where $\gamma = \sum_{\alpha \in R_+} k(\alpha)$ and $m = \dim V$ is the rank of R. The weight function ω_k is given by

$$\omega_k(y) = \prod_{\alpha \in R_+} \langle \alpha, y \rangle^{k(\alpha)}$$

C. Donati-Martin et al. (eds.), *Séminaire de Probabilités XLII*,
Lecture Notes in Mathematics 1979, DOI 10.1007/978-3-642-01763-6_6,
© Springer-Verlag Berlin Heidelberg 2009

and D_k^W is the generalized Bessel function. Thus, $\mathscr{L}_k$ may be written as

$$\mathscr{L}_k u(x) = \frac{1}{2}\Delta u(x) + \langle \nabla u(x), \nabla \log \omega_k(x)\rangle. \tag{2}$$

2 Motivation

Several reasons motivated us to investigate radial Dunkl processes associated with dihedral root systems. First, the dihedral group is a Coxeter, yet non Weyl in general, reflection group and covers an exceptional Weyl group known in the literature as G_2 which is of a particular interest ([1]). Second, the study of the Dunkl operators associated with dihedral root systems revealed a close relation to Gegenbauer and Jacobi polynomials which have interesting geometrical interpretations as harmonics and eigenfunctions of the radial part of the Laplacian on the sphere ([3], [11], [14]). The latter operator is a particular case of the Jacobi operator which generates a diffusion known as the Jacobi process that may be represented, up to a random time change, by means of two independent Bessel processes ([21]). Since the norm of the radial Dunkl process is a Bessel process, we wanted to gather all these materials in the present work and to see how do they interact. The last reason is that [7] and [8] emphasize the irreducible root systems of types A, B, C, D, which, together with dihedral root systems, exhaust the infinite families of irreducible root systems associated with finite Coxeter groups.

The remaining part of the paper consists of five sections. In order to be self-contained, some needed facts on dihedral systems are collected in the next section. Then, we write down the semi-group density, via a detailed analysis of the so-called spherical motion (see below). As a by-product, we deduce the generalized Bessel function. Once this is done, we express the W-invariant counterparts of the generalized Hermite polynomials ([20]) as products of univariate Laguerre and Jacobi polynomials. Next, we give a skew product decomposition of the radial Dunkl process using only independent Bessel processes. This mainly follows from the skew-product decomposition of the Jacobi process derived in [21]. Finally, we compute the tail distribution of the first hitting time of ∂C.

3 Dihedral Groups and Dihedral Systems

The dihedral group, denoted by $\mathcal{D}_2(n)$ for $n \geqslant 3$, is defined as the group of orthogonal transformations that preserve a regular n-sided polygon in $V = \mathbb{R}^2$ centered at the origin. Without loss of generality, one may assume that the y-axis is a mirror for the polygon. It contains n rotations through multiples of $2\pi/n$ and n reflections about the diagonals of the polygon. By a diagonal, we mean a line joining two opposite vertices or two midpoints of opposite sides if n is even, or a vertex to the midpoint of the opposite side if n is odd. The corresponding dihedral root system, $I_2(n)$, is characterized by its positive and simple systems given by:

$$R_+ = \{-ie^{i\pi l/n} := -ie^{i\theta_l},\, 1 \leqslant l \leqslant n\}, \quad S = \{e^{i\pi/n}e^{-i\pi/2}, e^{i\pi/2}\}$$

so that the Weyl chamber is a wedge of angle π/n. The reader can check that, for instance, $I_2(3)$ (equilateral triangle-preserving) is isomorphic to $R = A_2$ and $I_2(4)$ (square-preserving) is nothing but $R = B_2$ (see [16] for the definitions of both root systems). However, it is a bit more delicate to see that $I_2(6)$ (hexagon-preserving) corresponds to the exceptional Weyl group G_2 ([1]). When $n = 2p$ for $p \geqslant 2$, there are two orbits so that $k = (k_0, k_1) \in \mathbb{R}_+^2$; otherwise, there is only one orbit and k takes only one positive value.

4 Semi Group Density

4.1 Spherical Motion

Recall the skew-product decomposition of the radial Dunkl process into a radial and a spherical parts ([5] p. 53):

$$(X_t)_{t \geqslant 0} = (|X_t|\Theta_{A_t})_{t \geqslant 0} \quad A_t = \int_0^t \frac{ds}{|X_s|^2},$$

where $|X|$ is a Bessel process of index γ ([18]) and Θ is the spherical motion of X valued in the intersection of the sphere of the Euclidean space V and the closure $\overline{C}$ of the positive Weyl chamber, and is independent of $|X|$. For dihedral systems, Θ is valued in the unit circle and may be written as $(\cos\theta, \sin\theta)$ for some process θ valued in the interval $[0, \pi/n]$.

4.2 Semi Group Density of θ

Let us first split $\mathscr{L}_k$ of X into a radial and a spherical part. To proceed, we use (2) together with the expression of ω_k in polar coordinates (up to a constant factor, [11] p. 205)

$$\omega_k(r, \theta) = r^{nk}(\sin n\theta)^k \qquad \text{if } n \text{ is odd,}$$
$$\omega_k(r, \theta) = r^{p(k_0+k_1)}[\sin(p\theta)]^{k_0}[\cos(p\theta)]^{k_1} \qquad \text{if } n = 2p.$$

Thus

$$\mathscr{L}_k = \frac{1}{2}\left[\partial_r^2 + \frac{2\gamma + 1}{r}\partial_r\right] + \frac{1}{r^2}\left[\frac{\partial_\theta^2}{2} + nk\cot(n\theta)\partial_\theta\right]$$

when n is odd, where $\gamma = nk$, and

$$\mathscr{L}_k = \frac{1}{2}\left[\partial_r^2 + \frac{2\gamma + 1}{r}\partial_r\right] + \frac{1}{r^2}\left[\frac{\partial_\theta^2}{2} + p(k_0\cot(p\theta) - k_1\tan(p\theta))\partial_\theta\right]$$

when $n = 2p$, where $\gamma = p(k_0 + k_1)$. It follows that the generator of θ, say $\mathscr{L}_k^\theta$, acts on smooth functions as

$$\mathscr{L}_k^\theta = \frac{\partial_\theta^2}{2} + nk\cot(n\theta)\partial_\theta \qquad\qquad \text{when } n \text{ is odd,}$$

$$\mathscr{L}_k^\theta = \frac{\partial_\theta^2}{2} + p(k_0\cot(p\theta) - k_1\tan(p\theta))\partial_\theta \qquad \text{when } n = 2p.$$

Now, it is easy to see that the process N defined by $N_t := n\theta_{t/n^2}$ satisfies

$$dN_t = dB_t + k\cot(N_t)dt$$

when n is odd, while $(M_t := p\theta_{t/p^2})_{t\geqslant 0}$ satisfies

$$dM_t = dB_t + [k_0\cot(M_t) - k_1\tan(M_t)]dt$$

when $n = 2p$, B being a real Brownian motion. Let us first investigate the case of even $n = 2p$. The generator of M has a discrete spectrum given by $\lambda_j = -2j(j + k_0 + k_1)$, $j \geqslant 0$ corresponding to the Jacobi polynomials $P_j^{k_1-1/2,k_0-1/2}(\cos(2\theta))$ (see [11], p. 201). It is known that this set of orthogonal eigenpolynomials is complete for the Hilbert space $L^2([0,\pi/2], \mu_k(\theta)d\theta)$ where

$$\mu_k(\theta) := c_k\sin(\theta)^{2k_0}\cos(\theta)^{2k_1}$$

for some constant c_k. Accordingly, M has a semi-group density, say $m_t^k(\phi,\theta)$, given by (we use orthonormal polynomials, [19] p. 29)

$$m_t^k(\phi,\theta) = \sum_{j\geqslant 0} e^{\lambda_j t} P_j^{k_1-1/2,k_0-1/2}(\cos(2\phi)) P_j^{k_1-1/2,k_0-1/2}(\cos(2\theta))\mu_k(\theta) \quad (3)$$

for $\phi,\theta \in [0,\pi/2]$. It follows that the semi-group density of θ, say $K_t^{k,p}$, is given by

$$K_t^{k,p}(\phi,\theta) = pm_{p^2t}^k(p\phi,p\theta), \quad \phi,\theta \in [0,\pi/(2p)].$$

A similar spectral description holds for odd n: the generator of N has a discrete spectrum given by $\lambda_j = -2j(j + k)$, $j \geqslant 0$ corresponding to $P_j^{-1/2,k-1/2}(\cos(2\theta))$.

4.3 Semi Group Density of X

Let $(r,\theta) \mapsto f(r,\theta)$ be a nice function and let $\mathbb{P}_{\rho,\phi}$ denote the law of X starting at $x = (\rho,\phi) \in C$. Then, using the independence of θ and $|X|$ together with Fubini's Theorem, one has

$$\mathbb{E}_{\rho,\phi}[f(|X_t|,\theta_{A_t})] = \mathbb{E}_{\rho,\phi}[\mathbb{E}_{\rho,\phi}[f(|X_t|,\theta_{A_t})|\sigma(|X_s|, s \leqslant t)]]$$

$$= \int_0^{\pi/(2p)} \sum_{j\geqslant 0} \mathbb{E}_\rho^\gamma[f(|X_t|,\theta)e^{\lambda_j p^2 A_t}]P_j^{l_1,l_0}(\cos(2p\phi))P_j^{l_1,l_0}(\cos(2p\theta))\mu_k(p\theta)d\theta$$

where $\mathbb{P}_\rho^\gamma$ is the law of the Bessel process $|X|$ starting at ρ and of index γ. Next, for every $\theta \in [0, \pi/(2p)]$

$$\mathbb{E}_\rho^\gamma[f(|X_t|, \theta)e^{\lambda_j p^2 A_t}] = \mathbb{E}_\rho^\gamma[\mathbb{E}_\rho^\gamma[f(|X_t|, \theta)e^{\lambda_j p^2 A_t} || X_t |]$$
$$= \int_0^\infty \mathbb{E}_\rho^\gamma[e^{\lambda_j p^2 A_t} || X_t | = r] f(r, \theta) q_t(\rho, r) dr$$

where $q_t(\rho, r)$ is the semi-group density of the Bessel process $|X|$ of index γ ([18]):

$$q_t(\rho, r) = \frac{1}{t} \left(\frac{r}{\rho}\right)^\gamma r e^{-(\rho^2 + r^2)/2t} I_\gamma\left(\frac{\rho r}{t}\right)$$

where I_γ is the modified Bessel function of index γ ([17] p. 108). Moreover (see [22] p. 80)

$$\mathbb{E}_\rho^\gamma[e^{\lambda_j p^2 A_t} || X_t | = r] = \frac{I_{\sqrt{\gamma^2 - 2\lambda_j p^2}}(\rho r/t)}{I_\gamma(\rho r/t)}, \quad \lambda_j = -2j(j + k_0 + k_1).$$

Thus, we proved that

Proposition 1 *The semi-group density of the radial Dunkl process associated with even dihedral groups* $\mathcal{D}_2(2p)$ *is given by*

$$p_t^k(\rho, \phi, r, \theta) = \frac{1}{c_k t}\left(\frac{r}{\rho}\right)^\gamma e^{-(\rho^2 + r^2)/2t} \sin^{2k_0}(p\phi) \cos^{2k_1}(p\theta)$$
$$\sum_{j \geq 0} I_{2jp+\gamma}\left(\frac{\rho r}{t}\right) P_j^{l_1, l_0}(\cos(2p\phi)) P_j^{l_1, l_0}(\cos(2p\theta))$$

with respect to $dr\, d\theta$, *where* c_k *is a normalizing constant,* $l_0 = k_0 - 1/2$, $l_1 = k_1 - 1/2$ *and* $\rho, r \geq 0$, $0 \leq \phi, \theta \leq \pi/2p$. *For odd dihedral systems, one has to substitute in the above formula* $k_1 = 0, k_0 = k, p = n$.

Remarks 1 *1/ The j-th Jacobi polynomial* $P_j^{k_1 - 1/2, k_0 - 1/2}(\cos(2p\theta))$ *can be replaced by the generalized Gegenbauer polynomial* $C_{2j}^{k_1, k_0}(\cos(p\theta))$ *(see [11], p. 27). For* $k_1 = 0$, $C_{2j}^{k_1, k_0}(\cos(p\theta))$ *reduces to the Gegenbauer polynomial* $C_{2j}^{k_0}(\cos(p\theta))$.

2/ The heat kernel corresponding to a planar Brownian motion starting at $x \in C$ *and reflected on* ∂C *corresponds to* $k \equiv 0$. *Using the above formula, one deduces*

$$p_t^0(\rho, \phi, r, \theta) = \frac{1}{c_0 t} e^{-(\rho^2 + r^2)/2t} \sum_{j \geq 0} I_{2jp}\left(\frac{\rho r}{t}\right) T_j(\cos(2p\phi)) T_j(\cos(2p\theta))$$

where T_j *is the orthonormal j-th Tchebycheff polynomial defined by*

$$T_j(\cos\theta) = \cos(j\theta), \quad j \geq 0.$$

Thus

$$p_t^0(\rho, \phi, r, \theta) = \frac{1}{c_0 t} e^{-(\rho^2 + r^2)/2t} \sum_{j \geqslant 0} I_{2jp}\left(\frac{\rho r}{t}\right) \cos(2jp\phi) \cos(2jp\theta).$$

For $k \equiv 1$, one recovers the Brownian motion conditioned to stay in a wedge of angle π/n which is the $h = \omega_1$-transform in Doob's sense of a planar Brownian motion killed when it first hits ∂C ([15]). More precisely, one has for $n = 2p$

$$p_t^1(\rho, \phi, r, \theta)$$
$$= \frac{\omega_1^2(r, \theta)}{c_1 t}\left(\frac{1}{r\rho}\right)^{2p} e^{-(\rho^2 + r^2)/2t} \sum_{j \geqslant 0} I_{2(j+1)p}\left(\frac{\rho r}{t}\right) U_j(\cos(2p\phi)) U_j(\cos(2p\theta)),$$

where U_j is the j-th Tchebycheff polynomial of the second kind defined by

$$U_j(\cos \theta) = \frac{\sin(j+1)\theta}{\sin \theta}, \quad j \geqslant 0.$$

and d_j is a normalizing constant such that $d_j U_j$ has unit norm. Keeping in mind that

$$\omega_1(r, \theta) = cr^{2p} \sin(2p\theta)$$

for some constant c, elementary computations yield

$$p_t^1(\rho, \phi, r, \theta)$$
$$= \frac{\omega_1(r, \theta)}{\omega_1(\rho, \phi)} \frac{e^{-(r^2 + \rho^2)/2t}}{c_1 t} \sum_{j \geqslant 0} I_{2(j+1)p}\left(\frac{\rho r}{t}\right) \sin[2p(j+1)\phi)] \sin[2(j+1)p\theta]$$

which agrees with the ω_1-transform property. Besides, one deduces that the semi-group density of a planar Brownian motion killed when it first hits ∂C is given by

$$p_t^C(\rho, \phi, r, \theta) = \frac{e^{-(r^2 + \rho^2)/2t}}{t} \sum_{j \geqslant 0} I_{2(j+1)p}\left(\frac{\rho r}{t}\right) \sin[2p(j+1)\phi)] \sin[2p(j+1)\theta].$$

The expression of p_t^C should be compared with Lemma 1 in [4].

Writing p_t^k as

$$p_t^k(\rho, \phi, r, \theta) = \frac{1}{c_k t^{\gamma+1}}\left(\frac{t}{r\rho}\right)^{\gamma} e^{-(\rho^2 + r^2)/2t} \omega_k^2(r, \theta)$$
$$\sum_{j \geqslant 0} I_{(2j+k_0+k_1)p}\left(\frac{\rho r}{t}\right) P_j^{l_1, l_0}(\cos(2p\phi)) P_j^{l_1, l_0}(\cos(2p\theta))$$

and considering (1), we are led to the by-product

Corollary 1 (Generalized Bessel function) *For even dihedral groups, the generalized Bessel function is given by*

$$D_k^W(\rho,\phi,r,\theta) = c_{p,k}\left(\frac{2}{r\rho}\right)^\gamma \sum_{j\geq0} I_{2jp+\gamma}(\rho r)P_j^{l_1,l_0}(\cos(2p\phi))P_j^{l_1,l_0}(\cos(2p\theta))$$

where $\gamma = p(k_0 + k_1)$. *For odd dihedral groups, one has*

$$D_k^W(\rho,\phi,r,\theta)$$
$$= c_{n,k}\left(\frac{2}{r\rho}\right)^\gamma \sum_{j\geq0} I_{2jn+\gamma}(\rho r)P_j^{-1/2,l_0}(\cos(2n\phi))P_j^{-1/2,l_0}(\cos(2n\theta))$$

where $\gamma = nk$ *and* $l = k - 1/2$. *The constant* $c_{p,k}$ *and* $c_{n,k}$ *are such that* $D_k^W(0,0,r,\theta) = |W|$.

5 W-invariant Generalized Hermite Polynomials

In this section, we shall express the W-invariant counterparts of the so-called generalized Hermite polynomials by means of univariate Laguerre and Jacobi polynomials. This is done in three steps.

5.1 Generalized Hermite Polynomials

Recall from [20] that the generalized Hermite polynomials $(H_\tau)_{\tau\in\mathbb{N}^m}$ are defined by

$$H_\tau(x) = [e^{-\Delta_k/2}\phi_\tau](x)$$

where Δ_k denotes the Dunkl Laplacian ([20]) and $(\phi_\tau)_{\tau\in\mathbb{N}^m}$ is a basis of homogeneous polynomials orthogonal with respect to the pairing inner product defined in [13] (see also [20]):

$$[p,q]_k = \int_V e^{-\Delta_k/2}p(x)e^{-\Delta_k/2}q(x)\omega_k^2(x)dx$$

for two polynomials p,q (up to a constant factor). The family $(H_\tau)_\tau$ is then said to be associated to the basis $(\phi_\tau)_\tau$. Their W-invariant counterparts are defined by

$$H_\tau^W(x) := \sum_{w\in W} H_\tau(wx).$$

5.2 Mehler-type Formula

It is known that $(H_\tau)_\tau$ satisfies a Mehler-type formula ([11] p. 246[1])

$$\sum_{\tau \in \mathbb{N}^m} H_\tau(x) H_\tau(y) r^{|\tau|} = \frac{1}{(1-r^2)^{\gamma+m/2}} \exp -\frac{r^2(|x|^2 + |y|^2)}{2(1-r^2)} D_k \left(x, \frac{r}{1-r^2} y \right)$$

for $0 < r < 1$, $x, y \in V$. An analogous formula is satisfied by $(H_\tau^W)_\tau$ and follows after summing twice over W and using $D_k(wx, w'y) = D_k(x, w^{-1}w'y)$ ([20]):

$$\sum_{\tau \in \mathbb{N}^m} H_\tau^W(x) H_\tau^W(y) r^{|\tau|}$$

$$= \frac{|W|}{(1-r^2)^{\gamma+m/2}} \exp -\frac{r^2(|x|^2 + |y|^2)}{2(1-r^2)} D_k^W \left(x, \frac{r}{1-r^2} y \right).$$

5.3 Dihedral Systems

Let us express D_k^W through the hypergeometric function $_0\mathscr{F}_1$. This is done via the relation ([17])

$$I_\nu(z) = \frac{1}{\Gamma(\nu+1)} z^\nu {}_0\mathscr{F}_1(\nu+1, z).$$

It follows that

$$D_k^W(\rho, \phi, r, \theta)$$

$$= c_{p,k} \sum_{j \geq 0} \frac{(\rho r/2)^{2jp}}{\Gamma(2jp+\gamma+1)} {}_0\mathscr{F}_1 \left(2jp+\gamma+1, \frac{\rho^2 r^2}{4} \right)$$

$$P_j^{l_1, l_0}(\cos(2p\phi)) P_j^{l_1, l_0}(\cos(2p\theta)).$$

Using the Mehler-type formula for univariate Laguerre polynomials ([2] p. 200):

$$\sum_{q \geq 0} \frac{q!}{(2jp+\gamma+1)_q} L_q^{2jp+\gamma}(\rho^2/2) L_q^{2jp+\gamma}(r^2/2) z^{2q}$$

$$= (1-z^2)^{-2jp-\gamma-1} e^{-z^2(\rho^2+r^2)/[2(1-z^2)]} {}_0\mathscr{F}_1 \left(2jp+\gamma+1, \frac{z^2\rho^2 r^2}{4(1-z^2)^2} \right),$$

valid for $|z| < 1$, one gets

$$(1-z^2)^{-\gamma-1} e^{-z^2(\rho^2+r^2)/[2(1-z^2)]} D_k^W \left(\rho, \phi, \frac{zr}{1-z^2}, \theta \right) =$$

$$c_{p,k} \sum_{j,q \geq 0} \frac{q!}{\Gamma(2jp+q+\gamma+1)} N_{j,p}^{k,W}(\rho, \phi) N_{j,p}^{k,W}(r, \theta) \left(\frac{\rho r}{2} \right)^{2jp} z^{2(q+jp)}$$

---

[1] We use a different normalization from the one used in [11].

where

$$N_{j,p}^{k,W}(\rho,\phi) := L_q^{2jp+\gamma}\left(\frac{\rho^2}{2}\right)P_j^{l_1,l_0}(\cos(2p\phi)).$$

This suggests that the W-invariant generalized Hermite polynomials are given by

$$H_{\tau_1,\tau_2}^W(\rho,\phi) = \sqrt{\frac{q!}{\Gamma(2jp+q+\gamma+1)}}\left(\frac{\rho^2}{2}\right)^{jp}N_{j,p}^{k,W}(\rho,\phi)$$

for $\tau_1 = 2q\,(q \geqslant 0), \tau_2 = 2jp\,(j \geqslant 0)$ and zero otherwise. An elegant proof of this claim was given to us (private communication) by Professor C. F. Dunkl and is as follows: the j-th W-invariant harmonic is given by (see Proposition 3.15 in [12])

$$h_j^W(\rho,\phi) = \rho^{2jp}P_j^{l_1,l_0}(\cos(2p\phi))$$

so that by Proposition 3.9 in [13],

$$e^{-\Delta_k/2}[\rho^{2q}h_j^W(\rho,\phi)]$$

$$= e^{-\Delta_k^W/2}[\rho^{2q}h_j^W(\rho,\phi)] = (-2)^j j! L_q^{2jp+\gamma}\left(\frac{\rho^2}{2}\right)P_j^{l_1,l_0}(\cos(2p\phi)).$$

Remark 1 *A similar result holds for odd dihedral systems with* $k_1 = 0$, $k_0 = k$, $p = n$.

6 Skew-product Decomposition

In this section, we derive a skew-product decomposition for X using only independent Bessel processes. This is done by relating the process θ to a Jacobi process. That is why some results on Jacobi processes are collected below.

6.1 Facts on Jacobi Processes

The Jacobi process J of parameters $d, d' \geqslant 0$ is a $[0, 1]$-valued process and is a solution, whenever it exists, of ([21])

$$dJ_t = 2\sqrt{J_t(1 - J_t)}dB_t + (d - (d + d')J_t)dt, \tag{4}$$

where B is a real Brownian motion. As for squared Bessel processes, (4) has a unique strong solution for all $t \geqslant 0$ and all $J_0 \in [0, 1]$ since the diffusion coefficient is $1/2$-Hölder and the drift term is Lipschitz ([18] p. 360). When $d \wedge d' \geqslant 2$ and $J_0 \in [0, 1]$, then J remains in $]0, 1[$ while when $d \wedge d' > 0$, J is valued in $[0, 1]$ ([10] p. 135). Besides, J has the skew-product decomposition below ([21]):

$$\left(\frac{Z_1^2(t)}{Z_1^2(t) + Z_2^2(t)}\right)_{t \geq 0} = (J_{F_t})_{t \geq 0}, \quad F_t = \int_0^t \frac{ds}{Z_1^2(s) + Z_2^2(s)}, \tag{5}$$

where Z_1, Z_2 are two independent Bessel processes of dimension d, d' respectively such that $d + d' \geq 2$. Moreover, J is independent from $Z_1^2 + Z_2^2$ (and thereby also from F).

6.2 Relating θ and J

Assume $d \wedge d' \geq 1$, and define

$$(H_t := -\cos 2M_t)_{t \geq 0}$$

where $(M_t = p\theta_{t/p^2})_{t \geq 0}$. Then an application of Itô's formula and of pathwise uniqueness for the above SDE shows that $(H_t)_{t \geq 0} = (Y_{2t})_{t \geq 0}$ where

$$dY_t = \sqrt{2}\sqrt{1 - Y_t^2}dB_t - [(k_1 - k_0) + (k_0 + k_1 + 1)Y_t]dt.$$

In fact, it is easy to see that $(1 - Y_{2t})/2 = (1 - H_t)/2 = \cos^2(M_t)$ is a Jacobi process of parameters $d = 2k_1 + 1$, $d' = 2k_0 + 1$. As a result, one gets

$$(\theta_{A_t})_{t \geq 0} = \left(\frac{1}{p}\arccos(\sqrt{J_{p^2 A_t}})\right)_{t \geq 0}$$

where J is independent from X, thereby from A.

6.3 Skew-product Decomposition

On the one hand, it is a well known fact ([18]) that the sum of two independent squared Bessel processes of dimensions $d = 2k_1 + 1$, $d' = 2k_0 + 1$ is again a squared Bessel process of dimension $d + d'$, thus $Z := Z_1^2 + Z_2^2$ is a squared Bessel process of index $k_0 + k_1$. On the other hand, for any conjugate numbers r, q and any Bessel process R_ν of index $\nu > -1/q$, there exists a Bessel process $R_{\nu q}$ of index νq and defined on the same probability space such that the following holds ([18])

$$q^2 R_\nu^{2/q} = R_{\nu q}^2\left(\int_0^\cdot R_\nu^{-2/r}(s)ds\right). \tag{6}$$

Specializing (6) with $\nu = k_0 + k_1$, $q = p$, $R_\nu = Z$, there exists a Bessel process $R_{\nu q}$ such that

$$Z_t^{2/p} = \frac{R_{\nu q}^2}{p^2}\left(\int_0^t \frac{ds}{Z_s^{2(p-1)/p}}\right) := \frac{1}{p^2}R_{\nu q}^2(\tau_t), \quad r = \frac{p}{p-1}. \tag{7}$$

Let J be the Jacobi process defined in (5) with $d = 2k_1 + 1$, $d' = 2k_0 + 1$ and define a radial Dunkl process X by

$$X := \left(R_{\nu q}, \frac{1}{p} \arccos \sqrt{J}_{\int_0^{\cdot} p^2 \frac{ds}{R_{\nu q}^2(s)}} \right) = (|X|, \theta_{A_t})$$

Let $L_t := \inf\{s, \tau_s > t\}$ be the inverse of τ so that $Z_L^{2/p} = (1/p^2)|X|^2$. Then

$$p^2 A = p^2 \int_0^{\cdot} \frac{ds}{|X_s|^2} = \int_0^{\cdot} \frac{ds}{Z_{L_s}^{2/p}} = \int_0^{L_{\cdot}} \frac{d\tau_s}{Z_s^{2/p}} = \int_0^{L_{\cdot}} \frac{ds}{Z_s^2} = F_{L_{\cdot}}. \qquad (8)$$

As a result, when $k_0, k_1 \geqslant 0$ and $\theta \in C$, one has

$$(\theta_{A_t})_{t \geqslant 0} = \left(\frac{1}{p} \arccos(\sqrt{J_{F_{L_t}}}) \right)_{t \geqslant 0} = \frac{1}{p} \left(\arccos \sqrt{\frac{Z_1^2}{Z_1^2 + Z_2^2}}(L_t) \right)_{t \geqslant 0}.$$

Finally

Proposition 2 *Let $k_0, k_1 \geqslant 0$ and define the time-change τ by*

$$\tau := \int_0^t \frac{ds}{[Z_1^2(s) + Z_2^2(s)]^{2(p-1)/p}}$$

where Z_1, Z_2 are two independent Bessel processes of dimensions $d = 2k_1 + 1$, $d' = 2k_0 + 1$ respectively. Then there exists a radial Dunkl process X associated with the even dihedral group $\mathcal{D}_2(2p)$ such that X_τ is realized as the two-dimensional process

$$\left[p(Z_1^2 + Z_2^2)^{1/2p}, \frac{1}{p} \arccos \sqrt{\frac{Z_1^2}{Z_1^2 + Z_2^2}} \right].$$

A similar representation holds when n is odd.

7 On the First Hitting Time of a Wedge

Let $X_0 = x \in C$ and let

$$T_0 := \inf\{t, X_t \in \partial C\}$$

be the first hitting time of ∂C. Recall that for the dihedral groups $\mathcal{D}_2(2p)$, C is a wedge of angle $\pi/(2p)$. Recall also from ([9]) that if the index function $l := k - 1/2$ takes one striclty negative value for some simple root α, then

$< \alpha, X >$ hits zero a.s. so that $T_0 < \infty$ a.s (see also [5]). For even dihedral systems, two cases are to be considered:

- $1/2 \leqslant k_0, k_1 \leqslant 1$ with either $k_1 > 1/2$ or $k_0 > 1/2$ or equivalently $0 \leqslant l_0, l_1 \leqslant 1/2$ with either $l_0 > 0$ or $l_1 > 0$: in that case, the radial Dunkl process with index function $-l$ hits ∂C a.s. and we will use results from [5].
- One and only one of the index values is strictly negative while the other value is positive: in that case, the radial Dunkl process of index function l hits ∂C. We will follow a different strategy based on our representation of the angular process θ as a Jacobi process and on results from ([10]) on Jacobi processes. This strategy applies to the first case too.

For odd dihedral systems, we can only have $1/2 < k \leqslant 1$ and computations are similar to the ones done in the first case for even dihedral systems.

7.1 Even Dihedral Groups: First Case

We give two different approaches: while the first one has the advantage to be short, the second approach is shown to be efficient for both cases. In fact, the first approach uses the absolute-continuity relations for radial Dunkl processes derived in [5] and we are met with a complicated exponential functional when dealing with the second case (see [7] for more details). The second approach focuses on the angular process θ which was identified with a Jacobi process and uses absolute-continuity relations for Jacobi processes from [10].

First approach: write $x = \rho e^{i\phi}$, $\rho > 0$, $0 < \phi < \pi/(2p)$ and let $1/2 \leqslant k_0, k_1 \leqslant 1$ with either $k_0 > 1/2$ or $k_1 > 1/2$. Using part (c) of Proposition 2. 15 in [5], p. 38, the tail distribution of T_0 is given by:

$$\mathbb{P}_x^{-l}(T_0 > t) = \mathbb{E}_x^l \left[\left(\prod_{\alpha \in R_+} \frac{< \alpha, X_t >}{< \alpha, x >} \right)^{-2l(\alpha)} \right],$$

where $\mathbb{P}_x^l$ ($\mathbb{E}_x^l$) denotes the law of a radial Dunkl process starting at $x \in C$ and of index l (the corresponding expectation). From (1), one gets

$$\mathbb{P}_x^{-l}(T_0 > t) = \frac{e^{-\rho^2/2t}}{c_k} \left(\frac{\rho}{\sqrt{t}} \right)^{2p(l_0+l_1)} \sin^{2l_0}(p\phi) \cos^{2l_1}(p\phi) g\left(\frac{\rho}{\sqrt{t}}, \phi \right),$$

where

$$g(\rho, \phi) = \int_0^\infty \int_0^{\pi/n} e^{-r^2/2} D_k^W(\rho, \phi, r, \theta) r^{2p+1} \sin(2p\theta) dr d\theta.$$

With regard to Corollary 1, it amounts to evaluate

$$S_1(j) = \int_0^\infty e^{-r^2/2} I_{b_j}(\rho r) r^{2p+1-\gamma} dr,$$

$$S_2(j) = \int_0^{\pi/2p} P_j^{k_1-1/2, k_0-1/2}(\cos(2p\theta)) \sin(2p\theta) d\theta$$

for every $j \geqslant 0$, where $b_j := 2jp + \gamma$. In order to evaluate S_1, we use the expansion ([17], p. 108)

$$I_{b_j}(\rho r) = \sum_{q \geqslant 0} \frac{1}{\Gamma(b_j + q + 1)} \left(\frac{\rho r}{2}\right)^{2q+b_j}$$

and exchange the order of integration to get

$$S_1(j) = 2^{(p-\gamma)/2} \frac{\Gamma(a_j + 1)}{\Gamma(b_j + 1)} \left(\frac{\rho}{\sqrt{2}}\right)^{b_j} {}_1\mathscr{F}_1\left(a_j + 1, b_j + 1, \frac{\rho^2}{2}\right)$$

where

$$a_j = \frac{(2j + k_0 + k_1)p + 2p - \gamma}{2} = (j+1)p.$$

Using the variable change $s = \cos(2p\theta)$, $S_2(j)$ transforms to

$$S_2(j) = \frac{1}{2p} \int_{-1}^{1} P_j^{k_1 - 1/2, k_0 - 1/2}(s) ds = \frac{1}{p} \int_0^1 P_j^{k_1 - 1/2, k_0 - 1/2}(2s - 1) ds$$

which is easily computed using the expansion p. 21 in [11]. As a result, the tail distribution is given by

Proposition 3

$$\mathbb{P}_x^{-l}(T_0 > t) = \frac{e^{-\rho^2/2t}}{c_k} \left(\frac{\rho}{\sqrt{t}}\right)^{\gamma - 2p} \sin^{2l_0}(p\phi) \cos^{2l_1}(p\phi)$$

$$\sum_{j \geqslant 0} S_2(j) \frac{\Gamma(a_j + 1)}{\Gamma(b_j + 1)} \left(\frac{\rho}{\sqrt{2t}}\right)^{2jp} {}_1\mathscr{F}_1\left(a_j + 1, b_j + 1, \frac{\rho^2}{2t}\right) P_j^{l_1, l_0}(\cos(2p\phi))$$

$$= \frac{1}{c_k} \left(\frac{\rho}{\sqrt{t}}\right)^{\gamma - 2p} \sin^{2l_0}(p\phi) \cos^{2l_1}(p\phi)$$

$$\sum_{j \geqslant 0} S_2(j) \frac{\Gamma(a_j + 1)}{\Gamma(b_j + 1)} \left(\frac{\rho}{\sqrt{2t}}\right)^{2jp} {}_1\mathscr{F}_1\left(b_j - a_j, b_j + 1, -\frac{\rho^2}{2t}\right) P_j^{l_1, l_0}(\cos(2p\phi))$$

by Kummer's transformation ([17]).

Remark 2 The value $k \equiv 1$ corresponds to the first exit time of a Brownian motion from a wedge and our result fits the one in [4]. Moreover $\gamma = 2p$, $b_j = 2a_j$ and one may use the duplication formula to simplify the above ratio of Gamma functions and use some argument simplifications for the confluent hypergeometric function ([17]).

Second approach: recall that

$$(\theta_{A_t})_{t \geqslant 0} = \left(\frac{1}{p} \arccos(\sqrt{J_{p^2 A_t}})\right)_{t \geqslant 0}$$

where J is a Jacobi process of parameters $d = 2k_1 + 1$, $d' = 2k_0 + 1$ and is independent from $|X|$ (thereby from A). Then, for an appropriate index function l (so that $T_0 < \infty$), one has

$$\mathbb{P}_x^l(T_0 > t) = \mathbb{P}_x^l(0 < \theta_{A_t} < \pi/(2p)) = \mathbb{P}_x^l(0 < J_{p^2 A_t} < 1) = \mathbb{P}_x^l(T_J > p^2 A_t)$$

where

$$T_J := \inf\{t, \, J_t = 0\} \wedge \inf\{t, \, J_t = 1\}$$

is the first exit time from the interval $[0, 1]$ by a Jacobi process. Now, recall the following absolute continuity relation between the laws of Jacobi processes of different set of parameters (Theorem 9.4.3. p.140 in [10] specialized to $m = 1$): let $\mathbb{P}_z^{d,d'}$ denote the probability law of a Jacobi process starting at $z \in]0, 1[$ and of parameters $d \wedge d' > 0$. Writing T for T_J, then

$$\mathbb{P}_z^{d_1,d_1'} \big|_{\mathscr{F}_t \cap \{T > t\}}$$

$$= \left(\frac{J_t}{z}\right)^{\kappa} \left(\frac{1 - J_t}{1 - z}\right)^{\beta} \exp - \int_0^t ds \left[c + \frac{u}{J_s} + \frac{v}{1 - J_s}\right] \mathbb{P}_z^{d_2,d_2'} \big|_{\mathscr{F}_t \cap \{T > t\}} \quad (9)$$

where $(\mathscr{F}_t)_t$ is the natural filtration of J, $d_i \wedge d_i' > 0, i = 1, 2$ and

$$\kappa = \frac{d_1 - d_2}{4}, \beta = \frac{d_1' - d_2'}{4},$$

$$u = \frac{d_1 - d_2}{4}\left(\frac{d_1 + d_2}{2} - 2\right), v = \frac{d_1' - d_2'}{4}\left(\frac{d_1' + d_2'}{2} - 2\right),$$

$$c = \frac{d_1 + d_1' - d_2 - d_2'}{4}\left(2 - \frac{d_1 + d_1' + d_2 + d_2'}{2}\right).$$

A corollary of the above theorem (Corollary 9.4.6. p. 140[2]) states that if $d := 2k_1 + 1 := 2(l_1 + 1)$, $d' := 2k_0 + 1 := 2(l_0 + 1)$ with $0 \leqslant l_1, l_0 < 1$, then

$$\mathbb{P}_z^{-l_1,-l_0}(T > t) = \mathbb{E}_z^{l_1,l_0}\left[e^{-ct}\left(\frac{z}{J_t}\right)^{l_1}\left(\frac{1 - z}{1 - J_t}\right)^{l_0}\right]. \quad (10)$$

where $\mathbb{P}_z^{l_1,l_0}$ ($\mathbb{E}_z^{l_1,l_0}$) denotes the probability law of a Jacobi process of indices l_1, l_0 and starting at z (the corresponding expectation).

To recover the result in Proposition 3, proceed as follows. Let $-1/2 \leqslant -l_0, -l_1 \leqslant 0$ so that at least one value is strictly negative. Using the semi-group density of $J = \cos^2 M$ which follows from (3):

$$p_t^J(z, s) = \frac{1}{2\sqrt{s(1 - s)}} m_t^k(\arccos\sqrt{z}, \arccos\sqrt{s})$$

$$= c_k\left[\sum_{j \geqslant 0} e^{\lambda_j t} P_j^{l_1,l_0}(2z - 1) P_j^{l_1,l_0}(2s - 1)\right] s^{l_1}(1 - s)^{l_0}, \lambda_j \quad (11)$$

$$= -2j(j + k_0 + k_1),$$

_____

[2] The exponential factor e^{-ct} is missing in [10].

for some constant c_k and $z, s \in]0, 1[$, together with the independence of J and A and (10), one gets

$$
\mathbb{P}_x^{-l}(T_0 > t) = \mathbb{E}_z^{l_1, l_0}\left[e^{-cp^2 A_t} \left(\frac{z}{J_{p^2 A_t}} \right)^{l_1} \left(\frac{1-z}{1-J_{p^2 A_t}} \right)^{l_0} \right]
$$

$$
= c_k z^{l_1}(1-z)^{l_0} \sum_{j \geq 0} \mathbb{E}_\rho^{\gamma'}[e^{-p^2(c-\lambda_j)A_t}] P_j^{l_1, l_0}(2z-1) \int_0^1 P_j^{l_1, l_0}(2s-1)ds,
$$

where $\mathbb{E}_\rho^{\gamma'}$ is the law of the Bessel process $|X|$ of index

$$
\gamma' = (2 - k_0 - k_1)p = 2p - \gamma,
$$

corresponding to $-l$ and $c = 2(l_0 + l_1) = 2(k_0 + k_1 - 1)$. Now, note that integral in the RHS is up to a constant $S_2(j)$ defined in the previous section and that $z^{l_1}(1-z)^{l_0} = \sin^{2l_0}(p\phi)\cos^{2l_1}(p\phi)$. Next, use the formula

$$
\mathbb{E}_\rho^{\gamma'}[e^{-(c-\lambda_j)p^2 A_t}||X_t| = r] = \frac{I_{\sqrt{\gamma'^2 + 2(c-\lambda_j)p^2}}(\rho r/t)}{I_{\gamma'}(\rho r/t)},
$$

and the semi-group density of $|X|$

$$
q_t(\rho, r) = \frac{1}{t}\left(\frac{r}{\rho}\right)^{\gamma'} r e^{-(\rho^2 + r^2)/2t} I_{\gamma'}\left(\frac{\rho r}{t}\right)
$$

to deduce after elemantary computations that

$$
\mathbb{E}_\rho^{\gamma'}[e^{-(c-\lambda_j)p^2 A_t}] = \frac{e^{-\rho^2/2t}}{t} \rho^{2p-\gamma} \int_0^\infty r^{2p-\gamma+1} e^{-r^2/2t} I_{2jp+\gamma}\left(\frac{\rho r}{t}\right) dr
$$

$$
= e^{-\rho^2/2t}\left(\frac{\rho}{\sqrt{t}}\right)^{\gamma-2p} \int_0^\infty r^{2p-\gamma+1} e^{-r^2/2} I_{2jp+\gamma}(\rho r) dr
$$

$$
= e^{-\rho^2/2t}\left(\frac{\rho}{\sqrt{t}}\right)^{\gamma-2p} S_1(j).
$$

In fact,

$$
\gamma'^2 - 2(c-\lambda_j)p^2 = [(2-k_0-k_1)^2 + 4[(k_0+k_1-1)+j(j+k_0+k_1)]]p^2
$$

$$
= [(k_0+k_1)^2 - 4(k_0+k_1-1) + 4(k_0+k_1-1)
$$

$$
+ 4j(j+k_0+k_1)]p^2
$$

$$
= (2j+k_0+k_1)^2 p^2 = (2jp+\gamma)^2.
$$

Finally, it only remains to relate the modified Bessel function I_ν and the hypergeometric function ${}_0\mathscr{F}_1$ via:

$$
I_\nu(z) = \frac{1}{\Gamma(\nu+1)} z^\nu {}_0\mathscr{F}_1(\nu+1, z).
$$

7.2 Even Dihedral Groups: Second Case

We use the second approach developed above and we suppose for instance that $k_1 < 1/2$ while $k_0 \geqslant 1/2$. Take $1 \leqslant d_1 = d < 2$, $d'_1 = d' \geqslant 2$ in (9) then perform the parameters change

$$d'_2 = d'_1 = d' \geqslant 2, \quad d_2 = 4 - d_1 = 4 - d = 3 - 2k_1 > 2$$

so that the indices corresponding to the new parameters d_2, d'_2 are positive, wherefrom $T = T_J = \infty$ a.s. Moreover, one has $\beta = u = v = 0$, which yields

$$\mathbb{P}^l_x(T_0 > t) = \mathbb{P}^{4-d,d'}_z \left[e^{-cp^2 A_t} \left(\frac{J_{p^2 A_t}}{z} \right)^\kappa \right]$$

where $\kappa = (d - d_2)/2 = d/2 - 1 = l_1 < 0$ and $c = -d'(d-2) = d'(2-d) > 0$. Since the parameter d_2 corresponds to the index value $1/2 - k_1 = -l_1 > 0$ and multiplicity function $1 - k_1 > 1/2$, we get

$$\mathbb{P}^l_x(T_0 > t) = c_k z^{1-2k_1}(1-z)^{l_0} \sum_{j \geqslant 0} \mathbb{E}^\gamma_\rho[e^{p^2(\lambda_j - c)A_t}] P_j^{1/2-k_1, k_0 - 1/2}(2z - 1) F(j)$$

where $\lambda_j = -2j(j + k_0 + 1 - k_1)$, $\gamma = (k_0 + k_1)p$ and

$$F(j) = \int_0^1 (1-s)^{l_0} P_j^{1/2-k_1, k_0 - 1/2}(2s - 1) ds.$$

We leave the computations to the interested reader. A similar result holds when $k_0 < 1/2$ and $k_1 \geqslant 1/2$: substitute k_1, l_1 by k_0, l_0 respectively and s, z by $1 - s$, $1 - z$ respectively.

Acknowedgments: the author would like to give a special thank for Professor C. F. Dunkl for his intensive reading of the paper, for his fruitful remarks and for pointing to him the references [12], [13]. The author also thanks Professor M. Bozejko for his remarks and for stimulating discussions at the Wrocław Institute of Mathematics.

References

1. J. C. Baez. The octonions. *Bull. Amer. Math. Soc. (N. S.)* **39**, no. 2. 2002, 145–205.
2. T. H. Baker, P. J. Forrester. The Calogero-Sutherland model and generalized classical polynomials. *Comm. Math. Phys.* **188**. 1997, 175–216.
3. D. Bakry. Remarques sur les semi-groupes de Jacobi. *Hommage à P. A. Meyer et J. Neveu. Astérisque* **236**, 1996, 23–39.
4. R. Bañuelos, R. G. Smits. Brownian motions in cones. *P. T. R. F.* **108**, 1997, 299–319.

5. O. Chybiryakov. Processus de Dunkl et relation de Lamperti. Thèse de doctorat, Université Paris VI, June 2005.

6. N. Demni, M. Zani. Large deviations for statistics of Jacobi process. To appear in S. P. A.

7. N. Demni. First hitting time of the boundary of a Weyl chamber by radial Dunkl processes. SIGMA Journal. 4, 2008, 074, 14 pages.

8. N. Demni. Generalized Bessel function of type D. SIGMA Journal. 4, 2008, 075, 7 pages.

9. N. Demni. Note on radial Dunkl processes. Submitted to Ann. I. H. P.

10. Y. Doumerc. Matrix Jacobi Process. Thèse de doctorat, Université Paul Sabatier, May 2005.

11. C. F. Dunkl, Y. Xu. Orthogonal Polynomials of Several Variables. Encyclopedia of Mathematics and Its Applications. Cambridge University Press. 2001.

12. C. F. Dunkl. Differential-difference operators associated to reflection groups. Trans. Amer. Math. Soc. 311. 1989, no. 1, 167–183.

13. C. F. Dunkl. Integral kernels with reflection group invariance. Canad. J. Math. 43. 1991, no. 6, 1213–1227.

14. C. F. Dunkl. Generating functions associated with dihedral groups. Special functions (Hong Kong 1999), World Sci. Publ. Rier Edge, NJ. 2000, 72–87.

15. D. J. Grabiner. Brownian motion in a Weyl chamber, non-colliding particles and random matrices. Ann. IHP. 35, 1999, no. 2. 177–204.

16. J. E. Humphreys. Reflections Groups and Coxeter Groups. Cambridge University Press. 29. 2000.

17. N. N. Lebedev. Special Functions and their Applications. Dover Publications, INC. 1972.

18. D. Revuz, M. Yor. Continuous Martingales and Brownian Motion, 3rd ed., Springer, 1999.

19. W. Schoutens. Stochastic Processes and Orthogonal Polynomials. Lecture Notes in Statistics, 146. Springer, 2000.

20. M. Rösler. Dunkl operator: theory and applications, orthogonal polynomials and special functions (Leuven, 2002). Lecture Notes in Math. Vol. 1817, Springer, Berlin, 2003, 93–135.

21. J. Warren, M. Yor. The Brownian Burglar: conditioning Brownian motion by its local time process. Sém. Probab. XXXII., 1998, 328–342.

22. M. Yor. Loi de l'indice du lacet brownien et distribution de Hartman-Watson, Zeit. Wahr. verw. Geb. 53, no.1, 1980, 71–95.

Matrix Valued Brownian Motion and a Paper by Pólya

Philippe Biane

CNRS, Laboratoire d'Informatique Institut Gaspard Monge, Université Paris-Est
5 bd Descartes, Champs-sur-Marne, 77454 Marne-la-Vallée cedex 2, France
e-mail: Philippe.Biane@univ-mlv.fr

1 Introduction

This paper has two parts which are largely independent. In the first one I recall some known facts on matrix valued Brownian motion, which are not so easily found in this form in the literature. I will study three types of matrices, namely Hermitian matrices, complex invertible matrices, and unitary matrices, and try to give a precise description of the motion of eigenvalues (or singular values) in each case. In the second part, I give a new look at an old paper of G. Pólya [14], where he introduces a function close to Riemann's ξ function, and shows that it satisfies Riemann's hypothesis. As put by Marc Kac in his comments on Pólya's paper [11], "Although this beautiful paper takes you within a hair's breadth of Riemann's hypothesis it does not seem to have inspired much further work and reference to it in the mathematical litterature are rather scant". My aim is to point out that the function considered by Pólya is related in a more subtle way to Riemann's ξ function than it looks at first sight. Furthermore the nature of this relation is probabilistic, since these functions have a natural interpretation involving Mellin transforms of first passage times for diffusions. By studying infinite divisibility properties of the distributions of these first passage times, we will see that they are generalized gamma convolutions, whose mixing measures are related to the considerations in the first part of this note.

2 Matrix Brownian Motions

We will study three types of matrix spaces, and in each of these spaces consider a natural Brownian motion, and show that the motion of eigenvalues (or singular values) of this Brownian motion has a simple geometric description, using Doob's transform. The following results admit analogues in more general complex symmetric spaces, but for the sake of simplicity, discussion will be restricted to type A symmetric spaces. Actually the interesting case for us

C. Donati-Martin et al. (eds.), *Séminaire de Probabilités XLII*,
Lecture Notes in Mathematics 1979, DOI 10.1007/978-3-642-01763-6_7,
© Springer-Verlag Berlin Heidelberg 2009

in the second part will be the simplest one, of rank one, but I think that this almost trivial case is better understood by putting it in the more general context. Some references for results in this section are [2], [4], [5], [7], [8], [9], [10], [13], [15], [16].

2.1 Hermitian Matrices

Consider the space of $n \times n$ Hermitian matrices, with zero trace, endowed with the quadratic form

$$\langle A, B \rangle = Tr(AB).$$

Let $M(t)$ be a Brownian motion with values in this space, which is simply a Gaussian process with covariance

$$E[Tr(AM(t))Tr(BM(s))] = Tr(AB)s \wedge t$$

for A, B traceless Hermitian matrices.

Let $\lambda_1(t) \geq \lambda_2(t) \geq \ldots \geq \lambda_n(t)$ be the eigenvalues of $M(t)$; they perform a stochastic process with values in the Weyl chamber

$$\mathcal{C} = \{(x_1, \ldots, x_n) \in \mathbf{R}^n \mid x_1 \geq x_2 \geq \ldots \geq x_n\} \cap H_n$$

where

$$H_n = \left\{ (x_1, \ldots, x_n) \in \mathbf{R}^n \ \middle| \ \sum_{i=1}^n x_i = 0 \right\}.$$

Let p_t^0 be the transition probability semi-group of Brownian motion killed at the boundary of the cone $\mathcal{C}$. This cone is a fundamental domain for the action of the symmetric group S_n, which acts by permutation of coordinates on $\mathbf{R}^n$. Using the reflexion principle, one shows easily that

$$p_t^0(x, y) = \sum_{\sigma \in S_n} \epsilon(\sigma) p_t(x, \sigma(y)) \qquad x, y \in \mathcal{C}$$

where $p_t(x, y) = (2\pi t)^{-(n-1)/2} e^{-|x-y|^2/2t}$ and $\epsilon(\sigma)$ is the signature of σ. Let h be the function

$$h(x) = \prod_{i>j}(x_i - x_j).$$

Proposition 1. *The function h is the unique (up to a positive multiplicative constant) positive harmonic function for the semigroup p_t^0, on the cone $\mathcal{C}$, which vanishes on the boundary.*

The harmonic function h corresponds to the unique point at infinity in the Martin compactification of $\mathcal{C}$. Consider now the Doob's transform of p_t^0, which is the semigroup given by

$$q_t(x,y) = \frac{h(y)}{h(x)}p_t^0(x,y).$$

It is a diffusion semigroup on $\mathcal{C}$ with infinitesimal generator

$$\frac{1}{2}\Delta + \langle \nabla \log h, \nabla \cdot \rangle.$$

Proposition 2. *The eigenvalue process of a traceless Hermitian Brownian motion is a Markov diffusion process in the cone $\mathcal{C}$, with semigroup q_t.*

We can summarize the last proposition by saying that the eigenvalue process is a Brownian motion in $\mathcal{C}$, conditioned (in Doob's sense) to exit the cone at infinity.

2.2 The Group $SL_n(\mathbf{C})$

This is the group of complex invertible matrices of size $n \times n$, with determinant 1. Its Lie algebra is the space $\mathfrak{sl}_n(\mathbf{C})$ of complex traceless matrices. Consider the Hermitian form

$$\langle A, B \rangle = Tr(AB^*)$$

on $\mathfrak{sl}_n(\mathbf{C})$ which is invariant by left and right action of the unitary subgroup $SU(n)$. This Hermitian form determines a unique Brownian motion with values in $\mathfrak{sl}_n(\mathbf{C})$. The Brownian motion g_t, on $SL_n(\mathbf{C})$, is the stochastic exponential of this Brownian motion, solution to the Stratonovich stochastic differential equation

$$dg_t = g_t dw_t$$

where w_t is a Brownian motion in $\mathfrak{sl}_n(\mathbf{C})$.

There are two remarkable decompositions of $SL_n(\mathbf{C})$, the Iwasawa and Cartan decompositions. The first one is $SL_n(\mathbf{C}) = NAK$ where K is the compact group $SU(n)$, A is the group of diagonal matrices with positive coefficients, and determinant one, and N is the nilpotent group of upper triangular matrices with 1's on the diagonal. Each matrix of $SL_n(\mathbf{C})$ has a unique decomposition as a product $g = nak$ of elements of the three subgroups N, A, K. This can be easily inferred from the Gram-Schmidt orthogonalization process. If g_t is a Brownian motion in $SL_n(\mathbf{C})$, one can consider its components n_t, a_t, k_t. In particular, denoting by $(e^{w_1(t)}, \dots, e^{w_n(t)})$ the diagonal components of a_t the following holds (cf [15]).

Proposition 3. *The process $(w_1(t), \dots, w_n(t))$ is a Brownian motion with a drift $\rho = (-n+1, -n+3, \dots, n-1)$ in the subspace H_n.*

The other decomposition is the Cartan decomposition $SL_n(\mathbf{C}) = KA^+K$, where A^+ is the part of A consisting of matrices with positive nonincreasing coefficients along the diagonal. In order to get the Cartan decomposition of a matrix $g \in SL_n(\mathbf{C})$, take its polar decomposition $g = ru$ with r positive Hermitian, and u unitary, then diagonalize r which yields $g = vav'$ with v and v' unitary and a diagonal, with positive real coefficients which can be put in nonincreasing order along the diagonal. These coefficients are the singular values of the matrix g. This decomposition is not unique since the diagonal subgroup of $SU(n)$ commutes with A, but the singular values are uniquely defined. Call $(e^{a_1(t)}, \ldots, e^{a_n(t)})$ the singular values of the Brownian motion g_t, with $a_1 \geq a_2 \geq \ldots \geq a_n$. They form a process with values in the cone $\mathcal{C}$. Let us mention that this stochastic process can also be interpreted as the radial part of a Brownian motion with values in the symmetric space $SL_n(\mathbf{C})/SU(n)$. We will now give for the motion of singular values a similar description as the one of eigenvalues of the Hermitian Brownian motion. For this, consider a Brownian motion in H_n, with drift ρ, killed at the exit of the cone $\mathcal{C}$. This process has a semigroup given by

$$p_t^{0,\rho}(x, y) = e^{\langle \rho, y-x \rangle - t\langle \rho, \rho \rangle / 2} p_t^0(x, y).$$

Proposition 4. *The function*

$$h^\rho(y) = \prod_{i>j}(1 - e^{2(y_j - y_i)})$$

is a positive harmonic function for the semigroup $p_t^{0,\rho}$, in the cone $\mathcal{C}$, and vanishes at the boundary of the cone.

It is not true that this function is the unique positive harmonic function on the cone; indeed the Martin boundary at infinity is now much larger and contains a point for each direction inside the cone, see [8]. The Doob-transformed semigroup

$$q_t^\rho(x, y) = \frac{h^\rho(y)}{h^\rho(x)} p_t^{0,\rho}(x, y)$$

is a Markov diffusion semigroup in the cone $\mathcal{C}$, with infinitesimal generator

$$\frac{1}{2}\Delta + \langle \rho, \cdot \rangle + \langle \nabla \log h^\rho, \nabla \cdot \rangle.$$

Note that it can also be expressed as

$$\frac{1}{2}\Delta + \langle \nabla \log \tilde{h}^\rho, \nabla \cdot \rangle$$

with

$$\tilde{h}^\rho(y) = \prod_{i>j} \sinh(y_j - y_i)$$

(see[10]).

Proposition 5. *The logarithms of the singular values of a Brownian motion in $SL_n(\mathbf{C})$ perform a diffusion process in the cone $\mathcal{C}$, with semigroup q_t^ρ.*

As in the preceding case, we can summarize by saying that the process of singular values is a Brownian motion with drift ρ in the cone $\mathcal{C}$, conditioned (in Doob's sense) to exit the cone at infinity, in the direction ρ.

2.3 Unitary Matrices

The Brownian motion with values in $SU(n)$ is obtained by taking the stochastic exponential of a Brownian motion in the Lie algebra of traceless anti-Hermitian matrices, endowed with the Hermitian form

$$\langle A, B \rangle = -Tr(AB).$$

Let $e^{i\theta_1}, \ldots, e^{i\theta_n}$ be the eigenvalues of a matrix in $SU(n)$, which can be chosen so that $\sum_i \theta_i = 0$, and $\theta_1 \geq \theta_2 \geq \ldots \geq \theta_n$, $\theta_1 - \theta_n \leq 2\pi$. These conditions determine a simplex Δ_n in H_n, which is a fundamental domain for the action of the affine Weyl group on H_n. Recall that the affine Weyl group $\tilde{W}$ is the semidirect product of the symmetric group S_n, which acts by permutation of coordinates in H_n, and of the group of translations by elements of the lattice $(2\pi \mathbf{Z})^n \cap H_n$.

One can use the reflexion principle again to compute the semigroup of Brownian moton in this simplex killed at the boundary. One gets an alternating sum over the elements of $\tilde{W}$,

$$p_t^0(\theta, \xi) = \sum_{w \in \tilde{W}} \epsilon(w) p_t(\theta, w(\xi)).$$

The infinitesimal generator is $1/2 \times$ the Laplacian in the simplex, with Dirichlet boundary conditions. It is well known that this operator has a compact resolvent, and its eigenvalue with smallest module is simple, with an eigenfunction which can be chosen positive. Consider the function

$$h^u(\theta) = \prod_{j > k} (e^{i\theta_j} - e^{i\theta_k}).$$

Proposition 6. *The function h^u is positive inside the simplex Δ_n, it vanishes on the boundary, and it is the eigenfunction corresponding to the Dirichlet eigenvalue with smallest module on Δ_n. This eigenvalue is $\lambda = (n - n^3)/6$.*

The Doob-transformed semigroup

$$q_t^u(x, y) = \frac{h^u(y)}{h^u(x)} e^{-\lambda t} p_t^0(x, y)$$

is a Markov diffusion semigroup in Δ_n, with infinitesimal generator

$$\frac{1}{2} \Delta + \langle \nabla \log h^u, \nabla \cdot \rangle - \lambda.$$

Proposition 7. *The process of eigenvalues of a unitary Brownian motion is a diffusion with values in Δ_n with probability transition semigroup q_t^u.*

Again a good summary of this situation is that the motion of eigenvalues is that of a Brownian motion in the simplex Δ_n conditioned to stay forever in this simplex.

2.4 The Case of Rank 1

In the next section we will need the simplest case, that of 2×2 matrices. Consider first the case of Hermitian matrices. The process of eigenvalues is essentially a Bessel process of dimension 3, with infinitesimal generator

$$\frac{1}{2}\frac{d^2}{dx^2} + \frac{1}{x}\frac{d}{dx} \qquad \text{on} \quad]0, +\infty[,$$

obtained from Brownian motion killed at zero, of infinitesimal generator

$$\frac{1}{2}\frac{d^2}{dx^2}$$

with Dirichlet boundary condition at 0, by a Doob transform with the positive harmonic function $h(x) = x$.

In the case of the group $SL_2(\mathbf{C})$, or the symmetric space $SL_2(\mathbf{C})/SU(2)$, which is the hyperbolic space of dimension 3, the radial process has infinitesimal generator

$$\frac{1}{2}\frac{d^2}{dx^2} + \coth x \frac{d}{dx}$$

obtained from Brownian motion with a drift

$$\frac{1}{2}\frac{d^2}{dx^2} + \frac{d}{dx}$$

with Dirichlet boundary condition at 0, by a Doob's transform with the function $1 - e^{-2x}$.

Finally the last case is Brownian motion in $SU(2)$, where the eigenvalue process takes values in $[0, \pi]$ and has an infinitesimal generator

$$\frac{1}{2}\frac{d^2}{dx^2} + \cot x \frac{d}{dx}$$

obtained by a Doob transform at the bottom of the spectrum from

$$\frac{1}{2}\frac{d^2}{dx^2}$$

on $[0, \pi]$ with Dirichlet boundary conditions at 0 and π, by the function $\sin(x)$.

For these last two examples, we shall write a spectral decomposition of the generator $L_i, i = 1, 2$, of the form

$$f(x) = \int \Phi_\lambda^i(x) \left[\int \Phi_\lambda^i(y) f(y) \, dm_i(y) \right] d\nu_i(\lambda) \qquad i = 1, 2 \qquad (1)$$

for every $f \in L^2(m_i)$, where m_i is measure for which L_i is selfadjoint in $L^2(m_i)$, and the functions Φ_λ^i are solutions to

$$L_i \Phi_\lambda^i + \lambda \Phi_\lambda^i = 0$$

and ν_i is a spectral measure for L_i.

For $L_1 = \frac{1}{2} \frac{d^2}{dx^2}$ on $[0, \pi]$ with Dirichlet boundary conditions, $m_1(dx)$ is Lebesgue measure on $[0, \pi]$, and L_1 is selfadjoint on $L^2([0, \pi])$. Furthermore

$$\Phi_\lambda^1(x) = \sin(\sqrt{2\lambda} x)$$

and

$$\nu_1(d\lambda) = \frac{1}{\pi} \sum_{n=1}^\infty \delta_{n^2/2}(d\lambda). \qquad (2)$$

For $L_2 = \frac{1}{2} \frac{d^2}{dx^2} + \frac{d}{dx}$ on $[0, +\infty[$, the measure $m_2(dx) = e^{2x} dx$, and

$$\Phi_\lambda^2(x) = e^{-x} \sin(\sqrt{2\lambda - 1} x) \qquad \lambda > 1/2.$$

The spectral measure is

$$\nu_2(d\lambda) = \frac{1}{\pi\sqrt{2\lambda - 1}} d\lambda \qquad \lambda > 1/2 \qquad (3)$$

on $[1, +\infty[$. Of course formulas (1), (2), (3) are immediate consequences of ordinary Fourier analysis.

Note that the spectral decompositions, and in particular the measures ν_i, depend on the normalisation of the fonctions Φ_λ. We have made a natural choice, but it does not coincide with the usual normalisation of Weyl-Titchmarsh-Kodaira theory, see [6].

3 MacDonald's Function and Riemann's ξ Function

3.1 Pólya's Paper

In his paper [14], Pólya starts from Riemann's ξ function

$$\xi(s) = s(s - 1)\pi^{-s/2} \Gamma(s/2) \zeta(s)$$

where ζ is Riemann's zeta function. Then ξ is an entire function whose zeros are exactly the nontrivial zeros of ζ.

Putting $s = 1/2 + iz$ yields

$$\xi(z) = 2 \int_0^\infty \Phi(u) \cos(zu) \, du$$

with

$$\Phi(u) = 2\pi e^{5u/2} \sum_{n=1}^{\infty} (2\pi e^{2u} n^2 - 3) n^2 e^{-\pi n^2 e^{2u}} \tag{4}$$

and the function Φ is even, as follows from the functional equation for Jacobi θ function; furthermore

$$\Phi(u) \sim 4\pi^2 e^{9u/2 - \pi e^{2u}} \qquad u \to +\infty$$

so that

$$\Phi(u) \sim 4\pi^2 (e^{9u/2} + e^{-9u/2}) e^{-\pi(e^{2u} + e^{-2u})} \qquad u \to \pm\infty.$$

This lead Pólya to define a "falsified" ξ function

$$\xi^*(z) = 8\pi^2 \int_0^{\infty} (e^{9u/2} + e^{-9u/2}) e^{-\pi(e^{2u} + e^{-2u})} \cos(zu) du.$$

The main result of [14] is

Theorem 1. *The function ξ^* is entire, its zeros are real and simple. Let $N(r)$, (resp. $N^*(r)$) denote the number of zeros of $\xi(z)$ (resp. $\xi^*(z)$) with real part in the interval $[0, r]$, then $N(r) - N^*(r) = O(\log r)$.*

Recall that the same assertion about the zeros of the function ξ (without the statement about simplicity, beware also that $s = 1/2 + iz$) is Riemann's hypothesis. Recall also the well known estimate

$$N(r) = \frac{r}{2\pi} \log(r/2\pi) - \frac{r}{2\pi} + O(1).$$

Pólya's results rely on the intermediate study of the function

$$\mathfrak{G}(z, a) = \int_{-\infty}^{\infty} e^{-a(e^u + e^{-u}) + zu} du$$

from which ξ^* is obtained by

$$\xi^*(z) = 2\pi^2 (\mathfrak{G}(iz/2 - 9/4, \pi) + \mathfrak{G}(iz/2 + 9/4, \pi)).$$

Pólya shows that $\mathfrak{G}(z, a)$ has only purely imaginary zeros, (as a function of z) and the number of these zeros with imaginary part in $[0, r]$ grows as $\frac{r}{\pi} \log \frac{r}{a} - \frac{r}{\pi} + O(1)$. The results on ξ^* are then deduced through a nice lemma which played a role in the history of statistical mechanics (the Lee-Yang theorem on Ising model), as revealed by M. Kac [11]. We shall now concentrate on $\mathfrak{G}(z, a)$. In particular, for $a = \pi$, the function $\tilde{\xi}(z) = \mathfrak{G}(iz/2, \pi)$ is another approximation of ξ which has many interesting structural properties.

3.2 MacDonald Functions

The function denoted $\mathfrak{G}(z,a)$ by Pólya is actually a Bessel function. Indeed, MacDonald's function, also called modified Bessel function (see e.g. [1]), given by

$$K_\mu(x) = \int_0^\infty t^{\mu-1} e^{-\frac{x}{2}(t+t^{-1})} dt \qquad x > 0,\ \mu \in \mathbf{C}.$$

satisfies $K_z(2x) = \mathfrak{G}(z,x)$. The function $\mathfrak{G}(z,a)$ is therefore essentially a MacDonald function, as noted by Pólya. MacDonald function is an even function of μ and satisfies

$$\frac{2\mu}{x} K_\mu(x) = K_{\mu+1}(x) - K_{\mu-1}(x) \tag{5}$$

$$-2\frac{d}{dx} K_\mu(x) = K_{\mu+1}(x) + K_{\mu-1}(x) \tag{6}$$

The first of these equations is used by Pólya in a very clever way to prove that the zeros (in z) of $\mathfrak{G}(z,x)$ are purely imaginary.

3.3 Spectral Interpretation of the Zeros

From (5), (6)

$$(\tfrac{\mu}{x} - \tfrac{d}{dx}) K_\mu = K_{\mu+1}$$

$$(-\tfrac{\mu}{x} - \tfrac{d}{dx}) K_\mu = K_{\mu-1}$$

from which one gets

$$K_\mu = (-\tfrac{\mu+1}{x} - \tfrac{d}{dx})(\tfrac{\mu}{x} - \tfrac{d}{dx}) K_\mu$$

$$= (\tfrac{d^2}{dx^2} + \tfrac{1}{x}\tfrac{d}{dx} - \tfrac{\mu^2}{x^2}) K_\mu \ .$$

This differential equation will give us a spectral interpretation of the zeros of $\mathfrak{G}(z,x)$. Change variable by $\psi_\mu(x) = K_\mu(e^x)$ to get

$$(-\frac{d^2}{dx^2} + e^{2x})\psi_\mu = -\mu^2\psi_\mu \tag{7}$$

Since K_μ vanishes exponentially at infinity, the spectral theory of Sturm-Liouville operators on the half-line (see e.g. [6], [12]) implies that the squares of the zeros of $\mu \mapsto \psi_\mu(y)$ are the eigenvalues of $\frac{d^2}{dx^2} - e^{2x}$ on the interval $[y, +\infty[$ with the Dirichlet boundary condition at y, the functions ψ_μ being the eigenfunctions. Since this operator is selfadjoint and negative the zeros are purely imaginary, and are simple.

This spectral interpretation of the zeros of MacDonald function is well known [17], I do not know why Pólya does not mention it.

3.4 $H = xp$

Equation (7) can be put into Dirac's form, indeed the equations

$$\left(\tfrac{d}{dx} + \tfrac{1}{2} + e^x\right) f = \gamma g$$

$$\left(-\tfrac{d}{dx} + \tfrac{1}{2} + e^x\right) g = \gamma f$$

imply

$$\left(-\frac{d^2}{dx^2} + e^{2x}\right) f = \left(\gamma^2 - \frac{1}{4}\right) f.$$

Using the change of variables $u = e^x$, we get

$$\left(u\tfrac{d}{du} + \tfrac{1}{2} + u\right) f = \gamma g$$

$$\left(-u\tfrac{d}{du} + \tfrac{1}{2} + u\right) g = \gamma f.$$

Remark that this Dirac system yields a perturbation of the Hamiltonian $H = xp$ considered by Berry et Keating [3], in relation with Riemann's zeta function.

3.5 Asymptotics of the Zeros

General results on Sturm-Liouville operators allow one to recover the asymptotic behaviour of the spectrum, thanks to a semiclassical analysis, see e.g. [12]. One can get a more precise result using the integral representation of $K_{i\mu}$. Pólya gives the asymptotic estimate

$$K_{x+iy}(2a) = \frac{1}{\sqrt{2\pi y}} e^{-\frac{\pi}{2}y + i\frac{\pi}{2}x} \left[\left(\frac{y}{a}\right)^x e^{i\Phi} + \left(\frac{y}{a}\right)^{-x} e^{-i\Phi}\right] + O(e^{-\frac{\pi}{2}y} y^{|x|-3/2})$$

in the strip $|x| \leq 1$ uniformly as $y \to \infty$, where

$$\Phi = y \log \frac{y}{a} - y - \frac{\pi}{4}.$$

This estimate can be obtained by the stationary phase method, writing

$$K_z(2a) = \int_{-\infty}^{\infty} e^{zt - 2a \cosh(t)} dt.$$

Making a contour deformation we get

$$K_z(2a) = \int_{-\infty}^{-A} e^{zt - 2a \cosh(t)} dt + i \int_0^{\pi/2} e^{z(-A+it) - 2a \cosh(-A+it)} dt$$
$$+ \int_{-A}^{A} e^{z(t+i\frac{\pi}{2}) - 2ai \sinh(t)} dt - i \int_0^{\pi/2} e^{z(A-it+i\frac{\pi}{2}) - 2a \cosh(A-it+i\frac{\pi}{2})} dt$$
$$+ \int_A^{\infty} e^{zt - 2a \cosh(t)} dt$$

and Pólya's estimate can be obtained by standard methods, which give also estimates for the derivatives of MacDonald's function. Finally the zeros of $y \to K_{iy}(2a)$ behave like the solutions to

$$y \log \frac{y}{a} - y - \frac{\pi}{4} = (n + \frac{1}{2})\pi \qquad n \text{ integer}$$

The number of zeros with imaginary part in $[0, T]$ is thus $\frac{T}{\pi} \log \frac{T}{a} - \frac{T}{\pi} + O(1)$.

4 Probabilistic Interpretations

We will now give interpretations of the functions ξ and $\tilde{\xi}$ using first passage times of diffusions.

4.1 Brownian Motion with a Drift

The first passage time at $x > 0$ of Brownian motion started at 0 follows a $1/2$ stable distribution i.e.,

$$P(T_x \in dt) = x \frac{e^{-\frac{x^2}{2t}}}{\sqrt{2\pi t^3}} dt$$

with Laplace transform

$$E[e^{-\lambda^2 T_x/2}] = e^{-\lambda x} \ .$$

Adding a drift $a > 0$ to the Brownian motion gives a first passage distribution

$$P^a(T_x \in dt) = x \frac{e^{-\frac{x^2}{2t}}}{\sqrt{2\pi t^3}} e^{ax - \frac{a^2 t}{2}} dt$$

with Laplace transform

$$E^a[e^{-\lambda^2 T_x/2}] = e^{-x\sqrt{\lambda^2 + a^2} + ax}.$$

This is a generalized inverse Gaussian distribution. In particular, its Mellin transform is

$$E^a[T_x^s] = (x/a)^s \frac{K_{-1/2+s}(ax)}{K_{-1/2}(ax)} = (x/a)^s \sqrt{\pi/ax}\, e^{ax} K_{-1/2+s}(ax)$$

which gives a probabilistic interpretation of MacDonald's function (as a function of s) as a Mellin transform of a probability distribution.

4.2 Three Dimensional Bessel Process

There exists a similar interpretation of the ξ function, which is discussed in details in [4], [5], for example, Consider the first passage time at $a > 0$ of a three dimensional Bessel process (i.e., the norm of a three dimensional Brownian motion) starting from 0. The Laplace transform of this hitting time is

$$E[e^{-\frac{\lambda^2}{2}S_a}] = \frac{\lambda a}{\sinh \lambda a} .$$

Let S'_a be an independent copy of S_a, and let

$$W_a = S_a + S'_a ;$$

then the density of the distribution of W_a is obtained by inverting the Laplace transform. One gets

$$P(W_a \in dx) = \sum_{n=1}^{\infty} (\pi^4 n^4 x/a^4 - 3\pi^2 n^2/a^2)e^{-\pi^2 n^2 x/2a^2} dx$$

from which one can compute the Mellin transform

$$E[W_a^s] = 2(2a^2/\pi)^s \xi(2s).$$

The function 2ξ thus has a probabilistic interpretation, as Mellin transform of $\sqrt{\frac{\pi}{2}W_1}$.

4.3 Infinite Divisibility

The distributions of T_x and W_a are infinitely divisible. Indeed

$$\log E^a[\exp(-\tfrac{\lambda^2}{2}T_x)] = -x\sqrt{\lambda^2 + a^2} + ax$$
$$= x \int_0^{\infty} (e^{-\frac{\lambda^2}{2}t} - 1)\frac{e^{-\frac{a^2}{2}t}}{\sqrt{2\pi t^3}} dt$$

which shows that T_x is a subordinator with Lévy measure

$$\frac{e^{-\frac{a^2}{2}t}}{\sqrt{2\pi t^3}} dt.$$

Similarly

$$\log E[\exp(-\tfrac{\lambda^2}{2}W_a)] = 2\log(\lambda a/\sinh(\lambda a))$$
$$= 2 \int_0^{\infty} (e^{-\frac{\lambda^2}{2}t} - 1)\sum_{n=1}^{\infty} e^{-\pi^2 n^2 t/a^2} dt$$

therefore the variable W_a has the distribution of a subordinator, with Lévy measure

$$2\sum_{n=1}^{\infty} e^{-\pi^2 n^2 t/a^2} dt,$$

taken at time 1. Observe however that the process $(W_a)_{a \geq 0}$ is not a subordinator.

4.4 Generalized Gamma Convolution

The gamma distributions are

$$P(\gamma_{\omega,c} \in dt) = \frac{c^{-\omega}}{\Gamma(\omega)} t^{\omega-1} e^{-t/c} dt = \Gamma_{\omega,c}(dt)$$

where ω and c are > 0 parameters. The Laplace transform is

$$E[e^{-\lambda\gamma_{\omega,c}}] = (1 + \lambda/c)^{-\omega}.$$

The gamma distributions form a convolution semigroup with respect to the parameter ω, i.e.,

$$\Gamma_{\omega_1,c} * \Gamma_{\omega_2,c} = \Gamma_{\omega_1+\omega_2,c}.$$

The Lévy exponent of the gamma semigroup is

$$\psi_c(\lambda) = \log(1 + \lambda/c) = \int_0^\infty (1 - e^{-\lambda t}) \frac{e^{-ct}}{t} dt$$

so that this is the semigroup of a subordinator with Lévy measure $e^{-ct}/t\, dt$.

The generalized gamma convolutions are the distributions of linear combinations, with positive coefficients, of independent gamma variables, and their weak limits.

One can also characterize the generalized gamma convolutions as the infinitely divisible distributions with a Lévy exponent of the form

$$\psi(\lambda) = \int_0^\infty \psi_c(\lambda) d\nu(c)$$

for some positive measure ν which integrates $1/c$ at ∞. This measure is called the Thorin measure of the generalized gamma distribution. The variables T_x and W_a of the preceding paragraph are generalized gamma convolutions. Indeed it is easy to check, using the computations of section 4.3, that W_a has a generalized gamma convolution as distribution, with Thorin measure

$$\nu(dc) = 2 \sum_{n=1}^\infty \delta_{n^2/a^2}(dc) ; \tag{8}$$

whereas T_x is distributed as a generalized gamma convolution with Thorin measure

$$\nu(dc) = \frac{dc}{\pi\sqrt{c - a^2/2}} \qquad c > a^2/2 \tag{9}$$

since

$$\frac{e^{-a^2 t/2}}{\sqrt{\pi t^3}} = \int_{a^2/2}^\infty e^{-ct} \frac{dc}{\pi\sqrt{c - a^2/2}}.$$

4.5 Final Remarks

We can now make a connection between the preceding considerations and those of the first part of the paper. Indeed, the Thorin measures associated with the variables T_x and W_a can be expressed as spectral measures associated with the generators of Brownian motion on matrix spaces. The hitting times of Brownian motion with drift are related with the radial part of Brownian motion in the symmetric space $SL_2(\mathbf{C})/SU(2)$, whereas the hitting times of the Bessel three process are related with the Brownian motion on the unitary group $SU(2)$. The precise relations are contained in formulas (2), (3), (8), (9). Thus the Riemann ξ function, which is the Mellin transform of a hitting time of the Bessel three process, as in section 4.2, and the Polya $\tilde{\xi}$ function from section 3.1, which appears as Mellin transform of hitting time of Brownian motion with drift, are related in this non obvious way.

References

1. Andrews, G. E.; Askey, R.; Roy, R.: Special functions. Encyclopedia of Mathematics and its Applications, 71. Cambridge University Press, Cambridge, 1999.
2. Babillot, M.: A probabilistic approach to heat diffusion on symmetric spaces. J. Theoret. Probab. 7 (1994), no. 3, 599–607
3. Berry, M. V.; Keating, J. P.: The Riemann zeros and eigenvalue asymptotics. SIAM Rev. 41 (1999), no. 2, 236–266.
4. Biane, P.: La fonction zêta de Riemann et les probabilités. La fonction zêta, 165–193, Ed. Éc. Polytech., Palaiseau, 2003.
5. Biane, P.; Pitman, J.; Yor, M.: Probability laws related to the Jacobi theta and Riemann zeta functions, and Brownian excursions. Bull. Amer. Math. Soc. (N.S.) 38 (2001), no. 4, 435–465.
6. Coddington, E. A.; Levinson, N.: Theory of ordinary differential equations. McGraw-Hill Book Company, Inc., New York-Toronto-London, 1955.
7. Dyson, F. J.: A Brownian-motion model for the eigenvalues of a random matrix. J. Mathematical Phys. 3 1962 1191–1198.
8. Guivarc'h, Y.; Ji, L.; Taylor, J. C.: Compactifications of symmetric spaces. Progress in Mathematics, 156. Birkhäuser Boston, Inc., Boston, MA, 1998.
9. Helgason, S.: Groups and geometric analysis. Integral geometry, invariant differential operators, and spherical functions. Corrected reprint of the 1984 original. Mathematical Surveys and Monographs, 83. American Mathematical Society, Providence, RI, 2000.
10. Jones, L.; O'Connell, N.: Weyl chambers, symmetric spaces and number variance saturation. ALEA Lat. Am. J. Probab. Math. Stat. 2 (2006), 91–118
11. Kac M.: Comments on [93] Bemerkung über die Intergraldarstellung der Riemannsche ξ-Funktion, in Pólya, G.: Collected papers. Vol. II: Location of zeros. Edited by R. P. Boas. Mathematicians of Our Time, Vol. 8. The MIT Press, Cambridge, Mass.-London, 1974.
12. Levitan, B. M.; Sargsjan, I. S.: Introduction to spectral theory: selfadjoint ordinary differential operators. Translated from the Russian by Amiel Feinstein.

Translations of Mathematical Monographs, Vol. 39. American Mathematical Society, Providence, R.I., 1975.

13. Malliavin, M.-P.; Malliavin, P.: Factorisations et lois limites de la diffusion horizontale au-dessus d'un espace riemannien symétrique. Théorie du potentiel et analyse harmonique, pp. 164–217. Lecture Notes in Math., Vol. 404, Springer, Berlin, 1974.

14. Pólya, G.: Bemerkung über die Integraldarstellung der Riemannschen ζ-Funktion. Acta Math. 48 (1926), no. 3-4, 305–317.

15. Taylor, J. C.: The Iwasawa decomposition and the limiting behaviour of Brownian motion on a symmetric space of noncompact type. Geometry of random motion (Ithaca, N.Y., 1987), 303–332, Contemp. Math., 73, Amer. Math. Soc., Providence, RI, 1988.

16. Taylor, J. C.: Brownian motion on a symmetric space of noncompact type: asymptotic behaviour in polar coordinates. Canad. J. Math. 43 (1991), no. 5, 1065–1085.

17. Titchmarsh, E. C.: Eigenfunction Expansions Associated with Second-Order Differential Equations. Oxford, at the Clarendon Press, 1946.

On the Laws of First Hitting Times of Points for One-dimensional Symmetric Stable Lévy Processes

Kouji Yano[1], Yuko Yano[2], and Marc Yor[3,4,2]

[1] Department of Mathematics, Graduate School of Science, Kobe University, Kobe, Japan.
  E-mail: kyano@math.kobe-u.ac.jp
[2] Research Institute for Mathematical Sciences, Kyoto University, Kyoto, Japan.
[3] Laboratoire de Probabilités et Modèles Aléatoires, Université Paris VI, Paris, France.
[4] Institut Universitaire de France.

Summary. Several aspects of the laws of first hitting times of points are investigated for one-dimensional symmetric stable Lévy processes. Itô's excursion theory plays a key role in this study.

Keywords: Symmetric stable Lévy process, excursion theory, first hitting times.

1 Introduction

For one-dimensional Brownian motion, the laws of several random times, such as first hitting times of points and intervals, can be expressed explicitly in terms of elementary functions. Moreover, these laws are infinitely divisible (abbrev. as (ID)), and in fact, self-decomposable (abbrev. as (SD)).

The aim of the present paper is to study various aspects of the laws of first hitting times of points and last exit times for one-dimensional symmetric stable Lévy processes. We shall put some special emphasis on the following objects: (i) the laws of the ratio of two independent gamma variables, which, as is usual, we call *beta variables of the second kind*; (ii) harmonic transform of Itô's measure of excursions away from the origin. The present study is motivated by a recent work [49] by the authors about penalisations of symmetric stable Lévy paths.

The organisation of the present paper is as follows. In Section 2, we recall several facts concerning beta and gamma variables and their variants. In Section 3, we briefly recall Itô's excursion theory and make some discussions

C. Donati-Martin et al. (eds.), *Séminaire de Probabilités XLII*,
Lecture Notes in Mathematics 1979, DOI 10.1007/978-3-642-01763-6_8,
© Springer-Verlag Berlin Heidelberg 2009

about last exit times. In Section 4, we consider harmonic transforms of symmetric stable Lévy processes, which play an important role in our study. In Section 5, we discuss the laws of first hitting times of single points and last exit times for symmetric stable Lévy processes. In Section 6, we discuss the laws of those random times for the absolute value of symmetric stable Lévy processes, which include reflecting Brownian motion as a special case.

2 Preliminaries: Several Important Random Variables

2.1 Generalized Gamma Convolutions

For $a > 0$, we write $\mathcal{G}_a$ for a gamma variable with parameter a:

$$P(\mathcal{G}_a \in dx) = \frac{1}{\Gamma(a)} x^{a-1} e^{-x} dx, \qquad x > 0. \tag{2.1}$$

As a rather general framework, we recall the class of *generalized gamma convolutions* (abbrev. as (GGC)), which is an important subclass of (SD); namely,

$$(\text{GGC}) \subset (\text{SD}) \subset (\text{ID}). \tag{2.2}$$

A nice reference for details is the monograph [7] by Bondesson. A recent survey can be found in James–Roynette–Yor [26].

A random variable X is said to be *of (GGC) type* if it is a weak limit of linear combinations of independent gamma variables with positive coefficients.

Theorem 2.1 (See, e.g., [7, Thm.3.1.1]) *A random variable X is of (GGC) type if and only if there exist a non-negative constant a and a non-negative measure $U(dt)$ on $(0, \infty)$ with*

$$\int_{(0,1]} |\log t| U(dt) < \infty \qquad and \qquad \int_{(1,\infty)} \frac{1}{t} U(dt) < \infty \tag{2.3}$$

such that

$$E[e^{-\lambda X}] = \exp\left\{ -a\lambda - \int_{(0,\infty)} \log(1 + \lambda/t) U(dt) \right\}, \qquad \lambda > 0. \tag{2.4}$$

In what follows we shall call $U(dt)$ the *Thorin measure* associated with X. By simple calculations, it follows that

$$E[e^{-\lambda X}] = \exp\left\{ -a\lambda + \int_0^\infty \left(\frac{1}{s+\lambda} - \frac{1}{s} \right) U((0,s)) ds \right\} \tag{2.5}$$

$$= \exp\left\{ -a\lambda - \int_0^\infty (1 - e^{-\lambda u}) \frac{1}{u} \left(\int_{(0,\infty)} e^{-ut} U(dt) \right) du \right\}. \tag{2.6}$$

In particular, the following holds: *The law of X is of (ID) type and its Lévy measure has a density given by* $n(u) := \frac{1}{u} \int_{(0,\infty)} e^{-ut} U(dt)$. *Since* $un(u)$ *is non-increasing, the law of X is of (SD) type.*

Theorem 2.2 (see, e.g., [7, Thm.4.1.1 and 4.1.4]) *Suppose that X is of (GGC) type and that $a = 0$ and $b := U((0, \infty)) < \infty$. Then X may be represented as $X \stackrel{\text{law}}{=} \mathcal{G}_b Y$ for some random variable Y independent of $\mathcal{G}_b$. The total mass of the Thorin measure is given by*

$$b = \sup \left\{ p \geq 0 : \lim_{x \to 0+} \frac{\rho(x)}{x^{p-1}} = 0 \right\} \tag{2.7}$$

where ρ is the density of the law of X with respect to the Lebesgue measure:

$$\rho(x) = \frac{1}{\Gamma(b)} x^{b-1} E\left[\frac{1}{Y^b} \exp\left(-\frac{x}{Y} \right) \right]. \tag{2.8}$$

Remark 2.3 *For a given X, the law of the variable Y which represents X as in Theorem 2.2 is unique; in fact, the gamma distribution is simplifiable (see [11, Sec.1.12]).*

Remark 2.4 *We do not know how to characterise explicitly the class of possible Y's which represent variables of (GGC) type as in Theorem 2.2. As a partial converse, Bondesson (see [7, Thm.6.2.1]) has introduced a remarkable class which is closed under multiplication of independent gamma variables.*

2.2 Beta and Gamma Variables

We introduce notations and recall several basic facts concerning the beta and gamma variables. See [11, Chap.4] for details. For $a, b > 0$, we write $\mathcal{B}_{a,b}$ for a beta variable (of the first kind) with parameters a, b:

$$P(\mathcal{B}_{a,b} \in dx) = \frac{1}{B(a,b)} x^{a-1} (1-x)^{b-1} dx, \qquad 0 < x < 1 \tag{2.9}$$

where $B(a, b)$ is the beta function:

$$B(a, b) = \frac{\Gamma(a)\Gamma(b)}{\Gamma(a+b)}. \tag{2.10}$$

Note that $\mathcal{B}_{a,b} \stackrel{\text{law}}{=} 1 - \mathcal{B}_{b,a}$ for $a, b > 0$ and that $\mathcal{B}_{a,1} \stackrel{\text{law}}{=} \mathcal{U}^{\frac{1}{a}}$ for $a > 0$ where $\mathcal{U}$ is a uniform variable on $(0, 1)$. The following identity in law is well-known: For any $a, b > 0$,

$$\left(\mathcal{G}_a, \widehat{\mathcal{G}_b} \right) \stackrel{\text{law}}{=} \left(\mathcal{B}_{a,b} \mathcal{G}_{a+b}, (1 - \mathcal{B}_{a,b}) \mathcal{G}_{a+b} \right), \tag{2.11}$$

or equivalently,

$$\left(\mathcal{G}_a + \widehat{\mathcal{G}_b}, \frac{\mathcal{G}_a}{\mathcal{G}_a + \widehat{\mathcal{G}_b}} \right) \stackrel{\text{law}}{=} \left(\mathcal{G}_{a+b}, \mathcal{B}_{a,b} \right) \tag{2.12}$$

where on the left hand side $\mathcal{G}_a$ and $\widehat{\mathcal{G}}_b$ are independent and on the right hand side $\mathcal{B}_{a,b}$ and $\mathcal{G}_{a+b}$ are independent. The proof is elementary; it can be seen in [11, (4.2.1)], and so we omit it.

Using formula (2.11), we obtain another expression of the Thorin measure of a variable of (GGC) type.

Theorem 2.5 *Under the same assumption as in Theorem 2.2, the total mass of the Thorin measure is given by*

$$b = \inf\left\{c \geq 0 : X \overset{\text{law}}{=} \mathcal{G}_c Y_c \text{ for some } Y_c \text{ independent of } \mathcal{G}_c\right\}. \qquad (2.13)$$

Proof. Let us write $\widetilde{b}$ for the right hand side of (2.13).

By Theorem 2.2, we have $X \overset{\text{law}}{=} \mathcal{G}_b Y$ for some random variable Y independent of $\mathcal{G}_b$. For any $c > b$, we have $\mathcal{G}_b \overset{\text{law}}{=} \mathcal{G}_c \mathcal{B}_{b,c-b}$ where $\mathcal{G}_c$ and $\mathcal{B}_{b,c-b}$ are independent, which implies that $c \geq \widetilde{b}$ for any such c. Hence we obtain $b \geq \widetilde{b}$.

Suppose that $b > \widetilde{b}$. Then we may take c with $b > c > \widetilde{b}$ such that $X \overset{\text{law}}{=} \mathcal{G}_c Z$ for some random variable Z independent of $\mathcal{G}_c$. Then we have another expression of the density $\rho(x)$ as

$$\rho(x) = \frac{1}{\Gamma(c)} x^{c-1} E\left[\frac{1}{Z^c}\exp\left(-\frac{x}{Z}\right)\right]. \qquad (2.14)$$

By the monotone convergence theorem, this implies that

$$\lim_{x \to 0+} \frac{\rho(x)}{x^{c-1}} = \frac{1}{\Gamma(c)} E\left[\frac{1}{Z^c}\right] > 0, \qquad (2.15)$$

which shows that $c \geq b$ by formula (2.7). This leads to a contradiction. Therefore we conclude that $b = \widetilde{b}$. $\square$

2.3 Beta Variables of the Second Kind

Consider the ratios of two independent gamma variables, which are sometimes called *beta variables of the second kind* or *beta prime variables*. By identity (2.11), the following is obvious: *For any $a, b > 0$,*

$$\frac{\mathcal{G}_a}{\widehat{\mathcal{G}}_b} \overset{\text{law}}{=} \frac{\mathcal{B}_{a,b}}{1 - \mathcal{B}_{a,b}}. \qquad (2.16)$$

The law of the ratio $\mathcal{G}_a/\widehat{\mathcal{G}}_b$ is given as follows: *For any $a, b > 0$,*

$$P\left(\frac{\mathcal{G}_a}{\widehat{\mathcal{G}}_b} \in dx\right) = \frac{1}{B(a,b)} \frac{x^{a-1}}{(1+x)^{a+b}} dx, \qquad x > 0. \qquad (2.17)$$

In spite of its simple statement, the following theorem is rather difficult to prove.

Theorem 2.6 (see, e.g., [7, Ex.4.3.1]) *For any $a, b > 0$, the ratio $\mathcal{G}_a/\widehat{\mathcal{G}}_b$ is of (GGC) type. Its Thorin measure has total mass a.*

For the proof, see [7]. We omit the details.

2.4 α-Cauchy Variables and Linnik Variables

It is well-known that the standard Cauchy distribution $\frac{1}{\pi}\frac{1}{1+x^2}dx$ and the bilateral exponential distribution $\frac{1}{2}e^{-|x|}dx$ satisfy the following relation:

(R) The characteristic function of any of these two distributions is proportional to the density of the other.

We shall introduce α-analogues of these two distributions which satisfy the relation (R).

Let us introduce the α-analogue for $\alpha > 1$ of the standard Cauchy variable $\mathcal{C}$, which, as just recalled, is given by

$$P(\mathcal{C} \in dx) = \frac{1}{\pi}\frac{1}{1+x^2}dx, \qquad x \in \mathbb{R}. \tag{2.18}$$

Define the α-*Cauchy variable* $\mathcal{C}_\alpha$ as follows:

$$P(\mathcal{C}_\alpha \in dx) = \frac{\sin(\pi/\alpha)}{2\pi/\alpha}\frac{1}{1+|x|^\alpha}dx, \qquad x \in \mathbb{R}. \tag{2.19}$$

Note that $\mathcal{C}_2 \overset{\text{law}}{=} \mathcal{C}$. By a change of variables, the following is easy to see: *For $\alpha > 1$, let $\gamma = 1/\alpha \in (0,1)$. Then it holds that*

$$\mathcal{C}_\alpha = \epsilon \left(\frac{\mathcal{G}_\gamma}{\widehat{\mathcal{G}}_{1-\gamma}}\right)^\gamma \tag{2.20}$$

where ϵ is a Bernoulli variable: $P(\epsilon = 1) = P(\epsilon = -1) = 1/2$ independent of $\mathcal{G}_\gamma$ and $\widehat{\mathcal{G}}_{1-\gamma}$. In particular,

$$\frac{\mathcal{G}_\gamma}{\widehat{\mathcal{G}}_{1-\gamma}} = |\mathcal{C}_\alpha|^\alpha. \tag{2.21}$$

Note that the law of a standard Cauchy variable $\mathcal{C}_2$ is of (SD) type. Moreover, the following theorem is known:

Theorem 2.7 (Bondesson [6]) *For $1 < \alpha \le 2$, the law of $|\mathcal{C}_\alpha|$ is of (ID) type.*

It is easy to see that

$$|\mathcal{C}_\alpha| \overset{\text{law}}{\longrightarrow} \mathcal{U} \qquad \text{as } \alpha \to \infty \tag{2.22}$$

where $\mathcal{U}$ is a uniform variable on $(0,1)$.

Theorem 2.8 (Thorin [42]) *For $p > 0$, the law of $\mathcal{U}^{-p}$, which is called the Pareto distribution of index p, is of (GGC) type.*

Remark 2.9 *The following problems still remain open:*
(i) Is it true that the law of C_α is of (SD) type (or of (ID) type at least)?
(ii) Is it true that the law of $|C_\alpha|$ is of (SD) type?
(iii) Is it true that the law of $|C_\alpha|^{-p}$ for $p > 0$ is of (SD) type (or of (ID) type at least)?

Remark 2.10 *Bourgade–Fujita–Yor ([8]) have proposed a new probabilistic method of computing special values of the Riemann zeta function $\zeta(2n)$ via the Cauchy variable. Fujita–Y. Yano–Yor [15] have recently generalized their method via the α-Cauchy variables and obtained a probabilistic method for computing special values of the complementary sum of the Hurwitz zeta function: $\zeta(2n, \gamma) + \zeta(2n, 1 - \gamma)$ for $0 < \gamma < 1$.*

Following [12], we introduce the *Linnik variable* Λ_α of index $0 < \alpha \le 2$ as follows:

$$E[e^{i\theta\Lambda_\alpha}] = \frac{1}{1 + |\theta|^\alpha}, \qquad \theta \in \mathbb{R}. \tag{2.23}$$

It is easy to see that

$$\Lambda_\alpha \overset{\text{law}}{=} X_\alpha(\mathbf{e}) \tag{2.24}$$

where $X_\alpha = (X_\alpha(t) : t \ge 0)$ is the symmetric stable Lévy process of index α starting from 0 such that

$$P[e^{i\theta X_\alpha(t)}] = e^{-t|\theta|^\alpha}, \qquad \theta \in \mathbb{R} \tag{2.25}$$

and $\mathbf{e}$ is a standard exponential variable independent of X_α. Hence the laws of Linnik variables are of (SD) type. A Lévy process $(\Lambda_\alpha(t))$ with $\Lambda_\alpha(1) \overset{\text{law}}{=} \Lambda_\alpha$ is called a *Linnik process*; its characteristic function is:

$$E\left[e^{i\theta\Lambda_\alpha(t)}\right] = \frac{1}{(1 + |\theta|^\alpha)^t}, \qquad \theta \in \mathbb{R}. \tag{2.26}$$

See James [25] for his study of Linnik processes. Note that the law of Λ_α has a continuous density $L_\alpha(x)$, i.e.,

$$P(\Lambda_\alpha \in dx) = L_\alpha(x)dx. \tag{2.27}$$

Proposition 2.11 *Suppose that $1 < \alpha < 2$. Then the α-Cauchy distribution and the Linnik distribution of index α satisfy the relation (**R**).*

Proof. Note that the identities (2.23) and (2.27) show that

$$\int_{-\infty}^{\infty} e^{i\theta x} L_\alpha(x)dx = \frac{1}{1 + |\theta|^\alpha}, \qquad \theta \in \mathbb{R}. \tag{2.28}$$

By Fourier inversion, we obtain:

$$L_\alpha(x) = \frac{1}{2\pi} \int_{-\infty}^{\infty} e^{-ix\theta} \frac{1}{1 + |\theta|^\alpha} d\theta, \qquad x \in \mathbb{R}. \tag{2.29}$$

Hence:

$$E[e^{i\theta C_\alpha}] = \frac{\sin(\pi/\alpha)}{2\pi/\alpha} L_\alpha(\theta), \qquad \theta \in \mathbb{R}. \tag{2.30}$$

Now the proof is complete. □

2.5 Log-gamma Processes and their Variants

We recall the classes of log-gamma processes, z-processes and Meixner processes.

It is well-known (see, e.g., [41]) that the law of the logarithm of a gamma variable $\log \mathcal{G}_a$ is of (SD) type. Let us introduce a Lévy process $(\eta_a(t) : t \geq 0)$ such that

$$\eta_a(1) \overset{\text{law}}{=} \log \mathcal{G}_a. \tag{2.31}$$

Following Carmona–Petit–Yor [10], we call the process $(\eta_a(t) : t \geq 0)$ the *log-gamma process*. Please be careful not to confuse with the convention that log-normal variables stand for exponentials of normal variables. In (2.31), we simply take the logarithm of a gamma variable.

The Lévy characteristics of $(\eta_a(t) : t \geq 0)$ are given as follows.

Theorem 2.12 (see [10] and also [19]) *For any $a > 0$, the log-gamma process is represented as*

$$\eta_a(t) \overset{\text{law}}{=} t\Gamma'(1) + \sum_{j=0}^{\infty} \left\{ \frac{t}{j+1} - \frac{\gamma^{(j)}(t)}{j+a} \right\} \tag{2.32}$$

where $\gamma^{(0)}, \gamma^{(1)}, \ldots$ are independent gamma processes. In particular, the Lévy exponent of $(\eta_a(t) : t \geq 0)$ defined by

$$E\left[e^{i\theta\eta_a(t)}\right] = \left(\frac{\Gamma(a+i\theta)}{\Gamma(a)}\right)^t = e^{t\phi_a(\theta)} \tag{2.33}$$

admits the representation

$$\phi_a(\theta) = \log \frac{\Gamma(a+i\theta)}{\Gamma(a)} \tag{2.34}$$

$$= i\theta\psi(a) + \int_{-\infty}^{0} \left(e^{i\theta u} - 1 - i\theta u\right) \frac{e^{-a|u|}}{|u|(1 - e^{-|u|})} du \tag{2.35}$$

where $\psi(z) = \Gamma'(z)/\Gamma(z)$ is called the digamma function.

Let $(\eta_a(t) : t \geq 0)$ and $(\widehat{\eta}_b(t) : t \geq 0)$ be independent log-gamma processes. Then the difference $(\eta_a(t) - \widehat{\eta}_b(t) : t \geq 0)$ is called a *generalized z-process* (see [18]). In particular, we have

$$\eta_a(1) - \widehat{\eta}_b(1) \overset{\text{law}}{=} \log \frac{\mathcal{G}_a}{\widehat{\mathcal{G}}_b} \tag{2.36}$$

and this law is called a *z-distribution*. Its characteristic function is given by

$$E\left[\exp\left\{i\theta \log \frac{\mathcal{G}_a}{\widehat{\mathcal{G}}_b}\right\}\right] = \frac{B(a+i\theta, b-i\theta)}{B(a,b)} \tag{2.37}$$

and the law itself is given by

$$P\left(\log \frac{\mathcal{G}_a}{\widehat{\mathcal{G}}_b} \in dx\right) = \frac{1}{B(a,b)} \frac{e^{ax}}{(1+e^x)^{a+b}} dx. \tag{2.38}$$

2.6 Symmetric z-Processes

We now consider a particular case of symmetric z-processes, i.e.,

$$\sigma_a(t) = \frac{1}{\pi}\{\eta_a(t) - \widehat{\eta}_a(t)\}, \qquad t \geq 0. \tag{2.39}$$

We introduce a subordinator given by

$$\Sigma_a(t) = \frac{2}{\pi^2} \sum_{j=0}^{\infty} \frac{\gamma^{(j)}(t)}{(j+a)^2} \tag{2.40}$$

where $\gamma^{(0)}, \gamma^{(1)}, \ldots$ are independent gamma processes. The Lévy measure of $(\Sigma_a(t) : t \geq 0)$ may be obtained from the following:

$$E\left[e^{-\lambda\Sigma_a(t)}\right] = \prod_{j=0}^{\infty} E\left[\exp\left\{-\lambda \frac{2\gamma^{(j)}(t)}{\pi^2(j+a)^2}\right\}\right] \tag{2.41}$$

$$= \prod_{j=0}^{\infty} \exp\left\{-t \int_0^{\infty} \left(1 - \exp\left(-\frac{2\lambda u}{\pi^2(j+a)^2}\right)\right) e^{-u} \frac{du}{u}\right\} \tag{2.42}$$

$$= \prod_{j=0}^{\infty} \exp\left\{-t \int_0^{\infty} \left(1 - e^{-\lambda u}\right) \exp\left(-\frac{\pi^2(j+a)^2}{2}u\right) \frac{du}{u}\right\} \tag{2.43}$$

$$= \exp\left\{-t \int_0^{\infty} \left(1 - e^{-\lambda u}\right) n(u) du\right\} \tag{2.44}$$

where

$$n(u) = \frac{1}{u} \sum_{j=0}^{\infty} \exp\left(-\frac{\pi^2(j+a)^2}{2}u\right) = \frac{1}{u} \int_{(0,\infty)} e^{-ut} U(dt) \qquad (2.45)$$

with

$$U = \sum_{j=0}^{\infty} \delta_{\pi^2(j+a)^2/2}. \qquad (2.46)$$

Hence we conclude that the law of $\Sigma_a(t)$ for fixed t is of (GGC) type.

The following theorem, due to Barndorff-Nielsen–Kent–Sørensen [1], connects the two Lévy processes Σ_a and σ_a:

Theorem 2.13 ([1]; see, e.g., [10]) *The process $(\sigma_a(t) : t \geq 0)$ may be obtained as the subordination of a Brownian motion $(\widehat{B}(u) : u \geq 0)$ with respect to the subordinator $(\Sigma_a(t) : t \geq 0)$:*

$$\sigma_a(t) \stackrel{\text{law}}{=} \widehat{B}(\Sigma_a(t)), \qquad t \geq 0. \qquad (2.47)$$

Proof. By (2.32), the process $(\sigma_a(t) : t \geq 0)$ is represented as

$$\sigma_a(t) = \frac{1}{\pi} \sum_{j=0}^{\infty} \frac{\widehat{\gamma}^{(j)}(t) - \gamma^{(j)}(t)}{j+a} \qquad (2.48)$$

where $\gamma^{(0)}, \gamma^{(1)}, \ldots, \widehat{\gamma}^{(0)}, \widehat{\gamma}^{(1)}, \ldots$ are independent gamma processes. Note that

$$\widehat{\gamma}^{(j)}(t) - \gamma^{(j)}(t) \stackrel{\text{law}}{=} \Lambda_2(t) \stackrel{\text{law}}{=} \sqrt{2}\widehat{B}(\gamma(t)) \qquad (2.49)$$

where $(\Lambda_2(t) : t \geq 0)$ is a Linnik process, i.e., a Lévy process such that $\Lambda_2(1) \stackrel{\text{law}}{=} \Lambda_2$. Now we obtain

$$\widehat{B}(\Sigma_a(t)) \stackrel{\text{law}}{=} \frac{\sqrt{2}}{\pi} \sum_{j=0}^{\infty} \frac{\widehat{B}_j(\gamma^{(j)}(t))}{j+a} \qquad (2.50)$$

$$\stackrel{\text{law}}{=} \frac{1}{\pi} \sum_{j=0}^{\infty} \frac{\widehat{\gamma}^{(j)}(t) - \gamma^{(j)}(t)}{j+a} \qquad (2.51)$$

$$\stackrel{\text{law}}{=} \sigma_a(t) \qquad (2.52)$$

where on the right hand side of (2.50) the $\widehat{B}_j$'s are independent of $\gamma^{(j)}$'s. The proof is complete. $\square$

The characteristic function of $\sigma_a(t)$ is given by

$$E[e^{i\theta\sigma_a(t)}] = \left(E\left[\exp\left\{i\frac{\theta}{\pi}\log\frac{\mathcal{G}_a}{\widehat{\mathcal{G}}_a}\right\}\right]\right)^t = e^{-t\Phi_a(\theta)} \qquad (2.53)$$

where

$$\Phi_a(\theta) = \int_{-\infty}^{\infty} \left(1 - e^{i\theta u}\right) \frac{e^{-a\pi|u|}}{|u|(1 - e^{-\pi|u|})} du \qquad (2.54)$$

$$= 2 \int_{0}^{\infty} \left(1 - \cos\theta u\right) \frac{e^{-a\pi u}}{u(1 - e^{-\pi u})} du. \qquad (2.55)$$

For $t = 1$, the law of $\sigma_a(1)$ is given by

$$P\left(\sigma_a(1) \in dx\right) = P\left(\frac{1}{\pi} \log \frac{\mathcal{G}_a}{\widehat{\mathcal{G}}_a} \in dx\right) = \frac{\pi}{B(a,a)} \frac{e^{a\pi x}}{(1 + e^{\pi x})^{2a}} dx. \qquad (2.56)$$

Example 2.14 *When $a = 1/2$, put*

$$C_t := \Sigma_{\frac{1}{2}}(t), \qquad \mathbb{C}_t := \widehat{B}(C_t) = \sigma_{\frac{1}{2}}(t). \qquad (2.57)$$

The law of $\mathbb{C}_1$ is called the hyperbolic cosine distribution:

$$E[e^{i\theta \mathbb{C}_1}] = E[e^{-\frac{1}{2}\theta^2 C_1}] = \frac{1}{\cosh\theta}, \qquad P\left(\mathbb{C}_1 \in dx\right) = \frac{1}{\cosh\pi x} dx. \qquad (2.58)$$

Consequently, $\mathbb{C}_1$ and $\pi\mathbb{C}_1$ satisfy the relation (R).

Example 2.15 *When $a = 1$, put*

$$S_t := \Sigma_1(t), \qquad \mathbb{S}_t := \widehat{B}(S_t) = \sigma_1(t). \qquad (2.59)$$

The law of $\mathbb{S}_1$ is called the logistic distribution:

$$E[e^{i\theta \mathbb{S}_1}] = E[e^{-\frac{1}{2}\theta^2 S_1}] = \frac{\theta}{\sinh\theta}, \qquad P\left(\mathbb{S}_1 \in dx\right) = \frac{\pi}{(\cosh\pi x)^2} dx. \qquad (2.60)$$

Consequently, $\mathbb{S}_1$ and $\pi\mathbb{C}_2$ satisfy the relation (R).

Let us introduce a subordinator (T_t) and then $(\mathbb{T}_t)$ such that

$$\mathbb{T}_t = \widehat{B}(T_t) \qquad (2.61)$$

and that

$$E[e^{i\theta \mathbb{T}_t}] = E[e^{-\frac{1}{2}\theta^2 T_t}] = \left(\frac{\tanh\theta}{\theta}\right)^t. \qquad (2.62)$$

It is well-known that the law of T_1 is of (ID) type and hence that such processes exist. Now it is obvious that

$$C_t \stackrel{\text{law}}{=} T_t + S_t, \qquad \mathbb{C}_t \stackrel{\text{law}}{=} \mathbb{T}_t + \mathbb{S}_t \qquad (2.63)$$

where (T_t) and (S_t) are independent and so are $(\mathbb{T}_t)$ and $(\mathbb{S}_t)$.

For further study of these processes C_t, S_t and T_t, see Pitman–Yor [32]. By taking Laplace inversion, the density of the law of T_1 can be obtained in terms of the theta function; see Knight [28, Cor.2.1] for details.

2.7 Meixner Processes

Let $\beta \in (-\pi, \pi)$ and let $(\mathcal{M}_\beta(t) : t \geq 0)$ be a Lévy process such that

$$\mathcal{M}_\beta(1) \overset{\text{law}}{=} \frac{1}{2\pi} \log \frac{\mathcal{G}_a}{\widehat{\mathcal{G}}_{1-a}} \qquad \text{where } \beta = (2a-1)\pi. \qquad (2.64)$$

The law of $\mathcal{M}_\beta(t)$ for fixed t is called a *Meixner distribution* because of its close relation to *Meixner–Pollaczek polynomials* (See [38], [39], [40] and [17]). The characteristic function of $\mathcal{M}_\beta(t)$ is given by

$$E\left[e^{i\theta \mathcal{M}_\beta(t)}\right] = \left(\frac{\cos \frac{\beta}{2}}{\cosh \frac{\theta - i\beta}{2}}\right)^t = e^{t\xi_\beta(\theta)} \qquad (2.65)$$

where

$$\xi_\beta(\theta) = \frac{i\theta}{2\pi}\left\{\psi\left(\frac{\pi+\beta}{2\pi}\right) - \psi\left(\frac{\pi-\beta}{2\pi}\right)\right\} \\ + \int_{-\infty}^{\infty} \left(e^{i\theta u} - 1 - i\theta u\right) \frac{e^{\beta u}}{u\sinh(\pi u)} du. \qquad (2.66)$$

The law of $\mathcal{M}_\beta(t)$ itself is given by

$$P(\mathcal{M}_\beta(t) \in dx) = \frac{\left(2\cos\frac{\beta}{2}\right)^t}{2\pi\Gamma(t)} e^{\beta x} \left|\Gamma\left(\frac{t}{2} + ix\right)\right|^2 dx \qquad (2.67)$$

$$= \frac{\left(2\cos\frac{\beta}{2}\right)^t}{2\pi\Gamma(t)} e^{\beta x} \Gamma(t/2)^2 e^{-\Phi_{t/2}(x)} dx \qquad (2.68)$$

$$= \frac{\left(2\cos\frac{\beta}{2}\right)^t B(t/2, t/2)}{2\pi} e^{\beta x - \Phi_{t/2}(x)} dx \qquad (2.69)$$

where $\Phi_a(x)$ has been defined in (2.55).

We simply write $\mathcal{M}_\beta$ for $\mathcal{M}_\beta(1)$. Remark that this Meixner distribution is identical to that of the log of an α-Cauchy variable:

$$\mathcal{M}_\beta \overset{\text{law}}{=} \frac{\alpha}{2\pi} \log |\mathcal{C}_\alpha| \qquad \text{where } \beta = \left(\frac{2}{\alpha} - 1\right)\pi. \qquad (2.70)$$

Remark also that the law of $\mathcal{M}_\beta$ is symmetric if and only if $\beta = 0$ (or $a = 1/2$). Then the corresponding Meixner distribution is identical to the hyperbolic cosine distribution, up to the factor $1/2$; precisely:

$$\mathcal{M}_0 \overset{\text{law}}{=} \frac{1}{\pi} \log |\mathcal{C}| \overset{\text{law}}{=} \frac{1}{2\pi} \log \frac{\mathcal{G}_{1/2}}{\widehat{\mathcal{G}}_{1/2}} \overset{\text{law}}{=} \frac{1}{2}\mathbb{C}_1. \qquad (2.71)$$

2.8 α-Rayleigh Distributions

For an exponential variable $\mathfrak{e}$, the random variable

$$\mathcal{R} = \sqrt{2\mathfrak{e}} \tag{2.72}$$

is sometimes called a *Rayleigh variable*. We shall introduce an α-analogue of the Rayleigh variable.

Let $0 < \alpha \le 2$ and let $t > 0$. By Fourier inversion, we obtain from (2.25) that

$$P(X_\alpha(t) \in \mathrm{d}x) = p_t^{(\alpha)}(x)\mathrm{d}x \tag{2.73}$$

where

$$p_t^{(\alpha)}(x) = \frac{1}{2\pi}\int_{-\infty}^{\infty} e^{-ix\xi}e^{-t|\xi|^\alpha}\,\mathrm{d}\xi = \frac{1}{\pi}\int_0^\infty \cos(x\xi)e^{-t\xi^\alpha}\,\mathrm{d}\xi. \tag{2.74}$$

Note that

$$p_1^{(\alpha)}(0) = \frac{\Gamma(1/\alpha)}{\alpha\pi}. \tag{2.75}$$

Lemma 2.16 *Let $0 < \alpha \le 2$. Then there exists a non-negative random variable $\mathcal{R}_\alpha$ such that*

$$P(\mathcal{R}_\alpha > x) = \frac{p_1^{(\alpha)}(x)}{p_1^{(\alpha)}(0)}, \qquad x > 0. \tag{2.76}$$

In particular, $\mathcal{R}_2 = \sqrt{2}\mathcal{R} = 2\sqrt{\mathfrak{e}}$.

We call $\mathcal{R}_\alpha$ an α-*Rayleigh variable* and its law the α-*Rayleigh distribution*.

For the proof of Lemma 2.16, we introduce some notations. For $0 < \alpha < 1$, denote by T_α the *unilateral α-stable distribution*:

$$E\left[e^{-\lambda T_\alpha}\right] = e^{-\lambda^\alpha}, \qquad \lambda \ge 0. \tag{2.77}$$

We denote by T'_α the h-size biased variable of T_α with respect to $h(x) = x^{-1/2}$:

$$E\left[f(T'_\alpha)\right] = \frac{E\left[(T_\alpha)^{-1/2}f(T_\alpha)\right]}{E\left[(T_\alpha)^{-1/2}\right]} \tag{2.78}$$

for any non-negative Borel function f. The following lemma proves Lemma 2.16.

Lemma 2.17 *Suppose $0 < \alpha < 2$. Then the variable $\mathcal{R}_\alpha$ is given by*

$$\mathcal{R}_\alpha = 2\sqrt{\mathfrak{e}T'_{\alpha/2}} \tag{2.79}$$

where the variables $\mathfrak{e}$ and $T'_{\alpha/2}$ are independent.

Proof of Lemma 2.17. Since we have

$$X_\alpha(1) \overset{\text{law}}{=} \sqrt{2}\widehat{B}(T_{\frac{\alpha}{2}}), \tag{2.80}$$

we obtain the following expression:

$$p_1^{(\alpha)}(x) = E\left[\frac{1}{2\sqrt{\pi T_{\frac{\alpha}{2}}}} \exp\left\{-\frac{x^2}{4T_{\frac{\alpha}{2}}}\right\}\right]. \tag{2.81}$$

Hence we obtain

$$\frac{p_1^{(\alpha)}(x)}{p_1^{(\alpha)}(0)} = E\left[\exp\left\{-\frac{x^2}{4T'_{\frac{\alpha}{2}}}\right\}\right]. \tag{2.82}$$

Using an independent exponential variable $\mathfrak{e}$, we have

$$\frac{p_1^{(\alpha)}(x)}{p_1^{(\alpha)}(0)} = E\left[\mathfrak{e} > \frac{x^2}{4T'_{\frac{\alpha}{2}}}\right] = E\left[2\sqrt{\mathfrak{e}T'_{\frac{\alpha}{2}}} > x\right]. \tag{2.83}$$

Now the proof is complete. □

Remark 2.18 *Lemmas 2.16 and 2.17 can also be found in Cordero [50, §1.2.2].*

3 Discussions from Excursion Theoretic Viewpoint

Recall Itô's excursion theory ([23] and [30]). See also the standard textbooks [22] and [35], as well as [33].

3.1 Itô's Measure of Excursions away from the Origin

We simply write $\mathbb{D}$ for the space $\mathbb{D}([0,\infty);\mathbb{R})$ of càdlàg paths equipped with the Skorokhod topology. Let $X = (X(t) : t \geq 0)$ be a strong Markov process with paths in $\mathbb{D}$ starting from 0. Suppose that the origin is regular, recurrent and an instantaneous state. Then it is well-known (see [4, Thm.V.3.13]) that there exists a local time at the origin, which we denote by $L = (L(t) : t \geq 0)$, subject to the normalization:

$$E\left[\int_0^\infty e^{-t}dL(t)\right] = 1. \tag{3.1}$$

This is a choice made in this section; but later, we may make another choice, which will be indicated as $L^{(\alpha)}(t)$, $L(t)$ being always subject to (3.1). The local time process $L = (L(t) : t \geq 0)$ is continuous and non-decreasing almost surely. Thus its right-continuous inverse process

$$\tau(l) = \inf\{t > 0 : L(t) > l\} \tag{3.2}$$

is strictly-increasing. By the strong Markov property of X, we see that $\tau(l)$ is a subordinator. Call D the (random) set of discontinuities of τ:

$$D = \{l > 0 : \tau(l) - \tau(l-) > 0\}. \tag{3.3}$$

It is obvious that D is a countable set. Now we define a point function $\boldsymbol{p}$ on D which takes values in $\mathbb{D}$ as follows: For $l \in D$,

$$\boldsymbol{p}(l)(t) = \begin{cases} X(t + \tau(l-)) & \text{if } 0 \leq t < \tau(l) - \tau(l-), \\ 0 & \text{otherwise.} \end{cases} \tag{3.4}$$

We call $\boldsymbol{p} = (\boldsymbol{p}(l) : l \in D)$ the *excursion point process*. Then the fundamental theorem of Itô's excursion theory is stated as follows.

Theorem 3.1 (Itô [23]; see also Meyer [30]) *The excursion point process $\boldsymbol{p}$ is a Poisson point process, i.e.:*
(i) *$\boldsymbol{p}$ is σ-discrete almost surely, i.e., for almost every sample path, there exists a sequence $\{U_n\}$ of disjoint measurable subsets of $\mathbb{D}$ such that $\mathbb{D} = \cup_n U_n$ and $\{l \in D : \boldsymbol{p}(l) \in U_n\}$ is a finite set for all n;*
(ii) *$\boldsymbol{p}$ is renewal, i.e., $\boldsymbol{p}(\cdot \wedge s)$ and $\boldsymbol{p}(\cdot + s)$ are independent for each $s > 0$.*

For a measurable subset U of $\mathbb{D}$, we define a point process $\boldsymbol{p}_U : D_U \to \mathbb{D}$ as

$$D_U = \{l \in D : \boldsymbol{p}(l) \in U\} \quad \text{and} \quad \boldsymbol{p}_U = \boldsymbol{p}|_{D_U}. \tag{3.5}$$

We call $\boldsymbol{p}_U = (\boldsymbol{p}_U(l) : l \in D_U)$ the restriction of $\boldsymbol{p}$ on U. The measure on $\mathbb{D}$ defined by

$$\boldsymbol{n}(U) = E\left[\sharp((0,1] \cap D_U)\right] \tag{3.6}$$

is called *Itô's measure* of excursions.

Corollary 3.2 (Itô [23]) *The following statements hold:*
(i) *Let $\{U_n\}$ be a sequence of disjoint measurable subsets of $\mathbb{D}$. Then the point processes $\{\boldsymbol{p}_{U_n}\}$ are independent;*
(ii) *Let U be a measurable subset of $\mathbb{D}$ such that $\boldsymbol{n}(U) < \infty$. Then $(0,l] \cap D_U$ is a finite set for all $l > 0$ a.s. Set*

$$D_U = \{0 < \kappa_1 < \kappa_2 < \cdots\}, \quad \boldsymbol{p}_U(\kappa_n) = u_n, \ n = 1, 2, \ldots. \tag{3.7}$$

Then:
(ii-a) *$\{\kappa_n - \kappa_{n-1}, u_n : n = 1, 2, \ldots\}$ are independent where $\kappa_0 = 0$;*
(ii-b) *For each n, $\kappa_n - \kappa_{n-1}$ is exponentially distributed with mean $1/\boldsymbol{n}(U)$, i.e., $P(\kappa_n - \kappa_{n-1} > l) = e^{-l\boldsymbol{n}(U)}$ for $l > 0$;*

(ii-c) *For each n, $P(u_n \in \cdot) = \boldsymbol{n}(\cdot \cap U)/\boldsymbol{n}(U)$;*

(iii) *Let $F(l, u)$ be a non-negative measurable functional on $(0, \infty) \times \mathbb{D}$. Then*

$$E\left[\exp\left\{-\sum_{l \in D} F(l, \boldsymbol{p}(l))\right\}\right] = \exp\left\{-\int \left(1 - \mathrm{e}^{-F(l,u)}\right) \mathrm{d}l \otimes \boldsymbol{n}(\mathrm{d}u)\right\}; \quad (3.8)$$

(iv) *Let $F(t, u)$ be a non-negative measurable functional on $(0, \infty) \times \mathbb{D}$. Then*

$$E\left[\sum_{l \in D} F(\tau(l-), \boldsymbol{p}(l))\right] = \int E[F(\tau(l), u)]\mathrm{d}l \otimes \boldsymbol{n}(\mathrm{d}u). \quad (3.9)$$

The proofs of Theorem 3.1 and Corollary 3.2 are also found in [22] and [35].

For $u \in \mathbb{D}$, define

$$\zeta(u) = \sup\{t \geq 0 : u(t) \neq 0\}. \quad (3.10)$$

For each excursion path $\boldsymbol{p}(l)$, $l \in D$, $\zeta(\boldsymbol{p}(l))$ is finite and called the *lifetime* of the path $\boldsymbol{p}(l)$. For a measurable subset U of $\mathbb{D}$, we set

$$\tau_U(l) = \sum_{k \in (0,l] \cap D_U} \zeta(\boldsymbol{p}_U(k)). \quad (3.11)$$

Note that $\tau_{\mathbb{D}}(l) = \tau(l)$, $l \geq 0$. By Corollary 3.2 (iii), we see that the process $(\tau_U(l) : l \geq 0)$ is a subordinator with Laplace transform $E[\mathrm{e}^{-\lambda \tau_U(l)}] = \mathrm{e}^{-l\psi_U(\lambda)}$ given by

$$\psi_U(\lambda) = \boldsymbol{n}\left[1 - \mathrm{e}^{-\lambda\zeta}; U\right], \qquad \lambda > 0. \quad (3.12)$$

Since $\psi(\lambda) := \psi_{\mathbb{D}}(\lambda) < \infty$, we have $\boldsymbol{n}(\zeta > t) < \infty$ for all $t > 0$; in particular, we see that the measure $\boldsymbol{n}$ is σ-finite.

3.2 Decomposition of First Hitting Time Before and After Last Exit Time

We denote the first hitting time of a closed set F for X by

$$T_F(X) = \inf\{t > 0 : X(t) \in F\}. \quad (3.13)$$

In particular, if $F = \{a\}$, the closed set consisting of a single point $a \in \mathbb{R}$, $T_F(X)$ is nothing but the first hitting time of *point* $a \in \mathbb{R}$ for X:

$$T_{\{a\}}(X) = \inf\{t > 0 : X(t) = a\}. \quad (3.14)$$

The hitting time $T_{\{a\}}(X)$ may be decomposed at the last exit time from 0:

$$T_{\{a\}}(X) = G_{\{a\}}(X) + \Xi_{\{a\}}(X) \tag{3.15}$$

where $G_{\{a\}}(X)$ is the last exit time from 0 before $T_{\{a\}}(X)$, and where $\Xi_{\{a\}}(X)$ is the remaining time after $G_{\{a\}}(X)$, i.e.,

$$G_{\{a\}}(X) = \sup\{t \leq T_{\{a\}}(X) : X(t) = 0\} \tag{3.16}$$

and

$$\Xi_{\{a\}}(X) = T_{\{a\}}(X) - G_{\{a\}}(X). \tag{3.17}$$

The joint law of the random times $G_{\{a\}}(X)$ and $\Xi_{\{a\}}(X)$ is characterised by the following proposition:

Proposition 3.3 *Let $a \neq 0$. Then the random times $G_{\{a\}}(X)$ and $\Xi_{\{a\}}(X)$ are independent. Moreover, the law of $G_{\{a\}}(X)$ is of* (ID) *type. The Laplace transforms of $G_{\{a\}}(X)$ and $\Xi_{\{a\}}(X)$ are given as*

$$E\left[e^{-\lambda G_{\{a\}}(X)}\right] = \left\{1 + \frac{n\left[1 - e^{-\lambda \zeta}; T_{\{a\}} > \zeta\right]}{n(T_{\{a\}} < \zeta)}\right\}^{-1} \tag{3.18}$$

and

$$E\left[e^{-\lambda \Xi_{\{a\}}(X)}\right] = \frac{n\left[e^{-\lambda T_{\{a\}}}; T_{\{a\}} < \zeta\right]}{n(T_{\{a\}} < \zeta)}. \tag{3.19}$$

Consequently, the Laplace transform of $T_{\{a\}}(X)$ is given as

$$E\left[e^{-\lambda T_{\{a\}}(X)}\right] = \frac{n\left[e^{-\lambda T_{\{a\}}}; T_{\{a\}} < \zeta\right]}{n\left[1 - \left(e^{-\lambda \zeta} \cdot 1_{\{T_{\{a\}} > \zeta\}}\right)\right]}. \tag{3.20}$$

Proof. Set

$$U_a = \{u \in \mathbb{D} : T_{\{a\}}(u) < \infty\}. \tag{3.21}$$

By Corollary 3.2 (i), we see that $p_{U_a^c}$ and p_{U_a} are independent. We remark that $n(U_a) < \infty$; in fact, if we supposed otherwise, then there would exist a sequence $\{t_n\}$ such that $t_n \to 0$ decreasingly and that $X(t_n) = a$, which contradicts $X(0+) = X(0) = 0$. Set

$$\kappa_a = \inf\{l > 0 : p(l) \in U_a\}. \tag{3.22}$$

Then, by Corollary 3.2 (ii), we see that κ_a and $p(\kappa_a)$ are independent. Since $\kappa_a = \inf D_{U_a}$ and $p(\kappa_a) = p_{U_a}(\kappa_a)$, they are measurable with respect to the σ-field generated by p_{U_a}. Hence we see that $\{p_{U_a^c}, \kappa_a, p(\kappa_a)\}$ are independent. Note that

$$G_{\{a\}}(X) = \tau_{U_a^c}(\kappa_a) \qquad \text{and} \qquad \Xi_{\{a\}}(X) = T_{\{a\}}(p(\kappa_a)). \qquad (3.23)$$

Thus we conclude that $G_{\{a\}}(X)$ and $\Xi_{\{a\}}(X)$ are independent. Moreover, we see that the law of $G_{\{a\}}(X)$ is of (ID) type; in fact, $\tau_{U_a^c}$ is a subordinator with Laplace exponent $\psi_{U_a^c}(\lambda)$ and κ_a is an exponential variable with mean $1/n(U_a)$ independent of $\tau_{U_a^c}$. The law of $\Xi_{\{a\}}(X)$ is given by

$$P(\Xi_{\{a\}}(X) \in \cdot) = \frac{n(u \in \mathbb{D} : T_{\{a\}}(u) \in \cdot)}{n(U_a)}. \qquad (3.24)$$

Now the proof is complete. $\square$

3.3 Excursion Durations

Consider the excursion straddling t. For a general study in the setup of linear diffusions, see [37].

Define the last exit time from 0 before t and the first hitting time of point 0 after t as follows:

$$G_t(X) = \sup\{s \leq t : X(s) = 0\}, \qquad D_t(X) = \inf\{s > t : X(s) = 0\}. \qquad (3.25)$$

Define

$$\Xi_t(X) = t - G_t(X), \qquad \Delta_t(X) = D_t(X) - G_t(X). \qquad (3.26)$$

Recall (see (3.1)) that $L = (L(t) : t \geq 0)$ denotes the local time at 0 of X and $\tau = (\tau(l) : l \geq 0)$ its right-continuous inverse. Then we have

$$G_t(X) = \tau(L(t)-), \qquad D_t(X) = \tau(L(t)) \qquad (3.27)$$

and

$$\Xi_t(X) = t - \tau(L(t)-), \qquad \Delta_t(X) = \tau(L(t)) - \tau(L(t)-). \qquad (3.28)$$

If the local time process L has the self-similarity property with index γ:

$$\left(\frac{L(ct)}{c^\gamma} : t \geq 0\right) \stackrel{\text{law}}{=} (L(t) : t \geq 0), \qquad c > 0, \qquad (3.29)$$

then we have

$$\left\{\left(\frac{L(ct)}{c^\gamma} : t \geq 0\right), \left(\frac{\tau(c^\gamma l)}{c} : l \geq 0\right)\right\} \stackrel{\text{law}}{=} \{(L(t) : t \geq 0), (\tau(l) : l \geq 0)\} \qquad (3.30)$$

for any $c > 0$; in particular, τ is a stable subordinator of index γ. Hence the index γ must be in $(0, 1)$.

We now state two explicit results, the proofs of which are postponed after commenting about these results.

Theorem 3.4 *Suppose that the local time process has the self-similarity property of index $0 < \gamma < 1$. Then*

$$(\Xi_1(X), \Delta_1(X)) \overset{\text{law}}{=} \left(\mathcal{B}_{1-\gamma,\gamma}, \frac{\mathcal{B}_{1-\gamma,\gamma}}{\mathcal{U}^{\frac{1}{\gamma}}} \right) \tag{3.31}$$

where $\mathcal{B}_{1-\gamma,\gamma}$ is a beta variable of index $(1 - \gamma, \gamma)$ and $\mathcal{U}$ is an independent uniform variable on $(0, 1)$.

The following is a special case of Winkel [45, Cor.1]:

Theorem 3.5 ([45]) *Suppose that the local time process has the self-similarity property of index $0 < \gamma < 1$. Let $\mathfrak{e}$ be an independent exponential time. Then*

$$(G_{\mathfrak{e}}(X), \Xi_{\mathfrak{e}}(X), \Delta_{\mathfrak{e}}(X)) \overset{\text{law}}{=} \left(\mathcal{G}_\gamma, \widehat{\mathcal{G}}_{1-\gamma}, \frac{\widehat{\mathcal{G}}_{1-\gamma}}{\mathcal{U}^{\frac{1}{\gamma}}} \right) \tag{3.32}$$

where $\mathcal{G}_\gamma$ and $\widehat{\mathcal{G}}_{1-\gamma}$, respectively, are independent gamma variables of indices γ and $1 - \gamma$, respectively, and $\mathcal{U}$ is an independent uniform variable.

Generalizing a self-decomposability result of Bondesson (see [7, Ex.5.6.3]), Bertoin–Fujita–Roynette–Yor [3, Thm.1.1] and Roynette–Vallois–Yor [36, Thm.5] have recently proved the following:

Theorem 3.6 ([3] and [36]) *For any $\gamma \in (0, 1)$, the laws*

$$\frac{\mathcal{G}_{1-\gamma}}{\mathcal{U}^{\frac{1}{\gamma}}} \quad \text{and} \quad \left(\frac{1}{\mathcal{U}^{\frac{1}{\gamma}}} - 1 \right) \mathcal{G}_{1-\gamma} \tag{3.33}$$

are both of (GGC) type with their Thorin measures having total mass $1 - \gamma$. Here $\mathcal{G}_{1-\gamma}$ is a gamma variable of index $1 - \gamma$ and $\mathcal{U}$ is an independent uniform variable.

Example 3.7 *For a symmetric stable Lévy process of index α, it is well-known (see Kesten [27] and Bretagnolle [9]) that the origin is regular for itself if and only if $1 < \alpha \leq 2$. Let $X_\alpha = (X_\alpha(t) : t \geq 0)$ be the symmetric stable Lévy process of index $1 < \alpha \leq 2$. Then its local time process is given as*

$$L(t) = \lim_{\varepsilon \to 0+} \frac{C}{2\varepsilon} \int_0^t 1_{\{|X_\alpha(s)| < \varepsilon\}} ds \tag{3.34}$$

for some constant C. Since X satisfies the self-similarity property with index $1/\alpha$, so does L with index $1 - 1/\alpha$, and hence Theorems 3.4 and 3.5 hold with $\gamma = 1 - 1/\alpha$.

Example 3.8 *For a Bessel process of dimension d, it is well-known that the origin is regular for itself if and only if $0 < d < 2$. Let $X = (X(t) : t \geq 0)$ be a reflecting Bessel process starting from 0 of dimension $d = 2 - 2\alpha$, $0 < d < 2$*

(or $0 < \alpha < 1$) which is scaled so that it has natural scale and speed measure $m(0, x) = x^{\frac{1}{\alpha}-1}$. Then its local time process is given as

$$L(t) = \lim_{\varepsilon \to 0+} \frac{C}{m(0, \varepsilon)} \int_0^t 1_{\{|X(s)| < \varepsilon\}} \mathrm{d}s \qquad (3.35)$$

for some constant C. Since X satisfies the self-similarity property with index α, so does L with the same index α, and hence Theorems 3.4 and 3.5 hold with $\gamma = \alpha$. For the relations among several choices in the literature, see [13].

Example 3.9 *Let $\alpha > 0$ and $0 < \beta < \min\{1, 1/\alpha\}$ and consider the process $X = X_{m^{(\alpha)}, j^{(\beta)}, 0, 0}$ given in [47, Ex.2.4.(b)]. Then X satisfies the self-similarity property with index α, but this property seems to have nothing to do with the local time L. Since its inverse local time process $\tau = \eta_{m^{(\alpha)}, j^{(\beta)}, 0, 0}$ satisfies the self-similarity property with index $1/(\alpha\beta)$, so does L with index $\alpha\beta$, and hence Theorems 3.4 and 3.5 hold with $\gamma = \alpha\beta$.*

Remark 3.10 *The identities in law $\Xi_1(X_\alpha) \overset{\text{law}}{=} \mathcal{B}_{\gamma, 1-\gamma}$ and $D_1(X_\alpha) - 1 \overset{\text{law}}{=} \mathcal{G}_\gamma / \widehat{\mathcal{G}}_{1-\gamma}$ are found in Feller [14, XIV.3] as the long-time limit laws of similar random variables derived from random walks.*

Let us prove Theorems 3.4 and 3.5 for completeness of this paper.

Proof of Theorem 3.4. Since $\tau(c^\gamma l) \overset{\text{law}}{=} c\tau(l)$ for $c, l > 0$, we have $\psi(c\lambda) = c^\gamma \psi(\lambda)$. Hence we obtain

$$n(\zeta \in \mathrm{d}t) = C \frac{\mathrm{d}t}{t^{\gamma+1}} \qquad (3.36)$$

for some constant C. For $t > 0$, the excursion straddling time t is $p(L(t))$. Hence we have

$$G_t = \tau(L(t)-), \qquad \Xi_t = t - \tau(L(t)-), \qquad \Delta_t = \zeta(p(L(t))). \qquad (3.37)$$

Let p, q, r be positive constants. Then

$$E\left[\int_0^\infty e^{-pt - q\Xi_t - r\Delta_t} \mathrm{d}t\right] \qquad (3.38)$$

$$= E\left[\sum_{l \in D} \int_{\tau(l-)}^{\tau(l)} e^{-pt - q\Xi_t - r\Delta_t} \mathrm{d}t\right] \qquad (3.39)$$

$$= E\left[\sum_{l \in D} e^{-p\tau(l-) - r\zeta(p(l))} \int_0^{\zeta(p(l))} e^{-pt - qt} \mathrm{d}t\right] \qquad (3.40)$$

$$= \int_0^\infty E\left[e^{-p\tau(l)}\right] \mathrm{d}l \int_{\mathbb{D}} n(\mathrm{d}u) e^{-r\zeta(u)} \int_0^{\zeta(u)} e^{-pt - qt} \mathrm{d}t \qquad (3.41)$$

$$= \int_0^\infty e^{-l\psi(p)} \mathrm{d}l \int_0^\infty C \frac{\mathrm{d}s}{s^{\gamma+1}} e^{-rs} \int_0^s e^{-pt - qt} \mathrm{d}t \qquad (3.42)$$

$$= \frac{C}{\psi(p)} \int_0^\infty \mathrm{d}t e^{-pt - qt} \int_t^\infty \frac{\mathrm{d}s}{s^{\gamma+1}} e^{-rs}. \qquad (3.43)$$

Note that

$$\frac{1}{\psi(p)} = \frac{1}{\psi(1)p^\gamma} = \frac{1}{\psi(1)\Gamma(\gamma)} \int_0^\infty t^{\gamma-1} e^{-pt} dt. \tag{3.44}$$

Hence we have

$$(3.43) = \frac{C}{\psi(1)\Gamma(\gamma)} \int_0^\infty dt e^{-pt} \int_0^t dv(t-v)^{\gamma-1} e^{-qv} \int_v^\infty \frac{ds}{s^{\gamma+1}} e^{-rs} \tag{3.45}$$

$$= \frac{C}{\psi(1)\Gamma(\gamma)} \int_0^\infty dt e^{-pt} \int_0^1 dv v^{-\gamma}(1-v)^{\gamma-1} e^{-qvt} \int_1^\infty \frac{ds}{s^{\gamma+1}} e^{-rsvt} \tag{3.46}$$

$$= C' \int_0^\infty dt e^{-pt} E \left[\exp\left\{ -q\mathcal{B}_{1-\gamma,\gamma} t - r \frac{\mathcal{B}_{1-\gamma,\gamma}}{\mathcal{U}^{\frac{1}{\gamma}}} t \right\} \right] \tag{3.47}$$

$$= C' E \left[\frac{1}{p + q\mathcal{B}_{1-\gamma,\gamma} + r\mathcal{B}_{1-\gamma,\gamma}\mathcal{U}^{-1/\gamma}} \right] \tag{3.48}$$

for some constant C'.

On the other hand, by the self-similarity property (3.30), we have $(\Xi_t, \Delta_t) \overset{\text{law}}{=} (t\Xi_1, t\Delta_1)$ for fixed $t > 0$, and hence we have

$$E \left[\int_0^\infty e^{-pt-q\Xi_t-r\Delta_t} dt \right] = E \left[\frac{1}{p + q\Xi_1 + r\Delta_1} \right]. \tag{3.49}$$

Letting $q, r \to 0+$ and comparing (3.48) and (3.49), we have $C' = 1$. Therefore we obtain the desired identity in law (3.31) by the uniqueness property of Stieltjes transform. $\square$

Proof of Theorem 3.5. Note that $(G_\epsilon, \Xi_\epsilon, \Delta_\epsilon) \overset{\text{law}}{=} (\epsilon G_1, \epsilon\Xi_1, \epsilon\Delta_1)$ by the self-similarity property (3.30). We also note that $(\epsilon(1 - \mathcal{B}_{1-\gamma,\gamma}), \epsilon\mathcal{B}_{1-\gamma,\gamma}) \overset{\text{law}}{=} (\mathcal{G}_\gamma, \widehat{\mathcal{G}}_{1-\gamma})$ by the identity in law (2.11). Therefore we obtain the desired identity in law (3.32) as an immediate consequence of Theorem 3.4. $\square$

4 Harmonic Transforms of Symmetric Stable Lévy Processes

We keep the notation $X_\alpha = (X_\alpha(t) : t \geq 0)$ for the symmetric stable Lévy process of index α such that

$$P[e^{i\theta X_\alpha(t)}] = e^{-t|\theta|^\alpha}, \qquad \theta \in \mathbb{R}. \tag{4.1}$$

Note that, with (4.1), we have $X_2(t) \overset{\text{law}}{=} \sqrt{2}B(t)$. We have

$$P(X_\alpha(t) \in dx) = p_t^{(\alpha)}(x) dx \tag{4.2}$$

where

$$p_t^{(\alpha)}(x) = \frac{1}{2\pi}\int_{-\infty}^{\infty} e^{-ix\xi}e^{-t|\xi|^\alpha}\,d\xi = \frac{1}{\pi}\int_0^{\infty}\cos(x\xi)e^{-t\xi^\alpha}\,d\xi. \qquad (4.3)$$

We suppose that $1 < \alpha \le 2$. Then the Laplace transform

$$u_q^{(\alpha)}(x) = \int_0^{\infty} e^{-qt}p_t^{(\alpha)}(x)\,dt = \frac{1}{\pi}\int_0^{\infty}\frac{\cos(x\xi)}{q+\xi^\alpha}\,d\xi \qquad (4.4)$$

is finite. Define

$$h_q^{(\alpha)}(x) = u_q^{(\alpha)}(0) - u_q^{(\alpha)}(x), \qquad q > 0, \; x \in \mathbb{R} \qquad (4.5)$$

and

$$h^{(\alpha)}(x) = \lim_{q\to 0+} h_q^{(\alpha)}(x) = \lim_{q\to 0+}\{u_q^{(\alpha)}(0) - u_q^{(\alpha)}(x)\}, \qquad x \in \mathbb{R}. \qquad (4.6)$$

Lemma 4.1 (See also [29, Sec.4.2]) *Suppose that $1 < \alpha \le 2$. Then the following assertions hold:*
(i) $u_q^{(\alpha)}(0) = u_1^{(\alpha)}(0)q^{\frac{1}{\alpha}-1}$ *for any $q > 0$ where*

$$u_1^{(\alpha)}(0) = \frac{1}{\alpha\pi}\Gamma\left(1 - \frac{1}{\alpha}\right)\Gamma\left(\frac{1}{\alpha}\right); \qquad (4.7)$$

(ii) $h^{(\alpha)}(x) = h^{(\alpha)}(1)|x|^{\alpha-1}$ *for any $x \in \mathbb{R}$ where*

$$h^{(\alpha)}(1) = \left\{2\Gamma(\alpha)\sin\frac{(\alpha-1)\pi}{2}\right\}^{-1}; \qquad (4.8)$$

(iii) $\displaystyle\lim_{q\to 0+}\frac{u_q^{(\alpha)}(x)}{u_q^{(\alpha)}(0)} = 1$ *for any $x \in \mathbb{R}$.*

Proof. The assertion (i) is obvious by definition. It is also obvious that

$$h^{(\alpha)}(x) = \frac{1}{\pi}\int_0^{\infty}\frac{1 - \cos(x\xi)}{\xi^\alpha}\,d\xi = h^{(\alpha)}(1)|x|^{\alpha-1}, \qquad x \in \mathbb{R}. \qquad (4.9)$$

For the computation:

$$h^{(\alpha)}(1) = \frac{1}{\pi}\int_0^{\infty}\frac{1 - \cos\xi}{\xi^\alpha}\,d\xi = \left\{2\Gamma(\alpha)\sin\frac{(\alpha-1)\pi}{2}\right\}^{-1}, \qquad (4.10)$$

see Proposition 7.1 in the Appendix. Hence we obtain (ii). Assertion (iii) is obtained by noting that

$$\frac{u_q^{(\alpha)}(x)}{u_q^{(\alpha)}(0)} = 1 - \frac{h_q^{(\alpha)}(x)}{u_q^{(\alpha)}(0)} \xrightarrow{q\to 0+} 1. \qquad (4.11)$$

$\square$

Let $(L^{(\alpha)}(t) : t \geq 0)$ be the unique local time process such that

$$L^{(\alpha)}(t) = \lim_{\varepsilon \to 0+} \frac{1}{2\varepsilon} \int_0^t 1_{\{|X_\alpha(s)| < \varepsilon\}} ds \qquad \text{a.s.} \tag{4.12}$$

Then it is well-known (see [2, Lemma V.1.3]) that

$$E\left[\int_0^\infty e^{-t} dL^{(\alpha)}(t) \right] = u_1^{(\alpha)}(0). \tag{4.13}$$

Let $n^{(\alpha)}$ denote Itô's measure for the process X_α corresponding to this normalisation of the local time $(L^{(\alpha)}(t) : t \geq 0)$. Remark that

$$L^{(\alpha)}(t) = u_1^{(\alpha)}(0) L(t) \quad (t \geq 0), \qquad n^{(\alpha)} = \frac{1}{u_1^{(\alpha)}(0)} n \tag{4.14}$$

where $(L(t) : t \geq 0)$ and n, respectively, are as defined by (3.1) and (3.6), respectively.

Theorem 4.2 ([49] and [48]) *Suppose that* $1 < \alpha \leq 2$. *Then*

$$n^{(\alpha)}[h^{(\alpha)}(X(t)); \zeta > t] = 1, \qquad t > 0. \tag{4.15}$$

Consequently, there exists a unique probability measure $P^{h^{(\alpha)}}$ *on* $\mathbb{D}$ *such that*

$$E^{h^{(\alpha)}}[Z_t] = n^{(\alpha)}[Z_t h^{(\alpha)}(X(t)); \zeta > t] \tag{4.16}$$

for any $t > 0$ *and for any non-negative or bounded* $\mathcal{F}_t$-*measurable functional* Z_t.

The proof of Theorem 4.2 can be found in [49, Thm.4.7], so we omit it. See [48, Thm.1.2] for the proof of Theorem 4.2 for a fairly general class of one-dimensional symmetric Lévy processes. Several aspects of the law of local time process will be discussed in Hayashi–K. Yano [20].

Example 4.3 *In the case where* $\alpha = 2$, *we have* $X_2(t) = \sqrt{2}B(t)$, *and we have the following formulae:*

$$p_t^{(2)}(x) = \frac{1}{2\sqrt{\pi t}} e^{-\frac{x^2}{4t}}, \qquad t > 0, \ x \in \mathbb{R}, \tag{4.17}$$

$$u_q^{(2)}(x) = \frac{1}{2\sqrt{q}} e^{-\sqrt{q}|x|}, \qquad q > 0, \ x \in \mathbb{R}, \tag{4.18}$$

$$h^{(2)}(x) = \frac{1}{2}|x|, \qquad x \in \mathbb{R}. \tag{4.19}$$

The process $(\frac{1}{\sqrt{2}} X(t) : t \geq 0)$ *under* $P^{h^{(2)}}$ *is nothing but the symmetrised 3-dimensional Bessel process starting from the origin.*

Theorem 4.4 ([48]) *Let $q > 0$. Then the following assertions are valid:*
(i) *Suppose that $1 < \alpha \leq 2$. Then it holds that*

$$\lim_{x \to 0} \frac{h_q^{(\alpha)}(x)}{h^{(\alpha)}(x)} = 1; \tag{4.20}$$

(iia) *Suppose that $1 < \alpha < 2$. Let $a \neq 0$. Then it holds that*

$$\lim_{x \to 0} \frac{u_q^{(\alpha)}(a - x) - u_q^{(\alpha)}(a)}{h^{(\alpha)}(x)} = 0; \tag{4.21}$$

(iib) *Suppose that $\alpha = 2$. Let $a \neq 0$. Then it holds that*

$$\lim_{x \to \pm 0} \frac{u_q^{(2)}(a - x) - u_q^{(2)}(a)}{h^{(2)}(x)} = \pm e^{-\sqrt{q}|a|}. \tag{4.22}$$

The proof of the claim (i) of Theorem 4.4 can be found in [48, Lem.4.4], so we omit it. The proof of the claim (iib) of Theorem 4.4 is immediate from formulae (4.18) and (4.19), so we omit it, too. The proof of the claim (iia) of Theorem 4.4 is immediate from the following estimate:

Lemma 4.5 ([48]) *Suppose that $1 < \alpha < 2$. Let $a, x \in \mathbb{R}$ with $0 < 2|x| < |a|$. Then there exists a constant $C_q^{(\alpha)}$ such that*

$$|u_q^{(\alpha)}(a - x) - u_q^{(\alpha)}(a)| \leq \frac{C_q^{(\alpha)}}{|a|}. \tag{4.23}$$

The proof of Lemma 4.5 can be found in [48, Lem.6.2, (i)] in a rather general setting, but we give it for convenience of the reader.
Proof of Lemma 4.5. Integrating by parts, we have

$$u_q^{(\alpha)}(a - x) - u_q^{(\alpha)}(a) = \frac{1}{\pi} \int_0^\infty \frac{\cos a\xi - \cos(a - x)\xi}{q + \xi^\alpha} d\xi \tag{4.24}$$

$$= \frac{1}{\pi} \int_0^\infty \{\varphi(a\xi) - \varphi((a - x)\xi)\} \frac{\alpha \xi^\alpha d\xi}{(q + \xi^\alpha)^2} \tag{4.25}$$

where

$$\varphi(x) = \frac{\sin x}{x} \quad (x \neq 0), \qquad \varphi(0) = 1. \tag{4.26}$$

Since $\varphi'(x) = \frac{\cos x}{x} - \frac{\sin x}{x^2}$ $(x \neq 0)$, we have

$$|\varphi(a\xi) - \varphi((a - x)\xi)| \leq \left| \int_{(a-x)\xi}^{a\xi} |\varphi'(y)| dy \right| \leq \left| \int_{(a-x)\xi}^{a\xi} \frac{2}{|y|} dy \right|. \tag{4.27}$$

We change variables: $y = u\xi$, then we have

$$|\varphi(a\xi) - \varphi((a-x)\xi)| \leq \left| \int_{a-x}^{a} \frac{2}{|u|} du \right| \leq \frac{4|x|}{|a|}. \tag{4.28}$$

Thus we have proved the estimate (4.23). □

Let us prove the claim (iia) of Theorem 4.4.

Proof of the claim (iia) *of Theorem 4.4.* Without loss of generality, we may suppose that $0 < 2|x| < |a|$. Using the estimate (4.23), we obtain

$$\left| \frac{u_q^{(\alpha)}(a-x) - u_q^{(\alpha)}(a)}{h^{(\alpha)}(x)} \right| \leq \frac{C_q^{(\alpha)}}{|a|} \cdot \frac{|x|^{2-\alpha}}{h^{(\alpha)}(1)}, \tag{4.29}$$

which tends to zero since $\alpha < 2$. Now the proof is complete. □

5 First Hitting Time of a Single Point for X_α

5.1 The Case of One-dimensional Brownian Motion

Let $B = (B(t) : t \geq 0)$ denote the one-dimensional Brownian motion starting from 0. We consider the first hitting time of $a \in \mathbb{R}$ for B:

$$T_{\{a\}}(B) = \inf\{t > 0 : B(t) = a\}. \tag{5.1}$$

It is well-known (see, e.g., [35, Prop.II.3.7]) that the law of the hitting time is of (SD) type where its Laplace transform is given as follows:

$$E\left[e^{i\theta \widehat{B}(T_{\{a\}}(B))} \right] = E\left[e^{-\frac{1}{2}\theta^2 T_{\{a\}}(B)} \right] = e^{-|a\theta|}, \qquad \theta \in \mathbb{R} \tag{5.2}$$

where $\widehat{B} = (\widehat{B}(t) : t \geq 0)$ stands for an independent copy of B. Identity (5.2) can be expressed as

$$\widehat{B}(T_{\{a\}}(B)) \overset{\text{law}}{=} |a|\mathcal{C}, \qquad T_{\{a\}}(B) \overset{\text{law}}{=} 2a^2 T_{\frac{1}{2}}. \tag{5.3}$$

Let $a > 0$. Consider the random times $G_{\{a\}}(B)$ and $\Xi_{\{a\}}(B)$. The following path decomposition is due to Williams (see [43] and [44]; see also Prop.VII.4.8 and Thm.VII.4.9 of [35]):

Theorem 5.1 ([43] and [44]) *The process* $(B(t) : 0 \leq t \leq T_{\{a\}}(B))$ *is identical in law to the process* $(Y(t) : 0 \leq t \leq T')$ *defined as follows:*

$$Y(t) = \begin{cases} B_1(t) & \text{for } 0 \leq t < T_{\{M\}}(B_1); \\ B_2(T-t) & \text{for } T_{\{M\}}(B_1) \leq t < T; \\ R(T+t) & \text{for } T \leq t \leq T' \end{cases} \tag{5.4}$$

where M, B_1, B_2 and R are independent, M is a uniform variable on $(0, a)$, B_1 and B_2 are both identical in law to B, R is a 3-dimensional Bessel process starting at 0, and T and T' are random times defined as follows:

$$T = T_{\{M\}}(B_1) + T_{\{M\}}(B_2), \qquad T' = T + T_{\{a\}}(R). \tag{5.5}$$

From this path decomposition, we may compute the Laplace transform of $G_{\{a\}}(B)$ as follows:

$$E\left[e^{-qG_{\{a\}}(B)}\right] = \int_0^a \frac{dm}{a} \left(E\left[e^{-qT_{\{m\}}(B)}\right]\right)^2 = \frac{1 - e^{-2\sqrt{2q}a}}{2\sqrt{2q}a}, \qquad q > 0. \tag{5.6}$$

We may also compute the Laplace transform of $\Xi_{\{a\}}(B)$ as follows:

$$E\left[e^{-q\Xi_{\{a\}}(B)}\right] = E\left[e^{-qT_{\{a\}}(R)}\right] = \frac{\sqrt{2q}a}{\sinh(\sqrt{2q}a)}, \qquad q > 0. \tag{5.7}$$

In other words, we have

$$\Xi_{\{a\}}(B) \stackrel{\text{law}}{=} T_{\{a\}}(R) \stackrel{\text{law}}{=} a^2 S_1. \tag{5.8}$$

Remark 5.2 *The laws of first hitting times are known to be of* (SD) *type also for Bessel processes with drift (see Pitman–Yor [31]) and of* (ID) *type for one-dimensional diffusion processes (see Yamazato [46] and references therein).*

5.2 The Law of $T_{\{a\}}(X_\alpha)$

Consider the first hitting time of point $a \in \mathbb{R}$ for X_α of index $1 < \alpha \le 2$:

$$T_{\{a\}}(X_\alpha) = \inf\{t > 0 : X_\alpha(t) = a\}. \tag{5.9}$$

It is well-known (see, e.g., [2, Cor.II.5.18]) that

$$E[e^{-qT_{\{a\}}(X_\alpha)}] = \frac{u_q^{(\alpha)}(a)}{u_q^{(\alpha)}(0)}, \qquad q > 0. \tag{5.10}$$

Let $\widehat{X}_\alpha = (\widehat{X}_\alpha(t) : t \ge 0)$ be an independent copy of X_α. The following is a generalization of formulae (5.2) and (5.3).

Theorem 5.3 (See also Cordero [50, §1.2.2]) *Suppose that $1 < \alpha \le 2$. Let $a \in \mathbb{R}$. Then*

$$E\left[e^{i\theta\widehat{X}_\alpha(T_{\{a\}}(X_\alpha))}\right] = E\left[e^{-|\theta|^\alpha T_{\{a\}}(X_\alpha)}\right] = \frac{\sin(\pi/\alpha)}{2\pi/\alpha} L_\alpha(a\theta) \tag{5.11}$$

and

$$\widehat{X}_\alpha(T_{\{a\}}(X_\alpha)) \stackrel{\text{law}}{=} |a|\mathcal{C}_\alpha, \qquad T_{\{a\}}(X_\alpha) \stackrel{\text{law}}{=} \frac{|a|^\alpha}{(\mathcal{R}_\alpha)^\alpha \mathcal{B}_{1-\gamma,\gamma}} \tag{5.12}$$

where $\gamma = 1/\alpha$.

We can recover (5.3) if we take $\alpha = 2$, noting that

$$T_{\{a\}}(B) \overset{\text{law}}{=} T_{\{\sqrt{2}a\}}(X_2) \overset{\text{law}}{=} \frac{2a^2}{(\mathcal{R}_2)^2 \mathcal{B}_{1/2,1/2}} \overset{\text{law}}{=} \frac{a^2}{2\mathfrak{e}\mathcal{B}_{1/2,1/2}} \overset{\text{law}}{=} \frac{a^2}{2\mathcal{G}_{1/2}} \overset{\text{law}}{=} 2a^2 T_{1/2}.$$
(5.13)

Proof of Theorem 5.3. If we take $q = |\theta|^\alpha$, then

$$u_q^{(\alpha)}(x) = \frac{1}{\pi} \int_0^\infty \frac{\cos(x\xi)}{|\theta|^\alpha + |\xi|^\alpha} d\xi = \frac{|\theta|^{1-\alpha}}{\pi} \int_0^\infty \frac{\cos(\theta x\xi)}{1 + |\xi|^\alpha} d\xi, \qquad x \in \mathbb{R}. \quad (5.14)$$

Hence, by formula (5.10), we obtain

$$E[e^{-|\theta|^\alpha T_{\{a\}}(X_\alpha)}] = E[\cos(\theta|a|\mathcal{C}_\alpha)] = E[e^{i\theta|a|\mathcal{C}_\alpha}]. \quad (5.15)$$

This shows (5.11) and the first identity of (5.12).

To prove the second identity of (5.12), it suffices to prove the claim when $a = 1$; in fact, by the self-similarity property, we have

$$T_{\{a\}}(X_\alpha) \overset{\text{law}}{=} |a|^\alpha T_{\{1\}}(X_\alpha). \quad (5.16)$$

Note that

$$\int_0^\infty e^{-qt} P(T_{\{1\}}(X_\alpha) < t) dt = \frac{1}{q} E[e^{-qT_{\{1\}}(X_\alpha)}] = \frac{u_q^{(\alpha)}(1)}{q u_q^{(\alpha)}(0)}. \quad (5.17)$$

Since $u_q^{(\alpha)}(0) = u_1^{(\alpha)}(0) q^{\gamma-1}$ where $\gamma = \frac{1}{\alpha}$, we have

$$\frac{1}{q u_q^{(\alpha)}(0)} = \frac{1}{u_1^{(\alpha)}(0)} q^{-\gamma} = \frac{1}{u_1^{(\alpha)}(0)\Gamma(\gamma)} \int_0^\infty y^{\gamma-1} e^{-qy} dy. \quad (5.18)$$

Hence, by Laplace inversion, we obtain

$$P(T_{\{1\}}(X_\alpha) < t) = \frac{1}{u_1^{(\alpha)}(0)\Gamma(\gamma)} \int_0^t (t-s)^{\gamma-1} p_s^{(\alpha)}(1) ds. \quad (5.19)$$

By the scaling property

$$p_s^{(\alpha)}(x) = \frac{1}{s^\gamma} p_1^{(\alpha)}\left(\frac{x}{s^\gamma}\right), \qquad s > 0, \quad (5.20)$$

we obtain

$$P(T_{\{1\}}(X_\alpha) < t) = \frac{\Gamma(1-\gamma) p_1^{(\alpha)}(0)}{u_1^{(\alpha)}(0)} \int_0^t \frac{(t-s)^{\gamma-1} s^{-\gamma}}{\Gamma(\gamma)\Gamma(1-\gamma)} \cdot \frac{p_1^{(\alpha)}(1/s^\gamma)}{p_1^{(\alpha)}(0)} ds \quad (5.21)$$

$$= \int_0^1 \frac{(1-s)^{\gamma-1} s^{-\gamma}}{\Gamma(\gamma)\Gamma(1-\gamma)} \cdot \frac{p_1^{(\alpha)}(1/(ts)^\gamma)}{p_1^{(\alpha)}(0)} ds \quad (5.22)$$

(from (2.75) and (i) of Lemma 4.1)

$$=P\left(\mathcal{R}_\alpha > \frac{1}{(t\mathcal{B}_{1-\gamma,\gamma})^\gamma}\right) \qquad \text{(from (2.76))} \qquad (5.23)$$

$$=P\left(\frac{1}{(\mathcal{R}_\alpha)^\alpha \mathcal{B}_{1-\gamma,\gamma}} < t\right). \qquad (5.24)$$

Now the proof is complete. □

5.3 Laplace Transform Formula for First Hitting Time of Two Points

For later use, we prepare several important formulae concerning Laplace transforms for first hitting time of two points.

Denote the symmetric α-stable process starting from $x \in \mathbb{R}$ by $X_\alpha^x(t) = x + X_\alpha(t)$. Suppose that $1 < \alpha \leq 2$. Recall that the Laplace transform of first hitting time of a single point is given by (see (5.10))

$$\varphi_{x\to a}^q := E\left[e^{-qT_{\{a\}}(X_\alpha^x)}\right] = \frac{u_q^{(\alpha)}(x-a)}{u_q^{(\alpha)}(0)}. \qquad (5.25)$$

The Laplace transform of first hitting time of two points, i.e., $T_{\{a\}}(X_\alpha^x) \wedge T_{\{b\}}(X_\alpha^x)$, is given by the following formula:

Proposition 5.4 *Suppose that $1 < \alpha \leq 2$. Let $x, a, b \in \mathbb{R}$. Then*

$$\varphi_{x\to a,b}^q := E\left[e^{-qT_{\{a\}}(X_\alpha^x)\wedge T_{\{b\}}(X_\alpha^x)}\right] = \frac{u_q^{(\alpha)}(x-a) + u_q^{(\alpha)}(x-b)}{u_q^{(\alpha)}(0) + u_q^{(\alpha)}(a-b)}. \qquad (5.26)$$

Proof. For any closed set F, put

$$T_F^x = T_F(X_\alpha^x) = \inf\{t > 0 : X_\alpha^x(t) \in F\}. \qquad (5.27)$$

Following [2, p.49], we introduce the *capacitary measure* as

$$\mu_F^q(A) = q\int E\left[e^{-qT_F^z}; X_\alpha^z(T_F^z) \in A\right]dz, \qquad A \in \mathcal{B}(\mathbb{R}). \qquad (5.28)$$

Now we apply Theorem II.2.7 of [2] and obtain

$$\int \mu_F^q(dy)\int_A u_q^{(\alpha)}(x-y)dx = \int_A E\left[e^{-qT_F^x}\right]dx, \qquad A \in \mathcal{B}(\mathbb{R}), \qquad (5.29)$$

where we have used the fact that the process considered is symmetric. This implies that

$$E\left[e^{-qT_F^x}\right] = \int u_q^{(\alpha)}(x-y)\mu_F^q(dy). \qquad (5.30)$$

By definition of μ_F^q, we obtain

$$E\left[e^{-qT_F^x}\right] = q\int E\left[e^{-qT_F^z}u_q^{(\alpha)}(x - X_\alpha^z(T_F^z))\right]dz. \tag{5.31}$$

Now we let $F = \{a, b\}$. Then we have $T_F^x = T_{\{a\}}^x \wedge T_{\{b\}}^x$. Noting that $X_\alpha^z(T_F^z) = a$ or b almost surely, we have

$$E\left[e^{-qT_F^x}\right] = C_{a\prec b}^q u_q^{(\alpha)}(x - a) + C_{b\prec a}^q u_q^{(\alpha)}(x - b) \tag{5.32}$$

where

$$C_{a\prec b}^q = q\int E\left[e^{-qT_F^z}; T_{\{a\}}^z < T_{\{b\}}^z\right]dz. \tag{5.33}$$

Since $T_F^a = T_F^b = 0$ almost surely, we have

$$1 = C_{a\prec b}^q u_q^{(\alpha)}(0) + C_{b\prec a}^q u_q^{(\alpha)}(a - b), \tag{5.34}$$

$$1 = C_{a\prec b}^q u_q^{(\alpha)}(b - a) + C_{b\prec a}^q u_q^{(\alpha)}(0). \tag{5.35}$$

Hence we obtain

$$C_{a\prec b}^q = C_{b\prec a}^q = \frac{1}{u_q^{(\alpha)}(0) + u_q^{(\alpha)}(a - b)}. \tag{5.36}$$

Combining this with (5.32), we obtain the desired result.   □

The Laplace transform of first hitting time of point a before hitting b is given by the following formula:

Proposition 5.5 *Suppose that $1 < \alpha \leq 2$. Let $x, a, b \in \mathbb{R}$ with $a \neq b$. Then*

$$\varphi_{x\to a\prec b}^q := E\left[e^{-qT_{\{a\}}(X_\alpha^x)}; T_{\{a\}}(X_\alpha^x) < T_{\{b\}}(X_\alpha^x)\right] \tag{5.37}$$

$$= \frac{u_q^{(\alpha)}(0)u_q^{(\alpha)}(x - a) - u_q^{(\alpha)}(a - b)u_q^{(\alpha)}(x - b)}{\{u_q^{(\alpha)}(0)\}^2 - \{u_q^{(\alpha)}(a - b)\}^2}. \tag{5.38}$$

Proof. Keep the notations in the proof of Proposition 5.4. Noting that

$$T_{\{a\}}^x = T_{\{b\}}^x + T_{\{a\}}^x \circ \theta_{T_{\{b\}}^x} \qquad \text{on } \{T_{\{a\}}^x > T_{\{b\}}^x\}, \tag{5.39}$$

we see, by the strong Markov property, that

$$E\left[e^{-qT_{\{a\}}^x}; T_{\{a\}}^x > T_{\{b\}}^x\right] = \varphi_{x\to b\prec a}^q \varphi_{b\to a}^q. \tag{5.40}$$

Thus we have

$$\varphi_{x\to a}^q = \varphi_{x\to a\prec b}^q + \varphi_{x\to b\prec a}^q \varphi_{b\to a}^q. \tag{5.41}$$

Combining this with the trivial identity

$$\varphi^q_{x\to a,b} = \varphi^q_{x\to a\prec b} + \varphi^q_{x\to b\prec a},$$ (5.42)

we obtain

$$\varphi^q_{x\to a\prec b} = \frac{\varphi^q_{x\to a} - \varphi^q_{b\to a}\varphi^q_{x\to a,b}}{1 - \varphi^q_{b\to a}}.$$ (5.43)

This proves the desired result. □

Remark 5.6 *Formula (5.38) can be written as*

$$E\left[e^{-qT_{\{a\}}(X^x_\alpha)}; T_{\{a\}}(X^x_\alpha) < T_{\{b\}}(X^x_\alpha)\right]$$ (5.44)

$$= \frac{u^{(\alpha)}_q(x-b)h^{(\alpha)}_q(a-b) + u^{(\alpha)}_q(0)\{h^{(\alpha)}_q(x-b) - h^{(\alpha)}_q(x-a)\}}{\{u^{(\alpha)}_q(0) + u^{(\alpha)}_q(a-b)\}h^{(\alpha)}_q(a-b)}.$$ (5.45)

Letting $q \to 0+$, we obtain

$$P\left(T_{\{a\}}(X^x_\alpha) < T_{\{b\}}(X^x_\alpha)\right) = \frac{1}{2}\left\{1 + \frac{|x-b|^{\alpha-1} - |x-a|^{\alpha-1}}{|a-b|^{\alpha-1}}\right\},$$ (5.46)

which is a special case of Getoor's formula [16, Thm.6.5]. See also [48, Thm.6.1] for its application to Itô's measure for symmetric Lévy processes.

Remark 5.7 *Let $a < x < b$. Then, as corollaries of Propositions 5.4 and 5.5, we recover the following well-known formulae (see, e.g., [24, Problem 1.7.6]) for the Brownian motion $(B^x(t) = x + B(t) : t \geq 0)$ starting from x:*

$$E\left[e^{-qT_{\{a\}}(B^x)\wedge T_{\{b\}}(B^x)}\right] = \frac{\cosh\left(\sqrt{2q}\left(x - \frac{b+a}{2}\right)\right)}{\cosh\left(\sqrt{2q}\cdot\frac{b-a}{2}\right)}$$ (5.47)

and

$$E\left[e^{-qT_{\{a\}}(B^x)}; T_{\{a\}}(B^x) < T_{\{b\}}(B^x)\right] = \frac{\sinh\sqrt{2q}(b-x)}{\sinh\sqrt{2q}(b-a)}.$$ (5.48)

5.4 The Laplace Transforms of $G_{\{a\}}(X_\alpha)$ and $\Xi_{\{a\}}(X_\alpha)$

The following theorem generalises formulae (5.6) and (5.7):

Theorem 5.8 *Suppose that $1 < \alpha \leq 2$. Let $a \neq 0$. Then it holds that*

$$E\left[e^{-qG_{\{a\}}(X_\alpha)}\right] = \frac{\{u^{(\alpha)}_q(0)\}^2 - \{u^{(\alpha)}_q(a)\}^2}{2h^{(\alpha)}(a)u^{(\alpha)}_q(0)}$$ (5.49)

and that

$$E\left[e^{-q\Xi_{\{a\}}(X_\alpha)}\right] = \frac{u^{(\alpha)}_q(a)}{u^{(\alpha)}_q(0)} \cdot \frac{2h^{(\alpha)}(a)u^{(\alpha)}_q(0)}{\{u^{(\alpha)}_q(0)\}^2 - \{u^{(\alpha)}_q(a)\}^2}.$$ (5.50)

Remark 5.9 *The left hand sides of (5.49) and (5.50) are functions of $q|a|^\alpha$ since*

$$G_{\{a\}}(X_\alpha) \stackrel{\text{law}}{=} |a|^\alpha G_{\{1\}}(X_\alpha) \qquad \text{and} \qquad \Xi_{\{a\}}(X_\alpha) \stackrel{\text{law}}{=} |a|^\alpha \Xi_{\{1\}}(X_\alpha). \quad (5.51)$$

We may check that so are also the right hand sides by the following formulae:

$$u_q^{(\alpha)}(0) = |a|^{\alpha-1} u_{q|a|^\alpha}^{(\alpha)}(0), \qquad u_q^{(\alpha)}(a) = |a|^{\alpha-1} u_{q|a|^\alpha}^{(\alpha)}(1) \quad (5.52)$$

and

$$h^{(\alpha)}(a) = |a|^{\alpha-1} h^{(\alpha)}(1). \quad (5.53)$$

For the proof of Theorem 5.8, we need the following proposition.

Proposition 5.10 *Suppose that $X = X_\alpha$ with $1 < \alpha \le 2$. Let $a \ne 0$ and $q, r > 0$. Then*

$$\boldsymbol{n}^{(\alpha)}\left[e^{-qT_{\{a\}} - r(\zeta - T_{\{a\}})}; T_{\{a\}} < \zeta\right] = \frac{u_r^{(\alpha)}(a)}{u_r^{(\alpha)}(0)} \cdot \frac{u_q^{(\alpha)}(a)}{\{u_q^{(\alpha)}(0)\}^2 - \{u_q^{(\alpha)}(a)\}^2}. \quad (5.54)$$

Consequently, it holds that

$$\boldsymbol{n}^{(\alpha)}\left[e^{-qT_{\{a\}}}; T_{\{a\}} < \zeta\right] = \frac{u_q^{(\alpha)}(a)}{\{u_q^{(\alpha)}(0)\}^2 - \{u_q^{(\alpha)}(a)\}^2} \quad (5.55)$$

and that

$$\boldsymbol{n}^{(\alpha)}(T_{\{a\}} < \zeta) = \frac{1}{2h^{(\alpha)}(a)}. \quad (5.56)$$

Proof. Let us only prove formula (5.54); in fact, from this formula one can obtain formulae (5.55) and (5.56) immediately by the limit (4.6) and Lemma 4.1.

Let $\varepsilon > 0$. By the strong Markov property of $\boldsymbol{n}^{(\alpha)}$, we have

$$\boldsymbol{n}^{(\alpha)}\left[e^{-qT_{\{a\}} - r(\zeta - T_{\{a\}})}; \varepsilon < T_{\{a\}} < \zeta\right] \quad (5.57)$$

$$= e^{-q\varepsilon} \boldsymbol{n}^{(\alpha)}\left[\left.(\varphi_{x \to a \prec 0}^q)\right|_{x=X(\varepsilon)} \varphi_{0 \to a}^r; \varepsilon < T_{\{a\}} \wedge \zeta\right] \quad (5.58)$$

$$= e^{-q\varepsilon} \varphi_{0 \to a}^r E^{h^{(\alpha)}}\left[\left.\left(\frac{\varphi_{x \to a \prec 0}^q}{h^{(\alpha)}(x)}\right)\right|_{x=X(\varepsilon)}; \varepsilon < T_{\{a\}}\right]. \quad (5.59)$$

Here we used Theorem 4.2. Note that

$$\frac{\varphi_{x \to a \prec 0}^q}{h^{(\alpha)}(x)} \cdot \left\{\{u_q^{(\alpha)}(0)\}^2 - \{u_q^{(\alpha)}(a)\}^2\right\}$$

$$= u_q^{(\alpha)}(a) \cdot \frac{h_q^{(\alpha)}(x)}{h^{(\alpha)}(x)} - u_q^{(\alpha)}(0) \cdot \frac{u_q^{(\alpha)}(x-a) - u_q^{(\alpha)}(a)}{h^{(\alpha)}(x)}. \quad (5.60)$$

Suppose that $1 < \alpha < 2$. Then we see that the right hand side of (5.60) converges to $u_q^{(\alpha)}(a)$ as $x \to 0$ by Theorem 4.4. Letting $\varepsilon \to 0+$ in identity (5.57)-(5.59), we obtain formula (5.54) by the dominated convergence theorem.

Suppose that $\alpha = 2$. We may assume without loss of generality that $a > 0$. Then we see that the quantity (5.59) is equal to

$$\frac{1}{2} e^{-q\varepsilon} \varphi_{0 \to a}^{r} E^{h^{(2)}} \left[\left. \left(\frac{\varphi_{x \to a \to 0}^{q}}{h^{(2)}(x)} \right) \right|_{x = X(\varepsilon)} ; \ \varepsilon < T_{\{a\}}, \ X(\varepsilon) > 0 \right]; \qquad (5.61)$$

In fact, $P^{h^{(2)}}$ is nothing but the law of the symmetrisation of the 3-dimensional Bessel process starting from the origin. We also see that the right hand side of (5.60) converges to $2u_q^{(\alpha)}(a)$ as $x \to +0$ by Theorem 4.4. Hence, letting $\varepsilon \to 0+$ in identity (5.57)-(5.61), we obtain formula (5.54) by the dominated convergence theorem. Now the proof is complete. $\square$

Now we proceed to prove Theorem 5.8.

Proof of Theorem 5.8. Using formulae (5.56) and (5.54) (with $r = q$), we have

$$\mathbf{n}^{(\alpha)} \left[1 - e^{-q\zeta}; \ T_{\{a\}} > \zeta \right] \qquad (5.62)$$

$$= \mathbf{n}^{(\alpha)} \left[1 - e^{-q\zeta} \right] - \mathbf{n}^{(\alpha)}(T_{\{a\}} < \zeta) + \mathbf{n}^{(\alpha)} \left[e^{-q\zeta}; \ T_{\{a\}} < \zeta \right] \qquad (5.63)$$

$$= \frac{1}{u_q^{(\alpha)}(0)} - \frac{1}{2h^{(\alpha)}(a)} + \frac{u_q^{(\alpha)}(a)}{u_q^{(\alpha)}(0)} \cdot \frac{u_q^{(\alpha)}(a)}{\{u_q^{(\alpha)}(0)\}^2 - \{u_q^{(\alpha)}(a)\}^2}. \qquad (5.64)$$

Hence we obtain

$$\frac{\mathbf{n}^{(\alpha)} \left[1 - e^{-q\zeta}; \ T_{\{a\}} > \zeta \right]}{\mathbf{n}^{(\alpha)}(T_{\{a\}} < \zeta)} = \frac{2h^{(\alpha)}(a)u_q^{(\alpha)}(0)}{\{u_q^{(\alpha)}(0)\}^2 - \{u_q^{(\alpha)}(a)\}^2} - 1. \qquad (5.65)$$

Combining this with formula (3.18), we obtain (5.49).

Using formulae (5.55) and (5.56), we have

$$\frac{\mathbf{n}^{(\alpha)} \left[e^{-qT_{\{a\}}}; T_{\{a\}} < \zeta \right]}{\mathbf{n}^{(\alpha)}(T_{\{a\}} < \zeta)} = \frac{2h^{(\alpha)}(a)u_q^{(\alpha)}(a)}{\{u_q^{(\alpha)}(0)\}^2 - \{u_q^{(\alpha)}(a)\}^2}. \qquad (5.66)$$

Combining this with formula (3.19), we obtain (5.50). $\square$

5.5 Overshoots at the First Passage Time of a Level

For comparison with the description of the law of a first hitting time, we recall the law of the overshoot at the first passage time of a level. Let $X_\alpha = (X_\alpha(t) : t \geq 0)$ denote the symmetric stable Lévy process of index $0 < \alpha \leq 2$ starting from the origin such that $E[e^{i\lambda X_\alpha(t)}] = e^{-t|\lambda|^\alpha}$.

Consider the *first passage time of level a > 0* for X_α:

$$T_{[a,\infty)}(X_\alpha) = \inf\{t > 0 : X_\alpha(t) \geq a\}. \tag{5.67}$$

The variable $X_\alpha(T_{[a,\infty)}(X_\alpha)) - a$ is the overshoot at the first hitting time of level a.

The following theorem is due to Ray [34], although he does not express his result like this:

Theorem 5.11 ([34]) *Suppose that $0 < \alpha \leq 2$. Let $a > 0$. Then*

$$X_\alpha(T_{[a,\infty)}(X_\alpha)) - a \overset{\text{law}}{=} a\frac{\mathcal{G}_{1-\frac{\alpha}{2}}}{\widehat{\mathcal{G}}_{\frac{\alpha}{2}}} \tag{5.68}$$

where $\mathcal{G}_{1-\frac{\alpha}{2}}$ and $\widehat{\mathcal{G}}_{\frac{\alpha}{2}}$ are independent gamma variables of indices $1 - \frac{\alpha}{2}$ and $\frac{\alpha}{2}$, respectively.

For its multidimensional analogue, see Blumenthal–Getoor–Ray [5].

6 First Hitting Time of a Single Point for $|X_\alpha|$

6.1 The Case of One-dimensional Reflecting Brownian Motion

We consider the first hitting time of $a > 0$ for the reflecting Brownian motion $|B| = (|B|(t))$:

$$T_{\{a\}}(|B|) = \inf\{t > 0 : |B(t)| = a\} = T_{\{a\}}(B) \wedge T_{\{-a\}}(B). \tag{6.1}$$

It is well-known (see, e.g., [35, Prop.II.3.7]) that the law of the hitting time is of (SD) type where its Laplace transforms is given as follows:

$$E\left[e^{i\theta\widehat{B}(T_{\{a\}}(|B|))}\right] = E\left[e^{-\frac{1}{2}\theta^2 T_{\{a\}}(|B|)}\right] = \frac{1}{\cosh(a\theta)}, \qquad \theta \in \mathbb{R}. \tag{6.2}$$

Identity (6.2) can be expressed as

$$\widehat{B}(T_{\{a\}}(|B|)) \overset{\text{law}}{=} a\mathbb{C}_1 \overset{\text{law}}{=} 2a\mathcal{M}_0, \qquad T_{\{a\}}(|B|) \overset{\text{law}}{=} a^2 C_1. \tag{6.3}$$

Noting that

$$\frac{1}{\cosh(a\theta)} = \frac{2e^{-a|\theta|}}{1 + e^{-2a|\theta|}} = 2e^{-a|\theta|}\sum_{n=0}^{\infty}(-1)^n e^{-2na|\theta|}, \tag{6.4}$$

we have the following expansion:

$$E\left[e^{-qT_{\{a\}}(|B|)}\right] = 2\sum_{n=0}^{\infty}(-1)^n E\left[e^{-qT_{\{(2n+1)a\}}(B)}\right], \qquad q > 0. \tag{6.5}$$

Consider the random times $G_{\{a\}}(|B|)$ and $\Xi_{\{a\}}(|B|)$. By means of random time-change, Williams' path decomposition (Theorem 5.1) is also valid for the reflecting Brownian motion $|B|$ instead of B. Hence we may compute the Laplace transforms of these variables as follows:

$$E\left[e^{-qG_{\{a\}}(|B|)}\right] = \int_0^a \frac{dm}{a}\left(E\left[e^{-qT_{\{m\}}(|B|)}\right]\right)^2 = \frac{\tanh(\sqrt{2q}a)}{\sqrt{2q}a}, \qquad q > 0 \tag{6.6}$$

and

$$E\left[e^{-q\Xi_{\{a\}}(|B|)}\right] = E\left[e^{-qT_{\{a\}}(R)}\right] = \frac{\sqrt{2q}|a|}{\sinh(\sqrt{2q}|a|)}, \qquad q > 0. \tag{6.7}$$

In other words, we have

$$G_{\{a\}}(|B|) \overset{\text{law}}{=} a^2 T_1, \qquad \Xi_{\{a\}}(|B|) \overset{\text{law}}{=} T_{\{a\}}(R) \overset{\text{law}}{=} a^2 S_1. \tag{6.8}$$

6.2 Discussions about the Laplace Transform of $T_{\{a\}}(|X_\alpha|)$

Consider the first hitting time of point $a > 0$ for $|X_\alpha|$ of index $1 < \alpha \le 2$:

$$T_{\{a\}}(|X_\alpha|) = \inf\{t > 0 : |X_\alpha(t)| = a\} = T_{\{a\}}(X_\alpha) \wedge T_{\{-a\}}(X_\alpha). \tag{6.9}$$

The following theorem generalises the Laplace transform formula (6.2) and the expansion (6.5).

Theorem 6.1 *Suppose that* $1 < \alpha \le 2$. *Let* $a \in \mathbb{R}$. *Then*

$$E\left[e^{-qT_{\{a\}}(|X_\alpha|)}\right] = \frac{2u_q^{(\alpha)}(a)}{u_q^{(\alpha)}(0) + u_q^{(\alpha)}(2a)} \tag{6.10}$$

$$= 2\sum_{n=0}^{\infty}(-1)^n E\left[e^{-q\{T_{\{a\}}(X_\alpha)+T_{\{2a\}}(X_\alpha^{(1)})+\cdots+T_{\{2a\}}(X_\alpha^{(n)})\}}\right] \tag{6.11}$$

where $X_\alpha^{(1)},\ldots,X_\alpha^{(n)},\ldots$ *are independent copies of* X_α.

Proof. Applying Proposition 5.4 with $x = 0$ and $b = -a$, we obtain the first identity (6.10). Expanding the right hand side, we have

$$E\left[e^{-qT_{\{a\}}(|X_\alpha|)}\right] = \frac{2u_q^{(\alpha)}(a)}{u_q^{(\alpha)}(0)}\sum_{n=0}^{\infty}(-1)^n\left\{\frac{u_q^{(\alpha)}(2a)}{u_q^{(\alpha)}(0)}\right\}^n. \tag{6.12}$$

Using formula (5.25), we may rewrite the identity as

$$E\left[e^{-qT_{\{a\}}(|X_\alpha|)}\right] = 2E\left[e^{-qT_{\{a\}}(X_\alpha)}\right]\sum_{n=0}^{\infty}(-1)^n\left\{E\left[e^{-qT_{\{2a\}}(X_\alpha)}\right]\right\}^n. \tag{6.13}$$

This is nothing but the second identity (6.11). $\square$

In the case of Brownian motion $B = (B(t))$ on one hand, we have

$$T_{\{a\}}(B) + T_{\{2a\}}(B^{(1)}) + \cdots + T_{\{2a\}}(B^{(n)}) \overset{\text{law}}{=} T_{\{(2n+1)a\}}(B) \qquad (6.14)$$

where $B^{(1)}, \ldots, B^{(n)}$ are independent copies of B. In the case of symmetric α-stable process $X_\alpha = (X_\alpha(t))$ for $1 < \alpha < 2$ on the other hand, however, the law of the sum

$$T_{\{a\}}(X_\alpha) + T_{\{2a\}}(X_\alpha^{(1)}) + \cdots + T_{\{2a\}}(X_\alpha^{(n)}) \qquad (6.15)$$

differs from that of $T_{\{(2n+1)a\}}(X_\alpha)$. In fact, we have the following theorem.

Theorem 6.2 *Suppose that $1 < \alpha < 2$. Let $a \in \mathbb{R}$. Then, for any $q > 0$ and $n \geq 1$,*

$$E\left[e^{-q\{T_{\{a\}}(X_\alpha) + T_{\{2a\}}(X_\alpha^{(1)}) + \cdots + T_{\{2a\}}(X_\alpha^{(n)})\}}\right] < E\left[e^{-qT_{\{(2n+1)a\}}(X_\alpha)}\right]. \qquad (6.16)$$

Proof. Set

$$D_n = E\left[e^{-qT_{\{(2n+1)a\}}(X_\alpha)}\right] - E\left[e^{-qT_{\{(2n-1)a\}}(X_\alpha)}\right] E\left[e^{-qT_{\{2a\}}(X_\alpha)}\right]. \qquad (6.17)$$

Then it suffices to prove that $D_n > 0$ for all $n \geq 1$.

Keep the notations in the proof of Proposition 5.4. Note that

$$D_n = \varphi^q_{0 \to (2n+1)a} - \varphi^q_{0 \to (2n-1)a} \varphi^q_{0 \to 2a}. \qquad (6.18)$$

Using formula (5.41) and translation invariance, we have

$$\varphi^q_{0 \to (2n+1)a} = \varphi^q_{0 \to (2n+1)a \prec (2n-1)a} + \varphi^q_{0 \to (2n-1)a \prec (2n+1)a} \varphi^q_{(2n-1)a \to (2n+1)a} \qquad (6.19)$$

$$= \varphi^q_{0 \to (2n+1)a \prec (2n-1)a} + \varphi^q_{0 \to (2n-1)a \prec (2n+1)a} \varphi^q_{0 \to 2a}. \qquad (6.20)$$

Using formula (5.41), translation invariance, and the symmetry, we have

$$\varphi^q_{0 \to (2n-1)a} = \varphi^q_{0 \to (2n-1)a \prec (2n+1)a} + \varphi^q_{0 \to (2n+1)a \prec (2n-1)a} \varphi^q_{0 \to 2a}. \qquad (6.21)$$

Hence we obtain

$$D_n = \varphi^q_{0 \to (2n+1)a \prec (2n-1)a}\left\{1 - (\varphi^q_{0 \to 2a})^2\right\}, \qquad (6.22)$$

which turns out to be positive because both $\varphi^q_{0 \to (2n+1)a \prec (2n-1)a}$ and $\varphi^q_{0 \to 2a}$ are positive and less than 1. Now the proof is complete. $\square$

Remark 6.3 *The consistency of the two formulae (6.18) and (6.22) can be confirmed by formulae (5.25) and (5.38) as follows:*

$$\varphi^q_{0\to(2n+1)a\prec(2n-1)a}\left\{1-\left(\varphi^q_{0\to2a}\right)^2\right\}$$ (6.23)

$$=\frac{u_q^{(\alpha)}(0)u_q^{(\alpha)}((2n+1)a)-u_q^{(\alpha)}(2a)u_q^{(\alpha)}((2n-1)a)}{\{u_q^{(\alpha)}(0)\}^2-\{u_q^{(\alpha)}(2a)\}^2}\cdot\left\{1-\left(\frac{u_q^{(\alpha)}(2a)}{u_q^{(\alpha)}(0)}\right)^2\right\}$$ (6.24)

$$=\frac{u_q^{(\alpha)}((2n+1)a)}{u_q^{(\alpha)}(0)}-\frac{u_q^{(\alpha)}((2n-1)a)}{u_q^{(\alpha)}(0)}\cdot\frac{u_q^{(\alpha)}(2a)}{u_q^{(\alpha)}(0)}$$ (6.25)

$$=\varphi^q_{0\to(2n+1)a}-\varphi^q_{0\to(2n-1)a}\varphi^q_{0\to2a}.$$ (6.26)

6.3 The Laplace Transforms of $G_{\{a\}}(|X_\alpha|)$ and $\Xi_{\{a\}}(|X_\alpha|)$

Since $|X_\alpha|=(|X_\alpha(t)|:t\geq0)$ is a strong Markov process, the arguments of Section 3.2 are valid for $X=|X_\alpha|$. Let us compute the Laplace transforms of $G_{\{a\}}(|X_\alpha|)$ and $\Xi_{\{a\}}(|X_\alpha|)$.

Theorem 6.4 *Suppose that $1<\alpha\leq2$. Let $a>0$. Then it holds that*

$$E\left[e^{-qG_{\{a\}}(|X_\alpha|)}\right]=\frac{2V_q^{(\alpha)}(a)}{\{u_q^{(\alpha)}(0)+u_q^{(\alpha)}(2a)\}\{4h^{(\alpha)}(a)-h^{(\alpha)}(2a)\}}$$ (6.27)

and that

$$E\left[e^{-q\Xi_{\{a\}}(|X_\alpha|)}\right]=\frac{u_q^{(\alpha)}(a)\{4h^{(\alpha)}(a)-h^{(\alpha)}(2a)\}}{V_q^{(\alpha)}(a)}$$ (6.28)

where

$$V_q^{(\alpha)}(a):=\{u_q^{(\alpha)}(0)\}^2+u_q^{(\alpha)}(0)u_q^{(\alpha)}(2a)-2\{u_q^{(\alpha)}(a)\}^2.$$ (6.29)

For the proof of Theorem 6.4, we need a certain Laplace transform formula for the first hitting time of three points. Avoiding unnecessary generality, we are satisfied with the following special case:

Proposition 6.5 *Suppose that $1<\alpha\leq2$. Let $x,a\in\mathbb{R}$. Then*

$$\varphi^q_{x\to0,a,-a}:=E\left[e^{-qT_{\{0,a,-a\}}(X^x_\alpha)}\right]$$ (6.30)

$$=C^q_{0\prec a,-a}u_q(x)+C^q_{a\prec0,-a}u_q(x-a)+C^q_{-a\prec0,a}u_q(x+a)$$ (6.31)

where

$$C^q_{0\prec a,-a}=\frac{u_q^{(\alpha)}(0)+u_q^{(\alpha)}(2a)-2u_q^{(\alpha)}(a)}{V_q^{(\alpha)}(a)}$$ (6.32)

and

$$C^q_{a\prec0,-a}=C^q_{-a\prec0,a}=\frac{u_q^{(\alpha)}(0)-u_q^{(\alpha)}(a)}{V_q^{(\alpha)}(a)}.$$ (6.33)

The proof of Proposition 6.5 is similar to that of Proposition 5.4 based on identity (5.31) with $F = \{0, a, -a\}$, so we omit it.

Proposition 6.6 *Suppose that $1 < \alpha \le 2$. Let $x, a \in \mathbb{R}$ with $a \ne 0$. Then*

$$\varphi^q_{x \to a, -a \prec 0} := E\left[e^{-qT_{\{a,-a\}}(X^x_\alpha)}; T_{\{a,-a\}}(X^x_\alpha) < T_{\{0\}}(X^x_\alpha)\right] \tag{6.34}$$

$$= \frac{\varphi^q_{x \to a, -a} - \varphi^q_{0 \to a, -a} \varphi^q_{x \to 0, a, -a}}{1 - \varphi^q_{0 \to a, -a}} \tag{6.35}$$

$$= \frac{u_q^{(\alpha)}(0)\left\{u_q^{(\alpha)}(x - a) + u_q^{(\alpha)}(x + a)\right\} - 2u_q^{(\alpha)}(a)u_q^{(\alpha)}(x)}{V_q^{(\alpha)}(a)}. \tag{6.36}$$

The proof of Proposition 6.6 is similar to that of Proposition 5.5, so we omit it.

Let $m^{(\alpha)}$ denote Itô's measure for $|X_\alpha|$ corresponding to the local time satisfying (4.12). The following proposition is crucial to the proof of Theorem 6.4.

Proposition 6.7 *Suppose that $1 < \alpha \le 2$. Let $a > 0$ and $q, r > 0$. Then*

$$m^{(\alpha)}\left[e^{-qT_{\{a\}} - r(\zeta - T_{\{a\}})}; T_{\{a\}} < \zeta\right] = \frac{u_r^{(\alpha)}(a)}{u_r^{(\alpha)}(0)} \cdot \frac{2u_q^{(\alpha)}(a)}{V_q^{(\alpha)}(a)}. \tag{6.37}$$

Consequently, it holds that

$$m^{(\alpha)}\left[e^{-qT_{\{a\}}}; T_{\{a\}} < \zeta\right] = \frac{2u_q^{(\alpha)}(a)}{V_q^{(\alpha)}(a)} \tag{6.38}$$

and that

$$m^{(\alpha)}(T_{\{a\}} < \zeta) = \frac{2}{4h^{(\alpha)}(a) - h^{(\alpha)}(2a)}. \tag{6.39}$$

Proof of Proposition 6.7. By definitions of $n^{(\alpha)}$ and $m^{(\alpha)}$, we have

$$m^{(\alpha)}\left[e^{-qT_{\{a\}} - r(\zeta - T_{\{a\}})}; T_{\{a\}} < \zeta\right]$$
$$= n^{(\alpha)}\left[e^{-qT_{\{a,-a\}} - r(\zeta - T_{\{a,-a\}})}; T_{\{a,-a\}} < \zeta\right]. \tag{6.40}$$

Let $\varepsilon > 0$. Then we have

$$n^{(\alpha)}\left[e^{-qT_{\{a,-a\}} - r(\zeta - T_{\{a,-a\}})}; \varepsilon < T_{\{a,-a\}} < \zeta\right] \tag{6.41}$$

$$= e^{-q\varepsilon} n^{(\alpha)}\left[\left.\left(\varphi^q_{x \to a, -a \prec 0}\right)\right|_{x = X(\varepsilon)} \cdot \varphi^r_{a \to 0}; \varepsilon < T_{\{a,-a\}} \wedge \zeta\right] \tag{6.42}$$

$$= e^{-q\varepsilon} \varphi^r_{a \to 0} E^{h^{(\alpha)}}\left[\left.\left(\frac{\varphi^q_{x \to a, -a \prec 0}}{h^{(\alpha)}(x)}\right)\right|_{x = X(\varepsilon)}; \varepsilon < T_{\{a,-a\}}\right]. \tag{6.43}$$

Here we used Theorem 4.2. Noting that, by Theorem 4.4, we have

$$\lim_{x \to 0} \frac{u_q^{(\alpha)}(a-x) + u_q^{(\alpha)}(a+x) - 2u_q^{(\alpha)}(a)}{h^{(\alpha)}(x)} = 0 \tag{6.44}$$

in whichever case where $1 < \alpha < 2$ or $\alpha = 2$. Hence, we use Proposition 6.6 and obtain

$$\lim_{x \to 0} \frac{\varphi_{x \to a, -a \prec 0}^q}{h^{(\alpha)}(x)} = \frac{2u_q^{(\alpha)}(a)}{V_q^{(\alpha)}(a)}. \tag{6.45}$$

Thus, letting $\varepsilon \to 0+$ in formula (6.43), we obtain (6.37) by dominated convergence. Letting $r \to 0+$ in formula (6.37), we obtain (6.38). Noting that

$$V_q^{(\alpha)}(a) = 2\left\{u_q^{(\alpha)}(0) + u_q^{(\alpha)}(a)\right\} h_q^{(\alpha)}(a) - u_q^{(\alpha)}(0)h_q^{(\alpha)}(2a), \tag{6.46}$$

we have

$$\lim_{q \to 0+} \frac{2u_q^{(\alpha)}(a)}{V_q^{(\alpha)}(a)} = \frac{2}{4h^{(\alpha)}(a) - h^{(\alpha)}(2a)}. \tag{6.47}$$

Hence, by letting $q \to 0+$ in formula (6.38), we obtain (6.39). Now the proof is complete.   □

The proof of Theorem 6.4 is now completely parallel to that of Theorem 5.8; thus we omit it.

7 Appendix: Computation of the Constant $h^{(\alpha)}(1)$

Proposition 7.1 *For $1 < \alpha < 3$, it holds that*

$$\frac{1}{\pi} \int_0^\infty \frac{1 - \cos x}{x^\alpha} dx = \frac{1}{2\Gamma(\alpha) \sin \frac{\pi(\alpha-1)}{2}}. \tag{7.1}$$

As a check, formula (7.1) in the case when $\alpha = 2$ is equivalent via integration by parts to the well-known formula:

$$\int_0^\infty \frac{\sin x}{x} dx = \frac{\pi}{2}. \tag{7.2}$$

The proof of Proposition 7.1 can be found in Feller [14, XVII.3 (g)], but we give it for convenience of the reader.

Proof. We start with the identity:

$$\int_0^\infty x^{\gamma-1} e^{-zx} dx = \Gamma(\gamma) z^{-\gamma} \tag{7.3}$$

for $\gamma > 0$ and $\mathrm{Re}\, z > 0$. For $0 < \alpha < 1$, $\varepsilon > 0$ and $\lambda \in \mathbb{R}$, we set $\gamma = 1 - \alpha$ and $z = \varepsilon - i\lambda$. Then we obtain

$$\int_0^\infty \frac{e^{i\lambda x}e^{-\varepsilon x}}{x^\alpha}dx = \Gamma(1-\alpha)(\varepsilon - i\lambda)^{\alpha-1}. \tag{7.4}$$

Using the identity $\Gamma(2-\alpha) = (1-\alpha)\Gamma(1-\alpha)$ and subtracting (7.4) for $\lambda = 0$ from (7.4) for $\lambda = \lambda$, we obtain

$$\int_0^\infty \frac{(1-e^{i\lambda x})e^{-\varepsilon x}}{x^\alpha}dx = \Gamma(2-\alpha) \cdot \frac{\varepsilon^{\alpha-1} - (\varepsilon - i\lambda)^{\alpha-1}}{1-\alpha}. \tag{7.5}$$

Rewriting the right hand side, we obtain

$$\int_0^\infty \frac{(1-e^{i\lambda x})e^{-\varepsilon x}}{x^\alpha}dx = \Gamma(2-\alpha)\int_\varepsilon^{\varepsilon - i\lambda} z^{\alpha-2}dz \tag{7.6}$$

where integration on the right hand side is taken over a segment from $\{\varepsilon - il : l \in \mathbb{R}\}$. Since both sides of (7.6) are analytic on $0 < \mathrm{Re}\,\alpha < 2$, we see, by analytic continuation, that identity (7.6) remains true for $0 < \alpha < 2$.

Let us restrict ourselves to the case when $1 < \alpha < 2$. Taking the limit $\varepsilon \to 0+$ on both sides of identity (7.6), we obtain

$$\int_0^\infty \frac{1-e^{i\lambda x}}{x^\alpha}dx = \Gamma(2-\alpha)\int_0^{-i\lambda} z^{\alpha-2}dz \tag{7.7}$$

$$= \Gamma(2-\alpha) \cdot \frac{(-i\lambda)^{\alpha-1}}{\alpha - 1} \tag{7.8}$$

where the branch of $f(w) = w^{\alpha-1}$ is chosen so that $f(1) = 1$. Hence we obtain

$$\int_0^\infty \frac{1-e^{i\lambda x}}{x^\alpha}dx = \Gamma(2-\alpha) \cdot \frac{\lambda^{\alpha-1}}{\alpha - 1}e^{-\frac{\pi(\alpha-1)i}{2}}. \tag{7.9}$$

Taking real parts on both sides, we obtain

$$\int_0^\infty \frac{1-\cos\lambda x}{x^\alpha}dx = \Gamma(2-\alpha) \cdot \frac{\lambda^{\alpha-1}}{\alpha - 1}\cos\frac{\pi(\alpha - 1)}{2}. \tag{7.10}$$

Letting $\lambda = 1$, we obtain

$$\frac{1}{\pi}\int_0^\infty \frac{1-\cos x}{x^\alpha}dx = \frac{\Gamma(2-\alpha)}{\pi(\alpha - 1)} \cdot \cos\frac{\pi(\alpha - 1)}{2}. \tag{7.11}$$

(We may find formula (7.11) also in [21, pp.88].) By a simple computation, we have

$$(\text{RHS of (7.11)}) = \frac{\Gamma(2-\alpha)}{\pi(\alpha - 1)} \cdot \frac{\sin\pi(\alpha - 1)}{2\sin\frac{\pi(\alpha-1)}{2}} \tag{7.12}$$

$$= \frac{1}{(\alpha - 1)\Gamma(\alpha - 1)} \cdot \frac{1}{2\sin\frac{\pi(\alpha-1)}{2}} \tag{7.13}$$

$$= \frac{1}{2\Gamma(\alpha)\sin\frac{\pi(\alpha-1)}{2}}. \tag{7.14}$$

Hence we have proved identity (7.1) when $1 < \alpha < 2$. By analytic continuation, identity (7.1) is proved to be valid also when $2 \leq \alpha < 3$. Therefore the proof is complete. $\square$

References

1. O. Barndorff-Nielsen, J. Kent, and M. Sørensen. Normal variance-mean mixtures and z distributions. *Internat. Statist. Rev.*, 50(2):145–159, 1982.
2. J. Bertoin. *Lévy processes*, volume 121 of *Cambridge Tracts in Mathematics*. Cambridge University Press, Cambridge, 1996.
3. J. Bertoin, T. Fujita, B. Roynette, and M. Yor. On a particular class of self-decomposable random variables: the durations of Bessel excursions straddling independent exponential times. *Probab. Math. Statist.*, 26(2):315–366, 2006.
4. R. M. Blumenthal and R. K. Getoor. *Markov processes and potential theory*. Pure and Applied Mathematics, Vol. 29. Academic Press, New York, 1968.
5. R. M. Blumenthal, R. K. Getoor, and D. B. Ray. On the distribution of first hits for the symmetric stable processes. *Trans. Amer. Math. Soc.*, 99:540–554, 1961.
6. L. Bondesson. On the infinite divisibility of the half-Cauchy and other decreasing densities and probability functions on the nonnegative line. *Scand. Actuar. J.*, (3-4):225–247, 1987.
7. L. Bondesson. *Generalized gamma convolutions and related classes of distributions and densities*, volume 76 of *Lecture Notes in Statistics*. Springer-Verlag, New York, 1992.
8. P. Bourgade, T. Fujita, and M. Yor. Euler's formulae for $\zeta(2n)$ and products of Cauchy variables. *Electron. Comm. Probab.*, 12:73–80 (electronic), 2007.
9. J. Bretagnolle. Résultats de Kesten sur les processus à accroissements indépendants. In *Séminaire de Probabilités, V (Univ. Strasbourg, année universitaire 1969-1970)*, pages 21–36. Lecture Notes in Math., Vol. 191. Springer, Berlin, 1971.
10. P. Carmona, F. Petit, and M. Yor. On the distribution and asymptotic results for exponential functionals of Lévy processes. In *Exponential functionals and principal values related to Brownian motion*, Bibl. Rev. Mat. Iberoamericana, pages 73–130. Rev. Mat. Iberoamericana, Madrid, 1997.
11. L. Chaumont and M. Yor. *Exercises in probability, A guided tour from measure theory to random processes, via conditioning*, volume 13 of *Cambridge Series in Statistical and Probabilistic Mathematics*. Cambridge University Press, Cambridge, 2003.
12. L. Devroye. A note on Linnik's distribution. *Statist. Probab. Lett.*, 9(4):305–306, 1990.
13. C. Donati-Martin, B. Roynette, P. Vallois, and M. Yor. On constants related to the choice of the local time at 0, and the corresponding Itô measure for Bessel processes with dimension $d = 2(1 - \alpha)$, $0 < \alpha < 1$. *Studia Sci. Math. Hungar.*, 45(2):207–221, 2008.
14. W. Feller. *An introduction to probability theory and its applications*. Vol. II. Second edition. John Wiley & Sons Inc., New York, 1971.
15. T. Fujita, Y. Yano, and M. Yor. *in preparation*.

16. R. K. Getoor. Continuous additive functionals of a Markov process with applications to processes with independent increments. *J. Math. Anal. Appl.*, 13:132–153, 1966.

17. B. Grigelionis. Processes of Meixner type. *Liet. Mat. Rink.*, 39(1):40–51, 1999.

18. B. Grigelionis. Generalized z-distributions and related stochastic processes. *Liet. Mat. Rink.*, 41(3):303–319, 2001.

19. B. Grigelionis. On the self-decomposability of Euler's gamma function. *Liet. Mat. Rink.*, 43(3):359–370, 2003.

20. M. Hayashi and K. Yano. On the laws of total local times for h-paths of stable Lévy processes. *in preparation.*

21. I. A. Ibragimov and Yu. V. Linnik. *Independent and stationary sequences of random variables.* Wolters-Noordhoff Publishing, Groningen, 1971. With a supplementary chapter by I. A. Ibragimov and V. V. Petrov, Translation from the Russian edited by J. F. C. Kingman.

22. N. Ikeda and S. Watanabe. *Stochastic differential equations and diffusion processes*, volume 24 of *North-Holland Mathematical Library*. North-Holland Publishing Co., Amsterdam, second edition, 1989.

23. K. Itô. Poisson point processes attached to Markov processes. In *Proceedings of the Sixth Berkeley Symposium on Mathematical Statistics and Probability (Univ. California, Berkeley, Calif., 1970/1971), Vol. III: Probability theory*, pages 225–239, Berkeley, Calif., 1972. Univ. California Press.

24. K. Itô and H. P. McKean, Jr. *Diffusion processes and their sample paths.* Springer-Verlag, Berlin, 1974. Second printing, corrected, Die Grundlehren der mathematischen Wissenschaften, Band 125.

25. L. F. James. Gamma tilting calculus for GGC and Dirichlet means with applications to Linnik processes and occupation time laws for randomly skewed Bessel processes and bridges. *preprint, arXiv:math/0610218*, 2006.

26. L. F. James, B. Roynette, and M. Yor. Generalized Gamma convolutions, Dirichlet means, Thorin measures with explicit examples. *Probab. Surv.*, 5:346–415, 2008.

27. H. Kesten. *Hitting probabilities of single points for processes with stationary independent increments.* Memoirs of the American Mathematical Society, No. 93. American Mathematical Society, Providence, R.I., 1969.

28. F. B. Knight. Brownian local times and taboo processes. *Trans. Amer. Math. Soc.*, 143:173–185, 1969.

29. M. B. Marcus and J. Rosen. *Markov processes, Gaussian processes, and local times*, volume 100 of *Cambridge Studies in Advanced Mathematics*. Cambridge University Press, Cambridge, 2006.

30. P. A. Meyer. Processus de Poisson ponctuels, d'après K. Ito. In *Séminaire de Probabilités, V (Univ. Strasbourg, année universitaire 1969–1970)*, pages 177–190. Lecture Notes in Math., Vol. 191. Springer, Berlin, 1971.

31. J. Pitman and M. Yor. Bessel processes and infinitely divisible laws. In *Stochastic integrals (Proc. Sympos., Univ. Durham, Durham, 1980)*, volume 851 of *Lecture Notes in Math.*, pages 285–370. Springer, Berlin, 1981.

32. J. Pitman and M. Yor. Infinitely divisible laws associated with hyperbolic functions. *Canad. J. Math.*, 55(2):292–330, 2003.

33. J. Pitman and M. Yor. Itô's excursion theory and its applications. *Jpn. J. Math.*, 2(1):83–96, 2007.

34. D. Ray. Stable processes with an absorbing barrier. *Trans. Amer. Math. Soc.*, 89:16–24, 1958.

35. D. Revuz and M. Yor. *Continuous martingales and Brownian motion*, volume 293 of *Grundlehren der Mathematischen Wissenschaften*. Springer-Verlag, Berlin, third edition, 1999.

36. B. Roynette, P. Vallois, and M. Yor. A family of generalized gamma convoluted variables. *to appear in Prob. Math. Stat.*, 2009.

37. P. Salminen. On last exit decompositions of linear diffusions. *Studia Sci. Math. Hungar.*, 33(1-3):251–262, 1997.

38. W. Schoutens. *Stochastic processes and orthogonal polynomials*, volume 146 of *Lecture Notes in Statistics*. Springer-Verlag, New York, 2000.

39. W. Schoutens. *Lévy processes in finance: Pricing financial derivatives*. John Wiley & Sons Inc., 2003.

40. W. Schoutens and J. L. Teugels. Lévy processes, polynomials and martingales. *Comm. Statist. Stochastic Models*, 14(1-2):335–349, 1998. Special issue in honor of Marcel F. Neuts.

41. D. N. Shanbhag and M. Sreehari. On certain self-decomposable distributions. *Z. Wahrscheinlichkeitstheorie und Verw. Gebiete*, 38(3):217–222, 1977.

42. O. Thorin. On the infinite divisibility of the Pareto distribution. *Scand. Actuar. J.*, (1):31–40, 1977.

43. D. Williams. Decomposing the Brownian path. *Bull. Amer. Math. Soc.*, 76: 871–873, 1970.

44. D. Williams. Path decomposition and continuity of local time for one-dimensional diffusions. I. *Proc. London Math. Soc. (3)*, 28:738–768, 1974.

45. M. Winkel. Electronic foreign-exchange markets and passage events of independent subordinators. *J. Appl. Probab.*, 42(1):138–152, 2005.

46. M. Yamazato. Topics related to gamma processes. In *Stochastic processes and applications to mathematical finance*, pages 157–182. World Sci. Publ., Hackensack, NJ, 2006.

47. K. Yano. Convergence of excursion point processes and its applications to functional limit theorems of markov processes on a half line. *Bernoulli*, 14(4): 963–987, 2008.

48. K. Yano. Excursions away from a regular point for one-dimensional symmetric Lévy processes without Gaussian part. *submitted. preprint*, arXiv:0805.3881, 2008.

49. K. Yano, Y. Yano, and M. Yor. Penalising symmetric stable Lévy paths. *J. Math. Soc. Japan*, to appear in 2009.

50. F. Cordero. Sur la théorie des excursions des processus de Lévy et quelques applications. *in preparation*.

Lévy Systems and Time Changes

P.J. Fitzsimmons and R.K. Getoor

Department of Mathematics, 0112; University of California San Diego
9500 Gilman Drive, La Jolla, CA 92093–0112, USA
e-mail: pfitzsim@ucsd.edu

Summary. The Lévy system for a Markov process X provides a convenient description of the distribution of the totally inaccessible jumps of the process. We examine the effect of time change (by the inverse of a not necessarily strictly increasing CAF A) on the Lévy system, in a general context. They key to our time-change theorem is a study of the "irregular" exits from the fine support of A that occur at totally inaccessible times. This permits the construction of a partial predictable exit system (à la Maisonneuve).

The second part of the paper is devoted to some implications of the preceding in a (weak, moderate Markov) duality setting. Fixing an excessive measure m (to serve as duality measure) we obtain formulas relating the "killing" and "jump" measures for the time-changed process to the analogous objects for the original process. These formulas extend, to a very general context, recent work of Chen, Fukushima, and Ying. The key to our development is the Kuznetsov process associated with X and m, and the associated moderate Markov dual process $\widehat{X}$. Using $\widehat{X}$ and some excursion theory, we exhibit a general method for constructing excessive measures for X from excessive measures for the time-changed process.

Key words and phrases: Lévy system, exit system, time change, Markov process, continuous additive functional, excessive measure, Kuznetsov process.

2000 Mathematics Subject Classification. Primary: 60J55, Secondary: 60J40.

1 Introduction

Let $X = (X_t, \mathbf{P}^x)$ be a right Markov process with state space E. The *Lévy system* of X describes the intensity with which X makes totally inaccessible jumps of specified types. It consists of a continuous additive functional H and a kernel N on $(E, \mathcal{E})$ such that

$$t \mapsto \sum_{s \leq t} \Phi(X^r_{s-}, X_s) - \int_0^t N(X_s, \Phi) \, \mathrm{d}H_s \tag{1.1}$$

C. Donati-Martin et al. (eds.), *Séminaire de Probabilités XLII*,
Lecture Notes in Mathematics 1979, DOI 10.1007/978-3-642-01763-6_9,
© Springer-Verlag Berlin Heidelberg 2009

is a $\mathbf{P}^x$-martingale for each $x \in E$, provided $\mathbf{P}^x \int_0^t N(X_s, |\varPhi|) \, dH_s < \infty$ for each $t > 0$. In (1.1), X_{s-}^r is the left limit of X at time $s > 0$ taken in a suitable Ray topology on E, $\varPhi$ is a product measurable function on $E \times E$ with $\varPhi(x, x) = 0$ for all $x \in E$, and $N(x, \varPhi) := \int_E N(x, dy) \varPhi(x, y)$. Intuitively, the rate at which jumps from $x \in E$ to $\varLambda \in \mathcal{E}$ occur, *relative to the clock* H, is $N(x, \varLambda)$. The notion of a Lévy system, which is a far-reaching generalization of the Itô-Lévy description of the jumps of a Lévy process, is due to S. Watanabe [28]. He constructed Lévy systems for Hunt processes satisfying Meyer's hypothesis (L) (= the existence of a reference measure). Lévy systems for general right (and Ray) processes, without (L), were constructed by Benveniste and Jacod in [1].

Suppose that in addition to the right process X we have a CAF A with right continuous inverse τ. Let F denote the fine support of A; thus A increases when (and only when) X is in F. It is well known that the time-changed process $\widetilde{X}_t := X_{\tau(t)}$, $t \geq 0$, is a right process with state space F. Our goal in this paper is to express the Lévy system $(\widetilde{N}, \widetilde{H})$ of $\widetilde{X}$ in terms of (N, H) and an *exit system* $(\mathbf{P}_{\mathrm{pr}}^\bullet, C)$; the latter describes the relevant ways in which X exits the fine support F. The need for this second ingredient stems from the fact that A need not be strictly increasing—some of the totally inaccessible jumps of $\widetilde{X}$ correspond to totally inaccessible jumps of X, while others are generated by the excursions of X from F. The first work in this direction of which we are aware is that of H. Gzyl [17]. The recent work of Chen, Fukushima, and Ying [3, 4] has been the direct inspiration for the present study. The effect of time change on symmetric Markov processes is considered in [3] (see also [21] for the case of "nearly symmetric" Hunt processes, and [14] for symmetric *diffusions*). The same issues are examined in [4] for a standard process in weak duality with a second standard process, under the condition that semipolar sets are m-polar (m being the duality measure). In trying to understand [4], we came to realize that neither duality nor a restriction on semipolar sets was crucial for the discussion. Rather, the key seemed to lie in coming to grips with the "irregular" exits of X from F that occur at totally inaccessible times.

In section 2 we describe the hypotheses that will be in force throughout the paper, and we recall the basic facts about exit systems and Lévy systems. In section 3, following Maisonneuve [22], we investigate the notion of *predictable* exit system, and construct a partial predictable exit system that is sufficient for our purposes. We also provide, in Theorem 3.4, several conditions equivalent to the existence of a complete predictable exit system. In a short section 4 we recall the definition of the time-changed process $\widetilde{X}$ induced by a continuous additive functional A of the basic process X. Section 5 contains one of the main results of the paper. Namely, the Lévy system of $\widetilde{X}$ is expressed in terms of the Lévy system of X and the partial predictable exit system describing the excursions of X away from the fine support F of A; see Theorem 5.2. In section 6 we assume the existence of an excessive measure m, and recall several constructs depending on m: In particular, the Kuznetsov measure $\mathbf{Q}_m$

and associated processes Y and Y^*. We introduce the "jump measure" $\mathcal{J}$ and the "killing measure" $\mathcal{K}$ associated with X and m, and we express them in terms of the Lévy system of X. Following [4] we then define the Feller measure Λ and the supplementary Feller measure δ associated with excursions from F. Formulas (6.24) and (6.25) relate Λ and δ explicitly to the partial predictable exit system. The main result of this section, Theorem 6.5, gives formulas for the jump measure $\widetilde{\mathcal{J}}$ and the killing measure $\widetilde{\mathcal{K}}$ of $\widetilde{X}$ in terms of $\mathcal{J}$, $\mathcal{K}$, Λ, and δ. Section 7 introduces the left-continuous moderate Markov process $\widetilde{X}$ in weak duality with X relative to m. Using this duality we extend some of the results about excursions from a regular point presented in [12] to excursions from a finely perfect nearly Borel set F for which a predictable exit system exists.

We close this introduction with a few words on notation. If $(F, \mathcal{F}, \mu)$ is a measure space, then $b\mathcal{F}$ (resp. $p\mathcal{F}$) denotes the class of bounded real-valued (resp. $[0, \infty]$-valued) $\mathcal{F}$-measurable functions on F. For $f \in p\mathcal{F}$ we may use $\mu(f)$ or $\langle \mu, f \rangle$ to denote the integral $\int_F f \, d\mu$; similarly, if $D \in \mathcal{F}$ then $\mu(f; D)$ denotes $\int_D f \, d\mu$. On the other hand $f\mu$ denotes the measure $f(x)\mu(dx)$ and $\mu|_D$ the restriction of μ to D. We write $\mathcal{F}^*$ for the universal completion of $\mathcal{F}$; that is $\mathcal{F}^* = \cap_\nu \mathcal{F}^\nu$, where $\mathcal{F}^\nu$ is the ν-completion of $\mathcal{F}$ and the intersection runs over all finite measures on $(F, \mathcal{F})$. If $(E, \mathcal{E})$ is a second measurable space and $K = K(x, dy)$ is a kernel from $(F, \mathcal{F})$ to $(E, \mathcal{E})$ (i.e., $F \ni x \mapsto K(x, A)$ is $\mathcal{F}$-measurable for each $A \in \mathcal{E}$ and $K(x, \cdot)$ is a measure on $(E, \mathcal{E})$ for each $x \in F$), then we write μK for the measure $A \mapsto \int_F \mu(dx)K(x, A)$ and Kf for the function $x \mapsto \int_E K(x, dy)f(y)$. We shall use $\mathcal{B}$ to denote the Borel subsets of the real line $\mathbf{R}$. If T is a stopping time, then $[\![T]\!]$ denotes the graph $\{(\omega, t) \in \Omega \times [0, \infty[: t = T(\omega)\}$.

2 Preliminaries

Throughout the paper $X = (\Omega, \mathcal{F}, \mathcal{F}_t, \theta_t, X_t, \mathbf{P}^x)$ will denote the canonical realization of a Borel right Markov process with state space $(E, \mathcal{E})$. We shall use the standard notation for Markov processes as found, for example, in [2], [15], [7] and [26]. In short, X is a strong Markov process with right continuous sample paths, the state space E (with Borel σ-field $\mathcal{E}$) is homeomorphic to a Borel subset of a compact metric space, and the transition semigroup $(P_t)_{t\geq 0}$ of X preserves the class $b\mathcal{E}$ of bounded $\mathcal{E}$-measurable functions. It follows that the resolvent operators $U^q := \int_0^\infty e^{-qt} P_t \, dt$, $q \geq 0$, also preserve Borel measurability. We shall write U for U^0. We allow the transition semigroup (P_t) to be subMarkovian: $P_t 1(x) \leq 1$ for all $x \in E$ and all $t \geq 0$. To allow for the possibility $P_t 1_E(x) < 1$, an absorbing cemetery state Δ is adjoined to E as an isolated point, and the process is sent to Δ at its lifetime ζ. Thus X takes values in $E_\Delta := E \cup \{\Delta\}$ (endowed with the σ-field $\mathcal{E}_\Delta := \mathcal{E} \vee \{\Delta\}$; until section 6, the cemetery state will play no special role.

We write $\mathcal{E}^e$ for the σ-algebra on E generated by the 1-excessive functions. Because the semigroup (P_t) is Borel, all 1-excessive functions are nearly Borel measurable; consequently, $\mathcal{E}^e$ is contained in the σ-algebra of nearly Borel sets.

One of our concerns will be the excursions induced by a CAF A; to this end we recall Maisonneuve's notion of optional exit system. The related notion of predictable exit system will be discussed in section 3. It is known that the stopping time $\tau(0) = \inf\{t : A_t > 0\}$ is equal a.s. to the hitting time $T_F := \inf\{t > 0 : X_t \in F\}$ of the *fine support* F of A, defined by

$$F := \{x \in E : \mathbf{P}^x[\tau(0) = 0] = 1\}. \tag{2.1}$$

The set F is $\mathcal{E}^e$-measurable and *finely perfect* in the sense that $F = F^r$, where $F^r := \{x \in E : \mathbf{P}^x[T_F = 0] = 1\}$ denotes the set of points regular for F. Consequently, the optional set $\{X \in F\} := \{(\omega, t) : X_t(\omega) \in F\}$ has ω-sections that are right closed and without isolated points, almost surely. Let M be the (optional) subset of $\Omega \times [0, \infty[$ with ω-section $M(\omega)$ equal to the closure in $[0, \infty[$ of the visiting set $\{t \geq 0 : X_t(\omega) \in F\}$, for each $\omega \in \Omega$. The complement of $M(\omega)$ comprises a countable union of disjoint open intervals. We write $G(\omega)$ for the collection of strictly positive left endpoints of these "contiguous intervals". The associated random set G is progressively measurable, but not in general optional. More precisely, the "regular" part of G, given by

$$G^r := \{(\omega, s) \in G : X_s(\omega) \in F\}, \tag{2.2}$$

has evanescent intersection with the graph of any stopping time, while the "irregular" part

$$G^i := \{(\omega, s) \in G : X_s(\omega) \notin F\} \tag{2.3}$$

is a countable union of graphs of stopping times.

According to Maisonneuve [22] there is an *optional exit system* consisting of an AF B with bounded 1-potential, and a kernel $\mathbf{P}^\bullet_{\mathrm{op}}$ from $(E_\Delta, \mathcal{E}^*_\Delta)$ to $(\Omega, \mathcal{F}^*)$ such that

$$\mathbf{P}^x \sum_{s \in G} Z_s \, \Phi \circ \theta_s = \mathbf{P}^x \int_0^\infty Z_s \, \mathbf{P}^{X_s}_{\mathrm{op}}[\Phi] \, \mathrm{d}B_s, \tag{2.4}$$

for all optional $Z \geq 0$, $\Phi \in p\mathcal{F}^*$, and $x \in E_\Delta$. ($\mathbf{P}^\Delta_{\mathrm{op}}$ is the point mass at the dead path $[\Delta]$.) We can (and do) take the continuous part B^c of B to be the dual predictable projection of the raw AF

$$t \mapsto \sum_{s \leq t, s \in G^r} [1 - \exp(-T_F)] \circ \theta_s + \int_0^t 1_F(X_s) \, \mathrm{d}s,$$

and the discontinuous part of B to be

$$B^d_t := \sum_{s \leq t, s \in G^i} \mathbf{P}^{X_s}[1 - \exp(-T_F)]. \tag{2.5}$$

Notice that B^c grows only when X is in F. In view of Motoo's theorem [26, (66.2)], there exists $\ell \in p\mathcal{E}^e$ such that

$$\int_0^t 1_F(X_s)\,\mathrm{d}s = \int_0^t \ell(X_s)\,\mathrm{d}B_s^c = \int_0^t \ell(X_s)\,\mathrm{d}B_s. \qquad (2.6)$$

The second equality in (2.6) holds because we can (and do) take $\ell + \mathbf{P}_{\mathrm{op}}^{\bullet}[1 - e^{-T_F}] = 1$ on E_Δ and $\ell = 0$ on $E_\Delta \setminus F$. Moreover,

$$\mathbf{P}_{\mathrm{op}}^x[\Phi] = \frac{\mathbf{P}^x[\Phi]}{\mathbf{P}^x[1 - e^{-T_F}]}, \qquad \forall x \in E_\Delta \setminus F.$$

This choice of $(\mathbf{P}_{\mathrm{op}}^{\bullet}, B)$ having been made, we have

$$U_B^1 1_{E_\Delta}(x) = \mathbf{P}^x[\exp(-T_F)], \qquad \forall x \in E_\Delta. \qquad (2.7)$$

A second key ingredient in our development is the *Lévy system* describing the totally inaccessible jumps of X. Recall that a stopping time T is totally inaccessible if $\mathbf{P}^x[T = S] = 0$ for all x and all predictable S. Let $(X_{t-}^r)_{t>0}$ denote the left limit process of X, the limits being taken in some Ray-Knight compactification $\overline{E}$ of E_Δ; see [26, §17–18]. The set

$$J := \{(\omega, t) : X_t^r - (\omega) \in E_\Delta, X_t^r - (\omega) \neq X_t(\omega)\} \qquad (2.8)$$

is the union $\cup_n [\![T_n]\!]$ of a sequence of totally inaccessible stopping times. Indeed, a stopping time T is totally inaccessible if and only if $[\![T]\!] \subset J$ up to evanescence; see [26, (44.5)]. Also, if we write X_{t-} for the left limit of X at time $t > 0$ (in the original topology of E) whenever it exists, then

$$J \subset \{(\omega, t) : X_{t-}(\omega) = X_{t-}^r(\omega)\} \qquad (2.9)$$

up to evanescence; see [26, (46.3)]. The Lévy system consists of a kernel N_Δ from $(E, \mathcal{E})$ to $(E_\Delta, \mathcal{E}_\Delta)$ such that $N_\Delta(x, \{x\}) = 0$ for all $x \in E$, and a CAF H, such that

$$\mathbf{P}^x \sum_{s \in J} Z_s \Psi(X_{s-}, X_s) = \mathbf{P}^x \int_0^\infty Z_s \int_{E_\Delta} \Psi(X_s, y)\, N_\Delta(X_s, \mathrm{d}y)\, \mathrm{d}H_s, \qquad (2.10)$$

for all predictable $Z \geq 0$, $\Psi \in p(\mathcal{E} \otimes \mathcal{E}_\Delta)$, and $x \in E_\Delta$. We will often write $N_\Delta(x, \Psi)$ for $\int_{E_\Delta} \Psi(x, y)\, N_\Delta(x, \mathrm{d}y)$; with this notation the right side of (2.10) collapses to $\mathbf{P}^x \int_0^\infty Z_s\, N_\Delta(X_s, \Psi)\, \mathrm{d}H_s$. Because X cannot jump out of Δ, we can (and do) assume that $H_t = H_\zeta$ for all $t > \zeta$. Because H is a (finite) CAF, there is a strictly positive function $g \in \mathcal{E}^e$ such that $\sup_x \mathbf{P}^x \int_0^\infty e^{-t} g(X_t)\, \mathrm{d}H_t < \infty$. Therefore, at the cost of replacing H_t by $\int_0^t g(X_s)\, \mathrm{d}H_s$ and $N_\Delta(x, y)$ by $g(x)^{-1} N_\Delta(x, \mathrm{d}y)$, we can arrange for H to have a bounded 1-potential.

3 Predictable Exit System

In section 9 of [22], Maisonneuve constructed a predictable exit system, assuming condition (iii) in Lemma 3.1 below. In what follows we shall refer to this as a *complete* predictable exit system. Our purposes are served by a modified construction, yielding a *partial* predictable exit system $(\mathbf{P}^{\bullet}_{\mathrm{pr}}, C)$ that is more broadly applicable to the problem of Lévy systems and time changes. Before proceeding to the construction we introduce a supplementary hypothesis under which the partial exit system becomes a complete predictable exit system.

The set G^i, defined in (2.3), of irregular left endpoints of the intervals contiguous to M can be expressed as the disjoint union $\cup_n [\![T_n]\!]$ of graphs of stopping times. It turns out that only $G^i \cap J$ is germane to the present study. The following result captures the situation in which all of the irregular exits from F occur at totally inaccessible times. To state it define

$$D_t := \inf\{s > t : X_s \in F\} = t + T_F \circ \theta_t, \qquad t \geq 0.$$

The process D is increasing and right continuous, and $M \setminus G = \{(\omega, t) : D_t(\omega) = t\}$.

Lemma 3.1 *The following conditions are equivalent:*
(i) $G^i \subset J$;
(ii) The dual predictable projection of the AF B^d (defined in (2.5)) is continuous;
(iii) The 1-potential $\varphi_1 : x \mapsto \mathbf{P}^x[\exp(-T_F)]$ is regular.
(iv) The process $t \mapsto D_t$ is quasi-left continuous.

Proof. (i)$\Longrightarrow$(ii). Let $B^{d,p}$ denote the dual predictable projection of B^d. If T is a predictable time, then

$$\mathbf{P}^x[B^{d,p}_T - B^{d,p}_{T-}, 0 < T < \infty] = \mathbf{P}^x\left[\int_{]0,\infty[} 1_{[\![T]\!]}(s)\, \mathrm{d}B^{d,p}_s\right]$$

$$= \mathbf{P}^x\left[\int_{]0,\infty[} 1_{[\![T]\!]}(s)\, \mathrm{d}B^d_s\right]$$

$$= \mathbf{P}^x[1 - \varphi_1(X_T); T \in G^i] = 0,$$

for all $x \in E_\Delta$.

(ii)$\Longrightarrow$(iii) The process $t \mapsto e^{-t}\varphi_1(X_t)$ is a positive right-continuous supermartingale. Indeed, from (2.7),

$$e^{-t}\varphi_1(X_t) = \mathbf{P}^x\left[\int_{]t,\infty[} e^{-s}\, \mathrm{d}B_s \middle| \mathcal{F}_t\right]$$

$$= \mathbf{P}^x\left[\int_{]t,\infty[} e^{-s}\, \mathrm{d}B^p_s \middle| \mathcal{F}_t\right], \qquad \forall x \in E_\Delta, \tag{3.1}$$

where B is the AF component of the optional exit system for F and B^p is the dual predictable projection of B. The hypothesis (ii) implies that B^p is continuous, which in turn implies that φ_1 is regular, because of (3.1).

(iii)$\Longrightarrow$(iv) Let (T_n) be an increasing sequence of stopping times with limit T, and set $\Upsilon := \uparrow \lim_n D_{T_n} \le D_T$. Then, for $x \in E_\Delta$,

$$\mathbf{P}^x[\exp(-\Upsilon)] = \lim_n \mathbf{P}^x[\exp(-D_{T_n})] = \lim_n \mathbf{P}^x[e^{-T_n}\varphi_1(X_{T_n})]$$
$$= \mathbf{P}^x[e^{-T}\varphi_1(X_T)] = \mathbf{P}^x[\exp(-D_T)]$$

the third equality resulting from the assumed regularity of φ_1. It follows that $\Upsilon = D_T$ almost surely.

(iv)$\Longrightarrow$(i) Let T be a stopping time with $[\![T]\!] \subset G^i$. Then, on $\{T < \infty\}$, we have $0 < T = D_{T-} < D_T$. The quasi-left-continuity of D now implies that T is totally inaccessible. Thus, by [26, (44.5)], $[\![T]\!] \subset J$. $\square$

Remark 3.2 In view of [26, (46.2)],

$$G_0 := G^r \cup (G^i \cap J) \subset \{(\omega, t) : X_{t-}(\omega) = X^r_{t-}(\omega) \in E_\Delta\} \tag{3.2}$$

up to evanescence. This observation will be used several times in the sequel.

Define

$$B^J_t := \sum_{s \le t, s \in G^i \cap J} \mathbf{P}^{X_s}[1 - \exp(-T_F)], \qquad t \ge 0. \tag{3.3}$$

As preparation for the construction of a (partial) predictable exit system, we express the dual predictable projection of B^J in terms of the Lévy system.

Lemma 3.3 We have, for predictable $Z \ge 0$, $\Psi \in p\mathcal{F}^*$, and $x \in E_\Delta$,

$$\mathbf{P}^x \sum_{t \in G^i \cap J} Z_t \Psi \circ \theta_t = \mathbf{P}^x \int_0^\infty 1_F(X_t) Z_t \int_{F^c} N_\Delta(X_t, dy) \mathbf{P}^y[\Psi] \, dH_t. \tag{3.4}$$

Proof. Define $I_t := \limsup_{s \uparrow\uparrow t} 1_F(X_s)$; this is a predictable process by the discussion on pp. 202–203 of [26] or by [6, T-IV90(a)]. Moreover, $G^i \cap J = J \cap \{(\omega, t) : I_t(\omega) = 1, X_t(\omega) \in F^c\}$, up to evanescence. Therefore

$$\mathbf{P}^\bullet \sum_{t \in G^i \cap J} Z_t \Psi \circ \theta_t = \mathbf{P}^\bullet \sum_{t \in G^i \cap J} Z_t \mathbf{P}^{X_t}[\Psi]$$
$$= \mathbf{P}^\bullet \sum_{t \in J} I_t Z_t 1_{F^c}(X_t) \mathbf{P}^{X_t}[\Psi]$$
$$= \mathbf{P}^\bullet \int_0^\infty I_t Z_t \int_{F^c} N_\Delta(X_t, dy) \mathbf{P}^y[\Psi] \, dH_t \tag{3.5}$$
$$= \mathbf{P}^\bullet \int_0^\infty 1_F(X_t) Z_t \int_{F^c} N_\Delta(X_t, dy) \mathbf{P}^y[\Psi] \, dH_t.$$

We have used the fact that $\{(\omega,t) : t > 0, I_t(\omega) = 1, X_t(\omega) \notin F\} \subset G$, which implies that the sets $\{t : I_t = 1\}$ and $\{t : X_t \in F\}$ differ by at most a countable set, almost surely. This difference is not charged by H. □

Define $\psi(x) := \mathbf{P}^x[1 - \exp(-T_F)]$ and then take $\Psi = \psi(X_0)$ in (3.5) to see that the dual predictable projection of B^J is

$$\int_0^t 1_F(X_s)\, N_\Delta(X_s, \psi)\, dH_s, \qquad t \geq 0. \tag{3.6}$$

Accordingly we define a CAF C by

$$C_t := B_t^c + \int_0^t 1_F(X_s)\, N_\Delta(X_s, \psi)\, dH_s, \qquad t \geq 0, \tag{3.7}$$

noting that the 1-potential of C is

$$U_C^1 1_{E_\Delta}(x) = \mathbf{P}^x \int_0^\infty e^{-t}\, dC_t \leq \mathbf{P}^x \int_0^\infty e^{-t}\, dB_t = \mathbf{P}^x[\exp(-T_F)], \tag{3.8}$$

for all $x \in E_\Delta$.

By Motoo's theorem there are positive $\mathcal{E}^e$-measurable functions b and h such that

$$B_t^c = \int_0^t b(X_s)\, dC_s \quad \text{and} \quad \int_0^t 1_F(X_s)\, N_\Delta(X_s, \psi)\, dH_s = \int_0^t h(X_s)\, dC_s. \tag{3.9}$$

We may suppose that $b + h = 1$ on F and that $b = h = 0$ on $E_\Delta \setminus F$. Finally, define

$$\mathbf{P}_{\mathrm{pr}}^x[\Phi] := b(x) \cdot \mathbf{P}_{\mathrm{op}}^x[\Phi] + h(x)1_F(x) \cdot \frac{\int_{F^c} \mathbf{P}^y[\Phi]\, N_\Delta(x, dy)}{N_\Delta(x, \psi)}, \tag{3.10}$$

the ratio on the right being taken to be 0 when the denominator vanishes. Notice that

$$\int_0^t 1_F(X_s)\, ds = \int_0^t \gamma(X_s)\, dC_s, \qquad \forall t \geq 0, \tag{3.11}$$

where $\gamma := \ell \cdot b$.

Theorem 3.4 (a) *The pair* $(\mathbf{P}_{\mathrm{pr}}^\bullet, C)$ *is a partial predictable exit system for* F, *in the sense that*

$$\mathbf{P}^\bullet \sum_{t \in G_0} Z_t \Psi \circ \theta_t = \mathbf{P}^\bullet \int_0^\infty Z_t \mathbf{P}_{\mathrm{pr}}^{X_t}[\Psi]\, dC_t, \tag{3.12}$$

for all predictable $Z \geq 0$ *and* $\Psi \in p\mathcal{F}^*$, *and* C *is a CAF with fine support contained in* F.

(b) *Suppose that* $G^i \subset J$. *Then* $(\mathbf{P}_{\mathrm{pr}}^\bullet, C)$ *is a (complete) predictable exit system for* F, *in the sense that the fine support of* C *is all of* F, *and (3.12) holds with* G_0 *replaced by* G.

(c) *Conversely, if there is a complete predictable exit system for* F, *then the conditions listed in (3.1) hold.*

Proof. (a) From (3.4) we see that

$$\mathbf{P}^\bullet \sum_{t\in G^i\cap J} Z_t\Psi\circ\theta_t = \mathbf{P}^\bullet \int_0^\infty Z_t\mathbf{P}_0^{X_t}[\Psi]1_F(X_t)N_\Delta(X_t,\psi)\,\mathrm{d}H_t, \tag{3.13}$$

where

$$\mathbf{P}_0^x[\Psi] := \frac{\int_{F^c} N_\Delta(x,\mathrm{d}y)\mathbf{P}^y[\Psi]}{N_\Delta(x,\psi)},$$

with the understanding that the ratio vanishes when the denominator is zero. Combining this with (2.4), (3.9), and (3.10), we obtain (3.12). It follows from (3.7) that the fine support of C is contained in F.

(b) Suppose that $G^i \subset J$. Clearly $G = G_0$ in this case. To see that C has fine support equal to F, we observe that the inequality in (3.8) becomes an equality, so $\mathbf{P}^x[\exp(-T_F)] = U_C^1 1_{E_\Delta}(x)$ for all x. Let $R_C := \inf\{t : C_t > 0\}$. Clearly $R_C \geq T_F$ because C is carried by F. On the other hand

$$\mathbf{P}^x[\exp(-T_F)] = U_C^1 1_{E_\Delta}(x) = \mathbf{P}^x \int_0^\infty \mathrm{e}^{-t}\,\mathrm{d}C_t$$

$$= \mathbf{P}^x \int_{R_C}^\infty \mathrm{e}^{-t}\,\mathrm{d}C_t \tag{3.14}$$

$$= \mathbf{P}^x \left[\exp(-R_C)U_C^1 1_{E_\Delta}(X_{R_C})\right]$$

$$\leq \mathbf{P}^x[\exp(-R_C)],$$

because $U_C^1 1_{E_\Delta} \leq 1_{E_\Delta}$. Together with the previously noted inequality $R_C \geq T_F$, (3.14) implies that $R_C = T_F$ almost surely.

(c) Suppose, conversely, that $(\mathbf{P}_{\mathrm{pr}}^\bullet, C)$ is a predictable exit system for F; that is, (3.12) holds with G_0 replaced by G. One readily checks that $\varphi_1 := \mathbf{P}^\bullet[\mathrm{e}^{-T_F}]$ is the 1-potential of the CAF $t \mapsto \int_0^t 1_F(X_s)\,\mathrm{d}s + \int_0^t \mathbf{P}_{\mathrm{pr}}^{X(s)}[1 - \mathrm{e}^{-T_F}]\,\mathrm{d}C_s$, which implies that φ_1 is regular. $\square$

4 Time Change

Recall from section 2 that A is a CAF of X with fine support F. Thus F is finely perfect and the closed visiting set M has ω-sections that are perfect (or empty) almost surely. Let $\tau = (\tau(t))_{t\geq 0}$ denote the right-continuous inverse of A:

$$\tau_t = \tau(t) := \inf\{s : A_s > t\}, \qquad t \geq 0. \tag{4.1}$$

Then τ is strictly increasing (while finite), and as t varies the path $t \mapsto \tau(t)$ traces out $M \setminus G$.

As is well known the time-changed process $\widetilde{X}$ defined by

$$\widetilde{X}_t := X_{\tau(t)}, \qquad t \geq 0, \tag{4.2}$$

(with the convention $\widetilde{X}_\infty = \Delta$) is a right process with state space F, though $\widetilde{X}$ need not be a *Borel* right process.

We note in passing that if a nearly Borel set $L \subset F$ is X-polar (that is, $\mathbf{P}^x[X_t \in L$ for some $t > 0] = 0$ for all $x \in E$) then L is also $\widetilde{X}$-polar. Conversely, if $L \subset F$ is $\widetilde{X}$-polar, then L is X-semipolar. In fact, if L is $\widetilde{X}$-polar, then the visiting set $\{(\omega, t) : X_t(\omega) \in L\}$ is contained in the graph of T_F, up to evanescence. Indeed, since $L \subset F$, it is clear that $\{(\omega, t) : X_t(\omega) \in L\} \subset [\![T_F, \infty [\![$. In view of the observation made at the end of the first paragraph of this section, the $\widetilde{X}$-polarity of L implies that $\{(\omega, t) : X_t(\omega) \in L\} \subset G$. In particular, $\{t > 0 : X_t(\omega) \in L\}$ is countable, a.s. Thus, if we fix an initial distribution μ, then

$$\mathbf{P}^\mu \sum_{s \in G} 1_L(X_s)[1 - \exp(-T_F \circ \theta_s)] = \mathbf{P}^\mu \sum_{s \in G^r} 1_L(X_s)[1 - \exp(-T_F \circ \theta_s)]$$

$$= \mathbf{P}^\mu \int_0^\infty 1_L(X_s) \mathbf{P}_{\mathrm{op}}^{X_s}[1 - \exp(-T_F)]\, \mathrm{d}B_s^c = 0,$$

because B^c is continuous. To see that, in general, L need not be X-polar, consider the example of X equal to uniform motion to the right on $\mathbf{R}$ with $F = [0, \infty[$ and $L = \{0\}$.

5 Lévy System for $\widetilde{X}$

In this section we give an explicit description of the Lévy system of $\widetilde{X}$ in terms of the Lévy system of X and the partial predictable exit system for F. The key observation is contained in Lemma 5.1 below. Before coming to its statement and proof, it is necessary to introduce some notation. Let $(\widetilde{\mathcal{F}}_t)$ denote the filtration of the time-changed process $\widetilde{X}$; that is, the usual augmentation of $\sigma\{\widetilde{X}_s : 0 \le s \le t\}$, $t \ge 0$. Let ρ be a metric on E_Δ compatible with the Ray topology induced there by X. When viewed as a process with values in the metric space (E_Δ, ρ), X is a right process; consequently, $\widetilde{X}$ is a right process when viewed as a process with state space (F_Δ, ρ), where $F_\Delta := F \cup \{\Delta\}$. The corresponding Ray-Knight compactification $\overline{F}$ of F_Δ (determined by $\widetilde{X}$) induces a topology on F_Δ; let $\widetilde{\rho}$ be a metric compatible with that topology. We shall write $\widetilde{X}_{t-}^{\widetilde{r}}$ for the left limit (in $\overline{F}$) of $\widetilde{X}$ at time $t > 0$. We write $\widetilde{J}$ for the set of totally inaccessible jumps of $\widetilde{X}$; thus,

$$\widetilde{J} = \{(\omega, t) : \widetilde{X}_{t-}^{\widetilde{r}}(\omega) \in F_\Delta, \widetilde{X}_{t-}^{\widetilde{r}}(\omega) \neq \widetilde{X}_t(\omega)\},$$

and $\widetilde{J}$ encompasses the totally inaccessible stopping times of the filtration $(\widetilde{\mathcal{F}}_t)$. As in sections 2 and 3, we use $\widetilde{X}_{t-}^r$ and X_{t-}^r to denote left limits in the ρ-topology; these limits exist in $\overline{E}$ (the Ray compactification of E_Δ induced by X) for all $t > 0$ almost surely. Finally, $\widetilde{X}_{t-}$ and X_{t-} denote left limits taken in the original topology of E_Δ, whenever those limits exist.

We write Λ^+ (resp. Λ^-) for the set of points of right (resp. left) increase of A:

$$\Lambda^+ := \{(\omega, t) : t \geq 0, A_t(\omega) < A_{t+\epsilon}(\omega), \forall \epsilon > 0\}, \tag{5.1}$$
$$\Lambda^- := \{(\omega, t) : t > 0, A_{t-\epsilon}(\omega) < A_t(\omega), \forall \epsilon > 0\}. \tag{5.2}$$

The set Λ^+ is progressively measurable; in fact,

$$\Lambda^+ = M \setminus G. \tag{5.3}$$

Consequently, by the strong Markov property and Blumenthal's zero-one law, if T is a stopping time, then $[\![T]\!] \subset \Lambda^+$ if and only if $X_T \in F$, almost surely. Meanwhile, with $I_t = \limsup_{s \uparrow\uparrow t} 1_{\{X_s \in F\}}$ as before,

$$\Lambda^- = \{(\omega, t) : I_t(\omega) = 1\}, \tag{5.4}$$

so that Λ^- is predictable.

Lemma 5.1 *(a) Defining*

$$J^\# := \{(\omega, \tau_t(\omega)) : (\omega, t) \in \tilde{J}, \tau_{t-}(\omega) = \tau_t(\omega)\}, \tag{5.5}$$

we have $J^\# = J \cap \Lambda^- \cap \Lambda^+$, *up to evanescence.*
(b) Recalling that $G_0 := G^r \cup (G^i \cap J)$, *we have*

$$\{(\omega, \tau_{t-}(\omega)) : (\omega, t) \in \tilde{J}, \tau_{t-}(\omega) < \tau_t(\omega)\} = \{(\omega, s) \in G_0 : X^r_{s-}(\omega) \neq X_{D_s}(\omega)\},$$

up to evanescence, where $D_s = s + T_F \circ \theta_s$ *as before.*

Proof. In what follows, equalities or inclusions between subsets of $\Omega \times [0, \infty[$ are understood to hold modulo evanescence. Also, if $\Gamma \subset \Omega \times [0, \infty[$ and $S : \Omega \to [0, \infty]$, then we sometimes write $S \in \Gamma$ instead of $[\![S]\!] \subset \Gamma$.

(a) It is clear that $J^\# \subset \Lambda^- \cap \Lambda^+$. Moreover, since $\{(\omega, t) \in \tilde{J} : \tau_{t-}(\omega) = \tau_t(\omega)\}$ is $(\mathcal{F}_{\tau(t)})$-optional and has countable sections, it can be expressed as the countable union $\cup_n [\![T_n]\!]$ of graphs of $(\mathcal{F}_{\tau(t)})$-stopping times. Fix $n \in \mathbf{N}$ and write T for T_n and define $S := \tau(T)$, so that $T = A_S$. In view of (2.9) applied to $\tilde{X}$,

$$\rho\text{-}\lim_{u \uparrow T} \tilde{X}_u = \tilde{\rho}\text{-}\lim_{u \uparrow T} \tilde{X}_u = \tilde{X}^{\tilde{r}}_{T-}.$$

But, because X has left limits in the ρ-topology and $\tau(T-) = \tau(T)$,

$$\rho\text{-}\lim_{u \uparrow T} \tilde{X}_u = \rho\text{-}\lim_{u \uparrow t} X_{\tau(u)}$$

$$= X^r_{\tau(T-)-} = X^r_{\tau(T)-} = X^r_{S-}.$$

Hence $X^r_{S-} = \tilde{X}^{\tilde{r}}_{T-} \neq \tilde{X}_T = X_S$, from which we deduce that $S \in J$. This proves that $J^\# \subset J \cap \Lambda^- \cap \Lambda^+$.

For the reverse containment we begin by observing that $\Lambda^+ = M \setminus G$ is $(\mathcal{F}_t)$-progressive, so $J \cap \Lambda^+ \cap \Lambda^-$ is an $(\mathcal{F}_t)$-progressively measurable subset of the $(\mathcal{F}_t)$-optional set $J \cap \Lambda^-$, the latter set having countable sections. Thus, by [6, T-IV.88], there is a sequence (S_n) of $(\mathcal{F}_t)$-stopping times such that $J \cap \Lambda^+ \cap \Lambda^- = \cup_n [\![S_n]\!]$ up to evanescence. The containment at issue will therefore be established once we show that if S is an $(\mathcal{F}_t)$-stopping time with $[\![S]\!] \subset J \cap \Lambda^+ \cap \Lambda^-$ then $[\![S]\!] \subset J^\#$. Fix such a stopping time S and define $T := A_S$. Notice that $\tau(T-) = \tau(T) = S$ almost surely on $\{S < \infty\} = \{S < \zeta\}$, because $[\![S]\!] \subset \Lambda^+ \cap \Lambda^-$. Thus, to complete the proof it suffices to show that $[\![T]\!] \subset \tilde{J}$. Now

$$\widetilde{X}^r_{T-} = \rho\text{-}\lim_{u \uparrow A_S} X_{\tau(u)} = X^r_{\tau(A_S)-} = X^r_{S-} = X_{S-} \neq X_S = \widetilde{X}_T,$$

in which (i) the second, third, and last equalities hold because $S \in \Lambda^- \cap \Lambda^+$ so that A_S is a continuity point of τ with $\tau(A_S) = S$, and (ii) the fourth equality and the inequality follow from (2.9) because $S \in J$. Consequently, T is a discontinuity time of $\widetilde{X}$ in the ρ-topology, and $\widetilde{X}^r_{T-} \in E_\Delta$. Applying [26, (46.2)] to $\widetilde{X}$ (viewed as a right process in the ρ-topology on the state space $F_\Delta := F \cup \{\Delta\}$), the sets $\{\widetilde{X}^r_- \text{ does not exist in } F_\Delta\}$ and $\{\widetilde{X}^r_- \text{ exists in } F_\Delta \text{ but } \widetilde{X}^r_- \neq \widetilde{X}^{\tilde{r}}_-\}$ are both predictably meager; that is, they are countable unions of graphs of $(\widetilde{\mathcal{F}}_t)$-predictable stopping times. To show that the intersection of $[\![T]\!]$ with the union of these sets is evanescent, it therefore suffices to show that $[\![T]\!]$ meets the graph of no $(\widetilde{\mathcal{F}}_t)$-predictable time. Suppose for the moment that this has been established. Then since $\widetilde{X}^r_{T-} \neq \widetilde{X}_T$, the only remaining possibility is that $\widetilde{X}^r_{T-}$ exists in F_Δ and is equal to $\widetilde{X}^{\tilde{r}}_{T-}$. But this forces $T \in \tilde{J}$.

It remains to show that if R is an $(\widetilde{\mathcal{F}}_t)$-predictable time then $[\![R]\!] \cap [\![T]\!]$ is evanescent. Fix such an R, and let (R_n) announce R. Then

$$\{\tau(R_n) < t\} = \{R_n < A_t\} = \cup_{q \in \mathbf{Q}} \{R_n < q < A_t\}$$
$$= \cup_{q \in \mathbf{Q}} \{R_n < q, \tau(q) < t\} \in \mathcal{F}_t,$$

since $\{R_n < q\} \in \widetilde{\mathcal{F}}_q \subset \mathcal{F}_{\tau(q)}$ and $\tau(q)$ is an $(\mathcal{F}_t)$-stopping time. Thus $\tau(R_n)$ is an $(\mathcal{F}_t)$-stopping time. But $t \mapsto \tau(t-)$ is strictly increasing on $]0, A_\zeta]$ and identically infinite on $[A_\zeta, \infty[$. Therefore the sequence $(\tau(R_n) \wedge n)$ of $(\mathcal{F}_t)$-stopping times increases to $\tau(R-)$ strictly from below. Consequently $\tau(R-)$ is a predictable $(\mathcal{F}_t)$-stopping time. Recalling that $T = A_S$, we see that on the event $\{R = T < \infty\}$ we have $\tau(R-) = S$ since $S \in \Lambda^- \cap \Lambda^+$. This implies that $\mathbf{P}^x[R = T < \infty] \leq \mathbf{P}^x[S = \tau(R-) < \infty] = 0$ for all $x \in E_\Delta$, because $[\![S]\!] \subset J$ and J meets the graph of no predictable $(\mathcal{F}_t)$-stopping time. Thus $[\![R]\!] \cap [\![T]\!]$ is evanescent, and the proof of assertion (a) of Lemma 5.1 is complete.

(b) Since $\{t \in \tilde{J} : \tau(t-) < \tau(t)\}$ is $(\mathcal{F}_{\tau(t)})$-optional and has countable sections, it can be expressed as the countable union of graphs of $(\mathcal{F}_{\tau(t)})$-stopping times T_n, $n \in \mathbf{N}$. (Only the $\mathcal{F}$-measurability of these times is relevant in the subsequent discussion.) Fix $n \in \mathbf{N}$ and abbreviate T_n to T. Then $T \in \tilde{J}$ and so $\tilde{X}_T \neq \tilde{X}_{T-}^{\tilde{r}} = \tilde{X}_{T-}^r = X_{\tau(T-)}^r$ with $X_{\tau(T-)}^r \in F_\Delta \subset E_\Delta$ by the equality of the second and fourth terms. Define $S := \tau(T-)$. Then $S \in G$ and $D(S) = D_S = \tau(T)$. From the previous string of equalities it follows that

$$X_{D(S)} = X_{\tau(T)} = \tilde{X}_T \neq X_{S-}^r,$$

and so it remains to show that $S \in G_0$. We consider two cases. First, if $X_{S-}^r \neq X_S$ then $S \in G \cap J = G^i \cap J$ because $X_{S-}^r \in F_\Delta \subset E_\Delta$. On the other hand, if $X_{S-}^r = X_S$, then $X_S \in F_\Delta$. If $X_S \in F$ then $S \in G^r$. To rule out the remaining possibility, suppose $X_S = \Delta$. Then $\tilde{X}_{T-}^{\tilde{r}} = X_S = \Delta$. But this implies that T is not a discontinuity of $\tilde{X}$ in its Ray topology, since X is constantly equal to Δ on $[S, \infty[$, contradicting $T \in \tilde{J}$.

For the opposite inclusion, recall that $D_t = t + T_F \circ \theta_t$ is an $(\mathcal{F}_t)$-stopping time for each $t \geq 0$ and that $t \mapsto D_t$ is increasing and right continuous with $G = \{t > 0 : D_{t-} = t < D_t\}$. Therefore G, and hence $G_0 = G^r \cup (G^i \cap J)$, is optional relative to the filtration $(\mathcal{F}_{D(t)})$ and has countable sections. Consequently, $G_0 = \cup_n [\![S_n]\!]$, where each S_n is an $(\mathcal{F}_{D(t)})$-stopping time. Fix $n \in \mathbf{N}$ and let S denote S_n. Then $S \in G_0$ and $\tau(A_S -) = S < D_S = \tau(A_S)$. But

$$\tilde{X}_{A(S)-}^r = X_{\tau(A(S)-)-}^r = X_{S-}^r \neq X_{D(S)} = \tilde{X}_{A(S)},$$

so A_S is a time of discontinuity of $\tilde{X}$ in the ρ-topology. To complete the proof we must show that $A_S \in \tilde{J}$. This will follow once we show that $\tilde{X}_{A(S)-}^{\tilde{r}} = \tilde{X}_{A(S)-}^r \in F_\Delta$, and (using [26, (46.2)] as in the proof of the second part of (a)) this will follow in turn once we show that $[\![A_S]\!]$ meets the graph of no $(\tilde{\mathcal{F}}_t)$-predictable time. Let R be such a predictable time and suppose that $\mathbf{P}^x[R = A_S < \infty] > 0$ for some $x \in F_\Delta$. Then exactly as in the last paragraph of the argument for (i), $\tau(R-)$ is an $(\mathcal{F}_t)$-predictable time, and $\tau(R-) = S \in G_0$ on $\{R = A_S\}$. This is a contradiction since G^r meets the graph of no $(\mathcal{F}_t)$-stopping time and J meets the graph of no $(\mathcal{F}_t)$-predictable time. This completes the proof of Lemma 5.1. $\square$

Define

$$\tilde{C}_t := C_{\tau(t)}, \qquad \tilde{H}_t^F := \int_0^{\tau(t)} 1_F(X_s) \, dH_s. \tag{5.6}$$

Because the fine supports of C and $t \mapsto \int_0^t 1_F(X_s) \, dH_s$ are contained in F, both $\tilde{C}$ and $\tilde{H}^F$ are CAFs of $\tilde{X}$. Now define

$$\tilde{H}_t := \tilde{C}_t + \tilde{H}_t^F, \qquad t \geq 0. \tag{5.7}$$

Another application of Motoo's theorem yields the existence of $\mathcal{E}^e$-measurable densities $\tilde{c}$ and $\tilde{h}$ (vanishing off F) such that

$$\tilde{C}_t = \int_0^t \tilde{c}(\tilde{X}_s) \, d\tilde{H}_s, \quad \text{and} \quad \tilde{H}_t^F = \int_0^t \tilde{h}(\tilde{X}_s) \, d\tilde{H}_s, \qquad \forall t \geq 0, \tag{5.8}$$

almost surely. Finally, define a kernel $\tilde{N}_\Delta$ on $(F_\Delta, \mathcal{E}_\Delta \cap F_\Delta)$ by

$$\tilde{N}_\Delta(x, dy) := 1_{F \times F_\Delta}(x, y) \left[\tilde{c}(x) 1_{\{x \neq y\}} \mathbf{P}^x_{\mathrm{pr}}[X_{T_F} \in dy] + \tilde{h}(x) N_\Delta(x, dy) \right]. \tag{5.9}$$

Theorem 5.2 *The pair* $(\tilde{N}_\Delta, \tilde{H})$ *is a Lévy system for the totally inaccessible jumps of* $\tilde{X}$.

Proof. Fix an $\tilde{X}$-predictable process $\tilde{Z} \geq 0$, and a positive Borel function Φ on the product space $F \times F_\Delta$ such that $\Phi(x, x) = 0$ for all $x \in F$. Using [6, (IV.67.1)] it is not hard to check that $s \mapsto \tilde{Z}_{A(s)}$ is X-predictable. Then, using Lemma 5.1(a) for the first equality,

$$\mathbf{P}^{\bullet} \sum_{t \in \tilde{J}, \tau(t-)=\tau(t)} \tilde{Z}_t \Phi(\tilde{X}_{t-}, \tilde{X}_t)$$

$$= \mathbf{P}^{\bullet} \sum_{s \in J} 1_{\Lambda^-}(s) \tilde{Z}_{A(s)} \Phi(X_{s-}, X_s) 1_{\Lambda^+}(s)$$

$$= \mathbf{P}^{\bullet} \sum_{s \in J} 1_{\Lambda^-}(s) \tilde{Z}_{A(s)} \Phi(X_{s-}, X_s) 1_F(X_s)$$

$$= \mathbf{P}^{\bullet} \int_0^\infty 1_{\Lambda^-}(s) \tilde{Z}_{A(s)} \int_F N_\Delta(X_s, dy) \Phi(X_s, y) \, dH_s$$

$$= \mathbf{P}^{\bullet} \int_0^\infty 1_F(X_s) \tilde{Z}_{A(s)} \int_F N_\Delta(X_s, dy) \Phi(X_s, y) \, dH_s.$$

The second equality above follows from the discussion just after (5.3) because J is the disjoint union of graphs of stopping times, while the final equality holds because Λ^- differs from $\{X \in F\}$ by a countable set not charged by H. Consequently,

$$\mathbf{P}^{\bullet} \sum_{t \in \tilde{J}, \tau(t-)=\tau(t)} \tilde{Z}_t \, \Phi(\tilde{X}_{t-}, \tilde{X}_t) = \mathbf{P}^{\bullet} \int_0^\infty \tilde{Z}_t \int_F N_\Delta(\tilde{X}_t, dy) \, \Phi(\tilde{X}_t, y) \, d\tilde{H}_t^F. \tag{5.10}$$

On the other hand, using Lemma 5.1(b),

$$\mathbf{P}^{\bullet} \sum_{t \in \tilde{J}, \tau(t-) < \tau(t)} \tilde{Z}_t \, \Phi(\tilde{X}_{t-}, \tilde{X}_t)$$

$$= \mathbf{P}^{\bullet} \sum_{s \in G_0} \tilde{Z}_{A(s)} \, \Phi(X_{s-}, X_{D_s})$$

$$= \mathbf{P}^{\bullet} \int_0^\infty \tilde{Z}_{A(s)} \int_\Omega \Phi(X_s, X_{T_F}(\omega)) \mathbf{P}^{X_s}_{\mathrm{pr}}(d\omega) \, dC_s \tag{5.11}$$

$$= \mathbf{P}^{\bullet} \int_0^\infty \tilde{Z}_t \int_\Omega \Phi(\tilde{X}_t, X_{T_F}(\omega)) \mathbf{P}^{\tilde{X}_t}_{\mathrm{pr}}(d\omega) \, d\tilde{C}_t.$$

Taken together, (5.10) and (5.11) imply that $(\tilde{N}_\Delta, \tilde{H})$ is a Lévy system for the totally inaccessible jumps of $\tilde{X}$. $\square$

6 Jump Measures and Feller Measures

We now fix an *excessive measure* m to serve as background measure. Thus m is a σ-finite measure on $(E, \mathcal{E})$ such that $mP_t(L) \leq m(L)$ for all $L \in \mathcal{E}$ and $t > 0$. Because (P_t) is a right semigroup, we then have $mP_t \uparrow m$ (setwise) as $t \downarrow 0$; see [7, XII 36-37]. Here and in the remainder of the paper the (absorbing) state Δ is viewed as a cemetery state; the stopping time $\zeta := \inf\{t : X_t = \Delta\}$ is the *lifetime* of X. Accordingly, functions (resp. measures) defined on E (resp. $\mathcal{E}$) are extended to E_Δ (resp. $\mathcal{E}_\Delta$) by letting the value at Δ (resp. $\{\Delta\}$) be 0.

Let $R = (R_t)_{t \geq 0}$ be a raw (i.e., not necessarily adapted) additive functional (RAF) of X. The Revuz measure of R, relative to m, is defined by the monotone limit

$$\nu_R^m(f) := \uparrow \lim_{t \downarrow 0} t^{-1} \mathbf{P}^m \int_0^t f(X_s) \, dR_s, \qquad f \in p\mathcal{E}^*. \tag{6.1}$$

If R is a CAF then the measure ν_R^m is σ-finite, and two CAFs with the same Revuz measure are $\mathbf{P}^m$-indistiguishable. See [15], [10], and [11] for more details. The "local" formula (6.1) defining ν_R^m has a "global" counterpart expressed in terms of the Kuznetsov process $((Y_t)_{t \in \mathbf{R}}, Q_m)$ associated with X and m. The sample space for Y is W, the space of all paths $w : \mathbf{R} \to E_\Delta := E \cup \{\Delta\}$ that are right continuous and E-valued on an open interval $]\alpha(w), \beta(w)[$ and take the value Δ outside of this interval. The dead path $[\Delta]$, constantly equal to Δ, corresponds to the interval being empty; by convention $\alpha([\Delta]) = +\infty$, $\beta([\Delta]) = -\infty$. The σ-algebra $\mathcal{G}^\circ$ on W is generated by the coordinate maps $Y_t(w) = w(t)$, $t \in \mathbf{R}$, and $\mathcal{G}_t^\circ := \sigma(Y_s : s \leq t)$. The Kuznetsov measure Q_m is the unique σ-finite measure on $\mathcal{G}^\circ$ not charging $\{[\Delta]\}$ such that, for $-\infty < t_1 < t_2 < \cdots < t_n < +\infty$,

$$\begin{aligned}
Q_m(Y_{t_1} &\in dx_1, Y_{t_2} \in dx_2, \ldots, Y_{t_n} \in dx_n) \\
&= m(dx_1) \, P_{t_2 - t_1}(x_1, dx_2) \cdots P_{t_n - t_{n-1}}(x_{n-1}, dx_n).
\end{aligned} \tag{6.2}$$

Because the only times appearing on the right side of (6.2) are the differences $t_k - t_{k-1}$, the measure Q_m is invariant with respect to the shift operators σ_t, $t \in \mathbf{R}$, defined by

$$\sigma_t w(s) = [\sigma_t w](s) := w(t + s), \qquad s \in \mathbf{R};$$

that is

$$Q_m[\Phi \circ \sigma_t] = Q_m[\Phi], \qquad \forall \Phi \in p\mathcal{G}^\circ, t \in \mathbf{R}.$$

It will be convenient to take $X = (X_t, \mathbf{P}^x)$ to be the realization of (P_t) described on p. 53 of [15]. The sample space for X is

$$\Omega := \{\alpha = 0, Y_{\alpha+} \text{ exists in } E\} \cup \{[\Delta]\}, \tag{6.3}$$

X_t is the restriction of Y_t to Ω for $t > 0$, and X_0 is the restriction of Y_{0+}. Moreover, $\mathcal{F}^{\circ} := \sigma(X_t : t \geq 0)$ is the trace of $\mathcal{G}^{\circ}$ on Ω.

To discuss the strong Markov property of Y, as well as the moderate markov property of Y when time is reversed, we recall the modified process Y^* of [15, (6.12)]. Let d be a totally bounded metric on E compatible with the topology of E, and let $\mathcal{D}$ be a countable uniformly dense subset of the d-uniformly continuous bounded real-valued functions on E. Given a strictly positive $h \in b\mathcal{E}$ with $m(h) < \infty$ define $W(h) \subset W$ by the conditions:

(i) $\alpha \in \mathbf{R}$;
(ii) $Y_{\alpha+} := \lim_{t \downarrow \alpha} Y_t$ exists in E;
(iii) $U^q g(Y_{\alpha+1/n}) \to U^q g(Y_{\alpha+})$ as $n \to \infty$, for all $g \in \mathcal{D}$ and all rationals $q > 0$;
(iv) $Uh(Y_{\alpha+1/n}) \to Uh(Y_{\alpha+})$ as $n \to \infty$.

Evidently $\sigma_t^{-1}(W(h)) = W(h)$ for all $t \in \mathbf{R}$, and $W(h) \in \mathcal{G}_{\alpha+}^{\circ}$ since E is a Lusin space. We now define

$$Y_t^*(w) = \begin{array}{l} Y_{\alpha+}(w), \text{ if } t = \alpha(w) \text{ and } w \in W(h), \\ Y_t(w), \quad \text{otherwise.} \end{array} \qquad (6.4)$$

(If h' is another function with the properties of h then $\mathbf{Q}_m(W(h)\Delta W(h')) = 0$.)

The process Y^* features in a maximal form of the the strong Markov property, recorded in Proposition 6.1 below; for a proof see [15, (6.15)]. (This process will also be used in section 7 to define the moderate Markov dual of X with respect to m.) A clean statement of this result requires the "truncated shift" operators θ_t, $t \in \mathbf{R}$ defined by

$$\theta_t w(s) = [\theta_t w](s) := \begin{array}{l} w(t+s), \ s > 0; \\ \Delta, \qquad s \leq 0. \end{array}$$

The filtration $(\mathcal{G}_t^m)_{t \in \mathbf{R}}$ is obtained by augmenting $(\mathcal{G}_t^{\circ})_{t \in \mathcal{R}}$ with the $\mathbf{Q}_m$ null sets in the usual way.

Proposition 6.1 Let T be a $(\mathcal{G}_t^m)$-stopping time. Then $\mathbf{Q}_m$ restricted to $\mathcal{G}_T^m \cap \{Y_T^* \in E\}$ is a σ-finite measure and

$$\mathbf{Q}_m(F \circ \theta_T \mid \mathcal{G}_T^m) = P^{Y_T^*}(F), \quad \mathbf{Q}_m\text{-a.e.} \quad \text{on} \quad \{Y_T^* \in E\} \qquad (6.5)$$

for all $F \in p\mathcal{F}^{\circ}$.

Now given an RAF R there is a uniquely determined (up to $\mathbf{Q}_m$ evanescence) homogeneous random measure (HRM) κ_R such that

$$\kappa_R]s,t] = R_{t-s}\circ\theta_s, \quad \text{on } \{\alpha < s < \beta\}, \mathbf{Q}_m\text{-a.s.},$$

for all real $s < t$. The global counterpart to (6.1) that was alluded to earlier is this:

$$\mathbf{Q}_m \int_R f(Y_t, t)\,\kappa_R(dt) = \int_E \int_{\mathbf{R}} f(x, t)\,dt\,\nu_R^m(dx), \qquad \forall f \in p(\mathcal{E} \otimes \mathcal{B}). \quad (6.6)$$

See [15, (8.21), (8.25)].

As an example, let us consider additive functionals related to the Lévy system (N_Δ, H) discussed at the end of section 2. As is customary, we now break N_Δ into two pieces

$$N(x, dy) := 1_E(y)N_\Delta(x, dy), \qquad n(x) := N_\Delta(x, \{\Delta\}), \quad (6.7)$$

and define the "killing rate" CAF K by

$$K_t := \int_0^t n(X_s)\,dH_s, \qquad t \ge 0. \quad (6.8)$$

Taking $Z = 1_{]0,t]}$ in (2.10) we find that

$$\mathbf{P}^x \sum_{s \in J, s \le t} \Psi(X_{s-}, X_s)1_{\{s < \zeta\}} = \mathbf{P}^x \int_0^t N(X_s, \Psi)\,dH_s, \qquad x \in E, t \ge 0,$$
$$(6.9)$$

and

$$\mathbf{P}^x[f(X_{\zeta-}); \zeta \le t, \zeta \in J] = \mathbf{P}^x \int_0^t f(X_s)\,dK_s, \quad (6.10)$$

for $\Psi \in p(\mathcal{E} \otimes \mathcal{E})$ and $f \in p\mathcal{E}$. It follows from this discussion and (6.1) that

$$\mathcal{J}(\Psi) :=\uparrow \lim_{t \downarrow 0} t^{-1}\mathbf{P}^m \sum_{s \in J, s \le t} \Psi(X_{s-}, X_s)1_{\{s < \zeta\}} = \nu_H^m(N\Psi), \quad (6.11)$$

and

$$\mathcal{K}(f) :=\uparrow \lim_{t \downarrow 0} t^{-1}\mathbf{P}^m[f(X_{\zeta-}); \zeta \le t, \zeta \in J] = \nu_K^m(f) = \nu_H^m(nf). \quad (6.12)$$

That is, the "jump measure" $\mathcal{J}$ is given on $E \times E$ by the formula

$$\mathcal{J}(dx, dy) = \nu_H^m(dx)N(x, dy), \quad (6.13)$$

while the "killing measure" $\mathcal{K}$ is given on E by

$$\mathcal{K}(dx) = n(x)\,\nu_H^m(dx). \quad (6.14)$$

Now let J^* denote the set of totally inaccessible jumps of Y (defined as J was for X). Paralleling (2.9) we have (employing the obvious notation regarding left limits)

$$J^* \subset \{(w, t) \in W \times \mathbf{R} : \alpha(w) < t < \beta(w), Y_{t-}(w) = Y_{t-}^r(w)\} \quad (6.15)$$

up to $\mathbf{Q}_m$-evanescence. Combining (6.6) with the version of (2.10) valid for Y we now obtain, for suitably measurable $f \geq 0$,

$$\mathbf{Q}_m \sum_{t \in J^*} f(t, Y_{t-}, Y_t) = \mathbf{Q}_m \int_{\mathbf{R}} \int_E f(t, Y_t, y) \, N(Y_t, dy) \, \kappa_H(dt)$$

$$= \int_{\mathbf{R}} dt \int_{E \times E} f(t, x, y) \, \mathcal{J}(dx, dy), \tag{6.16}$$

and

$$\mathbf{Q}_m[f(\beta, Y_{\beta-}); \beta \in J^*] = \mathbf{Q}_m \int_{\mathbf{R}} f(t, Y_t) n(Y_t) \, \kappa_H(dt)$$

$$= \int_{\mathbf{R}} dt \int_E f(t, x) \, \mathcal{K}(dx). \tag{6.17}$$

We record the analogous results for the optional and predictable exit systems for F. The "optional" version comes from [13] (see also [15, (11.6)]); the "predictable" version is proved in a similar manner. Let F be a finely perfect nearly Borel set. For $w \in W$ let $M^*(w)$ be the closure in $\mathbf{R}$ of $\{t \in \mathbf{R} : Y_t(w) \in F\}$ and let $G^*(w)$ be the set of left endpoints (in $]\alpha(w), \infty[$) of the contiguous intervals of $M^*(w)$. It is readily verified that if $\alpha < s < t$ then $t \in M^*$ if and only if $t - s \in M \circ \theta_s$, and likewise for G^* and G_0^*, where G_0^* is defined over Y in analogy to G_0.

Proposition 6.2 *Let* $(\mathbf{P}_{op}^\bullet, B)$ *and* $(\mathbf{P}_{pr}^\bullet, C)$ *be optional and partial predictable exit systems for* F, *respectively. Let* $\nu_B = \nu_B^m$ *and* $\nu_C = \nu_C^m$ *be the corresponding Revuz measures, with respect to* m. *Then* ν_B *and* ν_C *are* σ-*finite, and*

$$\mathbf{Q}_m \sum_{s \in G^*} f(s, Y_s, \theta_s) = \int_{\mathbf{R}} dt \int_E \nu_B(dx) \, \mathbf{P}_{op}^x[f(t, x, \cdot)], \tag{6.18}$$

$$\mathbf{Q}_m \sum_{s \in G_0^*} f(s, Y_{s-}, \theta_s) = \int_{\mathbf{R}} dt \int_E \nu_C(dx) \, \mathbf{P}_{pr}^x[f(t, x, \cdot)], \tag{6.19}$$

provided $f \in p(\mathcal{B} \otimes \mathcal{E}_\Delta^* \otimes \mathcal{F}^*)$.

Following [4] (see also [14] and [3]) we define the *Feller measure*

$$\Lambda(\Gamma) := \uparrow \lim_{t \downarrow 0} t^{-1} \mathbf{P}^m \sum_{s \in G_0} 1_{]0,t]}(s) 1_\Gamma(X_{s-}, X_{D_s}) 1_{\{D_s < \infty\}}, \qquad \Gamma \in \mathcal{E} \otimes \mathcal{E}, \tag{6.20}$$

and the *supplementary Feller measure*

$$\delta(L) := \uparrow \lim_{t \downarrow 0} t^{-1} \mathbf{P}^m \sum_{s \in G_0} 1_{]0,t]}(s) 1_L(X_{s-}) 1_{\{D_s = \infty\}}, \qquad L \in \mathcal{E}. \tag{6.21}$$

Since $X_{D_s} = X_{T_F} \circ \theta_s$, the exit system formula (3.12) implies that

$$\mathbf{P}^m \sum_{s \in G_0} 1_{]0,t]}(s) 1_\Gamma(X_{s-}, X_{D_s}) 1_{\{D_s < \infty\}} = \mathbf{P}^m \int_0^t \phi(X_s) \, dC_s, \tag{6.22}$$

where
$$\phi(x) := \mathbf{P}_{\mathrm{pr}}^x[1_\Gamma(x, X_{T_F}); T_F < \infty], \qquad x \in E. \tag{6.23}$$

Formula (6.22) yields the existence of the monotone limit in (6.20) and even identifies the limit as $\nu_C(\phi)$. Hence,

$$\Lambda(\Gamma) = \nu_C(\phi) = \int_F \nu_C(\mathrm{d}x) \, \mathbf{P}_{\mathrm{pr}}^x[1_\Gamma(x, X_{T_F}); T_F < \infty]. \tag{6.24}$$

Similar considerations lead to the identification of the limit in (6.21) as

$$\delta(L) = \int_L \nu_C(\mathrm{d}x) \, \mathbf{P}_{\mathrm{pr}}^x[T_F = \infty]. \tag{6.25}$$

The next two formulas follow immediately from Proposition 6.2 and formulas (6.24) and (6.25).

Proposition 6.3 *For $\Phi \in p(\mathcal{B} \otimes \mathcal{E} \otimes \mathcal{E})$ and $f \in p(\mathcal{B} \otimes \mathcal{E})$,*

$$\mathbf{Q}_m \sum_{t \in G_0^*} \Phi(t, Y_{t-}, Y_{D_t}) 1_{\{D_t < \infty\}} = \int_{\mathbf{R}} \mathrm{d}t \int_{F \times F} \Phi(t, x, y) \Lambda(\mathrm{d}x, \mathrm{d}y), \tag{6.26}$$

and

$$\mathbf{Q}_m \sum_{t \in G_0^*} f(t, Y_{t-}) 1_{\{D_t = \infty\}} = \int_{\mathbf{R}} \int_F f(t, x) \, \mathrm{d}t \, \mathbf{P}_{\mathrm{pr}}^x[T_F = \infty] \nu_C(\mathrm{d}x). \tag{6.27}$$

Recall from section 5 the CAF A, its inverse τ, and the time-changed process $\widetilde{X} = X_\tau$. We are now going to exhibit formulas for the jump and killing measures of $\widetilde{X}$, in terms of the corresponding measures for X and the exit system for F. In the course of the proof we shall use the following result taken from section 6 of [9].

Proposition 6.4 *Let $\widetilde{m}$ denote ν_A^m, and let Z be a CAF of X with fine support contained in F (That is, $Z_{T_F} = 0$, a.s.) Then $\widetilde{m}$ is $\widetilde{X}$-excessive, the time change $\widetilde{Z}_t := Z_{\tau(t)}$, $t \geq 0$, defines a CAF of $\widetilde{X}$, and the Revuz measure $\widetilde{\nu}_{\widetilde{Z}}^{\widetilde{m}}$ of $\widetilde{Z}$, relative to $\widetilde{m}$, is equal to ν_Z^m.*

The following result extends [4, Thm. 5.6] to the context of this paper. We write $(F \times F)_0$ for $\{(x, y) \in F \times F : x \neq y\}$.

Theorem 6.5 *The jump measure $\widetilde{\mathcal{J}}$ and the killing measure $\widetilde{\mathcal{K}}$ for the time-changed process $\widetilde{X}$, with respect to the $\widetilde{X}$-excessive measure $\widetilde{m} := \nu_A^m$, are given respectively by the formulas*

$$\widetilde{\mathcal{J}} = 1_{(F \times F)_0}(\mathcal{J} + \Lambda), \tag{6.28}$$
$$\widetilde{\mathcal{K}} = 1_F \mathcal{K} + \delta. \tag{6.29}$$

Proof. We prove only (6.28), as the proof of (6.29) is quite similar. Let Γ be a Borel measurable subset of $(F \times F)_0$, and let us begin with

$$\mathbf{P}^x \sum_{s \in \tilde{J}, s \leq t} 1_\Gamma(\tilde{X}^r_{s-}, \tilde{X}_s) = \mathbf{P}^x \int_0^t \tilde{N}(\tilde{X}_s, 1_\Gamma) \, d\tilde{H}_s, \qquad x \in F. \tag{6.30}$$

In view of (5.6)–(5.9), the right side of (6.30) is equal to

$$\mathbf{P}^x \int_0^t N(\tilde{X}_s, 1_\Gamma) \, d\tilde{H}^F_s + \mathbf{P}^x \int_0^t \phi(\tilde{X}_s) \, d\tilde{C}_s,$$

where $\phi(x) := \mathbf{P}^x_{\mathrm{pr}}[1_\Gamma(x, X_{T_F}); T_F < \infty]$. By (5.6) and Proposition 6.4 the Revuz measure of $\tilde{H}^F$ (relative to $\tilde{m} := \nu^m_A$) is $1_F \nu^m_H$, while that of $\tilde{C}$ is ν^m_C. Therefore

$$\begin{aligned} \tilde{J}(\Gamma) &= \tilde{\nu}^{\tilde{m}}_H(\tilde{N} 1_\Gamma) \\ &= \int_\Gamma \nu^m_H(dx) \, N(x, dy) + \int_\Gamma \nu_C(dx) \, \mathbf{P}^x_{\mathrm{pr}}[X_{T_F} \in dy, T_F < \infty]. \end{aligned} \tag{6.31}$$

$\square$

7 Entrance Law

Because of the time-symmetry of the Markov property, the process $(Y_t, \mathbf{Q}_m)$ is a Markov process with respect to the reverse filtration $\tilde{\mathcal{G}}_t := \sigma\{Y_s : s \geq t\}$, $t \in \mathbf{R}$. Unlike the situation in "forward" time, this process need not be a strong Markov process, but it is a *moderate* Markov process. To make this precise we define

$$\hat{\Omega} := \{\beta = 0\} \subset W; \tag{7.1}$$

$$\hat{X}_t(\hat{\omega}) := Y^*_{-t}(\hat{\omega}), \qquad t > 0, \hat{\omega} \in \hat{\Omega} \tag{7.2}$$

$$\hat{\mathcal{F}}_t := \sigma\{\hat{X}_s : 0 < s \leq t\}, t > 0, \quad \hat{\mathcal{F}} := \sigma\{\hat{X}_s : s > 0\} \tag{7.3}$$

$$\check{\theta}_t w(s) := \begin{cases} w(t-s), s > 0; \\ \Delta, \qquad s \leq 0. \end{cases} \tag{7.4}$$

Then there is a Borel measurable family $\{\hat{\mathbf{P}}^x, x \in E\}$ of probability measures on $(\hat{\Omega}, \hat{\mathcal{F}})$ under which $(\hat{X}_t)_{t>0}$ has the moderate Markov property:

$$\hat{\mathbf{P}}^x[f(\hat{X}_{T+s})|\hat{\mathcal{F}}_{T-}] = \hat{\mathbf{P}}^{\hat{X}_T}[f(\hat{X}_s)], \qquad s > 0, f \in b\mathcal{E}, \tag{7.5}$$

whenever T is an $(\hat{\mathcal{F}}_t)$-predictable stopping time. (As a matter of convention, $\mathbf{P}^x[\hat{X}_0 = x] = 1$ and $\hat{\mathcal{F}}_{0-} = \{\emptyset, \hat{\Omega}\}$.) The measures $\hat{\mathbf{P}}^x$ are uniquely determined modulo an m-polar set. (A set $L \in \mathcal{E}^e$ is m-polar provided $\mathbf{P}^m[T_L < \infty] = 0$.) The link between Y and $\widehat{X}$ is this: If $T : W \to [-\infty, \infty]$ is $(\widehat{\mathcal{G}}_t)$-predictable, then for $\varPhi \in p\widehat{\mathcal{F}}$,

$$\mathbf{Q}_m[\varPhi \circ \check{\theta}_T | \widehat{\mathcal{G}}_{T-}] = \widehat{\mathbf{P}}^{Y_T^*}[\varPhi], \qquad \text{on } \{Y_T^* \in E\}, \tag{7.6}$$

the σ-finiteness of $\mathbf{Q}_m$ on $\widehat{\mathcal{G}}_{T-} \cap \{Y_T^* \in E\}$ being part of the assertion. For more details see [16, §2], [8, §4], and [23].

It follows easily from (7.6) (with T a fixed time) that the transition semi-group $(\widehat{P}_t)$ of $\widehat{X}$, defined by $\widehat{P}_t f = \widehat{\mathbf{P}}^\bullet[f(\widehat{X}_t)]$ is in duality with (P_t) with respect to m:

$$(P_t f, g) = (f, \widehat{P}_t g), \qquad f, g \in \mathcal{E}^*, t > 0, \tag{7.7}$$

in which $(f, g) := \int_E fg \, dm$ provided the integral exists. Likewise, defining the associated resolvent

$$\widehat{U}^\lambda f = \int_0^\infty e^{-\lambda t} \widehat{P}_t f \, dt = \widehat{\mathbf{P}}^\bullet \int_0^\infty e^{-\lambda t} f(\widehat{X}_t) \, dt, \tag{7.8}$$

we have

$$(U^\lambda f, g) = (f, \widehat{U}^\lambda g), \qquad f, g \in \mathcal{E}^*, \lambda > 0. \tag{7.9}$$

We usually omit the hat $\widehat{}$ in those places where it is obviously required. For example, we write $\widehat{\mathbf{P}}^\bullet[f(X_t)]$ in place of $\widehat{\mathbf{P}}^\bullet[f(\widehat{X}_t)]$.

Before proceeding, we collect some facts about the moderate Markov dual process. Recall that a set $L \in \mathcal{E}^e$ is m-semipolar provided the visiting set $\{t > 0 : X_t \in L\}$ is $\mathbf{P}^m$-a.s. at most countable. Also, property $P(x)$ depending on $x \in E$ is said to hold m-quasi-everywhere (m-q.e.) provided $\{x \in E : P(x)$ fails$\}$ is m-polar. Define

$$\widehat{T}_F := \inf\{t \in]0, \widehat{\zeta}[: \widehat{X}_t \in F\}. \tag{7.10}$$

Lemma 7.1 *Let F be a finely perfect nearly Borel subset of E.*
(i) $\{x \in F : \widehat{\mathbf{P}}^x[T_F = 0] < 1\}$ is m-semipolar.
(ii) $\{x \in E \setminus F : \widehat{\mathbf{P}}^x[T_F = 0] > 0\}$ is m-semipolar.
(iii) $t \mapsto \widehat{X}_t$ has right limits in $\overline{E}$ (with respect to the Ray topology) on $[0, \infty[$, $\widehat{\mathbf{P}}^x$-a.s. for m-q.e. $x \in E$.

Remark 7.2 Neither statement (i) nor statement (ii) of the lemma can be improved, even if X is continuous with a strong Markov continuous dual process $\widehat{X}$, as the example of uniform motion on $\mathbf{R}$ shows.

Proof. (i) Let μ be a finite measure on E not charging m-semipolar sets. Then there is a diffuse optional copredictable HRM κ with Revuz measure μ; see

[8, (5.22)] or [12, (3.10)]. Let ϕ be a strictly positive Borel function on $\mathbf{R}$ with $\int_{\mathbf{R}} \phi(t)\, dt = 1$. Since κ is copredictable,

$$\mathbf{Q}_m \int_{\mathbf{R}} \phi(t) 1_F(Y_t^*) 1_{\{\widehat{T}_F \circ \breve{\theta}_t = 0\}}\, \kappa(dt)$$

$$= \mathbf{Q}_m \int_{\mathbf{R}} \phi(t) 1_F(Y_t^*) \widehat{P}^{Y^*(t)}[\widehat{T}_F = 0]\, \kappa(dt) \qquad (7.11)$$

$$= \int_F \widehat{\mathbf{P}}^x[T_F = 0]\, \mu(dx).$$

Let Z denote the closure of $\{t \in]\alpha, \beta[: Y_t \in F\}$. Then

$$Z \cap \{t : \widehat{T}_F \circ \breve{\theta}_t > 0\} = Z \cap \{t : \exists \epsilon > 0,\ ]t - \epsilon, t[\cap Z = \emptyset\}.$$

Hence, $Z \cap \{t : \widehat{T}_F \circ \breve{\theta}_t > 0\}$ is contained in the set of right endpoints of the contiguous intervals of Z. But there are only countably many such intervals, and κ is diffuse, so

$$\mathbf{Q}_m \int_{\mathbf{R}} \phi(t) 1_F(Y_t^*) 1_{\{\widehat{T}_F \circ \breve{\theta}_t = 0\}}\, \kappa(dt) = \mu(F). \qquad (7.12)$$

It follows that $\{x \in F : \widehat{\mathbf{P}}^x[T_F = 0] < 1\}$ has μ-measure equal to 0. Since μ was an arbitrary finite measure not charging m-semipolars, a result of Dellacherie [5, p. 70] tells us that $\{x \in F : \widehat{\mathbf{P}}^x[T_F = 0] < 1\}$ is m-semipolar.

(ii) Using the notation established in the proof of point (i),

$$\mathbf{Q}_m \int_{\mathbf{R}} \phi(t) 1_{\{\widehat{T}_F \circ \breve{\theta}_t = 0\}} 1_{E \setminus F}(Y_t^*)\, \kappa(dt) = \int_{E \setminus F} \widehat{\mathbf{P}}^x[T_F = 0]\, \mu(dx). \qquad (7.13)$$

If $\widehat{T}_F \circ \breve{\theta}_t = 0$ then for every sufficiently small $\eta > 0$ the interval $]t - \eta, t[$ contains times at which Y is in F; if also $Y_t \in E \setminus F$ then t is an element of G^* because $E \setminus F$ is finely open. Since κ is diffuse, the above displayed integrals must vanish. Point (ii) now follows as did (i).

(iii) By considering $f(\widehat{X}_t)$ as f runs through a countable dense subset of $C(\overline{E})$ one sees that the set of $(\widehat{\omega}, t)$ such that $s \mapsto \widehat{X}_s(\widehat{\omega})$ fails to have a right limit in $\overline{E}$ at t is $(\widehat{\mathcal{F}}_t^P)_{t \geq 0}$-progressively measurable. Here P is an arbitrary probability measure on $(\widehat{\Omega}, \widehat{\mathcal{F}}^\circ)$, and $(\widehat{\mathcal{F}}_t^P)$ is the usual right-continuous completion of $(\widehat{\mathcal{F}}_t^\circ)$. See, for example, [6, IV-90]. It follows that the projection Π of the above-described set onto $\widehat{\Omega}$ is an element of $\widehat{\mathcal{F}}^* := \cap_P \widehat{\mathcal{F}}^P$. Note that Π is the set of $\widehat{\omega}$ for which $s \mapsto \widehat{X}_s(\widehat{\omega})$ fails to have a right limit in $\overline{E}$ at some $t > 0$. Hence, $f(x) := \widehat{\mathbf{P}}^x[\Pi]$ is $\mathcal{E}^*$-measurable, and then f is coexcessive. Since $\widehat{X}_s \circ \breve{\theta}_t = Y_{t-s}$ for $s > 0$, the set $\breve{\theta}_t^{-1} \Pi$ is contained in the set of $w \in W$ such that $r \mapsto Y_r(w)$ fails to have a left limit in $\overline{E}$ at some $r \in]-\infty, t[$, and so $\mathbf{Q}_m[\breve{\theta}_t^{-1} \Pi] = 0$ for all $t \in \mathbf{R}$. Now

$$m(f) = \mathbf{Q}_m\left[\widehat{\mathbf{P}}^{Y(0)}[\Pi]\right] = \mathbf{Q}_m[\breve{\theta}_0^{-1} \Pi, \alpha < 0 < \beta] = 0.$$

Hence $f = 0$, m-a.e., and therefore $\widehat{\mathbf{P}}^x[\Pi] = f(x) = 0$ for m-q.e. $x \in E$; see [16, (2.11)].   □

We assume, for the remainder of this section, that $G^i \subset J$.

This means that the partial predictable exit system constructed in section 3 is in fact "complete" in the sense of Theorem 3.4(b).

Define, for $\lambda \geq 0$ and $f \in p\mathcal{E}^*$,

$$\widehat{P}_F^\lambda f(x) := \widehat{\mathbf{P}}^x[e^{-\lambda T_F} f(X_{T_F})] \tag{7.14}$$

$$\widehat{P}_{F+}^\lambda f(x) := \widehat{\mathbf{P}}^x[e^{-\lambda T_F} f(\widehat{X}_{\widehat{T}_F+}^r)], \tag{7.15}$$

with the understanding that $\exp(-0 \cdot \infty) = 0$ and $f(\Delta) = 0$, so that $\widehat{P}_F f :=$ $\widehat{P}_F^0 f = \widehat{\mathbf{P}}^x[f(X_{T_F}) : T_F < \infty]$. Here $\widehat{X}_{\widehat{T}_F+}^r$ denotes the right limit (in $\overline{E}$ with its Ray topology) of $t \mapsto \widehat{X}_t$ at $\widehat{T}_F$. In (7.15), f is extended to all of $\overline{E}$ by declaring $f(x) = 0$ for $x \in \overline{E} \setminus E$. In the light of Lemma 7.1(iii), $\widehat{X}_{\widehat{T}_F+}^r$ exists $\widehat{\mathbf{P}}^x$-a.s. on $\{\widehat{T}_F < \infty\}$ for m-q.e. $x \in E$. Thus both $\widehat{P}_{F+}^\lambda f$ and $\widehat{P}_F^\lambda f$ are uniquely determined m-q.e. and are $\mathcal{E}^*$-measurable.

Recall the optional and predictable exit systems $(\mathbf{P}_{\mathrm{op}}^\bullet, B)$ and $(\mathbf{P}_{\mathrm{pr}}^\bullet, C)$ for F. Since the measure m will remain fixed in the sequel, we shall write ν_B for ν_B^m and ν_C for ν_C^m. The *balayage* of m on F is the excessive measure $R_F m$ defined by

$$R_F m(f) := \mathbf{Q}_m[f(Y_t); T_F < t], \qquad f \in p\mathcal{E}. \tag{7.16}$$

Here $T_F := \inf\{t \in]\alpha, \beta[: Y_t \in F\}$ extends the previously defined hitting time of F to all of W. Upon noting that $f(Y_t) 1_{\{T_F < t\}} = [f(Y_0) 1_{\{T_F < 0\}}] \circ \sigma_t$, it becomes clear that the right side of (7.16) does not depend on t. Moreover, because $\{T_F < 0\} = \check{\theta}_0^{-1}\{\widehat{T}_F < \widehat{\zeta}\}$, we see from (7.16) that

$$R_F m(f) = \mathbf{Q}_m\left[f(Y_0)\widehat{\mathbf{P}}^{Y_0}[\widehat{T}_F < \widehat{\zeta}]\right] = (\widehat{P}_F 1, f),$$

because $\widehat{\mathbf{P}}^x[T_F < \zeta] = \widehat{P}_F 1(x)$ for all $x \in E$. It is important to note at this stage that

$$\widehat{P}_{F+} 1(x) = \widehat{P}_F 1(x), \qquad \text{for } m\text{-a.e.} x \in E. \tag{7.17}$$

(In fact, for m-q.e. x, but the m-a.e. assertion is sufficient for our purposes.) To see this fix a strictly positive $f \in b\mathcal{E}$ with $m(f) < \infty$, and define $g_0 := \sup\{t \leq 0 : Y_t \in F\}$. Observe that $0 \leq \widehat{P}_F 1(x) - \widehat{P}_{F+} 1(x) = \widehat{\mathbf{P}}^x[T_F < \zeta, X_{T_F+} \notin E]$. Using (7.6) with $T = 0$ we have

$$\int_E f(x) \left[\widehat{P}_F 1(x) - \widehat{P}_{F+} 1(x)\right] m(dx) = \mathbf{Q}_m[f(Y_0); \alpha < g_0, Y_{g_0-}^r \notin E]$$

$$= \mathbf{Q}_m[f(Y_0); \alpha < g_0, Y_{g_0-} = \Delta]$$

$$= 0,$$

the second equality following from Remark 3.2 reinterpreted for Y, and the third from the fact that Δ is isolated in E_Δ.

The following decomposition of $R_F m$ (for general Borel F) appears in section 6 of [13]:

$$(\widehat{P}_F 1, f) = R_F m(f) = \nu_B(\ell f + V^0_{\mathrm{op}} f), \qquad f \in p\mathcal{E}, \tag{7.18}$$

where $V^0_{\mathrm{op}} f(x) := \mathbf{P}^x_{\mathrm{op}} \int_0^{T_F} f(X_t) \, dt$. Our principal goal in the remainder of this section is to generalize this decomposition and to obtain its predictable analog.

Notation 7.3. $(f, g)_0 := \int_{E \setminus F} fg \, dm$.

Theorem 7.4 *If $f, g \in p\mathcal{E}^*$, then*
   (i) $(\widehat{P}^\lambda_F f, g)_0 = \nu_B(f \cdot V^\lambda_{\mathrm{op}} g)$, *and*
   (ii) $(\widehat{P}^\lambda_{F+} f, g)_0 = \nu_C(f \cdot V^\lambda_{\mathrm{pr}} g)$.
Here $V^\lambda_{\mathrm{op}} f = \mathbf{P}^\bullet_{\mathrm{op}} \int_0^{T_F} e^{-\lambda t} f(X_t) \, dt$ and $V^\lambda_{\mathrm{pr}} f$ is defined analogously with $\mathbf{P}^\bullet_{\mathrm{pr}}$ replacing $\mathbf{P}^\bullet_{\mathrm{op}}$.

Proof. We shall prove only (ii), as the proof of (i) is similar but easier since no Ray limit is involved. From Lemma 7.1(ii), $\widehat{\mathbf{P}}^x[T_F > 0] = 1$, m-a.e. on $E \setminus F$. Thus

$$(\widehat{P}^\lambda_{F+} f, g)_0 = \mathbf{Q}_m[\widehat{P}^\lambda_{F+} f(Y_0) g(Y_0); Y_0 \in E \setminus F]$$
$$= \mathbf{Q}_m\left[e^{-\lambda \widehat{T}_F \circ \check{\theta}_0} f(\widehat{X}^r(\widehat{T}_{F+})) \circ \check{\theta}_0 \, g(Y_0); \widehat{T}_F \circ \check{\theta}_0 > 0, Y_0 \in E \setminus F \right],$$

because $\widehat{T}_F \circ \check{\theta}_0 > 0$, $\mathbf{Q}_m$-a.e. on the event $\{Y_0 \in E \setminus F\}$. If $s = -\widehat{T}_F \circ \check{\theta}_0 > \alpha$, then $s \in G^*$ (defined below (6.17)), and $]s, s + T_F \circ \theta_s[$ is the unique interval contiguous to M^* that contains 0. If $s \in G^*$, then $Y^r_{s-} = Y_{s-}$ by Remark 3.2. Therefore, using (6.19) for the second equality below,

$$(\widehat{P}^\lambda_{F+} f, g)_0 = \mathbf{Q}_m \sum_{s \in G^*, s < 0} e^{\lambda s} f(Y_{s-}) g(X_{-s}) \circ \theta_s \, 1_{\{s + T_F \circ \theta_s > 0\}}$$
$$= \int_0^\infty e^{-\lambda t} \, dt \int_F \nu_C(dx) f(x) \mathbf{P}^x_{\mathrm{pr}}[g(X_t); t < T_F]$$
$$= \nu_C(f \cdot V^\lambda_{\mathrm{pr}} g).$$

$\square$

Let $(Q_t)_{t \geq 0}$ and $(V^\lambda)_{\lambda \geq 0}$ denote the semigroup and resolvent for (X, T_F), the process X killed at time T_F:

$$Q_t f(x) := \mathbf{P}^x[f(X_t); t < T_F],$$
$$V^\lambda f(x) := \mathbf{P}^x \int_0^{T_F} e^{-\lambda t} f(X_t) \, dt = \int_0^\infty e^{-\lambda t} Q_t f(x) \, dt.$$

Let $(\widehat{X}, \widehat{T}_F)$ denote $\widehat{X}$ killed at $\widehat{T}_F$, with corresponding semigroup $(\widehat{Q}_t)_{t \geq 0}$ and resolvent $(\widehat{V}^\lambda)_{\lambda \geq 0}$. As is customary, $V := V^0$ and $\widehat{V} := \widehat{V}^0$. It is known that (X, T_F) and $(\widehat{X}, \widehat{T}_F)$ are dual processes, in the sense that $(V^\lambda f, g)_0 = (f, \widehat{V}^\lambda g)_0$ for all $f, g \in p\mathcal{E}^*$ and $\lambda \geq 0$. See [12, (A.7)]. We write $Q_t^{\mathrm{op}} f := \mathbf{P}_{\mathrm{op}}^\bullet[f(X_t); t < T_F]$, with an analogous definition for Q_t^{pr}.

Corollary 7.5 *Fix $f \in bp\mathcal{E}^*$. Then the formulas*

$$\eta^f(g) := (\widehat{P}_{F+}f, g)_0 \quad \text{and} \quad \xi^f(g) := (\widehat{P}_F f, g)_0, \qquad g \in p\mathcal{E}, \qquad (7.19)$$

define (σ-finite) purely excessive measures η^f and ξ^f for (X, T_F). Moreover, $\eta_t^f := (f\nu_C)Q_t^{\mathrm{pr}}$ and $\xi_t^f := (f\nu_B)Q_t^{\mathrm{op}}$, $t > 0$, define entrance laws for (X, T_F) such that

$$\eta^f = \int_0^\infty \eta_t^f \, dt \quad \text{and} \quad \xi^f = \int_0^\infty \xi_t^f \, dt.$$

If, in addition, $\nu_C(f) < \infty$ (resp. $\nu_B(f) < \infty$), then η_t^f (resp. ξ_t^f) is a finite measure for each $t > 0$.

Proof. Clearly η^f and ξ^f are σ-finite measures on $E \setminus F$. If $t > 0$ then $\widehat{Q}_t \widehat{P}_{F+}f = \widehat{\mathbf{P}}^\bullet[f \circ X_{\widehat{T}_{F+}}; t < \widehat{T}_F < \infty] \leq \widehat{P}_{F+}f$, and because of Lemma 7.1(ii) we have $\widehat{Q}_t \widehat{P}_{F+}f \uparrow \widehat{P}_{F+}f$, m-a.e. on $E \setminus F$. Since f is bounded, $\widehat{Q}_t \widehat{P}_{F+}f \to 0$ as $t \to \infty$. Hence η^f is a purely excessive measure for (X, T_F). It follows from Theorem 7.4(ii) that $\eta^f = \int_0^\infty \eta_t^f \, dt$. Using the fact that $(X_t)_{t>0}$ under $\mathbf{P}_{\mathrm{pr}}^x$ is Markovian with transition semigroup (P_t), one easily checks that $\eta_{t+s}^f = \eta_t^f Q_s$ for $t, s > 0$. Recall that $\mathbf{P}_{\mathrm{pr}}^\bullet[1 - \exp(-T_F)] \leq 1$; see (3.10) and the sentence following (2.6). Now $(1 - e^{-t}) \leq (1 - e^{-T_F})$ on $\{t < T_F\}$, and ν_C is σ-finite. This implies that η_t^f is a countable sum of finite measures for each $t > 0$. Fix $g \in p\mathcal{E}$ with $0 < g \leq 1$ on $E \setminus F$ and $\eta^f(g) < \infty$. Then $Vg > 0$ on $E \setminus F$, and we may use the Fubini theorem to conclude that

$$\eta_t^f(Vg) = \int_t^\infty \eta_s^f(g) \, ds \leq \eta^f(g) < \infty.$$

Therefore η_t^f is in fact σ-finite for each $t > 0$. Consequently, $(\eta_t^f)_{t>0}$ is an entrance law for (X, T_F). If, in addition, $\nu_C(f) < \infty$, then $\eta_t^f(1) \leq \nu_C(f) < \infty$. The treatment of $(\xi_t^f)_{t>0}$ is similar. $\square$

Corollary 7.6 *For $f, g \in \mathcal{E}^*$,*
(i) $(\widehat{P}_F^\lambda f, g) = \nu_B(\ell f g) + \nu_B(f V_{\mathrm{op}}^\lambda g) = \nu_{B^c}(\ell f g) + \nu_B(f V_{\mathrm{op}}^\lambda g)$;
(ii) $(\widehat{P}_{F+}^\lambda f, g) = \nu_C(\gamma g \widehat{P}_{0+}f) + \nu_C(f V_{\mathrm{pr}}^\lambda g)$.
Here ℓ comes from (2.6), B^c is the continuous part of B, γ is defined just below (3.11), and $\widehat{P}_{0+}f := \widehat{\mathbf{P}}^\bullet[f(\widehat{X}_{0+}^r)]$.

Proof. Since $\sigma_t \mathbf{Q}_m = \mathbf{Q}_m$ for all $t \in \mathbf{R}$,

$$\int_F \widehat{P}_F^\lambda f(x)g(x)\,m(dx) = \mathbf{Q}_m[\widehat{P}_F^\lambda f(Y_0)g(Y_0)1_F(Y_0)]$$

$$= \int_0^1 \mathbf{Q}_m[\widehat{P}_F^\lambda f(Y_t)g(Y_t)1_F(Y_t)]\,dt.$$

Also, (2.6) implies that $1_F(Y_t)\,dt = \ell(Y_t)\,\kappa_B(dt)$, where κ_B is the HRM of Y that extends B; notice that κ_B has Revuz measure ν_B. See, for example, the discussion on pages 89–91 of [15]. Therefore, by [15, (8.21)],

$$\int_F \widehat{P}_F^\lambda f(x)g(x)\,m(dx) = \mathbf{Q}_m \int_0^1 \widehat{P}_F^\lambda f(Y_t)g(Y_t)\ell(Y_t)\,\kappa_B(dt)$$

$$= \nu_B(\ell g \widehat{P}_F^\lambda f).$$

But $\ell = 0$ on $E \setminus F$ and $\ell \nu_B = \ell \nu_{B^c}$. In view of Lemma 7.1(i), $\widehat{\mathbf{P}}^x[T_F = 0] = 1$, ν_{B^c}-a.e. because ν_{B^c} doesn't charge m-semipolars. Hence $\nu_B(\ell g \widehat{P}_F^\lambda f) = \nu_{B^c}(\ell g f)$, since $\widehat{\mathbf{P}}^x[\widehat{X}_0 = x] = 1$ by convention. Combining this with Theorem 7.4(i) yields Corollary 7.6(i). A similar argument shows that

$$\int_F \widehat{P}_{F+}^\lambda f(x)g(x)\,m(dx) = \nu_C(\gamma g \widehat{P}_{F+}^\lambda f) = \nu_C(\gamma g \widehat{P}_{0+}f),$$

establishing Corollary 7.6(ii).   □

Proposition 7.7 *If* $f, g \in p\mathcal{E}^*$, *then*

$$(\widehat{P}_{F+}^\lambda f, P_F g) = \nu_C(\gamma g \widehat{P}_{0+}f) + \nu_C(f V_{\mathrm{pr}}^\lambda P_F g) = (\widehat{P}_{F+}f, P_F^\lambda g). \tag{7.20}$$

Proof. The first equality is an immediate consequence of Corollary 7.6(ii) since $P_F g = g$ on F. For the second equality, arguing as in the proof of Corollary 7.6, we have

$$\int_F \widehat{P}_{F+}f \cdot P_F^\lambda g\,dm = \nu_C(\gamma g \widehat{P}_{0+}f).$$

Also, as in the proof of Theorem 7.4,

$$(\widehat{P}_{F+}f, P_F^\lambda g)_0 = \mathbf{Q}_m \sum_{s \in G^*, s<0} f(Y_{s-})e^{-\lambda(s+T_F \circ \theta_s)}g(X_{T_F}) \circ \theta_s 1_{\{s+T_F \circ \theta_s > 0\}}$$

$$= \int_F \nu_C(dx)f(x)\mathbf{P}_{\mathrm{pr}}^x \int_{-\infty}^0 e^{-\lambda(s+T_F)}g(X_{T_F})1_{\{T_F > -s\}}\,ds$$

$$= \int_F \nu_C(dx)f(x)\mathbf{P}_{\mathrm{pr}}^x \int_0^{T_F} e^{-\lambda(T_F-s)}g(X_{T_F})\,ds$$

$$= \int_F \nu_C(dx)f(x)\mathbf{P}_{\mathrm{pr}}^x \int_0^{T_F} e^{-\lambda u}g(X_{T_F})\,du$$

$$= \int_F \nu_C(dx)f(x)\mathbf{P}_{\mathrm{pr}}^x \int_0^{T_F} e^{-\lambda u}g(X_{T_F}) \circ \theta_u\,du$$

$$= \nu_C(f V_{\mathrm{pr}}^\lambda P_F g).$$

Combining these observations yields the second equality in (7.20).   □

In the same way one has

$$(\widehat{P}_F^\lambda f, P_F g) = \nu_{B^c}(\ell f g) + \nu_B(f V_{\text{op}}^\lambda g) = (\widehat{P}_F f, P_F^\lambda g). \tag{7.21}$$

Formulas (7.20) and (7.21) reduce to (3.9) of [12] when F is a singleton.

Let us suppose in this paragraph that $R_F m = m$ (otherwise, replace m with $R_F m$.) Let us make the special choice $A = C$ in the preceding discussion. Then by (7.17) and Corollary 7.6(ii) with $\lambda = 0$ and $f = 1$,

$$m(g) = \nu_C^m(\gamma g + V_{\text{pr}} g), \qquad g \in p\mathcal{E}, \tag{7.22}$$

where $V_{\text{pr}} = V_{\text{pr}}^0$. Notice that the right side of (7.22) depends on m only through the Revuz measure ν_C^m, which is excessive for the time-changed process $\widetilde{X}$. Following up on earlier work ([18, 19, 20, 12, 27] we use this formula to construct an excessive measure for X, given an excessive measure for $\widetilde{X}$.

Proposition 7.8 *Suppose that* $\mathbf{P}^x[T_F < \infty] > 0$ *for all* $x \in E$. *Let* ν *be an excessive measure for* $\widetilde{X}$. *Then*

$$\eta(g) := \nu(\gamma g + V_{\text{pr}} g), \qquad g \in p\mathcal{E} \tag{7.23}$$

defines an excessive measure for X *such that (i)* $R_F \eta = \eta$ *and (ii)* $\nu_C^\eta = \nu$. *The measure* η *is uniquely determined by these two conditions.*

Proof. According to [9, (5.12), (5.13)], under the hypothesis of the proposition there is a uniquely determined X-excessive measure $m = m^\nu$ such that $R_F m = m$ and $\nu_C^m = \nu$. By Corollary 7.6(ii) we have

$$m(g) = R_F m(g) = (\widehat{P}_{F+} 1, g) = \nu_C^m(\gamma g + V_{\text{pr}} g) = \nu(\gamma g + V_{\text{pr}} g).$$

The right side of (7.23) therefore defines an excessive measure for X with the stated properties.   □

Remark 7.9 If the measure ν is conservative for $\widetilde{X}$, then (5.17) of [9] implies that η is conservative (hence invariant) for X. In particular, this is the case if ν is a finite invariant measure for $\widetilde{X}$. This extends a result of H. Kaspi [20] to all Borel right processes.

Define a measure Θ on $F \times F \times]0, \infty[$ by

$$\Theta(dx, dy, dt) := \nu_C(dx) \mathbf{P}_{\text{pr}}^x[X_{T_F} \in dy, T_F \in dt]. \tag{7.24}$$

Because ν_C is σ-finite and $0 < \mathbf{P}_{\text{pr}}^\bullet[1 - \exp(-T_F)] \leq 1$, it is easy to check that Θ is σ-finite. Notice that the Feller measure Λ is related to Θ by

$$\Lambda(dx, dy) = \Theta(dx, dy,]0, \infty[).$$

Intuitively, $\Theta(dx, dy, dt)$ is the rate at which excursions from F of duration t originate from x and terminate at y.

Proposition 7.10 *Suppose $f, g \in p(\mathcal{E}^* \cap F)$ with $\nu_C(f) < \infty$ and g bounded. Then $\Theta(f, g, dt)$ is a σ-finite measure on $]0, \infty[$, and for $\lambda > 0$,*

$$\lambda(\widehat{P}_{F+}^{\lambda} f, P_F g) = \lambda \nu_C(\gamma g \widehat{P}_{0+} f) + \int_0^{\infty} (1 - e^{-\lambda t}) \Theta(f, g, dt). \tag{7.25}$$

Proof. It is immediate that $\int_0^{\infty} (1 - e^{-t}) \Theta(f, g, dt)$ is dominated by $\|g\|_{\infty} \nu_C(f)$. Consequently, $\Theta(f, g, \cdot)$ is σ-finite. Then

$$\int_0^{\infty} (1 - e^{-\lambda t}) \Theta(f, g, dt) = \int_F f(x) \mathbf{P}_{\mathrm{pr}}^x [g(X_{T_F})(1 - e^{-\lambda T_F})] \nu_C(dx)$$

$$= \lambda \int_F f(x) \nu_C(dx) \mathbf{P}_{\mathrm{pr}}^x \int_0^{T_F} P_F g(X_t) e^{-\lambda t} \, dt$$

$$= \lambda \nu_C(f V_{\mathrm{pr}}^{\lambda} P_F g),$$

and combining this with Corollary 7.6(ii) we obtain (7.25). □

Remark 7.11 Analogous results hold for the optional exit system. For example, employing the obvious notation,

$$\lambda(\widehat{P}_F^{\lambda} f, P_F g) = \lambda \nu_{B^c}(\gamma f g) + \int_0^{\infty} \Theta_{\mathrm{op}}(f, g, dt). \tag{7.26}$$

As pointed out earlier, $\eta_t^f(g) = \nu_C(f Q_t^{\mathrm{pr}} g)$ and $\xi_t^f(g) = \nu_B(f Q_t^{\mathrm{op}} g))$ are entrance laws for (X, T_F) provided $f \in bp\mathcal{E}^*$. If $f \equiv 1$ then $\widehat{P}_{F+}^{\lambda} 1 = \widehat{P}_F^{\lambda} 1 = \widehat{\mathbf{P}}^{\bullet}[e^{-\lambda T_F}]$, m-a.e. by the obvious variant of (7.17). Let us write $\widehat{\varphi}$ for this last function when $\lambda = 0$. Theorem 7.4 implies that $\nu_C V_{\mathrm{pr}}^{\lambda} = \nu_B V_{\mathrm{op}}^{\lambda}$ as measures on $E \setminus F$ for all $\lambda \geq 0$. In particular, $\eta_t^1 = \xi_t^1$ since either entrance law integrates to the measure $\widehat{\varphi} m_0$, which is purely excessive for (X, T_F). (Here $m_0 := m|_{E \setminus F}$.)

We conclude by recording some additional extensions to the present context of some formulas obtained in [12] in the context of excursions from a point. First recall the definition of the energy functional L^0 of the killed process $X^0 := (X, T_F)$; see [15, §3]. If ξ is an X^0-excessive measure and f is an X^0-excessive function, then

$$L^0(\xi, f) := \sup\{\mu(f) : \mu V \leq \xi\}, \tag{7.27}$$

in which μ ranges over the σ-finite measures on F. If ξ is purely excessive for X^0, then [15, (3.6)]

$$L^0(\xi, f) = \lim_{\lambda \to \infty} \lambda \langle \xi - \lambda \xi V^{\lambda}, f \rangle = \lim_{t \downarrow 0} t^{-1} \langle \xi - \xi Q_t, f \rangle, \tag{7.28}$$

where $\langle \mu, f \rangle := \int f \, d\mu$. (Both of the limits in (7.28) are monotone increasing.) Define

$$\psi := 1 - P_F 1 = \mathbf{P}^{\bullet}[T_F = \infty]. \tag{7.29}$$

It is easily checked that ψ is X^0-excessive. We fix $f \in bp\mathcal{E}$. Then $\eta^f := \widehat{P}_{F+} f \cdot m_0 = \int_0^{\infty} \eta_t^f \, dt$ is purely excessive for X^0.

Theorem 7.12 *(i) If g is X^0-excessive then $L^0(\eta^f, g) = \lim_{t\downarrow 0} \eta_t^f(g)$.*
(ii) $L^0(\eta^f, \psi) = \int_F f(x) \mathbf{P}_{\mathrm{pr}}^x[T_F = \infty, \zeta > 0] \nu_C(\mathrm{d}x)$.
(iii) (Recall the definition (6.21) of the supplementary Feller measure δ.)

$$\delta(f) = \int_F f(x) \mathbf{P}_{\mathrm{pr}}^x[T_F = \infty] \nu_C(\mathrm{d}x)$$

$$= L^0(\eta^f, \psi) + \int_F f(x) \mathbf{P}_{\mathrm{pr}}^x[\zeta = 0]\, \nu_C(\mathrm{d}x). \tag{7.30}$$

Proof. Abbreviate $\eta = \eta^f$ and $\eta_t = \eta_t^f$ during this proof. Suppose first that $g \in p\mathcal{E}^*$ with $\eta(g) < \infty$. Then

$$\langle \eta - \eta Q_t, g \rangle = \eta(g) - \eta Q_t g = \int_0^t \eta_s(g)\, \mathrm{d}s.$$

The extreme terms in this display are positive measures in g, so for general $g \in p\mathcal{E}^*$ we deduce that

$$\langle \eta - \eta Q_t, g \rangle = \int_0^t \eta_s(g)\, \mathrm{d}s. \tag{7.31}$$

If g is X^0-excessive then $\eta_{t+s}(g) = \eta_t(Q_s g) \uparrow \eta_t(g)$ as $s \downarrow 0$. Thus $t \mapsto \eta_t(g)$ is right continuous and decreasing on $]0, \infty[$. In particular, $\uparrow \lim_{t\downarrow 0} \eta_t(g)$ exists, though it may equal $+\infty$. Therefore, by (7.31),

$$L^0(\eta, g) = \lim_{t\downarrow 0} t^{-1} \int_0^t \eta_s(g)\, \mathrm{d}s = \lim_{t\downarrow 0} \int_0^1 \eta_{tu}(g)\, \mathrm{d}u$$

$$= \lim_{t\downarrow 0} \eta_t(g),$$

by monotone convergence, establishing (i).

Next, $\eta_t(\psi) = \nu_C(f Q_t^{\mathrm{pr}} \psi)$ and $Q_t^{\mathrm{pr}} \psi = \mathbf{P}_{\mathrm{pr}}^\bullet[T_F = \infty; t < T_F \wedge \zeta] \uparrow \mathbf{P}_{\mathrm{pr}}^\bullet[T_F = \infty, \zeta > 0]$ as $t \downarrow 0$. Hence $L^0(\eta, \psi) = \mathbf{P}_{\mathrm{pr}}^{f\nu_C}[T_F = \infty, \zeta > 0]$. But $\mathbf{P}_{\mathrm{pr}}^{f\nu_C}[T_F = \infty, \zeta = 0] = \mathbf{P}_{\mathrm{pr}}^{f\nu_C}[\zeta = 0]$, proving both (ii) and (iii). $\square$

Remark 7.13 Intuitively, $\delta(f) = \mathbf{P}_{\mathrm{pr}}^{f\nu_C}[T_F = \infty]$ represents the rate (weighted by f) at which a final excursion of infinite length occurs, terminating M. Theorem 7.12 indicates that $\mathbf{P}_{\mathrm{pr}}^{f\nu_C}[\zeta = 0]$ is the weighted rate at which the process X is killed while in F; $L^0(\eta, \psi)$ is the corresponding rate of occurrence of an excursion in which the process wanders away from F, never to return. Exactly the same argument establishes the analogous facts in the optional case.

References

1. Benveniste, A. and Jacod, J.: Systèmes de Lévy des processus de Markov, *Invent. Math.* **21** (1973) 183–198.
2. Blumenthal, R.M. and Getoor, R.K.: *Markov Processes and Potential Theory.* Academic Press, New York, 1968.

3. Chen, Z., Fukushima, M., and Ying, J.: Traces of symmetric Markov processes and their characterizations, *Ann. Probab.* **34** (2006) 1052–1102.

4. Chen, Z., Fukushima, M., and Ying, J.: Entrance law, exit system and Lévy system of time changed processes, *Illinois J. Math.* **50** (2006) 269–312.

5. Dellacherie, C.: Autour des ensembles semi-polaires. In *Seminar on Stochastic Processes, 1987*, pp. 65–92. Birkhäuser Boston, 1988.

6. Dellacherie, C. and Meyer, P.-A.: *Probabilités et Potentiel. Chapitres I à IV.* Hermann, Paris, 1978.

7. Dellacherie, C. and Meyer, P.-A.: *Probabilités et Potentiel. Chapitres XII–XVI.* Hermann, Paris, 1987. Théorie du potentiel associée à une résolvante. Théorie des processus de Markov.

8. Fitzsimmons, P.J.: Homogeneous random measures and a weak order for the excessive measures of a Markov process. *Trans. Amer. Math. Soc.* **303** (1987) 431–478.

9. Fitzsimmons, P.J. and Getoor, R.K.: Revuz measures and time changes, *Math. Zeit.* **199** (1988) 233–256.

10. Fitzsimmons, P.J. and Getoor, R.K.: Smooth measures and continuous additive functionals of right Markov processes. In *Itô's Stochastic Calculus and Probability Theory*. Springer, Tokyo, 1996, pp. 31–49.

11. Fitzsimmons, P.J. and Getoor, R.K.: Homogeneous random measures and strongly supermedian kernels of a Markov process, *Electronic Journal of Probability* **8** (2003), Paper 10, 54 pages.

12. Fitzsimmons, P.J. and Getoor, R.K.: Excursion theory revisited, *Illinois J. Math.* **50** (2006) 413–437.

13. Fitzsimmons, P.J. and Maisonneuve, B.: Excessive measures and Markov processes with random birth and death, *Probab. Th. Rel. Fields* **72** (1986) 319–336.

14. Fukushima, M., He, P., and Ying, J.: Time changes of symmetric diffusions and Feller measures. *Ann. Probab.* **32** (2004) 3138–3166.

15. Getoor, R.K.: *Excessive Measures*. Birkhäuser, Boston, 1990.

16. Getoor, R.K.: Measure perturbations of Markovian semigroups. *Potential Anal.* **11** (1999) 101–133.

17. Gzyl, H.: Lévy systems for time-changed processes, *Ann. Probab.* **5** (1977) 565–570.

18. Harris, T.E.: The existence of stationary measures for certain Markov processes, In *Proceedings of the Third Berkeley Symposium on Mathematical Statistics and Probability, 1954–1955*, vol. II, pp. 113–124, Berkeley, 1956.

19. Kaspi, H.: Excursions of Markov processes: an approach via Markov additive processes, *Z. Wahrsch. verw. Gebiete* **64** (1983) 251–268.

20. Kaspi, H.: On invariant measures and dual excursions of Markov processes, *Z. Wahrsch. verw. Gebiete* **66** (1984) 185–204.

21. Le Jan, Y.: Balayage et formes de Dirichlet, *Z. Wahrsch. verw. Gebiete* **37** (1977) 297–319.

22. Maisonneuve, B.: Exit systems, *Ann. Probab.*, **3** (1975) 399–411.

23. Maisonneuve, B.: Processus de Markov: naissance, retournement, régénération. In *Springer Lecture Notes in Math.* **1541**, pp. 263–292. Springer, berlin, 1993.

24. Motoo, M.: The sweeping-out of additive functionals and processes on the boundary, *Ann. Inst. Statist. Math.* **16** (1964) 317–345.

25. Motoo, M.: Application of additive functionals to the boundary problem of Markov processes. Lévy's system of U-processes. In *Proc. Fifth Berkeley Sympos. Math. Statist. and Probability*, vol. II, part II, pp. 75–110, Berkeley, 1966.

26. Sharpe, M.J.: *General Theory of Markov Processes*. Academic Press, Boston, 1988.

27. Silverstein, M.: Classification of coharmonic and coinvariant functions for a Lévy process, *Ann. Probab.* **8** (1980) 539–575.

28. Watanabe, S.: On discontinuous additive functionals and Lévy measures of a Markov process, *Japan. J. Math.* **34** (1964) 53–70.

Self-Similar Branching Markov Chains

Nathalie Krell

Laboratoire de Probabilités et Modèles Aléatoires, Université Paris 6,
175 rue du Chevaleret, 75013 Paris, France.
e-mail: nathalie.krell@upmc.fr

Summary. The main purpose of this work is to study self-similar branching Markov chains. First we construct such a process. Using the theory of self-similar Markov processes, we show a limit theorem concerning a tagged individual. Finally, we get other results in particular a $L^p(\mathbb{P})$ limit theorem on the convergence of the empirical measure associated to the size of the fragment of the branching chain.

Key words: Branching process, Self-similar Markov process, Tree of generations, Limit Theorems.

MSC 2000: 60J80, 60G18, 60F25, 60J27.

1 Introduction

This work is a contribution to the study of a special type of branching Markov chains. We will construct a continuous time branching chain **X** which has a self-similar property and which takes its values in the space of finite point measures of $\mathbb{R}_+^*$. This type of process is a generalization of a self-similar fragmentation (see [4]), which may apply to cases where the size models non additive quantities as e.g. surface energy in aerosols. We will focus on the case where the self-similarity index α is non-negative, which means that bigger individuals reproduce faster than smaller ones. There is no loss of generality by considering this model, as the map $x \to x^{-1}$ on atoms in $\mathbb{R}_+^*$ transforms a self-similar process with index α into another one with index $-\alpha$ (and preserves the Markov property).

In this article we choose to construct the process by bare hand. We extend the method used in [4] to deal with more general processes where we allow an individual to have a mass bigger than that of its parent. We will explain in the sequel which difficulties this new set-up entails. There are closely related articles about branching processes, among others [18], [19] from Kyprianou and [12], [13] from Chauvin. However notice that the time of splitting of the process depends on the size of the atoms of the process.

C. Donati-Martin et al. (eds.), *Séminaire de Probabilités XLII,*
Lecture Notes in Mathematics 1979, DOI 10.1007/978-3-642-01763-6_10,
© Springer-Verlag Berlin Heidelberg 2009

More precisely we will first introduce a branching Markov chain as a marked tree and we will obtain a process indexed by generations (it is simply a random mark on the tree of generation, see Section 2). Using a martingale associated to the latter and the theory of random stopping lines on a tree of generation, we will define the process indexed by time. After having constructed the process, we will study the evolution of a randomly chosen branch of the chain, from which we shall deduce some Limit Theorems, relying on the theory of self-similar Markov processes. In an appendix we will consider the intrinsic process and give some properties in the spirit of the article of Jagers [15]. On the way we will show properties about the earlier martingale.

2 The Marked Tree

In this part we will introduce a branching Markov chain as a marked tree, which gives a genealogic description of the process that we will construct. This terminology comes from Neveu in [21] even if here the marked tree we consider is slightly different. First we introduce some notations and definitions.

A finite point measure on $\mathbb{R}_+^*$ is a finite sum of Dirac point masses $\mathbf{s} = \sum_{i=1}^n \delta_{s_i}$, where the s_i are called the atoms of $\mathbf{s}$ and $n \geqslant 0$ is an arbitrary integer. We shall often write $\sharp\mathbf{s} = n = \mathbf{s}(\mathbb{R}_+^*)$ for the number of atoms of $\mathbf{s}$, and $\mathcal{M}_p(\mathbb{R}_+^*)$ for the space of finite point measures on $\mathbb{R}_+^*$. We also define for $f : \mathbb{R}_+^* \to \mathbb{R}$ a measurable function and $\mathbf{s} \in \mathcal{M}_p(\mathbb{R}_+^*)$

$$\langle f, \mathbf{s} \rangle := \sum_{i=1}^{\sharp s} f(s_i),$$

by taking the sum over the atoms of $\mathbf{s}$ repeated according to their multiplicity; and we will sometimes use the slight abuse of notation

$$\langle f(x), \mathbf{s} \rangle := \sum_{i=1}^{\sharp s} f(s_i)$$

when f is defined as a function depending on the variable x. We endow the space $\mathcal{M}_p(\mathbb{R}_+^*)$ with the topology of weak convergence, which means that $\mathbf{s}_n$ converge to $\mathbf{s}$ if and only if $\langle f, \mathbf{s}_n \rangle$ converge to $\langle f, \mathbf{s} \rangle$ for all continuous bounded functions f.

Let $\alpha \geqslant 0$ be an index of self-similarity and ν be some **probability measure** on $\mathcal{M}_p(\mathbb{R}_+^*)$. The aim of this work is to construct a branching Markov chain $\mathbf{X} = ((\sum_{i=1}^{\sharp \mathbf{X}(t)} \delta_{X_i(t)})_{t \geqslant 0})$ with values in $\mathcal{M}_p(\mathbb{R}_+^*)$, which is self-similar with index α and has reproduction law ν. The index of self-similarity will play a part in the rate at which an individual will reproduce and the reproduction law ν will specify the distribution of the offspring. We stress that our setting includes the case when

$$\nu(\exists i \ : \ s_i > 1) > 0, \tag{1}$$

which means that with positive probability the size of a daughter can exceed that of her mother.

To do that, exactly as described in Chapter 1 section 1.2.1 of [4], we will construct a marked tree.

We consider the Ulam Harris labelling system

$$\mathcal{U} := \cup_{n=0}^{\infty} \mathbb{N}^n,$$

with the notation $\mathbb{N} = \{1, 2, \dots\}$ and $\mathbb{N}^0 = \{\emptyset\}$. In the sequel the elements of $\mathcal{U}$ are called nodes (or sometimes also individuals) and the distinguished node $\emptyset$ the root. For each $u = (u_1, \dots, u_n) \in \mathcal{U}$, we call n the generation of u and write $|u| = n$, with the obvious convention $|\emptyset| = 0$. When $n \geqslant 0$, $u = (u_1, \dots, u_n) \in \mathbb{N}^n$ and $i \in \mathbb{N}$, we write $ui = (u_1, \dots, u_n, i) \in \mathbb{N}^{n+1}$ for the i-th child of u. We also define for $u = (u_1, \dots, u_n)$ with $n \geqslant 2$,

$$mu = (u_1, \dots, u_{n-1})$$

the mother of u, $mu = \emptyset$ if $u \in \mathbb{N}$. If $v = m^n u$ for some $n \geqslant 0$ we write $v \preceq u$ and say that u stems from v. Additionally for M a set of $\mathcal{U}$, $M \preceq v$ means that $u \preceq v$ for some $u \in M$. Generally we write $M \preceq L$ if all $x \in L$ stem from M.

Here it will be convenient to identify the point measure $\mathbf{s}$ with the infinite sequence $(s_1, \dots, s_n, 0, \dots)$ obtained by aggregation of infinitely many 0's to the finite sequence of the atoms of $\mathbf{s}$.

In particular we say that a random infinite sequence $(\xi_i, i \in \mathbb{N})$ has law ν, if there is a (random) index n such that $\xi_i = 0 \Leftrightarrow i > n$ and the finite point measure $\sum_{i=1}^{n} \delta_{\xi_i}$ has law ν.

Definition 1. Let $(\overline{\xi}_u, u \in \mathcal{U})$ and $(\mathbf{e}_u, u \in \mathcal{U})$ be two independent families of i.i.d. variables indexed by the nodes of the tree, where for each $u \in \mathcal{U}$, $\overline{\xi}_u = (\tilde{\xi}_{ui})_{i \in \mathbb{N}}$ is distributed according to the law ν, and $(\mathbf{e}_{ui})_{i \in \mathbb{N}}$ is a sequence of i.i.d. exponential variables with parameter 1. Define recursively for some fixed $x > 0$

$$\xi_{\emptyset} := x, \qquad a_{\emptyset} := 0, \qquad \zeta_{\emptyset} := x^{-\alpha} \mathbf{e}_{\emptyset},$$

and for $u \in \mathcal{U}$ and $i \in \mathbb{N}$:

$$\xi_{ui} := \tilde{\xi}_{ui} \xi_u, \qquad a_{ui} := a_u + \zeta_u, \qquad \zeta_{ui} := \xi_{ui}^{-\alpha} \mathbf{e}_{ui}.$$

To each node u of the tree $\mathcal{U}$, we associate the mark (ξ_u, a_u, ζ_u) where ξ_u is the size, a_u the birth-time and ζ_u the lifetime of the individual with label u. We call

$$T_x = ((\xi_u, a_u, \zeta_u)_{u \in \mathcal{U}})$$

a marked tree with root of size x, and the associated law is denoted by $\mathbb{P}_x$. Let $\overline{\Omega}$ be the set of all possible marked trees.

The size of the individuals $(\xi_u, u \in \mathcal{U})$ defines a multiplicative cascade (see the references in Section 3 of [5]). However the latter is not sufficient to construct the process $\mathbf{X}$, in fact we also need the information given by $((a_u, \zeta_u), u \in \mathcal{U})$.

Another useful concept is that of a *line*. A subset $L \subset \mathcal{U}$ is a line if for every $u, v \in L$, $u \preceq v \Rightarrow u = v$. The *pre-L-sigma algebra* is

$$\mathcal{H}_L := \sigma(\tilde{\xi}_u, \mathbf{e}_u; \exists l \in L : u \preceq l).$$

A random set of individuals

$$\mathcal{J} : \bar{\Omega} \to \mathcal{P}(\mathcal{U})$$

is *optional* if $\{\mathcal{J} \preceq L\} \in \mathcal{H}_L$ for every line $L \subset \mathcal{U}$, where $\mathcal{P}(\mathcal{U})$ is the power set of $\mathcal{U}$. An *optional line* is a random line which is optional. For any optional set $\mathcal{J}$ we define the pre-$\mathcal{J}$-algebra by:

$$A \in \mathcal{H}_{\mathcal{J}} \Leftrightarrow \forall L \text{ line } \subset \mathcal{U} : A \cap \{\mathcal{J} \preceq L\} \in \mathcal{H}_L.$$

The first result is:

Lemma 1. *The marked tree constructed in Definition 1 satisfies the strong Markov branching property: for $\mathcal{J}$ an optional line and $\varphi_u : \bar{\Omega} \to [0,1]$, $u \in \mathcal{U}$, measurable functions, we have*

$$\mathbb{E}_1 \left(\prod_{u \in \mathcal{J}} \varphi_u \circ T^{\xi_u} \middle| \mathcal{H}_{\mathcal{J}} \right) = \prod_{u \in \mathcal{J}} \mathbb{E}_{\xi_u}(\varphi_u),$$

where T^{ξ_u} is the marked tree extracted from T_1 at the node (ξ_u, a_u, ζ_u). More precisely

$$T^{\xi_u} = ((\xi_{uv}, a_{uv} - a_u, \zeta_{uv})_{v \in \mathcal{U}}).$$

Proof. Thanks to the i.i.d. properties of the random variables $(\tilde{\xi}_u, u \in \mathcal{U})$ and $(\mathbf{e}_u, u \in \mathcal{U})$, the Markov property for lines is of course easily checked. In order to get the result for a more general optional line, we use Theorem 4.14 of [15]. Indeed, the tree we have constructed is a special case of the tree constructed by Jagers in [15]. In our case, Jagers's notation ϱ_u, τ_u and σ_u are such that the type ϱ_u of $u \in \mathcal{U}$ is the mass ξ_u of u, the birth time σ_u is a_u and τ_u is here equal to ζ_{mu} (because a mother dies when giving birth to her daughters). We notice that sisters always have the same birth time, which means that for all $u \in \mathcal{U}$ and all $i \in \mathbb{N}$, we have that τ_{ui} is here equal to ζ_u. $\square$

3 Malthusian Hypotheses and the Intrinsic Martingale

We introduce some notations to formulate the fundamental assumptions of this work:

$$\underline{p} := \inf\left\{p \in \mathbb{R} : \int_{\mathcal{M}_p(\mathbb{R}_+^*)} \langle x^p, \mathbf{s}\rangle \nu(ds) < \infty\right\},$$

and

$$p_\infty := \inf\left\{p > \underline{p} : \int_{\mathcal{M}_p(\mathbb{R}_+^*)} \langle x^p, \mathbf{s}\rangle \nu(ds) = \infty\right\}$$

(with the convention $\inf \emptyset = \infty$) and then for every $p \in (\underline{p}, p_\infty)$:

$$\kappa(p) := \int_{\mathcal{M}_p(\mathbb{R}_+^*)} (1 - \langle x^p, \mathbf{s}\rangle)\,\nu(ds).$$

Note that κ is a continuous and concave function (but not necessarily a strictly increasing function) on $(\underline{p}, p_\infty)$, as $p \to \int_{\mathcal{M}_p(\mathbb{R}_+^*)} \langle x^p, \mathbf{s}\rangle \nu(ds)$ is a convex application. By concavity, the equation $\kappa(p) = 0$ has at most two solutions on $(\underline{p}, p_\infty)$. When a solution exists, we denote by $p_0 := \inf\{p \in (\underline{p}, p_\infty) : \kappa(p) = 0\}$ the smallest, and call p_0 the Malthusian exponent.

We now make the fundamental:

Malthusian Hypotheses. *We suppose that the Malthusian exponent p_0 exists, that $p_0 > 0$, and that*

$$\kappa(p) > 0 \quad \text{for some} \ \ p > p_0. \tag{2}$$

Furthermore we suppose that the integral

$$\int_{\mathcal{M}_p(\mathbb{R}_+^*)} (\langle x^{p_0}, \mathbf{s}\rangle)^p\, \nu(ds) \tag{3}$$

is finite for some $p > 1$.

Throughout the rest of this article, these hypotheses will always be taken for granted.

Note that (2) always holds when $\nu(s_i \leqslant 1$ for all $i) = 1$ (fragmentation case). We stress that κ may not be strictly increasing, and may not be negative when p is sufficiently large (see Subsection 6.1 for a consequence of this fact.)

We will give one example based on the Dirichlet process (see Kingman's book [16]). Fix $n \geqslant 2$, $(v_1, \ldots, v_n)$ n positive real numbers and $v = \sum_{i=1}^n v_i$. Define the simplex Δ_n by

$$\Delta_n := \left\{(p_1, p_2, \ldots, p_n) \in \mathbb{R}_+^n, \ \sum_{j=1}^n p_i = 1\right\}.$$

The Dirichlet distribution of parameter $(v_1, \ldots, v_n)$ over the simplex Δ_n has density (with respect to the $(n-1)$-dimensional Lebesgue measure on Δ_n):

$$f(p_1, \ldots, p_n) = \frac{\Gamma(v)}{\Gamma(v_1)\ldots\Gamma(v_n)} p_1^{v_1-1} \ldots p_n^{v_n-1}.$$

Let $a := v(v+1)/(\sum_{i=1}^{n} v_i(v_i+1))$. Note that a is strictly larger than 1. Let the reproduction measure be the law of $(aX_1, \ldots, aX_n)$, where $(X_1, \ldots, X_n)$ is a random vector with Dirichlet distribution of parameter $(v_1, \ldots, v_n)$. Therefore

$$\kappa(p) = a^p \frac{\Gamma(v)}{\Gamma(v+p)} \sum_{i=1}^{n} \frac{\Gamma(p+v_i)}{\Gamma(v_i)},$$

$\underline{p} = -v$, $p_0 = 1$ and the Malthusian hypotheses are verified.

In this article we will call *extinction* the event that for some $n \in \mathbb{N}$, all nodes u at the n-th generation have zero size, and *non-extinction* the complementary event. The probability of extinction is always strictly positive whenever $\nu(s_1 = 0) > 0$, and equals zero if and only if $\nu(s_1 = 0) = 0$ (since we have supposed (3); see p.28 [4]).

After these definitions, we introduce a fundamental martingale associated to $(\xi_u, u \in \mathcal{U})$.

Theorem 1. *The process*

$$M_n := \sum_{|u|=n} \xi_u^{p_0}, \quad n \in \mathbb{N}$$

is a martingale in the filtration $(\mathcal{H}_{L_n})$, with L_n the line associated to the n-th generation (i.e., $L_n := \{u \in \mathcal{U} : |u| = n\}$). This martingale is bounded in $L^p(\mathbb{P})$ for some $p > 1$, and in particular uniformly integrable.

Moreover, conditionally on non-extinction, the terminal value M_∞ is strictly positive a.s.

Remark 1. As κ is concave, the equation $\kappa(p) = 0$ may have a second root $p_+ := \inf\{p > p_0, \ \kappa(p) = 0\})$. This second root is less interesting: even though

$$M_n^+ := \sum_{|u|=n} \xi_u^{p_+}, \quad n \in \mathbb{N},$$

is also a martingale, it is easy to check that for all $p > 1$ the p-variation of M_n^+ is infinite, i.e., $\mathbb{E}(\sum_{n=0}^{\infty} |M_{n+1} - M_n|^p) = \infty)$.

We can notice that for all $p \in (p_0, p_+)$, $(M_n^{(p)})_{n \in \mathbb{N}} := (\sum_{|u|=n} \xi_u^p)_{n \in \mathbb{N}}$ is a supermartingale.

Assumption (3) actually means that $\mathbb{E}(M_1^p) < \infty$.

Proof. • We will use the fact that the empirical measure of the logarithm of the sizes of fragments

$$Z^{(n)} := \sum_{|u|=n} \delta_{\log \xi_u} \tag{4}$$

can be viewed as a branching random walk (see the article of Biggins [8]) and use Theorem 1 of [8]. In order to do that we first introduce some notation: for $\theta > \underline{p}$, we define

$$m(\theta) := \mathbb{E}\left(\int \exp(\theta x) Z^{(1)}(dx)\right) = \mathbb{E}\left(\sum_{|u|=1} \xi_u^\theta\right) = 1 - \kappa(\theta)$$

and

$$W^{(n)}(\theta) := m(\theta)^{-n} \int \exp(\theta x) Z^{(n)}(dx) = (1 - \kappa(\theta))^{-n} \sum_{|u|=n} \xi_u^\theta.$$

Notice that $M_n = W^{(n)}(p_0)$. Therefore in order to apply Theorem 1 of [8] and to get convergence almost sure and in pth mean for some $p > 1$, it is enough to show that

$$\mathbb{E}(W^{(1)}(p_0)^\gamma) < \infty$$

for some $\gamma \in (1, 2]$ and

$$m(p p_0)/|m(p_0)|^p < 1$$

for some $p \in (1, \gamma]$. The first condition is a consequence of the Malthusian assumption. Moreover the second follows from the identities

$$m(p p_0)/|m(p_0)|^p = (1 - \kappa(p p_0))/|1 - \kappa(p_0)|^p = 1 - \kappa(p p_0)$$

which, by the definition of p_0, is smaller than 1 for $p > 1$ well chosen.

• Finally, let us now check that $M_\infty > 0$ a.s. conditionally on non-extinction. Define $q = \mathbb{P}(M_\infty = 0)$, therefore as $\mathbb{E}(M_\infty) = 1$ we get $q < 1$. Moreover, an application of the branching property yields

$$\mathbb{E}(q^{Z_n}) = q,$$

where Z_n is the number of individuals with positive size at the n-th generation. Notice that $Z_n = \langle Z^{(n)}, 1 \rangle$. By construction of the marked tree and as ν is a probability measure, $(Z_n, n \in \mathbb{N})$ is of course a Galton-Watson process and it follows that q is its probability of extinction. Since $M_\infty = 0$ conditionally on the extinction, the two events coincide a.s. □

4 Evolution of the Process in Continuous Time

After having defined the process indexed by generation and having shown that the martingale M_n is $L^p(\mathbb{P})$ bounded, we are now able to properly define the main objet of this paper. In order to do this, when an individual labelled by u has positive size $\xi_u > 0$, call $I_u := [a_u, a_u + \zeta_u)$ the time period during which this individual is alive. Otherwise, i.e., when $\xi_u = 0$, we decide that $I_u = \emptyset$. With this definition, we set:

Definition 2. *Define the process* $\mathbf{X} = (\mathbf{X}(t), t \geq 0)$ *by*

$$\mathbf{X}(t) = \sum_{u \in \mathcal{U}} \mathbb{1}_{\{t \in I_u\}} \delta_{\xi_u}, t \geq 0. \tag{5}$$

In particular one has for $f : \mathbb{R}_+ \to \mathbb{R}$ *any measurable function*

$$\langle f, \mathbf{X}(t) \rangle = \sum_{u \in \mathcal{U}} f(\xi_u) \mathbb{1}_{\{t \in I_u\}}.$$

For every $x > 0$, let $\mathbb{P}_x$ be the law of the process $\mathbf{X}$ starting from a single individual with size x. And for simplicity, write $\mathbb{P}$ for $\mathbb{P}_1$, and let $(\mathcal{F}_t)_{t \geq 0}$ be the natural filtration of the process $(\mathbf{X}(t), t \geq 0)$. We use the notation $(X_1(t), \ldots, X_{\sharp \mathbf{X}(t)}(t))$ for the sequence of atoms of $\mathbf{X}(t)$. In the following we will show that this sequence is almost surely finite. Of course the set $(X_1(t), \ldots, X_{\sharp \mathbf{X}(t)}(t))$ is the same as the set $((\xi_u); t \in I_u)$; but sometimes it will be clearer to use the notation $(X_i(t))$.

Define for $u \in \mathbb{R}_+$:

$$F(u) := \int_{\mathcal{M}_p(\mathbb{R}_+^*)} u^{\sharp \mathbf{s}} \nu(d\mathbf{s}).$$

Notice that $F(u)$ is the generating function of the Galton-Watson process $(Z_n, n \geq 0) = (\sharp \{u \in \mathcal{U} : \xi_u > 0 \text{ and } |u| = n\}, n \geq 0)$.

From now on, we will suppose that for every $\varepsilon > 0$

$$\int_{1-\varepsilon}^{1} \frac{du}{F(u) - u} = \infty. \tag{6}$$

Of course if $F'(1) = \mathbb{E}(Z_1) < \infty$ this assumption is fulfilled. Therefore we get the first theorem about the continous time process:

Theorem 2. *The process $\mathbf{X}$ takes its values in the set $\mathcal{M}_p(\mathbb{R}_+^*)$. It is a branching Markov chain, more precisely the conditional distribution of $\mathbf{X}(t+r)$ given that $\mathbf{X}(r) = \mathbf{s}$ is the same as that of the sum $\sum \mathbf{X}^{(i)}(t)$, where for each index i, $\mathbf{X}^{(i)}(t)$ is distributed as $\mathbf{X}(t)$ under $\mathbb{P}_{s_i}$ and the variables $\mathbf{X}^{(i)}(t)$ are independent.*

The process $\mathbf{X}$ also has the scaling property, namely for every $c > 0$, the distribution of the rescaled process $(c\mathbf{X}(c^\alpha t), t \geq 0)$ under $\mathbb{P}_1$ is $\mathbb{P}_c$.

In the fragmentation case, the fact that the size of the fragments decreases with time entails that the process of the fragments of size larger than or equal to ε is Markovian, and this easily leads to Theorem 2. This property is lost in the present case.

Proof. • First we will check that for all $t \geq 0$, $\mathbf{X}(t)$ is a (random) finite point measure. By Theorem 1 and Doob's L^p-inequality we get that for some $p > 1$:

$$\sup_{n \in \mathbb{N}} M_n = \sup_{n \in \mathbb{N}} \sum_{|u|=n} \xi_u^{p_0} \in L^p(\mathbb{P}).$$

As a consequence:

$$\sup_{u\in\mathcal{U}} \xi_u^{p_0} \in L^p(\mathbb{P}) \tag{7}$$

and then by the definition of the process $\mathbf{X}$, writing $X_1(t),\dots$ for the (possibly infinite) sequence of atoms of $\mathbf{X}(t)$

$$\sup_i \sup_{t\in\mathbb{R}_+} X_i(t)^{p_0} \in L^p(\mathbb{P}).$$

Recall that $p_0 > 0$ by assumption. Fix some arbitrarily large $m > 0$. We now work conditionally on the event that the size of all individuals is bounded by m, and we will show that the number of the individuals alive at time t is almost surely finite for all $t \geqslant 0$.

As we are conditioning on the event $\{\sup_{u\in\mathcal{U}} \xi_u \leqslant m\}$, by construction of the marked tree, we get that the life time of an individual can be stochastically bounded from below by an exponential variable of parameter m^α. Therefore we can bound the number of individuals present at time t by the number of individuals of a continuous time branching process (denoted by GW) in which each individual lives for a random time whose law is exponential of parameter m^α and the probability distribution of the offspring is the law of $\sharp s \vee 1$ under ν (we have taken the supremum with 1 to ensure the absence of death). For the Markov branching process GW, we are in the temporally homogeneous case and, we notice that

$$\int_{\mathcal{M}_p(\mathbb{R}_+^*)} u^{(n_\mathbf{s})\vee 1}\nu(d\mathbf{s}) = (f(u) - u)\nu(n_\mathbf{s} \neq 0) + u,$$

therefore as we have supposed (6), we can use Theorem 1 p.105 of the book of Athreya and Ney [3] (proved in Theorem 9 p.107 of the book of Harris [14]) and get that GW is non-explosive. As the number of the individuals is bounded by that of GW we get that the number of individuals at time t is a.s. finite.

Therefore, conditioning on the event $\{\sup_{u\in\mathcal{U}} \xi_u \leqslant m\}$, we have that for all $t \geqslant 0$, the number of individuals at time t is a.s. finite, i.e., $\mathbf{X}(t)$ is a finite point measure.

• Second we will show the Markov property. Fix $r \in \mathbb{R}_+$. Let τ_r be equal to $\{u \in \mathcal{U} : \ r \in I_u\}$. Notice that τ_r is an optional line. In fact for all lines $L \subset \mathcal{U}$ we have that

$$\{\tau_r \preceq L\} = \{r < a_u + \zeta_u \ \forall u \in L\} \in \mathcal{H}_L.$$

By definition, we have the identity

$$\sum_{i=1}^{\sharp \mathbf{X}(t+r)} \mathbb{1}_{\{X_j(t+r)>0\}}\delta_{X_j(t+r)} = \sum_{u\in\mathcal{U}} \mathbb{1}_{\{t+r\in I_u\}}\delta_{\xi_u}.$$

Let $\mathbf{X}(r) = \sum_{i=1}^{n} \delta_{\xi_{v_n}} \in \mathcal{M}_p(\mathbb{R}_+^*)$ with $n = \sharp\mathbf{X}(r)$ and $(v_1, \ldots, v_n)$ the nodes of $\mathcal{U}$. Define for all $i \leqslant n$,

$$\tilde{T}^{(i)} := ((\xi_{v_i u}, a_{v_i u} - a_{v_i}, \zeta_{v_i u} - \mathbb{1}_{\{u=\emptyset\}}(r - a_{v_i}))_{u \in \mathcal{U}}) = ((\tilde{\xi}_u^{(i)}, \tilde{a}_u^{(i)}, \tilde{\zeta}_u^{(i)})_{u \in \mathcal{U}}),$$

$$\tilde{I}_u^{(i)} := [\tilde{a}_u^{(i)}, \tilde{a}_u^{(i)} + \tilde{\zeta}_u^{(i)}[\text{ and }$$

$$\mathbf{X}^{(i)}(t) = \sum_{u \in \mathcal{U}} \mathbb{1}_{\{t \in \tilde{I}_u^{(i)}\}} \delta_{\tilde{\xi}_u^{(i)}}.$$

Then

$$\mathbf{X}(t+r) = \sum_{i=1}^{n} \mathbf{X}^{(i)}(t).$$

By lack of memory of the exponential variable, we have that for $u \in \mathcal{U}$, given $s \in I_u$ the law of the marked tree $\tilde{T}^{(i)}$ is the same as that of

$$T^{\xi_{v_i}} := ((\xi_{v_i u}, a_{v_i u} - a_{v_i}, \zeta_{v_i u})_{u \in \mathcal{U}}) := ((\xi_u^i, a_u^i, \zeta_u^i)_{u \in \mathcal{U}}).$$

Thus we have the equality in law:

$$\sum_{u \in \mathcal{U}} \mathbb{1}_{\{t \in \tilde{I}_u^{(i)}\}} \delta_{\tilde{\xi}^{(i)}} \overset{(d)}{=} \sum_{u \in \mathcal{U}} \mathbb{1}_{\{t \in I_u^i\}} \delta_{\xi_u^i},$$

with $I_u^i := [a_u^i, a_u^i + \zeta_u^i[$.

Let $\tau_r^i := \{v_i u \in \mathcal{U} : \ r \in I_u^i\}$. Moreover for all lines $L \in \mathcal{U}$ we have that

$$\{\tau_r^i \preceq L\} = \{r < a_{v_i u} + \zeta_{v_i u} \ \forall v_i u \in L\} \in \mathcal{H}_L.$$

Therefore τ_r^i is an optional line and by applying Lemma 1 for the optional line τ_s^i, we have that the conditional distribution of the point measure

$$\sum_{u \in \mathcal{U}} \mathbb{1}_{\{t + r \in I_u^i\}} \delta_{\xi_u^i}$$

given $\mathcal{H}_{\tau_r}$ is the law of $\mathbf{X}(t)$ under $\mathbb{P}_{x_i}$. Notice that $\mathcal{H}_{\tau_s} = \sigma(\tilde{\xi}_u, \mathbf{e}_u : a_u \leqslant s)$ is the same filtration as $\mathcal{F}_s = \sigma(\mathbf{X}(s') : \ s' \leqslant s)$. Therefore $(\mathbf{X}^{(1)}, \mathbf{X}^{(2)}, \ldots, \mathbf{X}^{(n)})$ is a sequence of independent random processes, where for each i, $\mathbf{X}^{(i)}(t)$ is distributed as $\mathbf{X}(t)$ under $\mathbb{P}_{x_i}$. We then have proven the Markovian property.

• The scaling property is an easy consequence of the definition of the tree T_x. □

Remark 2. For every measurable function $g : \mathbb{R}_+^* \to \mathbb{R}_+^*$, define a multiplicative functional such that for every $\mathbf{s} = \sum_{i=1}^{\sharp s} \delta_{s_i} \in \mathcal{M}_p(\mathbb{R}_+^*)$,

$$\phi_g(\mathbf{s}) = \exp(-\langle g, \mathbf{s} \rangle) = \exp(-\sum_{i=1}^{\sharp s} g(s_i)).$$

Then the generator G of the Markov process $\mathbf{X}(t)$ fulfills for every $\mathbf{y} = \sum_{i=1}^{\sharp\mathbf{y}} \delta_{y_i} \in \mathcal{M}_p(\mathbb{R}_+^*)$:

$$G\phi_g(\mathbf{y}) = \sum y_i^\alpha \exp(-\sum_{j\neq i} g(y_j)) \int_{\mathcal{M}_p(\mathbb{R}_+^*)} (e^{-\langle g(xy_i),\mathbf{s}\rangle} - e^{-g(y_i)})\nu(d\mathbf{s}).$$

The intrinsic martingale M_n is indexed by the generations; it will also be convenient to consider its analogue in continuous time, i.e.,

$$M(t) := \langle x^{p_0}, \mathbf{X}(t)\rangle = \sum_{u\in\mathcal{U}} \mathbb{1}_{\{t\in I_u\}} \xi_u^{p_0}.$$

It is straightforward to check that $(M(t), t \geqslant 0)$ is again a martingale in the natural filtration $(\mathcal{F}_t)_{t\geqslant 0}$ of the process $(\mathbf{X}(t), t \geqslant 0)$; and more precisely, the argument Proposition 1.5 in [4] gives:

Corollary 1. *The process $(M(t), t \geqslant 0)$ is a martingale, and more precisely*

$$M(t) = \mathbb{E}(M_\infty|\mathcal{F}_t),$$

where M_∞ is the terminal value of the intrinsic martingale $(M_n, n \in \mathbb{N})$. In particular $M(t)$ converges in $L^p(\mathbb{P})$ to M_∞ for some $p > 1$.

Proof. We will use the same argument as in the proof of Proposition 1.5 of [4]. But we have to deal here with the fact that $\sup_{u\in\mathcal{U}} \xi_u$ may be larger than 1. Therefore we will have to condition. We know that M_n converges in $L^p(\mathbb{P})$ to M_∞ as n tends to ∞, so

$$\mathbb{E}(M_\infty|\mathcal{F}_t) = \lim_{n\to\infty} \mathbb{E}(M_n|\mathcal{F}_t).$$

By Theorem 1, as already seen in (7), we have $\sup_{u\in\mathcal{U}} \xi_u^{p_0} \in L^p(\mathbb{P})$, so, fixing $m > 0$, we now work on the event $B_m := \{\sup_{u\in\mathcal{U}} \xi_u \leqslant m\}$.

By applying the Markov property at time t we easily get

$$\mathbb{E}(M_n|\mathcal{F}_t) = \sum_{i=1}^{\sharp\mathbf{X}(t)} X_i^{p_0}(t)\mathbb{1}_{\{\varrho(X_i(t))\leqslant n\}} + \sum_{|u|=n} \xi_u^{p_0}\mathbb{1}_{\{a_u+\zeta_u<t\}} \tag{8}$$

where $\varrho(\xi_v)$ stands for the generation of the individual v (i.e., $\varrho(\xi_v) = |v|$), and $a_u + \zeta_u$ is the instant when the individual corresponding to the node u reproduces. We can rewrite the latter as

$$a_u + \zeta_u = \xi_{m^{|u|}u}^{-\alpha}\mathbf{e}_0 + \xi_{m^{|u|-1}u}^{-\alpha}\mathbf{e}_1 + ... + \xi_u^{-\alpha}\mathbf{e}_{|u|}$$

where $\mathbf{e}_0,...$ is a sequence of independent exponential variables with parameter 1, which is also independent of ξ_u.

As α is nonnegative, and as we are working on the event B_m: $\xi_{m^i u}^{-\alpha} \geqslant m^{-\alpha}$ we have that for each fixed node $u \in \mathcal{U}$, $a_u + \zeta_u$ is bounded from below by the sum of $|u| + 1$ independent exponential variables with parameter m^α which are independent of ξ_u. Thus

$$\lim_{n \to \infty} \mathbb{E}\left(\sum_{|u|=n} \xi_u^{p_0} \mathbb{1}_{\{a_u + \zeta_u < t\}} \mathbb{1}_{\{B_m\}} \right) = 0,$$

and therefore by (8) on the event $\{B_m\}$, we get that for all $m > 0$: $\mathbb{E}(M_\infty | \mathcal{F}_t) \mathbb{1}_{\{B_m\}} = M(t) \mathbb{1}_{\{B_m\}}$. The result is then obtained by then by letting m tend to ∞. $\square$

5 A Randomly Tagged Leaf

We will here (as in [4]) use a tagged leaf to define a tagged individual.

We call *leaf* of the tree $\mathcal{U}$ an infinite sequence of integers $l = (u_1, \ldots)$. For each n, $l^n := (u_1, \ldots, u_n)$ is the ancestor of l at the generation n. We enrich the probabilistic structure by adding the information about a so called tagged leaf, chosen at random as follows. Let H_n be the space of bounded functionals Φ which depend on the mark M and of the leaf l up to the n-th first generation, i.e., such that $\Phi(M, l) = \Phi(M', l')$ if $l^n = l'^n$ and $M(u) = M'(u)$ whenever $|u| \leqslant n$. For such functionals, we use the slightly abusing notation $\Phi(M, l) = \Phi(M, l^n)$. As in [4] for a pair (M, λ) where $M : \mathcal{U} \to [0, 1] \times \mathbb{R}_+ \times \mathbb{R}_+$ is a random mark on the tree and λ is a random leaf of $\mathcal{U}$, the joint distribution denoted by $\mathbb{P}^*$ (and by $\mathbb{P}^*_x$ if the size of the first mark is x instead of 1) can be defined unambiguously by

$$\mathbb{E}^*(\Phi(M, \lambda)) = \mathbb{E}\left(\sum_{|u|=n} \Phi(M, u) \xi_u^{p_0} \right), \quad \Phi \in H_n.$$

Moreover since the intrinsic martingale $(M_n, n \in \mathbb{Z}_+)$ is uniformly integrable (cf. Theorem 1), the first marginal of $\mathbb{P}^*$ is absolutely continuous with respect to the law of the random mark M under $\mathbb{P}$, with density M_∞.

Let λ_n be the node of the tagged leaf at the n-th generation. We denote by $\chi_n := \xi_{\lambda_n}$ the size of the individual corresponding to the node λ_n and by $\chi(t)$ the size of the tagged individual alive at time t, viz.

$$\chi(t) := \chi_n \quad \text{if } a_{\lambda_n} \leqslant t < a_{\lambda_n} + \zeta_{\lambda_n},$$

because in the case under consideration $\sup_{n \in \mathbb{N}} a_{\lambda_n} = \infty$. We stress that, in general the process $\chi(t)$ is not monotonic. However as in [4], Lemma 1.4 there becomes:

Lemma 2. *Let $k : \mathbb{R}_+ \to \mathbb{R}_+$ be a measurable function such that $k(0) = 0$. Then we have for every $n \in \mathbb{N}$*

$$\mathbb{E}^*(k(\chi_n)) = \mathbb{E}\left(\sum_{|u|=n} \xi_u^{p_0} k(\xi_u) \right),$$

and for every $t \geqslant 0$

$$\mathbb{E}^*\big(k(\chi(t))\big) = \mathbb{E}\big(\langle x^{p_0} k(x), X(t)\rangle\big).$$

Proposition 1.6 of [4] becomes:

Proposition 1. *Under $\mathbb{P}^*$,*

$$S_n := \ln \chi_n, \quad n \in \mathbb{Z}_+$$

is a random walk on $\mathbb{R}$ with step distribution

$$\mathbb{P}(\ln \chi_n - \ln \chi_{n+1} \in dy) = \widetilde{\nu}(dy),$$

where the probability measure $\widetilde{\nu}$ is defined by

$$\int_{]0,\infty[} k(y)\widetilde{\nu}(dy) = \int_{\mathcal{M}_p(\mathbb{R}_+^*)} \langle x^{p_0} k(\ln(x)), \mathbf{s}\rangle \nu(d\mathbf{s}).$$

Equivalently, the Laplace transform of the step distribution is given by

$$\mathbb{E}^*(\exp(pS_1)) = \mathbb{E}^*(\chi_1^p) = 1 - \kappa(p + p_0), \quad p \geqslant 0.$$

Moreover, conditional on $(\chi_n, n \in \mathbb{Z}_+)$, the sequence of lifetimes $(\zeta_{\lambda_0}, \zeta_{\lambda_1}, \ldots)$ along the tagged leaf is a sequence of independent exponential variables with respective parameters $\chi_0^\alpha, \chi_1^\alpha, \ldots$

We now see that we can use this proposition to obtain the description of $\chi(t)$ using a Lamperti transformation. Let

$$\eta_t := S \circ N_t, \quad t \geqslant 0,$$

with N a Poisson process with parameter 1 which is independent of the random walk S; for probabilities and expectations related to η we use the notation P and E. The process $(\chi(t), t \geqslant 0)$ is Markovian and enjoys a scaling property. More precisely under $\mathbb{P}_x^*$ we get

$$\chi(t) \overset{(d)}{=} \exp(\eta_{\tau(tx^{-\alpha})}), \quad t \geqslant 0, \tag{9}$$

where η is the compound Poisson defined above and τ the time-change defined implicitly by

$$t = \int_0^{\tau(t)} \exp(\alpha \eta_s) ds, \quad t \geqslant 0. \tag{10}$$

6 Asymptotic Behaviors

6.1 The Convergence of the Size of a Tagged Individual

Let

$$\kappa'(p_0) = -\int_{\mathcal{M}_p(\mathbb{R}_+^*)} \langle x^{p_0} \ln(x), \mathbf{s} \rangle \nu(d\mathbf{s})$$

denote the derivative of κ at the Malthusian parameter p_0.

In this part we focus on the asymptotic behavior of the size of a tagged individual. In this direction, the quantity $\varpi_t = \exp(\alpha \eta_t)$ plays an important role, as it appears at the time change of the Lamperti transformation (see (10)), as shown by the next proposition:

Proposition 2. *Suppose that $\alpha > 0$, that the support of ν is not a discrete subgroup $r\mathbb{Z}$ for any $r > 0$ and that $0 < \kappa'(p_0) < \infty$. Then for every $y > 0$, under $\mathbb{P}_y^*$, $t^{1/\alpha}\chi(t)$ converges in law as $t \to \infty$ to a random variable Y whose law is specified by*

$$E(k(Y^\alpha)) = \frac{1}{\alpha m_1} E(k(I)I^{-1}),$$

for every measurable function $k : \mathbb{R}_+ \to \mathbb{R}_+$, with $I := \int_0^\infty \exp(\alpha \eta_s)ds$ and $m_1 := E(\eta_1) = -\kappa'(p_0)$.

Proof. As $-\kappa'(p_0)$ is the mean of the step distribution of the random walk S_n (see Proposition 1), therefore $\kappa'(p_0) > 0$ imply that $E(-\eta_1) > 0$ thus the assumption of Theorem 1 in the work of Bertoin and Yor [7] is fulfilled by the self-similar Markov process $\chi(t)^{-1}$, which gives the result.   □

We could also try to use the same method as in [6] for which we need Proposition 1.7 [4]. But in the latter we needed $\mathbb{E}(\langle x^p, X(t) \rangle)$ to be finite for large p, and its derivative to be completely monotone. But here neither of these requirements is necessarily true as κ is not necessarily positive when p is large. This explains why we have to use a different method.

Remark 3. In the case $\kappa'(p_0) = 0$ we can extend this proposition. More precisely, suppose that $\int_{\mathcal{M}_p(\mathbb{R}_+^*)} \langle x^{p_0} | \ln(x)|, \mathbf{s} \rangle \nu(d\mathbf{s}) < \infty$,

$$J := \int_1^\infty \frac{x\nu^-((x,\infty))dx}{1 + \int_0^x dy \int_y^\infty \nu^-((-\infty,-z))dz} < \infty,$$

(where ν^- is the image of $\tilde{\nu}$ by the map $u \to -u$ and $\tilde{\nu}$ is defined in Proposition 1) and $E\left(\log^+ \int_0^{T_1} \exp(-\eta_s)ds\right) < \infty$ (with $T_z := \inf\{t : -\eta_t \geqslant z\}$) hold; then, for any $y > 0$, under $\mathbb{P}_y^*$, $t^{1/\alpha}\chi(t)$ converges in law as $t \to \infty$ to a random variable $\tilde{Y}$ whose law is specified by

for any bounded, continuous function k, $\mathbb{E}\left(k(\tilde{Y}^\alpha)\right) = \lim_{\lambda \to 0} \frac{1}{\lambda} E\left(I_\lambda^{-1} k(I_\lambda)\right)$,

where $I_\lambda = \int_0^\infty \exp(\alpha \eta_s - \lambda s)ds$.

The proof is the same as the previous one, but uses Theorem 1 and Theorem 2 from the work of Caballero and Chaumont [11] instead of [7].

6.2 Convergence of the Mean Measure and L^p-convergence

We encode the configuration of masses $X(t) = \{(X_i(t))_{1 \leqslant i \leqslant \sharp\mathbf{X}(t)}\}$ by the weighted empirical measure

$$\sigma_t := \sum_{i=1}^{\sharp\mathbf{X}(t)} X_i^{p_0}(t)\delta_{t^{1/\alpha}X_i(t)}$$

which has total mass $M(t)$.

The associated mean measure σ_t^* is defined by the formula

$$\int_0^\infty k(x)\sigma_t^*(dx) = \mathbb{E}\left(\int_0^\infty k(x)\sigma_t(dx)\right)$$

which is required to hold for all compactly supported continuous functions k. Since $M(t)$ is a martingale, σ_t^* is a probability measure. We are interested in the convergence of this measure. This convergence was already established in the case of binary conservative fragmentation (see the results of Brennan and Durrett [9] and [10]). A very useful tool for this is the renewal theorem, for which they needed the fact that the process $\chi(t)$ is decreasing; here we no longer have such a monotonicity property. See also Theorem 2 and 5 of [6], Theorem 1.3 of [4] and Proposition 4 of [17] for results about empirical measures which have the property $\nu(s_i \leqslant 1 \ \forall i \in \mathbb{N}) = 1$.

Nonetheless, with Proposition 2 and Lemma 2, we easily get:

Corollary 2. *With the assumptions of Proposition 2 we get:*
1. *The measures σ_t^* converge weakly, as $t \to \infty$, to the distribution of Y i.e., for any continuous bounded function $k : \mathbb{R}_+ \to \mathbb{R}_+$, we have:*

$$\mathbb{E}\left(\langle x^{p_0}k(t^{1/\alpha}x), X(t)\rangle\right) \xrightarrow[t\to\infty]{} \mathbb{E}\left(k(Y)\right).$$

2. *For all $p_+ > p > p_0$:*

$$t^{(p-p_0)/\alpha}\mathbb{E}\left(\langle x^p, X(t)\rangle\right) \xrightarrow[t\to\infty]{} \mathbb{E}(Y^{p-p_0}).$$

We now formulate a more precise convergence result concerning the empirical measure:

Theorem 3. *Under the same assumptions as in Proposition 2 we get that for every bounded continuous function k:*

$$L^p - \lim_{t\to\infty} \int_0^\infty k(x)\sigma_t(dx) = M_\infty\mathbb{E}\left(k(Y)\right) = \frac{M_\infty}{\alpha m}E\left(k(I)I^{-1}\right),$$

for some $p > 1$.

Remark 4. A slightly different version of Corollary 2 and Theorem 3 exists also under the assumptions in Remark 3.

See also Asmussen and Kaplan [1] and [2] for a closely related result.

Proof. We follow the same method as Section 1.4. in [4] and in this direction we use Lemma 1.5 therefrom: for $(\lambda(t))_{t \geqslant 0} = (\lambda_i(t), i \in \mathbb{N})_{t \geqslant 0}$ a sequence of non-negative random variables such that for fixed $p > 1$

$$\sup_{t \geqslant 0} \mathbb{E}\left(\left(\sum_{i=1}^{\infty} \lambda_i(t)\right)^p\right) < \infty \quad \text{and} \quad \lim_{t \to \infty} \mathbb{E}\left(\sum_{i=1}^{\infty} \lambda_i(t)\right) = 0,$$

and for $(Y_i(t), i \in \mathbb{N})$ a sequence of random variables which are independent conditionally on $\lambda(t)$, we assume that there exists a sequence $(\bar{Y}_i, i \in \mathbb{N})$ of i.i.d. variables in $L^p(\mathbb{P})$, which is independent of $\lambda(t)$ for each fixed t, and such that $|Y_i(t)| \leqslant \bar{Y}_i$ for all $i \in \mathbb{N}$ and $t \geqslant 0$.

Then we know from Lemma 1.5 in [4] that

$$\lim_{t \to \infty} \sum_{i=1}^{\infty} \lambda_i(t)\left[Y_i(t) - \mathbb{E}\big(Y_i(t)\,\big|\,\lambda(t)\big)\right] = 0. \tag{11}$$

Now, let k be a continuous function bounded by 1 and let

$$A_t := \langle x^{p_0} k(t^{1/\alpha} x), X(t) \rangle.$$

The Markov property at time t for A_{t+s} and the self-similarity property of the process $\mathbf{X}$ allow to rewrite A_{t+s} as

$$\sum_{i=1}^{\sharp \mathbf{X}(t)} \lambda_i(t) Y_i(t, s)$$

where $\lambda_i(t) := X_i^{p_0}(t)$ and

$$Y_i(t, s) := \langle x^{p_0} k((t + s)^{1/\alpha} X_i(t) x), \mathbf{X}_{i,.}(s) \rangle,$$

with $\mathbf{X}_{1,.}, \mathbf{X}_{2,.,}\ldots$ a sequence of i.i.d. copies of $\mathbf{X}$ which is independent of $\mathbf{X}(t)$.

By Theorem 1 we get that

$$\sup_{t \geqslant 0} \mathbb{E}\left(\left(\sum_{i=1}^{\sharp \mathbf{X}(t)} \lambda_i(t)\right)^p\right) < \infty.$$

By the last corollary we also obtain that

$$\mathbb{E}\left(\sum_{i=1}^{\sharp \mathbf{X}(t)} \lambda_i^p(t)\right) \sim t^{-\frac{p-1}{\alpha} p_0} \mathbb{E}\big(\chi^{(p-1)p_0}(1)\big) \to 0,$$

as $t \to \infty$.

Moreover the variables $Y_i(t,s)$ are uniformly bounded by

$$Y_i = \sup_{s \geqslant 0} \langle x^{p_0}, \mathbf{X}_{i,.}(s) \rangle,$$

which are i.i.d. variables and also bounded in $L^p(\mathbb{P})$ thanks to Doob's inequality (as $\langle x_{p_0}, \mathbf{X}_{i,.}(s) \rangle$ is a martingale bounded in $L^p(\mathbb{P})$).

Thus we may apply (11), which reduces the study to that of the asymptotic behavior of:

$$\sum_{i=1}^{\sharp\mathbf{X}(t)} \lambda_i(t)\mathbb{E}(Y_i(t,s)|\mathbf{X}(t)),$$

as t tends to ∞. On the event $\{X_i(t) = y\}$, we get

$$\mathbb{E}\big(Y_i(t,s) \,\big|\, \mathbf{X}(t)\big) = \mathbb{E}\big(\langle x^{p_0} k((t+s)^{1/\alpha}yx), \mathbf{X}(s)\rangle\big).$$

Then by Lemma 2:

$$\mathbb{E}\big(\langle x^{p_0} k((t+s)^{1/\alpha}yx), \mathbf{X}_{i,.}(s)\rangle\big) = \mathbb{E}^*\big(k((t+s)^{1/\alpha}y\chi(s))\big).$$

With Proposition 2, we obtain

$$\lim_{s \to \infty} \mathbb{E}^*\big(k((t+s)^{1/\alpha}y\chi(s))\big) = \mathbb{E}\big(k(Y)\big).$$

Moreover recall from Corollary 1 that $\sum_{i=1}^{\sharp\mathbf{X}(t)} \lambda_i(t)$ converges to M_∞ in $L^p(\mathbb{P})$. Therefore we finally get that when t goes to infinity:

$$\sum_{i=1}^{\sharp\mathbf{X}(t)} \lambda_i(t)\mathbb{E}(Y_i(t,s)|X(t)) \sim \mathbb{E}(k(Y)) \sum_{i=1}^{\sharp\mathbf{X}(t)} \lambda_i(t) \sim \mathbb{E}(k(Y)) M_\infty. \qquad \square$$

Appendix: Further Results about the Intrinsic Process

We will give more general properties about the intrinsic process $\{M_Q, Q \subset \mathcal{U}\}$, $M_Q = \sum_{u \in Q} \xi_u^{p_0}$. By abuse of notation, let M_n stand for the process M_{L_n}, with $L_n = \{u \in \mathcal{U} : |u| = n\}$ the labels of the n-th generation. We introduce new definitions: we say that a line Q *covers* L, if $Q \succeq L$ and any individual stemming from L either stems from Q or has progeny in Q. If Q covers the ancestor it may simply be called *covering*. Let $\mathcal{C}_0$ be the class of covering lines with finite maximal generation. We denote the generation of Q: $|Q| = \sup_{u \in Q} |u|$. The origin of the intrinsic martingale comes from real time martingale of Nerman [20].

Also for $r \in \mathbb{R}_+^*$, let ϑ_r be the structural measure:

$$\vartheta_r(B) := \mathbb{E}_r(\sharp\{u \in \mathcal{U} : \xi_u \in B\}) = \sum_{i=1}^{\infty} \nu(rs_i \in B) \quad \text{for } B \subset \mathcal{B},$$

where $\mathcal{B}$ is the Borel algebra on $\mathbb{R}_+^*$. Let the reproduction measure μ on the sigma-field $\mathcal{B} \otimes \mathcal{B}$ be such that for every $r \geqslant 0$:

$$\mu(r, dv \times du) := r^\alpha \exp(-r^\alpha u) du \vartheta_r(dv)$$

and for any $\lambda \in \mathbb{R}$

$$\mu_\lambda(r, dv \times du) := \exp(-\lambda u) \mu(r, du \times dv).$$

The composition operation $*$ denotes the Markov transition on the size space $\mathbb{R}_+$ and convolution on the time space $\mathbb{R}_+$, so that: for all $A \in \mathcal{B}$ and $B \in \mathcal{B}$,

$$\mu^{*2}(s, A \times B) = \mu * \mu(s, A \times B) = \int_{\mathbb{R}_+ \times \mathbb{R}_+} \mu(r, A \times (B - u)) \mu(s, dr \times du).$$

With the convention that the $*$-power 0 is $\mathbb{1}_{\{A \times B\}}(s, 0)$ which gives full mass to $(s, 0)$. Define the renewal measure as

$$\psi_\lambda := \sum_0^\infty \mu_\lambda^{*n}.$$

Let

$$\alpha' := \inf\{\lambda : \quad \psi_\lambda(r, \mathbb{R}_+ \times \mathbb{R}_+) < \infty \text{ for some } r \in \mathbb{R}_+\}.$$

Moreover as

$$\mu_\lambda(r, \mathbb{R}_+ \times \mathbb{R}_+) = \begin{cases} mr^\alpha/(r^\alpha + \lambda) & \text{if } \lambda > -r^\alpha \\ \infty & \text{else,} \end{cases}$$

thus

$$\psi_\lambda(r, \mathbb{R}_+ \times \mathbb{R}_+) < \infty \text{ if and only if } \lambda < (r/(m-1))^{1/\alpha}$$

therefore we get $\alpha' = 0$. For $A \in \mathcal{B}$, let

$$\pi(A) := \lim_{n \to \infty} \mu^{*n}(1, A \times \mathbb{R}_+) \tag{12}$$

which is well defined as $\mu^{*n}(1, A \times \mathbb{R}_+)$ is a decreasing function in n and non-negative. Let $h(s) := s^{p_0}$ for all $s \in \mathbb{R}_+$ and $\beta := 1$. These objects correspond to those defined in [15].

Recall that the Galton-Watson process $(Z_n, n \geqslant 0))$ is equal to $(\#\{u \in \mathcal{U} : \xi_u > 0 \text{ and } |u| = n\}, n \geqslant 0)$.

We suppose that

$$m := \mathbb{E}(Z_1) < \infty,$$

i.e., $\int_{\mathcal{M}_p(\mathbb{R}_+^*)} \sharp s \nu(ds) < \infty$. This assumption is slightly stronger than (6), therefore we get

Proposition 3. *1. If $L \preceq Q$ are lines, then*

$$\mathbb{E}(M_Q|\mathcal{H}_L) \leqslant M_L.$$

If Q verifies $|Q| < \infty$ and covers L, then

$$\mathbb{E}(M_Q|\mathcal{H}_L) = M_L.$$

2. For all $s > 0$, $\{M_L;\ L \in \mathcal{C}_0\}$ is uniformly $\mathbb{P}_s$-integrable.
3. There is a random variable $M \geqslant 0$ such that for π-almost all $s > 0$

$$M_L = \mathbb{E}_s(M|\mathcal{H}_L)$$

and $M_L \overset{L^1(\mathbb{P}_s)}{\to} M$, as $L \in \mathcal{C}_0$ filters ($\preceq$). If $\varsigma_n \preceq \varsigma_{n+1} \in \mathcal{C}_0$ and to any $x \in \mathcal{U}$ there is an ς_n such that x has progeny in ς_n, then $M_{\varsigma_n} \to M$, as $n \to \infty$, also $\mathbb{P}_s$-a.s.

A consequence of the first and second points applied for $L_n = \{u \in \mathcal{U} : |u| = n\}$ and $L_m = \{u \in \mathcal{U} : |u| = m\}$ with $m \geqslant n \geqslant 0$, is that M_n is a martingale and the uniform $\mathbb{P}_s$-integrability of this martingale. The third point applied for the lines τ_t gives convergence of $M(t)$ in $L^1(\mathbb{P}_s)$ and a.s.

Proof. First the conditions of Malthusian population, as defined by Jagers in [15], are fulfilled, thus by Theorem 5.1 therein we get the first point.

Let $\bar{\xi} := \int_{\mathbb{R}_+ \times \mathbb{R}_+} h(s)r^\alpha \exp(-tr^\alpha)dt\vartheta_1(ds) = \sum_{|u|=1} \xi_u^{po}$ and $\mathbb{E}_\pi$ be the expectation with respect to $\int_{\mathbb{R}_+} \mathbb{P}_s(dw)\pi(ds)$. Therefore,

$$\mathbb{E}_\pi(\bar{\xi} \log^+ \bar{\xi}) = \int_{\mathbb{R}_+} \mathbb{E}_x\left(\sum_{i=1}^\infty \xi_i^{po}\left(\log^+ \sum_{j=1}^\infty \xi_j^{po}\right)\right)\pi(dx),$$

and it follows readily from the Malthusian hypotheses and the fact that $\sum_{|u|=n} \xi_u^{ppo}$ is a supermartingale, that this quantity is finite. So the assumption of Theorem 6.1 of [15] hold, which gives by Theorem 6.1 of [15] the second point and by Theorem 6.3 of [15] we get the third point. $\square$

Acknowledgements: I wish to thank J. Bertoin for his help and suggestions. I also wish to thank the anonymous referees of an earlier draft for their detailed comments and suggestions.

References

1. Asmussen S. and Kaplan N.: Branching random walks. I. Stochastic Process. Appl. 4, no. 1, 1-13 (1976).
2. Asmussen S. and Kaplan N.: Branching random walks. II. Stochastic Process. Appl. 4, no. 1, 15-31 (1976).

3. Athreya K. B. and Ney P. E.: Branching processes. Springer-Verlag Berlin Heidelberg (1972).

4. Bertoin J.: Random fragmentation and coagulation processes. Cambridge Univ. Pr. (2006).

5. Bertoin J.: Different aspects of a random fragmentation model. Stochastic Process. Appl. 116, 345-369, (2006).

6. Bertoin J. and Gnedin A. V.: Asymptotic laws for nonconservative self-similar fragmentations. Electron. J. Probab. 9, No. 19, 575-593, (2004).

7. Bertoin J. and Yor M.: The entrance laws of self-similar Markov processes and exponential functionals of Lévy processes. Potential Analysis 17 389-400, (2002).

8. Biggins J. D.: Uniform convergence of martingales in the branching random walk. Ann. Probab. 20, No. 1, 131-151, (1992).

9. Brennan M. D. and Durrett R.: Splitting intervals. Ann. Probab. 14, No. 3, 1024-1036, (1986).

10. Brennan M. D. and Durrett R.: Splitting intervals. II. Limit laws for lengths. Probab. Theory Related Fields. 75 No. 1, 109-127, (1987).

11. Caballero M.E. and Chaumont L.: Weak convergence of positive self-similar Markov processes and overshoots of Lévy processes. Ann. Probab. 34, No. 3, 1012-1034, (2006).

12. Chauvin B.: Arbres et processus de Bellman-Harris. Ann. Inst. Henri Poincaré. 22, No. 2, 209-232, (1986).

13. Chauvin B.: Product martingales and stopping lines for branching Brownian motion. Ann. Probab. 19, No. 3, 1195-1205, (1991).

14. Harris T. E.: The theory of branching processes. Springer (1963).

15. Jagers P.: General branching processes as Markov fields. Stochastic Process. Appl. 32, 183-212, (1989).

16. Kingman J. F. C.: Poisson processes. Oxford Studies in Probability, 3. Oxford Science Publications. The Clarendon Press, Oxford University Press (1993).

17. Krell N.: Multifractal spectra and precise rates of decay in homogeneous fragmentations. To appear in Stochastic Process. Appl. (2008).

18. Kyprianou A. E.: A note on branching Lévy processes. Stochastic Process. Appl. 82, No. 1, 1-14, (1999).

19. Kyprianou A. E.: Martingale convergence and the stopped branching random walk. Probab. Theory Related Fields 116, no. 3, 405-419, (2000).

20. Nerman O.: The growth and composition of supercritical branching populations on general type spaces. Technical report, Dept. Mathematics, Chalmers Univ. Technology and Goteborg Univ. (1984).

21. Neveu J.: Arbres et processus de Galton-Watson. Ann. Inst. H. Poincaré Probab. Statist. 22, No. 2, 199-207, (1986).

A Spine Approach to Branching Diffusions with Applications to $\mathcal{L}^p$-convergence of Martingales

Robert Hardy and Simon C. Harris

Department of Mathematical Sciences, University of Bath
Claverton Down, Bath, BA2 7AY, UK
E-mail: S.C.Harris@bath.ac.uk

Summary. We present a modified formalization of the 'spine' change of measure approach for branching diffusions in the spirit of those found in Kyprianou [40] and Lyons *et al.* [44, 43, 41]. We use our formulation to interpret certain 'Gibbs-Boltzmann' weightings of particles and use this to give an intuitive proof of a general 'Many-to-One' result which enables expectations of sums over particles in the branching diffusion to be calculated purely in terms of an expectation of one 'spine' particle. We also exemplify spine proofs of the $\mathcal{L}^p$-convergence ($p \geq 1$) of some key 'additive' martingales for three distinct models of branching diffusions, including new results for a multi-type branching Brownian motion and discussion of left-most particle speeds.

1 Introduction

Consider a branching Brownian motion (BBM) with constant branching rate r and offspring distribution A, which is a branching process where particles diffuse independently according to a standard Brownian motion and at any moment undergo fission at a rate r to be replaced by a random number of offspring, $1 + A$, where A is an independent random variable with distribution

$$P(A = i) = p_i, \qquad i \in \{0, 1, \dots\},$$

such that $m := P(A) = \sum_{i=0}^{\infty} i\, p_i < \infty$. Offspring move off from their parent's point of fission, and continue to evolve independently as above, and so on.

Let the configuration of this BBM at time t be given by the $\mathbb{R}$-valued point process $\mathbb{X}_t := \{X_u(t) : u \in N_t\}$ where N_t is the set of individuals alive at time t. Let the probabilities for this process be $\{P^x : x \in \mathbb{R}\}$, where P^x is the law starting from a single particle at position x, and let $(\mathcal{F}_t)_{t \geq 0}$ be the natural filtration. It is well known that for any $\lambda \in \mathbb{R}$,

$$Z_\lambda(t) := \sum_{u \in N_t} e^{-rmt} e^{\lambda X_u(t) - \frac{1}{2}\lambda^2 t} = \sum_{u \in N_t} e^{\lambda X_u(t) - E_\lambda t} \tag{1}$$

C. Donati-Martin et al. (eds.), *Séminaire de Probabilités XLII*,
Lecture Notes in Mathematics 1979, DOI 10.1007/978-3-642-01763-6_11,
© Springer-Verlag Berlin Heidelberg 2009

where $E_\lambda := -\lambda c_\lambda := \frac{1}{2}\lambda^2 + rm$, defines a *positive* martingale, so $Z_\lambda(\infty) :=$ $\lim_{t\to\infty} Z_\lambda(t)$ exists and is finite almost surely under each P^x. See Neveu [46], for example.

One of the central elements of the spine approach is to interpret the behaviour of a branching process under a certain change of measure. Chauvin and Rouault [9] showed that changing measure for BBM with the Z_λ martingale leads to the following 'spine' construction:

Theorem 1.1 *If we define the measure $\mathbb{Q}_\lambda^x$ via*

$$\left.\frac{d\mathbb{Q}_\lambda^x}{dP^x}\right|_{\mathcal{F}_t} = \frac{Z_\lambda(t)}{Z_\lambda(0)} = e^{-\lambda x} Z_\lambda(t), \tag{2}$$

then under $\mathbb{Q}_\lambda^x$ the point process $\mathbb{X}_t$ can be constructed as follows:
- *starting from position x, the original ancestor diffuses according to a Brownian motion on $\mathbb{R}$ with drift λ;*
- *at an accelerated rate $(1+m)r$ the particle undergoes fission producing $1+\tilde{A}$ particles, where the distribution of $\tilde{A}$ is independent of the past motion but is size-biased:*

$$\mathbb{Q}_\lambda(\tilde{A} = i) = \frac{(i+1)p_i}{m+1}, \qquad i \in \{0, 1, \ldots\}.$$

- *with equal probability, one of these offspring particles is selected;*
- *this chosen particle repeats stochastically the behaviour of the parent with the size-biased offspring distribution;*
- *each other particle initiates, from its birth position, an independent copy of a P branching Brownian motion with branching rate r and offspring distribution given by A (which is without the size-biasing).*

The chosen line of descent in such pathwise constructions of the measure, here $\mathbb{Q}_\lambda$, has come to be known as the *spine* as it can be thought of as the backbone of the branching process $\mathbb{X}_t$ from which all particles are born. The phenomena of size-biasing along the spine is a common feature of such measure changes when random offspring distributions are present.

Although Chauvin and Rouault's work on the measure change continued in a paper co-authored with Wakolbinger [10], where the new measure is interpreted as the result of building a conditioned tree using the concepts of Palm measures, it wasn't until the so-called 'conceptual proofs' of Lyons, Kurtz, Peres and Pemantle published around 1995 ([44, 43, 41]) that the spine approach really began to crystalize. These papers laid out a formal basis for spines using a series of new measures on two underlying spaces of sample trees with and without distinguished lines of descent (spines). Of particular interest is the paper by Lyons [43] which gave a spine-based proof of the $\mathcal{L}^1$-convergence of the well-known martingale for the Galton-Watson process. Here we first saw the *spine decomposition* of the martingale as the key to using the intuition provided by Chauvin and Rouault's pathwise construction of the new measure – Lyons used this together with a previously known measure-theoretic result on Radon-Nikodym derivatives that allows us to deduce the behaviour of the change-of-measure martingale under the original measure by investigating its behaviour under the second measure. Similar ideas have recently been used by Kyprianou [40] to investigate the $\mathcal{L}^1$-convergence of the BBM martingale (1), by Biggins and Kyprianou [4] for multi-type branching processes in discrete time, by Hu and Shi [33] for the minimal position in a branching random walk, by Geiger [16, 17] for Galton-Watson processes, by Georgii and Baake [19] to study ancestral type behaviour in a continuous time branching Markov chain, as well as Olofsson [47] for general branching processes. Also see Athreya [2], Geiger [15, 18], Iksanov [34], Rouault and Liu [42] and Waymire and Williams [49], to name just a few other papers where spine and size-biasing techniques have already proved extremely useful in branching process situations. For applications of spines in branching in random media see, for example, the survey by Engländer [13].

In this article[1], we present a modified formalization of the spine approach that attempts to improve on the schemes originally laid out by Lyons et al. [44, 43, 41] and later for BBM by Kyprianou [40]. Although the set-up costs of our spine formalization are quite large, at least in terms of definitions and notation, the underlying ideas are all extremely simple and intuitive. One advantage of this approach is that it has facilitated the development of further spine techniques, for example, in Hardy & Harris [23, 22], Git et al. [20] and J.W.Harris & S.C.Harris [27] where a number of technical problems and difficult non-linear calculations are by-passed with spine calculations enabling their reduction to relatively straightforward classical one-particle situations; this article also serves as a foundation for these and other works.

The basic concept of our approach is quite straightforward: given the original branching process, we first create an extended probability measure by enriching the process through (carefully) choosing at random one of the particles to be the so-called *spine*. Now, on this enriched process, changes of measure can easily be applied that *only* affect the behaviour along the path of this single distinguished 'spine' particle; in our examples, we add a drift to the spine's motion, increase rate of fission along the path of the spine and size-bias the spine's offspring distribution. However, projecting this new enriched and changed measure back onto the original process filtration (that is, without any knowledge of the distinguished spine) brings the fundamental 'additive' martingales into play as a Radon-Nikodym derivative. The four probability measures, various martingales, extra filtrations and clear process constructions afforded by our setup, together with some other useful properties and tricks, such as the *spine decomposition*, provide a very elegant, intuitive and powerful set of techniques for analysing the process.

---

[1] Based on the **arXiv** articles [24, 25]

The reader who is familiar with the work of Lyons *et al.* [44] or Kyprianou [40] will notice significant similarities as well as differences in our approach. In the first instance our modifications correct our perceived weakness in the Lyons *et al.* scheme where one of the measures they defined had a time-dependent mass and could not be normalized to be a probability measure in a natural way, hence lacked a clear interpretation in terms of any direct process construction; an immediate consequence of this improvement is that here *all* measure changes are carried out by *martingales* and we regain a clear intuitive construction. Another difference is in our use of filtrations and sub-filtrations, where Lyons *et al.* instead used marginalizing. As we shall show, this brings substantial benefits since it allows us to relate the spine and the branching diffusion through the conditional expectation operation, and in this way gives us a proper methodology for building *new* martingales for the branching diffusion based on known single particle martingales for the spine.

The conditional expectation approach also leads directly to simple proofs of some key results for branching diffusions. The first of these concerns the relation that becomes clear between the spine and the 'Gibbs-Boltzmann' weightings for the branching particles. Such weightings are well known in the theory of branching process, for example see Chauvin & Rouault [7], or Harris [30] which also studies the continuous-typed branching diffusion example introduced later. In our formulation these weightings can be interpreted as a conditional expectation of a spine event, and we can use them to immediately obtain a new interpretation of the additive operations previously seen only within the context of the Kesten-Stigum theorem and related problems. Our approach also leads to a substantially easier proof of a more general form of the *Many-to-One* theorem that is so often useful in branching processes applications; for example, in Champneys *et al.* [5] or Harris and Williams [28], special cases of this theorem were a key tool in their more classical approaches to branching diffusions.

As another application of spine techniques, we will analyze the $\mathcal{L}^p$-convergence properties (for $p \geq 1$) of some fundamental positive 'additive' martingales for three different models of branching diffusions.

Consider first the branching Brownian motion (BBM) with random family sizes. We recall that Kyprianou [40] used spine techniques to give necessary and sufficient conditions for $\mathcal{L}^1$-convergence of the Z_λ martingales:

Theorem 1.2 *Let* $\tilde{\lambda} := -\sqrt{2rm}$ *so that* $c_\lambda := -E_\lambda/\lambda$ *attains local maximum at* $\tilde{\lambda}$. *For each* $x \in \mathbb{R}$, *the limit* $Z_\lambda(\infty) := \lim_{t \to \infty} Z_\lambda(t)$ *exists* P^x-*almost surely where:*

- *if* $\lambda \leq \tilde{\lambda}$ *then* $Z_\lambda(\infty) = 0$ P^x-*almost surely;*
- *if* $\lambda \in (\tilde{\lambda}, 0]$ *and* $P(A \log^+ A) = \infty$ *then* $Z_\lambda(\infty) = 0$ P^x-*almost surely;*
- *if* $\lambda \in (\tilde{\lambda}, 0]$ *and* $P(A \log^+ A) < \infty$ *then* $Z_\lambda(t) \to Z_\lambda(\infty)$ *almost surely and in* $\mathcal{L}^1(P^x)$.

(Note, without loss of generality (by symmetry) we will suppose $\lambda \leq 0$ throughout this article.)

In fact, in many cases where the martingale has a non-trivial limit, the convergence will also be much stronger than merely in $\mathcal{L}^1(P^x)$, as indicated by the following $\mathcal{L}^p$-convergence result:

Theorem 1.3 *For each $x \in \mathbb{R}$, and for each $p \in (1, 2]$:*

- $Z_\lambda(t) \rightarrow Z_\lambda(\infty)$ *almost surely and in* $\mathcal{L}^p(P^x)$ *if* $p\lambda^2 < 2mr$ *and* $P(A^p) < \infty$
- Z_λ *is unbounded in* $\mathcal{L}^p(P^x)$, *that is* $\lim_{t\to\infty} P^x(Z_\lambda(t)^p) = \infty$, *if* $p\lambda^2 > 2mr$ *or* $P(A^p) = \infty$.

We shall give a spine-based proof of this $\mathcal{L}^p$-convergence theorem, but also see Neveu [46] for sufficient conditions in the special case of binary branching at unit rate using more classical techniques. Also see Harris [29] for further discussion of martingale convergence in BBM and applications. Iksanov [34] also uses similar spine techniques in the study of the branching random walk.

For our second model, we look at a finite-type BBM model where the type of each particle controls the rate of fission, the offspring distribution and the spatial diffusion. First, we will extend Kyprianou's [40] approach to give the analogous $\mathcal{L}^1$-convergence result for this multi-type BBM model. We will also briefly discuss the rate of convergence of the martingales to zero and the speed of the spatially left-most particle within the process. Next, we give a new result on $\mathcal{L}^p$-convergence criteria, extending our earlier spine based proof developed for the single-type BBM case.

The third model we consider has a *continuous* type-space where the type of each particle moves independently as an Ornstein-Uhlenbeck process on $\mathbb{R}$. This branching diffusion was first introduced in Harris and Williams [28] and has also been investigated in Harris [30], Git *et al.* [20] and Kyprianou and Engländer [12].

Proofs for each of these models run along similar lines and the techniques are quite general, and it is a powerful feature of the spine approach that this is possible. For example, they have since been extended to more general branching diffusions in Engländer *et al.* [14] and to fragmentation processes in Harris *et al.* [31]. More classical techniques based on the expectation semigroup are simply not able to generalize easily, since they often require either some *a priori* bounds on the semigroup or involve difficult estimates – for example, in Harris and Williams [28] their important bound of a non-linear term is made possible only by the existence of a good $\mathcal{L}^2$ theory for their operator, and this is not generally available.

Of course, to prove martingale convergence in $\mathcal{L}^p$ for some $p > 1$ we use Doob's theorem, and therefore need only show that the martingale is *bounded* in $\mathcal{L}^p$. The *spine decomposition* is an excellent tool here for showing boundedness of the martingale since it reduces difficult calculations over the whole collection of branching particles to just the single spine process. We find the same conditions are also *necessary* for $\mathcal{L}^p$-boundedness of the martingale when $p > 1$ by just considering the contributions along the spine at times of fission and observing when these are unbounded. Otherwise, to determine whether the martingale is merely $\mathcal{L}^1$-convergent or has an almost-surely zero limit, we determine whether the martingale is almost-surely bounded or not under its own change of measure – this was Kyprianou's [40] approach and relies on a measure-theoretic result that has become standard in the spine methodology since the important work of Lyons *et al.* [44, 43, 41].

There are a number of reasons why we may be interested in knowing about the $\mathcal{L}^p$ convergence of a martingale: in Neveu's original article [46] it was a means to proving $\mathcal{L}^1$-convergence of martingales which can then be used to represent (non-trivial) travelling-wave solutions to the FKPP reaction-diffusion equation as well as in understanding the growth and spread of the BBM, whilst Git *et al.* [20] and Asmussen and Hering [1] have used it to deduce the almost-sure rate of convergence of the martingale to its limit. Of equal importance are the *techniques* that we use here. The convergence of other additive martingales can be determined with similar techniques, for example, see an application to a BBM with inhomogeneous breeding potential in J.W. Harris and S.C. Harris [27]. Similar ideas have also been used in proving a lower bound for a number of problems in the large-deviations theory of branching diffusions – we have used the spine decomposition with Doob's submartingale inequality to get an upper-bound for the growth of the martingale under the new measure which then leads to a lower-bound on the probability that one of the diffusing particles follows an unexpected path. See Hardy and Harris [23] for a spine-based proof of a path large deviation result for branching Brownian motion, and see Hardy and Harris [22] for a proof of a lower bound in the model that we consider in Section 11.

The layout of this paper is as follows. In Section 2, we will introduce the branching models, describing a binary branching multi-type BBM that we will frequently use as an example, before describing a more general branching Markov process model with random family sizes. In Section 3, we introduce the *spine* of the branching process as a distinguished infinite line of descent starting at the initial ancestor, we describe the underlying space for the branching Markov process with spine and we also introduce various fundamental filtrations. In Section 4, we define some fundamental probability spaces, including a probability measure for the branching process with a randomly chosen spine. In Section 5, various martingales are introduced and discussed. In particular, we see how to use filtrations and conditional expectation to build 'additive' martingales for the branching process out of the product of three simpler 'one-particle' martingales that only depend on the behaviour along the path of the spine; used as changes of measure, one martingale will increase the fission rate along the path of the spine, another will size-bias the offspring distribution along the spine, whilst the other one will change the motion of the spine. Section 6 discusses changes of measure with these martingales and gives very important and useful intuitive constructions for the branching process with spine under both the original measure $\tilde{P}$ and the changed measure $\tilde{Q}$. Another extremely useful tool in the spine approach is the *spine decomposition* that we prove in Section 7; this gives an expression for the expectation of the 'additive' martingale under the new measure $\tilde{Q}$ conditional on knowing the behaviour all along the path of the spine (including the spine's motion, the times of fission along the spine and number of offspring at each of the spine's fissions). In Section 8, we use the spine formulation to derive an interpretation for certain Gibbs-Boltzmann weights of $\tilde{Q}$, discussing links with theorems of Kesten-Stigum and Watanabe, in addition to proving a 'Many-to-One' theorem. Finally, in sections 9, 10, and 11 we will prove the martingale convergence results for BBM, finite-type BBM and the continuous-type BBM models, respectively.

2 Branching Markov Models

Before we present the underlying constructions for spines, it will be useful to give the reader a further idea of the branching-diffusion models that we have in mind for applications. We first briefly introduce a finite-type branching diffusion (which will often serve as a useful example), before presenting a more general model that shall be used as the basis of our spine constructions in the following sections.

2.1 A Finite-type Branching Diffusion

Let θ be a strictly positive constant that can be considered as a temperature parameter. For some fixed $n \in \mathbb{N}$, define the finite type-space $I := \{1, \ldots, n\}$ and suppose that we are given two sets of positive constants $a(1), \ldots, a(n)$ and $R(1), \ldots, R(n)$.

A Single Particle Motion. Consider the process $(\xi_t, \eta_t)_{t \geq 0}$ moving on $J := \mathbb{R} \times I$ as follows:

(i) The *type* location, η_t, of the particle moves as an irreducible, time-reversible Markov chain on the finite type-space I with Q-matrix θQ and invariant measure $\pi = (\pi_1, \ldots, \pi_n)$;

(ii) the *spatial* location, ξ_t, moves as a driftless Brownian motion on $\mathbb{R}$ with diffusion coefficient $a(y) > 0$ whenever η_t is in state y, that is,

$$d\xi_t = a(\eta_t)^{\frac{1}{2}} \, dB_t, \qquad \text{where } B_t \text{ a Brownian motion.} \tag{3}$$

The formal generator of this process (ξ_t, η_t) is therefore:

$$\mathcal{H}F(x, y) = \frac{1}{2} a(y) \frac{\partial^2 F}{\partial x^2} + \theta \sum_{j \in I} Q(y, j) F(x, j), \qquad (F : J \to \mathbb{R}). \tag{4}$$

A Typed Branching Brownian Motion. Consider a branching diffusion where individual particles move independently according to the *single particle motion* as described above, and any particle currently of type y will undergo *binary* fission at rate $R(y)$ to be replaced by two particles at the same spatial and type positions as the parent. These offspring particles then move off independently, repeating stochastically the parent's behaviour, and so on.

Let the configuration of the whole branching diffusion at time t be given by the J-valued point process $\mathbb{X}_t = \{(X_u(t), Y_u(t)) : u \in N_t\}$, where N_t is the set of individuals alive at time t. Suppose probabilities for this process are given by $\{P^{x,y} : (x,y) \in J\}$ defined on the natural filtration, $(\mathcal{F}_t)_{t\geq 0}$, where $P^{x,y}$ is the law of the typed BBM process starting with one initial particle of type y at spatial position x.

This finite-type branching diffusion (with general offspring distribution) is investigated in Section 10 in this article, also see Hardy [21]. For now, we briefly introduce two fundamental *positive* martingales used to understand this model, the first based on the whole branching diffusion and the second based only on the single-particle model:

$$Z_\lambda(t) := \sum_{u \in N_t} v_\lambda(Y_u(t)) e^{\lambda X_u(t) - E_\lambda t}, \tag{5}$$

$$\zeta_\lambda(t) := e^{\int_0^t R(\eta_s)\, ds} v_\lambda(\eta_t) e^{\lambda \xi_t - E_\lambda t}, \tag{6}$$

where v_λ and E_λ satisfy

$$\left(\frac{1}{2}\lambda^2 A + \theta Q + R\right) v_\lambda = E_\lambda v_\lambda,$$

where $A := \operatorname{diag}(a(y) : y \in I)$ and $R := \operatorname{diag}(R(y) : y \in I)$. That is, v_λ is the (Perron-Frobenius) eigenvector of the matrix $\frac{1}{2}\lambda^2 A + \theta Q + R$, with eigenvalue E_λ. These two martingales should be compared with the corresponding martingales (1) and $e^{\lambda B_t - \frac{1}{2}\lambda^2 t}$ for BBM and a single Brownian motion respectively.

2.2 A General Branching Markov Process

The spine constructions in our formulation can be applied to a much more general branching Markov model, and we shall base the presentation on the following model, where particles move independently in a general space J as a stochastic copy of some given Markov process Ξ_t, and at a location-dependent rate undergo fission to produce a location-dependent random number of offspring that each carry on this branching behaviour independently.

Definition 2.1 (A General Branching Markov Process) *We   suppose that three initial elements are given to us:*

- *a Markov process Ξ_t in a measurable space $(J, \mathcal{B})$,*
- *a measurable function $R : J \to [0, \infty)$,*
- *for each $x \in J$ we are given a random variable $A(x)$ whose probability distribution on the numbers $\{0, 1, \ldots\}$ is $P(A(x) = k) = p_k(x)$, with mean $m(x) := \sum_{k=0}^{\infty} k p_k(x) < \infty$.*

From these ingredients we can build a branching process in J according to the following recipe:

- *Each particle of the branching process will live, move and die in this space* $(J, \mathcal{B})$, *and if an individual u is alive at time t we refer to its location in J as* $X_u(t)$. *Therefore the time-t configuration of the branching process is a J-valued point process* $\mathbb{X}_t := \{X_u(t) : u \in N_t\}$ *where* N_t *denotes the collection of all particles alive at time t.*
- *For each individual u, the stochastic behaviour of its motion in J is an independent copy of the given process* Ξ_t.
- *The function* $R : J \to [0, \infty)$ *determines the rate at which each particle dies: given that u is alive at time t, its probability of dying in the interval* $[t, t + dt)$ *is* $R(X_u(t))dt + o(dt)$.
- *If a particle u dies at location* $x \in J$ *it is replaced by* $1 + A_u$ *particles all positioned at x, where* A_u *is an independent copy of the random variable* $A(x)$. *All particles, once born, progress independently of each other.*

We suppose that the probabilities of this branching process are $\{P^x : x \in J\}$ *where under* P^x *one initial ancestor starts out at x.*

We shall first give a formal construction of the underlying probability space, made up of the sample trees of the branching process $\mathbb{X}_t$ in which the spines are the distinguished lines of descent. Once built, this space will be filtered in a natural way by the underlying family relationships of each sample tree, the diffusing branching particles and the diffusing spine, and then in section 4 we shall explain how we can define new probability measures $\tilde{P}^x$ that extend each P^x up to the finest filtration that contains all information about the spine and the branching particles. Much of the notation that we use for the underlying space of trees, the filtrations and the measures is closely related to that found in Kyprianou [40].

Although we do not strive to present our spine approach in the greatest possible generality, our model already covers many important situations whilst still being able to clearly demonstrate all the key spine ideas. In particular, in all our models, new offspring always inherit the position of their parent, although the same spine methods should also readily adapt to situations with random dispersal of offspring.

For greater clarity, we often use the finite-type branching diffusion of Section 2.1 to introduce the ideas before following up with the general formulation. For example, in this finite-type model we would take the process Ξ_t to be the single-particle process (ξ_t, η_t) which lives in the space $J := \mathbb{R} \times I$ and has generator $\mathcal{H}$ given by (4). The birth rate in this model at location $(x, y) \in J$ will be independent of x and given by the function $R(y)$ for all $y \in I$ and, since only binary branching occurs in this case, we also have $P(A(x, y) = 1) = 1$ for all $(x, y) \in J$.

3 The Underlying Space for Spines

3.1 Marked Galton-Watson Trees with Spines

The set of Ulam-Harris labels is to be equated with the set Ω of finite sequences of strictly-positive integers:

$$\Omega := \{\emptyset\} \cup \bigcup_{n \in \mathbb{N}} (\mathbb{N})^n,$$

where we take $\mathbb{N} = \{1, 2, \ldots\}$. For two words $u, v \in \Omega$, uv denotes the concatenated word ($u\emptyset = \emptyset u = u$), and therefore Ω contains elements like '213' (or '$\emptyset$213'), which represents 'the 3rd child of the 1st child of the 2nd child of the initial ancestor $\emptyset$'. For two labels $v, u \in \Omega$ the notation $v < u$ means that v is an *ancestor* of u, and $|u|$ denotes the length of u. The set of all ancestors of u is equally given by

$$\{v : v < u\} = \{v : \exists w \in \Omega \text{ such that } vw = u\}.$$

Collections of labels, ie. subsets of Ω, will therefore be groups of individuals. In particular, a subset $\tau \subset \Omega$ will be called a *Galton-Watson tree* if:

1. $\emptyset \in \tau$,
2. if $u, v \in \Omega$, then $uv \in \tau$ implies $u \in \tau$,
3. for all $u \in \tau$, there exists $A_u \in 0, 1, 2, \ldots$ such that $uj \in \tau$ if and only if $1 \leq j \leq 1 + A_u$, (where $j \in \mathbb{N}$).

That is just to say that a Galton-Watson tree:

1. has a single initial ancestor $\emptyset$,
2. contains all ancestors of any of its individuals v,
3. has the $1 + A_u$ children of an individual u labelled in a consecutive way,

and is therefore just what we imagine by the picture of a family tree descending from a single ancestor. Note that the '$1 \leq j \leq 1 + A_u$' condition in 3 means that each individual has *at least* one child, so that in our model we are insisting that Galton-Watson trees *never die out*.

The set of all Galton-Watson trees will be called $\mathbb{T}$. Typically we use the name τ for a particular tree, and whenever possible we will use the letters u or v or w to refer to the labels in τ, which we may also refer to as *nodes of τ* or *individuals in τ* or just as *particles*.

Each individual should have a *location* in J at each moment of its *lifetime*. Since a Galton-Watson tree $\tau \in \mathbb{T}$ in itself can express only the *family* structure of the individuals in our branching random walk, in order to give them these extra features we suppose that each individual $u \in \tau$ has a mark (X_u, σ_u) associated with it which we read as:

- $\sigma_u \in \mathbb{R}^+$ is the *lifetime* of u, which determines the *fission time* of particle u as $S_u := \sum_{v \leq u} \sigma_v$ (with $S_\emptyset := \sigma_\emptyset$). The times S_u may also be referred to as the *death* times;

- $X_u : [S_u - \sigma_u, S_u) \to J$ gives the *location* of u at time $t \in [S_u - \sigma_u, S_u)$.

To avoid ambiguity, it is always necessary to decide whether a particle is in existence or not at its death time.

Remark 3.1 *Our convention throughout will be that a particle u dies 'just before' its death time S_u (which explains why we have defined $X_u : [S_u - \sigma_u, S_u) \to \cdot$ for example). Thus at the time S_u the particle u has disappeared, replaced by its $1 + A_u$ children which are all alive and ready to go.*

We denote a single marked tree by (τ, X, σ) or (τ, M) for shorthand, and the set of all marked Galton-Watson trees by $\mathcal{T}$:

- $\mathcal{T} := \left\{ (\tau, X, \sigma) : \tau \in \mathbb{T} \text{ and for each } u \in \tau, \sigma_u \in \mathbb{R}^+, X_u : [S_u - \sigma_u, S_u) \to J \right\}.$

- For each $(\tau, X, \sigma) \in \mathcal{T}$, the set of particles that are alive at time t is defined as $N_t := \left\{ u \in \tau : S_u - \sigma_u \leq t < S_u \right\}$.

Where we want to highlight the fact that these values depend on the underlying marked tree we write e.g. $N_t((\tau, X, \sigma))$ or $S_u((\tau, M))$.

Any particle $u \in \tau$ that comes into existence creates a *subtree* made up from the collection of particles (and all their marks) that have u as an ancestor – and u is the original ancestor of this subtree.

- $(\tau, X, \sigma)_j^u$, or $(\tau, M)_j^u$ for shorthand, is defined as the *subtree* growing from individual u's jth child uj, where $1 \leq j \leq 1 + A_u$.

This subtree is a marked tree itself, but when considered as a part of the original tree we have to remember that it comes into existence at the space-time location $(X_u(S_u - \sigma_u), S_u - \sigma_u)$ – which is just the space-time location of the death of particle u (and therefore the space-time location of the birth of its child uj).

Before moving on there is a further useful extension of the notation: for any particle u we extend the definition of X_u from the time interval $[S_u - \sigma_u, S_u)$ to allow all earlier times $t \in [0, S_u)$:

Definition 3.2 *Each particle u is alive in the time interval $[S_u - \sigma_u, S_u)$, but we extend the concept of its path in J to all earlier times $t < S_u$:*

$$X_u(t) := \begin{cases} X_u(t) \text{ if } S_u - \sigma_u \leq t < S_u \\ X_v(t) \text{ if } v < u \text{ and } S_v - \sigma_v \leq t < S_v \end{cases}$$

Thus particle u inherits the path of its unique line of ancestors, and this simple extension will allow us to later write expressions like $\exp\{\int_0^t f(s)\,\mathrm{d}X_u(s)\}$ whenever $u \in N_t$, without worrying about the birth time of u.

For any given marked tree $(\tau, M) \in \mathcal{T}$ we can identify distinguished lines of descent from the initial ancestor: $\emptyset, u_1, u_2, u_3, \ldots \in \tau$, in which u_3 is a child of u_2, which itself is a child of u_1 which is a child of the original ancestor $\emptyset$. We'll call such a subset of τ a *spine*, and will refer to it as ξ:

- a *spine* ξ is a subset of nodes $\{\emptyset, u_1, u_2, u_3, \ldots\}$ in the tree τ that make up a unique line of descent. We use ξ_t to refer to the unique node in ξ that that is alive at time t.

In a more formal definition, which can for example be found in the paper by Rouault and Liu [42], a spine is thought of as a point on $\partial\tau$ the boundary of the tree – in fact the boundary is *defined* as the set of all infinite lines of descent. This explains the notation $\xi \in \partial\tau$ in the following definition: we augment the space $\mathcal{T}$ of marked trees to become

- $\tilde{\mathcal{T}} := \left\{(\tau, M, \xi) : (\tau, M) \in \mathcal{T} \text{ and } \xi \in \partial\tau\right\}$ is the set of *marked trees with distinguished spines*.

It is natural to speak of the *position of the spine at time t* which we think of as the position of the unique node that is in the spine and alive at time t:

- we define the time-t position of the spine as $\xi_t := X_u(t)$, where $u \in \xi \cap N_t$.

By using the notation ξ_t to refer to both the node in the tree and that node's spatial position we are introducing potential ambiguity. However, in practice the context will usually make clear which we intend, although if this is not the case we shall give the node a longer name:

- $\mathrm{node}_t((\tau, M, \xi)) := u$ if $u \in \xi$ is the node in the spine alive at time t,

which may also be written as $\mathrm{node}_t(\xi)$.

Finally, it will later be important to know how many fission times there have been in the spine, or what is the same, to know which generation of the family tree the node ξ_t is in (where the original ancestor $\emptyset$ is considered to be the 0th generation)

Definition 3.3 *We define the counting function*

$$n_t = \left|\mathrm{node}_t(\xi)\right|,$$

which tells us which generation the spine node is in, or equivalently how many fission times there have been on the spine. For example, if $\xi_t = (\emptyset, u_1, u_2)$ then both $\emptyset$ and u_1 have died and so $n_t = 2$.

3.2 Filtrations

The reader who is already familiar with the Lyons *et al.* [41, 43, 44] papers will recall that they used two separate underlying spaces of marked trees *with* and *without* the spines, then marginalized out the spine when wanting to deal only with the branching particles as a whole. Instead, we are going to use the single underlying space $\tilde{\mathcal{T}}$, but define *four* filtrations of it that will encapsulate different knowledge.

Filtration $(\mathcal{F}_t)_{t \geq 0}$

We define a filtration of $\tilde{\mathcal{T}}$ made up of the σ-algebras:

$$\mathcal{F}_t := \sigma\Big((u, X_u, \sigma_u) : S_u \leq t \ ; \ (u, X_u(s) : s \in [S_u - \sigma_u, t]) : t \in [S_u - \sigma_u, S_u)\Big).$$

Then, $\mathcal{F}_t$ *knows everything that has happened to all the branching particles up to the time t, but does not know which one is the spine.* Each of these σ-algebras will be a subset of the limit defined as

$$\mathcal{F}_\infty := \sigma\left(\bigcup_{t\geq0} \mathcal{F}_t\right).$$

Filtration $(\tilde{\mathcal{F}}_t)_{t\geq0}$

In order to know about the spine, we make this filtration finer, defining $\tilde{\mathcal{F}}_t$ by adding into $\mathcal{F}_t$ the knowledge of which node is the spine at time t:

$$\tilde{\mathcal{F}}_t := \sigma\big(\mathcal{F}_t, \mathrm{node}_t(\xi)\big), \qquad \tilde{\mathcal{F}}_\infty := \sigma\left(\bigcup_{t\geq0} \tilde{\mathcal{F}}_t\right).$$

Consequently, $\tilde{\mathcal{F}}_t$ *knows* everything *about the branching process and* everything *about the spine up to time t*, including which nodes make up the spine, when they were born, when they died (ie. the fission times S_u), and their family sizes.

Filtration $(\mathcal{G}_t)_{t\geq0}$

We define a filtration of $\tilde{\mathcal{T}}$, $\{\mathcal{G}_t\}_{t\geq0}$, which is generated by *only* the spatial motion of the spine by:

$$\mathcal{G}_t := \sigma\big(\xi_s : 0 \leq s \leq t\big), \qquad \mathcal{G}_\infty := \sigma\left(\bigcup_{t\geq0} \mathcal{G}_t\right),$$

Then, $\mathcal{G}_t$ *knows only about the spine's motion in J up to time t*, but does not actually know which line of descent in the family tree makes up the spine or anything about births along the spine.

Filtration $(\tilde{\mathcal{G}}_t)_{t\geq0}$

We augment $\mathcal{G}_t$ by adding in information on the nodes that make up the spine (as we did from $\mathcal{F}_t$ to $\tilde{\mathcal{F}}_t$), as well as the knowledge of when the fission times occurred on the spine and how big the families were that were produced:

$$\tilde{\mathcal{G}}_t := \sigma\big(\mathcal{G}_t, (\mathrm{node}_s(\xi) : s \leq t), (A_u : u < \mathrm{node}_t(\xi))\big), \qquad \tilde{\mathcal{G}}_\infty := \sigma\left(\bigcup_{t\geq0} \tilde{\mathcal{G}}_t\right).$$

Then, $\tilde{\mathcal{G}}_t$ *knows about* everything *along the spine up until time t*.

We note the obvious relationships between these filtrations of $\tilde{\mathcal{T}}$ that $\mathcal{F}_t \subset \tilde{\mathcal{F}}_t$ and $\mathcal{G}_t \subset \tilde{\mathcal{G}}_t \subset \tilde{\mathcal{F}}_t$. Trivially, we also note that $\mathcal{G}_t \not\subset \mathcal{F}_t$, since the filtration $\mathcal{F}_t$ does not know *which* line of descent makes up the spine.

4 Probability Measures

Having now carefully defined the underlying space for our probabilities, we remind ourselves of the probability measures:

Definition 4.1 *For each $x \in J$, let P^x be the measure on $(\tilde{T}, \mathcal{F}_\infty)$ such that the filtered probability space $(\tilde{T}, \mathcal{F}_\infty, (\mathcal{F}_t)_{t \geq 0}, P)$ is the canonical model for $\mathbb{X}_t$, the branching Markov process described in Definition 2.1.*

For details of how the measures P^x are formally constructed on the underlying space of trees, we refer the reader to the work of Neveu [45] and Chauvin [8, 6]. Note, we could equally think of P^x as a measure on $(T, \mathcal{F}_\infty)$, but it is convenient to use the enlarged sample space $\tilde{T}$ for all our measure spaces, varying only the filtrations.

Our spine approach relies first on building a measure $\tilde{P}^x$ under which the spine is a single genealogical line of descent chosen from the underlying tree. If we are given a sample tree (τ, M) for the branching process, it is easy to verify that, if at each fission we make a uniform choice amongst the offspring to decide which line of descent continues the spine ξ, when $u \in \tau$ we have

$$\text{Prob}(u \in \xi) = \prod_{v < u} \frac{1}{1 + A_v}. \tag{7}$$

In the binary-branching case, for example, $\text{Prob}(A_v = 1) = 1$ and then $\text{Prob}(u \in \xi) = 2^{-|u|}$. This simple observation is the key to our method for extending the measures, and for this we make use of the following representation found in Lyons [43].

Theorem 4.2 *If $f \in m\tilde{\mathcal{F}}_t$, that is f is an $\tilde{\mathcal{F}}_t$-measurable function, then we can write:*

$$f = \sum_{u \in N_t} f_u \mathbf{1}_{(\xi_t = u)} \tag{8}$$

where $f_u \in m\mathcal{F}_t$.

As a simple example of this, in the case of the finite-typed branching diffusion of Section 2.1, such a representation would be:

$$e^{\int_0^t R(\eta_s)\,ds} v_\lambda(\eta_t)\, e^{\lambda \xi_t - E_\lambda t} = \sum_{u \in N_t} e^{\int_0^t R(Y_u(s))\,ds} v_\lambda(Y_u(t))\, e^{\lambda X_u(t) - E_\lambda t} \mathbf{1}_{(\xi_t = u)}. \tag{9}$$

Definition 4.3 *Given the measure P^x on $(\tilde{T}, \mathcal{F}_\infty)$ we extend it to the probability measure $\tilde{P}^x$ on $(\tilde{T}, \tilde{\mathcal{F}}_\infty)$ by defining*

$$\int_{\tilde{T}} f\, d\tilde{P}^x := \int_{\tilde{T}} \sum_{u \in N_t} f_u \prod_{v < u} \frac{1}{1 + A_v}\, dP^x, \tag{10}$$

for each $f \in m\tilde{\mathcal{F}}_t$ with representation like (8).

The previous approach to spines, exemplified in Lyons [43], used the idea of *fibres* to get a measure analogous to our $\tilde{P}$ that could measure the spine. However, a perceived weakness in this approach was that the corresponding measure had time-dependent total mass and could not be normalized to become a probability measure with an intuitive construction, unlike our $\tilde{P}$. Our idea of using the down-weighting term of (7) in the definition of $\tilde{P}$ is crucial in ensuring that we get a natural *probability* measure (look ahead to Lemma 4.9), and leads to the very useful situation in which *all* measure changes in our formulation are carried out by *martingales*.

Theorem 4.4 *This measure $\tilde{P}^x$ is an extension of P^x in that $P = \tilde{P}|_{\mathcal{F}_\infty}$.*

Proof: If $f \in m\mathcal{F}_t$ then the representation (8) is trivial and therefore by definition

$$\int_{\tilde{T}} f \, d\tilde{P} = \int_{\tilde{T}} f \times \left(\sum_{u \in N_t} \prod_{v < u} \frac{1}{1 + A_v} \right) dP.$$

However, it can be seen that $\sum_{u \in N_t} \prod_{v < u} \frac{1}{1+A_v} = 1$ by retracing the sum back through the lines of ancestors to the original ancestor $\emptyset$, factoring out the product terms as each generation is passed. Thus

$$\int_{\tilde{T}} f \, d\tilde{P} = \int_{\tilde{T}} f \, dP. \quad \square$$

Definition 4.5 *The filtered probability space $(\tilde{T}, \tilde{\mathcal{F}}_\infty, (\tilde{\mathcal{F}}_t)_{t \geq 0}, \tilde{P})$ with $(\mathbb{X}_t, \xi_t)$ will be referred to as the **canonical model with spines**.*

In the single-particle model of section 2.1 we assumed the existence of a separate measure $\mathbb{P}$ and a process (ξ_t, η_t) that behaved stochastically like a 'typical' particle in the typed branching diffusion $\mathbb{X}_t$. In our formalization the *spine* is exactly the single-particle model:

Definition 4.6 *We define the measure $\mathbb{P}$ on $(\tilde{T}, \mathcal{G}_\infty)$ as the restriction of $\tilde{P}$:*

$$\mathbb{P}|_{\mathcal{G}_t} := \tilde{P}|_{\mathcal{G}_t}.$$

Under the measure $\mathbb{P}$ the spine process ξ_t has exactly the same law as Ξ_t.

Definition 4.7 *The filtered probability space $(\tilde{T}, \mathcal{G}_\infty, (\mathcal{G}_t)_{t \geq 0}, \mathbb{P})$ together with the spine process ξ_t will be referred to as the **single-particle model**.*

4.1 An Intuitive Construction of $\tilde{P}$

As the name suggests, we should be able to think of the spine as the backbone of the branching process. This is made precise by the following decomposition:

Theorem 4.8 *The measure $\tilde{P}$ on $\tilde{\mathcal{F}}_t$ can be decomposed as:*

$$d\tilde{P}(\tau, M, \xi) = dP(\xi)d\mathbb{L}^{(R(\xi))}(n)\left(\prod_{v<\xi_t}\frac{1}{1+A_v}\right)\left(\prod_{v<\xi_t}p_{A_v}(\xi_{S_v})\prod_{j=1}^{A_v}dP((\tau, M)_j^v)\right)$$
(11)

where $\mathbb{L}^{(R(\xi))}$ is the law of the Poisson (Cox) process with rate $R(\xi_t)$ at time t, and we recall that $n = (n_t : t \geq 0)$ is the counting process of fission times along the spine.

We can summarise a clear intuitive picture of this decomposition in the following lemma:

Lemma 4.9 *The decomposition of measure $\tilde{P}$ at (11) enables the following construction:*
- *the spine's motion is determined by the single-particle measure $\mathbb{P}$;*
- *the spine undergoes fission at time t at rate $R(\xi_t)$;*
- *at the fission time of node v on the spine, the single spine particle is replaced by $1 + A_v$ children, with A_v being chosen independently and distributed according to the location-dependent random variable $A(\xi_{S_v})$ with probabilities $(p_k(\xi_{S_v}) : k = 0, 1, \ldots)$;*
- *the spine is chosen uniformly from the $1+A_v$ children at the fission point v;*
- *each of the remaining A_v children gives rise to the independent subtrees $(\tau, M)_j^v$, for $1 \leq j \leq A_v$, which are not part of the spine and which are each determined by an independent copy of the original measure P shifted to their point and time of creation.*

5 Martingales

Starting with the single Markov process Ξ_t that lives in $(J, \mathcal{B})$ we have built $(\mathbb{X}_t, \xi_t)$, a branching Markov process with spines, in which the spine ξ_t behaves stochastically like the given Ξ_t. In this section we are going to show how *any* given martingale for the spine leads to a corresponding additive martingale for the whole branching model.

We have already seen an example of this for the finite-type model of section 2.1, when we introduced the two martingales:

$$Z_\lambda(t) := \sum_{u \in N_t} v_\lambda(Y_u(t))e^{\lambda X_u(t) - E_\lambda t}, \qquad \zeta_\lambda(t) := e^{\int_0^t R(\eta_s)\,ds}v_\lambda(\eta_t)e^{\lambda\xi_t - E_\lambda t}.$$

Just from their very form it has always been clear that they are closely related. What we shall later be demonstrating in full generality in this section is that the key to their relationship comes through generalising the following $\tilde{\mathcal{F}}_t$-measurable martingale for the multi-type BBM model:

Definition 5.1 *We define an $\tilde{\mathcal{F}}_t$-measurable martingale:*

$$\tilde{\zeta}_\lambda(t) := \prod_{u < \xi_t} (1 + A_u) \times v_\lambda(\eta_t) e^{\lambda \xi_t - E_\lambda t}. \tag{12}$$

An important result that we show in this article (Lemma 5.7) is that $Z_\lambda(t)$ and $\zeta_\lambda(t)$ are simply conditional expectations of this new martingale $\tilde{\zeta}_\lambda$. We emphasize that this relationship is only *possible* because of the construction of $\tilde{P}$ as a *probability* measure and using filtrations to capture the different knowledge generated by the spine and the branching particles. This idea of projection is also used in random fragmentation theory where it corresponds to the notion of tagged fragment, see Bertoin [3], for example.

Furthermore, in the general form that we present below it provides a consistent methodology for using well-known martingales for a single process ξ_t to get new additive martingales for the related branching process. In Hardy and Harris [23, 22] we use these powerful ideas to give substantially easier proofs of large-deviations problems in branching diffusions than have previously been possible.

Suppose that $\zeta(t)$ is a strictly positive $(\tilde{T}, (\mathcal{G}_t)_{t \geq 0}, \tilde{P})$-martingale, which is to say that it is a $\mathcal{G}_t$-measurable function that is a martingale with respect to the measure $\tilde{P}$. For example, in the case of our finite-type branching diffusion this could be the martingale $\zeta_\lambda(t)$ which is $\mathcal{G}_t$-measurable since it refers only to the spine process (ξ_t, η_t).

Definition 5.2 *We shall call $\zeta(t)$ a **single-particle martingale**, since it is $\mathcal{G}_t$-measurable and thus depends only to the spine ξ.*

Any such single-particle martingale can be used to define an additive martingale for the whole branching process via the representation (8):

Definition 5.3 *Suppose that we can represent the martingale $\zeta(t)$ as*

$$\zeta(t) = \sum_{u \in N_t} \zeta_u(t) \mathbf{1}_{(\xi_t = u)}, \tag{13}$$

for $\zeta_u(t) \in m\mathcal{F}_t$, as at (8). We can then define an $\mathcal{F}_t$-measurable process $Z(t)$ as

$$Z(t) := \sum_{u \in N_t} e^{-\int_0^t m(X_u(s)) R(X_u(s)) \, ds} \zeta_u(t),$$

*and refer to $Z(t)$ as the **branching-particle martingale**.*

The martingale property $Z(t)$ will be established in Lemma 5.7 after first building another martingale, $\tilde{\zeta}(t)$, from the single-particle martingale $\zeta(t)$. First, for clarity, we take a moment to discuss this definition of the additive martingale and the terms like $\zeta_u(t)$.

If we return to our familiar martingales (5) and (6), it is clear that

$$\zeta_\lambda(t) = e^{\int_0^t R(\eta_s)\,ds} v_\lambda(\eta_t) e^{\lambda\xi_t - E_\lambda t} = \sum_{u \in N_t} e^{\int_0^t R(Y_u(s))\,ds} v_\lambda(Y_u(t)) e^{\lambda X_u(t) - E_\lambda t} \mathbf{1}_{(\xi_t = u)}.$$

(14)

The 'ζ_u' terms of (13) could be here replaced with a more descriptive notation $\zeta_\lambda[(X_u, Y_u)](t)$, where

$$\zeta_u(t) = \zeta_\lambda[(X_u, Y_u)](t) := e^{\int_0^t R(Y_u(s))\,ds} v_\lambda(Y_u(t)) e^{\lambda X_u(t) - E_\lambda t},$$

can be seen to essentially be a functional of the space-type path $(X_u(t), Y_u(t))$ of particle u. In this way the original single-particle martingale ζ_λ would be understood as a functional of the space-type path (ξ_t, η_t) of the spine itself and we could write

$$\zeta_\lambda(t) = \zeta_\lambda[(\xi, \eta)](t) = \sum_{u \in N_t} \zeta_\lambda[(X_u, Y_u)](t) \mathbf{1}_{(\xi_t = u)}.$$

This is the idea behind the representation (13), and in those typical cases where the single-particle martingale is essentially a functional of the paths of the spine ξ_t, as is the case for our $\zeta_\lambda(t)$, we should just think of ζ_u as being that same functional but evaluated over the path $X_u(t)$ of particle u rather than the spine ξ_t. The representation (13) can also be used as a more general way of treating other martingales that perhaps are not such a simple functional of the spine path.

Finally, from (14) it is clear that the additive martingale being defined by definition 5.3 is our familiar $Z_\lambda(t)$:

$$Z_\lambda(t) = \sum_{u \in N_t} e^{-\int_0^t R(Y_u(s))\,ds} \zeta_\lambda[(X_u, Y_u)](t) = \sum_{u \in N_t} v_\lambda(Y_u(t)) e^{\lambda X_u(t) - E_\lambda t}.$$

Although definition 5.3 will work in general, in the main the spine approach is interested in martingales that can act as Radon-Nikodym derivatives between probability measures, and therefore we suppose from now on that $\zeta(t)$ is *strictly positive*, and therefore that the additive martingale $Z(t)$ is strictly positive.

The work of Lyons *et al.* [43, 41, 44], that of Chauvin and Rouault [9] and more recently of Kyprianou [40] suggests that when a change of measure is carried out with a branching-diffusion additive martingale like $Z(t)$ it is typical to expect three changes: the spine will gain a drift, its fission times will be increased and the distribution of its family sizes will be size-biased. In section 6.1 we shall confirm this, but we first take a separate look at the martingales that could perform these changes, and which we shall combine to obtain a martingale $\tilde{\zeta}(t)$ that will ultimately be used to change the measure $\tilde{P}$.

Theorem 5.4 *The expression*

$$\prod_{v<\xi_t} \left(1+m(\xi_{S_v})\right) e^{-\int_0^t m(\xi_s)R(\xi_s)\,\mathrm{d}s}$$

is a $\tilde{P}$-martingale that will increase the rate at which fission times occur along the spine from $R(\xi_t)$ to $(1+m(\xi_t))R(\xi_t)$:

$$\frac{\mathrm{d}\mathbb{L}_t^{((1+m(\xi))R(\xi))}}{\mathrm{d}\mathbb{L}_t^{(R(\xi))}} = \prod_{v<\xi_t} \left(1+m(\xi_{S_v})\right) e^{-\int_0^t m(\xi_s)R(\xi_s)\,\mathrm{d}s}$$

where $\mathbb{L}^{(R(\xi))}$ is the law of the Poisson (Cox) process with rate $R(\xi_t)$ at time t.

Theorem 5.5 *The term*

$$\prod_{v<\xi_t} \frac{1+A_v}{1+m(\xi_{S_v})}$$

is a $\tilde{P}$-martingale that will change the measure by size-biasing the family sizes born from the spine:

$$\text{if } v < \xi_t, \quad \text{then} \quad \text{Prob}(A_v = k) = \frac{(1+k)p_k(\xi_{S_v})}{1+m(\xi_{S_v})}.$$

The proof of these two results is left as an easy exercise for the reader. The product of these two martingales with the single-particle martingale $\zeta(t)$ will simultaneously perform the three changes mentioned above:

Definition 5.6 *We define a $\tilde{\mathcal{F}}_t$-measurable martingale as*

$$\tilde{\zeta}(t) := \prod_{v<\xi_t} (1+A_v) e^{-\int_0^t m(\xi_s)R(\xi_s)\,\mathrm{d}s} \times \zeta(t)$$

$$= \prod_{u<\xi_t} \frac{1+A_u}{1+m(\xi_{S_u})} \times \prod_{v<\xi_t} \left(1+m(\xi_{S_v})\right) e^{-\int_0^t m(\xi_s)R(\xi_s)\,\mathrm{d}s} \times \zeta(t). \quad (15)$$

Significantly, *only* the motion of the spine and the behaviour along the immediate path of the spine will be affected by any change of measure using this martingale. Also note, this martingale is the general form of $\tilde{\zeta}_\lambda(t)$ that we defined at (12) for our finite-type model.

The real importance of the size-biasing and fission-time-increase operations is that they introduce the correct terms into $\tilde{\zeta}(t)$ so that the following key relationships hold:

Lemma 5.7 *Both $Z(t)$ and $\zeta(t)$ are projections of $\tilde{\zeta}(t)$ onto their filtrations: for all $t \geq 0$,*

- $Z(t) = \tilde{P}\big(\tilde{\zeta}(t)\,|\,\mathcal{F}_t\big),$
- $\zeta(t) = \tilde{P}\big(\tilde{\zeta}(t)\,|\,\mathcal{G}_t\big).$

Proof: We use the representation (8) of $\tilde{\zeta}(t)$:

$$\tilde{\zeta}(t) = \sum_{u \in N_t} \prod_{v<u} (1 + A_v) e^{-\int_0^t m(X_u(s))R(X_u(s))\,ds} \zeta_u(t) \mathbf{1}_{(\xi_t=u)}. \qquad (16)$$

Since $\tilde{P}\big(\mathbf{1}_{(\xi_t=u)}|\mathcal{F}_t\big) = \mathbf{1}_{(u \in N_t)} \times \prod_{v<u}(1+A_v)^{-1}$, it follows that

$$\tilde{P}\big(\tilde{\zeta}(t)|\mathcal{F}_t\big) = \sum_{u \in N_t} e^{-\int_0^t m(X_u(s))R(X_u(s))\,ds} \zeta_u(t) \times \prod_{v<u}(1+A_v)\,\tilde{P}\big(\mathbf{1}_{(\xi_t=u)}|\mathcal{F}_t\big)$$

$$= \sum_{u \in N_t} e^{-\int_0^t m(X_u(s))R(X_u(s))\,ds} \zeta_u(t) = Z(t).$$

On the other hand, the martingale terms in (15) imply

$$\tilde{P}\big(\tilde{\zeta}(t)|\mathcal{G}_t\big) = \zeta(t) \times \tilde{P}\Big(\prod_{v<\xi_t}(1+A_v)e^{-\int_0^t m(\xi_s)R(\xi_s)\,ds}\Big|\mathcal{G}_t\Big) = \zeta(t).$$

$\square$

6 Changing the Measures

For the finite type model, the single-particle martingale $\zeta_\lambda(t)$ defined at (6) can be used to define a new measure for the single-particle model (as in [21]), via

$$\frac{d\mathbb{P}_\lambda}{d\mathbb{P}}\Big|_{\mathcal{G}_t} = \frac{\zeta_\lambda(t)}{\zeta(0)}.$$

We have now seen the close relationships between the three martingales ζ_λ, Z_λ and $\tilde{\zeta}_\lambda$:

$$Z_\lambda(t) = \tilde{P}\big(\tilde{\zeta}_\lambda(t)\,|\,\mathcal{F}_t\big), \qquad \zeta_\lambda(t) = \tilde{P}\big(\tilde{\zeta}_\lambda(t)\,|\,\mathcal{G}_t\big),$$

and in this section we show in a more general form how these close relationships mean that a new measure $\tilde{\mathbb{Q}}_\lambda$ defined in terms of $\tilde{P}$ as

$$\frac{d\tilde{\mathbb{Q}}_\lambda}{d\tilde{P}}\Big|_{\tilde{\mathcal{F}}_t} = \frac{\tilde{\zeta}_\lambda(t)}{\tilde{\zeta}_\lambda(0)},$$

will induce measure changes on the sub-filtrations $\mathcal{G}_t$ and $\mathcal{F}_t$ of $\tilde{\mathcal{F}}_t$ whose Radon-Nikodym derivatives are given by $\zeta_\lambda(t)$ and $Z_\lambda(t)$ respectively. We will also give a useful intuitive construction of the measures $\tilde{P}$ and $\tilde{\mathbb{Q}}$.

Definition 6.1 *A measure $\tilde{\mathbb{Q}}$ on $(\tilde{T}, \tilde{\mathcal{F}}_\infty)$ is defined via its Radon-Nikodym derivative with respect to $\tilde{P}$:*

$$\frac{d\tilde{\mathbb{Q}}}{d\tilde{P}}\Big|_{\tilde{\mathcal{F}}_t} = \frac{\tilde{\zeta}(t)}{\tilde{\zeta}(0)}.$$

As we did for the measures P and $\mathbb{P}$ in Section 4, we can restrict $\tilde{Q}$ to the sub-filtrations:

Definition 6.2 *We define the measure Q on $(\tilde{T}, \mathcal{F}_\infty, (\mathcal{F}_t)_{t\geq 0})$ via*

$$Q := \tilde{Q}|_{\mathcal{F}_\infty}.$$

Definition 6.3 *We define the measure $\hat{\mathbb{P}}$ on $(\tilde{T}, \mathcal{G}_\infty, (\mathcal{G}_t)_{t\geq 0})$ via*

$$\hat{\mathbb{P}} := \tilde{Q}|_{\mathcal{G}_\infty}.$$

A consequence of our new formulation in terms of filtrations and the equalities of Lemma 5.7 is that the changes of measure are carried out by $Z(t)$ and $\zeta(t)$ on their subfiltrations:

Theorem 6.4

$$\left.\frac{dQ}{dP}\right|_{\mathcal{F}_t} = \frac{Z(t)}{Z(0)}, \quad and \quad \left.\frac{d\hat{\mathbb{P}}}{d\mathbb{P}}\right|_{\mathcal{G}_t} = \frac{\zeta(t)}{\zeta(0)}.$$

Proof: These two results actually follow from a more general observation that if $\tilde{\mu}_1$ and $\tilde{\mu}_2$ are two measures defined on a measure space $(\Omega, \tilde{S})$ with Radon-Nikodym derivative

$$\frac{d\tilde{\mu}_2}{d\tilde{\mu}_1} = f,$$

and if S is a sub-σ-algebra of $\tilde{S}$, then the two measures $\mu_1 := \tilde{\mu}_1|_S$ and $\mu_2 := \tilde{\mu}_2|_S$ on (Ω, S) are related by the conditional expectation operation:

$$\frac{d\mu_2}{d\mu_1} = \tilde{\mu}_1(f|S).$$

Applying this general result and using the relationships between the general martingales given in Lemma 5.7 concludes the proof. □

6.1 Understanding the Measure $\tilde{Q}$

This decomposition of $\tilde{P}_t$ given at (11) will allow us to interpret the measure $\tilde{Q}$ if we appropriately factor the components of the change-of-measure martingale $\tilde{\zeta}(t)$ across this representation. On $\tilde{\mathcal{F}}_t$,

$$d\tilde{Q} = \tilde{\zeta}(t)\,d\tilde{P}$$

$$= \tilde{\zeta}(t) \times e^{-\int_0^t R(\xi_s)ds} \prod_{u<\xi_t}\left(1+m(\xi_{S_u})\right) \times \prod_{v<\xi_t}\frac{1+A_v}{1+m(\xi_{S_v})} \times d\tilde{P}$$

$$= d\hat{\mathbb{P}}(\xi)\,d\mathbb{L}^{((1+m(\xi))R(\xi))}(n)$$

$$\times \prod_{u<\xi_t}\frac{1}{1+A_u} \prod_{v<\xi_t}\frac{1+A_v}{1+m(\xi_{S_v})} p_{A_v}(\xi_{S_v}) \prod_{j=1}^{A_v} dP((\tau, M)_j^v). \quad (17)$$

Just as we did for $\tilde{P}$, we can offer a clear interpretation of this decomposition:

Lemma 6.5 *Under the measure* $\tilde{\mathbb{Q}}$,

- *the spine process* ξ_t *moves as if under the changed measure* $\hat{\mathbb{P}}$;
- *the fission times along the spine occur at an accelerated rate* $(1 + m(\xi_t))R(\xi_t)$;
- *at the fission time of node* v *on the spine, the single spine particle is replaced by* $1 + A_v$ *children, with* A_v *being chosen as an independent copy of the random variable* $\tilde{A}(y)$ *which has the size biased offspring distribution* $((1 + k)p_k(y)/(1 + m(y)) : k = 0, 1, \ldots)$, *where* $y = \xi_{S_v} \in J$ *is the spine's location at the time of fission;*
- *the spine is chosen uniformly from the* $1 + A_v$ *particles at the fission point* v;
- *each of the remaining* A_v *children gives rise to the independent subtrees* $(\tau, M)_j^v$, *for* $1 \le j \le A_v$, *which are not part of the spine and evolve as independent processes determined by the measure* P *shifted to their point and time of creation.*

Such an interpretation of the measure $\tilde{\mathbb{Q}}$ was first given by Chauvin and Rouault [9] in the context of BBM, allowing them to come to the important conclusion that under the new measure $\mathbb{Q}$ the branching diffusion remains largely unaffected, except that the Brownian particles of a single (random) line of descent in the family tree are given a changed motion, with an accelerated birth rate – although they did not have random family sizes, so the size-biasing aspect was not seen. Size-biasing has been known for a long time in the study of branching populations, and in the context of spines, it was introduced in the Lyons *et al.* papers [43, 41, 44]. Kyprianou [40] presented the decomposition of equation (17) and the construction of $\mathbb{Q}$ at Lemma 6.5 for BBM with random family sizes, but did not follow our natural approach of starting with the *probability* measure $\tilde{P}$.

7 The Spine Decomposition

One of the most important results introduced in Lyons [43] was the so-called *spine decomposition*, which in the case of the additive martingale

$$Z_\lambda(t) = \sum_{u \in N_t} v_\lambda(Y_u(t))e^{\lambda X_u(t) - E_\lambda t},$$

from the finite-type branching diffusion would be:

$$\tilde{\mathbb{Q}}_\lambda\big(Z_\lambda(t)|\tilde{\mathcal{G}}_\infty\big) = \sum_{u < N_t} v_\lambda(\eta_{S_u})e^{\lambda \xi_{S_u} - E_\lambda S_u} + v_\lambda(\eta_t)e^{\lambda \xi_t - E_\lambda t}. \tag{18}$$

To prove this we start by decomposing the martingale as

$$Z_\lambda(t) = \sum_{u \in N_t, u \notin \xi} v_\lambda(Y_u(t))e^{\lambda X_u(t) - E_\lambda t} + v_\lambda(\eta_t)e^{\lambda \xi_t - E_\lambda t},$$

which is clearly true since one of the particles $u \in N_t$ must be in the line of descent that makes up the spine ξ. Recalling that the σ-algebra $\tilde{\mathcal{G}}_\infty$ contains all information about the line of nodes that makes up the spine, all about the spine diffusion (ξ_t, η_t) for all times t, and also contains all information regarding the fission times and number of offspring along the spine, it is useful to partition the particles $v \in \{u \in N_t, u \notin \xi\}$ into the distinct subtrees $(\tau, M)^u$ that were born at the fission times S_u from the particles that made up the spine before time t, or in other words those nodes in the $\{u < \xi_t\}$ of ancestors of the current spine node ξ_t. Thus:

$$Z_\lambda(t) = \sum_{u < \xi_t} e^{\lambda \xi_{S_u} - E_\lambda S_u} \left\{ \sum_{v \in N_t, v \in (\tau, M)^u} v_\lambda(Y_v(t)) e^{\lambda(X_u(t) - \xi_{S_u}) - E_\lambda(t - S_u)} \right\}$$
$$+ v_\lambda(\eta_t) e^{\lambda \xi_t - E_\lambda t}.$$

If we now take the $\tilde{\mathbb{Q}}_\lambda$-conditional expectation of this, we find

$$\tilde{\mathbb{Q}}_\lambda(Z_\lambda(t) | \tilde{\mathcal{G}}_\infty)$$
$$= \sum_{u < \xi_t} e^{\lambda \xi_{S_u} - E_\lambda S_u} \tilde{\mathbb{Q}}_\lambda \left(\sum_{v \in N_t, v \in (\tau, M)^u} v_\lambda(Y_v(t)) e^{\lambda(X_u(t) - \xi_{S_u}) - E_\lambda(t - S_u)} \Big| \tilde{\mathcal{G}}_\infty \right)$$
$$+ v_\lambda(\eta_t) e^{\lambda \xi_t - E_\lambda t}.$$

We know from the decomposition (17) that the under the measure $\tilde{\mathbb{Q}}_\lambda$ the subtrees coming off the spine evolve as if under the measure P, and therefore

$$\tilde{\mathbb{Q}}_\lambda \left(\sum_{v \in N_t, v \in (\tau, M)^u} v_\lambda(Y_v(t)) e^{\lambda(X_u(t) - \xi_{S_u}) - E_\lambda(t - S_u)} \Big| \tilde{\mathcal{G}}_\infty \right)$$
$$= \tilde{P} \left(\sum_{v \in N_t, v \in (\tau, M)^u} v_\lambda(Y_v(t)) e^{\lambda(X_u(t) - \xi_{S_u}) - E_\lambda(t - S_u)} \Big| \tilde{\mathcal{G}}_\infty \right) = v_\lambda(\eta_{S_u}),$$

since the additive expression being evaluated on the subtree is just a shifted form of the martingale Z_λ itself.

This concludes the proof of (18), but before we go move on to give a similar proof for the general case, for easier reference through the cumbersome-looking general proof it is worth recalling that

$$\zeta_\lambda(t) = e^{\int_0^t R(\eta_s) \, ds} v_\lambda(\eta_t) e^{\lambda \xi_t - E_\lambda t},$$

and therefore noting that (18) can alternatively be written as

$$\tilde{\mathbb{Q}}_\lambda(Z_\lambda(t) | \tilde{\mathcal{G}}_\infty) = \sum_{u < N_t} e^{-\int_0^{S_u} R(\eta_s) \, ds} \zeta_\lambda(S_u) + e^{-\int_0^t R(\eta_s) \, ds} \zeta_\lambda(t).$$

Also, in the general model we are supposing that each particle u in the spine will give birth to a total of A_u subtrees that go off from the spine – the one remaining other offspring is used to continue the line of descent that makes up the spine. This explains the appearance of A_u in the general decomposition.

Theorem 7.1 (Spine decomposition) *We have the following **spine decomposition** for the additive branching-particle martingale:*

$$\tilde{Q}\big(Z(t)|\tilde{\mathcal{G}}_\infty\big) = \sum_{u<\xi_t} A_u\, e^{-\int_0^{S_u} m(\xi_s)R(\xi_s)\,ds}\zeta(S_u) + e^{-\int_0^t m(\xi_s)R(\xi_s)\,ds}\zeta(t).$$

Proof: In each sample tree one and only one of the particles alive at time t is the spine and therefore:

$$Z(t) = \sum_{u\in N_t} e^{-\int_0^t m(X_u(s))R(X_u(s))\,ds}\zeta_u(t),$$

$$= e^{-\int_0^t m(\xi_s)R(\xi_s)\,ds}\zeta(t) + \sum_{u\in N_t,\,u\neq\xi_t} e^{-\int_0^t m(X_u(s))R(X_u(s))\,ds}\zeta_u(t).$$

The other individuals $\{u \in N_t, u \neq \xi_t\}$ can be partitioned into subtrees created from fissions along the spine. That is, each node u in the spine ξ_t (so $u < \xi_t$) has given birth at time S_u to one offspring node uj (for some $1 \leq j \leq 1+A_u$) that was chosen to continue the spine whilst the other A_u individuals go off to make the subtrees $(\tau, M)_j^u$. Therefore,

$$Z(t) = e^{-\int_0^t m(\xi_s)R(\xi_s)\,ds}\zeta(t) + \sum_{u<\xi_t} e^{-\int_0^{S_u} m(\xi_s)R(\xi_s)\,ds} \sum_{\substack{j=1,\ldots,1+A_u\\ uj\notin\xi}} Z_{uj}(S_u;t),$$

$$\tag{19}$$

where for $t \geq S_u$,

$$Z_{uj}(S_u;t) := \sum_{v\in N_t,\,v\in(\tau,M)_j^u} e^{-\int_{S_u}^t m(X_v(s))R(X_v(s))\,ds}\zeta_v(t),$$

is, conditional on $\tilde{\mathcal{G}}_\infty$, a $\tilde{P}$-martingale on the subtree $(\tau, M)_j^u$, and therefore

$$\tilde{P}\big(Z_{uj}(S_u;t)|\tilde{\mathcal{G}}_\infty\big) = \zeta(S_u).$$

Thus taking $\tilde{Q}$-conditional expectations of (19) gives

$$\tilde{Q}\big(Z(t)|\tilde{\mathcal{G}}_\infty\big) = e^{-\int_0^t m(\xi_s)R(\xi_s)\,ds}\zeta(t)$$

$$+ \sum_{u<\xi_t} e^{-\int_0^{S_u} m(\xi_s)R(\xi_s)\,ds}\tilde{P}\bigg(\sum_{\substack{j=1,\ldots,1+A_u\\ uj\notin\xi}} Z_{uj}(S_u;t)\,\Big|\,\tilde{\mathcal{G}}_\infty\bigg)$$

$$= e^{-\int_0^t m(\xi_s)R(\xi_s)\,ds}\zeta(t) + \sum_{u<\xi_t} e^{-\int_0^{S_u} m(\xi_s)R(\xi_s)\,ds} A_u\,\zeta(S_u),$$

which completes the proof. □

This representation was first used in the Lyons *et al.* [43, 41, 44] papers and has become the standard way to investigate the behaviour of Z under the measure $\tilde{\mathbb{Q}}$. We also observe that the two measures $\tilde{P}$ and $\tilde{\mathbb{Q}}$ for the general model are equal when conditioned on $\tilde{\mathcal{G}}_\infty$ since this factors out their differences in the spine diffusion ξ_t, the family sizes born from the spine and the fission times on the spine. That is, $\tilde{P}\big(Z(t)|\tilde{\mathcal{G}}_\infty\big) = \tilde{\mathbb{Q}}\big(Z(t)|\tilde{\mathcal{G}}_\infty\big)$.

8 Spine Results

Having covered the formal basis for our spine approach, we now present some results that follow from our spine formulation: the Gibbs-Boltzmann weights, conditional expectations, and a simpler proof of the improved Many-to-One theorem.

8.1 The *Gibbs-Boltzmann* Weights of $\tilde{\mathbb{Q}}$

The Gibbs-Boltzmann weightings in branching processes are well-known, for example see Chauvin and Rouault [7] where they consider random measures on the boundary of the tree, and Harris [30] which gives convergence results for Gibbs-Boltzmann random measures. They have previously been considered via the individual terms of the additive martingale Z, but the following theorem gives a new interpretation of these weightings in terms of the spine. We recall that

$$Z(t) = \sum_{u \in N_t} e^{-\int_0^t m(X_u(s))R(X_u(s))\,ds} \zeta_u(t).$$

Theorem 8.1 *Let $u \in \Omega$ be a given and fixed label. Then*

$$\tilde{\mathbb{Q}}\big(\xi_t = u | \mathcal{F}_t\big) = \mathbf{1}_{(u \in N_t)} \frac{e^{-\int_0^t m(X_u(s))R(X_u(s))\,ds} \zeta_u(t)}{Z(t)}.$$

Proof: Suppose $F \in \mathcal{F}_t$. We aim to show:

$$\int_F \mathbf{1}_{(\xi_t = u)}\, d\tilde{\mathbb{Q}}(\tau, M, \xi) = \int_F \mathbf{1}_{(u \in N_t)} \frac{e^{-\int_0^t m(X_u(s))R(X_u(s))\,ds} \zeta_u(t)}{Z(t)}\, d\tilde{\mathbb{Q}}(\tau, M, \xi).$$

First of all we know that $d\tilde{\mathbb{Q}}/d\tilde{P} = \tilde{\zeta}(t)$ on $\mathcal{F}_t$ and therefore,

$$\text{LHS} = \int_F \mathbf{1}_{(\xi_t = u)} \prod_{v < \xi_t}(1 + A_v)e^{-\int_0^t m(\xi_s)R(\xi_s)\,ds}\zeta(t)\, d\tilde{P}(\tau, M, \xi),$$

by definition of $\tilde{\zeta}(t)$ at (15). The definition 4.3 of the measure $\tilde{P}$ requires us to express the integrand with a representation like (8):

$$\mathbf{1}_{(\xi_t=u)} \prod_{v<\xi_t} (1+A_v) e^{-\int_0^t m(\xi_s)R(\xi_s)\,ds} \zeta(t)$$

$$= \mathbf{1}_{(\xi_t=u)} \mathbf{1}_{(u\in N_t)} \prod_{v<u} (1+A_v) e^{-\int_0^t m(X_u(s))R(X_u(s))ds} \zeta_u(t),$$

and therefore

$$\text{LHS} = \int_F \mathbf{1}_{(u\in N_t)} \prod_{v<u} (1+A_v) e^{-\int_0^t m(X_u(s))R(X_u(s))\,ds} \zeta_u(t) \mathbf{1}_{(\xi_t=u)}\ d\tilde{P}(\tau,M,\xi),$$

$$= \int_F \mathbf{1}_{(u\in N_t)} e^{-\int_0^t m(X_u(s))R(X_u(s))\,ds} \zeta_u(t)\ dP(\tau,M,\xi),$$

by definition 4.3. Since $Z(t) > 0$ a.s., we know that on $\mathcal{F}_t$, $dP/d\mathbb{Q} = 1/Z(t)$, so

$$\text{LHS} = \int_F \mathbf{1}_{(u\in N_t)} e^{-\int_0^t m(X_u(s))R(X_u(s))\,ds} \zeta_u(t)\ \frac{1}{Z(t)} d\mathbb{Q}(\tau,M,\xi),$$

and the proof is concluded.   □

The above result combines with the representation (8) to show how we take conditional expectations under the measure $\tilde{\mathbb{Q}}$.

Theorem 8.2 *If $f(t) \in m\tilde{\mathcal{F}}_t$, and $f = \sum_{u\in N_t} f_u(t)\mathbf{1}_{(\xi_t=u)}$, with $f_u(t) \in m\mathcal{F}_t$ then*

$$\tilde{\mathbb{Q}}(f(t)|\mathcal{F}_t) = \sum_{u\in N_t} f_u(t)\frac{e^{-\int_0^t m(X_u(s))R(X_u(s))\,ds}\zeta_u(t)}{Z(t)}. \tag{20}$$

Proof: It is clear that

$$\tilde{\mathbb{Q}}(f(t)|\mathcal{F}_t) = \sum_{u\in N_t} f_u(t)\tilde{\mathbb{Q}}(\xi_t = u|\mathcal{F}_t),$$

and the result follows from Theorem 8.1.   □

A corollary to this useful result also appears to go a long way towards obtaining the Kesten-Stigum result in more general models:

Corollary 8.3 *If $g(\cdot)$ is a Borel function on J then*

$$\sum_{u\in N_t} g(X_u(t))\, e^{-\int_0^t m(X_u(s))R(X_u(s))\,ds}\zeta_u(t) = \tilde{\mathbb{Q}}(g(\xi_t)|\mathcal{F}_t) \times Z(t). \tag{21}$$

Proof: We can write $g(\xi_t) = \sum_{u\in N_t} g(X_u(t))\mathbf{1}(\xi_t = u)$, and now the result follows from the above corollary.   □

The classical Kesten-Stigum theorems of [37, 36, 38] for multi-dimensional Galton-Watson processes give conditions under which an operation like the left-hand side of (21) converges as $t \to \infty$, and it is found that when it exists the limit is a multiple of the martingale limit $Z(\infty)$. Also see Lyons *et al.* [41] for a more recent proof of this based on other spine techniques. Our spine formulation apparently gives a previously unknown but simple meaning to this operation in terms of a conditional expectation and, as we hope to pursue in further work, in many cases we would intuitively expect that $\tilde{\mathbb{Q}}(g(\xi_t)|\mathcal{F}_t)/\tilde{\mathbb{Q}}(g(\xi_t)) \to 1$ a.s., leading to alternative spine proofs of both Kesten-Stigum like theorems and Watanabe's theorem in the case of BBM.

8.2 The *Full* Many-to-One Theorem

A very useful tool in the study of branching processes is the *Many-to-One* result that enables expectations of sums over particles in the branching process to be calculated in terms of an expectation of a single particle. In the context of the finite-type branching diffusion of section 2.1, the Many-to-One theorem would be stated as follows:

Theorem 8.4 *For any measurable function* $f : J \to \mathbb{R}$ *we have*

$$P^{x,y}\left(\sum_{u \in N_t} f(X_u(t), Y_u(t))\right) = \mathbb{P}^{x,y}\left(e^{\int_0^t R(\eta_s)\,\mathrm{d}s} f(\xi_t, \eta_t)\right).$$

Intuitively it is clear that the up-weighting term $e^{\int_0^t R(\eta_s)\,\mathrm{d}s}$ incorporates the notion of the population growing at an exponential rate, whilst the idea of $f(\xi_t, \eta_t)$ being the 'typical' behaviour of $f(X_u(t), Y_u(t))$ is also reasonable.

Existing results tend to apply only to functions of the above form that depend only on *the time-t location* of the spine and existing proofs do not lend themselves to covering functions that depend on the entire *path history* of the spine up to time t.

With the spine approach we have the benefit of being able to give a much less complicated proof of the stronger version that covers the most general path-dependent functions.

Theorem 8.5 (Many-to-One) *If* $f(t) \in m\tilde{\mathcal{F}}_t$ *has the representation*

$$f(t) = \sum_{u \in N_t} f_u(t)\mathbf{1}_{(\xi_t=u)},$$

where $f_u(t) \in m\mathcal{F}_t$, *then*

$$P\left(\sum_{u \in N_t} f_u(t)e^{-\int_0^t m(X_u(s))R(X_u(s))\,\mathrm{d}s}\zeta_u(t)\right) = \tilde{P}\big(f(t)\,\tilde{\zeta}(t)\big) = \zeta(0)\,\tilde{\mathbb{Q}}\big(f(t)\big).$$

$$(22)$$

In particular, if $g(t) \in m\mathcal{G}_t$ *with* $g(t) = \sum_{u \in N_t} g_u(t)\mathbf{1}_{(\xi_t=u)}$ *where* $g_u(t) \in m\mathcal{F}_t$, *then*

$$P\left(\sum_{u \in N_t} g_u(t)\right) = \mathbb{P}\left(e^{\int_0^t m(\xi_s)R(\xi_s)\mathrm{d}s} g(t)\right) = \hat{\mathbb{P}}\left(\frac{g(t)\,\zeta(0)}{e^{-\int_0^t m(\xi_s)R(\xi_s)\mathrm{d}s}\,\zeta(t)}\right). \quad (23)$$

Proof: Let $f(t) \in m\tilde{\mathcal{F}}_t$ with the given representation. The tower property together with Theorem 8.2 gives

$$\tilde{\mathbb{Q}}(f(t)) = \tilde{\mathbb{Q}}\left(\tilde{\mathbb{Q}}(f(t)|\mathcal{F}_t)\right) = \mathbb{Q}\left(\tilde{\mathbb{Q}}(f(t)|\mathcal{F}_t)\right)$$

$$= \mathbb{Q}\left(\frac{1}{Z(t)} \sum_{u \in N_t} f_u(t) e^{-\int_0^t m(X_u(s))R(X_u(s))\,ds} \zeta_u(t)\right).$$

From Theorem 6.4,

$$\left.\frac{d\mathbb{Q}}{dP}\right|_{\mathcal{F}_t} = \frac{Z(t)}{Z(0)},$$

and therefore we have

$$\tilde{\mathbb{Q}}(f(t)) = P\left(Z(0)^{-1} \sum_{u \in N_t} f_u(t) e^{-\int_0^t m(X_u(s))R(X_u(s))\,ds} \zeta_u(t)\right).$$

On the other hand,

$$\left.\frac{d\tilde{\mathbb{Q}}}{d\tilde{P}}\right|_{\tilde{\mathcal{F}}_t} = \frac{\tilde{\zeta}(t)}{\tilde{\zeta}(0)},$$

we have

$$\tilde{\mathbb{Q}}(f(t)) = \tilde{P}\left(f(t) \times \tilde{\zeta}(t)\,\tilde{\zeta}(0)^{-1}\right).$$

Trivially noting $Z(0) = \zeta(0) = \tilde{\zeta}(0)$ as there is only one initial ancestor, we can combine these expressions to obtain (22). For the second part, given $g(t) \in m\mathcal{G}_t$, we can define

$$f(t) := e^{\int_0^t m(\xi_s)R(\xi_s)\,ds} g(t) \times \zeta(t)^{-1},$$

which is clearly $\mathcal{G}_t$-measurable and satisfies $f(t) = \sum_{u \in N_t} f_u(t) \mathbf{1}_{(\xi_t = u)}$ with

$$f_u(t) = g_u(t) e^{\int_0^t m(X_u(s))R(X_u(s))\,ds} \zeta_u(t)^{-1} \in m\mathcal{F}_t.$$

When we use this $f(t)$ in equation (22) and recall Lemma 5.7, that $\mathbb{P} := \tilde{P}|_{\mathcal{G}_\infty}$ from Definition 4.6 and that $\hat{\mathbb{P}} := \tilde{\mathbb{Q}}|_{\mathcal{G}_\infty}$ from Definition 6.3, we arrive at the particular case given at (23) in the theorem. $\square$

In the further special case in which $g = g(\xi_t)$ for some Borel-measurable function $g(\cdot)$, the trivial representation

$$g(\xi_t) = \sum_{u \in N_t} g(X_u(t)) \mathbf{1}_{(\xi_t = u)}$$

leads immediately to the weaker version of the Many-to-One result that was utilised and proven, for example, in Harris and Williams [28] and Champneys *et al.* [5] using resolvents and the Feynman-Kac formula, expressed in terms of our more general branching Markov process $\mathbb{X}_t$:

Corollary 8.6 *If* $g(\cdot) : J \to \mathbb{R}$ *is* $\mathcal{B}$-*measurable then*

$$P\left(\sum_{u \in N_t} g(X_u(t))\right) = \mathbb{P}\left(e^{\int_0^t R(\xi_s)\,ds} g(\xi_t)\right).$$

9 Branching Brownian Motion

We now return to the original BBM model where particles move as standard Brownian motions, branching at rate r with offspring distribution A, as in Section 1. Under the measure $\tilde{P}^x$, the spine diffusion ξ_t is a Brownian motion that starts at x and we note that the martingale Z_λ can be obtained as in Sections 5 & 6 by starting with the spine $\tilde{P}^x$-martingale

$$\tilde{\zeta}_\lambda(t) := e^{-mrt}(1+m)^{n_t} \times \prod_{v < \xi_t} \left(\frac{1 + A_v}{1 + m} \right) \times e^{\lambda \xi_t - \frac{1}{2}\lambda^2 t}.$$

That is, we define the measure $\tilde{\mathbb{Q}}_\lambda$ on $(\tilde{T}, \tilde{\mathcal{F}}_\infty)$ by

$$\left. \frac{\mathrm{d}\tilde{\mathbb{Q}}_\lambda^x}{\mathrm{d}\tilde{P}^x} \right|_{\tilde{\mathcal{F}}_t} := \frac{\tilde{\zeta}_\lambda(t)}{\tilde{\zeta}_\lambda(0)} = e^{\lambda(\xi_t - x) - E_\lambda t} \prod_{v < \xi_t} (1 + A_v) \tag{24}$$

then, under $\tilde{\mathbb{Q}}_\lambda^x$, the process $\mathbb{X}_t$ can be constructed as follows:

- starting from x, the spine ξ_t diffuses according to a Brownian motion with drift λ on $\mathbb{R}$;
- at accelerated rate $(1 + m)r$ the spine undergoes fission producing $1 + \tilde{A}$ particles, where $\tilde{A}$ is independent of the spine's motion with size-biased distribution $\{(1 + k)p_k/(1 + m) : k \geq 0\}$;
- with equal probability, one of the spine's offspring particles is selected to continue the path of the spine, repeating stochastically the behaviour of its parent;
- the other particles initiate, from their birth position, independent copies of a $P^\cdot$ branching Brownian motion with branching rate r and family-size distribution given by A, that is, $\{p_k : k \geq 0\}$.

Further, ignoring information identifying the spine by setting $\mathbb{Q}_\lambda^x := \tilde{\mathbb{Q}}_\lambda^x|_{\mathcal{F}_\infty}$, we find

$$\left. \frac{\mathrm{d}\mathbb{Q}_\lambda^x}{\mathrm{d}P^x} \right|_{\mathcal{F}_t} = \frac{Z_\lambda(t)}{Z_\lambda(0)} = \sum_{u \in N_t} e^{\lambda(X_u(t) - x) - E_\lambda t}. \tag{25}$$

Of course, this is all in full agreement with the equivalent definition of $\mathbb{Q}_\lambda$ initially introduced in Theorem 1.1 via its pathwise construction.

9.1 Proof of Theorem 1.3

Just before we proceed to the proof we recall the naturally occurring eigenvalue $E_\lambda := \frac{1}{2}\lambda^2 + mr$, noting that under the symmetry assumption that $\lambda \leq 0$ and for $p \in (1, 2]$:

$$pE_\lambda - E_{p\lambda} > 0 \quad \Leftrightarrow \quad c_\lambda > c_{p\lambda} \quad \Leftrightarrow \quad p\lambda^2 < 2mr$$

and that this always holds for some $p > 1$ whenever $\lambda \in (\tilde{\lambda}, 0]$, that is, when λ lies between the minimum of c_λ found at $\tilde{\lambda}$ and the origin.

Proof of Part 1:

We are going to prove that for every $p \in (1,2]$ the martingale Z_λ is $\mathcal{L}^p(P)$-convergent if $pE_\lambda - E_{p\lambda} > 0$. Furthermore, since $P^x(Z_\lambda(t)^p) = e^{p\lambda x} P^0(Z_\lambda(t)^p)$ we do not lose generality supposing that $x = 0$; from now on this is implicit if we drop the superscript by simply writing P.

From the change of measure at (25) it is clear that

$$P(Z_\lambda(t)^p) = P(Z_\lambda(t)^{p-1}Z_\lambda(t)) = \mathbb{Q}_\lambda(Z_\lambda(t)^q),$$

where $q := p - 1$. Our aim is to prove that $\mathbb{Q}_\lambda(Z_\lambda(t)^q)$ is bounded in t, since then $Z_\lambda^p(t)$ must be bounded in $\mathcal{L}^p(P)$ and Doob's theorem will then imply that Z_λ is convergent in $\mathcal{L}^p(P)$.

As we know from Theorem 7.1, the algebra $\tilde{\mathcal{G}}_\infty$ gives us the very important *spine-decomposition* of the martingale Z_λ:

$$\tilde{\mathbb{Q}}_\lambda\big(Z_\lambda(t)|\tilde{\mathcal{G}}_\infty\big) = \sum_{k=1}^{n_t} A_k e^{\lambda\xi_{S_k} - E_\lambda S_k} + e^{\lambda\xi_t - E_\lambda t}, \tag{26}$$

where A_k is the number of new particles produced from the fission at time S_k along the path of the spine, and the sum is taken to equal 0 if $n_t = 0$. The intuition is quite clear: since the particles that do not make up the spine grow to become independent copies of $\mathbb{X}_t$ distributed *as if under* P, the fact that Z_λ is a P-martingale on these subtrees implies that their contributions to the above decomposition are just equal to their *immediate* contribution on being born at time S_k at location ξ_{S_k}. Note, we emphasize that here we must use $\tilde{\mathbb{Q}}_\lambda$, since $\mathbb{Q}_\lambda$ cannot measure the algebra $\tilde{\mathcal{G}}_\infty \not\subseteq \mathcal{F}_\infty$.

We can now use the conditional form of Jensen's inequality followed by the spine decomposition of (26) coupled with the simple inequality,

Proposition 9.1 *If $q \in (0,1]$ and $u, v > 0$ then $(u + v)^q \leq u^q + v^q$,*

to obtain,

$$\tilde{\mathbb{Q}}_\lambda\big(Z_\lambda(t)^q|\tilde{\mathcal{G}}_\infty\big) \leq \tilde{\mathbb{Q}}_\lambda\big(Z_\lambda(t)|\tilde{\mathcal{G}}_\infty\big)^q \tag{27}$$

$$\leq \sum_{k=1}^{n_t} A_k^q e^{q\lambda\xi_{S_k} - qE_\lambda S_k} + e^{q\lambda\xi_t - qE_\lambda t}. \tag{28}$$

With the tower property of conditional expectations and noting that $\mathbb{Q}_\lambda$ and $\tilde{\mathbb{Q}}_\lambda$ agree on $\mathcal{F}_t$,

$$\mathbb{Q}_\lambda(Z_\lambda(t)^q) = \tilde{\mathbb{Q}}_\lambda(Z_\lambda(t)^q) = \tilde{\mathbb{Q}}_\lambda\left(\tilde{\mathbb{Q}}_\lambda\big(Z_\lambda(t)^q|\tilde{\mathcal{G}}_\infty\big)\right) \tag{29}$$

$$\leq \tilde{\mathbb{Q}}_\lambda\left(\sum_{k=1}^{n_t} A_k^q e^{q\lambda\xi_{S_k} - qE_\lambda S_k}\right) + \tilde{\mathbb{Q}}_\lambda\left(e^{q\lambda\xi_t - qE_\lambda t}\right),$$

$$\tag{30}$$

and the proof of $\mathcal{L}^p(P)$-boundedness will be complete once we show this is bounded in t.

As written, (30) is made up of two terms, and since they play a central role from here on we name them explicitly: on the far right we have the **spine term** $\tilde{\mathbb{Q}}_\lambda\big(e^{q\lambda\xi_t - qE_\lambda t}\big)$, the other being the **sum term** $\tilde{\mathbb{Q}}_\lambda\big(\sum_{k=1}^{n_t} A_k^q e^{q\lambda\xi_{S_k} - qE_\lambda S_k}\big)$.

The Spine Term: Changing from $\tilde{P}$ to $\tilde{\mathbb{Q}}_\lambda$ gives the spine a drift of λ, and therefore the change-of-measure for just the spine's motion (i.e. on the algebra $\mathcal{G}_t$) is carried out by the martingale $e^{\lambda\xi_t - \frac{1}{2}\lambda^2 t}$, so

$$
\tilde{\mathbb{Q}}_\lambda\left(e^{q\lambda\xi_t - qE_\lambda t}\right) = \tilde{P}\left(e^{q\lambda\xi_t - qE_\lambda t} \times e^{\lambda\xi_t - \frac{1}{2}\lambda^2 t}\right)
$$

$$
= e^{\{\frac{1}{2}(p\lambda)^2 - \frac{1}{2}\lambda^2\}t - qE_\lambda t} \tilde{P}\left(e^{p\lambda\xi_t - \frac{1}{2}(p\lambda)^2 t}\right)
$$

$$
= e^{-(pE_\lambda - E_{p\lambda})t} \tilde{\mathbb{Q}}_{p\lambda}(1) = e^{-(pE_\lambda - E_{p\lambda})t} \tag{31}
$$

since the second-line term $e^{p\lambda\xi_t - \frac{1}{2}(p\lambda)^2 t}$ is also a $\tilde{P}$-martingale and $\frac{1}{2}(p\lambda)^2 - \frac{1}{2}\lambda^2 = E_{p\lambda} - E_\lambda$.

The Sum Term: Conditioning on the motion of the spine (without knowledge of the fission times or family sizes) and appealing to intuitive results from Poisson process theory (see [35] for example) yields

$$
\tilde{\mathbb{Q}}_\lambda\left(\sum_{k=1}^{n_t} A_k^q e^{q\lambda\xi_{S_k} - qE_\lambda S_k} \,\Big|\, \mathcal{G}_t\right) = \int_0^t (1+m)r\, \tilde{\mathbb{Q}}_\lambda(\tilde{A}^q)\, e^{q\lambda\xi_s - qE_\lambda s}\, ds \tag{32}
$$

Taking expectations of both sides of (32) and using Fubini's theorem then gives

$$
\tilde{\mathbb{Q}}_\lambda\left(\sum_{k=1}^{n_t} A_k^q e^{q\lambda\xi_{S_k} - qE_\lambda S_k}\right) = (1+m)r\, \tilde{\mathbb{Q}}_\lambda(\tilde{A}^q) \int_0^t \tilde{\mathbb{Q}}_\lambda\left(e^{q\lambda\xi_s - qE_\lambda s}\right) ds
$$

$$
= (1+m)r\, \tilde{\mathbb{Q}}_\lambda(\tilde{A}^q) \int_0^t e^{-(pE_\lambda - E_{p\lambda})s}\, ds, \qquad \text{using (31)}.
$$

Thus we have found an explicit upper-bound (if $pE_\lambda \neq E_{p\lambda}$):

$$
P^x(Z_\lambda(t)^p) \leq e^{p\lambda x} \left(\frac{(1+m)r}{pE_\lambda - E_{p\lambda}}\left[1 - e^{-(pE_\lambda - E_{p\lambda})t}\right]\tilde{\mathbb{Q}}_\lambda(\tilde{A}^q) + e^{-(pE_\lambda - E_{p\lambda})t}\right).
$$

$$\tag{33}$$

Finally, we also observe that

Lemma 9.2 *If $p \in (1,2]$ and $q := p - 1$, $\tilde{\mathbb{Q}}_\lambda(\tilde{A}^q) < \infty$ if and only if $P(A^p) < \infty$*

since

$$\tilde{Q}_\lambda(\tilde{A}^q) = \sum_{i=1}^{\infty} i^q \frac{i+1}{m+1} p_i = \frac{P(A^p) + P(A^q)}{m+1} \leq \frac{2P(A^p)}{m+1}.$$

Hence, if we have $pE_\lambda - E_{p\lambda} > 0$ in addition to $P(A^p) < \infty$, this implies that $P^x(Z_\lambda(t)^p)$ will remain bounded as $t \to \infty$, which together with Doob's theorem will complete the proof of the first part of Theorem 1.3.   □

Proof of Part 2:

We seek to show that Z_λ is unbounded in $\mathcal{L}^p(P^x)$ if either $pE_\lambda - E_{p\lambda} < 0$ or $P(A^p) = \infty$.

Note that *if Z_λ is $\mathcal{L}^p(P^x)$ bounded* then

$$P^x(Z_\lambda(\infty)^p) = \lim_{t \to \infty} P^x(Z_\lambda(t)^p) < \infty$$

hence, $\tilde{Q}_\lambda^x(Z_\lambda(\infty)^q) < \infty$ and $Z_\lambda(\infty)^q$ is a uniformly integrable $\tilde{Q}_\lambda^x$-submartingale. In particular, for any stopping time T, $\tilde{Q}_\lambda^x(Z_\lambda(\infty)^q | \mathcal{F}_T) \geq Z_\lambda(T)^q$ hence $\tilde{Q}_\lambda^x(Z_\lambda(\infty)^q) \geq \tilde{Q}_\lambda^x(Z_\lambda(T)^q)$.

First, by considering only the contribution of the spine $Z_\lambda(t) \geq e^{\lambda \xi_t - E_\lambda t}$ for all $t \geq 0$ and recalling (31), we see that

$$\tilde{Q}_\lambda^x(Z_\lambda(t)^q) \geq \tilde{Q}_\lambda^x(e^{q\lambda \xi_t - q E_\lambda t}) = e^{q\lambda x - (pE_\lambda - E_{p\lambda})t}$$

and Z_λ is therefore unbounded in $\mathcal{L}^p(P^x)$ if $pE_\lambda - E_{p\lambda} < 0$.

Now, let T be any fission time along the path of the spine, then

$$Z_\lambda(T) \geq (1 + \tilde{A})e^{\lambda \xi_T - E_\lambda T}$$

where $\tilde{A}$ is the number of additional offspring produced at the time of fission. Then,

$$\tilde{Q}_\lambda^x(Z_\lambda(T)^q) \geq \tilde{Q}_\lambda^x\left((1 + \tilde{A})^q\right) e^{q\lambda x} \tilde{Q}_{p\lambda}^x(e^{-(pE_\lambda - E_{p\lambda})T})$$

and so Z_λ is unbounded in $\mathcal{L}^p(P^x)$ if $\tilde{Q}_\lambda^x\left((1 + \tilde{A})^q\right) = \infty$, which is true iff $P(\tilde{A}^p) = \infty$.   □

10 A Typed Branching Diffusion

We move on to consider a general offspring distribution version of the typed branching diffusion introduced in Section 2.1. We will follow a similar notation and setup as before, but leave some details to the reader.

Recall the *single particle motion* $(X_t, Y_t)_{t\geq0}$ from Section 2.1, where the type Y_t evolves as a Markov chain on $I := \{1, \ldots, n\}$ with Q-matrix θQ and the *spatial* location, X_t, moves as a driftless Brownian motion on $\mathbb{R}$ with diffusion coefficient $a(y) > 0$ whenever η_t is in state y.

Consider a *typed branching Brownian motion* where individual particles move independently according to the single particle motion as above, and any particle currently of type y will undergo fission at rate $R(y)$ to be replaced by a random number of offspring, $1 + A(y)$, where $A(y) \in \{0, 1, 2, \ldots\}$ is an independent RV with distribution

$$P\big(A(y) = i\big) = p_i(y), \qquad i \in \{0, 1, \ldots\},$$

and mean $M(y) := P\big(A(y)\big) < \infty$ for all $y \in I$. At birth, offspring inherit the parent's spatial and type positions and then move off independently, repeating stochastically the parent's behaviour, and so on. We gather together the mean number of offspring in matrix $M := \mathrm{diag}[M(1), \ldots, M(n)]$ and also recall that $R := \mathrm{diag}[R(1), \ldots, R(n)]$ and $A := \mathrm{diag}[a(1), \ldots, a(n)]$.

As usual, let the configuration of the whole branching diffusion at time t be given by the J-valued point process $\mathbb{X}_t = \big\{\big(X_u(t), Y_u(t)\big) : u \in N_t\big\}$, where N_t is the set of individuals alive at time t. Let the probabilities for this process be given by $\big\{P^{x,y} : (x, y) \in J\big\}$ defined on the natural filtration, $(\mathcal{F}_t)_{t \geq 0}$, where $P^{x,y}$ is the law of the typed BBM process starting with one initial particle of type y at spatial position x. Recall, under the extended measures $\big\{\tilde{P}^{x,y} : (x, y) \in J\big\}$ where we identify a distinguished infinite line of decent starting from the initial particle, this *spine* $(\xi_t, \eta_t)_{t \geq 0}$ will simply move like the *single particle motion* above.

It should be noted that the condition of time-reversibility on the Markov chain is not absolutely necessary, and is really just a simplifying assumption that gives us an easier $\mathcal{L}^2$ theory for the matrices and eigenvectors; our aim is really to show how the spine techniques work – lessening the geometric complexity of the model serves a good purpose.

Note, the special case of the 2-type BBM model was considered in Champneys *et al.* [5] by different means. Also, in our model, at the time of fission a type-y individual can produce only type-y offspring. This is not the same as the case in which a type-y individual may produce a random collection of particles of *different* types – as considered in T.E. Harris's classic text [32], for example. Other forms of typed branching processes have also been dealt with by spine techniques, for example, see Lyons *et al.* [41] or Athreya [2] for discrete-time models in which a particle's type does not change during its life but a type-w individual can give offspring of any type according to some distribution. See also the remarkable work of Georgii and Baake [19] that uses spine techniques to study ancestral type behaviour in a continuous time branching Markov chain where particles can give birth to across all types. In principle, our spine methods will be robust enough to extend to all these other type behaviours (with added spatial diffusion).

10.1 The Martingale

Via the many-to-one Lemma 8.5, it is easy to see that for any $\lambda \in \mathbb{R}$, any function (vector) $v_\lambda : I \to \mathbb{R}$ and any number $E_\lambda \in \mathbb{R}$, the expression

$$Z_\lambda(t) := \sum_{u \in N(t)} v_\lambda(Y_u(t)) \, e^{\lambda X_u(t) - E_\lambda t},$$

will be a martingale if and only if v_λ and E_λ satisfy:

$$\left(\frac{1}{2}\lambda^2 A + \theta Q + MR\right)v_\lambda = E_\lambda v_\lambda. \tag{34}$$

That is, v_λ must be an eigenvector of the matrix $\frac{1}{2}\lambda^2 A + \theta Q + MR$, with eigenvalue E_λ.

Definition 10.1 *For two vectors u, v on I, we define*

$$\langle u, v \rangle_\pi := \sum_{i=1}^{n} u_i v_i \pi_i,$$

which gives us a Hilbert space which we refer to as $\mathcal{L}^2(\pi)$. We suppose that the eigenvector v_λ is normalized so that $\|v_\lambda\|_\pi := \langle v_\lambda, v_\lambda \rangle_\pi = 1$.

The fact that the Markov chain is time-reversible implies that the matrix $\frac{1}{2}\lambda^2 A + \theta Q + MR$ is self-adjoint with respect to this inner product. This in itself is enough to guarantee the existence of eigenvectors in $\mathcal{L}^2(\pi)$, but the fact that we are dealing with a finite-state Markov chain means that we also have the *Perron-Frobenius* theory to hand, which allows us to suppose that v_λ is a *strictly positive* eigenvector whose eigenvalue E_λ is real and the farthest to the right of all the other eigenvalues – see Seneta [48] for details. This implies a useful representation for the eigenvalue:

Theorem 10.2

$$E_\lambda = \sup_{\|v\|_\pi = 1} \langle ((\lambda^2/2)A + \theta Q + MR) \, v, v \rangle_\pi, \tag{35}$$

since it is the rightmost eigenvalue.

A proof can be found in Kreyzig [39]. From this it is not difficult to show that E_λ is a strictly-convex function of λ. Interestingly, it will be seen in our proofs that it is the geometry of the eigenvalue E_λ that determines the interval that gives rise to martingales $Z_\lambda(t)$ that are $\mathcal{L}^p$-convergent.

Corollary 10.3 *As a function of λ, E_λ is strictly-convex and infinitely differentiable with*

$$E_\lambda' = \lambda \langle A v_\lambda, v_\lambda \rangle_\pi. \tag{36}$$

If we define the speed function

$$c_\lambda := -E_\lambda/\lambda, \tag{37}$$

then on $(-\infty, 0)$ the function c_λ has just one minimum at a single point $\tilde{\lambda}(\theta)$, either side of which c_λ is strictly increasing to $+\infty$ as either $\lambda \downarrow -\infty$ or $\lambda \uparrow 0$. In particular, for each $\lambda \in (\tilde{\lambda}(\theta), 0]$ there is some $p > 1$ such that $c_\lambda > c_{p\lambda}$; on the other hand, if $\lambda < \tilde{\lambda}(\theta)$ there is no such $p > 1$.

We refer to the function c_λ as the speed function since it relates to the asymptotic speed of the travelling waves associated with the martingale $Z_\lambda(t)$; see Harris [29] or Champneys *et al.* [5] for details of the relationship between branching-diffusion martingales and travelling waves.

Since $Z_\lambda(t)$ is a strictly-positive martingale it is immediate that $Z_\lambda(\infty) := \lim_{t \to \infty} Z_\lambda(t)$ exists and is finite almost-surely under $P^{x,y}$. As before, *by symmetry we shall assume that $\lambda \leq 0$ and, without loss of generality, we also suppose that $P(A(y) = 0) = 1$ whenever $r(y) = 0$ to simplify statements.* We shall prove necessary and sufficient conditions for $\mathcal{L}^1$-convergence of the Z_λ martingales:

Theorem 10.4 *For each $x \in \mathbb{R}$, the limit $Z_\lambda(\infty) := \lim_{t \to \infty} Z_\lambda(t)$ exists $P^{x,y}$-a.s. where:*

- *if $\lambda \leq \tilde{\lambda}(\theta)$ then $Z_\lambda(\infty) = 0$ $P^{x,y}$-almost surely;*
- *if $\lambda \in (\tilde{\lambda}(\theta), 0]$ and $P(A(y) \log^+ A(y)) = \infty$ for some $y \in I$, then $Z_\lambda(\infty) = 0$ $P^{x,y}$-a.s.;*
- *if $\lambda \in (\tilde{\lambda}(\theta), 0]$ and $P(A(y) \log^+ A(y)) < \infty$ for all $y \in I$, then $Z_\lambda(t) \to Z_\lambda(\infty)$ almost surely and in $\mathcal{L}^1(P^{x,y})$.*

Once again, in many cases where the martingale has a non-trivial limit, the convergence will be much stronger than merely in $\mathcal{L}^1(P^{x,y})$, as indicated by the following new $\mathcal{L}^p$-convergence result that we will prove by extending our earlier new spine approach:

Theorem 10.5 *For each $x \in \mathbb{R}$, and for each $p \in (1, 2]$:*

- *$Z_\lambda(t) \to Z_\lambda(\infty)$ a.s. and in $\mathcal{L}^p(P^{x,y})$ if $pE_\lambda - E_{p\lambda} > 0$ and $P(A(y)^p) < \infty$ for all $y \in I$.*
- *Z_λ is unbounded in $\mathcal{L}^p(P^{x,y})$, that is $\lim_{t \to \infty} P^{x,y}(Z_\lambda(t)^p) = \infty$, if either $pE_\lambda - E_{p\lambda} < 0$ or $P(A(y)^p) = \infty$ for some $y \in I$.*

Note, when $\lambda \leq 0$, the inequality $pE_\lambda - E_{p\lambda} > 0$ is equivalent to $c_\lambda > c_{p\lambda}$ and holds for some $p \in (1, 2]$ if and only if $\lambda \in (\tilde{\lambda}(\theta), 0]$.

10.2 New Measures for the Typed BBM

As usual, we can define a measure $\tilde{\mathbb{Q}}_\lambda$ via a Radon-Nikodym derivative with respect to $\tilde{P}$ by combining three simpler changes of measures that only affect behaviour along the spine.

First, we observe that for $\lambda \in \mathbb{R}$,

$$v_\lambda(\eta_t)\, e^{\int_0^t MR(\eta_s)\,\mathrm{d}s}\, e^{\lambda\xi_t - E_\lambda t}$$

is a $\tilde{P}$-martingale. This fact is easy to confirm with some classical 'one-particle' calculations, for example, using the Feynman-Kac formula, the generator (4) and noting the relation (34).

We can obtain the Z_λ martingale as in Sections 5 & 6 by using the $\tilde{P}$-martingale

$$\tilde{\zeta}_\lambda(t) := e^{-\int_0^t MR(\eta_s)\,\mathrm{d}s}\prod_{u<\xi_t}(1 + A_u) \times v_\lambda(\eta_t)\, e^{\int_0^t MR(\eta_s)\,\mathrm{d}s}\, e^{\lambda\xi_t - E_\lambda t}.$$

That is, for each $\lambda \in \mathbb{R}$ we define a measure $\tilde{\mathbb{Q}}_\lambda^{x,y}$ on $(\tilde{\mathcal{T}}, \tilde{\mathcal{F}}_\infty)$ via

$$\left.\frac{\mathrm{d}\tilde{\mathbb{Q}}_\lambda^{x,y}}{\mathrm{d}\tilde{P}^{x,y}}\right|_{\tilde{\mathcal{F}}_t} := \frac{\tilde{\zeta}_\lambda(t)}{\tilde{\zeta}_\lambda(0)} = \frac{v_\lambda(\eta_t)}{v_\lambda(y)}\, e^{\lambda(\xi_t - x) - E_\lambda t}\prod_{v<\xi_t}(1 + A_v) \qquad (38)$$

and then ignoring information about the spine by defining $\mathbb{Q}_\lambda^{x,y} := \tilde{\mathbb{Q}}_\lambda^{x,y}|_{\mathcal{F}_\infty}$, we find that

$$\left.\frac{\mathrm{d}\mathbb{Q}_\lambda^{x,y}}{\mathrm{d}P^{x,y}}\right|_{\mathcal{F}_t} = \frac{Z_\lambda(t)}{Z_\lambda(0)} = v_\lambda(y)^{-1}\sum_{u\in N(t)}v_\lambda(Y_u(t))\, e^{\lambda(X_u(t) - x) - E_\lambda t}. \qquad (39)$$

We emphasise that, starting with the three simple 'spine' martingales, we have actually shown that Z_λ must, in fact, be a martingale. This route offers a simple way of getting general 'additive' martingales for the branching process.

10.3 The Spine Process (ξ_t, η_t) Under $\tilde{\mathbb{Q}}_\lambda$

It remains to identify the behaviour of the spine under the change of measure. In the BBM model it was clear to see that the spine ξ_t received a drift under the measure $\tilde{\mathbb{Q}}_\lambda$, and something similar happens here:

Lemma 10.6 *Under $\tilde{\mathbb{Q}}_\lambda$ the spine process (ξ_t, η_t) has generator:*

$$\mathcal{H}_\lambda F(x,y) := \frac{1}{2}a(y)\frac{\partial^2 F}{\partial x^2} + a(y)\lambda\frac{\partial F}{\partial x} + \sum_{j\in I}\theta Q_\lambda(y,j)F(x,j), \qquad (40)$$

where Q_λ is an honest Q-matrix:

$$\theta Q_\lambda(i,j) = \begin{cases} \theta Q(i,j)\frac{v_\lambda(j)}{v_\lambda(i)} & \text{if } i \neq j \\ \theta Q(i,i) + \frac{\lambda^2}{2}a(i) - E_\lambda + r(i) & \text{if } i = j \end{cases}$$

That is, under $\tilde{\mathbb{Q}}_\lambda$, ξ_t is a Brownian motion with instantaneous variance $a(\eta_t)$ and instantaneous drift $a(\eta_t)\lambda$, and η_t is a Markov chain on I with Q-matrix θQ_λ and invariant measure $\pi_\lambda = v_\lambda^2\pi$.

The form of this above generator $\mathcal{H}_\lambda$ can be obtained from the theory of Doob's h-transforms, due to the fact that on the algebra $\mathcal{G}_t$ the change of measure is given by:

$$\left.\frac{\mathrm{d}\mathbb{Q}_\lambda^{x,y}}{\mathrm{d}P^{x,y}}\right|_{\mathcal{G}_t} = \frac{1}{v_\lambda(y)e^{\lambda x}}\, v_\lambda(\eta_t)\, e^{\int_0^t MR(\eta_s)\,\mathrm{d}s}\, e^{\lambda\xi_t - E_\lambda t}. \tag{41}$$

The long-term behaviour under $\tilde{\mathbb{Q}}_\lambda$ of the spine diffusion ξ_t can now be retrieved from the generator (40) and the properties of E_λ stated in Lemma 10.3:

Corollary 10.7 *Almost surely under* $\tilde{\mathbb{Q}}_\lambda^{x,y}$, *the long-term drift of the spine is given explicitly as*

$$\lim_{t\to\infty} t^{-1}\xi_t = E_\lambda'$$

and hence

$$\xi_t + c_\lambda t \to \begin{cases} \infty & \text{if } \lambda \in (\tilde{\lambda}(\theta), 0] \\ -\infty & \text{if } \lambda < \tilde{\lambda}(\theta) \end{cases} \tag{42}$$

whereas, if $\lambda = \tilde{\lambda}$ *the process* $\xi_t + c_\lambda t$ *will be recurrent on* $\mathbb{R}$ *under* $\tilde{\mathbb{Q}}_\lambda$.

Proof: From the generator stated at (40) we can write:

$$\xi_t = B\left(\int_0^t a(\eta_s)\mathrm{d}s\right) + \lambda \int_0^t a(\eta_s)\mathrm{d}s,$$

where $B(t)$ is a $\tilde{\mathbb{Q}}_\lambda$-Brownian motion. Then by the ergodic theorem and the fact that $\pi_\lambda = v_\lambda^2 \pi$:

$$t^{-1}\xi_t \to \lambda \sum_{y\in I} a(y)\pi_\lambda(y) = \lambda \sum_{y\in I} a(y)v_\lambda^2(y)\pi(y) = \lambda\langle Av_\lambda, v_\lambda\rangle_\pi = E_\lambda'.$$

Direct calculation from (37) gives $E_\lambda' = -c_\lambda - \lambda c_\lambda'$, and therefore $t^{-1}(\xi_t + c_\lambda t) \to -\lambda c_\lambda'$, whence whether we are to the left or right of the local minimum of c_λ found at $\tilde{\lambda}$ determines the behaviour of $\xi_t + c_\lambda t$, as is required. Lastly, when $\lambda = \tilde{\lambda}$, with the laws of the iterated logarithm in mind, it is not difficult to see that both $B(\int_0^t a(\eta_s)\,\mathrm{d}s)$ and $\int_0^t (\lambda a(\eta_s) + c_\lambda)\,\mathrm{d}s$ will fluctuate about the origin, hence $\xi_t + c_\lambda t$ will be recurrent under $\tilde{\mathbb{Q}}_\lambda$. $\square$

10.4 Construction of the Process under $\tilde{\mathbb{Q}}_\lambda$

Drawing together the elements from this section, we now present the pathwise construction of the new measure $\tilde{\mathbb{Q}}_\lambda^{x,y}$:

Theorem 10.8 *Under* $\tilde{Q}_\lambda^{x,y}$, *the process* X_t *evolves as follows:*

- *starting from* (x,y), *the spine* (ξ_t, η_t) *evolves as a Markov process with generator* $\mathcal{H}_\lambda$, *that is,* η_t *evolves as Markov chain on* I *with Q-matrix* θQ_λ *and* ξ_t *moves as a Brownian motion on* $\mathbb{R}$ *with variance coefficient* $a(\eta_t)$ *and drift* $a(\eta_t)\lambda$.
- *whenever the type of the spine* η *is in state* $y \in I$, *the spine undergoes fission at an accelerated rate* $(1 + m(y))r$, *producing* $1 + \tilde{A}(y)$ *particles where* $\tilde{A}(y)$ *is independent of the spine's motion with size-biased distribution* $\{(1+k)p_k(y)/(1+m(y)) : k \geq 0\}$;
- *with equal probability, one of the spine's offspring particles is selected to continue the path of the spine, repeating stochastically the behaviour of its parent;*
- *the other particles initiate, from their birth position, independent copies of* $P^{\cdot,\cdot}$ *typed branching Brownian motions.*

10.5 Proof of Theorem 10.4

The following proof is an extension of that given for BBM by Kyprianou [40]. The second part of the following theorem is the key element in using the measure change (38) to determine properties of the martingale Z_λ:

Theorem 10.9 *Suppose that* P *and* Q *are two probability measures on a space* $(\Omega, \mathcal{F}_\infty)$ *with filtration* $(\mathcal{F}_t)_{t\geq 0}$, *such that for some positive martingale* Z_t,

$$\left.\frac{dQ}{dP}\right|_{\mathcal{F}_t} = Z_t.$$

The limit $Z_\infty := \limsup_{t\to\infty} Z_t$ *therefore exists and is finite almost surely under* P. *Furthermore, for any* $F \in \mathcal{F}_\infty$

$$Q(F) = \int_F Z_\infty \, dP + Q(F \cap \{Z_\infty = \infty\}), \tag{43}$$

and consequently

$$\text{(a)} \quad P(Z_\infty = 0) = 1 \iff Q(Z_\infty = \infty) = 1 \tag{44}$$
$$\text{(b)} \quad P(Z_\infty) = 1 \iff Q(Z_\infty < \infty) = 1 \tag{45}$$

A proof of the decomposition (43) can be found in Durrett [11], at page 241.

Suppose that $\lambda \leq \tilde{\lambda} < 0$. Ignoring all contributions except for the spine, it is immediate that

$$Z_\lambda(t) = \sum_{u \in N_t} v_\lambda(Y_u(t)) \, e^{\lambda X_u(t) - E_\lambda t} \geq v_\lambda(\eta_t) \, e^{\lambda(\xi_t + c_\lambda t)}$$

where, from Corollary 10.7, under the measure $\tilde{Q}_\lambda$ the spine satisfies $\liminf\{\xi_t + c_\lambda t\} = -\infty$ a.s. and $v_\lambda > 0$, hence $\limsup_{t\to\infty} Z_\lambda(t) = \infty$ almost surely under $\tilde{Q}_\lambda$, yielding $P(Z_\lambda(\infty) = 0) = 1$.

Note that, for $y \in I$, $P(A(y) \log^+ A) < \infty \iff \sum_{k \geq 1} P(\log^+ \tilde{A}(y) > ck) < \infty$ for any $c > 0$, where recall that $\tilde{A}(y)$ has the size-biased distribution $\{(i+1)p_k(y)/(1+m(y)) : k \geq 0\}$. Then for an IID sequence $\{\tilde{A}_n(y)\}$ of copies of $\tilde{A}(y)$, Borel-Cantelli reveals that, P almost surely,

$$\limsup_{t \to \infty} n^{-1} \log^+ \tilde{A}_n(y) = \begin{cases} 0 & \text{if } P(A(y) \log^+ A(y)) < \infty, \\ \infty & \text{if } P(A(y) \log^+ A(y)) = \infty. \end{cases} \qquad (46)$$

Now suppose that $\lambda \in (\tilde{\lambda}, 0]$ and $P(A(y) \log^+ A(y)) = \infty$ for some $y \in I$ (with $r(y) > 0$). Let S_k be the time of the kth fission along the spine producing $\tilde{A}_k(\eta_{S_k})$ additional particles, then

$$Z_\lambda(S_k) \geq \tilde{A}_k(\eta_{S_k}) v_\lambda(\eta_{S_k}) e^{\lambda(\xi_{S_k} + c_\lambda S_k)}$$

where $(\xi_t + c_\lambda t)/t \to -\lambda c'_\lambda > 0$, η_t is ergodic so the event $\{\eta_{S_k} = y\}$ will occur for infinitely many k since $r(y) > 0$, and $n_t/t \to < Rv_\lambda, v_\lambda >_\pi$ so $S_k/k \to < Rv_\lambda, v_\lambda >_\pi^{-1}$, hence the super-exponential growth for $\tilde{A}_k(y)$ from (46) gives $\limsup_{t \to \infty} Z_\lambda(t) = \infty$ $\tilde{\mathbb{Q}}_\lambda$-almost surely which then implies that $P(Z_\lambda(\infty) = 0) = 1$.

Finally, suppose that $\lambda \in (\tilde{\lambda}, 0]$ and $P(A(y) \log^+ A(y)) < \infty$ for all $y \in I$. Recall from (26):

$$\tilde{\mathbb{Q}}_\lambda(Z_\lambda(t) | \tilde{\mathcal{G}}_\infty) = \sum_{k=1}^{n_t} \tilde{A}_k(\eta_{S_k}) v_\lambda(\eta_{S_k}) e^{\lambda(\xi_{S_k} + c_\lambda S_k)} + v_\lambda(\eta_t) e^{\lambda(\xi_t + c_\lambda t)}. \qquad (47)$$

In this case, the facts that $(\xi_t + c_\lambda t)/t \to -\lambda c'_\lambda > 0$ and $S_k/k \to < Rv_\lambda, v_\lambda >_\pi^{-1}$ together with the moment conditions and (46) implying that the $\tilde{A}_k(y)$'s all have sub-exponential growth means that

$$\limsup_{t \to \infty} \tilde{\mathbb{Q}}_\lambda(Z_\lambda(t) | \tilde{\mathcal{G}}_\infty) < \infty \qquad \tilde{\mathbb{Q}}_\lambda\text{-a.s.}$$

Fatou's lemma then gives $\liminf_{t \to \infty} Z_\lambda(t) < \infty$, $\tilde{\mathbb{Q}}_\lambda$-a.s., hence also $\mathbb{Q}_\lambda$-a.s. In addition, since $Z_\lambda(t)^{-1}$ is a positive $\mathbb{Q}_\lambda$-martingale (recall (39)) with an almost sure limit, this means that $\lim_{t \to \infty} Z_\lambda(t) < \infty$, $\mathbb{Q}_\lambda$-a.s. and then (45) yields that $P(Z_\lambda(\infty)) = 1$ and so $Z_\lambda(t)$ converges almost surely and in $\mathcal{L}^1(P)$. $\square$

Discussion of Rate of Convergence to Zero and Left-most Particle Speed.

Alternatively, when $\lambda < \tilde{\lambda}$ we can readily obtain the rate of convergence to zero with the following simple argument, adapted from Git *et al.* [20]. By Proposition 9.1,

$$Z_\lambda(t)^q \leq \sum_{u \in N(t)} v_\lambda(Y_u(t))^q e^{q\lambda(X_u(t) + c_{q\lambda})} e^{q\lambda(c_\lambda - c_{q\lambda})t} \leq K Z_{q\lambda}(t) e^{q\lambda(c_\lambda - c_{q\lambda})t}$$

where $K := \max_{y \in I} v_\lambda^q(y)/v_{q\lambda}(y) < \infty$ since I is finite and $v_\lambda > 0$. Recall that c_λ has a minimum over $\lambda \in (-\infty, 0]$ at $\tilde{\lambda}$ with $c_{\tilde{\lambda}} = -E_{\tilde{\lambda}} = -\tilde{\lambda} < Av_{\tilde{\lambda}}, v_{\tilde{\lambda}} >_\pi$. Then, since $Z_{q\lambda}(t)$ is a convergent martingale, we can choose q such that $q\lambda = \tilde{\lambda}$ giving $Z_\lambda(t)$ decaying exponentially to zero at least at rate $\lambda(c_\lambda - c_{\tilde{\lambda}})$.

Further, once we know that P and $\mathbb{Q}_\lambda$ are equivalent for every $\lambda \in (\tilde{\lambda}, 0]$, since the spine moves such that $\xi_t/t \to -c_\lambda - \lambda c_\lambda'$ under $\mathbb{Q}_\lambda$, the left-most particle $L(t) := \inf_{u \in N(t)} X_u(t)$ must satisfy $\liminf_t L(t)/t \leq -c_{\tilde{\lambda}}$, P-a.s. On the other hand, the convergence of the Z_λ P-martingales quickly gives the same upper bound on the fastest speed of any particle, leading to $L(t)/t \to -c_{\tilde{\lambda}}$, P-a.s. This result also reveals that the rate of exponential decay found above is actually best possible.

10.6 Proof of Theorem 10.5

Proof of Part 1:

Suppose $p \in (1, 2]$, then with $q := p - 1$ a slight modification of the BBM proof arrives at

$$P^{x,y}(Z_\lambda(t)^p) = e^{\lambda x} v_\lambda(y) \tilde{\mathbb{Q}}_\lambda^{x,y}(Z_\lambda(t)^q)$$

$$\leq e^{\lambda x} v_\lambda(y) \tilde{\mathbb{Q}}_\lambda^{x,y} \left(\sum_{k=1}^{n_t} A_k^q v_\lambda(\eta_{S_k})^q e^{q\lambda \xi_{S_k} - qE_\lambda S_k} \right)$$

$$+ e^{\lambda x} v_\lambda(y) \tilde{\mathbb{Q}}_\lambda^{x,y} \left(v_\lambda(\eta_t)^q e^{q\lambda \xi_t - qE_\lambda t} \right)$$

and the proof of $\mathcal{L}^p$-boundedness will be complete once we show that this RHS expectation is bounded in t.

The Spine Term. Since I is finite we note that $\langle v_\lambda^p, v_{p\lambda} \rangle_\pi < \infty$. It is always useful to first focus on the spine term, since we can change the measure with (41) to get

$$\tilde{\mathbb{Q}}_\lambda^{x,y} \left(v_\lambda(\eta_t)^q e^{q\lambda \xi_t - qE_\lambda t} \right) = \tilde{P}^{x,y} \left(v_\lambda(\eta_t)^q e^{q\lambda \xi_t - qE_\lambda t} \cdot \frac{v_\lambda(\eta_t) e^{\int_0^t MR(\eta_s)\,ds} e^{\lambda \xi_t - E_\lambda t}}{v_\lambda(y) e^{\lambda x}} \right)$$

$$= e^{q\lambda x} \frac{v_{p\lambda}(y)}{v_\lambda(y)} g_t(y) e^{-(pE_\lambda - E_{p\lambda})t} \tag{48}$$

where, for all $y \in I$

$$g_t(y) := \tilde{\mathbb{Q}}_{p\lambda}^{0,y} \left(\frac{v_\lambda^p}{v_{p\lambda}}(\eta_t) \right) \to \langle v_\lambda^p, v_{p\lambda} \rangle_\pi$$

as $t \to \infty$ and $\langle g_t v_{p\lambda}, v_{p\lambda} \rangle_\pi = \langle v_\lambda^p, v_{p\lambda} \rangle_\pi$ for all $t \geq 0$, since η_t is a finite-state irreducible Markov chain under $\tilde{\mathbb{Q}}_\mu$ with invariant distribution $\pi_\mu(y) = v_\mu(y)^2 \pi(y)$. It follows that the long term the growth or decay of the spine term is determined by the sign of $pE_\lambda - E_{p\lambda}$.

The Sum Term. We now assume that $pE_\lambda - E_{p\lambda} > 0$. We know that under $\tilde{\mathbb{Q}}_\lambda$ and conditional on knowing η, the fission times $\{S_k : k \geq 0\}$ on the spine occur as a Poisson process of rate $(1+m(\eta_s))r(\eta_s)$ with the kth fission yielding an additional A_k offspring, each A_k being an independent copy of $\tilde{A}(y)$ which has the size-biased distribution $\{(1 + k)p_k(y)/(1 + m(y)) : k \geq 0\}$ where $y = \eta_{S_k}$ is the type at the time of fission. We also recall from Lemma 9.2 that

$$M_q(y) := \tilde{\mathbb{Q}}_\lambda(\tilde{A}^q(y)) < \infty \iff P(A^p(y)) < \infty.$$

Therefore, if we condition on $\mathcal{G}_t$ which knows about (ξ_s, η_s) at all times $0 \leq s \leq t$ we can transform the sum into an integral, use Fubini's theorem and the change of measure used in (48):

$$\tilde{\mathbb{Q}}_\lambda^{x,y}\left(\sum_{k=1}^{n_t} A_k^q v_\lambda(\eta_{S_k})^q e^{q\lambda\xi_{S_k} - qE_\lambda S_k}\right)$$

$$= \tilde{\mathbb{Q}}_\lambda^{x,y}\left(\tilde{\mathbb{Q}}_\lambda^{x,y}\left(\sum_{k=1}^{n_t} A_k^q v_\lambda(\eta_{S_k})^q e^{q\lambda\xi_{S_k} - qE_\lambda S_k}\Big|\mathcal{G}_t\right)\right)$$

$$= \tilde{\mathbb{Q}}_\lambda^{x,y}\left(\int_0^t (1 + m(\eta_s))r(\eta_s)\, M_q(\eta_s)v_\lambda(\eta_s)^q e^{q\lambda\xi_s - qE_\lambda s}\, ds\right)$$

$$= \int_0^t \tilde{\mathbb{Q}}_\lambda^{x,y}\left((1 + m(\eta_s))r(\eta_s)M_q(\eta_s)v_\lambda(\eta_s)^q e^{q\lambda\xi_s - qE_\lambda s}\right) ds$$

$$= e^{q\lambda x}\frac{v_{p\lambda}(y)}{v_\lambda(y)}\int_0^t h_s(\eta_s)e^{-(pE_\lambda - E_{p\lambda})s}\, ds$$

$$= e^{q\lambda x}\frac{v_{p\lambda}(y)}{v_\lambda(y)} \times \frac{k_t(y)}{pE_\lambda - E_{p\lambda}}$$

where

$$h_s(y) := \tilde{\mathbb{Q}}_{p\lambda}^{0,y}\left(\tilde{r}(\eta_s)M_q(\eta_s)\frac{v_\lambda^p}{v_{p\lambda}}(\eta_s)\right), \qquad \tilde{r}(y) := (1 + m(y))r(y),$$

$$\text{and} \qquad k_t(y) := \mathbb{E}(h_U(y); U \leq t)$$

with U an independent exponential of rate $(pE_\lambda - E_{p\lambda}) > 0$. Note that, for all $y \in I$, $h_s(y) \to \langle \tilde{r}M_q v_\lambda^p, v_{p\lambda}\rangle$ and $k_t(y) \uparrow k_\infty(y)$ as $t \to \infty$, where $\langle k_t v_{p\lambda}, v_{p\lambda}\rangle = \langle \tilde{r}M_q v_\lambda^p, v_{p\lambda}\rangle \mathbb{P}(U \leq t) \uparrow \langle k_\infty v_{p\lambda}, v_{p\lambda}\rangle = \langle \tilde{r}M_q v_\lambda^p, v_{p\lambda}\rangle$. Then, since $M_q(w) < \infty \iff P(A(w)^p) < \infty$, and I is finite, we are guaranteed that $k_\infty(y) < \infty$ for all $y \in I$ as long as $P(A(w)^p) < \infty$ for all $w \in I$.

Having dealt with both the spine term and the sum term, we have obtained the upper-bound

$$P^{x,y}\left(Z_\lambda(t)^p\right) \leq \frac{e^{p\lambda x}v_{p\lambda}(y)}{(pE_\lambda - E_{p\lambda})}\left(k_t(y) + g_t(y)\,(pE_\lambda - E_{p\lambda})\,e^{-(pE_\lambda - E_{p\lambda})t}\right)$$

and since $Z_\lambda(t)^p$ is a P-submartingale, we find that

$$P^{x,y}\big(Z_\lambda(t)^p\big) \leq \frac{e^{p\lambda x}v_{p\lambda}(y)}{(pE_\lambda - E_{p\lambda})}k_\infty(y) \qquad (\forall t \geq 0)$$

and $Z_\lambda(t)$ will be bounded in $\mathcal{L}^p(P^{x,y})$ if we have both $pE_\lambda - E_{p\lambda} > 0$ and $P(A^p(w)) < \infty$ for all $w \in I$. □

Proof of Part 2:

The earlier proof for BBM goes through with minor modification. Exactly as in the BBM case, looking only at the contribution of the spine means that Z_λ is unbounded in $\mathcal{L}^p(P^{x,y})$ if $pE_\lambda - E_{p\lambda} < 0$. In addition, letting T be any fission time along the path of the spine,

$$Z_\lambda(T) \geq (1 + \tilde{A}(\eta_T))v_\lambda(\eta_T)e^{\lambda \xi_T - E_\lambda T}$$

where $\tilde{A}(\eta_T)$ is the number of additional offspring produced at the time of fission. Then, with $m_q(y) := \tilde{\mathbb{Q}}((1 + \tilde{A}(y))^q) < \infty \iff P(A^p(y)) < \infty$,

$$\tilde{\mathbb{Q}}_\lambda^{x,y}(Z_\lambda(T)^q) \geq e^{q\lambda x}\, \tilde{\mathbb{Q}}_\lambda^{x,y}(m_q(\eta_T)v_\lambda(\eta_T)^q e^{q\lambda \xi_T - qE_\lambda T})$$

$$= e^{q\lambda x}\, \tilde{\mathbb{Q}}_{p\lambda}^{x,y}\left(m_q(\eta_T)\frac{v_\lambda(\eta_T)^p}{v_{p\lambda}(\eta_T)}\, e^{-(pE_\lambda - E_{p\lambda})T}\right)$$

and so Z_λ will also be unbounded in $\mathcal{L}^p(P^{x,y})$ if $m_q(y) = \infty \iff P(A^p(y)) = \infty$ for any $y \in I$ (taking a fission time when also in state y). □

Remarks on Signed Martingales and Kesten-Stigum Type Theorems

In the multi-typed BBM, for each λ there will be other (signed) additive martingales corresponding to the different eigenvectors and eigenvalues obtained from solving (34); the Z_λ martingale simply corresponds to the Perron-Frobenius, or ground-state, eigenvalue E_λ and (strictly positive) eigenvector v_λ. Since $|u + v|^q \leq (|u| + |v|)^q \leq |u|^q + |v|^q$ for all $u, v \in \mathbb{R}$, the above proof will also adapt to give convergence results for signed martingales. In fact, when there is a complete orthonormal set of eigenvectors, a Kesten-Stigum like theorem would then swiftly follow (for example, see Harris [30] in the context of the continuous-type model of the next section).

11 A Continuous-Typed Branching-Diffusion

The previous finite-type model was originally inspired by the model that we now turn to, originally laid out in Harris and Williams [28]. In this model the type moves on the real line as an Orstein-Uhlenbeck process associated with the generator

$$Q_\theta := \frac{\theta}{2}\left(\frac{\partial^2}{\partial y^2} - y\frac{\partial}{\partial y}\right), \qquad \text{with } \theta > 0 \text{ considered as the } \textit{temperature,}$$

which has the standard normal density as its invariant distribution:

$$\pi(y) := (2\pi)^{-\frac{1}{2}}e^{-\frac{1}{2}y^2}.$$

The spatial movement of a particle of type y is a driftless Brownian motion with instantaneous variance

$$A(y) := ay^2, \qquad \text{for some fixed } a > 0,$$

and fission of a particle of type y occurs at a rate

$$R(y) := ry^2 + \rho, \qquad \text{where } r, \rho > 0 \text{ are fixed,}$$

to produce two particles at the same type-space location as the parent (we consider only binary splitting). The model has very different behaviour for low temperature values (i.e. low θ), but most studies have considered the high temperature regime where $\theta > 8r$. Also, the parameter λ must be restricted to an interval $(\lambda_{\min}, 0)$ in order for some of the model's parameters to remain in $\mathbb{R}$, where

$$\lambda_{\min} := -\sqrt{\frac{\theta - 8r}{4a}}.$$

Generally, *unboundedness* in a model's rates is a serious obstacle to classical proofs since they often depend on the expectation semigroup of the branching process, and unbounded rates tend to lead to unbounded *eigenfunctions*. Here this is the case, but the existence of a spectral theory for their particular expectation operator allowed Harris and Williams to get a sufficiently good bound in particular for a *non-linear* term (see Theorem 5.1 of [28]), and therefore to prove $\mathcal{L}^p$-convergence of the martingale. Other convergence results for various martingales and weighted sums over particles for this model also appear in Harris [30], again using more classical methods and requiring 'non-linear' calculations. The spine approach we again adopt here is both simple and more generic in nature; requiring no such special 'non-linear' calculations, it elegantly produces very good estimates that only involve easy one-particle calculations.

We use the same notation as previously, $\mathbb{X}_t = \{(X_u(t), Y_u(t)) : u \in N_t\}$ to denote the point process of space-type locations in $\mathbb{R} \times \mathbb{R}$, and suppose that the measures $\{\tilde{P}^{x,y} : (x, y) \in \mathbb{R}^2\}$ on the natural filtration with a spine $(\tilde{\mathcal{F}}_t)_{t \geq 0}$ are such that the initial ancestor starts at (x, y) and $(\mathbb{X}_t, (\xi_t, \eta_t))$ becomes the above-described branching diffusion with a spine.

11.1 The Measure Change

Although there are some significant differences, this model is similar in flavour to our finite-type model. There is a strictly-positive martingale Z_λ defined as

$$Z_\lambda(t) := \sum_{u \in N_t} v_\lambda(Y_u(t)) e^{\lambda X_u(t) - E_\lambda t}$$

where v_λ and E_λ are the eigenvector and eigenvalue associated with the self-adjoint (in $\mathcal{L}^2(\pi)$) operator:

$$Q_\theta + \frac{1}{2} \lambda^2 A(y) + R(y).$$

The eigenfunction v_λ is normalizable against the $\mathcal{L}^2(\pi)$ norm, and can be found explicitly as

$$v_\lambda(y) = e^{\psi_\lambda^- y^2}$$

where

$$\psi_\lambda^- := \frac{1}{4} - \frac{\mu_\lambda}{2\theta}, \quad \mu_\lambda := \frac{1}{2}\sqrt{\theta^2 - \theta(8r + 4a\lambda^2)},$$

are both positive for all $\lambda \in (\lambda_{\min}, 0)$; another important parameter is $\psi_\lambda^+ := \frac{1}{4} + \frac{\mu_\lambda}{2\theta}$. The eigenvalue E_λ is then given by

$$E_\lambda = \rho + \theta \psi_\lambda^-.$$

We again define the speed function $c_\lambda := -E_\lambda/\lambda$, and $\tilde{\lambda}(\theta) < 0$ is the unique point (on the negative axis) at which c_λ hits its minimum $\tilde{c}(\theta)$ – further details are given in Harris and Williams [28]. We are going to use spines to prove the following result, in which the critical case of $\lambda = \tilde{\lambda}$ and the necessary conditions for $\mathcal{L}^p(P)$-convergence are new results:

Theorem 11.1 *Suppose that $\lambda \in (\lambda_{\min}, 0)$.*

1. *Let $p \in (1, 2]$. The martingale Z_λ is $\mathcal{L}^p(P)$-bounded if both $pE_\lambda - E_{p\lambda} > 0$ and $p\psi_\lambda^- < \psi_{p\lambda}^+$. In particular, for all $\lambda \in (\tilde{\lambda}(\theta), 0]$, Z_λ is a uniformly-integrable martingale.*
2. *Z_λ is unbounded in $\mathcal{L}^p(P)$ if either $pE_\lambda - E_{p\lambda} < 0$ or $p\psi_\lambda^- > \psi_{p\lambda}^+$.*
3. *Almost surely under P, $Z_\lambda(\infty) = 0$ if $\lambda \leq \tilde{\lambda}(\theta)$.*

Once again, for each $\lambda \leq 0$ we define a measure $\tilde{\mathbb{Q}}_\lambda^{x,y}$ on $(\tilde{\mathcal{T}}, \tilde{\mathcal{F}}_\infty)$ via

$$\frac{d\tilde{\mathbb{Q}}_\lambda^{x,y}}{d\tilde{P}^{x,y}}\bigg|_{\tilde{\mathcal{F}}_t} := \frac{1}{v_\lambda(y) e^{\lambda x}} 2^{n_t} v_\lambda(\eta_t) e^{\lambda \xi_t - E_\lambda^- t}, \tag{49}$$

so that with $\mathbb{Q}_\lambda := \tilde{\mathbb{Q}}_\lambda|_{\mathcal{F}_\infty}$ we have

$$\frac{d\mathbb{Q}_\lambda^{x,y}}{dP^{x,y}}\bigg|_{\mathcal{F}_t} = \frac{Z_\lambda(t)}{Z_\lambda(0)} = \frac{Z_\lambda(t)}{v_\lambda(y) e^{\lambda x}}.$$

The facts are that under $\tilde{\mathbb{Q}}_\lambda$:

- the spine diffusion ξ_t has instantaneous drift $a\eta_t^2\lambda$;
- the type process η_t has generator $\frac{\theta}{2}\left(\frac{\partial^2}{\partial y^2} - \frac{2\mu_\lambda}{\theta}y\frac{\partial}{\partial y}\right)$ and an invariant probability measure $\pi_\lambda := \langle v_\lambda, v_\lambda\rangle_\pi^{-1}v_\lambda^2\pi$, corresponding to a normal distribution, $N(0, \frac{\theta}{2\mu_\lambda})$;
- fission times on the spine occur at the accelerated rate of $2R(\eta_t)$;
- all particles not in the spine behave as if under the original measure P.

We briefly comment that, along similar lines as discussed for the finite-typed BBM case, we could now give a straightforward spine proof that the asymptotic right-most particle speed in this continuous typed BBM model is almost surely $\tilde{c}(\theta)$.

11.2 Proof of Theorem 11.1

Proof of Part 1: Suppose $p \in (1, 2]$. Then using the spine decomposition with Jensen's inequality and Proposition 9.1 we find,

$$P^{x,y}(Z_\lambda^-(t)^p) \leq e^{\lambda x}v_\lambda(y)\tilde{\mathbb{Q}}_\lambda^{x,y}\left(\sum_{u<\xi_t} v_\lambda(\eta_{S_u})^q e^{q\lambda\xi_{S_u}-qE_\lambda S_u}\right)$$
$$+ e^{\lambda x}v_\lambda(y)\tilde{\mathbb{Q}}_\lambda^{x,y}\left(v_\lambda(\eta_t)^q e^{q\lambda\xi_t-qE_\lambda t}\right).$$

Assume that $pE_\lambda - E_{p\lambda} > 0$ and $p\psi_\lambda^- < \psi_{p\lambda}^+$. As seen in Harris and Williams [28], we can do many calculations explicitly in this model, largely due to the fact that under $\tilde{\mathbb{Q}}_{p\lambda}^{0,y}$

$$\eta_s \sim N\left(e^{-\mu_{p\lambda}s}y, \frac{\theta(1-e^{-2\mu_{p\lambda}s})}{2\mu_{p\lambda}}\right) \to N\left(0, \frac{\theta}{2\mu_{p\lambda}}\right)$$

and the eigenfunctions v_λ have such simple exponential form. For example,

$$\tilde{\mathbb{Q}}_{p\lambda}^{0,y}\left(\frac{v_\lambda^p}{v_{p\lambda}}(\eta_s)\right) = \tilde{\mathbb{Q}}_{p\lambda}^{0,y}\left(e^{(p\psi_\lambda^- - \psi_{p\lambda}^-)\eta_s^2}\right) \tag{50}$$

can easily be seen to be finite and bounded for all $s \geq 0$ if and only if $p\psi_\lambda^- - \psi_{p\lambda}^- - \frac{\mu_{p\lambda}}{\theta} = p\psi_\lambda^- - \psi_{p\lambda}^+ < 0$, and just as readily calculated explicitly.

In fact, more 'natural' conditions for $\mathcal{L}^p$ convergence of the martingales would be that

$$\langle RM_q v_\lambda^p, v_{p\lambda}\rangle_\pi < \infty, \qquad \langle v_\lambda^p, v_{p\lambda}\rangle_\pi < \infty, \qquad \text{and} \qquad pE_\lambda - E_{p\lambda} < 0,$$

where $M_q(y) := \tilde{\mathbb{Q}}(\tilde{A}^q(y))$ with $\tilde{A}$ the size-biased offspring distribution (here, binary splitting means $\tilde{A}(y) \equiv 1$), and we present arguments below that are more generic in nature, at least in terms of adapting to other 'suitably' ergodic type motions and random family sizes. Note, the last condition above is related to the natural convexity of E_λ and, in our specific model, both integrability conditions are guaranteed by $p\psi_\lambda^- - \psi_{p\lambda}^+ < 0$.

The Spine Term. On the algebra $\mathcal{G}_t$ the change of measure takes the form

$$\left.\frac{d\tilde{\mathbb{Q}}_\lambda^{x,y}}{d\tilde{P}^{x,y}}\right|_{\mathcal{G}_t} = \frac{v_\lambda(\eta_s)}{v_\lambda(y)} \exp\left(\int_0^t R(\eta_s)\,ds + \lambda(\xi_t - x) - E_\lambda^- t\right),$$

which we can use on the spine term to arrive at

$$f_t(x,y) := e^{\lambda x} v_\lambda(y) \tilde{\mathbb{Q}}_\lambda^{x,y}\left(v_\lambda(\eta_t)^q e^{q\lambda\xi_t - qE_\lambda t}\right) = e^{p\lambda x} v_{p\lambda}(y) g_t(y) e^{-(pE_\lambda - E_{p\lambda})t} \tag{51}$$

with $g_t(y) := \tilde{\mathbb{Q}}_{p\lambda}^{0,y}\left(v_\lambda^p(\eta_t)/v_{p\lambda}(\eta_t)\right)$. Under the assumption that $p\psi_\lambda^- < \psi_{p\lambda}^+$, it easy to check that $\langle v_\lambda^p, v_{p\lambda}\rangle_\pi < \infty$, that is $v_\lambda^p/v_{p\lambda} \in L^1(\pi_{p\lambda})$ from which it follows that $g_t \in L^1(\pi_{p\lambda})$ for all $t \geq 0$. Since η has equilibrium $\pi_{p\lambda}$ under $\tilde{\mathbb{Q}}_{p\lambda}$, we find $\langle g_t v_{p\lambda}, v_{p\lambda}\rangle_\pi = \langle v_\lambda^p, v_{p\lambda}\rangle_\pi < \infty$ and $g_t(y) \to \langle v_\lambda^p, v_{p\lambda}\rangle_\pi \langle v_{p\lambda}, v_{p\lambda}\rangle_\pi^{-1} < \infty$ as $t \to \infty$ for all $y \in \mathbb{R}$.

We also note that since $g_t \in L^1(\pi_{p\lambda})$, we have $f_t \in L^1(\tilde{\pi}_{p\lambda})$ where $\tilde{\pi}_\mu := \langle 1, v_\mu\rangle_\pi^{-1} v_\mu \pi$ and then

$$\int_{y\in\mathbb{R}} \tilde{\pi}_{p\lambda}(y) f_t(x,y)\,dy = e^{p\lambda x} \frac{\langle v_\lambda^p, v_{p\lambda}\rangle_\pi}{\langle 1, v_{p\lambda}\rangle_\pi} e^{-(pE_\lambda - E_{p\lambda})t}.$$

The Sum Term. Note that under the parameter assumptions we have $\langle Rv_\lambda^p, v_{p\lambda}\rangle_\pi < \infty$. As for the finite-type model the fission times S_u on the spine occur as a Cox process and therefore

$$g_t(x,y) := e^{\lambda x} v_\lambda(y) \tilde{\mathbb{Q}}_\lambda^{x,y}\left(\sum_{u<\xi_t} v_\lambda(\eta_{S_u})^q e^{q\lambda\xi_{S_u} - qE_\lambda S_u}\right)$$

$$= e^{\lambda x} v_\lambda(y) \int_0^t \tilde{\mathbb{Q}}_\lambda^{x,y}\left(2R(\eta_s) v_\lambda(\eta_s)^q e^{q\lambda\xi_s - qE_\lambda s}\right) ds$$

$$= e^{\lambda x} v_\lambda(y) \int_0^t \tilde{\mathbb{Q}}_{p\lambda}^{0,y}\left(2R(\eta_s) \frac{v_\lambda^p}{v_{p\lambda}}(\eta_s)\right) e^{-(pE_\lambda - E_{p\lambda})s} ds$$

$$= e^{p\lambda x} v_{p\lambda}(y) k_t(y)$$

where

$$k_t(y) := \int_0^t h_s(y) e^{-(pE_\lambda - E_{p\lambda})s}\,ds, \qquad h_s(y) := \tilde{\mathbb{Q}}_{p\lambda}^{0,y}\left(2R(\eta_s)\frac{v_\lambda^p}{v_{p\lambda}}(\eta_s)\right)$$

and $h_t, k_t \in L^1(\pi_{p\lambda})$. Note, $k_t(y) \uparrow k_\infty(y) \in L^1(\pi_{p\lambda})$ as $t \to \infty$ where

$$\frac{\langle k_t v_{p\lambda}, v_{p\lambda}\rangle_\pi}{\langle v_{p\lambda}, v_{p\lambda}\rangle_\pi} = \langle 2Rv_\lambda^p, v_{p\lambda}\rangle_\pi \frac{(1 - e^{-(pE_\lambda - E_{p\lambda})t})}{(pE_\lambda - E_{p\lambda})}$$

$$\uparrow \frac{\langle 2Rv_\lambda^p, v_{p\lambda}\rangle_\pi}{(pE_\lambda - E_{p\lambda})} = \langle k_\infty v_{p\lambda}, v_{p\lambda}\rangle_\pi < \infty.$$

Note, $k_t \in L^1(\pi_{p\lambda})$ implies $g_t \in L^1(\tilde{\pi}_{p\lambda})$, with an explicit calculation again possible.

Bringing together the results for the sum and spine terms, we have an upper bound

$$P^{x,y}(Z_\lambda(t)^p) \leq e^{p\lambda x} v_{p\lambda}(y) \left\{ g_t(y) e^{-(pE_\lambda - E_{p\lambda})t} + k_t(y) \right\} \in L^1(\tilde{\pi}_{p\lambda}) \quad (52)$$

and hence the submartingale property reveals that

$$P^{x,y}(Z_\lambda(t)^p) \leq e^{p\lambda x} v_{p\lambda}(y) k_\infty(y) < \infty$$

for all $t \geq 0$ and all $y \in \mathbb{R}$.   □

Proof of Part 2: We need only dominate the martingale by the spine at time t, yielding

$$\tilde{\mathbb{Q}}_\lambda^{x,y}(Z_\lambda(t)^q) \geq \tilde{\mathbb{Q}}_\lambda^{x,y}(v_\lambda(\eta_t)^q e^{q\lambda\xi_t - qE_\lambda t})$$

$$= e^{q\lambda x} \frac{v_{p\lambda}(y)}{v_\lambda(y)} \tilde{\mathbb{Q}}_{p\lambda}^y \left(\frac{v_\lambda^p}{v_{p\lambda}}(\eta_t) \right) e^{-(pE_\lambda - E_{p\lambda})t}.$$

Hence Z_λ is unbounded in $\mathcal{L}^p(P^x)$ if either $pE_\lambda - E_{p\lambda} < 0$ or $\langle v_\lambda^p, v_{p\lambda}\rangle_\pi = \infty$.   □

Proof of Part 3: The proof that we have seen in the finite-type model will work here with little change: under $\tilde{\mathbb{Q}}_\lambda$ the spatial motion is

$$\xi_t = B\left(\int_0^t a(\eta_s)ds\right) + \lambda\int_0^t a(\eta_s)ds,$$

and the type process η_s has invariant distribution $N(0, \frac{\theta}{2\mu_\lambda})$, whence $t^{-1}\xi_t \to \lambda a\theta/\mu_\lambda = E'_\lambda$. Therefore it follows that under $\tilde{\mathbb{Q}}_\lambda$ the diffusion $\xi_t + c_\lambda t$ drifts off to $-\infty$ if $\lambda < \tilde{\lambda}(\theta)$. When $\lambda = \tilde{\lambda}$, it is also simple to check that $\xi_t + c_\lambda t$ is recurrent, so has $\liminf\{\xi_t + c_\lambda t\} = -\infty$. Whence, in either case, bounding Z_λ below by the spine's contribution as done before, we have $Z_\lambda(t) \geq v_\lambda(\eta_t)e^{\lambda(\xi_t + c_\lambda t)}$ and since $v_\lambda > 0$ and η_t recurrent, we see that $\limsup_{t\to\infty} Z_\lambda(t) = \infty$ almost surely under $\tilde{\mathbb{Q}}_\lambda^{x,y}$.   □

Acknowledgement. We would like to thank the referees for their careful reading of the original articles and for their helpful suggestions.

References

1. S. Asmussen and H. Hering, *Strong limit theorems for general supercritical branching processes with applications to branching diffusions*, Z. Wahrschein-lichkeitstheorie und Verw. Gebiete **36** (1976), no. 3, 195–212.

2. K. B. Athreya, *Change of measures for Markov chains and the $L \log L$ theorem for branching processes*, Bernoulli **6** (2000), no. 2, 323–338.5

3. J. Bertoin, Random fragmentation and coagulation processes, Cambridge University Press, 2006.

4. J. D. Biggins and A. E. Kyprianou, *Measure change in multitype branching*, Adv. in Appl. Probab. **36** (2004), no. 2, 544–581.

5. A. Champneys, S. Harris, J. Toland, J. Warren, and D. Williams, *Algebra, analysis and probability for a coupled system of reaction-diffusion equations*, Philosophical Transactions of the Royal Society of London **350** (1995), 69–112.

6. B. Chauvin, *Arbres et processus de Bellman-Harris*, Ann. Inst. H. Poincaré Probab. Statist. **22** (1986), no. 2, 209–232.

7. B. Chauvin and A. Rouault, *Boltzmann-Gibbs weights in the branching random walk*, Classical and modern branching processes (Minneapolis, MN, 1994), IMA Vol. Math. Appl., vol. 84, Springer, New York, 1997, pp. 41–50.

8. B. Chauvin, *Product martingales and stopping lines for branching Brownian motion*, Ann. Probab. **19** (1991), no. 3, 1195–1205.

9. B. Chauvin and A. Rouault, *KPP equation and supercritical branching Brownian motion in the subcritical speed area. Application to spatial trees*, Probab. Theory Related Fields **80** (1988), no. 2, 299–314.

10. B. Chauvin, A. Rouault, and A. Wakolbinger, *Growing conditioned trees*, Stochastic Process. Appl. **39** (1991), no. 1, 117–130.

11. R. Durrett, *Probability: Theory and examples*, 2nd ed., Duxbury Press, 1996.

12. J. Engländer and A. E. Kyprianou, *Local extinction versus local exponential growth for spatial branching processes*, Ann. Probab. **32** (2004), no. 1A, 78–99.

13. J. Engländer, *Branching diffusions, superdiffusions and random media*, Probab. Surveys Vol. **4** (2007) 303-364.

14. J. Engländer, S. C. Harris, and A. E. Kyprianou, *Laws of Large numbers for spatial branching processes*, Annales de l'Institut Henri Poincaré (B) Probability and Statistics, (2009), to appear.

15. J. Geiger, *Size-biased and conditioned random splitting trees*, Stochastic Process. Appl. **65** (1996), no. 2, 187–207.

16. J. Geiger, *Elementary new proofs of classical limit theorems for Galton-Watson processes*, J. Appl. Probab. **36** (1999), no. 2, 301–309.

17. J. Geiger, *Poisson point process limits in size-biased Galton-Watson trees*, Electron. J. Probab. **5** (2000), no. 17, 12 pp. (electronic).

18. J. Geiger and L. Kauffmann, *The shape of large Galton-Watson trees with possibly infinite variance*, Random Structures Algorithms **25** (2004), no. 3, 311–335.

19. H. O. Georgii and E. Baake, *Supercritical multitype branching processes: the ancestral types of typical individuals*, Adv. in Appl. Probab. **35** (2003), no. 4, 1090–1110.

20. Y. Git, J. W. Harris, and S. C. Harris, *Exponential growth rates in a typed branching diffusion*, Ann. App. Probab., **17** (2007), no. 2, 609–653. doi:10.1214/105051606000000853

21. R. Hardy, *Branching diffusions*, Ph.D. thesis, University of Bath Department of Mathematical Sciences, 2004.
22. R. Hardy and S. C. Harris, *Some path large deviation results for a branching diffusion*, (2007), submitted.
23. R. Hardy and S. C. Harris, *A conceptual approach to a path result for branching Brownian motion*, Stoch. Proc. and Applic., **116** (2006), no. 12, 1992–2013. doi:10.1016/j.spa.2006.05.010
24. R. Hardy and S. C. Harris, *A new formulation of the spine approach for branching diffusions*, (2006). arXiv:math.PR/0611054
25. R. Hardy and S. C. Harris, *Spine proofs for $\mathcal{L}^p$-convergence of branching-diffusion martingales*, (2006). arXiv:math.PR/0611056
26. J. W. Harris, S. C. Harris, and A. E. Kyprianou, *Further probabilistic analysis of the Fisher-Kolmogorov-Petrovskii-Piscounov equation: one-sided travelling waves*, Ann. Inst. H. Poincaré Probab. Statist. **42** (2006), no. 1, 125–145.
27. J. W. Harris and S. C. Harris, *Branching Brownian motion with an inhomogeneous breeding potential*, Ann. Inst. H. Poincaré Probab. Statist. (2008), to appear.
28. S. C. Harris and D. Williams, *Large deviations and martingales for a typed branching diffusion. I*, Astérisque (1996), no. 236, 133–154, Hommage à P. A. Meyer et J. Neveu.
29. S. C. Harris, *Travelling-waves for the FKPP equation via probabilistic arguments*, Proc. Roy. Soc. Edinburgh Sect. A **129** (1999), no. 3, 503–517.
30. S. C. Harris, *Convergence of a "Gibbs-Boltzmann" random measure for a typed branching diffusion*, Séminaire de Probabilités, XXXIV, Lecture Notes in Math., vol. 1729, Springer, Berlin, 2000, pp. 239–256.
31. S. C. Harris, R. Knobloch, and A.E. Kyprianou, *Strong Law of Large Numbers for Fragmentation Processes*, (2008) arXiv:0809.2958v1, submitted.
32. T. E. Harris, *The theory of branching processes*, Dover ed., Dover, 1989.
33. Y. Hu and Z. Shi, *Minimal position and critical martingale convergence in branching random walks, and directed polymers on disordered trees*, (2008) Ann. App. Probab., to appear.
34. A. M. Iksanov, *Elementary fixed points of the BRW smoothing transforms with infinite number of summands*, Stochastic Process. Appl. **114** (2004), no. 1, 27–50.
35. O. Kallenberg, *Foundations of modern probability*, Springer-Verlag, 2002.
36. H. Kesten and B. P. Stigum, *Additional limit theorems for indecomposable multidimensional Galton-Watson processes*, Ann. Math. Stat. **37** (1966), 1463–1481.
37. H. Kesten and B. P. Stigum, *A limit theorem for multidimensional Galton-Watson processes*, Ann. Math. Stat. **37** (1966), 1211–1223.
38. H. Kesten and B. P. Stigum, *Limit theorems for decomposable multi-dimensional Galton-Watson processes*, J. Math. Anal. Applic. **17** (1967), 309–338.
39. E. Kreyszig, *Introductory functional analysis with applications*, John Wiley and Sons, 1989.
40. A. E. Kyprianou, *Travelling wave solutions to the K-P-P equation: alternatives to Simon Harris' probabilistic analysis*, Ann. Inst. H. Poincaré Probab. Statist. **40** (2004), no. 1, 53–72.
41. T. Kurtz, R. Lyons, R. Pemantle, and Y. Peres, *A conceptual proof of the Kesten-Stigum theorem for multi-type branching processes*, Classical and modern

branching processes (Minneapolis, MN, 1994), IMA Vol. Math. Appl., vol. 84, Springer, New York, 1997, pp. 181–185.

42. Q. Liu and A. Rouault, *On two measures defined on the boundary of a branching tree*, Classical and modern branching processes (Minneapolis, MN, 1994), IMA Vol. Math. Appl., vol. 84, Springer, New York, 1997, pp. 187–201.

43. R. Lyons, *A simple path to Biggins' martingale convergence for branching random walk*, Classical and modern branching processes (Minneapolis, MN, 1994), IMA Vol. Math. Appl., vol. 84, Springer, New York, 1997, pp. 217–221.

44. R. Lyons, R. Pemantle, and Y. Peres, *Conceptual proofs of $L \log L$ criteria for mean behavior of branching processes*, Ann. Probab. **23** (1995), no. 3, 1125–1138.

45. J. Neveu, *Arbres et processus de Galton-Watson*, Ann. Inst. H. Poincaré Probab. Statist. **22** (1986), no. 2, 199–207.

46. J. Neveu, *Multiplicative martingales for spatial branching processes*, Seminar on Stochastic Processes (E. Çinlar, K.L.Chung, and R.K.Getoor, eds.), Birkhäuser, 1987, pp. 223–241.

47. P. Olofsson, *The $x \log x$ condition for general branching processes*, J. Appl. Probab. **35** (1998), no. 3, 537–544.

48. E. Seneta, *Non-negative matrices and Markov chains*, Springer-Verlag, 1981.

49. E. C. Waymire and S. C. Williams, *A general decomposition theory for random cascades*, Bull. Amer. Math. Soc. (N.S.) **31** (1994), no. 2, 216–222.

Penalisation of the Standard Random Walk by a Function of the One-sided Maximum, of the Local Time, or of the Duration of the Excursions

Pierre Debs

Institut Élie Cartan Nancy
B.P. 239, 54506 Vandœuvre-lès-Nancy Cedex, France
E-mail: Pierre.Debs@iecn.u-nancy.fr

Summary. Call $(\Omega, \mathcal{F}_\infty, \mathbb{P}, X, \mathcal{F})$ the canonical space for the standard random walk on $\mathbb{Z}$. Thus, Ω denotes the set of paths $\phi : \mathbb{N} \to \mathbb{Z}$ such that $|\phi(n+1) - \phi(n)| = 1$, $X = (X_n, n \geqslant 0)$ is the canonical coordinate process on Ω; $\mathcal{F} = (\mathcal{F}_n, n \geqslant 0)$ is the natural filtration of X, $\mathcal{F}_\infty$ the σ-field $\bigvee_{n \geqslant 0} \mathcal{F}_n$, and $\mathbb{P}_0$ the probabilitiy on $(\Omega, \mathcal{F}_\infty)$ such that under $\mathbb{P}_0$, X is the standard random walk started form 0, i.e., $\mathbb{P}_0 (X_{n+1} = j \mid X_n = i) = \frac{1}{2}$ when $|j - i| = 1$.

Let $G : \mathbb{N} \times \Omega \to \mathbb{R}^+$ be a positive, adapted functional. For several types of functionals G, we show the existence of a positive $\mathcal{F}$-martingale $(M_n, \ n \geqslant 0)$ such that, for all n and all $\Lambda_n \in \mathcal{F}_n$,

$$\frac{\mathbb{E}_0[\mathbb{1}_{\Lambda_n} G_p]}{\mathbb{E}_0[G_p]} \longrightarrow \mathbb{E}_0[\mathbb{1}_{\Lambda_n} M_n] \qquad \text{when} \quad p \to \infty.$$

Thus, there exists a probability Q on $(\Omega, \mathcal{F}_\infty)$ such that $Q(\Lambda_n) = \mathbb{E}_0[\mathbb{1}_{\Lambda_n} M_n]$ for all $\Lambda_n \in \mathcal{F}_n$. We describe the behavior of the process $(\Omega, X, \mathcal{F})$ under Q.

The three sections of the article deal respectively with the three situations when G is a function:

- of the one-sided maximum;
- of the sign of X and of the time spent at zero;
- of the length of the excursions of X.

1 Introduction

Let $\left\{ \Omega, (X_t, \mathcal{F}_t)_{t \geqslant 0}, \mathcal{F}_\infty, \mathbb{P}_x \right\}$ be the canonical one-dimensional Brownian motion. For several types of positive functionals $\Gamma : \mathbb{R}^+ \times \Omega \to \mathbb{R}^+$, B. Roynette, P. Vallois and M. Yor show in [RVY06] that, for fixed s and for all $\Lambda_s \in \mathcal{F}_s$,

$$\lim_{t \to \infty} \frac{\mathbb{E}_x[\mathbb{1}_{\Lambda_s} \Gamma_t]}{\mathbb{E}_x[\Gamma_t]}$$

C. Donati-Martin et al. (eds.), *Séminaire de Probabilités XLII*,
Lecture Notes in Mathematics 1979, DOI 10.1007/978-3-642-01763-6_12,
© Springer-Verlag Berlin Heidelberg 2009

exists and has the form $\mathbb{E}_x[\mathbb{1}_{\Lambda_s} M_s^x]$, where $(M_s^x, s \geqslant 0)$ is a positive martingale. This enables them to define a probability Q_x on $(\Omega, \mathcal{F}_\infty)$ by:

$$\forall \Lambda_s \in \mathcal{F}_s \qquad Q_x(\Lambda_s) = \mathbb{E}_x[\mathbb{1}_{\Lambda_s} M_s^x];$$

moreover, they precisely describe the behavior of the canonical process X under Q_x. This they do for numerous functionals Γ, for instance a function of the one-sided maximum, or of the local time, or of the age of the current excursion (cf. [RVY06], [RVY]).

Our purpose is to study a discrete analogue of their results. More precisely, let Ω denote the set of all functions ϕ from $\mathbb{N}$ to $\mathbb{Z}$ such that $|\phi(n+1) - \phi(n)| = 1$, let $X = (X_n, n \geqslant 0)$ be the process of coordinates on that space, $\mathcal{F} = (\mathcal{F}_n, n \geqslant 0)$ the canonical filtration, $\mathcal{F}_\infty$ the σ-field $\bigvee_{n \geqslant 0} \mathcal{F}_n$, and $\mathbb{P}_x$ ($x \in \mathbb{N}$) the family of probabilities on $(\Omega, \mathcal{F}_\infty)$ such that under $\mathbb{P}_x$ X is the standard random walk started at x. For notational simplicity, we often write $\mathbb{P}$ for $\mathbb{P}_0$. Our aim is to establish that for several types of positive, adapted functionals $G : \mathbb{N} \times \Omega \to \mathbb{N}$,
i) for each $n \geqslant 0$ and each $\Lambda_n \in \mathcal{F}_n$,

$$\frac{\mathbb{E}_0[\mathbb{1}_{\Lambda_n} G_p]}{\mathbb{E}_0[G_p]},$$

tends to a limit when p tends to infinity;
ii) this limit is equal to $\mathbb{E}_0[\mathbb{1}_{\Lambda_n} M_n]$, for some $\mathcal{F}$-martingale M such that $M_0 = 1$.

Call $Q(\Lambda_n)$ this limit. Assuming i) and ii), Q is a probability on each σ-field $\mathcal{F}_n$; it extends in a unique way to a probability (still called Q) on the σ-field $\mathcal{F}_\infty$. This can be seen either by applying Kolmogorov's theorem on projective limits (knowing Q on the $\mathcal{F}_n$ amounts to knowing the finite marginal laws of the process X), or directly, since every finitely additive probability on the Boolean algebra $\mathcal{A} = \bigcup_n \mathcal{F}_n$ extends to a σ-additive probability on $\mathcal{F}_\infty$ (a Cantorian diagonal argument shows that every decreasing sequence (A_k) in $\mathcal{A}$ with limit $\bigcap_k A_k = \varnothing$ is stationary; hence every finitely additive probability on $\mathcal{A}$ is σ-additive on $\mathcal{A}$). In short, Q is the unique probability on $(\Omega, \mathcal{F}_\infty)$ such that

$$\forall n \in \mathbb{N} \quad \forall \Lambda_n \in \mathcal{F}_n \qquad Q(\Lambda_n) = \mathbb{E}_0[\mathbb{1}_{\Lambda_n} M_n].$$

We will also study the process X under Q.
1) In the first section, G is a function of the one-sided maximum, i.e.,

$$G_p = \varphi(S_p)$$

where $S_p = \sup\{X_k, k \leqslant p\}$ and where φ is a function from $\mathbb{N}$ to $\mathbb{R}^+$ satisfying

$$\sum_{k=0}^{\infty} \varphi(k) = 1$$

We will also need the function $\Phi : \mathbb{N} \longrightarrow \mathbb{R}^+$ given by

$$\Phi(k) := \sum_{j=k}^{\infty} \varphi(j).$$

The results of Section 1 are summarized in the following statement:

Theorem 1.   *1.   a) For each $n \geqslant 0$ and each $\Lambda_n \in \mathcal{F}_n$, one has*

$$\lim_{p \to \infty} \frac{\mathbb{E}[\mathbb{1}_{\Lambda_n} \varphi(S_p)]}{\mathbb{E}[\varphi(S_p)]} = \mathbb{E}[\mathbb{1}_{\Lambda_n} M_n^\varphi],$$

where $M_n^\varphi := \varphi(S_n)(S_n - X_n) + \Phi(S_n)$.

b) $(M_n^\varphi, n \geqslant 0)$ is a positive martingale, with $M_0^\varphi = 1$, non uniformly integrable; in fact, M_n^φ tends a.s. to 0 when $n \to \infty$.

2. Call Q^φ the probability on $(\Omega, \mathcal{F}_\infty)$ characterized by

$$\forall n \in \mathbb{N}, \Lambda_n \in \mathcal{F}_n, \quad Q^\varphi(\Lambda_n) = \mathbb{E}[\mathbb{1}_{\Lambda_n} M_n^\varphi].$$

Then

a) S_∞ is finite Q^φ-a.s. and satisfies for every $k \in \mathbb{N}$:

$$Q^\varphi(S_\infty = k) = \varphi(k). \tag{1}$$

b) Under Q^φ, the r.v. $T_\infty := \inf\{n \geqslant 0, X_n = S_\infty\}$ (which is not a stopping time in general) is a.s. finite and

i. $(X_{n \wedge T_\infty}, n \geqslant 0)$ and $(S_\infty - X_{T_\infty + n}, n \geqslant 0)$ are two independent processes;

ii. conditional on the r.v. S_∞, the process $(X_{n \wedge T_\infty}, n \geqslant 0)$ is a standard random walk stopped when it first hits the level S_∞;

iii. $(S_\infty - X_{T_\infty + n}, n \geqslant 0)$ is a 3-Bessel walk started from 0.

3. Put $R_n = 2S_n - X_n$. Under Q^φ, $(R_n, n \geqslant 0)$ is a 3-Bessel walk independent of S_∞.

The proofs of the second and third parts of this theorem rest largely upon a theorem due to Pitman (cf. [Pit75]) and on the study of the large p asymptotics of $\mathbb{P}(\Lambda_n | S_p = k)$ for $\Lambda_n \in \mathcal{F}_n$.

We must now explain the precise meaning of the '3-Bessel walk' mentioned in the theorem and further in this article. In fact, two processes, which we call the 3-Bessel walk and the 3-Bessel* walk, will play a role in this work; they are identical up to a one-step space shift.

The 3-Bessel walk is the Markov chain $(R_n, n \geqslant 0)$, with values in $\mathbb{N} = \{0, 1, 2, \ldots\}$, whose transition probabilities from $x \geqslant 0$ are given by

$$\pi(x, x+1) = \frac{x+2}{2x+2}; \qquad \pi(x, x-1) = \frac{x}{2x+2}. \tag{2}$$

The 3-Bessel* walk is the Markov chain $(R_n^*, n \geqslant 0)$, valued in $\mathbb{N}^* = \{1, 2, \ldots\}$, such that $R^* - 1$ is a 3-Bessel walk. So its transition probabilities from $x \geqslant 1$ are

$$\pi^*(x, x+1) = \frac{x+1}{2x}; \qquad \pi^*(x, x-1) = \frac{x-1}{2x}.$$

2) In the second section, the functional G_p will be a function of the local time at 0 of the random walk. The local time is the process $(L_n, n \geqslant 0)$ such that L_n is the number of times that X was null *strictly* before time n. In other words,

$$L_n = \sum_{m \geqslant 0} \mathbb{1}_{m<n} \mathbb{1}_{X_m=0}.$$

Observe that L_n is also the sum of the number of up-crossings from 0 to 1 and of the number of down-crossings from 0 to -1, up to time n. Given two functions h^+ and h^- from $\mathbb{N}^*$ to $\mathbb{R}^+$ such that

$$\frac{1}{2} \sum_{k=1}^{\infty} (h^+(k) + h^-(k)) = 1,$$

we consider the penalisation functional

$$G_p := h^+(L_p) \mathbb{1}_{X_p>0} + h^-(L_p) \mathbb{1}_{X_p<0}.$$

Putting

$$\Theta(x) = \frac{1}{2} \sum_{k=x+1}^{\infty} (h^+(k) + h^-(k)),$$

we obtain the following penalisation theorem.

Theorem 2.   1.   a) For each $n \geqslant 0$ and each $\Lambda_n \in \mathcal{F}_n$, one has

$$\lim_{p \to \infty} \frac{\mathbb{E}[\mathbb{1}_{\Lambda_n} G_p]}{\mathbb{E}[G_p]} = \mathbb{E}[\mathbb{1}_{\Lambda_n} M_n^{h^+, h^-}], \tag{3}$$

where $M_n^{h^+, h^-} := X_n^+ h^+(L_n) + X_n^- h^-(L_n) + \Theta(L_n).$

b) $M_n^{h^+,h^-}$ is a positive, non uniformly integrable martingale ; indeed, it tends to 0 when n tends to infinity.

2. Call Q^{h^+,h^-} the probability on $\mathcal{F}_\infty$ whose restriction to $\mathcal{F}_n$ is given by

$$\forall \Lambda_n \in \mathcal{F}_n, \qquad Q^{h^+,h^-}(\Lambda_n) = \mathbb{E}[\mathbb{1}_{\Lambda_n} M_n^{h^+,h^-}].$$

This Q^{h^+,h^-} has the following properties:

a) L_∞ is Q^{h^+,h^-}-a.s. finite and satisfies

$$\forall k \in \mathbb{N}^*, \quad Q^{h^+,h^-}(L_\infty = k) = \frac{1}{2}\left(h^+(k) + h^-(k)\right).$$

b) The r.v. $g := \sup\{n \geqslant 0, X_n = 0\}$ is Q^{h^+,h^-}-a.s. finite and, under Q^{h^+,h^-},

i. The processes $(X_{g+u}, u \geqslant 0)$ and $(X_{u \wedge g}, u \geqslant 0)$ are independent.

ii. With probability $\frac{1}{2}\sum_{k=1}^\infty h^+(k)$, the process $(X_{g+u}, u \geqslant 1)$ is a 3-Bessel* walk started from 1.
With probability $\frac{1}{2}\sum_{k=1}^\infty h^-(k)$, the process $(-X_{g+u}, u \geqslant 1)$ is a 3-Bessel* walk started from 1.

iii. Conditional on $L_\infty = l$, the process $(X_{u \wedge g}, u \geqslant 0)$ is a standard random walk stopped at its l-th passage at 0.

Our unusual choice for the definition of the local time at 0 will be helpful when proving the first point. The second part of the proof of this theorem rests essentially on an article by Le Gall (cf [LeG85]) which enables us to assess, under specific conditions, that a 3-Bessel* walk for $\mathbb{P}$ is is still a 3-Bessel* walk for Q^{h^+,h^-}.

3) In the third part, the penalisation functional G_p will be a function of the longest excursion completed until time p. Set $g_n := \sup\{k \leqslant n, X_k = 0\}$, $d_n := \inf\{k \geqslant n, X_k = 0\}$, and $\Sigma_n := \sup\{d_k - g_k, d_k \leqslant n\}$; for $n \geqslant 0$, Σ_n is the duration of the longest excursion completed until time n.

Fix an even integer $x \geqslant 0$, and consider the penalisation functional

$$G_p := \mathbb{1}_{\Sigma_p \leqslant x}.$$

To study penalisation by this G, we must also introduce $A_n := n - g_n$, which is the age of the current excursion, and $A_n^* := \sup_{k \leqslant n} A_k$, which is the longest duration of a (complete or incomplete) excursion until n. We also call $\tau = \inf\{n > 0, X_n = 0\}$ the first return time to 0, and we put

$$\theta(x) := \mathbb{E}_0\big[|X_x| \,\big|\, \tau > x\big].$$

Theorem 3.   *1.* *a) For each $n \geqslant 0$ and each $\Lambda_n \in \mathcal{F}_n$:*

$$\lim_{p \to \infty} \frac{\mathbb{E}_0[\mathbb{1}_{\Lambda_n} \mathbb{1}_{\Sigma_p \leqslant x}]}{\mathbb{P}_0[\Sigma_p \leqslant x]} = \mathbb{E}_0[\mathbb{1}_{\Lambda_n} M_n], \tag{4}$$

where

$$M_n := \left\{ \frac{|X_n|}{\theta(x)} + \tilde{\mathbb{P}}_{X_n}(\tilde{T}_0 \leqslant x - A_n) \, \mathbb{1}_{A_n \leqslant x} \right\} \mathbb{1}_{\Sigma_n \leqslant x}.$$

(In this expression and in similar ones, the meaning of $\tilde{\mathbb{P}}$ and $\tilde{T}_0$ is to be interpreted as follows: $\tilde{\mathbb{P}}_{X_n}(\tilde{T}_0 \leqslant x - A_n)$ stands for $f(X_n, x - A_n)$, with $f(y, z) = \mathbb{P}_y(T_0 \leqslant z)$.)

b) Moreover, $(M_n, n \geqslant 0)$ is a positive martingale, non uniformly inte-grable; indeed, $\lim_{n \to \infty} M_n = 0$ $\mathbb{P}$-a.s.

2. Call Q^x the probability on $\mathcal{F}_\infty$ whose restriction to $\mathcal{F}_n$ is defined by

$$\forall \Lambda_n \in \mathcal{F}_n, \qquad Q^x(\Lambda_n) = \mathbb{E}\left[\mathbb{1}_{\Lambda_n} M_n\right].$$

Under Q^x, one has:

a) $\Sigma_\infty \leqslant x$ a.s. and satisfies for all $y \leqslant x$:

$$Q^x(\Sigma_\infty > y) = 1 - \frac{\mathbb{P}(\tau > x)}{\mathbb{P}(\tau > y)}.$$

b) $A^*_\infty = \infty$ a.s.

c) The r.v. $g := \sup\{n \geqslant 0, X_n = 0\}$ is a.s. finite. Moreover, if $p = 2l$ or $2l + 1$ with $l \geqslant 0$,

$$Q^x(g > p) = \left(\frac{1}{2}\right)^l \sum_{k=0}^{l \wedge \frac{x}{2}} C_{2l-2k}^{l-k} C_{2k}^{k} \left(1 - \frac{\mathbb{P}(\tau > x)}{\mathbb{P}(\tau > 2k)}\right).$$

d) For y such that $0 \leqslant y \leqslant x$,

i. $(A_n, n \leqslant T_y^A)$ has the same law under $\mathbb{P}$ and Q^x.

ii. $(A_n, n \leqslant T_y^A)$ and $X_{T_y^A}$ are independent under $\mathbb{P}$ and under Q^x.

iii. Under Q^x, the law of $X_{T_y^A}$ is given by

$$Q^x(X_{T_y^A} = k) = \left\{ \frac{|k|}{\theta(x)} + \mathbb{P}_k(T_0 \leqslant x - y) \right\} \mathbb{P}(X_y = k \mid \tau > y).$$

iv.

$$Q^x \left(g > T_y^A \right) = 1 - \frac{\mathbb{P}\left(\tau > x\right)}{\mathbb{P}\left(\tau > y\right)}.$$

v. Under Q^x, $\left(A_n, \, n \leqslant T_y^A\right)$ is independent of $\left\{ g > T_y^A \right\}$.

3. Under Q^x,

a) The processes $(X_{n \wedge g}, \, n \geqslant 0)$ and $(X_{g+n}, \, n \geqslant 0)$ are independent.

b) With probability $\frac{1}{2}$, the process $(X_{g+n}, n \geqslant 0)$ is a 3-Bessel* walk and with probability $\frac{1}{2}$, the process $(-X_{g+n}, n \geqslant 0)$ is a 3-Bessel* walk.

c) Conditional on $L_\infty = l$, the process $(X_{n \wedge g}, \, n \geqslant 0)$ is a standard random walk stopped at its l-th return time to 0 and conditioned by $\{\Sigma_{\tau_l} \leqslant x\}$, where τ_l is the l-th return time to 0.

The proof of the first point of this theorem rests largely on a Tauberian theorem (cf [Fel50]) which gives the large p asymptotics of $\mathbb{P}\left(\Sigma_p \leqslant x\right)$. And the study of the process X under Q^x rests on arguments similar to those used in the proof of Theorem 2.

2 Principle of Penalisation

Penalisation can intuitively be interpreted as a generalisation of conditioning by a null event.

Consider the event $A_\infty := \{S_\infty \leqslant a\}$, where $a \in \mathbb{N}$. By recurrence of the standard walk, A_∞ is a $\mathbb{P}$-null event. One way of conditioning by A_∞, which involves the filtration $(\mathcal{F}_n)$, is to consider the sequence of events $A_p := \{S_p \leqslant a\}$ and to study the limit

$$\lim_{p \to \infty} \frac{\mathbb{E}\left[\mathbb{1}_{A_n \cap \{S_p \leqslant a\}}\right]}{\mathbb{E}\left[\mathbb{1}_{\{S_p \leqslant a\}}\right]}, \tag{5}$$

for each $n \in \mathbb{N}$ and each $A_n \in \mathcal{F}_n$.

Simple arguments show that the limit in (5) exists and equals

$$\mathbb{E}\left[\mathbb{1}_{\{A_n, \, S_n \leqslant a\}} \frac{a + 1 - X_n}{a + 1}\right].$$

Put $M_n := \mathbb{1}_{\{S_n \leqslant a\}} \frac{a+1-X_n}{a+1}$. The process M is the martingale $\frac{a+1-X}{a+1}$ stopped when S first hits $a + 1$; so it is a positive $\mathbb{P}_0$-martingale. Since $M_0 = 1$ and $M_\infty = 0$ a.s., M is not uniformly integrable. But a probability $Q_{(n)}$ can be defined on $\mathcal{F}_n$ by

$$\frac{dQ_{(n)}}{d\mathbb{P}}\bigg|_{\mathcal{F}_n} = M_n;$$

moreover, for $m < n$, $Q_{(m)}$ and $Q_{(n)}$ agree on $\mathcal{F}_m$. By Kolmogorov's existence theorem (cf [Bil] pp. 430-435), there exists a probability Q on $(\Omega, \mathcal{F}_\infty)$ whose restriction to each $\mathcal{F}_n$ is the corresponding $Q_{(n)}$; in other words, Q is characterized by

$$Q(\Lambda_n) := \mathbb{E}\left[\mathbb{1}_{\{\Lambda_n, S_n \leqslant a\}} \frac{a + 1 - X_n}{a + 1}\right]$$

for all $n \in \mathbb{N}$ and $\Lambda_n \in \mathcal{F}_n$.

When studying the behavior of $(X_n, n \geqslant 0)$ under the new probability Q, one obtains that S_∞ is a.s. finite and uniformly distributed on $[0, a]$. A more detailed study shows that:

- $(X_{n \wedge T_\infty}, n \geqslant 0)$ and $(S_\infty - X_{T_\infty + n}, n \geqslant 0)$ are two independent processes.
- Conditional on $\{S_\infty = k\}$, $(X_{n \wedge T_\infty}, n \geqslant 0)$ is a standard random walk stopped when it reaches the value k.
- $(S_\infty - X_{T_\infty + n}, n \geqslant 0)$ is a 3-Bessel walk started from 0, independent from (S_∞, T_∞).

This raises several natural questions: What happens when $\mathbb{1}_{\{S_n \leqslant a\}}$ is replaced with a more complicated function of the supremum? In that case, what does the limit (5) become? Can one still define a probability Q, and how is the behavior of $(X_n, n \geqslant 0)$ under Q influenced by this modification?

This simple idea of replacing the indicator by a more complex function is the essence of penalisation. All this is evidently not limited to the case of the one-sided maximum, but extends to many other increasing, adapted functionals tending $\mathbb{P}$-a.s. to $+\infty$. There exist various examples of penalisation, and also a general principle (cf [Deb07]) but this article is only devoted to three examples of penalisation functionals: the one-sided maximum, the local time at 0 and the maximal duration of the completed excursions.

3 Penalisation by a Function of the One-sided Maximum: Proof of Theorem 1

1) We start by recalling a few facts.

The next result is classical (cf. [Fel50] p. 75):

Lemma 1. For $k \in \mathbb{Z}$ and $n \in \mathbb{N}$,

$$\mathbb{P}_0(X_n = k) = \left(\frac{1}{2}\right)^n C_n^{\frac{n+k}{2}}.$$

Remark 1. In the sequel, we put

$$p_{n,k} := \mathbb{P}_0(X_n = k);$$

observe that $p_{n,k} \neq 0$ if and only if n and k have the same parity and $|k| \leqslant n$.

Lemma 2. *For k in $\mathbb{Z}$ and n and r in $\mathbb{N}$, one has*

$$\mathbb{P}_0(X_n = k, \; S_n \geqslant r) = \begin{cases} \mathbb{P}(X_n = k) & \text{if } k > r; \\ \mathbb{P}(X_n = 2r - k) & \text{if } k \leqslant r. \end{cases} \tag{6}$$

Proof. This formula is trivial when $k > r$; when $k \leqslant r$, it is Désiré André's well-known reflection principle (see for instance [Fel50] p. 72 and pp. 88-89). $\square$

From Lemma 2 and Remark 1, one easily derives the law of S:

Lemma 3. *For n and r in $\mathbb{N}$, one has*

$$\mathbb{P}_0(S_n = r) = p_{n,r} + p_{n,r+1} = p_{n,r} \vee p_{n,r+1}. \tag{7}$$

Proof. Summing (6) over all $k \in \mathbb{Z}$ gives

$$\mathbb{P}(S_n \geqslant r) = \sum_{k>r} \mathbb{P}(X_n = k) + \sum_{k \leqslant r} \mathbb{P}(X_n = 2r - k) = \mathbb{P}(X_n > r) + \mathbb{P}(X_n \geqslant r).$$

Consequently,

$$\mathbb{P}(S_n = r) = \mathbb{P}(S_n \geqslant r) - \mathbb{P}(S_n \geqslant r+1) = \mathbb{P}(X_n = r+1) + \mathbb{P}(X_n = r),$$

and (7) follows by definition of $p_{n,k}$ and by Remark 1. $\square$

2) We start showing point 1 of Theorem 1.

Lemma 4. *For each $k \geqslant 0$, the ratio*

$$\frac{\mathbb{P}(S_n = k)}{\mathbb{P}(S_n = 0)}$$

is majorized by 1 for all $n \geqslant 0$ and tends to 1 when $n \to +\infty$.

Proof. The denominator is minorated by $\mathbb{P}(X_1 = \ldots = X_n = -1) = 2^{-n}$; so it does not vanish. Observe that, for even n and even $k \geqslant 2$,

$$\frac{\mathbb{P}(S_n = k-1)}{\mathbb{P}(S_n = 0)} = \frac{\mathbb{P}(S_n = k)}{\mathbb{P}(S_n = 0)} = \frac{p_{n,k}}{p_{n,0}} = \left(\frac{n-k+2}{n+2}\right)\left(\frac{n-k+4}{n+4}\right)\cdots\left(\frac{n}{n+k}\right);$$

and for odd n and odd $k \geqslant 1$,

$$\frac{\mathbb{P}(S_n = k-1)}{\mathbb{P}(S_n = 0)} = \frac{\mathbb{P}(S_n = k)}{\mathbb{P}(S_n = 0)} = \frac{p_{n,k}}{p_{n,1}} = \left(\frac{n-k+2}{n+1}\right)\left(\frac{n-k+4}{n+3}\right)\cdots\left(\frac{n+1}{n+k}\right).$$

Clearly, these products are smaller than 1 and tend to 1 when n goes to infinity. $\square$

Lemma 5. *For all $x \in \mathbb{N}$ and $y \in \mathbb{Z}$ such that $y \leqslant x$, the ratio*

$$\frac{\mathbb{E}[\varphi(x \vee (y+S_n))]}{\mathbb{P}(S_n = 0)}$$

is majorized for all $n \in \mathbb{N}$ by $(x-y)\varphi(x)+\Phi(x)$ and tends to $(x-y)\varphi(x)+\Phi(x)$ when n tends to infinity.

Proof. Write

$$\frac{\mathbb{E}[\varphi(x \vee (y+S_n))]}{\mathbb{P}(S_n = 0)} = \varphi(x) \frac{\mathbb{P}(y + S_n < x)}{\mathbb{P}(S_n = 0)} + \sum_{k \geqslant x} \varphi(k) \frac{\mathbb{P}(y + S_n = k)}{\mathbb{P}(S_n = 0)}$$

$$= \varphi(x) \sum_{k < x-y} \frac{\mathbb{P}(S_n = k)}{\mathbb{P}(S_n = 0)} + \sum_{k \geqslant x} \varphi(k) \frac{\mathbb{P}(S_n = k - y)}{\mathbb{P}(S_n = 0)}.$$

By Lemma 4, this sum is majorized by $(x - y)\varphi(x) + \sum_{k \geqslant x} \varphi(k)$ and tends to this value by dominated convergence. □

To establish point 1 of Theorem 1, observe first that

$$M_n^\varphi = \varphi(S_n)(S_n - X_n) + \Phi(S_n)$$

is a positive martingale. Positivity is obvious: φ, Φ, and $S - X$ are positive. To see that M^φ is a martingale, consider $M_{n+1}^\varphi - M_n^\varphi$.

If $S_{n+1} = S_n$, the only thing that varies in the expression of M^φ when n is changed to $n + 1$ is X; so, in that case,

$$M_{n+1}^\varphi - M_n^\varphi = -\varphi(S_n)(X_{n+1} - X_n).$$

On the other hand, if $S_{n+1} \neq S_n$, one has $S_{n+1} = S_n + 1$ because each step of S is 0 or 1; one also has $X_{n+1} = S_{n+1}$ because S can increase only when pushed up by X, and $X_n = S_n$ because X_n must simultaneously be $\leqslant S_n$ and at distance 1 from X_{n+1}. So $S_{n+1} - X_{n+1} = S_n - X_n = 0$, giving

$$M_{n+1}^\varphi - M_n^\varphi = \Phi(S_{n+1}) - \Phi(S_n) = \Phi(S_n + 1) - \Phi(S_n)$$
$$= -\varphi(S_n) = -\varphi(S_n)(X_{n+1} - X_n).$$

All in all, the equality $M_{n+1}^\varphi - M_n^\varphi = -\varphi(S_n)(X_{n+1} - X_n)$ holds everywhere; this entails that M^φ is a martingale, verifying

$$|M_n^\varphi - M_0^\varphi| \leqslant n; \tag{8}$$

and since $M_0^\varphi = \Phi(0) = 1$, one has $\mathbb{E}[M_n^\varphi] = 1$.

We now proceed to prove 1.a of Theorem 1. For $0 \leqslant n \leqslant p$, one can write $S_p = S_n \vee (X_n + \widetilde{S}_{p-n})$, where $\widetilde{S}$ is the maximal process of the standard random walk $(X_{n+k} - X_n)_{k \geqslant 0}$, which is independent from $\mathcal{F}_n$. Hence

$$\mathbb{E}[\varphi(S_p) \mid \mathcal{F}_n] = \widetilde{\mathbb{E}}[\varphi(S_n \vee (X_n + \widetilde{S}_{p-n}))],$$

where $\widetilde{\mathbb{E}}$ integrates over $\widetilde{S}_{p-n}$ only, S_n and X_n being kept fixed. So, for $\Lambda_n \in \mathcal{F}_n$,

$$\frac{\mathbb{E}[\mathbb{1}_{\Lambda_n} \varphi(S_p)]}{\mathbb{P}(S_{p-n} = 0)} = \mathbb{E}\left[\mathbb{1}_{\Lambda_n} \frac{\widetilde{\mathbb{E}}[\varphi(S_n \vee (X_n + \widetilde{S}_{p-n}))]}{\widetilde{\mathbb{P}}(\widetilde{S}_{p-n} = 0)}\right].$$

When p tends to infinity, Lemma 5 says that the ratio in the right-hand side tends to M_n^φ and is dominated by M_n^φ, which is integrable by (8). Consequently,

$$\frac{\mathbb{E}[\mathbb{1}_{\Lambda_n} \varphi(S_p)]}{\mathbb{P}(S_{p-n} = 0)} \begin{cases} \text{is majorated by } \mathbb{E}[\mathbb{1}_{\Lambda_n} M_n^\varphi] \text{ for all } p \geqslant n \\ \text{and tends to } \quad \mathbb{E}[\mathbb{1}_{\Lambda_n} M_n^\varphi] \quad \text{when } p \to \infty. \end{cases} \tag{9}$$

Taking in particular $\Lambda_n = \Omega$, one also has

$$\frac{\mathbb{E}[\varphi(S_p)]}{\mathbb{P}(S_{p-n} = 0)} \to \mathbb{E}[M_n^\varphi] = 1 \qquad \text{when } p \to \infty,$$

and to establish 1.a of Theorem 1, it suffices to take the ratio of these two limits.

Half of 1.b is already proven: we have seen above that M^φ is a positive martingale, with $M_0^\varphi = 1$. The proof that $M_n^\varphi \to 0$ a.s. is postponed; we first establish 2.a.

The set-function Q^φ defined on the Boolean algebra $\bigcup_n \mathcal{F}_n$ by $Q^\varphi(\Lambda_n) = \mathbb{E}[\mathbb{1}_{\Lambda_n} M_n^\varphi]$ if $\Lambda_n \in \mathcal{F}_n$, is a probability on each σ-field $\mathcal{F}_n$. As recalled in the introduction, Q^φ automatically extends to a probability (still called Q^φ) on the σ-field $\mathcal{F}_\infty$.

For k and n in $\mathbb{N}$, the event $\{S_n \geqslant k\}$ is equal to $\{T_k \leqslant n\}$, where $T_k = \inf\{m : X_m \geqslant k\} = \inf\{m : S_m \geqslant k\}$. Now, by Doob's stopping theorem,

$$Q^\varphi(S_n \geqslant k) = Q^\varphi(T_k \leqslant n) = \mathbb{E}[\mathbb{1}_{T_k \leqslant n} M_n^\varphi]$$
$$= \mathbb{E}[\mathbb{1}_{T_k \leqslant n} M_{n \wedge T_k}^\varphi] = \mathbb{E}[\mathbb{1}_{T_k \leqslant n} M_{T_k}^\varphi].$$

But $\mathbb{P}_0$-a.s., $X_{T_k} = S_{T_k} = k$ and $M_{T_k}^\varphi = \Phi(k)$; wherefrom

$$Q^\varphi(S_n \geqslant k) = \Phi(k) \, \mathbb{P}(S_n \geqslant k).$$

Fixing k, let now n tend to infinity. The events $\{S_n \geqslant k\}$ form an increasing sequence, with limit $\{S_\infty \geqslant k\}$; hence

$$Q^\varphi(S_\infty \geqslant k) = \Phi(k) \, \mathbb{P}(S_\infty \geqslant k) = \Phi(k).$$

This implies that S_∞ is Q^φ-a.s. finite, with

$$Q^\varphi(S_\infty = k) = \Phi(k) - \Phi(k+1) = \varphi(k);$$

so 2.a is established.

This also implies that the $\mathbb{P}$-a.s. limit M_∞^φ of M^φ is null, by the following argument. Using Fatou's lemma, one writes

$$\mathbb{E}[\mathbb{1}_{S_\infty \geqslant k} M_\infty^\varphi] = \mathbb{E}\big[\lim_n (\mathbb{1}_{S_n \geqslant k} M_n^\varphi)\big]$$
$$\leqslant \liminf_n \mathbb{E}[\mathbb{1}_{S_n \geqslant k} M_n^\varphi]$$
$$= \liminf_n Q^\varphi(S_n \geqslant k) = Q^\varphi(S_\infty \geqslant k) = \Phi(k);$$

then, by dominated convergence, one has

$$\mathbb{E}[\mathbb{1}_{S_\infty = \infty} M_\infty^\varphi] = \mathbb{E}\big[\lim_k (\mathbb{1}_{S_\infty \geqslant k} M_\infty^\varphi)\big] = \lim_k \mathbb{E}[\mathbb{1}_{S_\infty \geqslant k} M_\infty^\varphi] \leqslant \lim_k \Phi(k) = 0,$$

and $\mathbb{P}(S_\infty = \infty) = 1$ now implies $\mathbb{E}[M_\infty^\varphi] = 0$. Point 1.b is proven.

3) Here are now a few facts on 3-Bessel walks, which will play an important role in the rest of the proof of Theorem 1.

Proposition 1. *Let $(R_n, n \geqslant 0)$ be a 3-Bessel walk; put $J_n = \inf_{m \geqslant n} R_m$.*
1. *Conditional on $\mathcal{F}_n^R$, the law of J_n is uniform on $\{0, 1, \ldots, R_n\}$.*
2. *Suppose now $R_0 = 0$ (therefore $J_0 = 0$ too).*
   a) *The process $(Z_n, n \geqslant 0)$ defined by $Z_n = 2J_n - R_n$ is a standard random walk, and its natural filtration $\mathcal{Z}$ is also the natural filtration of the 2-dimensional process (R, J).*
   b) *If T is a stopping time for $\mathcal{Z}$ such that $R_T = J_T$, then the process $(R_{T+n} - R_T, n \geqslant 0)$ is a 3-Bessel walk started from 0 and independent of $\mathcal{Z}_T$.*

Proof. 1. By the Markov property, it suffices to show that if $R_0 = k$, the r.v. J_0 is uniformly distributed on $\{0, \ldots, k\}$. The function $f(x) = 1/(1+x)$ defined for $x \geqslant 0$ is bounded and verifies for $x \geqslant 1$

$$f(x) = \pi(x, x-1) f(x-1) + \pi(x, x+1) f(x+1),$$

where π is the transition kernel of the 3-Bessel walk, given by (2). Thus f is π-harmonic except at $x = 0$, and $f(R_{n \wedge \sigma_0})$ is a bounded martingale, where σ_x denotes the hitting time of x by R. (This result is due to [LeG85] p. 449.) For $0 \leqslant a \leqslant k$, by stopping, $\mu_n^a = f(R_{n \wedge \sigma_a})$ is also a bounded martingale. A Borel-Cantelli argument shows that the paths of R are a.s. unbounded; hence $\liminf_{n \to \infty} f(R_n) = 0$ and $\mu_\infty^a = f(a) \mathbb{1}_{J_0 \leqslant a}$. The martingale equality $f(a) \mathbb{P}(J_0 \leqslant a) = \mathbb{E}[\mu_\infty^a] = \mathbb{E}[\mu_0^a] = f(k)$ yields $\mathbb{P}(J_0 \leqslant a) = (a+1)/(k+1)$, so the law of J_0 is uniform on $\{0, \ldots, k\}$.

Part 2 of Proposition 1 depends only on the law of the process R, so we need not prove it for *all* 3-Bessel walks started at 0, it suffices to prove it for *some particular* 3-Bessel walk started at 0. Given a standard random walk Z' with $Z_0' = 0$ and its past maximum $S_n' = \sup_{m \leqslant n} Z_m'$, Pitman's theorem [Pit75] says that the process $R = 2S' - Z'$ is a 3-Bessel walk started from 0, with future minimum $J_n = \inf_{m \geqslant n} R_m$ given by $J = S'$. We shall prove 2.a and 2.b for this particular 3-Bessel walk R.

The process $Z = 2J' - R = Z'$, so it is a standard random walk. Both $J = S'$ and $R = 2S' - Z'$ are adapted to the filtration of Z; conversely, $Z = 2J - R$ is adapted to the filtration generated by R and J. This proves 2.a.

To show 2.b, let T be $\mathcal{Z}$-stopping time such that $R_T = J_T$. One has

$$Z'_T = 2J_T - R_T = J_T = S'_T.$$

The process $\widetilde{Z}$ defined by $\widetilde{Z}_n = Z'_{T+n} - Z'_T$ is a standard random walk independent of $\mathcal{Z}_T$, started from 0, with past maximum

$$\widetilde{S}_n = \sup_{m \leqslant n} \widetilde{Z}_m = S'_{T+n} - Z'_T = S'_{T+n} - S'_T.$$

By Pitman's theorem, $\widetilde{R} = 2\widetilde{S} - \widetilde{Z}$ is a 3-Bessel walk, and it is independent of $\mathcal{Z}_T$ because so is $\widetilde{Z}$. Now,

$$\widetilde{R}_n = 2\widetilde{S}_n - \widetilde{Z}_n = 2(S'_{T+n} - S'_T) - (Z'_{T+n} - Z'_T) = R_{T+n} - R_T;$$

thus 2.b holds and Proposition 1 is established.   □

4) The next step is the proof of point 3 in Theorem 1. We start with a small computation:

Lemma 6. Let a r.v. U be uniformly distributed on $\{0, .., r\}$. Then

$$\mathbb{E}\big[\varphi(U)(r - U) + \Phi(U)\big] = 1$$

Proof. It suffices to write

$$(r+1)\,\mathbb{E}\big[1 - \Phi(U)\big] = \sum_{i=0}^{r}(1 - \Phi(i)) = \sum_{i=0}^{r}\sum_{j=0}^{i-1}\varphi(j) = \sum_{j=0}^{r-1}\sum_{i=j+1}^{r}\varphi(j)$$

$$= \sum_{j=0}^{r-1}(r - j)\varphi(j) = (r+1)\,\mathbb{E}\big[(r - U)\varphi(U)\big]. \quad \Box$$

The next proposition proves the first half of point 3 in Theorem 1.

Proposition 2. Under Q^φ, the process $(R_n, n \geqslant 0)$ given by $R_n = 2S_n - X_n$ is a 3-Bessel started from 0.

Proof. According to Pitman's theorem [Pit75], under the probability $\mathbb{P}$, the process $(R_n, n \geqslant 0)$ is a 3-Bessel walk with future infimum $J_n = S_n$. Call $\mathcal{R}$ the natural filtration of R. By Proposition 1.1, the conditional law of S_n given $\mathcal{R}_n$ is uniform on $\{0, \ldots, R_n\}$; consequently Lemma 6 gives

$$\mathbb{E}[M_n^\varphi | \mathcal{R}_n] = \mathbb{E}[\varphi(S_n)(R_n - S_n) + \Phi(S_n) \mid \mathcal{R}_n] = 1.$$

Now, let f be any bounded function on $\mathbb{N}^{n+1}$. One has

$$\mathbb{E}^{Q^\varphi}[f(R_0,\ldots,R_n)] = \mathbb{E}[f(R_0,\ldots,R_n)M_n^\varphi]$$
$$= \mathbb{E}\big[f(R_0,\ldots,R_n)\,\mathbb{E}[M_n^\varphi|\mathcal{R}_n]\big] = \mathbb{E}[f(R_0,\ldots,R_n)].$$

As n and f were arbitrary, R has the same law under Q^φ as under $\mathbb{P}$, that is, Q^φ also makes R a 3-Bessel walk. □

To finish proving point 3, it remains to establish that R is independent of S_∞ under Q^φ. This will easily follow from the next lemma, which decomposes Q^φ as a sum of measures carried by the level sets of S_∞.

Lemma 7. *Call $Q^{(k)}$ the probability Q^φ for $\varphi = \delta_k$, that is, $\varphi(k) = 1$ and $\varphi(x) = 0$ for $x \neq k$. Then $Q^{(k)}$ is supported by the event $\{S_\infty = k\}$, and, for a general φ and for all $\Lambda \in \mathcal{F}_\infty$ one has*

$$Q^\varphi(\Lambda) = \sum_{k \geqslant 0} \varphi(k)\, Q^{(k)}(\Lambda);$$

$$Q^\varphi(\Lambda \mid S_\infty = k) = Q^{(k)}(\Lambda) \qquad \text{for all } k \text{ such that } \varphi(k) > 0.$$

Proof. For $\Lambda_n \in \mathcal{F}_n$, one can use formula (9) twice and write

$$Q^\varphi(\Lambda_n) = \lim_p \frac{\mathbb{E}[\mathbb{1}_{\Lambda_n}\,\varphi(S_p)]}{\mathbb{P}(S_{p-n} = 0)} = \lim_p \sum_k \varphi(k)\,\frac{\mathbb{P}(\Lambda_n \cap \{S_p = k\})}{\mathbb{P}(S_{p-n} = 0)}$$

$$= \sum_k \varphi(k)\, \lim_p \frac{\mathbb{P}(\Lambda_n \cap \{S_p = k\})}{\mathbb{P}(S_{p-n} = 0)} = \sum_k \varphi(k)\, Q^{(k)}(\Lambda_n),$$

where $\lim$ and Σ commute by dominated convergence, owing to the majoration in (9). So the probabilities Q^φ and $\sum_k \varphi(k)\, Q^{(k)}$ coincide on $\bigcup_n \mathcal{F}_n$; therefore they also coincide on $\mathcal{F}_\infty$.

Applying now equation (1) with $\varphi = \delta_k$ gives $Q^{(k)}(S_\infty = k) = 1$, that is, $Q^{(k)}$ is supported by $\{S_\infty = k\}$.

Consequently, for any $\Lambda \in \mathcal{F}_\infty$, one has $Q^\varphi(\Lambda \cap \{S_\infty = k\}) = \varphi(k)\, Q^{(k)}(\Lambda)$ because all other terms in the series vanish. Using (1) again, one may replace $\varphi(k)$ with $Q^\varphi(S_\infty = k)$; this proves $Q^\varphi(\Lambda \mid S_\infty = k) = Q^{(k)}(\Lambda)$ whenever $\varphi(k) > 0$. □

The proof of independence in Theorem 1.3 is now a child's play: Proposition 2 says that the law of R under Q^φ is always the law of the 3-Bessel walk, whatever the choice of φ. We may in particular take $\varphi = \delta_k$, so it is also true under $Q^{(k)}$. Since $Q^{(k)}$ is also the conditioning of Q^φ by $\{S_\infty = k\}$, under Q^φ the law of R conditional on $\{S_\infty = k\}$ does not depend upon k, thus R is independent of S_∞.

5) So far, all of Theorem 1 has been established, except 2.b, to which the rest of the proof will be devoted. Finiteness of T_∞ is due to X being integer-valued and its supremum S_∞ being finite.

Put $U_n = X_{n \wedge T_\infty}$ and $V_n = S_\infty - X_{T_\infty + n}$. To prove 2.b.i and 2.b.iii we have to show that under Q^φ the process V is a 3-Bessel walk independent of the process U. Call ν the law of the 3-Bessel walk. For bounded functionals F and G, we must prove that

$$\mathbb{E}^{Q^\varphi}[F{\circ}U \, G{\circ}V] = \mathbb{E}^{Q^\varphi}[F{\circ}U] \int G(v) \, \nu(\mathrm{d}v).$$

Replacing now Q^φ by $\sum_k \varphi(k) \, Q^{(k)}$ (see Lemma 7), it suffices to show it when $\varphi = \delta_k$. Similarly, 2.b.ii only refers to a conditional law given S_∞; by Lemma 7 again, we may replace Q^φ by $Q^{(k)}$. Finally, *when proving 2.b, we may suppose $\varphi = \delta_k$ and $Q^\varphi = Q^{(k)}$ for a fixed $k \geqslant 0$.* Hence the random time T_∞ becomes the stopping time $T_k = \inf \{n \geqslant 0, \; X_n = k\}$, and it remains to show that

- $(X_{n \wedge T_k}, \; n \geqslant 0)$ is a standard random walk stopped when it first hits the level k;
- $(2k - X_{T_k + n}, \; n \geqslant 0)$ is a 3-Bessel walk started at 0;
- These two processes are independent.

By point 3 of Theorem 1, we know that $R = 2S - X$ is a 3-Bessel walk; and as we are now working under $Q^{(k)}$, we have $S_\infty = k$ a.s. Put $J_n = \inf_{m \geqslant n} R_m$.

We shall first show that the processes J and S are equal on the interval $[0, T_k]$. Given n, call τ the first time $p \geqslant n$ when $X_p = S_n$, and observe that on the event $\{T_k \geqslant n\}$, τ is finite because $X_n \leqslant S_n \leqslant k = X_{T_k}$. For all $m \geqslant n$, one has $R_m = S_m + (S_m - X_m) \geqslant S_n + 0$, with equality for $m = \tau$; thus $J_n = S_n$ on $\{\tau < \infty\}$ and a fortiori on $\{T_k \geqslant n\}$.

We shall now apply Proposition 1.2 to the 3-Bessel walk $R = 2S - X$ and its future infimum J. Part 2.a of this proposition says that $Z = 2J - R$ is a standard random walk. We just saw that $J = S$ on the random time-interval $[0, T_k]$; consequently, on this interval, $Z = 2S - R = X$. And as T_k is the first time when $X = k$, it is also the first time when $Z = k$. This proves that $(X_{n \wedge T_k}, \; n \geqslant 0)$ is a standard random walk stopped at level k, and also that the $\mathcal{Z}$-stopping time T_k satisfies $\mathcal{Z}_{T_k} = \mathcal{F}_{T_k}$, where $\mathcal{Z}$ is the filtration of Z.

Remarking that $R_{T_k} = J_{T_k} = k$, part 2.b of proposition 1 can be applied to T_k; it says that $(R_{T_k + n} - k, \; n \geqslant 0)$ is a 3-Bessel walk independent of $\mathcal{F}_{T_k}$, and hence also of the process $(X_{n \wedge T_k}, \; n \geqslant 0)$. But $R_{T_k + n} = 2S_{T_k + n} - X_{T_k + n} = 2k - X_{T_k + n}$ since $S_{T_k} = k = S_\infty$; so this 3-Bessel walk is nothing but $(k - X_{T_k + n}, \; n \geqslant 0)$. This concludes the proof of Theorem 1.

4 Penalisation by a Function of the Local Time: Proof of Theorem 2

Definition 1. *Recall that the 3-Bessel* walk is the Markov chain $(R_n^*, n \geqslant 0)$, valued in $\mathbb{N}^* = \{1, 2, \ldots\}$, such that $R^* - 1$ is a 3-Bessel walk. So its transition probabilities from $x \geqslant 1$ are*

$$\pi^*(x, x+1) = \frac{x+1}{2x}; \qquad \pi^*(x, x-1) = \frac{x-1}{2x}.$$

1) We now prove point 1 of Theorem 2. First, $(M_n^{h^+,h^-}, n \geqslant 0)$ is a positive martingale. Positivity is obvious from the definitions of h, h^- and Θ. To see that M^{h^+,h^-} is a martingale, we shall verify that the increment $M_{n+1}^{h^+,h^-} - M_n^{h^+,h^-}$ has the form $(X_{n+1} - X_n) K_n$, where K_n is $\mathcal{F}_n$-measurable and $|K_n| \leqslant 1$. There are three cases, depending on the value of X_n.

If $X_n > 0$, then $X_{n+1} \geqslant 0$, so $X_n^+ = X_n$, $X_{n+1}^+ = X_{n+1}$, and $L_{n+1} = L_n$. Consequently, in that case, $M_{n+1}^{h^+,h^-} - M_n^{h^+,h^-} = (X_{n+1} - X_n) h^+(L_n)$.

Similarly, if $X_n < 0$, one has $X_n^- = -X_n$, $X_{n+1}^- = -X_{n+1}$, $L_{n+1} = L_n$ and $M_{n+1}^{h^+,h^-} - M_n^{h^+,h^-} = -(X_{n+1} - X_n) h^-(L_n)$.

Last, if $X_n = 0$, then $L_{n+1} = L_n + 1$ and $X_{n+1} = \pm 1$. In that case,

$$M_{n+1}^{h^+,h^-} - M_n^{h^+,h^-} = \mathbb{1}_{\{X_{n+1}=1\}} h^+(L_n+1) + \mathbb{1}_{\{X_{n+1}=-1\}} h^-(L_n+1)$$
$$+\Theta(L_n+1) - \Theta(L_n)$$

$$= h^{\mathrm{sgn}(X_{n+1}-X_n)}(L_n+1) - \frac{1}{2}\big(h^+(L_n+1) + h^-(L_n+1)\big)$$

$$= (X_{n+1} - X_n)\frac{1}{2}\big(h^+(L_n+1) - h^-(L_n+1)\big).$$

This establishes the claim; consequently, M^{h^+,h^-} is a martingale which satisfies

$$\big|M_n^{h^+,h^-} - M_0^{h^+,h^-}\big| \leqslant n$$

and, as $M_0^{h^+,h^-} = 1$, one has $\mathbb{E}\big[M_n^{h^+,h^-}\big] = 1$.

To finish the proof of point 1 in Theorem 2, it remains to show formula (3). This will use the following lemma.

Lemma 8. *For each integer k such that $0 < k < \lfloor \frac{n}{2} \rfloor$,*

$$\frac{\mathbb{P}(L_n = k)}{\mathbb{P}(S_n = 0)}$$

is bounded above by 2 and tends to 1 when $n \to \infty$.

Remark 2. In the sequel, for $h : \mathbb{N} \to \mathbb{R}^+$ such that $\sum_{k=1}^{\infty} h(k) < \infty$, we put $M_n^{h,0} = X_n^+ h(L_n) + \Theta(L_n)$ for $n \geqslant 0$. When $\sum_{k=1}^{\infty} h(k) = 1$, this notation is consistent with the one used so far; in general, $M^{h,0}$ is a martingale too, for dividing it by the constant $\Theta(0) = \sum_{k=1}^{\infty} h(k)$ reduces it to the previous case.

Lemma 9. *Let $h : \mathbb{N} \longrightarrow \mathbb{R}^+$ be such that $\sum_{k=1}^{\infty} h(k) < \infty$. For $a \geqslant 0$ and $x \in \mathbb{Z}$,*

$$\frac{\mathbb{E}_x[h(L_n + a) \mathbb{1}_{X_n > 0}]}{\mathbb{P}(S_n = 0)}$$

is bounded above by $2(h(a)x^+ + \frac{1}{2}\sum_{k \geqslant a+1} h(k))$ and converges to $h(a)x^+ + \frac{1}{2}\sum_{k \geqslant a+1} h(k)$ when $n \to \infty$.

Proof of Lemma 8. Call $\gamma_n = |\{p \leqslant n, X_p = 0\}|$ the number of visits to 0 up to time n. Clearly, $\gamma_n = L_{n+1}$ and

$$\mathbb{P}(L_n = k) = \mathbb{P}(\gamma_{n-1} = k).$$

We shall study the law of γ_n. Define a sequence $(V_n, \ n \geqslant 0)$ by

$$\begin{cases} V_0 &= 0 \\ V_{n+1} &= \inf\{k > 0, \ X_{V_n+k} = 0\} \end{cases}$$

and put $(X_n^{(k)})_{n \geqslant 0} = (X_{V_k+n})_{n \geqslant 0}$ and $T_i^{(k)} = \inf\{n \geqslant 0, \ X_n^{(k)} = i\}$.

Owing to the symmetry of the random walk and the Markov property,

$$\forall i \geqslant 1 \qquad \mathbb{P}(V_i = k) = \mathbb{P}(T_1^{(i-1)} = k - 1).$$

So $\forall i \geqslant 1$, $V_i \overset{\mathcal{L}}{=} T_1^{(i-1)} + 1$. Moreover, according to the strong Markov property, $(X_n^{(2)}, n \geqslant 0)$ is independent of $\mathcal{F}_{V_1}$ and hence

$$V_1 + V_2 \overset{\mathcal{L}}{=} T_1^{(0)} + T_1^{(1)} + 2.$$

Wherefrom, by induction,

$$V_1 + V_2 + ... + V_k \overset{\mathcal{L}}{=} T_1^{(0)} + T_1^{(1)} + ... + T_1^{(k-1)} + k.$$

So

$$\mathbb{P}(\gamma_n = k) = \mathbb{P}(V_1 + ... + V_{k-1} \leqslant n < V_1 + ... + V_k)$$
$$= \mathbb{P}(T_1^{(0)} + T_1^{(1)} + ... + T_1^{(k-2)} + k - 1 \leqslant n < T_1^{(0)} + T_1^{(1)} + ... + T_1^{(k-1)} + k)$$
$$= \mathbb{P}(T_{k-1} + k - 1 \leqslant n < T_k + k) = \mathbb{P}(S_{n-k+1} \geqslant k - 1, S_{n-k} < k)$$
$$= \mathbb{P}(S_{n-k+1} = k - 1) + \mathbb{P}(T_k = n - k + 1).$$

Taking inspiration from the proof of Lemma 4, it is easy to see that

$$\frac{\mathbb{P}(S_{n-k} = k - 1)}{\mathbb{P}(S_n = 0)}$$

is majorated by 1 and tends to 1 when n tends to infinity.

According to [Fel50] p. 89,

$$\mathbb{P}(T_r = n) = \frac{r}{n} C_n^{\frac{n+r}{2}} \left(\frac{1}{2}\right)^n.$$

Appealing again to the proof of Lemma 4, it is easy to show that

$$\frac{\mathbb{P}(T_k = n - k)}{\mathbb{P}(S_n = 0)}$$

is majorated by 1 and tends to 0 when n goes to infinity. The proof is over.

$\square$

Remark 3. From the preceding result, one easily sees that

$$\frac{\mathbb{P}_x\left(L_n = k, X_n > 0\right)}{\mathbb{P}\left(S_n = 0\right)}$$

is majorated by 1 and tends to $\frac{1}{2}$ when $n \to \infty$.

Proof of Lemma 9. Start from

$$\mathbb{E}_x[h(L_n + a)\, \mathbb{1}_{X_n > 0}] = \mathbb{E}_x[h(L_n + a)\mathbb{1}_{X_n > 0}\,(\mathbb{1}_{T_0 > n} + \mathbb{1}_{T_0 \leqslant n})]$$

One has

$$h(L_n + a)\, \mathbb{1}_{X_n > 0}\mathbb{1}_{T_0 > n} = \begin{cases} 0 & si\ x \leqslant 0 \\ h(a)\, \mathbb{1}_{T_0 > n} & si\ x > 0 \end{cases}$$

According to Lemma 4,

$$\frac{h(a)\, \mathbb{1}_{x > 0}\mathbb{P}_x(T_0 > n)}{\mathbb{P}(S_n = 0)}$$

is majorated by $x^+ h(a)$ and converges to $x^+ h(a)$.

Write

$$\frac{\mathbb{E}_x[h(L_n + a)\, \mathbb{1}_{\{X_n > 0, T_0 \leqslant n\}}]}{\mathbb{P}(S_n = 0)} = \sum_{k \geqslant 1} \frac{\mathbb{P}_x\left(L_n = k, X_n > 0\right)}{\mathbb{P}\left(S_n = 0\right)}\, h(k + a)$$

By Lemma 8,this sum is majorated by $\sum_{k \geqslant 1} h(k + a)$ and converges to $\frac{1}{2}\sum_{k \geqslant 1} h(k + a)$ when $n \to \infty$. $\square$

We shall now prove point 1.a in Theorem 2. For each $0 \leqslant n \leqslant p$, one has $L_p = L_n + \tilde{L}_{p-n}$ where $\tilde{L}$ is the local time at 0 of the standard random walk $(X_{n+k})_{k \geqslant 0}$ which, given X_n, is independent of $\mathcal{F}_n$. So

$$\mathbb{E}\left[h(L_p)\, \mathbb{1}_{X_p > 0} \mid \mathcal{F}_n\right] = \tilde{\mathbb{E}}_{X_n}\left[h(L_n + \tilde{L}_{p-n})\, \mathbb{1}_{\tilde{X}_{p-n} > 0}\right]$$

where $\tilde{\mathbb{E}}$ integrates only $\tilde{L}_{p-n}$ and $\tilde{X}_{p-n}$ and where L_n and X_n are fixed. Then, for all $\Lambda_n \in \mathcal{F}_n$,

$$\frac{\mathbb{E}\left[h(L_p)\, \mathbb{1}_{X_p > 0, \Lambda_n}\right]}{\mathbb{P}(S_{p-n} = 0)} = \mathbb{E}\left[\mathbb{1}_{\Lambda_n} \frac{\tilde{\mathbb{E}}_{X_n}\left[h(L_n + \tilde{L}_{p-n})\, \mathbb{1}_{\tilde{X}_{p-n} > 0}\right]}{\mathbb{P}(S_{p-n} = 0)}\right]$$

When $p \to \infty$, Lemma 9 says that the ratio in the right-hand side tends to $M_n^{h,0}$ and is dominated by $2M_n^{h,0}$, which is integrable. Consequently, when $p \to \infty$,

$$\frac{\mathbb{E}[h(L_p)\, \mathbb{1}_{X_p > 0, \Lambda_n}]}{\mathbb{P}\left(S_{p-n} = 0\right)} \to \mathbb{E}[\mathbb{1}_{\Lambda_n} M_n^{h,0}],$$

and taking $\Lambda_n = \Omega$, one has

$$\frac{\mathbb{E}[h(L_p)\,\mathbb{1}_{X_p>0}]}{\mathbb{P}(S_{p-n}=0)} \to \mathbb{E}[M_n^{h,0}].$$

Taking the ratio of these two limits yields

$$\frac{\mathbb{E}[h(L_p)\,\mathbb{1}_{X_p>0,\Lambda_n}]}{\mathbb{E}[h(L_p)\,\mathbb{1}_{X_p>0}]} \to \frac{\mathbb{E}[\Lambda_n M_n^{h,0}]}{\mathbb{E}[M_n^{h,0}]}.$$

To finalize the proof of point 1.a, it now suffices to use the symmetry of the standard random walk and the fact that $\mathbb{E}[M_n^{h^+,h^-}] = 1$.

2) Let us now show point 2 in Theorem 2. Put $\tau_l = \inf\{k \geqslant 0, \gamma_k = l\}$. Then

$$Q^{h^+,h^-}(L_n \geqslant l) = Q^{h^+,h^-}(\tau_l \leqslant n-1)$$
$$= \mathbb{E}[\mathbb{1}_{\tau_l \leqslant n-1} M_{\tau_l}^{h^+,h^-}] = \Theta(l-1)\mathbb{P}(\tau_l \leqslant n-1).$$

For fixed l, the sequence of events $\{L_n \geqslant l\}$ is increasing and tends to $\{L_\infty \geqslant l\}$; so

$$Q^{h^+,h^-}(L_\infty \geqslant l) = \Theta(l-1)\mathbb{P}(\tau_l \leqslant \infty) = \Theta(l-1).$$

Hence L_∞ is Q^{h^+,h^-}-a.s. finite, with

$$Q^{h^+,h^-}(L_\infty = l) = \Theta(l-1) - \Theta(l) = \frac{1}{2}\left(h^+(l) + h^-(l)\right)$$

and 2.a is established.

To show that the $\mathbb{P}$-a.s. limit $M_\infty^{h^+,h^-}$ of M^{h^+,h^-} is null, it suffices to apply the same method as for M^φ, with L instead of S and M^{h^+,h^-} instead of M^φ.

The study of the process $(X_n, n \geqslant 0)$ under Q^{h^+,h^-} starts with the next three lemmas.

Lemma 10. *Under $\mathbb{P}_1$ and conditional on the event $\{T_p < T_0\}$, the process $(X_n, 0 \leqslant n \leqslant T_p)$ is a 3-Bessel* walk started from 1 and stopped when it first hits the level p (cf. [LeG85]).*

For typographical simplicity, call $T_{p,n} := \inf\{k > n,\ X_k = p\}$ the time of the first visit to p after n, and $\mathcal{H}_l := \{T_{p,\tau_l} < \tau_{l+1}, X_{\tau_l+1}=1\}$, the event that the l-th excursion is positive and reaches level p.

Lemma 11. *Under the law Q^{h^+,h^-} and conditional on the event $\mathcal{H}_l$, the process $(X_{n+\tau_l}, 1 \leqslant n \leqslant T_{p,\tau_l} - \tau_l)$ is a 3-Bessel* walk started from 1 and stopped when it first hits the level p.*

Lemma 12. *Put $\Gamma^+ := \{X_{n+g} > 0, \forall n > 0\}$ and $\Gamma^- := \{X_{n+g} < 0, \forall n > 0\}$. Then:*

$$Q^{h^+,h^-}(\Gamma^+) = 1 - Q^{h^+,h^-}(\Gamma^-) = \frac{1}{2}\sum_{k=1}^{\infty} h^+(k)$$

Proof of Lemma 11. Let G be a function from $\mathbb{Z}^n$ to $\mathbb{R}^+$. Then, according to the definition of the probability Q^{h^+,h^-} and owing to Doob's stopping theorem,

$$
\begin{aligned}
\mathcal{K} &:= Q^{h^+,h^-}\left[G(X_{\tau_l+1},\dots,X_{\tau_l+n})\,\mathbb{1}_{n+\tau_l\leqslant T_{p,\tau_l}}\;\middle|\;\mathcal{H}_l\right] \\
&= \frac{Q^{h^+,h^-}\left[G(X_{\tau_l+1},\dots,X_{\tau_l+n})\,\mathbb{1}_{\tau_l+n\leqslant T_{p,\tau_l}<\tau_{l+1},X_{\tau_l+1}=1}\right]}{Q^{h^+,h^-}(\mathcal{H}_l)} \\
&= \frac{\mathbb{E}\left[G(X_{\tau_l+1},\dots,X_{\tau_l+n})\,\mathbb{1}_{\tau_l+n\leqslant T_{p,\tau_l}<\tau_{l+1},X_{\tau_l+1}=1}M^{h^+,h^-}_{\tau_{l+1}}\right]}{\mathbb{E}\left[\mathbb{1}_{\mathcal{H}_l}M^{h^+,h^-}_{\tau_{l+1}}\right]}.
\end{aligned}
$$

Replacing $M^{h^+,h^-}_{\tau_{l+1}}$ by the constant $\Theta(l)$ and using the Markov property, one gets

$$
\begin{aligned}
\mathcal{K} &= \frac{\mathbb{E}\left[G(X_{\tau_l+1},\dots,X_{\tau_l+n})\,\mathbb{1}_{\tau_l+n\leqslant T_{p,\tau_l}<\tau_{l+1},X_{\tau_l+1}=1}\right]}{\mathbb{P}(\mathcal{H}_l)} \\
&= \frac{\mathbb{E}_1\left[G(X_0,\dots,X_{n-1})\,\mathbb{1}_{n-1\leqslant T_p<T_0}\right]}{\mathbb{P}_1(T_p<T_0)} \\
&= \mathbb{E}_1\left[G(X_0,\dots,X_{n-1})\,\mathbb{1}_{n-1\leqslant T_p}\;\middle|\;T_p<T_0\right]. \qquad \square
\end{aligned}
$$

Remark 4. By letting $p\to\infty$, one deduces therefrom that, conditional on $\{g=\tau_l,X_{\tau_l+1}=1\}$, $(X_{n+g},n\geqslant 1)$ is a 3-Bessel* walk under Q^{h^+,h^-}.

Proof of Lemma 12. As g is Q^{h^+,h^-}-a.s. finite and as $X_n\neq 0$ for $n>g$, one has

$$
Q^{h^+,h^-}(\Gamma^+) = \lim_{n\to\infty} Q^{h^+,h^-}(X_n>0). \tag{10}
$$

Now,

$$
Q^{h^+,h^-}(X_n>0) = \mathbb{E}[\mathbb{1}_{X_n>0}M^{h^+,h^-}_n] = \mathbb{E}[\mathbb{1}_{X_n>0}\Theta(L_n)+X_n^+h^+(L_n)].
$$

Since $\mathbb{1}_{X_n>0}\,\Theta(L_n)\leqslant\Theta(L_n)\leqslant 1$, the dominated convergence theorem gives

$$
\mathbb{E}[\mathbb{1}_{X_n>0}\Theta(L_n)]\overset{n\to\infty}{\longrightarrow}0.
$$

We already know that $M^{h^+,0}$ is a martingale. Consequently,

$$
\mathbb{E}[M^{h^+,0}_n] = \mathbb{E}[M^{h^+,0}_0] = \frac{1}{2}\sum_{k=1}^\infty h^+(k),
$$

wherefrom

$$
\mathbb{E}[X_n^+h^+(L_n)] = \frac{1}{2}\mathbb{E}\left[\sum_{k=1}^{L_n}h^+(k)\right]\leqslant\frac{1}{2}\sum_{k=1}^\infty h^+(k).
$$

By dominated convergence again,

$$\lim_{n\to\infty} \mathbb{E}[X_n^+ h^+(L_n)] = \frac{1}{2}\sum_{k=1}^{\infty} h^+(k),$$

and so, according to (10), $Q^{h^+,h^-}(\Gamma^+) = \frac{1}{2}\sum_{k=1}^{\infty} h^+(k)$.   □

For $F : \mathbb{Z}^n \to \mathbb{R}^+$,

$$\mathbb{E}^Q\big[F(X_{g+1},\dots,X_{g+n})\,\mathbb{1}_{X_{g+1}=1}\big] = \sum_{l\geqslant 1} \mathbb{E}^Q\big[F(X_{g+1},\dots,X_{g+n})\,\mathbb{1}_{g=\tau_l,X_{g+1}=1}\big]$$

$$= \sum_{l\geqslant 1} \mathbb{E}^Q\big[F(X_{g+1},\dots,X_{g+n})\mid g=\tau_l, X_{\tau_l+1}=1\big]\,Q^{h^+,h^-}(g=\tau_l,X_{\tau_l+1}=1)$$

$$= \mathbb{E}_1\big[F(X_0,\dots,X_{n-1})\mid T_0=\infty\big] \sum_{l\geqslant 1} Q^{h^+,h^-}(g=\tau_l,X_{\tau_l+1}=1)$$

$$= \mathbb{E}_1\big[F(X_0,\dots,X_{n-1})\mid T_0=\infty\big]\,Q^{h^+,h^-}(\Gamma^+).$$

This shows half of point 2.b.ii. The other half, when $X_{g+1}=-1$, is easily obtained using the symmetry of the walk.

To end of the proof of Theorem 2, we shall show that, conditional on $\{L_\infty=l\}$ and under the law Q^{h^+,h^-}, the process $(X_u,\ u<g)$ is a standard random walk stopped at its l-th passage at 0.

Let F be a function from $\mathbb{Z}^n$ to $\mathbb{R}^+$ and l an element of $\mathbb{N}^*$. From the definition of Q^{h^+,h^-} and the optional stopping theorem,

$$\mathbb{E}^Q\big[F(X_1,\dots,X_n)\,\mathbb{1}_{n<\tau_l}\mid L_\infty=l\big] = \frac{\mathbb{E}^Q\big[F(X_1,\dots,X_n)\,\mathbb{1}_{n<\tau_l<\infty}\mathbb{1}_{\tau_{l+1}=\infty}\big]}{Q^{h^+,h^-}(L_\infty=l)}$$

$$= \frac{\mathbb{E}^Q\big[F(X_1,\dots,X_n)\,\mathbb{1}_{n<\tau_l<\infty}\big] - \mathbb{E}^Q\big[F(X_1,\dots,X_n)\,\mathbb{1}_{n<\tau_l<\tau_{l+1}<\infty}\big]}{Q^{h^+,h^-}(L_\infty=l)}$$

$$= \frac{\mathbb{E}\big[F(X_1,\dots,X_n)\,\mathbb{1}_{n<\tau_l}M_{\tau_l}\big] - \mathbb{E}^Q\big[F(X_1,\dots,X_n)\,\mathbb{1}_{n<\tau_l}M_{\tau_{l+1}}\big]}{Q^{h^+,h^-}(L_\infty=l)}$$

$$= \frac{\mathbb{E}\big[F(X_1,\dots,X_n)\,\mathbb{1}_{n<\tau_l}\big]\,(\Theta(l-1)-\Theta(l))}{\frac{1}{2}(h^+(l)+h^-(l))} = \mathbb{E}\big[F(X_1,\dots,X_n)\,\mathbb{1}_{n<\tau_l}\big]. \quad □$$

5 Penalisation by the Length of the Excursions

5.1 Notation

For $n\geqslant 0$, call g_n (respectively d_n) the last zero before n (respectively after n):

$$g_n := \sup\{k\leqslant n,\ X_k=0\}$$
$$d_n := \inf\{k>n,\ X_k=0\}$$

Thus $d_n - g_n$ is the duration of the excursion that straddles n. Put

$$\Sigma_n = \sup\{d_k - g_k, \ d_k \leqslant n\},$$

so Σ_n is the longest excursion before g_n; remark that

$$\Sigma_n = \Sigma_{g_n}. \tag{11}$$

Define $(A_n, \ n \geqslant 0)$, the "age process", by

$$A_n = n - g_n,$$

and call $\mathcal{A}_n = \sigma\,(A_n, \ n \geqslant 0)$ the filtration generated by A. Set

$$A_n^* = \sup_{k \leqslant n} A_k, \tag{12}$$

and observe that

$$A_n^* = (\Sigma_n - 1) \vee (n - g_n),$$

wherefrom

$$A_{g_n}^* = \Sigma_{g_n} - 1. \tag{13}$$

In the sequel, $\gamma_l := \sum_{k=0}^{n} \mathbb{1}_{\{X_k = 0\}}$ is the number of passage times at 0 up to time n, $\tau = \inf\{n > 0, \ X_n = 0\}$ is the first return time to 0 and a function θ is defined by

$$\mathbb{E}\left[\|X_x\| \mid \tau > x\right] =: \theta(x).$$

5.2 Proof of Theorem 3

1) We start with point 1 of Theorem 3. To show formula (4), we need:

Proposition 3.

$$\mathbb{P}(\Sigma_k \leqslant x) \underset{k \to \infty}{\sim} \left(\frac{2}{\pi k}\right)^{\frac{1}{2}} \theta(x).$$

To establish this Proposition, we will use the following lemma:

Lemma 13. *For every* $f : \mathbb{Z} \to \mathbb{R}^+$, *every* $n \geqslant 0$ *and every* $k > 0$,

$$\mathbb{E}\left[f(X_n) \mid A_n = k\right] = \mathbb{E}\left[f(X_k) \mid \tau > k\right].$$

and a Tauberian Theorem:

Theorem 4 (Cf. [Fel71] p. 447). *Given* $q_n \geqslant 0$, *suppose that the series*

$$S(s) = \sum_{n=0}^{\infty} q_n s^n$$

converges for $0 \leqslant s < 1$. *If* $0 < p < \infty$ *and if the sequence* $\{q_n\}$ *is monotone, then the two relations:*

$$S(s) \underset{s \to 1^-}{\sim} \frac{1}{(1-s)^p} C$$

and

$$q_n \underset{n \to \infty}{\sim} \frac{1}{\Gamma(p)} n^{p-1} C$$

where $0 < C < \infty$, are equivalent.

Proof of Lemma 13. By the Markov property,

$$\mathbb{E}\left[f(X_n) \mid A_n = k\right] = \mathbb{E}\left[f(X_n) \mid n - g_n = k\right]$$
$$= \mathbb{E}\left[f(X_n) \mid X_{n-k} = 0, X_{n-k+1} \neq 0, \ldots, X_n \neq 0\right] = \mathbb{E}\left[f(X_k) \mid \tau > k\right]. \quad \square$$

Proof of Proposition 3. Let δ_β be a geometric r.v. with parameter β, where $0 < \beta < 1$, and such that δ_β is independent of the walk X. Then

$$\mathbb{P}\left(\Sigma_{\delta_\beta} \leqslant x\right) = \sum_{k=1}^\infty \mathbb{P}\left(\delta_\beta = k\right) \mathbb{P}\left(\Sigma_k \leqslant x\right) = \sum_{k=1}^\infty (1-\beta)^{k-1} \beta \, \mathbb{P}\left(\Sigma_k \leqslant x\right).$$

Now, from (11) and (13),

$$\mathbb{P}(\Sigma_{\delta_\beta} \leqslant x) = \mathbb{P}(\Sigma_{g_{\delta_\beta}} \leqslant x) = \mathbb{P}(A^*_{g_{\delta_\beta}} \leqslant x) = \mathbb{P}(T^A_x \geqslant g_{\delta_\beta})$$
$$= \mathbb{P}(\delta_\beta \leqslant d_{T^A_x}) = 1 - \mathbb{P}(\delta_\beta > d_{T^A_x}) = 1 - \mathbb{E}\left[(1-\beta)^{d_{T^A_x}}\right]$$
$$= 1 - \mathbb{E}\left[(1-\beta)^{T^A_x} (1-\beta)^{T_0 \circ \theta_{T^A_x}}\right]$$
$$= 1 - \mathbb{E}\left[(1-\beta)^{T^A_x} \mathbb{E}_{X_{T^A_x}}\left[(1-\beta)^{T_0}\right]\right]. \qquad \square$$

$$(14)$$

Definition 2. A stopping time T is said to be X-standard if T is a.s. finite and if the stopped process $(X_{n \wedge T}, n \geqslant 0)$ is uniformly integrable.

According to [ALR04], if T is X-standard and if T is independent of X_T, then

$$\forall \alpha \in \mathbb{R} \qquad \mathbb{E}\left[\text{ch}(\alpha)^{-T}\right] = \mathbb{E}\left[\exp(\alpha X_T)\right]^{-1}. \qquad (15)$$

Recall that $\text{Arg ch}(\alpha) = \ln\left(\alpha + \sqrt{\alpha^2 - 1}\right)$. When $\text{ch}\,\alpha = (1-\beta)^{-1}$,

$$\alpha = \text{Arg ch}\left(\frac{1}{1-\beta}\right) = \ln\left(\frac{1}{1-\beta} + \sqrt{\frac{1}{(1-\beta)^2} - 1}\right) = \ln\left(\frac{1 + \sqrt{2\beta - \beta^2}}{1-\beta}\right).$$

According to [ALR04], T_k and T^A_x satisfy these properties, hence

$$\mathbb{E}_k\left[(1-\beta)^{T_0}\right] = \mathbb{E}_0\left[(1-\beta)^{T_k}\right] = \left(\frac{1 + \sqrt{2\beta - \beta^2}}{1-\beta}\right)^{-k}$$

$$\mathbb{E}\left[(1-\beta)^{T^A_x}\right] = \mathbb{E}\left[\left(\frac{1 + \sqrt{2\beta - \beta^2}}{1-\beta}\right)^{X_{T^A_x}}\right]^{-1}$$

So, owing to the independence of T_x^A et $X_{T_x^A}$ and the above formulae,

$$\mathbb{P}\left(\Sigma_{\delta_\beta} \leqslant x\right) = 1 - \mathbb{E}\left[(1-\beta)^{T_x^A}\right]\mathbb{E}\left[\left(\frac{1+\sqrt{2\beta-\beta^2}}{1-\beta}\right)^{-|X_{T_x^A}|}\right]$$

$$= \frac{\frac{1}{2}\left[\mathbb{E}\left[\left(\frac{1+\sqrt{2\beta-\beta^2}}{1-\beta}\right)^{|X_{T_x^A}|}\right] - \mathbb{E}\left[\left(\frac{1+\sqrt{2\beta-\beta^2}}{1-\beta}\right)^{-|X_{T_x^A}|}\right]\right]}{\mathbb{E}\left[\left(\frac{1+\sqrt{2\beta-\beta^2}}{1-\beta}\right)^{X_{T_x^A}}\right]}.$$

For all $k \in \mathbb{N}$,

$$\left[\frac{1+\sqrt{2\beta-\beta^2}}{1-\beta}\right]^k \underset{\beta\to0+}{\sim} 1 + k\sqrt{2\beta},$$

and consequently $\mathbb{P}\left(\Sigma_{\delta_\beta} \leqslant x\right) \underset{\beta\to0+}{\sim} \mathbb{E}\left[|X_{T_x^A}|\right]\sqrt{2\beta}$.

Thus we have obtained

$$\sum_{k=1}^\infty (1-\beta)^k \mathbb{P}\left(\Sigma_k \leqslant x\right) \underset{\beta\to0+}{\sim} \sqrt{\frac{2}{\beta}}(1-\beta)\mathbb{E}\left[|X_{T_x^A}|\right].$$

In order to apply Theorem 4, put $\alpha = 1 - \beta$. This gives

$$\sum_{k=1}^\infty \alpha^k \mathbb{P}\left(\Sigma_k \leqslant x\right) \underset{\alpha\to1-}{\sim} \frac{\sqrt{2}}{\sqrt{1-\alpha}}\mathbb{E}\left[|X_{T_x^A}|\right],$$

and now Theorem 4 with $p = \frac{1}{2}$ and $C = \sqrt{2}\mathbb{E}\left[|X_{T_x^A}|\right]$ gives

$$\mathbb{P}(\Sigma_k \leqslant x) \underset{\alpha\to1-}{\sim} \frac{1}{\Gamma\left(\frac{1}{2}\right)}k^{\frac{1}{2}-1}C = \left(\frac{2}{\pi k}\right)^{\frac{1}{2}}\mathbb{E}\left[|X_{T_x^A}|\right].$$

By Lemma 13,

$$\mathbb{E}\left[|X_{T_x^A}|\right] = \mathbb{E}\left[|X_{T_x^A}| \mid A_{T_x^A} = x\right] = \mathbb{E}\left[|X_x| \mid \tau > x\right] = \theta(x).$$

It is now possible to finalise the proof of point 1.a. Let $\tilde{T}_0$ be the hitting time of 0 by the walk $(X_{n+k})_{k\geqslant0}$, and Σ' be the maximal length of the excursions of the walk $(X_{k+n+\tilde{T}_0})_{k\geqslant0}$.

$$\mathbb{E}\left[\mathbb{1}_{\Lambda_n,\Sigma_p\leqslant x}\right] = \mathbb{E}\left[\mathbb{1}_{\Lambda_n,\Sigma_n\leqslant x,T_0\circ\theta_n>p-n}\right]$$
$$+ \mathbb{E}\left[\mathbb{1}_{\Lambda_n,\Sigma_n\leqslant x,T_0\circ\theta_n\leqslant(p-n)\wedge(x-A_n),\Sigma'_{p-n-T_0\circ\theta_n}\leqslant x}\right] = (1) + (2)$$

Call $\tilde{\mathbb{P}}$ the measure associated to the walk $(X_{n+k})_{k\geqslant 0}$, X_n and A_n being fixed. Then

$$(1) = \mathbb{E}\left[\mathbb{1}_{A_n, \Sigma_n \leqslant x}\tilde{\mathbb{P}}_{X_n}\left(\tilde{T}_0 > p - n\right)\right] \underset{p\to\infty}{\sim} \mathbb{E}\left[\mathbb{1}_{A_n, \Sigma_n \leqslant x}\left(\frac{2}{\pi p}\right)^{\frac{1}{2}}|X_n|\right]$$

Call also $\mathbb{P}'$ the measure associated to the walk $(X_{k+n+\tilde{T}_0})_{k\geqslant 0}$, $\tilde{T}_0$ being fixed. For $p > n + x$, $(p - n) \wedge x - A_n = x - A_n$; consequently

$$(2) \underset{p\to\infty}{\sim} \mathbb{E}\left[\mathbb{1}_{A_n, \Sigma_n \leqslant x, A_n \leqslant x}\tilde{\mathbb{P}}_{X_n}(\tilde{T}_0 \leqslant x - A_n)\mathbb{P}'\left(\Sigma'_{p-n-\tilde{T}_0} \leqslant x\right)\right]$$

$$\underset{p\to\infty}{\sim} \mathbb{E}\left[\mathbb{1}_{A_n, \Sigma_n \leqslant x, A_n \leqslant x}\tilde{\mathbb{P}}_{X_n}(\tilde{T}_0 \leqslant x - A_n)\left(\frac{2}{\pi p}\right)^{\frac{1}{2}}\theta(x)\right].$$

One derives therefrom

$$\lim_{p\to\infty}\frac{\mathbb{E}[\mathbb{1}_{A_n}\mathbb{1}_{\Sigma_p \geqslant x}]}{\mathbb{E}[\mathbb{1}_{\Sigma_p \geqslant x}]} = \mathbb{E}\left[\mathbb{1}_{A_n}\left\{\frac{|X_n|}{\theta(x)} + \tilde{\mathbb{P}}_{X_n}(\tilde{T}_0 \leqslant x - A_n)\mathbb{1}_{A_n \leqslant x}\right\}\mathbb{1}_{\Sigma_n \leqslant x}\right].$$

Remark 5. These notations $\tilde{\mathbb{P}}$ et $\tilde{T}_0$, or similar ones, will frequently occur in the sequel. We have not been completely rigorous when defining them; a rigorous definition is possible as follows: $\tilde{\mathbb{P}}_{\tilde{X}_n}(\tilde{T}_0 \leqslant x - A_n)$ stands for $f(X_n, x - A_n)$ where $f(y, z) = \mathbb{P}_y(T_0 \leqslant z)$.

We shall now see that $(M_n, n \geqslant 0)$ is indeed a martingale. The parity of $n + 1$ comes into play, so we shall consider two cases.

Suppose first that $n+1$ is odd. In that case, $\Sigma_{n+1} = \Sigma_n$ and $A_{n+1} = A_n+1$. Recall that $x \to |x|$ is harmonic except at 0 for the symmetric random walk. Hence, on the event $\{X_n \neq 0\}$, the only relevant term is

$$\mathcal{C}_{n+1} := \mathbb{1}_{\{A_{n+1} \leqslant x, \Sigma_n \leqslant x\}}\tilde{\mathbb{P}}_{X_{n+1}}(\tilde{T}_0 \leqslant x - A_{n+1}),$$

and on $X_n = 0$, it sufices to verify that, when conditioned by $\mathcal{F}_n$, this quantity equals $\left(1 - \frac{1}{\theta}\right)\mathbb{1}_{\Sigma_n \leqslant x}$.

By the Markov property, if $X_n \neq 0$,

$$\tilde{\mathbb{P}}_{X_n}(\tilde{T}_0 \leqslant x - A_n) = \frac{1}{2}(\tilde{\mathbb{P}}_{X_{n+1}}(\tilde{T}_0 \leqslant x - A_n - 1) + \tilde{\mathbb{P}}_{X_{n-1}}(\tilde{T}_0 \leqslant x - A_n - 1)).$$

So

$$\mathbb{E}[\mathbb{1}_{X_n \neq 0}\mathcal{C}_{n+1}|\mathcal{F}_n] = \mathbb{E}[\mathbb{1}_{X_n \neq 0}(\mathbb{1}_{X_{n+1}=X_n+1} + \mathbb{1}_{X_{n+1}=X_n-1})\mathcal{C}_{n+1}|\mathcal{F}_n]$$

$$= \mathbb{1}_{X_n \neq 0, \Sigma_n \leqslant x, A_n \leqslant x-1}\frac{1}{2}[\tilde{\mathbb{P}}_{X_n+1}(\tilde{T}_0 \leqslant x - A_n - 1) + \tilde{\mathbb{P}}_{X_n-1}(\tilde{T}_0 \leqslant x - A_n - 1)]$$

$$= \mathbb{1}_{X_n \neq 0, \Sigma_n \leqslant x, A_n \leqslant x-1}\tilde{\mathbb{P}}_{X_n}(\tilde{T}_0 \leqslant x - A_n)$$

And, as $\mathbb{1}_{X_n \neq 0, A_n = x} \tilde{\mathbb{P}}_{X_n}(\tilde{T}_0 \leqslant x - A_n) = 0$, one has

$$\mathbb{E}[\mathbb{1}_{X_n \neq 0} C_{n+1}|\mathcal{F}_n] = \mathbb{1}_{X_n \neq 0} C_n.$$

It remains to show that

$$\mathbb{E}[\mathbb{1}_{X_n = 0} C_{n+1}|\mathcal{F}_n] = \mathbb{1}_{X_n = 0, \Sigma_n \leqslant x} \left(1 - \frac{1}{\theta}\right). \tag{16}$$

This will use the classical result ([Fel50] pp. 73-77)

$$\mathbb{P}(X_1 > 0, \ldots, X_{2n-1} > 0, X_{2n} = 2r) = \frac{1}{2}(p_{2n-1,2r-1} - p_{2n-1,2r+1}). \tag{17}$$

where $p_{n,r} = \frac{1}{2^n} C_n^{\frac{n+r}{2}}$.

Using formula (17) with $x = 2n$, one can write

$$\mathbb{P}(\tau > x)\theta(x) = \mathbb{P}(\tau > x)\mathbb{E}[|X_x| \mid \tau > x] = \mathbb{E}\left[|X_x|\mathbb{1}_{\{\tau > x\}}\right]$$

$$= \mathbb{E}\left[X_x \mathbb{1}_{\{\tau > x, X_x > 0\}}\right] - \mathbb{E}\left[X_x \mathbb{1}_{\{\tau > x, X_x < 0\}}\right] = 2\mathbb{E}\left[X_x \mathbb{1}_{\{\tau > x, X_x > 0\}}\right]$$

$$= 2 \sum_{k>0, k \text{ even}}^{x} k\,\mathbb{P}(X_x = k, \tau > x) = 4 \sum_{\ell > 0}^{n} \ell\,\mathbb{P}(X_{2n} = 2\ell, \tau > 2n)$$

$$= 2 \sum_{\ell > 0}^{n} \ell\,(p_{2n-1,2\ell-1} - p_{2n-1,2\ell+1}) = \left(\frac{1}{2}\right)^{2n-2} \sum_{\ell > 0}^{n} \ell\,(C_{2n-1}^{n+\ell-1} - C_{2n-1}^{n+\ell}).$$

Now, $\sum_{\ell=1}^{n} \ell\,(C_{2n-1}^{n+\ell-1} - C_{2n-1}^{n+\ell}) = \sum_{\ell=0}^{n-1} C_{2n-1}^{n+\ell} = 2^{2n-2}$; so we obtain

$$\theta(x)\,\mathbb{P}(\tau > x) = 1. \tag{18}$$

On the other hand,

$$\mathbb{E}[\mathbb{1}_{X_n=0}C_{n+1}|\mathcal{F}_n] = \mathbb{1}_{X_n=0, \Sigma_n \leqslant x}\frac{1}{2}\left(\mathbb{P}_1\,(T_0 \leqslant x - 1) + \mathbb{P}_{-1}\,(T_0 \leqslant x - 1)\right)$$

$$= \mathbb{1}_{X_n=0, \Sigma_n \leqslant x}\mathbb{P}(\tau \leqslant x) = \mathbb{1}_{X_n=0, \Sigma_n \leqslant x}\,(1 - \mathbb{P}(\tau > x)); \tag{19}$$

hence, considering (18) and (19), formula (16) is established.

We now consider the case that $n + 1$ is even. In that case, $\{A_n \leqslant x\} = \{A_n \leqslant x - 1\}$. Indeed, $A_n = n - g_n$ is odd and x is even by hypothesis, so the event $\{A_n = x\}$ is null. Moreover, if $|X_n| \geqslant 3$, on a $\Sigma_{n+1} = \Sigma_n$. Last, if $|X_n| = 1$, there are two cases. Either $X_{n+1} \neq 0$ and one always has $\Sigma_{n+1} = \Sigma_n$, or $X_{n+1} = 0$ and we must see that in that case

$$\{\Sigma_{n+1} \leqslant x\} = \{\Sigma_n \leqslant x, n + 1 - g_n \leqslant x\} = \{\Sigma_n \leqslant x, A_n \leqslant x - 1\}.$$

So, one is always on the event $\{\Sigma_n \leqslant x, A_n \leqslant x - 1\}$, and the same argument as when $n + 1$ was odd and $X_n \neq 0$ shows that, conditional on $\mathcal{F}_n$, M_{n+1} is equal to M_n. This shows that M is a martingale; positivity is immediate. The proof that M is not uniformly integrable is postponed until later in this section.

2) We now start studying the process Σ under Q^x. We shall first show that, for all $y \leqslant x$, $Q^x (\Sigma_\infty > y) = 1 - \frac{\mathrm{P}(\tau > x)}{\mathrm{P}(\tau > y)}$.

Put $T_y^\Sigma := \inf \{n \geqslant 0, \ \Sigma_n > y\}$. Clearly, $X_{T_y^\Sigma} = 0$ and hence

$$Q^x (\Sigma_p > y) = Q^x (T_y^\Sigma \leqslant p) = \mathbb{E}\left[\mathbb{1}_{T_y^\Sigma \leqslant p} M_{T_y^\Sigma} \right]$$

$$= \mathbb{E}\left[\mathbb{1}_{T_y^\Sigma \leqslant p} \left\{ \frac{|X_{T_y^\Sigma}|}{\theta(x)} + \tilde{\mathbb{P}}_{X_{T_y^\Sigma}} \left(\tilde{T}_0 \leqslant x - A_{T_y^\Sigma} \right) \mathbb{1}_{A_{T_y^\Sigma} \leqslant x} \right\} \mathbb{1}_{\Sigma_{T_y^\Sigma} \leqslant x} \right]$$

$$= \mathrm{P}\left[T_y^\Sigma \leqslant p, \Sigma_{T_y^\Sigma} \leqslant x \right].$$

Letting p go to infinity, we obtain that $Q^x (\Sigma_\infty > y) = \mathrm{P}(\Sigma_{T_y^\Sigma} \leqslant x)$. For $y \leqslant x$, $\{\Sigma_{T_y^A} \leqslant x\}$ is a full event; so

$$\{\Sigma_{T_y^\Sigma} \leqslant x\} = \{\Sigma_{T_y^A} \leqslant x\} \cap \{T_0 \circ \theta_{T_y^A} + y \leqslant x\} = \{T_0 \circ \theta_{T_y^A} + y \leqslant x\}.$$

By the Markov property and Lemma 13,

$$Q^x (\Sigma_\infty > y) = \mathbb{E}\left[\mathbb{E}\left[\mathbb{1}_{T_0 \circ \theta_{T_y^A} + y \leqslant x} \mid \mathcal{A}_{T_y^A} \right] \right] = \mathbb{E}\left[\tilde{\mathbb{P}}_{X_{T_y^A}} \left(\tilde{T}_0 \leqslant x - y \right) \right]$$

$$= \mathbb{E}\left[\tilde{\mathbb{P}}_{X_y} \left(\tilde{T}_0 \leqslant x - y \right) \mid \tau > y \right] = 1 - \frac{\mathbb{E}\left[\tilde{\mathbb{P}}_{X_y} \left(\tilde{T}_0 > x - y \right) \mathbb{1}_{\tau > y} \right]}{\mathrm{P}(\tau > y)}$$

$$= 1 - \frac{\mathbb{E}\left[\mathbb{1}_{T_0 \circ \theta_y > x - y, \tau > y} \right]}{\mathrm{P}(\tau > y)} = 1 - \frac{\mathrm{P}(\tau > x)}{\mathrm{P}(\tau > y)}.$$

On the other hand, for all $n \geqslant 0$, one has $Q^x (\Sigma_n \leqslant x) = 1$. According to the definition of the probability Q^x,

$$Q^x (\Sigma_n \leqslant x) = \lim_{p \to \infty} \frac{\mathrm{P}(\Sigma_n \leqslant x, \Sigma_p \leqslant x)}{\mathrm{P}(\Sigma_p \leqslant x)} = \lim_{p \to \infty} \frac{\mathrm{P}(\Sigma_p \leqslant x)}{\mathrm{P}(\Sigma_p \leqslant x)} = 1.$$

3) We shall now describe several properties of g and $(A_n, n \geqslant 0)$ under Q^x.

a) We first show that g is Q^x-a.s. finite; this implies that $A_\infty = \infty$ Q^x-a.s.

Lemma 14. *For all $n \geqslant 0$ and $k \geqslant 0$,*

$$\mathbb{P}(A_{2n} = 2k) = \mathbb{P}(A_{2n+1} = 2k + 1) = C_{2n-2k}^{n-k} C_{2k}^k \left(\frac{1}{2} \right)^n.$$

Proof. According to [Fel50] p. 79, "Arcsin law for last visit",

$$\mathbb{P}(g_{2n} = 2k) = C_{2n-2k}^{n-k} C_{2k}^k \left(\frac{1}{2} \right)^n.$$

Therefore

$$\mathbb{P}(A_{2n} = 2k) = \mathbb{P}(2n - g_{2n} = 2k) = \mathbb{P}(g_{2n} = 2n - 2k) = C_{2n-2k}^{n-k} C_{2k}^k \left(\frac{1}{2} \right)^n;$$

and as $A_{2n+1} = A_{2n} + 1$, the proof is over.   □

The next lemma is instrumental in the sequel.

Lemma 15. *For each $p > 0$,*

$$Q^x (g > p \mid \mathcal{F}_p) = \tilde{\mathbb{P}}_{X_p}(\tilde{\tau} \leqslant x - A_p) \frac{1}{M_p}.$$

Proof. Recall that $T_{0,p} := \inf \{n > p, X_n = 0\}$ is the first zero after p, and remark that $\Sigma_{T_{0,p}} = \Sigma_p \vee \{A_p + \tau \circ \theta_p\}$. Recall also that under Q^x, the event $\{\Sigma_p \leqslant x\}$ is almost sure. So, for every $\Lambda_p \in \mathcal{F}_p$,

$$Q^x (\{\Lambda_p\} \cap \{g > p\}) = Q^x (\{\Lambda_p\} \cap \{T_{0,p} < \infty\})$$
$$= \mathbb{E}[\mathbb{1}_{\Lambda_p} M_{T_{0,p}}] = \mathbb{E}[\mathbb{1}_{\Lambda_p} \mathbb{1}_{\Sigma_{T_{0,p}} \leqslant x}] = \mathbb{E}[\mathbb{1}_{\Lambda_p, \Sigma_p \leqslant x} \tilde{\mathbb{P}}_{X_p} [\tilde{\tau} \leqslant x - A_p]]$$
$$= \mathbb{E}\left[\mathbb{1}_{\Lambda_p} \frac{\tilde{\mathbb{P}}_{X_p} [\tilde{\tau} \leqslant x - A_p]}{M_p} M_p\right] = \mathbb{E}^{Q^x}\left[\mathbb{1}_{\Lambda_p} \frac{\tilde{\mathbb{P}}_{X_p} [\tilde{\tau} \leqslant x - A_p]}{M_p}\right],$$

and consequently one has

$$Q^x (g > p \mid \mathcal{F}_p) = \tilde{\mathbb{P}}_{X_p}(\tilde{\tau} \leqslant x - A_p) \frac{1}{M_p}. \qquad \square$$

We now suppose that $p = 2l$ where $l \geqslant 0$; when $p = 2l+1$ the computation is similar, we won't give it (see Lemma 14). According to Lemma 15,

$$Q^x(g > p) = \mathbb{E}^{Q^x}\left[\mathbb{E}^{Q^x}[\mathbb{1}_{g > p} \mid \mathcal{F}_p]\right] = \mathbb{E}^{Q^x}\left[\tilde{\mathbb{P}}_{X_p}(\tilde{\tau} \leqslant x - A_p) \frac{1}{M_p}\right]$$

$$= \mathbb{E}[\tilde{\mathbb{P}}_{X_p}(\tilde{\tau} \leqslant x - A_p)] = \sum_{k=0}^{l \wedge \frac{x}{2}} \mathbb{E}[\tilde{\mathbb{P}}_{X_p}(\tilde{\tau} \leqslant x - A_p) \mathbb{1}_{A_p = 2k}]$$

$$= \sum_{k=0}^{l \wedge \frac{x}{2}} \mathbb{E}[\tilde{\mathbb{P}}_{X_p}(\tilde{\tau} \leqslant x - A_p) \mid A_p = 2k] \, \mathbb{P}(A_p = 2k)$$

$$= \sum_{k=0}^{l \wedge \frac{x}{2}} \mathbb{E}[\tilde{\mathbb{P}}_{X_{2k}}(\tilde{\tau} \leqslant x - 2k) \mid \tau > 2k] \, \mathbb{P}(A_p = 2k)$$

$$= \sum_{k=0}^{l \wedge \frac{x}{2}} \frac{\mathbb{E}[\tilde{\mathbb{P}}_{X_{2k}}(\tilde{\tau} \leqslant x - 2k) \mathbb{1}_{\tau > 2k}]}{\mathbb{P}(\tau > 2k)} \, \mathbb{P}(A_p = 2k)$$

$$= \sum_{k=0}^{l \wedge \frac{x}{2}} \left[1 - \frac{\mathbb{P}(\tau > x)}{\mathbb{P}(\tau > 2k)}\right] \mathbb{P}(A_p = 2k)$$

$$= \sum_{k=0}^{l \wedge \frac{x}{2}} C_{2l-2k}^{l-k} C_{2k}^{k} \left(\frac{1}{2}\right)^l \left(1 - \frac{\mathbb{P}(\tau > x)}{\mathbb{P}(\tau > 2k)}\right).$$

This gives the law of g under Q^x. Then, for $p > 2$, $Q^x(g > p) \leqslant \mathbb{E}[\mathbb{1}_{A_p} \leqslant x]$. Now, A_p tends to infinity $\mathbb{P}$-a.s.; consequently,

$$Q^x(g = \infty) = \lim_{p \to \infty} Q^x(g > p) \leqslant \lim_{p \to \infty} \mathbb{P}(A_p \leqslant x) = 0,$$

and g is Q^x-a.s. finite.

Remark 6. It is now easy to see that M is not uniformly integrable. Indeed, as g is finite, so is also L_∞, and the argument given earlier for M^φ and S immediately adapts to M and L.

b) To establish 2.d.i et 2.d.ii., we shall need:

Lemma 16. *For all $y \leqslant x$, one has*

$$\mathbb{E}\big[M_{T_y^A}\big] = 1$$

Proof of Lemma 16. Recall that the event $\{\Sigma_{T_y^A} \leqslant x\}$ has probability 1. By formula (18) and the proof of point 2.a,

$$\mathbb{E}\big[M_{T_y^A}\big] = \mathbb{E}\Big[\frac{|X_{T_y^A}|}{\theta(x)} + \tilde{\mathbb{P}}_{X_{T_y^A}}(\tilde{T}_0 \leqslant x - y)\Big]$$

$$= \frac{\theta(y)}{\theta(x)} + \mathbb{E}\big[\tilde{\mathbb{P}}_{X_{T_y^A}}(\tilde{T}_0 \leqslant x - y)\big] = \frac{\mathbb{P}(\tau > x)}{\mathbb{P}(\tau > y)} + 1 - \frac{\mathbb{P}(\tau > x)}{\mathbb{P}(\tau > y)}.$$

Let F be a positive functional and $G : \mathbb{R} \to \mathbb{R}^+$. Recall that after [ALR04], $X_{T_y^A}$ and $\mathcal{A}_{T_y^A}$ are independent under $\mathbb{P}$. On the other hand, as $M_{T_y^A}$ is a function of $X_{T_y^A}$, one has

$$\mathbb{E}^{Q^x}\big[F\big(A_n, n \leqslant T_y^A\big) G(X_{T_y^A})\big] = \mathbb{E}\big[F\big(A_n, n \leqslant T_y^A\big) G(X_{T_y^A}) M_{T_y^A}\big]$$

$$= \mathbb{E}\big[F\big(A_n, n \leqslant T_y^A\big)\big] \mathbb{E}\big[G(X_{T_y^A}) M_{T_y^A}\big]. \quad (20)$$

So, taking $G \equiv 1$ and using Lemma 16, one has

$$\mathbb{E}^{Q^x}\big[F\big(A_n, n \leqslant T_y^A\big)\big] = \mathbb{E}\big[F\big(A_n, n \leqslant T_y^A\big)\big],$$

which shows that $(A_n, n \leqslant T_y^A)$ has the same law under $\mathbb{P}$ and Q^x. Using again formula (20), one obtains

$$\mathbb{E}^{Q^x}\big[F\big(A_n, n \leqslant T_y^A\big) G(X_{T_y^A})\big] = \mathbb{E}^{Q^x}\big[F\big(A_n, n \leqslant T_y^A\big)\big] \mathbb{E}^{Q^x}\big[G(X_{T_y^A})\big];$$

this shows that $(A_n, n \leqslant T_y^A)$ and $X_{T_y^A}$ are independent under Q^x.

c) The rest of the proof of point 2 is quite easy, taking into account what has already been done:

$$\mathbb{E}^{Q^x}\left[G(X_{T_y^A})\right] = \mathbb{E}\left[G(X_{T_y^A})M_{T_y^A}\right] = \mathbb{E}\left[\mathbb{E}\left[G(X_{T_y^A})M_{T_y^A} \mid \mathcal{A}_{T_y^A}\right]\right]$$

$$= \mathbb{E}\left[\mathbb{E}\left[G(X_y)\left\{\frac{|X_y|}{\theta(x)} + \tilde{\mathbb{P}}_{X_y}(\tilde{T}_0 \leqslant x - y)\right\} \mid \tau > y\right]\right]$$

$$= \mathbb{E}\left[G(X_y)\left\{\frac{|X_y|}{\theta(x)} + \tilde{\mathbb{P}}_{X_y}(\tilde{T}_0 \leqslant x - y)\right\} \mid \tau > y\right]$$

$$= \mathbb{E}\left[G(X_y)\left\{\frac{|X_y|}{\theta(x)} + \tilde{\mathbb{P}}_{X_y}(\tilde{T}_0 \leqslant x - y)\right\} \mid \tau > y\right]$$

$$= \sum_k G(k)\left\{\frac{|k|}{\theta(x)} + \mathbb{P}_k(\tilde{T}_0 \leqslant x - y)\right\} \mathbb{P}(X_y = k \mid \tau > y).$$

Consequently, the law of $X_{T_y^A}$ under Q^x satisfies

$$Q^x(X_{T_y^A} = k) = \left\{\frac{|k|}{\theta(x)} + \mathbb{P}_k(\tilde{T}_0 \leqslant x - y)\right\} \mathbb{P}(X_y = k \mid \tau > y).$$

(The quantity $\mathbb{P}(X_y = k \mid \tau > y)$ is explicitly given in [Fel50] p. 77).
    We now compute $Q^x(g > T_y^A)$:

$$Q^x(g > T_y^A) = \mathbb{E}^{Q^x}\left[\mathbb{E}^{Q^x}\left[\mathbb{1}_{g>T_y^A} \mid \mathcal{F}_{T_y^A}\right]\right] = \mathbb{E}^{Q^x}\left[\frac{\tilde{\mathbb{P}}_{X_{T_y^A}}(\tilde{\tau} \leqslant x - y)}{M_{T_y^A}}\right]$$

$$= \mathbb{E}\left[\tilde{\mathbb{P}}_{X_{T_y^A}}(\tilde{\tau} \leqslant x - y)\right] = \mathbb{E}\left[\tilde{\mathbb{P}}_{X_y}(\tilde{\tau} \leqslant x - y) \mid \tau > y\right] = 1 - \frac{\mathbb{P}(\tau > x)}{\mathbb{P}(\tau > y)}.$$

Last, we now show that $(A_n, n \leqslant T_y^A)$ and $\{g > T_y^A\}$ are independent under Q^x; we use again the independence of $X_{T_y^A}$ and $A_{T_y^A}$ under $\mathbb{P}$.

$$\mathbb{E}^{Q^x}\left[F(A_n, n \leqslant T_y^A)\mathbb{1}_{g>T_y^A}\right] = \mathbb{E}^{Q^x}\left[F(A_n, n \leqslant T_y^A)\mathbb{E}^{Q^x}\left[\mathbb{1}_{g>T_y^A} \mid \mathcal{A}_{T_y^A}\right]\right]$$

$$= \mathbb{E}^{Q^x}\left[\frac{F(A_n, n \leqslant T_y^A)\tilde{\mathbb{P}}_{X_{T_y^A}}(\tilde{\tau} \leqslant x - y)}{M_{T_y^A}}\right]$$

$$= \mathbb{E}\left[F(A_n, n \leqslant T_y^A)\right]\mathbb{E}\left[\tilde{\mathbb{P}}_{X_{T_y^A}}(\tilde{\tau} \leqslant x - y)\right]$$

$$= \mathbb{E}^{Q^x}\left[F(A_n, n \leqslant T_y^A)\right]Q^x(g > T_y^A).$$

4) To study the process $(X_n, n \geqslant 0)$ under Q^x, we start with the law of the process $(X_n, n \geqslant g)$. Recall that $\Gamma^+ = \{X_n > 0, n > g\}$ and $\Gamma^- = \{X_n < 0, n > g\}$; these events Γ^+ and Γ^- are symmetric under Q_0^x:

Lemma 17.

$$Q^x(\Gamma^+) = Q^x(\Gamma^-) = \frac{1}{2}.$$

Proof. First remark that

$$Q^x(\Gamma^+) = \lim_{n\to\infty} Q^x(X_n > 0), \qquad Q^x(\Gamma^-) = \lim_{n\to\infty} Q^x(X_n < 0).$$

By definition of Q^x,

$$Q^x(X_n > 0) = \mathbb{E}\Big[\mathbb{1}_{X_n > 0}\Big\{\frac{|X_n|}{\theta(x)} + \tilde{\mathbb{P}}_{X_n}\big(\tilde{T}_0 \leqslant x - A_n\big)\mathbb{1}_{A_n \leqslant x}\Big\}\mathbb{1}_{\Sigma_n \leqslant x}\Big].$$

Owing to the symmetry of the walk under $\mathbb{P}$, one has

$$Q^x(X_n > 0) = \mathbb{E}\Big[\mathbb{1}_{X_n < 0}\Big\{\frac{|X_n|}{\theta(x)} + \tilde{\mathbb{P}}_{X_n}\big(\tilde{T}_0 \leqslant x - A_n\big)\mathbb{1}_{A_n \leqslant x}\Big\}\mathbb{1}_{\Sigma_n \leqslant x}\Big]$$

$$= Q^x(X_n < 0).$$

One also has $\lim_{n\to\infty} Q^x(X_n = 0) = 0$ because g is Q^x-a.s. finite; and as

$$Q^x(X_n > 0) + Q^x(X_n < 0) + Q^x(X_n = 0) = 2Q^x(X_n > 0) + Q^x(X_n = 0) = 1,$$

taking limits when n tends to infinity, on obtains

$$Q^x(\Gamma^+) + Q^x(\Gamma^-) = 2Q^x(\Gamma^+) = 1. \qquad \square$$

We now describe the behavior of $(X_{n+g},\ n > 0)$ under Q^x on Γ^+ (the other case is completely similar). Take $a \in \mathbb{N}^*$ and $p \geqslant x$, and set $q_{a,a+1} := Q(X_{n+1} = a + 1 | X_n = a, n > g)$.

$$q_{a,a+1} = Q(X_{n+1} = a + 1 | X_n = a, \forall i \leqslant p\ X_{n+i} > 0)$$

$$= \frac{Q(X_{n+1} = a + 1,\ X_n = a,\ \forall i \leqslant p\ X_{n+i} > 0)}{Q(X_n = a,\ \forall i \leqslant p\ X_{n+i} > 0)}$$

$$= \frac{\mathbb{E}\big[\mathbb{1}_{X_{n+1}=a+1,\, X_n=a,\, \forall i \leqslant p\ X_{n+i} > 0}\, M_{p+n}\big]}{\mathbb{E}\big[\mathbb{1}_{X_n=a,\, \forall i \leqslant p\ X_{n+i} > 0}\, M_{p+n}\big]}.$$

Here $M_{p+n} = \frac{X_{p+n}}{\theta(x)}\mathbb{1}_{\Sigma_n \leqslant x}$; hence we can condition the numerator (resp. the denominator) by $\mathcal{F}_{n+1}$ (resp. $\mathcal{F}_n$). The Markov property gives

$$q_{a,a+1} = \frac{\mathbb{E}\big[\mathbb{1}_{X_{n+1}=a+1,\, X_n=a,\, \Sigma_n \geqslant x}\mathbb{E}_{a+1}\big[X_p\mathbb{1}_{X_i>0,\forall i\leqslant p-1}\big]\big]}{\mathbb{E}\big[\mathbb{1}_{X_n=a,\, \Sigma_n \geqslant x}\mathbb{E}_a\big[X_p\mathbb{1}_{X_i>0,\forall i\leqslant p}\big]\big]}.$$

Clearly, $(X_p\mathbb{1}_{X_i>0,\forall i\leqslant p})_{p\geqslant 0}$ is a martingale, wherefrom

$$q_{a,a+1} = \frac{(a+1)\mathbb{E}\big[\mathbb{1}_{X_{n+1}=a+1,\, X_n=a,\, \Sigma_n \geqslant x}\big]}{a\mathbb{E}\big[\mathbb{1}_{X_n=a,\, \Sigma_n \geqslant x}\big]}.$$

Last, conditioning the numerator by $\mathcal{F}_n$ one gets

$$q_{a,a+1} = \frac{a+1}{2a},$$

the transition probability of a 3-Bessel* walk.

Recall the following notation:

$$\gamma_n := |\{k \leqslant n, X_k = 0\}| \,, \quad \gamma_\infty := \lim_{n \to \infty} \gamma_n$$

$$\tau_1 := T_0 \,, \quad \forall n \geqslant 2, \ \tau_n := \inf\{k > \tau_{n-1}, X_k = 0\}$$

It remains to show that, conditional on $\{\gamma_\infty = l\}$, $(X_u, \ u \leqslant g)$ is a standard random walk stopped at τ_l and conditioned by $\Sigma_{\tau_l} \leqslant x$.

Let F be a functional on $\mathbb{Z}^n$.

$$\mathbb{E}^{Q^x}\big[F(X_1,\ldots,X_n)\,\mathbb{1}_{n\leqslant\tau_l} \mid \gamma_\infty = l\big] = \frac{\mathbb{E}^{Q^x}\big[F(X_1,\ldots,X_n)\,\mathbb{1}_{n\leqslant\tau_l}\mathbb{1}_{\gamma_\infty=l}\big]}{\mathbb{E}^{Q^x}[\gamma_\infty = l]}$$

$$= \frac{\mathbb{E}^{Q^x}\big[F(X_1,\ldots,X_n)\,\mathbb{1}_{n\leqslant\tau_l<\infty}\big] - \mathbb{E}^{Q^x}\big[F(X_1,\ldots,X_n)\mathbb{1}_{n\leqslant\tau_l<\tau_{l+1}<\infty}\big]}{\mathbb{E}^{Q^x}\big[\mathbb{1}_{\tau_l<\infty}\mathbb{1}_{\tau_{l+1}=\infty}\big]}$$

$$= \frac{\mathbb{E}^{Q^x}\big[F(X_1,\ldots,X_n)\,\mathbb{1}_{n\leqslant\tau_l<\infty}\big] - \mathbb{E}^{Q^x}\big[F(X_1,\ldots,X_n)\mathbb{1}_{n\leqslant\tau_l<\tau_{l+1}<\infty}\big]}{\mathbb{E}^{Q^x}\big[\mathbb{1}_{\tau_l<\infty}\big] - \mathbb{E}^{Q^x}\big[\mathbb{1}_{\tau_{l+1}<\infty}\big]}$$

$$= \frac{\mathbb{E}\big[F(X_1,\ldots,X_n)\,\mathbb{1}_{n\leqslant\tau_l<\infty}\,M_{\tau_l}\big] - \mathbb{E}\big[F(X_1,\ldots,X_n)\mathbb{1}_{n<\tau_{l+1}<\infty}\,M_{\tau_{l+1}}\big]}{\mathbb{E}\big[\mathbb{1}_{\tau_l<\infty}\,M_{\tau_l}\big] - \mathbb{E}\big[\mathbb{1}_{\tau_{l+1}<\infty}\,M_{\tau_{l+1}}\big]}.$$

Under $\mathbb{P}$, $\{\tau_l < \infty\}$ has probability 1, and so

$$M_{\tau_l} - M_{\tau_{l+1}} = \mathbb{1}_{\Sigma_{\tau_l}\leqslant x}(1 - \mathbb{1}_{\tau_{l+1}-\tau_l\leqslant x}) = \mathbb{1}_{\Sigma_{\tau_l}\leqslant x,\,\tau_{l+1}-\tau_l>x}.$$

As $\tau_{l+1} - \tau_l$ is independent of $\mathcal{F}_{\tau_l}$, one gets

$$\mathbb{E}^{Q^x}\big[F(X_1,\ldots,X_n)\,\mathbb{1}_{n\leqslant\tau_l} \mid \gamma_\infty = l\big] = \frac{\mathbb{E}\big[F(X_1,\ldots,X_n)\,\mathbb{1}_{n\leqslant\tau_l}(M_{\tau_l} - M_{\tau_{l+1}})\big]}{\mathbb{E}\big[M_{\tau_l} - M_{\tau_{l+1}}\big]}$$

$$= \frac{\mathbb{E}\big[F(X_1,\ldots,X_n)\,\mathbb{1}_{\{n\leqslant\tau_l,\Sigma_{\tau_l}\leqslant x,\tau_{l+1}-\tau_l>x\}}\big]}{\mathbb{E}\big[\mathbb{1}_{\{\Sigma_{\tau_l}\leqslant x,\tau_{l+1}-\tau_l>x\}}\big]}$$

$$= \frac{\mathbb{E}\big[F(X_1,\ldots,X_n)\,\mathbb{1}_{n\leqslant\tau_l,\Sigma_{\tau_l}\leqslant x}\big]\mathbb{E}\big[\mathbb{1}_{\{\tau_{l+1}-\tau_l>x\}}\big]}{\mathbb{E}\big[\mathbb{1}_{\Sigma_{\tau_l}\leqslant x}\big]\mathbb{E}\big[\mathbb{1}_{\tau_{l+1}-\tau_l>x}\big]}$$

$$= \frac{\mathbb{E}\big[F(X_1,\ldots,X_n)\,\mathbb{1}_{\{n\leqslant\tau_l,\Sigma_{\tau_l}\leqslant x\}}\big]}{\mathbb{E}\big[\mathbb{1}_{\Sigma_{\tau_l}\leqslant x}\big]} = \mathbb{E}\big[F(X_1,\ldots,X_n)\,\mathbb{1}_{n\leqslant\tau_l} \mid \Sigma_{\tau_l} \leqslant x\big].$$

References

[ALR04] C. Ackermann, G. Lorang, and B. Roynette, *Independence of time and position for a random walk*, Revista Matematica Iberoamericana **20** (2004), no. 3, pp. 915–917.

[Bil] P. Billingsley, *Probability measures*.

[Deb07] Pierre Debs, *Pénalisations de marches aléatoires*, Thèse de Doctorat, Institut Élie Cartan, 2007.

[Fel50]   Feller, *An Introduction to Probability Theory and its Applications*, vol. 1, 1950.

[Fel71]   ———, *An Introduction to Probability Theory and its Applications*, vol. 2, 1966-1971.

[LeG85]   J.F. LeGall, *Une approche élémentaire des théorèmes de décomposition de Williams*, Lecture Notes in Mathematics, Séminaire de Probabilités XX, 1984-1985, pp. 447–464.

[Pit75]   J. Pitman, *One-dimensional Brownian motion and the three-dimensional Bessel process*, Advances in Appl. Probability, vol. 7, 1975, pp. p. 511–526.

[RVY]   B. Roynette, P. Vallois, and M. Yor, *Brownian penalisations related to excursion lengths*, submitted to Annales de l'Institut Henri Poincaré.

[RVY06]   ———, *Limiting laws associated with Brownian motion perturbed by its maximum, minimum and local time*, Studia sci. Hungarica Mathematica **43** (2006), no. 3.

Canonical Representation for Gaussian Processes

M. Erraoui[1] and E.H. Essaky[2]

[1] Université Cadi Ayyad, Faculté des Sciences Semlalia,
   Département de Mathématiques, B.P. 2390, Marrakech, Maroc.
   E-mail: erraoui@ucam.ac.ma
[2] Université Cadi Ayyad, Faculté Poly-disciplinaire,
   Département de Mathématiques et d'Informatique, B.P 4162, Safi, Maroc.
   E-mail: essaky@ucam.ac.ma

Summary. We give a canonical representation for a centered Gaussian process which has a factorizable covariance function with respect to a positive measure. We also investigate this representation in order to construct a stochastic calculus with respect to this Gaussian process.

Mathematics Subject Classification (2000): 60G15, 60H05, 60H07, 46E22.
Keywords: Volterra processes; Stochastic integrals; Reproducing kernel Hilbert space; Malliavin calculus; Girsanov theorem.

1 Introduction

Let $\{X_t,\ t \in [0,1]\}$ be a centered Gaussian process with covariance $K(t,s) = \mathbb{E}(X_t X_s)$ and $\mathcal{F}_t^X$ be the σ-field generated by $\{X_s;\ s \leq t\}$. It is well-known that the law of a centered Gaussian process is uniquely determined by the covariance function $K(t,s)$. The Gaussian process which has been studied most extensively is, of course, the Wiener process $\{W_t,\ t \in [0,1]\}$. Therefore it is natural to seek a representation of centered Gaussian process in terms of Wiener processes. It is known that if a centered Gaussian process has a factorizable covariance function of the form

$$K(t,s) = \int_0^1 k(s,u)k(t,u)du,$$

then X_t has a stochastic integral representation

$$X_t \stackrel{law}{=} \int_0^1 k(t,u)dW_u. \tag{1}$$

The fractional Brownian motion is the most important example of such processes, cf [AMN01], [DU99] and the references therein.

Most representations for Gaussian processes of type (1) are in law; then, no comparison between $\mathcal{F}_t^X$ and $\mathcal{F}_t^W$ is given. So, a natural question arises: can this representation hold strongly? In another term, is it possible to construct a Brownian motion such that $\mathcal{F}_t^X = \mathcal{F}_t^W$? This question is closely related to the problem of canonical representation for Gaussian processes, which means that the process X has the representation (1) and moreover $\mathcal{F}_t^X = \mathcal{F}_t^W$ for each t. The theory of the canonical and noncanonical representations for Gaussian processes was initiated by Lévy [Lev65]. Since then this theory has been extented in several directions, see, for example, Hida [Hid60], Cramér [Cra61], Hitsuda [His68], Hibino-Hitsuda-Muraoka [HHM97] and Shepp [She66]. Nevertheless, it seems interesting to investigate the above question in order to construct a stochastic calculus with respect to a Gaussian process.

An important class of Gaussian processes is the Gaussian-Markov ones. It is well known, see for example [RY91] p. 81, that if K is continuous and strictly positive then there exists a continuous function u and a continuous strictly increasing function v such that $K(t, s) = u(t)u(s)v(t \wedge s)$. It follows that there exits a positive measure ϱ given by $\varrho([0, t]) = v(t)$ such that X has the representation in law $X_t = u(t)W_{\rho([0,t])}$, where $\{W_{\varrho([0,s])}, \ s \in [0, 1]\}$ is the time-changed Brownian motion. Moreover the Gaussian space generated by X is isomorphic to the space $L^2(\varrho)$. For example when X is the standard Brownian bridge on $[0, 1]$, then $u(t) = 1 - t$; $v(t) = t/(1 - t)$; when X is the Ornstein-Uhlenbeck process of parameter $\lambda > 0$, $u(t) = \exp(\lambda t)$; $v(t) = (1 - \exp(-2\lambda t))/2\lambda$. It then seems interesting to consider Gaussian processes whose covariance functions are factorizable with respect to a positive measure.

In the sequel, we study a class of Gaussian processes having a factorizable covariance function of the form

$$K(t, s) = \int_0^1 k(s, u)k(t, u)\varrho(du), \qquad (2)$$

where ϱ is a probability measure on $[0, 1]$. Let $\{X_t, \ t \in [0, 1]\}$ be a centered Gaussian process with covariance of the form (2). Some natural questions are: is it possible to represent X by means of linear transformations of the Wiener process? If so, can this representation be canonical? In answering those questions we construct a Brownian motion W such that X has the following representation

$$X_t = \int_0^1 k(t, u)dW_{\varrho([0,s])}; \qquad (3)$$

moreover, $\mathcal{F}_{C_t}^X = \mathcal{F}_t^W \ \forall t \in [0, 1]$, with $C_t = \inf\{s \geq 0, \ A_s := \varrho([0, s]) > t\} \wedge 1$ and, as usual, $\inf(\emptyset) = +\infty$.

It should be pointed out that, for $\varrho(du) = du$ on $[0, 1]$, we recover the result of [Hul03] where the representation (1) is given only in law.

The paper is organized as follows. In section 2, we give a description of the reproducing kernel Hilbert space associated with the kernel $k(t, .)$ which will be useful in the sequel. A canonical representation of the Gaussian process X is given in Section 3. Section 4 is devoted to a stochastic calculus of variations with respect to a Gaussian process X. The divergences associated with the processes X and W are related by the formula $\delta^X(u) = \delta^W((\mathcal{K}^*u)(C.))$, where $\mathcal{K}^*$ is the adjoint of the operator $\mathcal{K}$ defined in Section 2. Finally, in Section 5, a Girsanov transformation is given.

2 Reproducing Kernel Hilbert Space

Let ϱ be a probability measure on $[0, 1]$ and K a kernel given by

$$K(t, s) = \int_0^1 k(s, u)k(t, u)\varrho(du), \qquad (4)$$

where $k : [0, 1] \times [0, 1] \longrightarrow \mathbb{R}$ satisfies

$$\sup_{t \in [0,1]} \|k(t, \cdot)\|_{L^2(\varrho)} < +\infty. \qquad (5)$$

Then K is a bounded reproducing kernel. Let $\mathcal{H}$ be the reproducing kernel Hilbert space associated with K. Now we give a description of $\mathcal{H}$.

Proposition 1. *(See Suquet [Suq95], Proposition 1.)*
Let E be the closed subspace of $L^2(\varrho)$ spanned by $\{k(t, \cdot), t \in [0, 1]\}$.
1. A function $h : [0, 1] \longrightarrow \mathbb{R}$ belongs to $\mathcal{H}$ if and only if there is a unique $g_h \in E$ such that

$$h(t) = \int_0^1 k(t, u)g_h(u)\varrho(du). \qquad (6)$$

2. The scalar product on $\mathcal{H}$ is given by

$$\langle h_1, h_2 \rangle_{\mathcal{H}} = \langle g_{h_1}, g_{h_2} \rangle_{L^2(\varrho)},$$

and its associated norm is denoted by $\|.\|_{\mathcal{H}}$.
3. The representation (6) defines a isometry of Hilbert spaces: $\Psi : \mathcal{H} \longrightarrow E$, $h \mapsto g_h$.

Now taking into account the integrability assumption (5) on k, we define a bounded operator in $L^2(\varrho)$ by

$$(\mathcal{K}h)(t) = \int_0^1 k(t, u)h(u)\varrho(du).$$

In particular, for $h = k(t, .)$ we have

$$(\mathcal{K}k(t,.))\,(s) = \int_0^1 k(t,u)k(s,u)\varrho\,(du) = K(t,s).$$

Having in mind that any $f \in \mathcal{H}$ has the representation (6), it follows that

$$\mathcal{H} = \mathcal{K}(E), \quad \text{and} \quad \mathcal{N}(\mathcal{K}) = E^\perp,$$

where $\mathcal{N}$ denotes the null space of $\mathcal{K}$. Note that as a vector space $\mathcal{H}$ is equal to $\mathcal{K}(E)$, but the norm on each of these spaces are different since for $h \in \mathcal{H}$

$$\|h\|_{\mathcal{H}} = \|g_h\|_{L^2(\varrho)} \quad \text{and} \quad \|h\|_{L^2(\varrho)} = \|\mathcal{K}g_h\|_{L^2(\varrho)}.$$

It also follows that $\mathcal{H}$ is the closed subspace spanned by $\{K(t, \cdot),\ t \in [0,1]\}$, for the norm $\|\cdot\|_{\mathcal{H}}$.

Moreover, for all $h \in E$, we have $\|\mathcal{K}h\|_{\mathcal{H}} = \|h\|_{L^2(\varrho)}$, and for all $h \in E^\perp$, we get $\mathcal{K}h = 0$, which means that $\mathcal{K}$ is partially isometric with the initial set $(E, \|\cdot\|_{L^2(\varrho)})$ and the final set $(\mathcal{H}, \|\cdot\|_{\mathcal{H}})$. This is equivalent to

$$\|\mathcal{K}h\|_{\mathcal{H}} = \|P_E h\|_{L^2(\varrho)},$$

where P_E is the orthogonal projection of $L^2(\varrho)$ on E. On other hand, we have $\mathcal{K}^* h = 0$, for $h \in (\mathcal{K}(E))^\perp$ and $\|\mathcal{K}^* h\|_{L^2(\varrho)} = \|h\|_{\mathcal{H}}$, for $h \in \mathcal{K}(E)$, where $\mathcal{K}^*$ is the adjoint operator of $\mathcal{K}$.

As a consequence we obtain the following relations:

(i) For $f \in L^2(\varrho)$, we have $\mathcal{K}^*\mathcal{K}(f) = P_E(f)$.

(ii) For $f \in \mathcal{H}$, we have $\mathcal{K}\mathcal{K}^*(f) = f$.

For more details on partially isometric operators see Chaleyat-Maurel and Jeulin [CJE85] and Kato [Kat95].

Moreover, if we assume that $\mathcal{K}$ is injective, which is equivalent to $E = L^2(\varrho)$, then $\mathcal{H} = \mathcal{K}(L^2(\rho))$ and

$$\langle h, \mathcal{K}f \rangle_{\mathcal{H}} = \langle \mathcal{K}^* h, f \rangle_{L^2(\varrho)},$$

for all $h \in \mathcal{H}$ and $f \in L^2(\varrho)$.

Hence the operator $\mathcal{K}^*$ is an isometry between $\mathcal{H}$ and $L^2(\varrho)$, that is,

$$\mathcal{H} = (\mathcal{K}^*)^{-1}\left(L^2(\varrho)\right). \tag{7}$$

In this case (i) and (ii) becomes:

(j) For $f \in L^2(\varrho)$, we have $\mathcal{K}^*\mathcal{K}(f) = f$.

(jj) For $f \in \mathcal{H}$, we have $\mathcal{K}\mathcal{K}^*(f) = f$.

3 Strong Representation of Gaussian Processes

Let $\Omega = C_0([0,1], \mathbb{R})$ be the Banach space of continuous functions, null at time 0, equipped with the sup-norm and the Borel σ-field $\mathcal{F}$. The canonical process $X = \{X_t : t \in [0,1]\}$ is defined by $X_t(\omega) = \omega(t)$ and the probability measure $\mathbb{P}$ on $(\Omega; \mathcal{F})$ is the unique measure such that X is centered Gaussian with covariance given by

$$\mathbb{E}(X_t X_s) = K(t,s) = \int_0^1 k(s,u)k(t,u)\varrho\,(du). \tag{8}$$

Denote by $\{\mathcal{F}_t^X : t \in [0,1]\}$ the natural filtration generated by X. Let H be the closure in $L^2(\Omega)$ of the space spanned by $\{X_t : t \in [0,1]\}$ equipped with the inner product $\langle \xi, \zeta \rangle_H = \mathbb{E}(\xi\zeta)$, for $\xi, \zeta \in H$. The reproducing kernel Hilbert space $\mathbf{H}$ associated with X is the space $R(H) = \{R(\xi) : \xi \in H\}$ where for any $\xi \in H$, $R(\xi)$ is the function $R(\xi)(t) = \langle \xi, X_t \rangle_H = \mathbb{E}(\xi X_t)$. $\mathbf{H}$ has inner product $\langle F, G \rangle_{\mathbf{H}} = \langle R^{-1}F, R^{-1}G \rangle_H$. Since $K(s,.) = R(X_s)$, we clearly have $K(s,.) \in \mathbf{H}$. Hence $span\{K(t,.), t \in [0,1]\} \subseteq \mathbf{H}$. More precisely, $\mathbf{H}$ is the closure of the space spanned by $\{K(t,.) : t \in [0,1]\}$ with respect to the norm associated to $\langle .,. \rangle_{\mathbf{H}}$. For more details on reproducing kernel Hilbert spaces we refer to Grenander [Gre81] or Janson [Jan97].

Remark 1. It should be pointed out that

$$\langle K(t,.), K(s,.) \rangle_{\mathbf{H}} = \langle K(t,.), K(s,.) \rangle_{\mathcal{H}} = K(t,s), \; \forall s,t \in [0,1],$$

and then $\mathbf{H} = \mathcal{H}$.

Let $\mathcal{J}_X$ be the isometry from $L^2([0,1], \varrho)$ onto H defined by $\mathcal{J}_X = R^{-1} \circ \mathcal{K}$. With this definition we see that $\mathcal{J}_X$ maps $k(t,.) \mapsto X(t)$, for more details see [Hul03].

Now, consider a deterministic function $k : [0,1] \times [0,1] \longrightarrow [0,+\infty)$ satisfying the following hypotheses:

($\mathbf{H_1}$) $k(0,s) = 0$ for all $s \in [0,1]$ and $k(t,s) = 0$ for $s > t$.

($\mathbf{H_2}$) There are constants $C, \alpha > 0$ such that for all $s,t \in [0,1]$

$$\int_0^1 ((k(t,r) - k(s,r))^2 \varrho(dr) \leq C|t-s|^\alpha.$$

($\mathbf{H_3}$) $\mathcal{K}$ is injective as a transformation of functions in $L^2(\varrho)$.

It is easily seen, using assumption ($\mathbf{H_2}$), that for sufficiently large p there exists a constant C_p such that

$$\mathbb{E}|X_t - X_s|^p \leq C_p \left(\int_0^1 (k(t,u) - k(s,u))^2 \varrho\,(du) \right)^{p/2}$$
$$\leq C_p|t-s|^{\alpha p/2} \leq C_p|t-s|^{1+\varepsilon},$$

for some $\varepsilon > 0$. Hence, by Kolmogorov's continuity criterion, X has a continuous modification. Henceforth, we will consider only this continuous modification.

3.1 Particular Case

In this subsection, we assume that $\varrho(ds) = ds$. Hult has shown in [Hul03] that, under hypotheses $(\mathbf{H}_1) - (\mathbf{H}_3)$, X has the representation

$$X_t \stackrel{law}{=} \int_0^1 k(t,u)dB_u,$$

where B is a standard Brownian motion.

Our aim is to prove that the above representation holds in the pathwise sense, with a fixed standard Brownian motion constructed on $(\Omega, \mathbb{P})$, and it is canonical. Following the work [Hul03], we define the Gaussian process $\{W_t : t \in [0,1]\}$ by $W_t := \mathcal{J}_X(1_{[0,t]}), t \in [0,1]$. Now, denote by H_t the closure in $L^2(\Omega)$ of the space spanned by $\{X_s : s \in [0,t]\}$ for $t \in [0,1]$. It follows that $\mathcal{J}_X$ is an isometry from $L^2([0,t],\varrho)$ onto H_t. Hence, for all $t \in [0,1]$, W_t is $\mathcal{F}_t^X$-measurable, therefore W is $\mathcal{F}_t^X$-adapted.

Lemma 1. *(i) For all $s < t$, we have*

$$\mathcal{J}_X(1_{[s,t]}) = W_t - W_s.$$

(ii) W_t is an $(\mathcal{F}_t^X, \mathbb{P})$- Brownian motion.

Proof. (i) Since $\mathcal{J}_X$ is linear we have, for all $s < t$

$$\mathcal{J}_X(1_{[s,t]}) = \mathcal{J}_X(1_{[0,t]}) - \mathcal{J}_X(1_{[0,s]}) = W_t - W_s.$$

(ii) First, $\mathbb{E}(W_t W_s) = \langle \mathcal{K}(1_{[0,t]}), \mathcal{K}(1_{[0,s]}) \rangle_{\mathcal{H}} = \langle 1_{[0,t]}, 1_{[0,s]} \rangle_{L^2(ds)} = s$, for all $s \leq t$.

To prove that W is a $\mathcal{F}_t^X$-Brownian motion, it is sufficient to show that $(W_t - W_s)$ is independent of $\mathcal{F}_s^X$, for all $t \geq s$. This is a consequence of the fact that

$$\mathbb{E}((W_t - W_s)X_r) = \langle \mathcal{K}(1_{[s,t]}), \mathcal{K}(k(r,.)) \rangle_{\mathcal{H}}$$
$$= \langle 1_{[s,t]}, k(r,.) \rangle_{L^2(ds)}$$
$$= \int_s^t k(r,u)du = 0, \forall r \leq s.$$

Remark 2. To prove that W_t is an $\mathcal{F}_t^X$- Brownian motion, it is not sufficient to show only that W is Gaussian process with covariance $t \wedge s$. For instance, it is clear that the process β defined by

$$\beta_t = B_t - \int_0^t \frac{B_u}{u} du = \int_0^t \left(1 + \ln(\frac{u}{t})\right) dB_u,$$

where B is a Brownian motion, is an $\mathcal{F}_t^B$-adapted Gaussian process with covariance $t \wedge s$. On other hand if it is an $\mathcal{F}_t^B$- Brownian motion then we have $\int_0^t \frac{B_u}{u} du = 0$ a.s. which is absurd. Moreover, it is easy to see that

$\mathbb{E}\left(\beta_s B_t\right) = 0$ for all $s \in [0,t]$. It follows that $\mathcal{F}_t^\beta \subsetneq \mathcal{F}_t^B$ for all $t > 0$. More precisely, for all $t > 0$, the Gaussian space Γ_t generated by $(\beta_s; s \leq t)$ is given by

$$\Gamma_t = \left\{ \int_0^t f(u)dB_u; \ f \in L^2\left([0,t], du\right), \quad \text{and} \quad \int_0^t f(u)du = 0 \right\}.$$

For more details on this example, see Jeulin and Yor [JY90] or Yor [Yor92].

The following theorem gives the representation of X.

Theorem 1. *The process X satisfies*

$$X_t = \int_0^t k(t,u)dW_u.$$

Furthermore

$$\{\mathcal{F}_t^X : t \in [0,1]\} = \{\mathcal{F}_t^W : t \in [0,1]\}.$$

Proof. Since W is a standard Brownian motion, the following limit holds in $L^2(\Omega)$

$$\int_0^t k(t,s)dWs = \lim_{n \to +\infty} \sum_{i=1}^n \left(n \int_{t_{i-1}}^{t_i} k(t,s)ds \right)(W_{t_i} - W_{t_{i-1}}); \qquad (9)$$

where $t_i = \frac{i}{n}$, $0 \leq i \leq n$. On the other hand,

$$n \sum_{i=1}^n \left(\int_{t_{i-1}}^{t_i} k(t,s)ds \right) 1_{[t_{i-1},t_i[}$$

converges in $L^2([0,1])$ to $k(t,s)$ as n goes to infinity. Applying Lemma 1, we get

$$\mathcal{J}_X \left(n \sum_{i=1}^n \left(\int_{t_{i-1}}^{t_i} k(t,s)ds \right) 1_{[t_{i-1},t_i[} \right)$$
$$= R^{-1} \circ \mathcal{K} \left(n \sum_{i=1}^n \left(\int_{t_{i-1}}^{t_i} k(t,s)ds \right) 1_{[t_{i-1},t_i[} \right)$$
$$= \sum_{i=1}^n \left(n \int_{t_{i-1}}^{t_i} k(t,s)ds \right)(W_{t_i} - W_{t_{i-1}})$$

Now, by continuity of $\mathcal{J}_X$ and equality (9) we have

$$X_t = \int_0^1 k(t,u)dW_u.$$

Then $\{\mathcal{F}_t^X : t \in [0,1]\} \subset \{\mathcal{F}_t^W : t \in [0,1]\}$. The equality follows from (ii) of Lemma 1.

Remark 3. By using arguments similar to the ones above, one can see that

$$\forall f \in L^2([0,1]), \ \mathcal{J}_X(f) = \int_0^1 f(s)dW_s.$$

For $h \in \mathcal{H}$, we have $\mathcal{K}^*h \in L^2([0,1])$. Then, it follows that

$$\mathcal{J}_X(\mathcal{K}^*h) = \int_0^1 \mathcal{K}^*h(s)dW_s.$$

In particular, for $h = K(t,.)$, we have

$$\mathcal{J}_X(\mathcal{K}^*(K(t,.))) = \mathcal{J}_X(k(t,.)) = \int_0^1 \mathcal{K}^*(K(t,.))(s)dW_s = \int_0^1 k(t,s)dW_s = X_t.$$

3.2 General Case

Our aim in this subsection is to give a canonical representation of X which has a factorizable covariance with respect to a probability measure ϱ.

Set $A_t = \varrho([0,t])$. It is well known that the variation of A corresponds to the total variation of ϱ and the function $C_t = \inf\{s : A_s > t\} \wedge 1$ for $t \in [0,1]$, is increasing and right-continuous. Moreover $A_{C_t} \geq t$, $C_{A_t} \geq t$, for every t and

$$A_t = \inf\{s : C_s > t\}.$$

For every $t \in [0,1]$ we have

$$\int_0^t h(s)\varrho\,(ds) = \int_0^t h(s)dA_s = \int_0^{A_t} h(C_s)ds.$$

Now we assume the following condition:
(**H₄**) A is continuous and $A_0 = 0$.
Note that, under condition (**H₄**), we have

$$A_{C_t} = t, \ \ \varrho([C_{t-}, C_t[) = 0 \ \text{ and } \ supp(\varrho) = \{s \in [0,1] : s = C_{A_s}\}. \tag{10}$$

Lemma 2. *The process* $W_t = \mathcal{J}_X(1_{[0,C_t]})$ *is a* $(\mathcal{F}^X_{C_t}, \mathbb{P})$*-Brownian motion. Moreover, for all $s < t$, we have*

$$\mathcal{J}_X(1_{[C_s,C_t]}) = W_t - W_s.$$

Proof. First, remark that $\mathcal{J}_X$ is an isometry from $L^2([0,C_t], \varrho)$ onto H_{C_t}, where H_{C_t} is the closure in $L^2(\Omega)$ of the space spanned by $\{X_s : s \in [0,C_t]\}$ for $t \in [0,1]$. Hence, for all $t \in [0,1]$, W_t is $\mathcal{F}^X_{C_t}$-measurable, therefore W is $\mathcal{F}^X_{C_t}$-adapted.

Moreover, we have

$$\mathbb{E}(W_t W_s) = \langle \mathcal{K}(1_{[0,C_t]}, \mathcal{K}(1_{[0,C_s]})) \rangle_{\mathcal{H}} = \langle 1_{[0,C_t]}, 1_{[0,C_s]} \rangle_{L^2(\varrho)} = A_{C_s} = s,$$

for all $s \leq t$. Since $\mathcal{J}_X$ is linear we have, for all $s < t$

$$\mathcal{J}_X(1_{[C_s,C_t]}) = \mathcal{J}_X(1_{[0,C_t]}) - \mathcal{J}_X(1_{[0,C_s]}) = W_t - W_s.$$

In order to prove that W is a $\mathcal{F}^X_{C_t}$-Brownian motion, it is sufficient to show that $(W_t - W_s)$ is independent of $\mathcal{F}^X_{C_s}$, for all $t \geq s$. This is a consequence of the fact that

$$\mathbb{E}((W_t - W_s)X_r) = \int_{C_s}^{C_t} k(r,u)\varrho(du) = 0, \forall r \leq C_s,$$

where we have used assumption $(\mathbf{H_1})$.

The main result of this subsection is the following.

Theorem 2. *The process X satisfies*

$$X_t = \int_0^1 k(t,s)dW_{A_s}.$$

Furthermore

$$\mathcal{F}^X_t \subseteq \mathcal{F}^W_{A_t} \subseteq \mathcal{F}^X_{C_{A_t}}, \quad \forall t \in [0,1],$$

which implies that

$$\mathcal{F}^X_{C_t} = \mathcal{F}^W_t, \quad \forall t \in [0,1].$$

Proof. Let $\tau_i = \frac{i}{n}$, $0 \leq i \leq n$, a subdivision of $[0,1]$. First remark that

$$\int_0^1 k(t,s)dW_{A_s} = \int_0^{A_t} k(t,C_s)dW_s = \int_0^1 k(t,C_s)dW_s.$$

Since W is a standard Brownian motion, the following limit holds in $L^2(\Omega)$

$$\int_0^1 k(t,C_s)dW_s = \lim_{n \to +\infty} \sum_{i=1}^n \left(\frac{1}{\tau_i - \tau_{i-1}} \int_{\tau_{i-1}}^{\tau_i} k(t,C_u)du \right)(W_{\tau_i} - W_{\tau_{i-1}}).$$

$$(11)$$

On the other hand,

$$\sum_{i=1}^n \frac{1}{\tau_i - \tau_{i-1}} \left(\int_{\tau_{i-1}}^{\tau_i} k(t,C_u)du \right) 1_{[\tau_{i-1},\tau_i[}(s),$$

converges in $L^2([0,1],ds)$ to $k(t,C_s)$ as n goes to infinity. That is,

$$\int_0^1 \left[\sum_{i=1}^n \frac{1}{\tau_i - \tau_{i-1}} \left(\int_{\tau_{i-1}}^{\tau_i} k(t,C_u)du \right) 1_{[\tau_{i-1},\tau_i[}(s) - k(t,C_s) \right]^2 ds \longrightarrow 0,$$

as n goes to infinity. On the other hand, using (10) we obtain

$$\int_0^1 \left[\sum_{i=1}^n \frac{1}{\tau_i - \tau_{i-1}} \left(\int_{\tau_{i-1}}^{\tau_i} k(t, C_u) du \right) 1_{[\tau_{i-1}, \tau_i[}(s) - k(t, C_s) \right]^2 ds$$

$$= \int_0^1 \left[\sum_{i=1}^n \frac{1}{\tau_i - \tau_{i-1}} \left(\int_{\tau_{i-1}}^{\tau_i} k(t, C_u) du \right) 1_{[\tau_{i-1}, \tau_i[}(A_s) - k(t, s) \right]^2 \rho(ds)$$

$$= \int_0^1 \left[\sum_{i=1}^n \frac{1}{\tau_i - \tau_{i-1}} \left(\int_{\tau_{i-1}}^{\tau_i} k(t, C_u) du \right) 1_{[C_{\tau_{i-1}}, C_{\tau_i}[}(s) - k(t, s) \right]^2 \rho(ds)$$

$$= \int_0^1 \left[\sum_{i=1}^n \frac{1}{\tau_i - \tau_{i-1}} \left(\int_{\tau_{i-1}}^{\tau_i} k(t, C_u) du \right) 1_{[C_{\tau_{i-1}}, C_{\tau_i}[}(s) - k(t, s) \right]^2 \rho(ds).$$

Hence $\left(\sum_{i=1}^n \frac{1}{\tau_i - \tau_{i-1}} \left(\int_{\tau_{i-1}}^{\tau_i} k(t, C_u) du \right) 1_{[C_{\tau_{i-1}}, C_{\tau_i}[}(s) \right)$ converges in $L^2(\varrho)$ to $k(t, s)$ as n goes to infinity. Applying Lemma 2, we get

$$\mathcal{J}_X \left(\sum_{i=1}^n \frac{1}{\tau_i - \tau_{i-1}} \left(\int_{\tau_{i-1}}^{\tau_i} k(t, C_u) du \right) 1_{[C_{\tau_{i-1}}, C_{\tau_i}[}(s) \right)$$

$$= R^{-1} \circ K \left(\sum_{i=1}^n \frac{1}{\tau_i - \tau_{i-1}} \left(\int_{\tau_{i-1}}^{\tau_i} k(t, C_u) du \right) 1_{[C_{\tau_{i-1}}, C_{\tau_i}[}(s) \right)$$

$$= \sum_{i=1}^n \left(\frac{1}{\tau_i - \tau_{i-1}} \int_{\tau_{i-1}}^{\tau_i} k(t, C_u) du \right) (W_{t_i} - W_{t_{i-1}}).$$

Now, by continuity of $\mathcal{J}_X$ and equality (11) we have

$$X_t = \int_0^1 k(t, C_s) dW_s.$$

Since

$$X_t = \int_0^1 k(t, C_s) dW_s = \int_0^{A_t} k(t, C_s) dW_s,$$

it follows that

$$\mathcal{F}_t^X \subseteq \mathcal{F}_{A_t}^W, \quad \forall t \in [0, 1].$$

On other hand, since W_t is a $\mathcal{F}_{C_t}^X$-Brownian motion, we have

$$\mathcal{F}_t^W \subseteq \mathcal{F}_{C_t}^X, \quad \forall t \in [0, 1],$$

and then

$$\mathcal{F}_{A_t}^W \subseteq \mathcal{F}_{C_{A_t}}^X, \quad \forall t \in [0, 1].$$

The proof is then finished by using (10).

Remark 4. By using a similar argument, one can see that

$$\forall f \in L^2(\varrho), \quad \mathcal{J}_X(f) = \int_0^1 f(C_s)dW_s.$$

For $h \in \mathcal{H}$, we have $\mathcal{K}^*h \in L^2(\varrho)$. Then, it follows that

$$\mathcal{J}_X(\mathcal{K}^*h) = \int_0^1 (\mathcal{K}^*h)(C_s)dW_s.$$

In particular, for $h = K(t, .)$, we have

$$\mathcal{J}_X(\mathcal{K}^*(K(t, .))) = \mathcal{J}_X(k(t, .)) = \int_0^1 \mathcal{K}^*(K(t, .))(C_s)dW_s$$

$$= \int_0^1 k(t, C_s)dW_s = X_t.$$

Remark 5. It should be noted that if A is continuous and strictly increasing, then C is also continuous and strictly increasing and we then have $C_{A_t} = A_{C_t} = t$. Hence

$$\mathcal{F}_t^X = \mathcal{F}_{A_t}^W, \quad \forall t \in [0, 1].$$

Example 1. • The fractional Brownian motion (fBm) with Hurst parameter $H \in [\frac{1}{2}, 1)$ is a centered Gaussian process B^H with covariance function

$$K^H(t, s) = \frac{1}{2}\left(t^{2H} + s^{2H} - |t - s|^{2H}\right).$$

It is known (see [DU99]) that B^H admits a canonical Volterra type representation

$$B_t^H = \int_0^t k^H(t, u)dW_u,$$

where W is a standard Brownian motion, the kernel k^H has the expression

$$k^H(t, s) = c_H \left(H - \frac{1}{2}\right) s^{\frac{1}{2} - H} \int_s^t (u - s)^{H - \frac{3}{2}} u^{H - \frac{1}{2}} \, du \, \mathbf{1}_{[0,t]}(s),$$

and c_H is a normalizing constant given by

$$c_H = \left[\frac{2H \, \Gamma\left(\frac{3}{2} - H\right)}{\Gamma\left(H + \frac{1}{2}\right)\Gamma(2 - 2H)}\right]^{1/2}.$$

We denote by $\mathbb{S}$ the set of step functions on $[0, 1]$. Let $\mathbb{H}_{\mathbb{S}}$ be the Hilbert space defined as the closure of $\mathbb{S}$ with respect to the scalar product

$$\langle \mathbf{1}_{[0,t]}, \mathbf{1}_{[0,s]} \rangle_{\mathbb{H}_{\mathbb{S}}} = K^H(t, s).$$

The mapping $1_{[0,t]} \mapsto B_t^H$ can be extended to an isometry between $\mathbb{H}_S$ and the Gaussian space associated to B^H. We will denote this isometry by $f \mapsto \int_0^1 f(s)dB_s^H$. Let $f : [0;1] \longrightarrow [0;1]$ be an absolutely continuous function such that $f > 0$ and f' is locally square integrable. It is known that both processes $\int_0^t f(s)dB_s^H$ and $B^H_{\int_0^t f(s)^{\frac{1}{H}}ds}$ are locally equivalent (see Baudoin and Nualart [BN03]). Moreover

$$B^H_{\int_0^t f(s)^{\frac{1}{H}}ds} = \int_0^t k_H\left(\int_0^t f(u)^{\frac{1}{H}}du, \int_0^s f(u)^{\frac{1}{H}}du\right)f(s)^{\frac{1}{2H}}dW_s.$$

Now, putting $A_s = \int_0^s f(u)^{\frac{1}{H}}du$, we find

$$B^H_{\int_0^t f(s)^{\frac{1}{H}}ds} = \int_0^t k_H\left(\int_0^t f(u)^{\frac{1}{H}}du, \int_0^s f(u)^{\frac{1}{H}}du\right)f(s)^{\frac{1}{2H}}dW_s$$

$$= \int_0^t k_H(A_t, A_s)f(s)^{\frac{1}{2H}}dW_s$$

$$= \int_{C_0}^{C_t} k_H(A_t, s)f(C_s)^{\frac{1}{2H}}dW_{C_s} = \int_0^1 k(t,s)dW_{\rho([0,s])},$$

with $k(t,s) = k_H(A_t,s)f(C_s)^{\frac{1}{2H}}1_{[C_0,C_t]}(s)$ and $\rho([0,s]) = C_s$.

- For each $\gamma > -1$, the weighted process is defined as a Gaussian process X with covariance function of the form $K(t,s) = s^\gamma t^\gamma (s \wedge t)$ and $X_0 = 0$. Observe that with $k(t,s) = t^\gamma 1_{[0,t]}(s)$ and $\varrho([0,s]) = s$, the covariance K takes the form

$$K(t,s) = \int_0^1 k(t,u)k(s,u)\varrho(du).$$

Thanks to Theorem 2 we have the canonical representation of X as follows

$$X_t = \int_0^1 k(t,s)dW_{\rho([0,s])} = t^\gamma W_t.$$

- Let X be a Gaussian-Markov process with continuous and strictly positive covariance function $K(t,s)$. It is well known that there exist a continuous process u and a continuous strictly increasing process v such that $K(t,s) = u(t)u(s)v(s \wedge t)$ (see [RY91], p. 81). It also follows from Theorem 4.3 in [RY91] that there exits a probability measure ϱ given by $\varrho([0,t]) = v(t)$. So, with $k(t,s) = u(t)1_{[0,t]}(s)$ and $\rho([0,s]) = v(s)$ we have

$$K(t,s) = \int_0^1 k(t,u)k(s,u)\varrho(du).$$

Then we obtain via Theorem 2 the following representation

$$X_t = \int_0^1 k(t,s)dW_{\rho([0,s])} = u(t)W_{v(t)}.$$

It should be pointed out that the above representation holds not only in law but strongly.

4 Malliavin Calculus

First recall that stochastic calculus of variations or Malliavin calculus is valid for an arbitrary Gaussian process (see Malliavin [Mal97] and Nualart [Nua95]). The first part of this section is devoted to the orthogonal chaos decomposition for square integrable functionals of our Gaussian process X. In the second part we establish relationships between derivation operators and divergences associated with the processes X and W. To the stochastic process $\{X_t,\ t \in [0,1]\}$ we associate the isonormal Gaussian process $\{X(f),\ f \in \mathcal{H}\}$, defined by $X(f) = \mathcal{J}_X(\mathcal{K}^*(f)), f \in \mathcal{H}$. Denote by D^X and δ^X the Malliavin derivative and the Skorohod integral associated with the process X.

Let S be the set of smooth and cylindrical random variables of the form

$$F = f\left(X\left(\varphi_1\right),\cdots,X\left(\varphi_n\right)\right) \tag{12}$$

where $n \geq 1$, $f \in C_b^\infty\left(\mathbb{R}^n\right)$ (f and all its derivatives are bounded), and $\varphi_1,\cdots,\varphi_n \in \mathcal{H}$. Given a random variable F of the form (12), we define its derivative as the $\mathcal{H}$-valued random variable given by

$$D^X F = \sum_{j=1}^n \frac{\partial f}{\partial x_j}\left(X\left(\varphi_1\right),\cdots,X\left(\varphi_n\right)\right)\varphi_j.$$

The derivative operator D^X is a closable unbounded operator from $L^p\left(\Omega\right)$ into $L^p\left(\Omega;\mathcal{H}\right)$ for any $p \geq 1$. In a similar way, the iterated derivative operator $D^{X,m}$ maps $L^p\left(\Omega\right)$ into $L^p\left(\Omega;\mathcal{H}^{\otimes m}\right)$. For any positive integer m and real $p \geq 1$, we denote by $\mathbb{D}_p^{X,m}$ the closure of S with respect to the norm defined by

$$\|F\|_{X,m,p}^p = \|F\|_{L^p(\Omega)}^p + \sum_{j=1}^m \left\|D^{X,j}F\right\|_{L^p(\Omega;\mathcal{H}^{\otimes j})}^p.$$

The domain of δ^X (denoted by $Dom\,\delta^X$) in $L^2(\Omega)$ is the set of elements $u \in L^2\left(\Omega;\mathcal{H}\right)$ such that there exists a constant c verifying

$$\left|\mathbb{E}\left\langle D^X F,u\right\rangle_{\mathcal{H}}\right| \leq c\|F\|_2,$$

for all $F \in S$. If $u \in Dom\,\delta^X$, $\delta^X\left(u\right)$ is the element in $L^2\left(\Omega\right)$ defined by the duality relationship

$$\mathbb{E}\left(\delta^X\left(u\right)F\right) = \mathbb{E}\left\langle D^X F,u\right\rangle_{\mathcal{H}}, \qquad F \in \mathbb{D}_2^{X,1}.$$

Let V be a separable Hilbert space. We can similarly define the spaces $\mathbb{D}_p^{X,m}\left(V\right)$ of V-valued random variables. Recall that the space $\mathbb{D}_2^{X,1}\left(\mathcal{H}\right)$ of $\mathcal{H}$-valued random variables is included in the domain of δ^X and for any element u in $\mathbb{D}_2^{X,1}\left(\mathcal{H}\right)$ we have

$$\mathbb{E}\left(\delta^X\left(u\right)^2\right) \leq \mathbb{E}\|u\|_{\mathcal{H}}^2 + \mathbb{E}\left\|D^X u\right\|_{\mathcal{H}\otimes\mathcal{H}}^2.$$

A random variable F of the form (12) is said to be a polynomial functional when f is an element of the set of real polynomials with n variables. We will denote by $\mathcal{P}$ the set of polynomial functionals. For a more complete presentation, see [Nua95].

Consider $\mathcal{P}_0 = \mathbb{R}$ and for $n \in \mathbb{N}^*$, define $\mathcal{P}_n$ as the closed space spanned in $L^2(\mathbb{P})$ by the elements of $\mathcal{P}$ of degree less than n. Set $\mathcal{C}_0 = \mathcal{P}_0$ and suppose that $\mathcal{C}_1, ..., \mathcal{C}_n$ are defined. Then, we define $\mathcal{C}_{n+1}$ as the orthogonal of $\mathcal{C}_1 \oplus ... \oplus \mathcal{C}_n$ in $\mathcal{P}_{n+1}$. As for all Gaussian spaces, we have the chaos decomposition:

Theorem 3.
$$L^2(\mathbb{P}) = \bigoplus_{n \geq 0} \mathcal{C}_n.$$

This means that every $\mathbb{P}$-square integrable functional from Ω to $\mathbb{R}$ can be written in a unique way as
$$F = \sum_{n \geq 0} J_n F,$$

where J_n is the orthogonal projection of $L^2(\mathbb{P})$ onto $\mathcal{C}_n$.

Henceforth we will denote by D^W, δ^W, $\mathbb{D}_p^{W,m}$ the operators and spaces associated with the Wiener process W. Now, remark from (jj) that for $\varphi \in \mathcal{H}$ and $t \in [0,1]$ we have

$$D_t^W(X(\varphi)) = D_t^W \int_0^1 (\mathcal{K}^*\varphi)(C_s)dW_s = (\mathcal{K}^*\varphi)(C_t).$$

Then, for $F = f(X(\varphi))$ and $t \in [0,1]$, we get

$$D_t^W F = f'(X(\varphi))(\mathcal{K}^*\varphi)(C_t).$$

It follows that

$$\|F\|_{W,1,2}^2 = \|F\|_{L^2(\Omega)}^2 + \|D^W F\|_{L^2(\Omega;L^2(dt))}^2$$
$$= \|F\|_{L^2(\Omega)}^2 + \mathbb{E}\left[(f'(X(\varphi)))^2 \int_0^1 [(\mathcal{K}^*\varphi)(C_s)]^2 \, ds\right]$$
$$= \|F\|_{L^2(\Omega)}^2 + \mathbb{E}\left[(f'(X(\varphi)))^2 \int_0^1 [(\mathcal{K}^*\varphi)(s)]^2 \, \varrho(ds)\right].$$

As a consequence we obtain from equality (7) that

$$\mathbb{D}_2^{X,m} = (\mathcal{K}^*)^{-1}(\mathbb{L}_2^{W,1}), \tag{13}$$

where $\mathbb{L}_2^{W,1} = \mathbb{D}_2^{W,m}(L^2(\varrho))$.

Proposition 2. *For any smooth random variable F and any $u \in L^2(\Omega;\mathcal{H})$:*

$$\mathbb{E}\langle D^X F, u\rangle_{\mathcal{H}} = \mathbb{E}\langle D^W F, (\mathcal{K}^* u)(C_.)\rangle_{L^2(dt)}$$

Proof. It is sufficient to consider F of the form $f(X(\mathcal{K}\varphi))$. In this case we have

$$
\begin{aligned}
\mathbb{E}\langle D^X F, u\rangle_{\mathcal{H}} &= \mathbb{E}\langle f'X(\mathcal{K}\varphi))\mathcal{K}\varphi, u\rangle_{\mathcal{H}} \\
&= \mathbb{E}\langle f'(X(\mathcal{K}\varphi))\varphi, \mathcal{K}^* u\rangle_{L^2(\varrho(dt))} \\
&= \mathbb{E}\int_0^1 f'(X(\mathcal{K}\varphi))\varphi(t)(\mathcal{K}^* u)(t)\varrho(dt) \\
&= \mathbb{E}\int_0^1 f'(X(\mathcal{K}\varphi))\varphi(C_t)(\mathcal{K}^* u)(C_t)\,dt \\
&= \mathbb{E}\langle D^W_. F, (\mathcal{K}^* u)(C.)\rangle_{L^2(dt)}.
\end{aligned}
$$

The above proposition and equality (13) have the following consequence:

Corollary 1. *1. For any $\mathcal{H}$-valued random variable u in $Dom\,\delta^X$, we have*

$$
\delta^X(u) = \delta^W((\mathcal{K}^* u)(C.))
$$

2. $(\mathcal{K}^*)^{-1}(\mathbb{L}_2^{W,1})$ is included in the domain of δ^X.

It should be noted that for $\varrho(ds) = ds$ we obtain

$$
\delta^X(u) = \delta^W(\mathcal{K}^* u),
$$

for any $\mathcal{H}$-valued random variable u in $Dom\,\delta^X$. On the other hand

$$
(\mathcal{K}^*)^{-1}\mathbb{L}_2^{W,1} = (\mathcal{K}^*)^{-1}\mathbb{D}_2^{W,m}(L^2(ds))
$$

is included in $Dom\,\delta^X$.

5 Girsanov Transformation

For $h \in \mathcal{H}$, we define

$$
\Lambda^h = \exp\left(\delta^X(h) - \frac{1}{2}\|h\|_{\mathcal{H}}^2\right).
$$

Let $h \in \mathcal{H}$ and $\tau_h(\mathbb{P})$ be the translate of $\mathbb{P}$ by h. In this section we look for the law of $\tau_h(\mathbb{P})$. Since $h \in \mathcal{H}$ there exists $g_h \in L^2(\varrho)$ such that

$$
h(t) = \int_0^1 k(t,s)g_h(s)\varrho(ds).
$$

So we define the following transformation

$$
(TX)_t = X_t + \int_0^1 k(t,u)g_h(u)\varrho(du),
$$

which has the law $\tau_h(\mathbb{P})$.

Proposition 3. *The Gaussian measure* τ_h $(\mathbb{P})$ *is equivalent to* $\mathbb{P}$ *and the density is equal to* Λ^h.

Proof. First we remark that $(TX)_\cdot$ has the representation

$$(TX)_t = \int_0^1 k(t,s)dW_{\varrho([0,s])} + \int_0^1 k(t,s)g_h(s)\varrho(ds).$$

Now we have

$$W_{\varrho([0,t])} + \int_0^t g_h(s)\varrho(ds) = \int_0^{A_t} dW_s + \int_0^{A_t} g_h(C_s)ds.$$

The classical Girsanov theorem asserts that the process

$$t \longrightarrow W_t + \int_0^t g_h(C_s)ds,$$

is a Brownian motion $\widetilde{W}$ under the law $\widetilde{\mathbb{P}}$ defined by

$$\frac{d\widetilde{\mathbb{P}}}{d\mathbb{P}}\bigg|_{\mathcal{F}_t^W} = \exp\bigg(\int_0^t g_h(C_s)dW_s - \frac{1}{2}\int_0^t (g_h(C_s))^2 ds\bigg).$$

It follows that the process $\left(\int_0^{A_t} dW_s + \int_0^{A_t} g_h(C_s)ds\right)$ is equivalent to $\widetilde{W}_{A_t}$ under the law

$$\frac{d\widetilde{\mathbb{P}}}{d\mathbb{P}}\bigg|_{\mathcal{F}_{A_t}^W} = = \exp\bigg(\int_0^{A_t} g_h(C_s)dW_s - \frac{1}{2}\int_0^{A_t} (g_h(C_s))^2 ds\bigg)$$

$$= \exp\bigg(\int_0^t g_h(s)dW_{A_s} - \frac{1}{2}\int_0^t (g_h(s))^2 dA_s\bigg) = \Lambda^h.$$

Corollary 2. *If* $\mathbb{E}\left(\Lambda^h\right) = 1$, *then the law of the process*

$$(TX)_t = \int_0^1 k(t,s)dW_{\varrho([0,s])} + \int_0^1 k(t,s)g_h(s)\varrho(ds),$$

under the probability

$$\frac{d\widetilde{\mathbb{P}}}{d\mathbb{P}}\bigg|_{\mathcal{F}_{A_t}^W} = \exp\bigg(\int_0^t g_h(s)dW_{\varrho([0,s])} - \frac{1}{2}\int_0^t (g_h(s))^2 \varrho(ds)\bigg),$$

is the same as the law of the canonical process X_t *under* $\mathbb{P}$.

Acknowledgments The anonymous referee is acknowledged for suggestions on improving the presentation of the paper.

References

[AMN01]   Alòs, E., Mazet, O., Nualart, D.: Stochastic calculus with respect to Gaussian process. The Annals of Probability **29** (2), 766-801 (2001).

[BN03]   Baudoin, F., Nualart, D.: Equivalence of Volterra processes. Stochastic Proc. Appl. 107, 327-350 (2003).

[CJE85]   Chaleyat-Maurel, M., Jeulin, T.: Grossissement gaussien de la filtration brownienne, Lecture Notes in Math. **1118**, Springer, 59-109 (1985).

[Cra61]   Cramér, H.: On some classes of non-stationary processes. Proc. 4th Berkeley Sympo. Math. Stat. and Prob. **2**, 57-77 (1961).

[DU99]   Decreusefond, L., Üstünel, A. S.: Stochastic analysis of the fractional Brownian motion. Potential Anal. **10** (2), 177–214 (1999).

[Gre81]   Grenander, U.: Abstract Inference. Wiley, New York (1981).

[Jan97]   Janson, S.: Gaussian Hilbert Spaces. Cambridge University Press, Cambridge (1997).

[JY90]   Jeulin, T., Yor, M.: Filtrations des ponts browniens, et équations différentielles stochastiques linéaires. Sém. Prob. XXIV, Lect. Notes in Maths 1426, Springer, Berlin, 227-265 (1990).

[HHM97]   Hibino, Y., Hitsuda, M., Muraoka, H.: Construction of noncanonical representations of a Brownian motion. Hiroshima Math. J. **27** , no. 3, 439-448 (1997).

[Hid60]   Hida, T.: Canonical representation of Gaussian processes and their applications. Mem. Coll. Sci. Univ. Kyoto Ser.A. **33**, 109-155 (1960).

[His68]   Hitsuda, M.: Representation of Gaussian processes equivalent to Wiener process. Osaka J. Math. 5, 299–312 (1968).

[Hul03]   Hult, H.: Approximating some Volterra type stochastic integrals with applications to parameter estimation. Stochastic Processes and their Applications 105, 1 – 32 (2003).

[Kat95]   Kato, T.: Perturbation Theory for Linear Operators. Reprint of the 1980 edition. Classics in Mathematics. Springer-Verlag, Berlin (1995).

[Lev65]   Lévy, P.: Processus Stochastiques et Mouvement Brownien. (1948, second edition). Gauthier-Villars, Paris (1965).

[Mal97]   Malliavin, P.: Stochastic Analysis. Springer, New York (1997).

[Nua95]   Nualart, D.: The Malliavin Calculus and Related Topics. Springer, Berlin (1995).

[She66]   Shepp, L.A.: Radon–Nikodym derivatives of Gaussian measures. Ann. Math. Statist. **37**, 321–354 (1966).

[Suq95]   Suquet, Ch.: Distances euclidiennes sur les mesures signées et application à des théorèmes de Berry-Esséen. Bull. Belg. Math. Soc. **2**, 161–181 (1995).

[RY91]   Revuz, D., Yor, M.: Continuous Martingales and Brownian Motion. Springer, Berlin (1991).

[Yor92]   Yor, M.: Some Aspects of Brownian Motion. Part I. Some Special Functionals, Lectures Notes in Math. ETH Zürich, Birkhäuser Verlag, Basel (1992).

Recognising Whether a Filtration is Brownian: a Case Study

Michel Émery

IRMA, Université de Strasbourg et C.N.R.S.
7 rue René Descartes, 67 084 Strasbourg Cedex, France
e-mail: emery@math.u-strasbg.fr

> [...] l'on s'introduit dans l'espace des signes.
>
> R. BARTHES. *L'empire des signes.*

Summary. A filtration on a probability space is said to be *Brownian* when it is generated by some Brownian motion started from 0. Recognising whether a given filtration $\mathcal{F} = (\mathcal{F}_t)_{t\geqslant 0}$ is Brownian may be a difficult problem; but when $\mathcal{F}$ is *Brownian after zero,* a necessary and sufficient condition for $\mathcal{F}$ to be Brownian is available, namely, the self-coupling property (ii) of Theorem 1 of [4]. ('Brownian after zero' means that for each $\varepsilon > 0$, the shifted filtration $\mathcal{F}^\varepsilon = (\mathcal{F}_{\varepsilon+t})_{t\geqslant 0}$ is generated by its initial σ-field $\mathcal{F}_\varepsilon$ and by some $\mathcal{F}^\varepsilon$-Brownian motion.) In all concrete examples where this self-coupling criterion has been used to establish Brownianity, another, more constructive proof was also available. The situation presented below is different. We are interested in a certain process, introduced in 1991 by Beneš, Karatzas and Rishel; the natural filtration of this process turns out to be also generated by some Brownian motion, but we have not been able to exhibit such a generating Brownian motion; the general, non constructive criterion is the only proof we know that this filtration is indeed Brownian.

The filtration to be studied is the one generated by the process $Z = (X, Y)$, where X and Y are two Brownian motions linked by the relation

$$\operatorname{sgn} X \, dX + \operatorname{sgn} Y \, dY = 0 . \tag{1}$$

This process was first considered by Beneš, Karatzas and Rishel [1], to solve a partially-observed stochastic control problem that turns out not to admit a strict-sense optimal law. I thank Ioannis Karatzas for bringing this reference to my attention and raising the question of the nature of the filtration generated by Z. I am also grateful to the Minerva Foundation, who supported my visit to Columbia University, where most of this work was done.

C. Donati-Martin et al. (eds.), *Séminaire de Probabilités XLII,*
Lecture Notes in Mathematics 1979, DOI 10.1007/978-3-642-01763-6_14,
© Springer-Verlag Berlin Heidelberg 2009

Notation and Conventions. A probability space $(\Omega, \mathcal{A}, \mathbb{P})$ is always complete; by a sub-σ-field of $\mathcal{A}$, we always mean an $(\mathcal{A}, \mathbb{P})$-complete sub-$\sigma$-field. A *raw filtration* $\mathcal{F}$ on $(\Omega, \mathcal{A}, \mathbb{P})$ is an increasing family $(\mathcal{F}_t)_{t \geqslant 0}$ of sub-σ-fields of $\mathcal{A}$ (so each of them is $(\mathcal{A}, \mathbb{P})$-complete); if furthermore $t \mapsto \mathcal{F}_t$ is right-continuous, that is, if $\mathcal{F}_t = \bigcap_{\varepsilon > 0} \mathcal{F}_{t+\varepsilon}$ for each t, then $\mathcal{F}$ is simply called a *filtration*. If $\mathcal{F}^\circ$ is a raw filtration, the filtration $\mathcal{F}$ generated by $\mathcal{F}^\circ$, i.e., the smallest filtration containing $\mathcal{F}^\circ$, is given by $\mathcal{F}_t = \bigcap_{\varepsilon > 0} \mathcal{F}^\circ_{t+\varepsilon}$.

If $\mathcal{F}$ and $\mathcal{G}$ are two raw filtrations, $\mathcal{F} \mathbin{\vee\!\!\!\!\vee} \mathcal{G}$ denotes the raw filtration generated by $\mathcal{F}$ and $\mathcal{G}$; it is given by $(\mathcal{F} \mathbin{\vee\!\!\!\!\vee} \mathcal{G})_t = \mathcal{F}_t \vee \mathcal{G}_t$ (the small circle is a reminder that $\mathcal{F} \mathbin{\vee\!\!\!\!\vee} \mathcal{G}$ is a raw filtration, not necessarily right-continuous). The filtration generated by $\mathcal{F}$ and $\mathcal{G}$, or equivalently by $\mathcal{F} \mathbin{\vee\!\!\!\!\vee} \mathcal{G}$, is denoted by $\mathcal{F} \vee \mathcal{G}$.

We will use the convention that a stochastic integral $\int V \, dU$ is always started from 0 (not from $V_0 U_0$). L^U will denote the local time at 0 of the continuous semimartingale U. Tanaka's well-known formula asserts that $\int \operatorname{sgn} U \, dU = |U| - |U_0| - L^U$; the process $|U| - L^U$ is the *Lévy transform* of U, we shall denote it by $\mathcal{T}U$. So (1) says that $\mathcal{T}X + \mathcal{T}Y$ is constant (and hence equal to $|X_0| + |Y_0|$).

If T is a stopping time, the process $U^{T]}$ is U stopped at T; its value at time t is $U_{t \wedge T}$. The process $U^{[T} = U - U^{T]}$ is null up to T and varies as U after T.

Before focusing on the filtration of the solution to (1), we shall first describe this process. The interesting case is when $X_0 = Y_0 = 0$, but it will be helpful to consider the more general case when the initial value Z_0 is an arbitrary point $z_0 = (x_0, y_0)$ in the plane.

Definition. A process $Z = (X, Y)$, defined on some filtered probability space $(\Omega, \mathcal{A}, \mathbb{P}, \mathcal{F})$ and taking its values in $\mathbb{R}^2$, is called a *BKR process* if X and Y are two $\mathcal{F}$-Brownian motions (not necessarily started from 0) linked by (1).

The next proposition is borrowed from [1]; when $z_0 \neq 0$, it entails existence and uniqueness in law of the BKR process Z and shows that the filtration of Z is Brownian.

Proposition 1 (Beneš, Karatzas and Rishel). *Fix a point $z_0 \neq 0$ in the plane. On some filtered probability space $(\Omega, \mathcal{A}, \mathbb{P}, \mathcal{F})$, let be given a BKR process $Z = (X, Y)$ started from $(X_0, Y_0) = z_0$. Define a Brownian motion B by*

$$dB = -\operatorname{sgn} Y \, dX = \operatorname{sgn} X \, dY \; ; \qquad B_0 = 0 \, . \tag{2}$$

The processes B and Z generate the same filtration. More precisely, there exists a functional Φ such that $(X, Y) = \Phi(z_0, B)$ whenever X, Y and B are any three $\mathcal{F}$-Brownian motions satisfying $(X_0, Y_0) = z_0$ and (2).

Conversely, given a Brownian motion B on some filtered probability space, with $B_0 = 0$, the process $Z = (X, Y)$ defined by $Z = \Phi(z_0, B)$ is a BKR process started from z_0 and satisfying (2).

Proof (borrowed from [1]). Observe first that $(2) \Rightarrow (1)$; more precisely, the second equality in (2) is equivalent to (1). (The convention for $\operatorname{sgn} 0$ is irrelevant here, since if U and V are two Brownian motions, $\mathbb{1}_{U=0}\, d\langle V, V\rangle = \mathbb{1}_{U=0}\, dt = 0$.)

The proof of the proposition consists in describing Φ as an algorithm yielding X and Y from the data x_0, y_0 and B. Here is the first step of this algorithm.

Put $S = \inf\{t \geqslant 0 \ : \ B_t = x_0 \operatorname{sgn} y_0 \text{ or } B_t = -y_0 \operatorname{sgn} x_0\}$; on $[\![0, S]\!]$, set $X = x_0 - \operatorname{sgn} y_0\, B$ and $Y = y_0 + \operatorname{sgn} x_0\, B$. Notice that S is also the first time when X or Y vanishes. On $[\![0, S]\!]$, (1) and (2) hold; conversely, if X, Y and B are three Brownian motions satisfying (2) and respectively started from x_0, y_0 and 0, one must have $X = x_0 - \operatorname{sgn} y_0\, B$ and $Y = y_0 + \operatorname{sgn} x_0\, B$ on $[\![0, S]\!]$.

For the second step, observe that one of X_S and Y_S vanishes, and put

$$T_0 = \begin{cases} S & \text{if } X_S = 0, \\ \inf\{t \geqslant S \ : \ \mathcal{J}(B^{[S]})_t = |X_S|\} & \text{if } X_S \neq 0 \text{ and } Y_S = 0; \end{cases}$$

on the interval $]\!]S, T_0]\!]$, set $Y_t = \operatorname{sgn} X_S\, B_t^{[S]}$ and $X_t = X_S - \operatorname{sgn} X_S\, \mathcal{J}(B^{[S]})_t$; remark that T_0 is also the first time after S that X vanishes. On the interval $]\!]S, T_0]\!]$ one has $dY = \operatorname{sgn} X_S\, dB$ and $dX = -\operatorname{sgn} X_S \operatorname{sgn} B^{[S}\, dB = -\operatorname{sgn} Y\, dB$, so (2) and (1) hold. Conversely, if X, Y and B satisfy (2), if T_0 denotes the first time from S on when $X = 0$, one must have $Y = \operatorname{sgn} X_S\, B^{[S}$ on $]\!]S, T_0]\!]$, and also $dX = -\operatorname{sgn} X_S \operatorname{sgn} B^{S]}\, dB = -\operatorname{sgn} X_S\, d\mathcal{J}(B^{[S]})$ on this interval, consequently T_0 is the first time that $\mathcal{J}(B^{[S]}) = |X_S|$ if $X_S \neq 0$.

The proof keeps proceeding the same way, on successive intervals: suppose that the algorithm manufacturing X and Y from B has been constructed up to some time T_{2n} such that $X_{T_{2n}} = 0$, and that the so-obtained X and Y have been shown to be the only Brownian motions started from x_0 and y_0 and satisfying (2) on $[\![0, T_{2n}]\!]$. Put

$$T_{2n+1} = \inf\{t \geqslant T_{2n} \ : \ \mathcal{J}(B^{[T_{2n}]})_t = |Y_{T_{2n}}|\} \ ;$$

and define X and Y on $]\!]T_{2n}, T_{2n+1}]\!]$ by

$$X = -\operatorname{sgn} Y_{T_{2n}}\, B^{[T_{2n}} \ ; \qquad Y = Y_{T_{2n}} - \operatorname{sgn} Y_{T_{2n}}\, \mathcal{J}(B^{[T_{2n}]}) \ .$$

The same arguments as before show that the validity and the uniqueness of the construction extend up to T_{2n+1}, which is the first time after T_{2n} such that $Y = 0$.

Define then

$$T_{2n+2} = \inf\{t \geqslant T_{2n+1} \ : \ \mathcal{J}(B^{[T_{2n+1}]})_t = |X_{T_{2n+1}}|\} \ ;$$

and on $]\!]T_{2n+1}, T_{2n+2}]\!]$ set

$$Y = \operatorname{sgn} X_{T_{2n+1}}\, B^{[T_{2n+1}} \ ; \qquad X = X_{T_{2n+1}} - \operatorname{sgn} X_{T_{2n+1}}\, \mathcal{J}(B^{[T_{2n+1}]}) \ .$$

This extends the construction to $[\![T_{2n+1}, T_{2n+2}]\!]$, and T_{2n+2} is the first time that X vanishes after T_{2n+1}.

To complete the proof, it remains to show that $T_\infty = \lim_n T_n$ is a.s. infinite. On the event $\{T_\infty < \infty\}$, one has $X_{T_\infty} = \lim X_{T_{2n}} = 0$ and similarly $Y_{T_\infty} = \lim Y_{T_{2n+1}} = 0$. This is impossible, because Tanaka's formula gives

$$d|X| + d|Y| = \operatorname{sgn} X \, dX + dL^X + \operatorname{sgn} Y \, dY + dL^Y = dL^X + dL^Y \; ;$$

so $|X| + |Y|$ is increasing, and consequently $|X_{T_\infty}| + |Y_{T_\infty}| \geqslant |x_0| + |y_0| > 0$ on $\{T_\infty < \infty\}$.   □

This proof is more informative than the statement of the proposition; it gives an intuitive description of how the BKR process behaves. The point Z_t lives on the sides of the random square $|x| + |y| = |x_0| + |y_0| + L_t^X + L_t^Y$, whose vertices are on the axes. When Z is on one side of this square, it moves Brownianly on this side, which remains fixed. The square inflates only when Z is at a vertex of the square, or equivalently when Z is on one of the axes; at those times, the square expands according to the local time spent by Z on this axis. And dB measures the movement of Z, when the square is given the same orientation as the trigonometric circle.

Our aim is to study BKR processes issued from the origin; but we shall first state and prove a few elementary lemmas concerning the behaviour of filtrations. These lemmas will be needed later, to establish existence and uniqueness in law of BKR processes.

Definition. If $\mathcal{F}$ and $\mathcal{G}$ are two raw filtrations on $(\Omega, \mathcal{A}, \mathbb{P})$, one says that $\mathcal{F}$ *is immersed in* $\mathcal{G}$, and one writes $\mathcal{F} \overset{m}{\subset} \mathcal{G}$, if ($\mathcal{F}_t \subset \mathcal{G}_t$ for each $t \geqslant 0$ and) every $\mathcal{F}$-martingale is a $\mathcal{G}$-martingale.

Lemma 1. *If $\mathcal{F}$ and $\mathcal{G}$ are two raw filtrations and if $\mathcal{F}_t \subset \mathcal{G}_t$ for all $t \geqslant 0$, the following four statements are equivalent:*

(i) $\mathcal{F}$ is immersed in $\mathcal{G}$;

(ii) for each $t \geqslant 0$, the operators of conditional expectation satisfy

$$\mathbb{E}^{\mathcal{G}_t} \mathbb{E}^{\mathcal{F}_\infty} = \mathbb{E}^{\mathcal{F}_t} \; ;$$

(iii) for each $t \geqslant 0$, the operators of conditional expectation satisfy

$$\mathbb{E}^{\mathcal{F}_\infty} \mathbb{E}^{\mathcal{G}_t} = \mathbb{E}^{\mathcal{F}_t} \; ;$$

(iv) for each $t \geqslant 0$, the σ-fields $\mathcal{F}_\infty$ and $\mathcal{G}_t$ are conditionally independent given $\mathcal{F}_t$.

Proof. First, (i) holds if and only if every uniformly integrable $\mathcal{F}$-martingale is a $\mathcal{G}$-martingale, that is, if and only if $\mathbb{E}^{\mathcal{G}_t}[J] = \mathbb{E}^{\mathcal{F}_t}[J]$ for each $J \in L^1(\mathcal{F}_\infty)$; this can be rewritten $\mathbb{E}^{\mathcal{G}_t} \mathbb{E}^{\mathcal{F}_\infty} = \mathbb{E}^{\mathcal{F}_t} \mathbb{E}^{\mathcal{F}_\infty}$, which is tantamount to (ii).

Equivalence between (ii), (iii) and (iv) stems immediately from the following classical exercise: *if three sub-σ-fields $\mathcal{B}$, $\mathcal{C}$ and $\mathcal{D}$ of $\mathcal{A}$ satisfy $\mathcal{D} \subset \mathcal{B} \cap \mathcal{C}$,* *a necessary and sufficient condition for* $\mathcal{B}$ *and* $\mathcal{C}$ *to be conditionally indepen-dent given* $\mathcal{D}$ *is* $\mathbb{E}^{\mathcal{B}}\mathbb{E}^{\mathcal{C}} = \mathbb{E}^{\mathcal{D}}$. Indeed, if conditional independence holds, for all $B \in L^{\infty}(\mathcal{B})$ and $C \in L^{\infty}(\mathcal{C})$ one has $\mathbb{E}^{\mathcal{D}}[BC] = \mathbb{E}^{\mathcal{D}}[B\,\mathbb{E}^{\mathcal{D}}[C]]$, whence $\mathbb{E}[BC] = \mathbb{E}[B\,\mathbb{E}^{\mathcal{D}}[C]]$, then $\mathbb{E}^{\mathcal{B}}[C] = \mathbb{E}^{\mathcal{D}}[C]$, and finally $\mathbb{E}^{\mathcal{B}}\mathbb{E}^{\mathcal{C}} = \mathbb{E}^{\mathcal{D}}\mathbb{E}^{\mathcal{C}} = \mathbb{E}^{\mathcal{D}}$. Conversely, from $\mathbb{E}^{\mathcal{B}}\mathbb{E}^{\mathcal{C}} = \mathbb{E}^{\mathcal{D}}$, one derives $\mathbb{E}^{\mathcal{D}}[BC] = \mathbb{E}^{\mathcal{D}}\mathbb{E}^{\mathcal{B}}[BC] = \mathbb{E}^{\mathcal{D}}[B\,\mathbb{E}^{\mathcal{B}}[C]] = \mathbb{E}^{\mathcal{D}}[B\,\mathbb{E}^{\mathcal{D}}[C]] = \mathbb{E}^{\mathcal{D}}[B]\,\mathbb{E}^{\mathcal{D}}[C]$.   □

Lemma 2. *Let* $\mathcal{F}^{\circ}$ *and* $\mathcal{G}^{\circ}$ *be two raw filtrations; call* $\mathcal{F}$ *and* $\mathcal{G}$ *the filtrations respectively generated by* $\mathcal{F}^{\circ}$ *and* $\mathcal{G}^{\circ}$. *If* $\mathcal{F}^{\circ}$ *is immersed in* $\mathcal{G}^{\circ}$, *then* $\mathcal{F}$ *is immersed in* $\mathcal{G}$.

Proof. By Lemma 1 (ii), one has $\mathbb{E}^{\mathcal{G}_t^{\circ}}\mathbb{E}^{\mathcal{F}_{\infty}} = \mathbb{E}^{\mathcal{F}_t^{\circ}}$ for all t. Replacing t by $t+\varepsilon$ and letting $\varepsilon \downarrow 0$ gives $\mathbb{E}^{\mathcal{G}_t}\mathbb{E}^{\mathcal{F}_{\infty}} = \mathbb{E}^{\mathcal{F}_t}$ by reverse martingale convergence. Then, by Lemma 1 (ii) again, one obtains $\mathcal{F} \overset{m}{\subset} \mathcal{G}$.   □

Lemma 3 (preservation of immersions by enlargements). *Let* $\mathcal{F}$, $\mathcal{G}$ *and* $\mathcal{S}$ *be three filtrations such that the final* σ-*fields* $\mathcal{S}_{\infty}$ *and* $\mathcal{G}_{\infty}$ *are condi-tionally independent given* $\mathcal{F}_{\infty}$. *If* $\mathcal{F}$ *is immersed in* $\mathcal{G}$, *the raw filtration* $\mathcal{F} \vee \mathcal{S}$ *is immersed in the raw filtration* $\mathcal{G} \vee \mathcal{S}$, *and the filtration* $\mathcal{F} \vee \mathcal{S}$ *is immersed in the filtration* $\mathcal{G} \vee \mathcal{S}$.

Proof. Every bounded, $\mathcal{G}_t$-measurable r.v. G_t is conditionally independent of $\mathcal{S}_{\infty}$ (or of $\mathcal{F}_{\infty} \vee \mathcal{S}_{\infty}$) given $\mathcal{F}_{\infty}$; consequently $\mathbb{E}^{\mathcal{F}_{\infty} \vee \mathcal{S}_{\infty}}[G_t] = \mathbb{E}^{\mathcal{F}_{\infty}}[G_t]$. In turn, $\mathbb{E}^{\mathcal{F}_{\infty}}[G_t] = \mathbb{E}^{\mathcal{F}_t}[G_t]$ by immersion of $\mathcal{F}$ in $\mathcal{G}$ and Lemma 1 (iii); so $\mathbb{E}^{\mathcal{F}_{\infty} \vee \mathcal{S}_{\infty}}[G_t] = \mathbb{E}^{\mathcal{F}_t}[G_t]$. If now S_t is any bounded, $\mathcal{S}_t$-measurable r.v., one has $\mathbb{E}^{\mathcal{F}_{\infty} \vee \mathcal{S}_{\infty}}[G_t\,S_t] = S_t\,\mathbb{E}^{\mathcal{F}_{\infty} \vee \mathcal{S}_{\infty}}[G_t] = S_t\,\mathbb{E}^{\mathcal{F}_t}[G_t]$; as this is measurable in $\mathcal{F}_t \vee \mathcal{S}_t$, one gets $\mathbb{E}^{\mathcal{F}_{\infty} \vee \mathcal{S}_{\infty}}[G_t\,S_t] = \mathbb{E}^{\mathcal{F}_t \vee \mathcal{S}_t}[G_t\,S_t]$. A monotone class argu-ment then gives $\mathbb{E}^{\mathcal{F}_{\infty} \vee \mathcal{S}_{\infty}}[H_t] = \mathbb{E}^{\mathcal{F}_t \vee \mathcal{S}_t}[H_t]$ for all H_t in $L^{\infty}(\mathcal{G}_t \vee \mathcal{S}_t)$. Hence $\mathbb{E}^{\mathcal{F}_{\infty} \vee \mathcal{S}_{\infty}}\mathbb{E}^{\mathcal{G}_t \vee \mathcal{S}_t} = \mathbb{E}^{\mathcal{F}_t \vee \mathcal{S}_t}\mathbb{E}^{\mathcal{G}_t \vee \mathcal{S}_t} = \mathbb{E}^{\mathcal{F}_t \vee \mathcal{S}_t}$, and $\mathcal{F} \vee \mathcal{S}$ is immersed in $\mathcal{G} \vee \mathcal{S}$ by Lemma 1 (iii). Immersion of $\mathcal{F} \vee \mathcal{S}$ in $\mathcal{G} \vee \mathcal{S}$ follows by Lemma 2.   □

Remark 1. Lemma 3 encaptures some situations where an immersion is pre-served by an enlargement, but not all such situations. A trivial counterexample is obtained by taking $\mathcal{S} = \mathcal{G}$ and $\mathcal{F}$ degenerate: $\mathcal{F}$ is then immersed in $\mathcal{G}$ and $\mathcal{F} \vee \mathcal{S}$ is immersed in $\mathcal{G} \vee \mathcal{S}$, but $\mathcal{S}$ and $\mathcal{G}$ are not independent.

Two particular cases of Lemma 3 are noteworthy:

Corollary 1. *Let* $\mathcal{F}$, $\mathcal{G}$ *and* $\mathcal{S}$ *be three filtrations such that* $\mathcal{S}_{\infty} \subset \mathcal{F}_{\infty}$. *If* $\mathcal{F}$ *is immersed in* $\mathcal{G}$, *the raw filtration* $\mathcal{F} \vee \mathcal{S}$ *is immersed in the raw filtration* $\mathcal{G} \vee \mathcal{S}$, *and the filtration* $\mathcal{F} \vee \mathcal{S}$ *is immersed in the filtration* $\mathcal{G} \vee \mathcal{S}$.

Proof. Given $\mathcal{F}_{\infty}$, the σ-field $\mathcal{S}_{\infty}$ is conditionally independent of anything. □

Corollary 2. *Let* $\mathcal{F}$, $\mathcal{G}'$ *and* $\mathcal{G}''$ *be three filtrations such that* $\mathcal{F}$ *is immersed in* $\mathcal{G}'$ *and in* $\mathcal{G}''$. *If the terminal* σ-*fields* $\mathcal{G}'_{\infty}$ *and* $\mathcal{G}''_{\infty}$ *are conditionally in-dependent given* $\mathcal{F}_{\infty}$, *the three filtrations* $\mathcal{F}$, $\mathcal{G}'$ *and* $\mathcal{G}''$ *are immersed in the filtration* $\mathcal{G}' \vee \mathcal{G}''$ *generated by* $\mathcal{G}'$ *and* $\mathcal{G}''$.

Proof. Lemma 3 with $\mathcal{S} = \mathcal{G}'$ and $\mathcal{G} = \mathcal{G}''$ gives $\mathcal{G}' \overset{m}{\subset} \mathcal{G}' \vee \mathcal{G}''$; similarly, one has $\mathcal{G}'' \overset{m}{\subset} \mathcal{G}' \vee \mathcal{G}''$, and $\mathcal{F} \overset{m}{\subset} \mathcal{G}' \vee \mathcal{G}''$ follows by transitivity of immersions. $\square$

We can now come back to the BKR process $Z = (X, Y)$ started from 0. The behaviour of $(|X|, |Y|)$ is an immediate consequence of a well known property of the Lévy transform:

Proposition 2. *If X and Y are two Brownian motions defined on some probability space $(\Omega, \mathcal{A}, \mathbb{P})$, started from 0 and satisfying (1), one has*

$$|X_t| = W_t - I_t \qquad and \qquad |Y_t| = S_t - W_t ,$$

where W is the Brownian motion defined by

$$W_t = \int_0^t \mathrm{sgn}\, X_s \, \mathrm{d}X_s = -\int_0^t \mathrm{sgn}\, Y_s \, \mathrm{d}Y_s ,$$

and where $I_t = \inf_{s \in [0,t]} W_s$ and $S_t = \sup_{s \in [0,t]} W_s$.

Moreover, the three processes $|X|$, $|Y|$, and W generate the same filtration.

Proof. The Lévy transform of X is the Brownian motion $W_t = \int_0^t \mathrm{sgn}\, X_s \, \mathrm{d}X_s = |X_t| - L_t^X$. Putting $g_t = \sup\{s \in [0, t] \,:\, X_s = 0\}$, one has for $s \in [0, t]$

$$W_s = |X_s| - L_s^X \geqslant 0 - L_t^X = |X_{g_t}| - L_{g_t}^X = W_{g_t} ,$$

wherefrom $I_t = W_{g_t} = -L_t^X$ and $W_t = |X_t| + I_t$. The proof of $W_t = S_t - |Y_t|$ is similar.

As was observed by Lévy, the process $W = S - |Y|$ is adapted to the filtration of $|Y|$ (see for instance Remark 2.25 in Chapter 6 of [6]); conversely, $|Y| = S - W$ is adapted to the filtration of W. Hence $|Y|$ and W generate the same filtration; so does also $|X|$ for a similar reason. $\square$

Proposition 2 describes the process $(|X|, |Y|)$ as a functional of the Brownian motion W, namely $(|X|, |Y|) = \Psi(I, W, S)$, where $\Psi(i, w, s) = (w-i, s-w)$. This implies in particular uniqueness in law of $(|X|, |Y|)$: if (X', Y') and (X'', Y'') are any two solutions to (1) started from $(0, 0)$, the processes $(|X'|, |Y'|)$ and $(|X''|, |Y''|)$ have the same law.

For fixed time t, the joint law of (I_t, W_t, S_t) can be explicitly written (see formula 1 1.15.8 in Borodin and Salminen [2]); the law of $(|X_t|, |Y_t|)$ can be derived therefrom.

Observe that Proposition 2 needs the Brownian motions X and Y to be linked by (1), but they do not have to form a BKR process, that is, to be Brownian motions for some common filtration. It is not difficult to see that all solutions (X, Y) to (1), started from $(0, 0)$ and not constrained to be Brownain motions for some common filtration, are obtained (in law) by the following procedure: First, construct $(|X|, |Y|)$ as $\Psi(I, W, S)$ for some Brownian motion

W started from 0. Then, conditionally on $(|X|, |Y|)$, construct X by choosing the signs of the excursions of X as an i.i.d. sequence uniform on $\{-, +\}$. Last, do the same for the signs of the excursions of Y. *Conditional on* $(|X|, |Y|)$, *each of the two sequences giving the signs of* X *and the signs of* Y *is i.i.d. and uniform on* $\{-, +\}$, *but their joint conditional law given* $(|X|, |Y|)$ *is arbitrary:* these two sequences may be correlated in any way. This assertion is an easy consequence of the following fact from excursion theory: *A process* U *is a Brownian motion if and only if* $|U|$ *is a Brownian motion reflected at the origin and, conditional on* $|U|$, *the signs of the excursions of* U *form an i.i.d. sequence uniform on* $\{-, +\}$.

To make this statement rigorous, we need a formal definition of the sequence of excursion signs of a process; for the sake of definiteness, we shall use the following one.

Definition. Fix once and for all a dense sequence (r_k) in $(0, \infty)$. If U is a Brownian motion, call J_k the (a.s. well defined) excursion interval of U straddling r_k. Define a sub-sequence (J'_n) by deleting from the sequence (J_k) any interval that already occurs earlier in that sequence; this sub-sequence is a.s. infinite. The *sequence of excursion signs of* U is $\varepsilon = (\varepsilon_n)$, where ε_n is the sign of U during J'_n.

Proposition 3. *On a suitable filtered probability space* $(\Omega, \mathcal{A}, \mathbb{P}, \mathcal{F})$, *there exists a BKR process started from the origin.*

Proof. Start with an independent triple (W, ε, η), where W is a Brownian motion started from 0, and ε and η are two i.i.d. sequences, uniform on $\{-, +\}$. Define I and S from W as in Proposition 2, and (with an abuse of notation) put $|X| = W - I$ and $|Y| = S - W$. Define X from $|X|$ by choosing the sequence of excursion signs of X equal to ε and similarly define Y so that the signs of its excursions are given by η.

As ε and η are independent of the reflected Brownian motions $|X|$ and $|Y|$, both X and Y are Brownian motions. As W and $W - I$ generate the same filtration, W is the martingale part of $|X|$, i.e., $dW = \operatorname{sgn} X \, dX$; similarly, $dW = -\operatorname{sgn} Y \, dY$, so (1) holds.

It remains to see that X and Y are $\mathcal{F}$-Brownian motions for some filtration $\mathcal{F}$. Call $\mathcal{W}$ (resp. $\mathcal{X}, \mathcal{Y}$) the natural filtration of W (resp. X, Y). The process $W = \int \operatorname{sgn} X \, dX$ is an $\mathcal{X}$-Brownian motion, whence $\mathcal{W} \overset{m}{\subset} \mathcal{X}$; similarly, $\mathcal{W} \overset{m}{\subset} \mathcal{Y}$. Now, $\mathcal{X}_\infty = \sigma(W, \varepsilon)$ and $\mathcal{Y}_\infty = \sigma(W, \eta)$ are conditionally independent given $\mathcal{W}_\infty = \sigma(W)$. So Corollary 2 applies, and $\mathcal{X}$ and $\mathcal{Y}$ are immersed in $\mathcal{X} \vee \mathcal{Y} = \mathcal{F}$; thus X and Y are $\mathcal{F}$-Brownian motions. $\qquad \square$

Uniqueness in law, established in Proposition 1 for BKR processes started from $z_0 \neq 0$, also holds when $z_0 = 0$. In other words, every solution to (1), where X and Y are $\mathcal{F}$-Brownian motions started from 0, can be obtained as in the proof of Proposition 3. This will be established in Proposition 4; we first state a small lemma on the signs of Brownian excursions.

Lemma 4. *If X is an $\mathcal{F}$-Brownian motion, then for $0 < s \leqslant t$ the conditional law of $\operatorname{sgn} X_t$ knowing $\mathcal{F}_s$ and the whole process $|X|$, is given by*

$$\mathcal{L}\big[\operatorname{sgn} X_t \mid \mathcal{F}_s \vee \sigma(|X|)\big] = \begin{cases} \delta_{\operatorname{sgn} X_s} & \text{if } X \text{ does not vanish between } s \text{ and } t, \\ \frac{1}{2}(\delta_- + \delta_+) & \text{if } X \text{ vanishes between } s \text{ and } t. \end{cases}$$

Proof. Observe that the event $\{X$ does not vanish between s and $t\}$ belongs to $\sigma(|X|)$. If $\mathcal{F} = \mathcal{X}$ where $\mathcal{X}$ is the natural filtration of X, the lemma is an easy consequence of the structure of X: given $|X|$, the excursion signs of X are i.i.d. and uniform on $\{-, +\}$.

In the general case, since X is an $\mathcal{F}$-Brownian motion, its natural filtration $\mathcal{X}$ is immersed in $\mathcal{F}$. Denote by $\mathcal{S}$ the constant filtration such that $\mathcal{S}_t = \sigma(|X|)$ for all t. Corollary 1 gives the immersion of $\mathcal{X} \vee \mathcal{S}$ in $\mathcal{F} \vee \mathcal{S}$, which entails $\mathcal{L}\big[\operatorname{sgn} X_t \mid \mathcal{F}_s \vee \sigma(|X|)\big] = \mathcal{L}\big[\operatorname{sgn} X_t \mid \mathcal{X}_s \vee \sigma(|X|)\big]$; so the result established for $\mathcal{X}$ carries over to $\mathcal{F}$. □

Proposition 4 (uniqueness in law of BKR processes started at 0). *On a filtered probability space $(\Omega, \mathcal{A}, \mathbb{P}, \mathcal{F})$, let (X, Y) be a BKR process started from the origin. The process $(|X|, |Y|)$ (whose law is given by Proposition 2), the sequence of excursion signs of X, and the sequence of excursion signs of Y are independent.*

Proof. Recall from Proposition 2 that the processes $|X|$, $|Y|$ and W generate the same filtration, and a fortiori the same σ-field. In particular, the excursion intervals of X and those of Y are $\sigma(W)$-measurable. To establish Proposition 3, it suffices to show that, conditional on $\sigma(W)$, the excursion signs of X and the excursion signs of Y are uniformly distributed on $\{-, +\}$ and independent. This amounts to verifying that, for each n, the conditional joint law $\mathcal{L}\big[\operatorname{sgn} X_{r_1}, \ldots, \operatorname{sgn} X_{r_n}, \operatorname{sgn} Y_{r_1}, \ldots, \operatorname{sgn} Y_{r_n} \mid W\big]$ makes all these signs uniform and independent or equal, with equality holding for, and only for, the signs of the same excursion of the same process.

By chronologically re-ordering the r_k and conditioning, it suffices to check that, for fixed $0 < s < t$, the conditional law $\mathcal{L}\big[(\operatorname{sgn} X_t, \operatorname{sgn} Y_t) \mid \mathcal{F}_s \vee \sigma(W)\big]$ is the uniform law on

- the singleton $\{(\operatorname{sgn} X_s, \operatorname{sgn} Y_s)\}$ if neither $|X|$ nor $|Y|$ vanishes between s and t;
- the doubleton $\{(-, \operatorname{sgn} Y_s), (+, \operatorname{sgn} Y_s)\}$ if $|X|$ vanishes between s and t, but $|Y|$ does not;
- the doubleton $\{(\operatorname{sgn} X_s, -), (\operatorname{sgn} X_s, +)\}$ if $|X|$ does not vanish between s and t, but $|Y|$ does;
- the 4-point set $\{-, +\}^2$ if both $|X|$ and $|Y|$ vanish between s and t.

The first case is trivial. Call g_t^X (resp. g_t^Y) the last zero of X (resp. of Y) on $[0, t]$; these random variables are $\sigma(W)$-measurable. Note that the event $\{g_t^X = g_t^Y\}$ is negligible, since W cannot simultaneously reach its current maximum and its current minimum. The second case deals with the

$\sigma(W)$-event $\{g_t^Y < s < g_t^X\}$. On this event, $\operatorname{sgn} Y_t = \operatorname{sgn} Y_s$, and the conditional law $\mathcal{L}\big[\operatorname{sgn} X_t \mid \mathcal{F}_s \vee \sigma(W)\big]$ is uniform on $\{-,+\}$ by Lemma 4; the claim follows. The third case is similar to the second one, by exchanging X and Y. The fourth case takes place on the event $\{g_t^X > s$ and $g_t^Y > s\}$. By symmetry, we may work on the event $\{s < g_t^Y < g_t^X\}$; and by considering all rational u, it suffices to work on $\{s < g_t^Y < u < g_t^X\}$; notice that this event is in $\sigma(W)$. To compute $\mathbb{E}^{\mathcal{F}_s \vee \sigma(W)}\big[f(\operatorname{sgn} X_t)\, g(\operatorname{sgn} Y_t)\big]$ on that event, replace $\mathbb{E}^{\mathcal{F}_s \vee \sigma(W)}$ by $\mathbb{E}^{\mathcal{F}_s \vee \sigma(W)}\mathbb{E}^{\mathcal{F}_u \vee \sigma(W)}$; this separates the operations on X and Y, and applying twice Lemma 4 finishes the proof. □

Another proof of Proposition 4 is possible: instead of establishing Lemma 4 in full generality, we only need to know it in the particular setting considered here, namely when $\mathcal{F}$ is generated by two Brownian motions linked by (1). As interest focuses on what happens between s and t, the behaviour at time 0 has no bearing on the result, so independence of the signs chosen at the beginning of all excursions can be deduced from Propositions 4 (existence) and 1 (uniqueness in law when $z_0 \neq 0$). We have preferred a more general-theoretic argument because we find Lemma 4 interesting in its own right, and because Lemma 3 was needed anyway in our proof of existence.

With Propositions 1, 3 and 4, we know existence and uniqueness in law of the BKR process started from any initial position in the plane.

Here is an immediate consequence of Proposition 4:

Corollary 3. *The law of the BKR process started from the origin is invariant under the eight transformations $(x,y) \mapsto (\pm x, \pm y)$ and $(x,y) \mapsto (\pm y, \pm x)$ (the planar isometries preserving or exchanging the axes).*

In particular, for fixed $t > 0$, the conditional law of (X_t, Y_t) given $(|X_t|, |Y_t|)$ is uniform on the four points $(\pm|X_t|, \pm|Y_t|)$.

Together with the remark, made after Proposition 2, that for fixed t the joint law of $(|X_t|, |Y_t|)$ can be explicitly computed, this corollary makes it possible to write the planar density of the joint law of (X_t, Y_t).

Other consequences of Proposition 4 are scaling invariance, time-inversion invariance, and the Markov property:

Corollary 4. *The BKR process started from the origin has the same scaling property as planar Brownian motion: for any real $\lambda \neq 0$, the law of the process is invariant under the space-time change $(x, y, t) \mapsto (\lambda x, \lambda y, \lambda^{-2} t)$.*

Proof. If $(X_t, Y_t)_{t \geqslant 0}$ is a BKR process started from $(0, 0)$, so is also the process $(\lambda X_{t/\lambda^2}, \lambda Y_{t/\lambda^2})_{t \geqslant 0}$; their laws are equal by Proposition 4. □

Corollary 5. *The law of the BKR process started from the origin is invariant under the space-time transformation $(x, y, t) \mapsto (tx, ty, 1/t)$.*

Proof. The processes $X_t' = t\, X_{1/t}$ and $Y_t' = t\, Y_{1/t}$ are Brownian motions, and $|X'| + |Y'|$ remains constant on any time-interval during which neither X' nor

Y' ever vanishes; so $\mathbb{1}_{\{X'Y'\neq0\}}(\operatorname{sgn} X'\, dX'+\operatorname{sgn} Y'\, dY')=0$. This entails that $Z'=(X',Y')$ is a BKR process since the amount of time $\mathbb{1}_{\{X'Y'=0\}}\, dt$ spent on the axes is null. Hence Z' has the same law as Z. $\square$

Corollary 6. *The BKR process Z is strong Markov: if T is a stopping time and f a bounded functional, $\mathbb{E}\big[f(Z_{T+t},\ t\geqslant 0)\ \big|\ \mathcal{F}_T\big]=\int f(w)\,\nu^{Z_T}(dw)$, where, for each $z\in\mathbb{R}^2$, ν^z denotes the law of the BKR process started from z.*

Proof. If X and Y are two $\mathcal{F}$-Brownian motions linked by (1), then $(X_{T+t})_{t\geqslant0}$ and $(Y_{T+t})_{t\geqslant0}$ are two $(\mathcal{F}_{T+t})_{t\geqslant0}$-Brownian motions linked by (1). $\square$

Corollary 7. *Let $\mathcal{F}$ be some filtration. A given process Z is a BKR process for $\mathcal{F}$ if and only if its law is that of a BKR process and its natural filtration is immersed in $\mathcal{F}$.*

Proof. Call $\mathcal{Z}$ the natural filtration of Z.

If Z is a BKR process for $\mathcal{F}$, the conditional law $\mathcal{L}[(Z_{t+h},\ h\geqslant 0)|\mathcal{F}_t]$ equals ν^{Z_t} by Corollary 6, hence $\mathcal{L}[Z|\mathcal{F}_t]$ is $\mathcal{Z}_t$-measurable, and, conditional on $\mathcal{Z}_t$, Z is independant of $\mathcal{F}_t$. So Lemma 1 (iv) says that $\mathcal{Z}$ is immersed in $\mathcal{F}$.

Conversely, suppose Z to have the law of a BKR process and $\mathcal{Z}$ to be immersed in $\mathcal{F}$. Then $\mathcal{L}[(X,Y)|\mathcal{F}_t]=\mathcal{L}[Z|\mathcal{Z}_t]$, hence X and Y are $\mathcal{F}$-Brownian motions; they satisfy Equation (1) which is a property of their joint law. $\square$

When the initial condition z_0 is not the origin, Proposition 1 says that the filtration generated by the BKR process Z is also generated by the Brownian motion B; so this filtration is Brownian. When $z_0=0$, formula (2) still defines a Brownian motion B, but B now contains strictly less information than Z; for instance, $-Z$ is another BKR process, but changing Z to $-Z$ does not change B. (More generally, the rotations with angle $k\pi/2$ preserve the law of Z and do not change B. There probably exist uncountably many other path transforms which preserve the law of Z without changing B, but describing them seems to be difficult.)

Anyway, our purpose is not to investigate the loss of information from Z to B, but to describe the filtration generated by Z when $z_0=0$. Call $\mathcal{F}$ this filtration; by hypothesis, X and Y are $\mathcal{F}$-Brownian motions. Proposition 1 and Corollary 6 imply that for each $s>0$, the shifted filtration $(\mathcal{F}_{s+t})_{t\geqslant0}$ is generated by the σ-field $\mathcal{F}_s$ and the Brownian motion $(B_{s+t}-B_s)_{t\geqslant0}$. With the vocabulary of [4], this means that the filtration $\mathcal{F}$ is *Brownian after zero*. This is weaker than being Brownian, i.e., being generated by a Brownian motion started from 0; for instance, Brownianity after zero does not even imply that $\mathcal{F}_0$ is degenerate ($\mathcal{F}_0=\bigcap_{t>0}\mathcal{F}_t$, since filtrations are right-continuous by definition). As it happens, Brownianity after zero plus degeneracy of $\mathcal{F}_0$ are not sufficient to imply Brownianity; this was discovered by Vershik [8] in the early seventies, in the framework of discrete time with time $\downarrow-\infty$. When adapted to the continuous setting with time $\downarrow0+$, Vershik's theory gives a necessary and sufficient condition for a filtration which is Brownian after zero to be Brownian. This condition is the *self-coupling criterion* (ii) from Theorem 1 of [4]; it is inspired from the concept of cosiness, introduced in 1997 by Tsirelson [7], and modified into I-cosiness in [5]. The precise phrasing of this criterion will be recalled in due time, in the proof of Proposition 5. We shall make use of the criterion to show that $\mathcal{F}$ is indeed Brownian; this will first need a *coupling lemma*, as the key ingredient. In other instances (see [5] and [3]), the coupling lemma opens the way to a constructive proof of Brownianity: it allows to exhibit a generating Brownian motion without resorting to the criterion. This does not seem to apply here; we have not been able to bypass the criterion and to give a rigorous and constructive proof.

Lemma 5 (coupling lemma). *Let (Z'_0, Z''_0) be a random vector in $\mathbb{R}^2 \times \mathbb{R}^2$, defined on some sufficiently rich probability space $(\Omega, \mathcal{A}, \mathbb{P})$.*

There exist on $(\Omega, \mathcal{A}, \mathbb{P})$ a filtration $\mathcal{H}$, two BKR processes Z' and Z'' for $\mathcal{H}$, respectively started from Z'_0 and Z''_0, and an $\mathcal{H}$-stopping time T such that $T < \infty$ a.s. and $Z' = Z''$ on $[\![T, \infty[\![$.

Proof. By "sufficiently rich", we mean that besides the vector (Z'_0, Z''_0), there also exist on $(\Omega, \mathcal{A}, \mathbb{P})$ a linear Brownian motion x started from 0, and two i.i.d. sequences η' and η'' uniform on $\{-, +\}$; these four ingredients are assumed to be independent. Put $Z'_0 = (X'_0, Y'_0)$ and $Z''_0 = (X''_0, Y''_0)$. Define a filtration $\mathcal{F}$ by

$$\mathcal{F}_t = \sigma\left(Z'_0, Z''_0, x^{t]}\right) .$$

The process $X' = X'_0 + x$ is a real $\mathcal{F}$-Brownian motion; so is also $X''_0 - x$. These two processes meet at time $S = \inf\{t : 2x_t = X''_0 - X'_0\}$; define X'' to equal $X''_0 - x$ on $[\![0, S]\!]$ and X' on $[\![S, \infty[\![$. This X'' is an $\mathcal{F}$-Brownian motion too, equal to X' from S on. Put $W' = \int \operatorname{sgn} X' \, dX'$ and $W'' = \int \operatorname{sgn} X'' \, dX''$, and set

$$R'_t = \left(|Y'_0| \vee \sup_{s \in [0,t]} W'_s\right) - W'_t \quad \text{and} \quad R''_t = \left(|Y''_0| \vee \sup_{s \in [0,t]} W''_s\right) - W''_t .$$

The processes R' and R'' are two 1-Bessel for $\mathcal{F}$; more precisely R' (resp. R'') is the 1-Bessel with martingale part $-W'$ (resp. $-W''$) and started from $|Y'_0|$ (resp. $|Y''_0|$). Remark that $dW' = dW''$ on $]\!]S, \infty[\![$; consequently, for some random t_0, one has

$$\forall t \geqslant t_0 \qquad W'_t = \sup_{s \in [0,t]} W'_s \quad \Leftrightarrow \quad W''_t = \sup_{s \in [0,t]} W''_s ,$$

and hence the $\mathcal{F}$-stopping time

$$T = \inf\{t : t \geqslant S \text{ and } W'_t = \sup_{s \leqslant t} W'_s \geqslant |Y'_0| \text{ and } W''_t = \sup_{s \leqslant t} W''_s \geqslant |Y''_0|\}$$

is a.s. finite. Observe that $R'_T = R''_T = 0$, and that $R' = R''$ on $[\![T, \infty[\![$. Call y' (resp. y'') the Brownian motion with absolute value R' (resp. R'')

and whose excursion signs are given by η' (resp. η''). By Lemma 4, $\mathcal{F}$ is immersed in the filtration $\mathcal{G}$ generated by $\mathcal{F}$ and y', and y' is a $\mathcal{G}$-Brownian motion; and by the same lemma, $\mathcal{G}$ is in turn immersed in the filtration $\mathcal{H}$ generated by $\mathcal{G}$ and y'', and y'' is an $\mathcal{H}$-Brownian motion. By immersion, X', X'' and y' are also $\mathcal{H}$-Brownian motions, and by definition of y' and y'' one has $\operatorname{sgn} y'\,dy' = -dW' = -\operatorname{sgn} X'\,dX'$ and $\operatorname{sgn} y''\,dy'' = -dW'' = -\operatorname{sgn} X''\,dX''$. To finish the proof, some innocuous modifications of y' and y'' are needed, because their initial values are $\pm|Y_0'|$ and $\pm|Y_0''|$ but not necessarily Y_0' and Y_0'', and because $y' = y''$ at time T but not on the whole interval $[\![T, \infty[\![$. It suffices to call τ' (resp. τ'') the first zero of y' (resp. y''), to observe that $\tau' \leqslant T$ and $\tau'' \leqslant T$ because $R_T' = R_T'' = 0$, and to put

$$Y' = \begin{cases} \operatorname{sgn}(y_0' Y_0')\, y' & \text{on } [\![0, \tau'[\![\,, \\ y' & \text{on } [\![\tau', \infty[\![\,; \end{cases} \qquad Y'' = \begin{cases} \operatorname{sgn}(y_0'' Y_0'')\, y'' & \text{on } [\![0, \tau''[\![\,, \\ y'' & \text{on } [\![\tau'', T[\![\,, \\ Y' & \text{on } [\![T, \infty[\![\,. \end{cases}$$

Y' and Y'' are $\mathcal{H}$-Brownian motions with the same absolute values as y' and y'', so $Z' = (X', Y')$ and $Z'' = (X'', Y'')$ are BKR processes for $\mathcal{H}$. One has $Z'' = Z'$ after T by definition of X'' and Y'', and the initial values of Z' and Z'' are the given vectors Z_0' and Z_0''. $\square$

Corollary 8. *Fix $\alpha < 1$. On a suitable filtered probability space $(\Omega, \mathcal{A}, \mathbb{P}, \mathcal{F})$, there exist two BKR processes Z' and Z'' for $\mathcal{F}$, such that*

- $Z_0' = Z_0'' = 0$;
- *the stopped processes $Z'^{\,1]}$ and $Z''^{\,1]}$ are independent;*
- *for some deterministic time $u > 0$, one has $\mathbb{P}\,[\forall t \geqslant u \quad Z_t' = Z_t''\,] > \alpha$.*

Proof. Run up to time 1 two independent BKR processes Z' and Z'' started from the origin; this yelds a random vector (Z_1', Z_1'') in $\mathbb{R}^2 \times \mathbb{R}^2$. Starting from this random vector and using Lemma 5, run Z' and Z'' after time 1 so that they couple at some finite time $T > 1$. The third property is obtained by choosing u large enough so that $\mathbb{P}\,[T \leqslant u] > \alpha$. $\square$

Corollary 9. *Fix $\alpha < 1$ and $\varepsilon > 0$. There exist a filtered probability space $(\Omega, \mathcal{A}, \mathbb{P}, \mathcal{F})$ and two BKR processes Z' and Z'' for $\mathcal{F}$, such that*

- $Z_0' = Z_0'' = 0$;
- *for some $s > 0$, the stopped processes $Z'^{\,s]}$ and $Z''^{\,s]}$ are independent;*
- $\mathbb{P}\,[\forall t \geqslant \varepsilon \quad Z_t' = Z_t''\,] > \alpha$.

Proof. Immediate from the previous corollary by simultaneously scaling the processes and the filtration (Corollary 4): it suffices to change the time-scale so as to transform u into ε, and 1 becomes some $s > 0$. $\square$

Proposition 5. *The filtration generated by a BKR process started from the origin is also generated by some real Brownian motion started from the origin.*

Proof. Let $Z = (X, Y)$ be a BKR process started from 0, and $\mathcal{F}$ its natural filtration. As in Proposition 1, define a real Brownian motion B by

$$B = -\int \operatorname{sgn} Y \, \mathrm{d}X = \int \operatorname{sgn} X \, \mathrm{d}Y$$

and, for $s > 0$, define the shifted process B^s by $B_t^s = B_{s+t} - B_s$. The latter is an $\mathcal{F}^s$-Brownian motion started from 0, where the shifted filtration $\mathcal{F}^s$ is defined by $\mathcal{F}_t^s = \mathcal{F}_{s+t}$. According to Proposition 1 and Corollary 6, the process $(Z_{s+t})_{t \geqslant 0}$ is equal to $\Phi(Z_s, B^s)$; consequently, the filtration $\mathcal{F}^s$ is generated by the σ-field $\mathcal{F}_s$ and by the $\mathcal{F}^s$-Brownian motion B^s. With the language of [4], one says that $\mathcal{F}$ is *Brownian after zero*.

For such a filtration, a (necessary and) sufficient condition to be Brownian is Condition (ii) in Theorem 1 of [4]. This criterion is stated in [4] for general filtrations; in the particular case considered here, using the fact that $\mathcal{F}$ is generated by a BKR process Z issued from 0, the statement of the criterion can be made slightly less obscure. Here it is: *For each integrable r.v. of the form $f \circ Z$ and each $\delta > 0$, there exists a filtered probability space $(\Omega, \mathcal{A}, \mathbb{P}, \mathcal{G})$ and two BKR processes Z' and Z'' for $\mathcal{G}$, such that*

- $Z_0' = Z_0'' = 0$;
- *for some $s > 0$, the stopped processes $Z'^{s]}$ and $Z''^{s]}$ are independent;*
- $\mathbb{E}\big[|f \circ Z' - f \circ Z''|\big] < \delta.$

Moreover, as remarked at the bottom of page 288 of [4], this need not be checked for all integrable functionals $f \circ Z$, but only for those belonging to some dense subset of $L^1(\sigma(Z))$. So we shall verify that the above statement holds true with $f \circ Z = g(Z_t, \, t \geqslant \varepsilon)$, for some $\varepsilon > 0$ and some bounded Borel functional g on continuous paths indexed by $[\varepsilon, \infty)$.

Fix such ε and g, as well as $\delta > 0$; let $M > 0$ be a bound for $|g|$. Corollary 9 with $\alpha = 1 - \delta/(2M)$ gives two BKR processes for some filtration, meeting the first two requirements of the criterion. The third one is met too, because the elementary estimate

$$\big|g(Z_t', \, t \geqslant \varepsilon) - g(Z_t'', \, t \geqslant \varepsilon)\big| \leqslant 2M \, \mathbb{1}_{\{\exists t \geqslant \varepsilon \ Z_t' \neq Z_t''\}}$$

and the third point in Corollary 9 imply

$$\mathbb{E}\big[\big|g(Z_t', \, t \geqslant \varepsilon) - g(Z_t'', \, t \geqslant \varepsilon)\big|\big] < 2M \, (1 - \alpha) = \delta \, . \qquad \square$$

Corollary 10 (zero-one law). *If Z is a BKR process with Z_0 deterministic and with natural filtration $\mathcal{F}$, the σ-field $\mathcal{F}_0$ is degenerate.*

Proof. By Proposition 1 when $Z_0 \neq 0$ and by Proposition 5 when $Z_0 = 0$, the natural filtration of Z is Brownian. $\qquad \square$

Proposition 5 asserts that a BKR process started from the origin has a Brownian filtration, but the proof does not explicitly exhibit any generating

Brownian motion. (Strictly speaking, the proof given in [4] is constructive, or rather can be made constructive; but from a practical point of view it is very far from effective.) A possible strategy to exhibit a generating Brownian motion would be to use the same ansatz as in Proposition 3 of [5] or in Theorem 8 of [3]; this approach needs a stronger version of the coupling lemma (Lemma 5), where one further demands that both BKR processes generate the same filtration.

Question. *Let z' and z'' be two points in the punctured plane $\mathbb{R}^2 \setminus \{0\}$. Do there exist two BKR processes Z' and Z'', generating the same filtration, started from $Z'_0 = z'$ and $Z''_0 = z''$, and such that their coupling time $\inf \{t : Z'_t = Z''_t\}$ is a.s. finite?*

If the answer is positive, a constructive proof of Brownianity, inspired from [5] and [3], can be derived from this result; the more effectively Z'' is constructed in the filtration of Z', the more explicit the generating Brownian motion will be.

References

1. Beneš, V.E., Karatzas, I., Rishel, R.W.: The separation principle for a Bayesian adaptive control problem with no strict-sense optimal law. In: Applied stochastic analysis (London, 1989), 121–156, Stochastics Monogr., 5, Gordon and Breach, New York (1991)
2. Borodin, A.N., Salminen, P.: Handbook of Brownian motion – Facts and Formulae. Birkhäuser, Basel (first edition, 1996, or second edition, 2002)
3. Brossard, J, Leuridan, C.: Transformations browniennes et compléments indépendants : résultats et problèmes ouverts. In: Séminaire de Probabilités XLI, 265–278, Lecture Notes in Math. 1934, Springer-Verlag, Berlin (2008)
4. Émery, M.: On certain almost Brownian filtrations. *Ann. Inst. H. Poincaré Probab. Statist.* **41**, 285–305 (2005)
5. Émery, M., Schachermayer, W.: A remark on Tsirelson's stochastic differential equation. In: Séminaire de Probabilités XXXIII, 291–303, Lecture Notes in Math. 1709, Springer-Verlag, Berlin (1999)
6. Karatzas, I., Shreve, S.E.: Brownian Motion and Stochastic Calculus. Second edition. Springer-Verlag, Berlin (1991)
7. Tsirelson, B.: Triple points: from non-Brownian filtrations to harmonic measures. *Geom. Funct. Anal.* **7**, 1096–1142 (1997)
8. Vershik, A.M.: Theory of decreasing sequences of measurable partitions. English version in *St. Petersburg Math. J.* **6**, 705–761 (1995)

Markovian Properties of the Spin-Boson Model

Ameur Dhahri

Ceremade, UMR CNRS 7534, Université Paris Dauphine
Place de Lattre de Tassigny, 75775 Paris Cedex 16, France
email: dhahri@ceremade.dauphine.fr

Summary. We systematically compare the Hamiltonian and Markovian approaches of quantum open system theory, in the case of the spin-boson model. We first give a complete proof of the weak coupling limit and we compute the Lindblad generator of this model. We study properties of the associated quantum master equation such as decoherence, detailed quantum balance and return to equilibrium at inverse temperature $0 < \beta \leq \infty$. We further study the associated quantum Langevin equation, its associated interaction Hamiltonian. We finally give a quantum repeated interaction model describing the spin-boson system where the associated Markovian properties are satisfied without any assumption.

1 Introduction

In the quantum theory of irreversible evolutions two different approaches have usually been considered by physicists as well as mathematicians: the Hamiltonian and the Markovian ones.

The Hamiltonian approach consists in giving a full Hamiltonian model for the interaction of a simple quantum system with a quantum field (particle gas, heat bath...) and to study the ergodic properties of the associated quantum dynamical system. The usual tools are then typically: modular theory of von Neumann algebras, KMS states...(cf [BR96], [DJP03], [JP96a], [JP96b]).

The Markovian approach consists in giving up the idea of modeling the environment and concentrating on the effective dynamics of the small system. This dynamics is supposed to be described by a (completely positive) semigroup and the studies concentrate on its Lindblad generator, or on the associated quantum Langevin equation (cf [F06], [F99], [F93], [FR06], [FR98], [P92], [HP84], [M95]).

In this article we systematically compare the two approaches in the case of the well-known spin-boson model. The first step in relating the Hamiltonian and Markovian models is to derive the Lindblad generator from the Hamiltonian description, by means of the weak coupling limit. We indeed give a

C. Donati-Martin et al. (eds.), *Séminaire de Probabilités XLII*,
Lecture Notes in Mathematics 1979, DOI 10.1007/978-3-642-01763-6_15,
© Springer-Verlag Berlin Heidelberg 2009

complete proof of the convergence of the Hamiltonian evolution to a Lindblad semigroup in the van Hove limit. We derive an explicit form for the generator in terms of Hamiltonian, this is treated in section 3.

In section 4, 5 and 6 we study the basic properties of the quantum master equation associated to the Lindbladian obtained in section 3. We investigate the quantum decoherence property. We show that the quantum detailed balance condition is satisfied with respect to the thermodynamical equilibrium state of the spin system and we prove the convergence to equilibrium in all cases.

In section 7 we consider the natural quantum Langevin equation associated to the Lindblad generator of the spin-boson system. We indeed introduce a natural unitary ampliation of the quantum master equation in terms of a Schrödinger equation perturbed by quantum noises. Such a quantum Langevin equation is actually a unitary evolution in the interaction picture, we compute the associated Hamiltonian which we compare to the initial Hamiltonian.

Finally, we give a quantum repeated interaction model which allows to prove that the Markovian properties of the spin-boson system are satisfied without assuming any hypothesis.

2 The Model

2.1 Spin-boson System

The model we shall consider all along this article is the spin-boson model, that is, a two level atom interacting with a reservoir modelled by a free Bose gas at thermal equilibrium for the temperature $T = \frac{1}{k\beta}$ (the case of zero temperature, i.e., $\beta = \infty$, is also treated). Let us start by defining the spin-boson system at positive temperature. We first introduce the isolated spin and the free reservoir, and we describe the coupled system.

The Hilbert space of the isolated spin is $\mathcal{K} = \mathbb{C}^2$ and its Hamiltonian is $h_S = \sigma_z$, where

$$\sigma_z = \begin{pmatrix} 1 & 0 \\ 0 & -1 \end{pmatrix}.$$

The associated eigenenergies are $e_\pm = \pm 1$ and we denote the corresponding eigenstates by $\Psi_\pm$. The algebra of observables of the spin is M_2, the algebra of all complex 2×2 matrix. At inverse temperature β, the equilibrium state of the spin is the normal state defined by the Gibbs Ansatz

$$\omega_S(A) = \frac{1}{Z} \mathrm{Tr}(\exp(-\beta\sigma_z)A), \quad \text{for all } A \in M_2,$$

where $Z = \mathrm{Tr}(\exp(-\beta\sigma_z))$.

The dynamics of the spin is defined as

$$\tau_S^t(A) = e^{it\sigma_z} A e^{-it\sigma_z}, \quad \text{for all } A \in M_2, \ t \in \mathbb{R}.$$

The free reservoir is modelled by a free Bose gas which is described by the symmetric Fock space $\Gamma_s(L^2(\mathbb{R}^3))$. If we call $w(k) = |k|$ the energy of a single boson with momentum $k \in \mathbb{R}^3$, then the Hamiltonian of the reservoir is given by the second differential quantization $d\Gamma(w)$ of w. In terms of the usual creation and annihilation operators $a^*(k)$, $a(k)$, we have

$$d\Gamma(w) = \int_{\mathbb{R}^3} w(k)a^*(k)a(k)dk.$$

The Weyl's operator associated to an element $f \in L^2(\mathbb{R}^3)$ is the operator

$$W(f) = \exp(i\varphi(f)),$$

where $\varphi(f)$ is the self-adjoint field operator defined by

$$\varphi(f) = \frac{1}{\sqrt{2}} \int_{\mathbb{R}^3} (a(k)\bar{f}(k) + a^*(k)f(k)) \, dk.$$

Call $\mathcal{D}_{loc}$ the space of $f \in L^2(\mathbb{R})$ with compactly supported Fourier transform. It follows from [JP96b] that the Weyl's algebra, $\mathcal{A}_{loc} = W(\mathcal{D}_{loc})$, the algebra generated by the set $\{W(f), f \in \mathcal{D}_{loc}\}$ is a natural minimal set of observables associated to the reservoir. The equilibrium state of the reservoir at inverse temperature β is given by

$$\omega_R(W(f)) = \exp\left[-\frac{\|f\|^2}{4} - \frac{1}{2}\int_{\mathbb{R}^3} |f(k)|^2 \rho(k) \, dk\right],$$

where $\rho(k)$ is related to $w(k)$ by Planck's radiation law

$$\rho(k) = \frac{1}{e^{\beta w(k)} - 1}.$$

The dynamics of the reservoir is generated by $H_b = [d\Gamma(w), .]$ and it induces a Bogoliubov transformation

$$\exp(itd\Gamma(w))W(f)\exp(-itd\Gamma(w)) = W(\exp(iwt)f).$$

The coupled system is described by the $\mathbb{C}^*$-algebra $M_2 \otimes \mathcal{A}_{loc}$. The free dynamics is given by

$$\tau_0^t(A) = \tau_S^t \otimes \tau_R^t(A), \quad \text{for all } A \in M_2 \otimes \mathcal{A}_{loc}.$$

2.2 Semistandard Representation

The semistandard representation of the coupled system (reservoir+spin) is the representation which is standard on its reservoir part, but not standard on the spin part (cf [DF06]). Now, let us introduce the Araki-Woods representation of the couple $(\omega_R, \mathcal{A}_{loc})$ which is the triple $(\mathcal{H}_R, \pi_R, \Omega_R)$, defined by

- $\mathcal{H}_R = l^2(\Gamma_s(L^2(\mathbb{R}^3))$, the space of Hilbert-Schmidt on $\Gamma_s(L^2(\mathbb{R}^3))$ which is naturally identified as $\Gamma_s(L^2(\mathbb{R}^3)) \otimes \overline{\Gamma_s(L^2(\mathbb{R}^3))}$ and equipped with the scalar product $(X,Y) = \text{Tr}(X^*Y)$,
- $\pi_R(W(f)) : X \longmapsto W((1+\rho)^{1/2}f)XW(\rho^{1/2}\bar{f})$ for all $X \in \mathcal{H}_R$,
- $\Omega_R = |\Omega\rangle\langle\Omega|$, where Ω is the vacuum vector of $\Gamma_s(L^2(\mathbb{R}^3))$.

Moreover a straightforward computation shows that

$$\omega_R(A) = (\Omega_R, \ \pi_R(A)\Omega_R),$$

and the relation

$$\pi_R(\exp(itd\Gamma(\omega))A\exp(-itd\Gamma(\omega))) = \exp(it[d\Gamma(\omega),.])\pi_R(A)\exp(-it[d\Gamma(\omega),.])$$

defines a dynamics on $\mathcal{M}_R = \pi_R(\mathcal{A}_{loc})''$ whose generator is the operator

$$L_R = [d\Gamma(\omega),.].$$

The free semi-Liouvillean associated to the semistandard representation of the spin-boson system is defined by

$$L_0^{semi} = \sigma_z \otimes 1 + 1 \otimes L_R.$$

The full semi-Liouvillean is the operator

$$L_\lambda^{semi} = L_0^{semi} + \lambda\sigma_x \otimes \varphi_{AW}(\alpha),$$

where $\lambda \in \mathbb{R}$, and where $\alpha \in L^2(\mathbb{R}^3)$ is called the test function (or cut-off function), $\varphi_{AW}(\alpha)$ is the field operator of the Araki-Woods representation which can be identified as follows

$$\varphi_{AW}(\alpha) \simeq \varphi((1+\rho)^{1/2}\alpha) \otimes 1 + 1 \otimes \varphi(\bar{\rho}^{1/2}\bar{\alpha})$$

(see [JP96b], [DJ03] for more details) and

$$\sigma_x = \begin{pmatrix} 0 & 1 \\ 1 & 0 \end{pmatrix}.$$

The following proposition follows from [JP96b].

Proposition 2.1 *If $(\omega+\omega^{-1})\alpha$ is in $L^2(\mathbb{R}^3)$, the operator L_λ^{semi} is essentially self-adjoint on $\mathbb{C}^2 \otimes D(d\Gamma(\omega)) \otimes D(d\Gamma(\omega))$ for all $\lambda \in \mathbb{R}$.*

An immediate consequence of the above proposition is that

$$\tau_\lambda^t(A) = e^{itL_\lambda^{semi}}Ae^{-itL_\lambda^{semi}}$$

defines a dynamics on $\mathcal{M} = M_2 \otimes \mathcal{M}_R$.

2.3 Reservoir 1-particle Space

After taking the Araki-Woods representation of the pair $(\omega_R, \mathcal{A}_{loc})$, we distinguish that the reservoir state is a non-Fock state (i.e., it cannot be represented as a pure state on a Fock space) and this case is more complicated to treat. By using the identifications given in [DJ03] and [JP96a], we see that this state can be represented as a pure state on a Fock space. Hence we have

$$\Gamma_s(L^2(\mathbb{R}^3)) \otimes \overline{\Gamma_s(L^2(\mathbb{R}^3))}$$
$$\simeq \Gamma_s(L^2(\mathbb{R}^3)) \otimes \Gamma_s(\overline{L^2(\mathbb{R}^3)}) \simeq \Gamma_s(L^2(\mathbb{R}^3) \oplus \overline{L^2(\mathbb{R}^3)}),$$
$$L_R \simeq d\Gamma(\omega \oplus -\overline{\omega}),$$
$$\varphi_{AW}(\alpha) \simeq \varphi((1+\rho)^{1/2}\alpha \oplus \overline{\rho}^{1/2}\overline{\alpha}),$$
$$\Omega_R \simeq \Omega \oplus \overline{\Omega}.$$

Therefore, it is obvious that ω_R is a pure state which is defined on the Fock space $\Gamma_s(L^2(\mathbb{R}^3) \oplus \overline{L^2(\mathbb{R}^3)})$. Moreover we have the Bogoliubov transformation

$$e^{itd\Gamma(\omega \oplus -\overline{\omega})}\varphi_{AW}(\alpha)e^{-itd\Gamma(\omega \oplus -\overline{\omega})} = \varphi_{AW}(e^{it\omega}\alpha).$$

This simplifies our formulation.

3 Weak Coupling Limit of the Spin-Boson System

3.1 Abstract Theory of the Weak Coupling Limit

Let $\mathcal{Y}$ be a Banach space and $\mathcal{X}$ its dual, i.e., $\mathcal{X} = \mathcal{Y}^*$. Let P be a projection on $\mathcal{X}$ and $e^{it\delta_0}$ a one parameter group of isometries on $\mathcal{X}$ which commutes with P. Put $E = P\delta_0$. It is clear that E is the generator of a one parameter group of isometries on $\mathrm{Ran}P$. Consider a perturbation Q of δ_0 such that $\mathcal{D}(Q) \supset \mathcal{D}(\delta_0)$.

We introduce the following assumptions:

(1) P is a w^*-continuous projection on $\mathcal{X}$ with norm is equal to one.
(2) $e^{it\delta_0}$ a one parameter group of w^*-continuous isometries (C_0^*-group) on $\mathcal{X}$,
(3) For $|\lambda| < \lambda_0$, $i\delta_\lambda = i\delta_0 + i\lambda Q$ is the generator of a one parameter C_0^*-semigroup of contractions.

Consider now the operator

$$K_\lambda(t) = i\int_0^{\lambda^{-2}t} e^{-is(E+\lambda PQP)}PQe^{is(1-P)\delta_\lambda(1-P)}QP\,ds.$$

For the proof of the following theorem we refer the interested reader to [DF06].

Theorem 3.1 *Suppose that assumptions (1), (2) and (3) are true. Assume that the following hypotheses are satisfied:*

(4) P is a finite range projection and $PQP = 0$,
(5) For all $t_1 > 0$, there exists a constant c such that

$$\sup_{|\lambda|<1} \sup_{0\leq t\leq t_1} \|K_\lambda(t)\| \leq c.$$

(6) There exists an operator K defined on $\mathrm{Ran}P$ such that

$$\lim_{\lambda\to 0} K_\lambda(t) = K$$

for all $0 < t < \infty$.
Put

$$K^\sharp = \sum_{e\in spE} \mathbb{1}_e(E) K \mathbb{1}_e(E) = \lim_{T\to\infty} \frac{1}{T} \int_0^T e^{itE} K e^{-itE} dt.$$

Then we have
i) $e^{itK^\sharp}$ is a semigroup of contractions,
ii) For all $t_1 > 0$,

$$\lim_{\lambda\to 0} \sup_{0\leq t\leq t_1} \|e^{-itE/\lambda^2} P e^{it(\delta_0 + \lambda Q)/\lambda^2} P - e^{itK^\sharp}\| = 0.$$

3.2 Application to the Spin-boson System

Recall that in the semistandard representation of the spin-boson system, the free semi-Liouvillean is the operator

$$L_0^{semi} = \sigma_z \otimes 1 + 1 \otimes L_R,$$

and the full semi-Liouvillean is given by

$$L_\lambda^{semi} = L_0^{semi} + \lambda\sigma_x \otimes \varphi_{AW}(\alpha).$$

Set $V = \sigma_x \otimes \varphi_{AW}(\alpha)$. Put

$$\delta_\lambda = [L_\lambda^{semi}, .] = \delta_0 + \lambda[V, .],$$

with $\delta_0 = [L_0^{semi}, .]$, the generator of the dynamics τ_λ^t. For $B \otimes C \in \mathcal{M}$, we define the projection P by

$$P(B \otimes C) = \omega_R(C) B \otimes 1_{\mathcal{H}_R}.$$

In particular we have

$$E = P\delta_0 = \delta_0 P = [\sigma_z, .]P \text{ and } P[V, .]P = 0.$$

Set $P_1 = 1 - P$. Then it follows that

$$K_\lambda(t) = i \int_0^{\lambda^{-2}t} e^{-isE} P[V,.] e^{isP_1[L_\lambda^{semi},.]P_1} [V,.] P \, ds.$$

Note that $P[V,.]P = 0$, P_1 commutes with $[L_0^{semi},.]$ and

$$e^{isP_1[L_0^{semi},.]P_1} = e^{is[L_0^{semi},.]}P_1 + P.$$

Thus, if we suppose that

$$K = i \int_0^\infty e^{-isE} P[V,.] e^{isP_1[L_0^{semi},.]P_1} [V,.] P \, ds$$

exists, we have

$$K = i \int_0^\infty e^{-isE} P[V,.] e^{is[L_0^{semi},.]} [V,.] P \, ds.$$

In the following we assume that $(\omega + \omega^{-1})\alpha \in L^2(\mathbb{R}^3)$ and we propose to show, under some conditions, that K exists and the operator K_λ converges to K when $\lambda \to 0$. Set

$$U_t^\lambda = e^{itP_1[L_\lambda^{semi},.]P_1}, \quad U_t = e^{itP_1[L_0^{semi},.]P_1}.$$

We thus have

$$U_t^\lambda = U_t + i\lambda \int_0^t U_{t-s} P_1[V,.] P_1 U_s^\lambda \, ds.$$

Hence, the operator $U_{-t}U_t^\lambda$ satisfies the equation

$$U_{-t}U_t^\lambda = 1 + i\lambda \int_0^t (U_{-s} P_1[V,.] P_1 U_s)(U_{-s}U_s^\lambda) \, ds.$$

Therefore, we get the following series of iterated integrals

$$U_{-t}U_t^\lambda = 1 + \sum_{n\geq 1}(i\lambda)^n \int_{0\leq t_n \leq ... \leq t_1 \leq t} (U_{-t_1} P_1[V,.] P_1 U_{t_1})...$$
$$(U_{-t_n} P_1[V,.] P_1 U_{t_n}) \, dt_n ... dt_1.$$

Note that the operator U_{t_k} commutes with P_1. So, if we put

$$Q_k = U_{-t_k}[V,.] U_{t_k},$$

then

$$U_{-t}U_t^\lambda = 1 + \sum_{n\geq 1}(i\lambda)^n \int_{0\leq t_n \leq .... \leq t_1 \leq t} (P_1 Q_1 P_1)...(P_1 Q_n P_1) \, dt_n ... dt_1,$$

and

$$K_\lambda(t) = i \int_0^{\lambda^{-2}t} e^{-isE} P[V,.]e^{isP_1[L_0^{semi},.]P_1}[V,.]P\, ds$$

$$+i\sum_{n\geq 1}(i\lambda)^n \int_{0\leq t_n\leq...\leq t_0\leq \lambda^{-2}t} e^{-it_0 E}P[V,.]U_{t_0}(P_1Q_1P_1)... \quad (1)$$

$$(P_1Q_nP_1)[V,.]P\, dt_n\,...\,dt_0.$$

Put

$$R_n(t) = \int_{0\leq t_n\leq...\leq t_0\leq t} e^{-it_0 E}P[V,.]U_{t_0}(P_1Q_1P_1)....(P_1Q_nP_1)[V,.]P\, dt_n\,...\,dt_0.$$

Recall that $PU_{-t_0} = P$. Hence, if we set $Q_{n+1} = U_{-t_{n+1}}[V,.]U_{t_{n+1}}$, with $t_{n+1} = 0$, we get

$$R_n(t) = \int_{0\leq t_n\leq...\leq t_0\leq t} e^{-it_0 E}PQ_0(P_1Q_1P_1)...(P_1Q_nP_1)Q_{n+1}P\, dt_n\,...\,dt_0. (2)$$

Lemma 3.2

$$R_n(t) = \int_{0\leq t_n\leq...\leq t_0\leq t} P[\sigma_{x,0}\otimes\varphi_{AW}(e^{-it_0\omega}\alpha),.]P_1...$$

$$P_1[\sigma_{x,n+1}\otimes\varphi_{AW}(e^{-it_{n+1}\omega}),.]P\, dt_n\,...\,dt_0,$$

where $t_{n+1} = 0$, $\sigma_{x,r} = e^{-it_r\sigma_z}\sigma_x e^{it_r\sigma_z}$.

Proof. Let us start by computing $P_1Q_rP_1$ for $r \geq 1$. We have

$$U_{t_r} = e^{it_r[\sigma_z,.]}e^{it_r[L_R,.]}P_1 + P,$$

and

$$U_{t_r}P_1 = e^{it_r[\sigma_z,.]}e^{it_r[L_R,.]}P_1.$$

Therefore, it follows that

$$P_1U_{-t_r}[V,.]U_{t_r}P_1 = P_1e^{-it_r[\sigma_z,.]}e^{-it_r[L_R,.]}[V,.]e^{it_r[\sigma_z,.]}e^{it_r[L_R,.]}P_1.$$

Furthermore we have

$$e^{-it_r[\sigma_z,.]}e^{-it_r[L_R,.]}[V,.]e^{it_r[\sigma_z,.]}e^{it_r[L_R,.]}(B\otimes C)$$

$$= [\sigma_{x,r}\otimes e^{-it_rL_R}\varphi_{AW}(\alpha)e^{it_rL_R},.](B\otimes C),$$

and

$$e^{-it_rL_R}\varphi_{AW}(\alpha)e^{it_rL_R} = \varphi_{AW}(e^{-it_r\omega}\alpha).$$

This gives

$$P_1Q_rP_1 = P_1[\sigma_{x,r}\otimes\varphi_{AW}(e^{-it_r\omega}\alpha),.]P_1.$$

Besides, $Pe^{-t_0[\sigma_z,\cdot]} = Pe^{-it_0[\sigma_z,\cdot]}e^{-it_0[L_R,\cdot]}$ and

$$
\begin{aligned}
e^{-it_0 E} P Q_0 P_1 &= Pe^{-it_0[\sigma_z,\cdot]}[V,\cdot]e^{it_0[\sigma_z,\cdot]}e^{it_0[L_R,\cdot]}P \\
&= Pe^{-it_0[\sigma_z,\cdot]}e^{-it_0[L_R,\cdot]}[V,\cdot]e^{it_0[\sigma_z,\cdot]}e^{it_0[L_R,\cdot]}P \\
&= P[\sigma_{x,0} \otimes \varphi_{AW}(e^{-it_0\omega}\alpha),\cdot]P_1.
\end{aligned}
$$

Thus from relation (2), the lemma holds.

Lemma 3.3

$$R_{2n+1}(t) = 0.$$

Proof. Note that

$$
\begin{aligned}
&P[\sigma_{x,0} \otimes \varphi_{AW}(e^{-it_0\omega}\alpha),\cdot]P_1....P_1[\sigma_{x,2n+2} \otimes \varphi_{AW}(e^{-it_{2n+2}\omega}\alpha),\cdot]P \\
&= P[\sigma_{x,0} \otimes \varphi_{AW}(e^{-it_0\omega}\alpha),\cdot](1-P)[\sigma_{x,1} \otimes \varphi_{AW}(e^{-it_1\omega}\alpha),\cdot](1-P)... \\
&...(1-P)[\sigma_{x,2n+2} \otimes \varphi_{AW}(e^{-it_{2n+2}\omega}\alpha),\cdot]P. \qquad (3)
\end{aligned}
$$

Therefore, if we expand the right-hand side of equation (3), we get a sum of terms each of which is a product of elements of the form

$$P[\sigma_{x,p_k} \otimes \varphi_{AW}(e^{-it_{p_k}\omega}\alpha),\cdot]....[\sigma_{x,p_m} \otimes \varphi_{AW}(e^{-it_{p_m}\omega}\alpha),\cdot]P,$$

where $0 \le p_k \le ... \le p_m \le ... \le 2n+2$. But, in each product there exists at least an element of the form

$$P[\sigma_{x,r_1} \otimes \varphi_{AW}(e^{-it_{r_1}\omega}\alpha),\cdot]....[\sigma_{x,r_{2p+1}} \otimes \varphi_{AW}(e^{-it_{r_{2p+1}}\omega}\alpha),\cdot]P,$$

where $0 \le r_1 \le ... \le r_{2p+1} \le ... \le r_{2n+2}$. Furthermore, it is easy to show that

$$[\sigma_{x,r_1} \otimes \varphi_{AW}(e^{-it_{r_1}\omega}\alpha),\cdot]....[\sigma_{x,r_{2p+1}} \otimes \varphi_{AW}(e^{-it_{r_{2p+1}}\omega}\alpha),\cdot]P(B \otimes C)$$

is a sum of terms each of which has a second component composed by $2p+1$ number product of vector fields. But the projection P acts uniquely in the second component and the Gibbs state ω_R of the reservoir is a quasi-free state (see [BR]). Then it follows that

$$P[\sigma_{x,r_1} \otimes \varphi_{AW}(e^{-it_{r_1}\omega}\alpha),\cdot]....[\sigma_{x,r_{2p+1}} \otimes \varphi_{AW}(e^{-it_{r_{2p+1}}\omega}\alpha),\cdot]P(B \otimes C) = 0,$$

and by Lemma 3.2, $R_{2n+1}(t) = 0$.

Remark 2: From the proof of Lemma 3.3 we can deduce that $R_{2n}(t)$ is a sum of 2^n terms each of which is a product containing only an even number of products of commutators of the form $[\sigma_{x,r} \otimes \varphi_{AW}(e^{-it_r\omega}\alpha),\cdot]$ between two successive projections P.

Theorem 3.4 *Suppose that the following assumptions hold:*
(i) $\|R_{2n}(t)\| \le c_n t^n$, *where the series* $\sum_{n\ge1} c_n t^n$ *has infinite radius of convergence.*

(ii) *There exists $0 < \varepsilon < 1$ and a sequence $d_n \geq 0$ such that*

$$\|R_{2n}(t)\| \leq d_n t^{n-\varepsilon}.$$

Then

$$\lim_{\lambda \to 0} \sum_{n \geq 1} (i\lambda)^n R_{2n}(\lambda^{-2}t) = 0.$$

Proof. The proof of this theorem is a straightforward application of Lebesgue's Theorem.

Now, the aim is to introduce some conditions which ensures that assumptions (i) and (ii) of the above theorem are satisfied. Set

$$h(t) = \langle e^{-itL_R}\varphi_{AW}(\alpha)e^{itL_R}\varphi_{AW}(\alpha)\Omega_R, \Omega_R \rangle.$$

Recall that

$$L_R = [d\Gamma(\omega), .] \simeq d\Gamma(\omega \oplus -\bar{\omega})$$

and

$$e^{-itL_R}\varphi_{AW}(\alpha)e^{itL_R} = \varphi_{AW}(e^{-it\omega}\alpha).$$

Therefore we get

$$h(t) = \langle \varphi_{AW}(e^{-it\omega}\alpha)\varphi_{AW}(\alpha)\Omega_R, \Omega_R \rangle.$$

Moreover, a straightforward computation shows that

$$h(t - s) = \langle \varphi_{AW}(e^{-it\omega}\alpha)\varphi_{AW}(e^{-is\omega}\alpha)\Omega_R, \Omega_R \rangle.$$

Now, for any integer n we define the set $\mathcal{P}_n$ of pairings as the set of permutations σ of $(1, ..., 2n)$ such that

$$\sigma(2r - 1) < \sigma(2r) \text{ and } \sigma(2r - 1) < \sigma(2r + 1)$$

for all r. Put

$$\langle \varphi_{AW}(\alpha_1)...\varphi_{AW}(\alpha_n) \rangle = \omega_R(\varphi_{AW}(\alpha_1)...\varphi_{AW}(\alpha_n))$$
$$= \langle \Omega_R, \varphi_{AW}(\alpha_1)...\varphi_{AW}(\alpha_n)\Omega_R \rangle.$$

If $n = 2$ then $\langle \varphi_{AW}(\alpha_1)\varphi_{AW}(\alpha_2) \rangle$ is called the two point correlations matrix. Besides, we have

$$\langle \varphi_{AW}(\alpha_1)...\varphi_{AW}(\alpha_{2n}) \rangle = \sum_{\sigma \in \mathcal{P}_n} \prod_{r=1}^{n} \langle \varphi_{AW}(\alpha_{\sigma(2r-1)})\varphi_{AW}(\alpha_{\sigma(2r)}) \rangle, \quad (4)$$

and

$$\langle \varphi_{AW}(\alpha_1)...\varphi_{AW}(\alpha_{2n+1}) \rangle = 0.$$

(see [BR96] P 40 for more details).
The proof of the following lemma is similar to the one of Lemma 3.2 in [D74].

Lemma 3.5 *If* $\|h\|_1 \leq \infty$, *then for any permutation* π *of* $(0, 1, ..., 2n+1)$ *we have*

$$\left| \sum_{\sigma \in \mathcal{P}_{(0,1,...,2n+1)}} \int_{0 \leq t_{2n} \leq ... \leq t_0 \leq t} \prod_{r=0}^{n} h(t_{\pi\sigma(2r)} - t_{\pi\sigma(2r+1)}) dt_{2n}...dt_0 \right|$$

$$\leq \frac{1}{2^{n+1}(n+1)!} \|h\|_1^{n+1} t^n,$$

with $t_{2n+1} = 0$.

We now prove the following.

Theorem 3.6 *If* $\|h\|_1 \leq \infty$ *then*

$$\|R_{2n}(t)\| \leq 2^{2n+1} \|h\|_1^{n+1} \frac{t^n}{(n+1)!}.$$

Proof. Put

$$\Phi_r = \varphi_{AW}(e^{-it_r \omega} \alpha), \ \Phi_r^L C = \Phi_r C, \ \Phi_r^R C = C\Phi_r,$$

$$\sigma_{x,r}^L B = \sigma_{x,r} B, \ \sigma_{x,r}^R B = B\sigma_{x,r},$$

$$\beta : \text{ a function from } \{0, 1, ..., 2n+1\} \text{ to } \{L, R\},$$

$$k_\beta = \sharp\{r \in \{0, 1, ..., 2n+1\} \text{ such that } \beta(r) = R\}.$$

In the sequel, we simplify the notation $\sigma_{x,r} \otimes \Phi_r$ into $\sigma_{x,r}\Phi_r$. With this notations we have

$$[\sigma_{x,r}\Phi_r, .] = \sigma_{x,r}^L \Phi_r^L - \sigma_{x,r}^R \Phi_r^R.$$

Recall that, from remark 2 and Lemma 3.2, $R_{2n}(t)$ is a sum of 2^n terms each of which is of the form

$$C_{2n,j}(t) = (-1)^j \int_{0 \leq t_{2n} \leq ... \leq t_0 \leq t} \sum_{\beta} (-1)^{k_\beta} P(\sigma_{x,0}^{\beta(0)} \Phi_0^{\beta(0)})(\sigma_{x,1}^{\beta(1)} \Phi_1^{\beta(1)})...$$

$$...(\sigma_{x,p_1-1}^{\beta(p_1-1)} \Phi_{p_1-1}^{\beta(p_1-1)}) P(\sigma_{x,p_1}^{\beta(p_1)} \Phi_{p_1}^{\beta(p_1)})...(\sigma_{x,p_j-1}^{\beta(p_j-1)} \Phi_{p_j-1}^{\beta(p_j-1)}) \times$$

$$P(\sigma_{x,p_j}^{p_j} \Phi_{p_j}^{\beta(p_j)})...(\sigma_{x,2n}^{\beta(2n)} \Phi_{2n}^{\beta(2n)})(\sigma_{x,2n+1}^{\beta(2n+1)} \Phi_{2n+1}^{\beta(2n+1)}) P \, dt_{2n} ... dt_0,$$

where $0 = p_0 < p_1 < p_2 < ... < p_j < p_{j+1} = 2n+2$, each p_k is an even number and $j = N - 2$, with N is the number of projections P, which appear in the expression of $C_{2n,j}(t)$.

Hence we have

$$\|C_{2n,j}(t)(B \otimes C)\|$$

$$\leq \|B \otimes C\| \sum_{\beta} \int_{0 \leq t_{2n} \leq ... \leq t_0 \leq t} \prod_{r=0}^{j} |\omega_R(\Phi_{p_r}^{\beta(p_r)}...\Phi_{p_{r+1}-1}^{\beta(p_{r+1}-1)})| \, dt_{2n} ... dt_0,$$

$$\leq \|B \otimes C\| \sum_{\beta} \int_{0 \leq t_{2n} \leq \ldots \leq t_0 \leq t} \prod_{r=0}^{j} |\langle \Phi_{p_r}^{\beta(p_r)} \ldots \Phi_{p_{r+1}-1}^{\beta(p_{r+1}-1)} \rangle| \, dt_{2n} \ldots dt_0,$$

$$\leq \|B \otimes C\| \sum_{\beta} \int_{\leq t_{2n} \leq \ldots \leq t_0 \leq t} \prod_{r=0}^{j} |\langle \Phi_{\pi(p_r)} \ldots \Phi_{\pi(p_{r+1}-1)} \rangle| \, dt_{2n} \ldots dt_0,$$

where π is a permutation which depends on β.

Thus from equation (4) and Lemma 3.5 we get

$$\|C_{2n,j}(t)\|$$

$$\leq \sum_{\beta} \sum_{\sigma \in P_{(0,1,\ldots,2n+1)}} \int_{0 \leq t_{2n} \leq \ldots \leq t_0 \leq t} \prod_{r=0}^{n} |\langle \Phi_{\pi(\sigma(2r))} \Phi_{\pi(\sigma(2r+1))} \rangle| dt_{2n} \ldots dt_0,$$

$$\leq 2^{2n+2} \|h\|_1^{n+1} \frac{t^n}{2^{n+1}(n+1)!},$$

Therefore, $C_{2n,j}$ is dominated uniformly in j. Finally, this proves that

$$\|R_{2n}(t)\| \leq 2^{2n+1} \|h\|_1^{n+1} \frac{t^n}{(n+1)!}.$$

The following theorem ensures that assumption (ii) of Theorem 3.4 holds.

Theorem 3.7 *If*

$$\int_0^{\infty} (1 + t^{\varepsilon}) |h(t)| dt < \infty$$

for some $0 < \varepsilon < 1$, *then there exists* $d_n > 0$ *such that*

$$\|R_{2n}(t)\| \leq d_n t^{n-\varepsilon}.$$

Proof. We have that $R_{2n}(t)$ is a sum of 2^n terms each of which takes the form of $C_{2n,j}$ which is defined previously. In order to prove this theorem we group those terms pairwise as follows:

$$(-1)^j \int_{0 \leq t_{2n} \leq \ldots \leq t_0 \leq t} \sum_{\beta} (-1)^{k_{\beta}} P(\sigma_{x,0}^{\beta(0)} \Phi_0^{\beta(0)}) \ldots$$

$$(\sigma_{x,p_1-1}^{\beta(p_1-1)} \Phi_{p_1-1}^{\beta(p_1-1)}) P \ldots P(\sigma_{x,p_j}^{\beta(p_j)} \Phi_{p_j}^{\beta(p_j)}) \ldots (\sigma_{x,2n-1}^{\beta(2n-1)} \Phi_{2n-1}^{\beta(2n-1)})(\sigma_{x,2n}^{\beta(2n)} \Phi_{2n}^{\beta(2n)})$$

$$(\sigma_{x,2n+1}^{\beta(2n+1)} \Phi_{2n+1}^{\beta(2n+1)}) P \, dt_{2n} \ldots dt_0$$

$$+(-1)^{(j+1)} \int_{0 \leq t_{2n} \leq \ldots \leq t_0 \leq t} \sum_{\beta} (-1)^{k_{\beta}} P(\sigma_{x,0}^{\beta(0)} \Phi_0^{\beta(0)}) \ldots (\sigma_{x,p_1-1}^{\beta(p_1-1)} \Phi_{p_1-1}^{\beta(p_1-1)}) P$$

$$\ldots P(\sigma_{x,p_j}^{\beta(p_j)} \Phi_{p_j}^{\beta(p_j)}) \ldots (\sigma_{x,2n-1}^{\beta(2n-1)} \Phi_{2n-1}^{\beta(2n-1)}) P(\sigma_{x,2n}^{\beta(2n)} \Phi_{2n}^{\beta(2n)})$$

$$(\sigma_{x,2n+1}^{\beta(2n+1)} \Phi_{2n+1}^{\beta(2n+1)}) P \, dt_{2n} \ldots dt_0$$

$$= (-1)^j \int_{0 \leq t_{2n} \leq \dots \leq t_0 \leq t} \sum_\beta (-1)^{k_\beta} P(\sigma_{x,0}^{\beta(0)} \Phi_0^{\beta(0)}) \dots (\sigma_{x,p_1-1}^{\beta(p_1-1)} \Phi_{p_1-1}^{\beta(p_1-1)}) P \dots$$

$$\dots \left\{ P(\sigma_{x,p_j}^{\beta(p_j)} \Phi_{p_j}^{\beta(p_j)}) \dots (\sigma_{x,2n-1}^{\beta(2n-1)} \Phi_{2n-1}^{\beta(2n-1)})(\sigma_{x,2n}^{\beta(2n)} \Phi_{2n}^{\beta(2n)})(\sigma_{x,2n+1}^{\beta(2n+1)} \Phi_{2n+1}^{\beta(2n+1)}) P \right.$$

$$- P(\sigma_{x,p_j}^{\beta(p_j)} \Phi_{p_j}^{\beta(p_j)}) \dots (\sigma_{x,2n-1}^{\beta(2n-1)} \Phi_{2n-1}^{\beta(2n-1)}) P(\sigma_{x,2n}^{\beta(2n)} \Phi_{2n}^{\beta(2n)})$$

$$\left. (\sigma_{x,2n+1}^{\beta(2n+1)} \Phi_{2n+1}^{\beta(2n+1)}) P \right\} dt_{2n} \dots dt_0.$$

Therefore, the right-hand side of the above equation is dominated by

$$\sum_\beta \int_{0 \leq t_{2n} \leq \dots \leq t_0 \leq t} \prod_{k=0}^{j-1} \left| \langle \Phi_{p_k}^{\beta(p_k)} \Phi_{p_k+1}^{\beta(p_k+1)} \dots \Phi_{p_{k+1}-1}^{\beta(p_{k+1}-1)} \rangle \right|$$

$$\times \left| \left\{ \langle \Phi_{p_j}^{\beta(p_j)} \dots \Phi_{2n}^{\beta(2n)} \Phi_{2n+1}^{\beta(2n+1)} \rangle \right. \right. \tag{5}$$

$$\left. \left. - \langle \Phi_{p_j}^{\beta(p_j)} \dots \Phi_{2n-1}^{\beta(2n-1)} \rangle \langle \Phi_{2n}^{\beta(2n)} \Phi_{2n+1}^{\beta(2n+1)} \rangle \right\} \right| dt_{2n} \dots dt_0.$$

Note that in the between bracket terms, there is no product of two point correlation matrix where $2n$ is paired with $(2n + 1)$. Moreover this term is equal to

$$\sum_{\sigma \in \mathcal{P}_{(p_j,\dots,2n+1)}} \prod_{r=\frac{1}{2}p_j}^n \langle \Phi_{\sigma(\pi(2r))} \Phi_{\sigma(\pi(2r+1))} \rangle,$$

where $2n$ is not paired with $(2n + 1)$ and π is a permutation which depends on β.
Thus the term in equation (5) is dominated by

$$\sum_\sigma \int_{0 \leq t_{2n} \leq \dots \leq t_0 \leq t} \prod_{r=0}^n \left| \langle \Phi_{\sigma(2r)} \Phi_{\sigma(2r+1)} \rangle \right| dt_{2n} \dots dt_0,$$

where $\sum_\sigma$ indicates the sum over all pairings of $\{0, 1 \dots, 2n + 1\}$ such that $2n$ is not paired with $(2n + 1)$, $(t_{2n+1} = 0)$.
But we have

$$\int_{0 \leq t_{2n} \leq \dots \leq t_0 \leq t} \prod_{r=0}^n \left| \langle \Phi_{\sigma(2r)} \Phi_{\sigma(2r+1)} \rangle \right| dt_{2n} \dots dt_0$$

$$= \int_{0 \leq t_{2n} \leq \dots \leq t_0 \leq t} \prod_{r=0}^n \left| h(t_{\sigma(2r)} - t_{\sigma(2r+1)}) \right| dt_{2n} \dots dt_0$$

$$\leq cst \, \|h\|_1^n t^k \int_0^t |h(s)| s^{n-k} ds$$

$$\leq cst \, \|h\|_1^n t^{n-\varepsilon} \int_0^t |h(s)| s^\varepsilon ds,$$

with $0 \leq k \leq n - 1$. This ends the proof of the above theorem.

All together applying relation (1), Lemma 3.3, Theorem 3.4 to 3.7, we have proved the following.

Theorem 3.8 *Suppose that the following assumptions are satisfied:*
(1) $(\omega + \omega^{-1})\alpha \in L^2(\mathbb{R}^3)$,
(2) $\int_0^\infty (1 + t^\varepsilon)|h(t)|dt < \infty$, for some $0 < \varepsilon < 1$,
then

$$\lim_{\lambda \to 0} K_\lambda(t) = K(t),$$

for all t. Moreover

$$K^\sharp = i \int_0^\infty \sum_{e \in sp([\sigma_z,.])} e^{-ise} P\mathbb{1}_e([\sigma_z,.])[V,.]e^{is[L_0^{semi},.]}[V,.]\mathbb{1}_e([\sigma_z,.])P \, ds.$$

3.3 Lindbladian of the Spin-boson System

Let

$$\mathcal{L} = iK^\sharp.$$

The aim of this subsection is to give an explicit formula of $\mathcal{L}$. Moreover, we prove that this operator has the form of a Lindblad generator (or Lindbladian). Let us introduce the well known formula of distribution theory

$$\int_0^\infty e^{\pm it\omega} \, dt = \frac{\pm i}{\omega \pm i0} = \pi\delta(\omega) \pm iV_p\left(\frac{1}{\omega}\right), \tag{6}$$

where

$$\frac{1}{x + i0} = \lim_{\varepsilon \to 0} \frac{1}{x + i\varepsilon},$$

$$\int f(x)\delta(x) \, dx = f(0),$$

$$\int f(x)V_p\left(\frac{1}{x}\right) dx = \lim_{\varepsilon \to 0} \int_{|x| \ge \varepsilon} \frac{f(x)}{x} \, dx = PP \int \frac{f(x)}{x} \, dx,$$

$$\int f(x)\frac{1}{x + i0} \, dx = \lim_{\varepsilon \to 0} \int f(x)\frac{1}{x + i\varepsilon} \, dx,$$

for all f, such that $\mathbb{R} \ni x \mapsto f(x)$ is a continuous function and provided the integrals on the right are well defined and the limits exist.

Note that the eigenvalues of $[\sigma_z, .]$ are 2, -2 and 0 where 2, -2 are non degenerate and 0 has multiplicity two. Besides, the corresponding eigenvectors are respectively given by $|\Psi_+\rangle\langle\Psi_-|$, $|\Psi_-\rangle\langle\Psi_+|$ and $|\Psi_+\rangle\langle\Psi_+|$, $|\Psi_-\rangle\langle\Psi_-|$.

Put

$$n_+ = \begin{pmatrix} 1 & 0 \\ 0 & 0 \end{pmatrix}, \quad n_- = \begin{pmatrix} 0 & 0 \\ 0 & 1 \end{pmatrix}, \quad \sigma_+ = \begin{pmatrix} 0 & 1 \\ 0 & 0 \end{pmatrix}, \quad \sigma_- = \begin{pmatrix} 0 & 0 \\ 1 & 0 \end{pmatrix},$$

$$n_+^L X = n_+ X, \quad n_+^R X = X n_+, \quad n_-^L X = n_- X, \quad n_-^R X = X n_-,$$

$$N(\omega) = \frac{1}{e^{\beta \omega(k)} - 1}.$$

It is easy to check that

$$\mathbb{1}_2([\sigma_z, .]) = n_+^L n_-^R,$$
$$\mathbb{1}_{-2}([\sigma_z, .]) = n_-^L n_+^R,$$
$$\mathbb{1}_0([\sigma_z, .]) = n_+^L n_+^R + n_-^L n_-^R.$$

The explicit formula of the Lindbladian associated to the spin-boson system is given as follows.

Theorem 3.9 *If the following assumptions are met:*
i) $\int_0^\infty |h(t)|\, dt < \infty$,
ii) α *is a* C^1 *function in a neighborhood of the sphere* $B(0,2) = \{k \in \mathbb{R}^3,\ |k| = 2\}$,
iii) $(1 + \omega)\alpha \in L^\infty(\mathbb{R}^3)$,
then for all $X \in M_2$,

$$
\begin{aligned}
\mathcal{L}(X) = \ & i\Big(\mathrm{Im}(\alpha, \alpha)_+^- - \mathrm{Im}(\alpha, \alpha)_-^+\Big)[n_+, X] \\
& + i\Big(\mathrm{Im}(\alpha, \alpha)_-^- - \mathrm{Im}(\alpha, \alpha)_+^+\Big)[n_-, X] \\
& + \mathrm{Re}(\alpha, \alpha)_-^+ \Big(2\sigma_+ X \sigma_- - \{n_+, X\}\Big) \\
& + \mathrm{Re}(\alpha, \alpha,)_-^- \Big(2\sigma_- X \sigma_+ - \{n_-, X\}\Big),
\end{aligned}
$$

where

$$\mathrm{Im}(\alpha, \alpha)_+^+ = \int_{\mathbb{R}^3} \frac{N(\omega) + 1}{\omega + 2} |\alpha(k)|^2\, dk,$$

$$\mathrm{Im}(\alpha, \alpha)_-^- = PP \int \frac{N(\omega)}{\omega - 2} |\alpha(k)|^2\, dk,$$

$$\mathrm{Im}(\alpha, \alpha)_-^+ = PP \int \frac{N(\omega) + 1}{\omega - 2} |\alpha(k)|^2\, dk,$$

$$\mathrm{Im}(\alpha, \alpha)_+^- = \int_{\mathbb{R}^3} \frac{N(\omega)}{\omega + 2} |\alpha(k)|^2\, dk,$$

$$\mathrm{Re}(\alpha, \alpha)_-^+ = \pi \frac{e^{2\beta}}{e^{2\beta} - 1} \int_{\mathbb{R}^3} |\alpha(k)|^2 \delta(\omega - 2)\, dk,$$

$$\mathrm{Re}(\alpha, \alpha)_-^- = \frac{\pi}{e^{2\beta} - 1} \int_{\mathbb{R}^3} |\alpha(k)|^2 \delta(\omega - 2)\, dk.$$

Proof. A straightforward computation shows that for all $X \in M_2$,

$$\mathbb{1}_2([\sigma_z, .])[V, .]e^{is[L_0^{semi}, .]}[V, .]\mathbb{1}_2([\sigma_z, .])PX$$
$$= \Big[\varphi_{AW}(\alpha)\varphi_{AW}(e^{is\omega}\alpha) + \varphi_{AW}(e^{is\omega}\alpha)\varphi_{AW}(\alpha)\Big] n_+Xn_-,$$

$$\mathbb{1}_{-2}([\sigma_z, .])[V, .]e^{is[L_0^{semi}, .]}[V, .]\mathbb{1}_{-2}([\sigma_z, .])PX$$
$$= \Big[\varphi_{AW}(\alpha)\varphi_{AW}(e^{is\omega}\alpha) + \varphi_{AW}(e^{is\omega}\alpha)\varphi_{AW}(\alpha)\Big] n_-Xn_+,$$

$$\mathbb{1}_0([\sigma_z, .])[V, .]e^{is[L_0^{semi}, .]}[V, .]\mathbb{1}_0([\sigma_z, .])PX$$
$$= \Big[e^{-2is}\varphi_{AW}(\alpha)\varphi_{AW}(e^{is\omega}\alpha) + e^{2is}\varphi_{AW}(e^{is\omega}\alpha)\varphi_{AW}(\alpha)\Big] n_+Xn_+$$
$$+ \Big[e^{2is}\varphi_{AW}(\alpha)\varphi_{AW}(e^{is\omega}\alpha) + e^{-2is}\varphi_{AW}(e^{is\omega}\alpha)\varphi_{AW}(\alpha)\Big] n_-Xn_-$$
$$- \Big[e^{-2is}\varphi_{AW}(\alpha)\varphi_{AW}(e^{is\omega}\alpha) + e^{2is}\varphi_{AW}(e^{is\omega}\alpha)\varphi_{AW}(\alpha)\Big] \sigma_+X\sigma_-$$
$$- \Big[e^{2is}\varphi_{AW}(\alpha)\varphi_{AW}(e^{is\omega}\alpha) + e^{-2is}\varphi_{AW}(e^{is\omega}\alpha)\varphi_{AW}(\alpha)\Big] \sigma_-X\sigma_+.$$

Hence, for all $X \in M_2$, we have

$$\sum_{e \in sp([\sigma_z, .])} e^{-ise}P\mathbb{1}_e([\sigma_z, .])[V, .]e^{is[L_0^{semi}, .]}[V, .]\mathbb{1}_e([\sigma_z, .])(X)$$
$$= \Big[e^{-2is}\langle\varphi_{AW}(\alpha)\varphi_{AW}(e^{is\omega}\alpha)\rangle + e^{-2is}\langle\varphi_{AW}(e^{is\omega}\alpha)\varphi_{AW}(\alpha)\rangle\Big] n_+Xn_-$$
$$+ \Big[e^{2is}\langle\varphi_{AW}(\alpha)\varphi_{AW}(e^{is\omega}\alpha)\rangle + e^{2is}\langle\varphi_{AW}(e^{is\omega}\alpha)\varphi_{AW}(\alpha)\rangle\Big] n_-Xn_+$$
$$- 2\mathrm{Re}\Big(e^{2is}\langle\varphi_{AW}(e^{is\omega}\alpha)\varphi_{AW}(\alpha)\rangle\Big)\Big[\sigma_+X\sigma_- - n_+Xn_+\Big]$$
$$- 2\mathrm{Re}\Big(e^{-2is}\langle\varphi_{AW}(e^{is\omega}\alpha)\varphi_{AW}(\alpha)\rangle\Big)\Big[\sigma_-X\sigma_+ - n_-Xn_-\Big].$$

It follows that

$$\mathcal{L}(X) = -\Big[\int_0^\infty e^{-2is}\Big(\langle\varphi_{AW}(\alpha)\varphi_{AW}(e^{is\omega}\alpha)\rangle + \langle\varphi_{AW}(e^{is\omega}\alpha)\varphi_{AW}(\alpha)\rangle\Big)ds\Big]$$
$$n_+Xn_-$$
$$- \Big[\int_0^\infty e^{2is}\Big(\langle\varphi_{AW}(\alpha)\varphi_{AW}(e^{is\omega}\alpha)\rangle + \langle\varphi_{AW}(e^{is\omega}\alpha)\varphi_{AW}(\alpha)\rangle\Big)ds\Big]$$
$$n_-Xn_+$$
$$+ 2\mathrm{Re}\Big(\int_0^\infty e^{2is}\langle\varphi_{AW}(e^{is\omega}\alpha)\varphi_{AW}(\alpha)\rangle ds\Big)\Big[\sigma_+X\sigma_- - n_+Xn_+\Big]$$
$$+ 2\mathrm{Re}\Big(\int_0^\infty e^{-2is}\langle\varphi_{AW}(e^{is\omega}\alpha)\varphi_{AW}(\alpha)\rangle ds\Big)\Big[\sigma_-X\sigma_+ - n_-Xn_-\Big].$$

But we have

$$\langle \varphi_{AW}(\alpha)\varphi_{AW}(e^{is\omega}\alpha)\rangle$$
$$= \int_{\mathbb{R}^3} e^{is\omega}(N(\omega)+1)|\alpha(k)|^2\, dk + \int_{\mathbb{R}^3} e^{-is\omega}N(\omega)|\alpha(k)|^2\, dk$$
$$= \langle \varphi_{AW}(e^{is\omega}\alpha)\varphi_{AW}(\alpha)\rangle.$$

Now, by assumptions i), ii) and iii) of the above theorem, we apply formula (6) to get

$$\int_0^\infty e^{-2is}\langle \varphi_{AW}(\alpha)\varphi_{AW}(e^{is\omega}\alpha)\rangle\, ds = \mathrm{Re}(\alpha,\alpha)_-^+ + i\mathrm{Im}(\alpha,\alpha)_-^+ - i\mathrm{Im}(\alpha,\alpha)_+^-,$$

$$\int_0^\infty e^{-2is}\langle \varphi_{AW}(e^{is\omega}\alpha)\varphi_{AW}(\alpha)\rangle\, ds = \mathrm{Re}(\alpha,\alpha)_-^- + i\mathrm{Im}(\alpha,\alpha,)_-^- - i\mathrm{Im}(\alpha,\alpha)_+^+,$$

$$\int_0^\infty e^{2is}\langle \varphi_{AW}(e^{is\omega}\alpha)\varphi_{AW}(\alpha)\rangle\, ds = \mathrm{Re}(\alpha,\alpha)_-^+ - i\mathrm{Im}(\alpha,\alpha)_-^+ + i\mathrm{Im}(\alpha,\alpha)_+^-,$$

$$\int_0^\infty e^{2is}\langle \varphi_{AW}(\alpha)\varphi_{AW}(e^{is\omega}\alpha)\rangle\, ds = \mathrm{Re}(\alpha,\alpha)_-^- + i\mathrm{Im}(\alpha,\alpha)_+^- - i\mathrm{Im}(\alpha,\alpha)_-^-.$$

Therefore we obtain

$$\mathcal{L}(X) = \Big\{ -\mathrm{Re}(\alpha,\alpha)_-^+ - \mathrm{Re}(\alpha,\alpha)_-^- + i\Big(\mathrm{Im}(\alpha)_+^- - \mathrm{Im}(\alpha,\alpha)_-^+\Big)$$
$$-i\Big(\mathrm{Im}(\alpha,\alpha)_-^- - \mathrm{Im}(\alpha,\alpha_+^+)\Big)\Big\} n_+ X n_- + \Big\{ -\mathrm{Re}(\alpha,\alpha)_-^+ - \mathrm{Re}(\alpha,\alpha)_-^-$$
$$-i\Big(\mathrm{Im}(\alpha,\alpha)_+^- - \mathrm{Im}(\alpha,\alpha)_-^+\Big) + i\Big(\mathrm{Im}(\alpha,\alpha)_-^- - \mathrm{Im}(\alpha,\alpha)_+^+\Big)\Big\} n_- X n_+$$
$$+2\mathrm{Re}(\alpha,\alpha)_-^+\big[\sigma_+ X\sigma_- - n_+ X n_+\big] + 2\mathrm{Re}(\alpha,\alpha)_-^-\big[\sigma_- X\sigma_+ - n_- X n_-\big].$$

Hence, we get the following

$$\mathcal{L}(X) = i\Big(\mathrm{Im}(\alpha,\alpha)_+^- - \mathrm{Im}(\alpha,\alpha)_-^+\Big)\big[n_+ X n_- - n_- X n_+\big]$$
$$+i\Big(\mathrm{Im}(\alpha,\alpha)_-^- - \mathrm{Im}(\alpha,\alpha)_+^+\Big)\big[n_- X n_+ - n_+ X n_-\big]$$
$$+\mathrm{Re}(\alpha,\alpha)_-^+\big[2\sigma_+ X\sigma_- - 2n_+ X n_+ - \big(n_+ X n_- + n_- X n_+\big)\big]$$
$$+\mathrm{Re}(\alpha,\alpha)_-^-\big[2\sigma_- X\sigma_+ - 2n_- X n_- - \big(n_+ X n_- + n_- X n_+\big)\big].$$

Note that we have

$$n_+ X n_- + n_- X n_+ = \{n_+, X\} - 2n_+ X n_+ = \{n_-, X\} - 2n_- X n_-,$$
$$n_+ X n_- - n_- X n_+ = [n_+, X],$$
$$n_- X n_+ - n_+ X n_- = [n_-, X].$$

This proves the theorem.

4 Properties of the Quantum Master Equation

In this section we state some properties of the quantum master equation associated to the spin-boson system, such as quantum decoherence and quantum
detailed balance condition. Note that the log-Sobolev inequality with explicit
computation of optimal constants are known in this context. We refer the
interested reader to [C04].

4.1 Quantum Master Equation

Let $\rho \in M_2$ be a density matrix. Then the quantum master equation of the
spin-boson system is given by

$$
\begin{aligned}
\frac{d\rho(t)}{dt} &= i\Big(\mathrm{Im}(\alpha,\alpha)^+_- - \mathrm{Im}(\alpha,\alpha)^-_+\Big)[n_+,\rho(t)] \\
&\quad + i\Big(\mathrm{Im}(\alpha,\alpha)^+_+ - \mathrm{Im}(\alpha,\alpha)^-_-\Big)[n_-,\rho(t)] \\
&\quad + \mathrm{Re}(\alpha,\alpha)^+_-\Big(2\sigma_-\rho(t)\,\sigma_+ - \{n_+,\rho(t)\}\Big) \\
&\quad + \mathrm{Re}(\alpha,\alpha)^-_-\Big(2\sigma_+\rho(t)\,\sigma_- - \{n_-,\rho(t)\}\Big).
\end{aligned}
$$

Put

$$
\rho(t) = \rho_{11}(t)\,n_+ + \rho_{12}(t)\,\sigma_+ + \rho_{21}(t)\,\sigma_- + \rho_{22}(t)\,n_-.
$$

Therefore, the above master equation is equivalent to the following system of
ordinary differential equations

$$
\frac{d}{dt}\rho_{11}(t) = 2\mathrm{Re}(\alpha,\alpha)^-_-\,\rho_{22}(t) - 2\mathrm{Re}(\alpha,\alpha)^+_-\,\rho_{11}(t)
$$

$$
\begin{aligned}
\frac{d}{dt}\rho_{12}(t) &= \Big[-i\Big(\mathrm{Im}(\alpha,\alpha)^+_+ - \mathrm{Im}(\alpha,\alpha)^-_-\Big) + i\Big(\mathrm{Im}(\alpha,\alpha)^+_- - \mathrm{Im}(\alpha,\alpha)^-_+\Big) \\
&\quad -\mathrm{Re}(\alpha,\alpha)^-_- - \mathrm{Re}(\alpha,\alpha)^+_-\Big]\rho_{12}(t)
\end{aligned}
$$

$$
\begin{aligned}
\frac{d}{dt}\rho_{21}(t) &= \Big[-i\Big(\mathrm{Im}(\alpha,\alpha)^+_- - \mathrm{Im}(\alpha,\alpha^-_+)\Big) + i\Big(\mathrm{Im}(\alpha,\alpha)^+_+ - \mathrm{Im}(\alpha,\alpha)^-_-\Big) \\
&\quad -\mathrm{Re}(\alpha,\alpha)^+_- - \mathrm{Re}(\alpha,\alpha)^-_-\Big]\rho_{21}(t)
\end{aligned}
$$

$$
\frac{d}{dt}\rho_{22}(t) = 2\mathrm{Re}(\alpha,\alpha)^+_-\,\rho_{11}(t) - 2\mathrm{Re}(\alpha,\alpha)^-_-\,\rho_{22}(t).
$$

Hence, it is straightforward to show that the thermodynamical equilibrium
state ρ_β of the spin system is the only solution of the above equation.

4.2 Quantum Decoherence of the Spin System

Definition 1 *We say that the dynamical evolution of a quantum system describes decoherence , if there exists an orthonormal basis of $\mathcal{H}_s$ such that the
off-diagonal elements of its time evolved density matrix in this basis vanish as
$t \to \infty$.*

From the system of ordinary differential equations introduced in the previous subsection, we have

$$\rho_{12}(t) = \rho_{12}(0) \exp\Big(-i(\mathrm{Im}(\alpha,\alpha)_+^- + \mathrm{Im}(\alpha,\alpha)_+^+ - \mathrm{Im}(\alpha,\alpha)_-^- - \mathrm{Im}(\alpha,\alpha)_-^+)t \Big)$$

$$\times \exp\Big(-(\mathrm{Re}(\alpha,\alpha)_-^- + \mathrm{Re}(\alpha,\alpha)_+^-)t \Big)$$

$$= \rho_{12}(0) \exp\Big(-i(\mathrm{Im}(\alpha,\alpha)_+^- + \mathrm{Im}(\alpha,\alpha)_+^+ - \mathrm{Im}(\alpha,\alpha)_-^- - \mathrm{Im}(\alpha,\alpha)_-^+)t \Big)$$

$$\times \exp\Big(-\pi\big(\frac{e^{2\beta}+1}{e^{2\beta}-1}\big) \int_{\mathbb{R}^3} |\alpha(k)|^2 \delta(\omega - 2) dk)t \Big),$$

$$\rho_{21}(t) = \rho_{21}(0) \exp\Big(-i(\mathrm{Im}(\alpha,\alpha)_-^+ + \mathrm{Im}(\alpha,\alpha)_-^- - \mathrm{Im}(\alpha,\alpha)_+^- - \mathrm{Im}(\alpha,\alpha)_+^+)t \Big)$$

$$\times \exp\Big(-(\pi\frac{e^{2\beta}+1}{e^{2\beta}-1} \int_{\mathbb{R}^3} |\alpha(k)|^2 \delta(\omega - 2) dk)t \Big).$$

Therefore, the spin system describes quantum decoherence if and only if

$$\int_{\mathbb{R}^3} |\alpha(k)|^2 \delta(\omega - 2) dk \neq 0.$$

Thus, the decoherence of the spin system is controlled by the cut-off function α.

4.3 Quantum Detailed Balance Condition

The following definition is taken from [AL87].

Definition 2 *Let Θ be a generator of a quantum dynamical semigroup written as*

$$\Theta = i\,[H, .] + \Theta_0,$$

where H is a self-adjoint operator. We say that Θ satisfies a quantum detailed balance condition with respect to a stationary state ρ if
  i) $[H, \rho] = 0$,
  ii) $\langle \Theta_0(A), B \rangle_\rho = \langle A, \Theta_0(B) \rangle_\rho$, for all $A, B \in D(\Theta_0)$,
     with $\langle A, B \rangle_\rho = \mathrm{Tr}(\rho A^* B)$.

Actually, we prove the following.

Theorem 4.1 *The generator $\mathcal{L}$ of the quantum dynamical semigroup $T_t = e^{itK^\sharp}$ satisfies a quantum detailed balance condition with respect to the thermo-dynamical equilibrium state of the spin system*

$$\rho_\beta = \frac{e^{-\beta\sigma_z}}{\mathrm{Tr}(e^{-\beta\sigma_z})}.$$

Proof. Note that

$$\mathcal{L}(A) = i [H, A] + \mathcal{L}_D(A),$$

with

$$H = \left(\mathrm{Im}(\alpha, \alpha)_+^- - \mathrm{Im}(\alpha, \alpha)_-^+ \right) n_+ + \left(\mathrm{Im}(\alpha, \alpha)_-^- - \mathrm{Im}(\alpha, \alpha)_+^+ \right) n_-,$$

and

$$\mathcal{L}_D(\rho) = \mathrm{Re}(\alpha, \alpha)_-^+ \left(2\sigma_+\rho\sigma_- - \{n_+, \rho\} \right) + \mathrm{Re}(\alpha, \alpha)_-^- \left(2\sigma_-\rho\sigma_+ - \{n_-, \rho\} \right).$$

Therefore, it is clear that H is a self-adjoint operator and $[H, \rho_\beta] = 0$. More-over it is straightforward to show that $\mathcal{L}_D$ is self-adjoint for the $\langle , \rangle_{\rho_\beta}$ scalar product.

5 Return to Equilibrium for the Spin-boson System

5.1 Hamiltonian Case

In this subsection we recall the results of return to equilibrium for the spin-boson system proved in [JP96b].

For $f \in L^2(\mathbb{R}^3)$ we define $\tilde{f}$ on $\mathbb{R} \times S^2$ by

$$\tilde{f}(s, \hat{k}) = \begin{cases} -|s|^{1/2}\bar{f}(|s|\hat{k}), & s < 0, \\ s^{1/2}f(s\hat{k}), & s \geq 0. \end{cases}$$

Put

$$\mathcal{C}(\delta) = \left\{ z \in \mathbb{C} \text{ s.t } |Imz| < \delta \right\},$$

$$H^2(\delta, \eta) = \left\{ f : \mathcal{C}(\delta) \to \eta \text{ s.t } \|f\|_{H^2(\delta,\eta)} = \sup_{|a|<\delta} \int_{-\infty}^{+\infty} \|f(x + ia)\|_\eta^2 dx < \infty \right\},$$

where η is a Hilbert space.

Definition 3 *Let $\mathcal{M}$ be a W^*-algebra, τ a dynamics on $\mathcal{M}$ and ω a faithful normal state on $\mathcal{M}$. We say that the triple $(\mathcal{M}, \tau, \omega)$ has the property of return to equilibrium if for all $A \in \mathcal{M}$ and all normal state μ, we have*

$$\lim_{t \to \infty} \mu(\tau^t(A)) = \omega(A).$$

Then, in the Hamiltonian approach of the spin-boson system, the following is proved in [JP96b].

Theorem 5.1 *Assume that the following assumptions are satisfied:*
(i) $(\omega + \omega^{-1})\alpha \in L^2(\mathbb{R}^3)$,
(ii) $\int_{S^2} |\alpha(2\hat{k})|^2 d\sigma(\hat{k}) > 0$, where $d\sigma$ is the surface measure on S^2,
(iii) There exists $0 < \delta < \frac{2\pi}{\beta}$ such that $\tilde{\alpha} \in H^2(\delta, L^2(S^2))$.

Then, for all $\beta > 0$ there exists a constant $\Lambda(\beta) > 0$ which depends only on the cut-off function α, such that the spin-boson system has the property of return to equilibrium for all $0 < |\lambda| < \Lambda(\beta)$.

Remark: In the above theorem the authors show that for any fixed temperature $\beta \in]0, +\infty[$, the spectrum of the full-Liouvillean L_λ associated to the spin-boson system is absolutely continuous uniformly on $\lambda \in]0, \Lambda(\beta)[$ and in particular for λ very small (weak coupling). Moreover they used the theory of perturbation of KMS-states for constructing the eigenvector of L_λ associated to the eigenvalue 0. Therefore, for any fixed $\beta \in]0, +\infty[$, the spin-boson system weakly coupled has the property of return to equilibrium.

5.2 Markovian Case

We shall compare the above conditions for the return to equilibrium to the one we obtain in the Markovian approach. Let $(T_t)_{t \geq 0}$ be a quantum dynamical semigroup on $\mathcal{B}(\eta)$ such that its generator has the form

$$\mathcal{L}(X) = G^* X + X G + \sum_{k \geq 1} L_k^* X L_k,$$

where $G = -\frac{1}{2} \sum_{k \geq 1} L_k^* L_k - iH$.
Put

$$\mathcal{A}(T) = \left\{ X \in \mathcal{B}(\eta) \text{ s.t } T_t(X) = X, \text{ for all } t \geq 0 \right\},$$

$$\mathcal{N}(T) = \left\{ X \in \mathcal{B}(\eta) \text{ s.t } T_t(X^* X) = T_t(X^*) T_t(X) \text{ and } \right.$$

$$\left. T_t(XX^*) = T_t(X) T_t(X^*), \text{ for all } t \geq 0 \right\}.$$

The following result is useful for the study of approach to equilibrium in the Markovian case.

Theorem 5.2 *(Frigerio-Verri)*
  If T has a faithful stationary state ρ and $\mathcal{N}(T) = \mathcal{A}(T)$, then

$$w^* - \lim_{t \to \infty} T_t(X) = T_\infty(X), \forall X \in \mathcal{B}(\eta),$$

where $X \to T_\infty(X)$ is a conditional expectation. In particular the quantum dynamical semigroup T has the property of return to equilibrium.

We state without proof the following result which is a special case of a theorem proved in [FR98].

Theorem 5.3 *Suppose that $(T_t)_t$ is a norm continuous quantum dynamical semigroup which has a faithful normal stationary state and H is a self-adjoint* *operator which has a pure point spectrum. Then $(T_t)_t$ has the property of return to equilibrium if and only if*

$$\left\{ L_k,\, L_k^*,\, H,\, k \geq 1 \right\}' = \left\{ L_k,\, L_k^*,\, k \geq 1 \right\}'.$$

Applying the above result, we now prove the following.

Theorem 5.4 *Suppose that the following assumptions are satisfied:*
i) $\mathrm{Im}(\alpha,\alpha)_{\mp}^{\pm}$ are given by real numbers,
ii) $\int_{S^2} |\alpha(2k)|^2\, dk > 0$.
Then the quantum dynamical semigroup of the spin-boson system at positive temperature has the property of return to equilibrium.

Proof. Set

$$H = \left(\mathrm{Im}(\alpha,\alpha)_{+}^{-} - \mathrm{Im}(\alpha,\alpha)_{-}^{+} \right) n_{+} + \left(\mathrm{Im}(\alpha,\alpha)_{-}^{-} - \mathrm{Im}(\alpha,\alpha)_{+}^{+} \right) n_{-},$$

$$L_1 = \left(2\mathrm{Re}(\alpha,\alpha)_{-}^{+} \right)^{1/2} \sigma_{-},$$

$$L_2 = \left(2\mathrm{Re}(\alpha,\alpha)_{-}^{-} \right)^{1/2} \sigma_{+}, \tag{7}$$

$$G = -\frac{1}{2} \sum_{k=1}^{2} L_k^* L_k - iH.$$

Then the Lindbladian of the spin-boson system takes the form

$$\mathcal{L}(X) = G^* X + XG + \sum_{k=1}^{2} L_k^* X L_k,$$

for all $X \in M_2$.

Note that the quantum dynamical semigroup T of the spin-boson system has the thermodynamical equilibrium state ρ_β of the spin system as a faithful normal stationary state. Moreover H is a self-adjoint bounded operator which has a pure point spectrum and it is clear that

$$\left\{ L_k,\, L_k^*,\, H,\, k = 1,2 \right\}' = \left\{ L_k,\, L_k^*,\, k = 1,2 \right\}' = \mathbb{C}I.$$

Thus from the previous theorem, the quantum dynamical semigroup of the spin-boson system has the property of return to equilibrium.

Note that compared to the Hamiltonian approach, we have in Theorem 5.4 a simplification of conditions for return to equilibrium of the spin-boson system. So in this theorem we need only that assumptions i) and ii) are satisfied. Hypothesis i) ensures that $\mathrm{Im}(\alpha,\alpha)_{\mp}^{\pm}$ exist and are finite, while if ii) holds, then $\mathrm{Re}(\alpha,\alpha)_{-}^{\pm}$ are not vanishing.

5.3 Spin-boson System at Zero Temperature

In the Hamiltonian case, if a quantum dynamical system which its Liouvillean L has a purely absolutely continuous spectrum, except for the simple eigenvalue 0, then this system has the property of return to equilibrium (cf [JP96b]). At inverse temperature β $(0 < \beta < \infty)$, by using the perturbation theory of KMS-states (cf [DJP03]), we can give an explicit formula of the eigenstate of L associated to the eigenvalue 0. But it is not the case for zero temperature $(\beta = \infty)$. On the other hand, the ground state of the spin system is not faithful and by Theorem 5.3 we cannot conclude. Let us describe the spin-boson system at zero temperature.

At zero temperature, the Hilbert space of the spin-boson system is

$$\mathcal{H} = \mathbb{C}^2 \otimes \Gamma_s(L^2(\mathbb{R}^3)).$$

The free Hamiltonian is defined as

$$h_0 = \sigma_z \otimes 1 + 1 \otimes d\Gamma(\omega),$$

and its full Hamiltonian with interaction is the operator

$$h_\lambda = h_0 + \lambda \sigma_x \otimes \varphi(\alpha),$$

where $\alpha \in L^2(\mathbb{R}^3)$ is a test function.

The zero temperature equilibrium state of the spin system is the vector state corresponding to the ground state of σ_z and it has a density matrix

$$\rho_\infty = |\Psi_-\rangle\langle\Psi_-|.$$

The weak coupling limit of the spin-boson system at zero temperature can be proved in the same way as for positive temperature. The associated Lindbladian can be deduced from the one at positive temperature by taking $\beta = \infty$ and it has the form

$$\mathcal{L}_\infty(X) = -i\nu_1[n_+, X] - i\nu_2[n_-, X] + \nu_3\Big(2\sigma_+ X\sigma_- - \{n_+, X\}\Big),$$

where

$$\nu_1 = \int_{\mathbb{R}^3} \frac{1}{\omega + 2}|\alpha(k)|^2 \, dk,$$

$$\nu_2 = PP \int \frac{1}{\omega - 2}|\alpha(k)|^2 \, dk,$$

$$\nu_3 = \pi \int_{\mathbb{R}^3} |\alpha(k)|^2 \delta(\omega - 2) \, dk.$$

Hence, for all density matrix $\rho \in M_2$, the associated quantum master equation is given by

$$\frac{d\rho(t)}{dt} = i\nu_1[n_+, \rho(t)] + i\nu_2[n_-, \rho(t)] + \nu_3\Big(2\sigma_-\rho(t)\,\sigma_+ - \{n_+, \rho(t)\}\Big) = \mathcal{L}_\infty^*(\rho(t)).$$

Now, in order to conclude the property of return to equilibrium for the quantum dynamical semigroup associated to the spin-boson system at zero temperature, we have to show it by direct computation.

Theorem 5.5 *Assume that:*

i) ν_2 is given by a real number,

ii) $\int_{S^2} |\alpha(2k)|^2 \, dk > 0.$

Then the spin-boson system at zero temperature has the property of return to equilibrium. Moreover we have

$$\lim_{t \to \infty} \mathrm{Tr}(e^{t\mathcal{L}_\infty^*} \rho A) = \mathrm{Tr}(\rho_\infty A),$$

for all $A \in M_2$ and all ρ be a given density matrix.

Proof. Consider the orthonormal basis of M_2 given by

$$\Big\{ |\Psi_+\rangle\langle\Psi_+|, \; |\Psi_+\rangle\langle\Psi_-|, \; |\Psi_-\rangle\langle\Psi_+|, \; |\Psi_-\rangle\langle\Psi_-| \Big\}.$$

Then in this basis we have

$$e^{t\mathcal{L}_\infty^*} = \begin{pmatrix} e^{-2t\nu_3} & 0 & 0 & 0 \\ 0 & e^{-t\nu_3}e^{it(\nu_1-\nu_2)} & 0 & 0 \\ 0 & 0 & e^{-t\nu_3}e^{-it(\nu_1-\nu_2)} & 0 \\ -e^{-2t\nu_3}+1 & 0 & 0 & 1 \end{pmatrix}.$$

Therefore we get

$$\lim_{t \to \infty} e^{t\mathcal{L}_\infty^*} = \Pi_\infty^*,$$

where

$$\Pi_\infty^* = \begin{pmatrix} 0 & 0 & 0 & 0 \\ 0 & 0 & 0 & 0 \\ 0 & 0 & 0 & 0 \\ 1 & 0 & 0 & 1 \end{pmatrix}.$$

A direct computation gives

$$\Pi_\infty^*(A) = \sigma_- A\sigma_+ + n_- An_-, \; \forall A \in M_2.$$

Consider a density matrix ρ of the form

$$\rho = \begin{pmatrix} \alpha & \beta \\ \bar{\beta} & 1-\alpha \end{pmatrix},$$

with $\alpha \in [0,1]$, $\beta \in \mathbb{C}$. We have

$$\Pi_\infty^*(\rho) = \begin{pmatrix} 0 & 0 \\ 0 & 1 \end{pmatrix} = |\Psi_-\rangle\langle\Psi_-| = \rho_\infty.$$

Therefore, it follows that

$$\lim_{t \to \infty} \mathrm{Tr}(e^{t\mathcal{L}_\infty^*} \rho A) = \mathrm{Tr}(\Pi_\infty^*(\rho)A) = \mathrm{Tr}(\rho_\infty A),$$

$\forall A \in M_2$. This proves our theorem.

6 Quantum Langevin Equation and Associated Hamiltonian

It is shown in [HP84] that any quantum master equation of a simple quantum system $\mathcal{H}_S$ can be dilated into a unitary quantum Langevin equation (quantum stochastic differential equation) on a larger space $\mathcal{H}_S \otimes \Gamma$ where Γ is a Fock space in which are naturally living quantum noises. Note that in the literature it is shown that natural quantum stochastic differential equations can be obtained by the stochastic limit of the full Hamiltonian system which is developed in [ALV02].

Now, let us introduce some notations that need in the sequel.

6.1 Basic Notations

Let $\mathcal{Z}$ be a Hilbert space for which we fix an orthonormal basis $\{z_k, \ k \in J\}$. We denote by $\Gamma_s(\mathbb{R}_+)$, the symmetric Fock space constructed over the Hilbert space $\mathcal{Z} \otimes L^2(\mathbb{R}_+)$. Therefore, from the following identification

$$\mathcal{Z} \otimes L^2(\mathbb{R}_+) \simeq L^2(\mathbb{R}_+, \mathcal{Z}) \simeq L^2(\mathbb{R}_+ \times J),$$

we get

$$\Gamma_s(\mathbb{R}_+) = \Gamma_{\text{sym}}(L^2(\mathbb{R}_+ \times J)).$$

The space $\mathcal{Z}$ is called the multiplicity space and $\dim \mathcal{Z}$ is called the multiplicity. The set J is equal to $\{1, ..., N\}$ in the case of finite multiplicity N and is equal to $\mathbb{N}$ in the case of infinite multiplicity.

Let us introduce another Hilbert space $\mathcal{H}$ called initial or system space and we identify the tensor product

$$\mathcal{K}(\mathbb{R}_+) = \mathcal{H} \otimes \Gamma_s(\mathbb{R}_+) = \mathcal{H} \otimes \bigoplus_{n=0}^{\infty} L^2(\mathbb{R}_+ \times J)^{\otimes n} = \bigoplus_{n=0}^{\infty} \mathcal{H} \otimes L^2(\mathbb{R}_+ \times J)^{\otimes n}$$

with the direct sum

$$\bigoplus_{n=0}^{\infty} \mathcal{H} \otimes L^2_{\text{sym}}((\mathbb{R}_+ \times J)^n) \simeq \bigoplus_{n=0}^{\infty} L^2_{\text{sym}}((\mathbb{R}_+ \times J)^n, \mathcal{H}),$$

consisting of the vectors $\Psi = (\Psi_n)_{n \geq 0}$ such that $\Psi_n \in L^2_{\text{sym}}((\mathbb{R}_+ \times J)^n, \mathcal{H})$ and

$$\|\Psi\|^2_{\mathcal{K}(\mathbb{R}_+)} = \sum_{n \geq 0} \frac{1}{n!} \|\Psi_n\|^2_{L^2_{\text{sym}}((\mathbb{R}_+ \times J)^n, \mathcal{H})} < \infty.$$

Note that for $f \in L^2(\mathbb{R}_+ \times J)$, we define its associated exponential vector by

$$\varepsilon(f) = \sum_{n \geq 0} \frac{f^{\otimes n}}{\sqrt{n!}}.$$

6.2 Hudson-Parthasarathy Equation

Let H, R_k and S_{kl}, k, $l \geq 1$ be bounded operators on $\mathcal{H}$ such that

$$H = H^*, \quad \sum_j S_{jk}^* S_{jl} = \sum_j S_{kj} S_{lj}^* = \delta_{kl}, \tag{8}$$

and the sum $\sum_k R_k^* R_k$ are assumed to be strongly convergent to a bounded operator. Through H, R_k and S_{kl} we define the following operators

$$S \in \mathcal{U}(\mathcal{H} \otimes \mathcal{Z}), \ R \in \mathcal{B}(\mathcal{H}, \mathcal{H} \otimes \mathcal{Z}), \ G \in \mathcal{B}(\mathcal{H}),$$

by

$$Ru = \sum_k (R_k u) \otimes z_k, \ \forall u \in \mathcal{H},$$

$$S = \sum_{kl} S_{kl} \otimes |z_k\rangle\langle z_l|,$$

$$G = -iH - \frac{1}{2} \sum_k R_k^* R_k = -iH - \frac{1}{2} R^* R.$$

The basic quantum noises are the processes

$$A_i(t) = A(\mathbb{1}_{(0,t)} \otimes z_i),$$
$$A_i^+(t) = A^+(\mathbb{1}_{(0,t)} \otimes z_i),$$
$$\Lambda_{ij}(t) = \Lambda(\pi_{(0,t)} \otimes |z_i\rangle\langle z_j|),$$

where i, $j \in J$, $\mathbb{1}_{(0,t)}$ is the indicator function over $(0,t)$, while $\pi_{(0,t)}$ is the multiplication operator by $\mathbb{1}_{(0,t)}$ in $L^2(\mathbb{R}_+)$.

The Hudson-Parthasarathy equation is defined as follows

$$(HP) \begin{cases} dU(t) = \left\{ \sum_k R_k dA_k^+(t) + \sum_{kl}(S_{kl} - \delta_{kl}) d\Lambda_{kl}(t) \right. \\ \left. - \sum_{kl} R_k^* S_{kl} dA_l(t) + Gdt \right\} U(t) \\ U(0) = 1. \end{cases}$$

Note that in order to have a unitary solution U of (HP), we need some conditions on the system operators. Actually the following theorem holds.

Theorem 6.1 *Suppose that the system operators H, R_k, S_{kl} satisfies (8). Then there exists a unique strongly continuous unitary adapted process $U(t)$ which satisfies equation (HP).*

Proof. For the proof of this theorem we refer the reader to [P92]. ∎

Now, in order to associate a group V to the solution U of (HP), we first introduce the one-parameter strongly continuous unitary group θ in $L^2(\mathbb{R}, \mathcal{Z})$ and its associated second quantization Θ in $\Gamma(\mathbb{R})$, defined by

$$\theta_t f(r) = f(r + t), \quad \forall f \in L^2(\mathbb{R}, \mathcal{Z}),$$
$$\Theta_t e(f) = e(\theta_t f), \quad \forall f \in L^2(\mathbb{R}, \mathcal{Z}). \tag{9}$$

Note that Θ and $U(t)$ can be extended to act on the space

$$\mathcal{K}(\mathbb{R}) = \mathcal{H} \otimes \Gamma_s(\mathbb{R}_+) \otimes \Gamma_s(\mathbb{R}_-) = \mathcal{K}(\mathbb{R}_+) \otimes \Gamma_s(\mathbb{R}_-) = \mathcal{H} \otimes \Gamma_s(\mathbb{R}),$$

by

$$\Theta_t = 1 \otimes \Theta_t \quad \text{in } \mathcal{H} \otimes \Gamma_s(\mathbb{R}),$$
$$U(t) = U(t) \otimes 1 \quad \text{in } \mathcal{K}(\mathbb{R}_+) \otimes \Gamma_s(\mathbb{R}_-).$$

Theorem 6.2 *Let Θ be the one-parameter strongly continuous group defined by (9) and U the solution of the EDSQ (HP) with system operators satisfying (8). Then*

$$U(t + s) = \Theta_s^* U(t) \Theta_s U(s), \quad \forall s, \, t \geq 0,$$

and the family $V = \{V_t\}_{t \in \mathbb{R}}$ such that

$$V_t = \begin{cases} \Theta_t U(t), & t \geq 0 \\ U^*(|t|)\Theta_t, & t \leq 0, \end{cases}$$

defines a one-parameter strongly continuous unitary group. Furthermore, the family of two-parameter unitary operators

$$U(t, s) = \Theta_t^* V_{t-s} \Theta_s = \Theta_s^* U(t - s) \Theta_s, \quad \forall s \leq t,$$

is strongly continuous in t and in s and satisfies the composition law

$$U(t, s) U(t, r) = U(t, r), \quad \forall r \leq s \leq t.$$

Proof. See [B06] for the proof of this theorem.

The group V defined as above, describes the reversible evolution of the small system plus the reservoir which is modelled by the free Bose gas. The free evolution of the reservoir is represented by the group Θ whose generator is formally given by

$$E_0 = d\Gamma(i \frac{\partial}{\partial x}).$$

Note that $U(t) = U(t, 0) = \Theta_t^* V_t$ is the evolution operator giving the dynamics state from time 0 to time t of the whole system in the interaction picture. Moreover by the Stone theorem

$$d\Theta_t = -iE_0\Theta_t dt,$$
$$dV_t = -iKV_t dt.$$

The operators H, E_0 represent respectively the energy associated to the small system and the reservoir. The operator K represents the total energy of the combined system in the interaction picture and the system operators R_j, S_{ij} control this interaction. Besides, if we take $R_j = 0$, $S_{ij} = \delta_{ij}$, then we get

$$U(t) = e^{itH}, \quad V_t = e^{-itE_0}e^{-itH},$$

and $K = E_0 + H$ which is self-adjoint operator defined on $\mathcal{H} \otimes D(E_0)$.

In [G01], Gregoratti give an essentially self-adjoint restriction of the Hamiltonian K which appears as a singular perturbation of $E_0 + H$.

6.3 Hamiltonian Associated to the Hudson-Parthasarathy Equation

Recall that the generators ϵ_0 and E_0 of the groups θ in $L^2(\mathbb{R}, \mathcal{Z})$ and Θ in $\mathcal{K}$ are self-adjoint unbounded operators. In order to explicit their domains we introduce the Sobolev space

$$H^\Sigma((\mathbb{R} \times J)^n, \mathcal{H})$$
$$= \Big\{u \in L^2((\mathbb{R} \times J)^n, \mathcal{H}) \text{ such that } \sum_{k=1}^n \partial_k u \in L^2((\mathbb{R} \times J)^n, \mathcal{H})\Big\},$$

where all the derivatives of u are in the sense of distributions in $(\mathbb{R} \times J)^n$ $(n \geq 1)$ and

$$H^\Sigma((\mathbb{R} \times J)^0, \mathcal{H}) = \mathcal{H}.$$

Furthermore $H^\Sigma((\mathbb{R} \times J)^n, \mathcal{H})$ is a Hilbert space with respect to the scalar product

$$\langle u, v\rangle_{H^\Sigma((\mathbb{R}\times J)^n, \mathcal{H})} = \langle u, v\rangle_{L^2((\mathbb{R}\times J)^n, \mathcal{H})} + \Big\langle \sum_{k=1}^n \partial_k u, \sum_{k=1}^n \partial_k v\Big\rangle_{L^2((\mathbb{R}\times J)^n, \mathcal{H})}.$$

Set

$$H^\Sigma_{\text{sym}}((\mathbb{R} \times J)^n, \mathcal{H}) = H^\Sigma((\mathbb{R} \times J)^n, \mathcal{H}) \cap L^2_{\text{sym}}((\mathbb{R} \times J)^n, \mathcal{H}).$$

We have

$$D(\epsilon_0) = H^1(\mathbb{R}, \mathcal{Z}), \quad \text{and} \quad \epsilon_0 u = iu',$$

Besides, the domain of E_0 is given by

$$D(E_0) = \Big\{\Phi \in \mathcal{K} \text{ s.t } \Phi_n \in H^\Sigma_{\text{sym}}((\mathbb{R} \times J)^n, \mathcal{H}), \ \forall n \text{ and}$$

$$\sum_{n \geq 1} \frac{1}{n!}\|\sum_{k=1}^n \partial_k \Phi_n\|^2 < \infty\Big\},$$

and this operator acts on its domain by $(E_0\Phi)_n = i\sum_{k=1}^n \partial_k \Phi_n$.

Set $\mathbb{R}_* = \mathbb{R} \setminus \{0\}$. Let us introduce the dense subspaces in $\mathcal{K}$ defined by

$$\mathcal{W} = \left\{ \Phi \in \mathcal{K} \ s.t \ \Phi_n \in H_{sym}^{\Sigma}((\mathbb{R}_* \times J)^n, \mathcal{H}), \ \forall n \ \text{ and } \right.$$

$$\left. \sum_{n \geq 1} \frac{1}{n!} \| \sum_{k=1}^{\infty} \partial_k \Phi_n \|_{L^2(\mathbb{R} \times J)^n, \mathcal{H}}^2 < \infty \right\},$$

$$\nu_s = \left\{ \Phi \in \mathcal{W} \ s.t \ \sum_{n \geq 0} \frac{1}{n!} \| \Phi_{n+1} |_{\{r_{n+1}=s\}} \|_{\mathcal{Z} \otimes L^2((\mathbb{R} \times J)^n, \mathcal{H})}^2 < \infty \right\},$$

$$\nu_{0\pm} = \nu_{0-} \cap \nu_{0+},$$

where $\Phi_{n+1}|_{\{r_{n+1}=s\}}$ is the trace (restriction) of the function Φ_{n+1} on the hyperplane $\{r_{n+1} = s\}$, for all $s \in \mathbb{R}_* \cup \{0^-, 0^+\}$. Clearly

$$\nu_{0\pm} \subseteq \mathcal{W}.$$

Define the trace operator $a(s) : \nu_s \to \mathcal{Z} \otimes \mathcal{K}$ such that

$$(a(s)\Phi)_n = \Phi_{n+1}|_{\{r_{n+1}=s\}}.$$

Note that $\varepsilon(H^1(\mathbb{R}^*, \mathcal{Z})) \subset \nu_s$ and

$$a(s)\Psi(u) \otimes h = u(s) \otimes \Psi(u) \otimes h, \quad \forall u \in H^1(\mathbb{R}_*, \mathcal{Z}), \ h \in \mathcal{H},$$

where

$$\Psi(u) = (1, u, u^{\otimes 2}, ..., u^{\otimes n}, ...).$$

Moreover $\mathcal{W} \supset D(E_0)$ and E_0 can be extended to a non-symmetric unbounded operator in $\mathcal{W}$ by

$$(E\Phi)_n = i \sum_{k=1}^n \partial_k \Phi_n.$$

The following theorem gives an essentially self-adjoint restriction of the Hamiltonian operator associated to (HP) and it is proved in [G01].

Theorem 6.3 *Let K be the Hamiltonian operator associated to the equation (HP) such that the system operators satisfying (8). Then*

(1) $D(K) \cap \nu_{0\pm} = \left\{ \Phi \in \nu_{0\pm} \ s.t \ a(0^-)\Phi = Sa(0^+)\Phi + R\Phi \right\}$,

(2) $K\Phi = \left(H + E - iR^*a(0^-) + \frac{i}{2}R^*R \right)\Phi, \ \forall \Phi \in D(K) \cap \nu_{0\pm}$,

(3) $K|_{D(K) \cap \nu_{0\pm}}$ is a essentially self-adjoint operator.

6.4 Hamiltonian Associated to the Stochastic Evolution of the Spin-boson System

Recall that the quantum Langevin equation of the spin-boson system is defined on $\mathbb{C}^2 \otimes \Gamma_s(L^2(\mathbb{R}_+, \mathbb{C}^2))$ by

$$\begin{cases} dU(t) = \Big\{ G dt + \sum_{k=1}^{2} L_k dA_k^+(t) - \sum_{k=1}^{2} L_k^* dA_k(t) \Big\} U(t) \\ U(0) = I, \end{cases}$$

where G, L_k, $k \in \{0,1\}$ are given by the relation (7).
Note that this equation satisfies the class of Hudson-Parthasarathy equation with $S_{ij} = \delta_{ij}$. Moreover we have

$S = I,$

$Ru = \big(2\mathrm{Re}(\alpha,\alpha)_-^+\big)^{1/2} \sigma_- u \otimes \Psi_+ + \big(2\mathrm{Re}(\alpha,\alpha)_-^-\big)^{1/2} \sigma_+ u \otimes \Psi_-, \ \forall u \in \mathbb{C}^2,$

$R^* u \otimes \varphi = \langle \Psi_+, \varphi \rangle \big(2\mathrm{Re}(\alpha,\alpha)_-^+\big)^{1/2} \sigma_+ u + \langle \Psi_-, \varphi \rangle \big(2\mathrm{Re}(\alpha,\alpha)_-^-\big)^{1/2} \sigma_- u,$

$\forall u, \ \varphi \in \mathbb{C}^2,$

$R^* R = 2\mathrm{Re}(\alpha,\alpha)_-^+ n_+ + 2\mathrm{Re}(\alpha,\alpha)_-^- n_-.$

Therefore we get

$$\nu_{0\pm} \cap D(K) = \Big\{ \Phi \in \nu_{0\pm} \ \text{s.t} \ a(0^-)\Phi = a(0^+)\Phi + R\Phi \Big\},$$

and

$$K\Phi = \Big(H + E - iR^*a(0^-) + i\big(\mathrm{Re}(\alpha,\alpha)_-^+ n_+ + \mathrm{Re}(\alpha,\alpha)_-^- n_-\big)\Big)\Phi,$$

for every $\Phi \in \nu_{0\pm} \cap D(K)$.

Recall that the associated energy of the reservoir is given by $E = d\Gamma(i\frac{\partial}{\partial x})$. Therefore, by using the spectral theorem, $i\frac{\partial}{\partial x}$ is a multiplication operator by a variable ω in $\mathbb{R}$. Thus we get

$$E = d\Gamma(\omega),$$

and E is the same as the usual Hamiltonian. On the other hand, the operator

$$H = \Big(\mathrm{Im}(\alpha/\alpha)_+^- - \mathrm{Im}(\alpha/\alpha)_-^+ \Big) n_+ + \Big(\mathrm{Im}(\alpha/\alpha)_-^- - \mathrm{Im}(\alpha/\alpha)_+^+ \Big) n_-,$$

describes the energy of the spin. Note that the constants $\mathrm{Im}(\alpha/\alpha)_{\pm}^{\pm}$ have an important physical interpretation. In some sense they contain all physical information on the original Hamiltonian of the spin. The free evolution of the combined system is described by $\mathcal{H}_f = H + E$ and the Hamiltonian K appears as a singular perturbation of H_f, where the operator R defined as above controls the interaction between the spin and the reservoir.

7 Repeated Quantum Interaction Model

In this section, we start by describing the repeated quantum interaction model (cf [AP06]). We prove that the quantum Langevin equation of the spin-boson system at zero temperature can be obtained as the continuous limit of an

Hamiltonian repeated interaction model. Moreover we compare the Lindbladian of the spin-boson system at positive temperature to the one obtained by using the method introduced in [AJ07].

Consider a small system $\mathcal{H}_0$ coupled with a piece of environment $\mathcal{H}$. The interaction between the two systems is described by the Hamiltonian H which is defined on $\mathcal{H}_0 \otimes \mathcal{H}$. The associated unitary evolution during the interval $[0, h]$ of times is

$$\mathbb{L} = e^{-ihH}.$$

After the first interaction, we repeat this time coupling the same $\mathcal{H}_0$ with a new copy of $\mathcal{H}$. Therefore, the sequence of the repeated interactions is described by the space

$$\mathcal{H}_0 \otimes \bigotimes_{\mathbb{N}^*} \mathcal{H}.$$

The unitary evolution of the small system in interaction picture with the $n-th$ copy of $\mathcal{H}$, denoted by $\mathcal{H}_n$, is the operator $\mathbb{L}_n$ which acts as $\mathbb{L}$ on $\mathcal{H}_0 \otimes \mathcal{H}_n$ and acts as the identity on the copies of $\mathcal{H}$ other than $\mathcal{H}_n$. The associated evolution equation of this model is defined on $\mathcal{H}_0 \otimes \bigotimes_{\mathbb{N}^*} \mathcal{H}$ by

$$\begin{cases} u_{n+1} = \mathbb{L}_{n+1} u_n \\ u_0 = I \end{cases} \tag{10}$$

Let $\{X_i\}_{i \in \Lambda \cup \{0\}}$ be an orthonormal basis of $\mathcal{H}$ with $X_0 = \Omega$ and let us consider the coefficients $(\mathbb{L}_j^i)_{i,j \in \Lambda \cup \{0\}}$ which are operators on $\mathcal{H}_0$ of the matrix representation of $\mathbb{L}$ in the basis $\{X_i\}_{i \in \Lambda \cup \{0\}}$.

Theorem 7.1 *If*

$$\mathbb{L}_0^0 = I - h(iH + \frac{1}{2} \sum_k L_k^* L_k) + h\omega_0^0,$$

$$\mathbb{L}_j^0 = \sqrt{h} L_j + \sqrt{h}\omega_j^0,$$

$$\mathbb{L}_0^i = -\sqrt{h} \sum_k L_k^* S_i^k + \sqrt{h}\omega_0^i,$$

$$\mathbb{L}_j^i = S_j^i + h\omega_j^i,$$

where H is a self-adjoint bounded operator, $(S_j^i)_{i,j}$ is a family of unitary operator, $(L_i)_i$ are operators on $\mathcal{H}_0$ and the terms ω_j^i converge to 0 when h tends to 0, then the solution $(u_n)_{n \in \mathbb{N}}$ of (10) is made of invertible operators which are locally uniformly bounded in norm. Moreover $u_{[t/h]}$ converges weakly to the solution $U(t)$ of the equation

$$\begin{cases} dU(t) = \sum_{i,j} L_j^i U(t) da_j^i(t) \\ U(0) = I \end{cases}$$

where

$$L_0^0 = -\left(iH + \frac{1}{2}\sum_k L_k^* L_k\right),$$

$$L_j^0 = L_j,$$

$$L_0^i = -\sum_k L_k^* S_i^k,$$

$$L_j^i = S_j^i - \delta_{ij} I.$$

Proof. See [AP06] for the proof of this theorem.

Now, let us put $\mathcal{H}_0 = \mathcal{H} = \mathbb{C}^2$ and consider the dipole interaction Hamiltonian defined on $\mathbb{C}^2 \otimes \mathbb{C}^2$ as

$$H = \sigma_z \otimes I + I \otimes H_R + \frac{1}{\sqrt{h}}(\sigma_- \otimes a^* + \sigma_+ \otimes a),$$

where

$H_R = \begin{pmatrix} 0 & 0 \\ 0 & 2 \end{pmatrix}$, is the Hamiltonian of the piece of the reservoir,

$V = \sigma_-,$

$a = \begin{pmatrix} 0 & 1 \\ 0 & 0 \end{pmatrix}$ and a^* is the adjoint of a.

Fix an orthonormal basis $\{\Omega, X\}$ of $\mathbb{C}^2$ such that

$$\Omega = \begin{pmatrix} 1 \\ 0 \end{pmatrix}, \quad X = \begin{pmatrix} 0 \\ 1 \end{pmatrix}.$$

The unitary evolution during the interval $[0, h]$ of time is $\mathbb{L} = e^{-ihH}$ such that

$$\mathbb{L}_0^0 = \langle \Omega, \mathbb{L}\Omega \rangle = I - ih\,\sigma_z - \frac{1}{2}h\sigma_+\sigma_- + o(h),$$

$$\mathbb{L}_0^1 = \langle \Omega, \mathbb{L}X \rangle = -i\sqrt{h}\,\sigma_+ + o(\sqrt{h}),$$

$$\mathbb{L}_1^0 = \langle X, \mathbb{L}\Omega \rangle = -i\sqrt{h}\,\sigma_- + o(\sqrt{h}),$$

$$\mathbb{L}_1^1 = \langle X, \mathbb{L}X \rangle = I - ih\sigma_z - ihI - \frac{1}{2}h\,\sigma_-\sigma_+ + o(h).$$

Therefore we obtain

$$\frac{\mathbb{L}_0^0 - I}{h} \xrightarrow{h \to 0} G_0 = -i\sigma_z - \frac{1}{2}\sigma_+\sigma_-,$$

$$\frac{\mathbb{L}_0^1}{\sqrt{h}} \xrightarrow{h \to 0} -L^* = -i\sigma_+,$$

$$\frac{\mathbb{L}_1^0}{\sqrt{h}} \xrightarrow{h \to 0} L = -i\sigma_-.$$

Thus by Theorem 7.1, the solution $(u_n)_{n\in\mathbb{N}}$ of the equation

$$\begin{cases} u_{n+1} = \mathbb{L}_{n+1}\, u_n \\ u_0 = I \end{cases}$$

is made of invertible operators which are locally uniformly bounded in norm and in particular $u_{[t/h]}$ converges weakly to the solution $U(t)$ of the equation

$$\begin{cases} dU(t) = \left\{ G_0\, dt + L\, dA^+(t) - L^*\, dA^-(t) \right\} U(t) \\ U(0) = I. \end{cases}$$

Theorem 7.2 *The quantum dynamical semigroup of the repeated quantum interaction model associated to the spin-boson system at zero temperature converges towards to equilibrium.*

Proof. The associated Lindbladian of the above equation is of the form

$$\mathcal{L}(X) = i[\sigma_z, X] + 2\sigma_+ X\sigma_- - \{n_+, X\},$$

and the proof is similar as the one of Theorem 5.5.

Now, at inverse temperature β, we suppose that the piece of the reservoir is described by the state

$$\rho = \frac{1}{1 + e^{-\beta}} e^{-\beta H_R} = \begin{pmatrix} \beta_0 & 0 \\ 0 & \beta_1 \end{pmatrix}.$$

The GNS representation of $(\mathbb{C}^2, \rho)$ is the triple $(\pi, \widetilde{\mathcal{H}}, \Omega_R)$, such that

- $\Omega_R = I$,
- $\widetilde{\mathcal{H}} = M_2$, the algebra of all complex 2×2 matrix which equipped by the scalar product

$$\langle A, B \rangle = \mathrm{Tr}(\rho A^* B),$$

- $\pi : M_2 \to \mathcal{B}(\widetilde{\mathcal{H}})$, such that $\pi(M)A = MA$, $\forall M,\, A \in M_2$.

Set

$$X_1 = \frac{1}{\sqrt{\beta_1}} \begin{pmatrix} 0 & 1 \\ 0 & 0 \end{pmatrix}, \quad X_2 = \frac{1}{\sqrt{\beta_0}} \begin{pmatrix} 0 & 0 \\ 1 & 0 \end{pmatrix}, \quad X_3 = \frac{1}{\sqrt{\beta_0\beta_1}} \begin{pmatrix} \beta_1 & 0 \\ 0 & -\beta_0 \end{pmatrix}.$$

It is easy to show that $(\Omega_R, X_1, X_2, X_3)$ is an orthonormal basis of M_2. Now, if we put $\widetilde{\mathbb{L}} = \pi(\mathbb{L})$ which is defined on $\mathbb{C}^2 \otimes M_2$, then a straightforward computation shows that the coefficients $(\widetilde{\mathbb{L}}^i_j)_{i,j}$, which are operators on $\mathbb{C}^2$, of the matrix representation of $\widetilde{\mathbb{L}}$, are given by

$$\tilde{\mathbb{L}}_0^0 = I - ih\sigma_z - ih\,\beta_1 I - \frac{1}{2}h\beta_0\sigma_+\sigma_- - \frac{1}{2}h\beta_1\sigma_-\sigma_+ + o(h^2),$$

$$\tilde{\mathbb{L}}_1^0 = -i\sqrt{\beta_1}\sqrt{h}\,\sigma_+ + o(h^{3/2}),$$

$$\tilde{\mathbb{L}}_2^0 = -i\sqrt{\beta_0}\sqrt{h}\,\sigma_- + o(h^{3/2}),$$

$$\tilde{\mathbb{L}}_3^0 = o(h),$$

$$\tilde{\mathbb{L}}_0^1 = -i\sqrt{\beta_1}\sqrt{h}\,\sigma_- + o(h^{3/2}),$$

$$\tilde{\mathbb{L}}_0^2 = -i\sqrt{\beta_0}\sqrt{h}\,\sigma_+ + o(h^{3/2}),$$

$$\tilde{\mathbb{L}}_0^3 = o(h),$$

$$\tilde{\mathbb{L}}_1^1 = I + o(h),$$

$$\tilde{\mathbb{L}}_2^2 = I + o(h),$$

$$\tilde{\mathbb{L}}_3^3 = I + o(h),$$

$$\tilde{\mathbb{L}}_1^2 = \tilde{\mathbb{L}}_2^1 = \tilde{\mathbb{L}}_1^3 = \tilde{\mathbb{L}}_3^1 = \tilde{\mathbb{L}}_3^2 = \tilde{\mathbb{L}}_3^2 = 0.$$

Hence we get

$$\frac{\tilde{\mathbb{L}}_0^0 - I}{h} \xrightarrow{h \to 0} L_0^0 = -i\sigma_z - i\beta_1 I - \frac{1}{2}\beta_0\,\sigma_+\sigma_- - \frac{1}{2}\beta_1\,\sigma_-\sigma_+,$$

$$\frac{\tilde{\mathbb{L}}_1^0}{\sqrt{h}} \xrightarrow{h \to 0} L_1^0 = -i\sqrt{\beta_1}\,\sigma_+,$$

$$\frac{\tilde{\mathbb{L}}_2^0}{\sqrt{h}} \xrightarrow{h \to 0} L_2^0 = -i\sqrt{\beta_0}\,\sigma_-,$$

$$\frac{\tilde{\mathbb{L}}_0^1}{\sqrt{h}} \xrightarrow{h \to 0} L_0^1 = -i\sqrt{\beta_1}\sigma_-,$$

$$\frac{\tilde{\mathbb{L}}_0^2}{\sqrt{h}} \xrightarrow{h \to 0} L_0^2 = \sqrt{\beta_0}\,\sigma_+,$$

and the other terms converges to 0 when h tends to 0. Thus the solution $(\tilde{u}_n)_{n\in\mathbb{N}}$ of the equation

$$\begin{cases} \tilde{u}_{n+1} = \tilde{\mathbb{L}}_{n+1}\tilde{u}_n \\ \tilde{u}_0 = I \end{cases}$$

is made of invertible operators which are locally uniformly bounded in norm and in particular $\tilde{u}_{[t/h]}$ converges weakly to the solution $\tilde{U}(t)$ of the equation

$$\begin{cases} d\tilde{U}(t) = \Big\{ -\Big(i\sigma_z + i\beta_1 I + \frac{1}{2}\beta_0\,\sigma_+\sigma_- + \frac{1}{2}\beta_1\,\sigma_-\sigma_+\Big)\,dt \\ -i\sigma_-\Big(\sqrt{\beta_1}\,da_0^1(t) + \sqrt{\beta_0}da_2^0(t)\Big) - i\sigma_+\Big(\sqrt{\beta_1}\,da_1^0(t) + \sqrt{\beta_0}\,da_0^2(t)\Big) \Big\}\tilde{U}(t) \\ \tilde{U}(0) = I. \end{cases}$$

Theorem 7.3 *The quantum dynamical semigroup of the repeated quantum interaction model associated to the spin-boson system converges towards the equilibrium.*

Proof. It suffices to observe that the associated Lindbladian of the above equation has the form

$$\mathcal{L}(X) = i[\sigma_z, X] + \frac{1}{2}\beta_0\left[2\sigma_- X\sigma_+ - \{n_-, X\}\right]$$
$$+ \frac{1}{2}\beta_1\left[2\sigma_+ X\sigma_- - \{n_+, X\}\right].$$

Remark: Note that by using the repeated quantum interaction model we can prove that the Markovian properties of the spin-boson system are satisfied without using any assumption.

References

[AK00]   L. Accardi, S. Kozyrev: Quantum interacting particle systems. Volterra International School (2000).

[AFL90] L. Accardi, A. Frigerio, Y.G. Lu: Weak coupling limit as a quantum functional central limit theorem. *Com. Math. Phys.* **131**, 537-570 (1990).

[ALV02] L. Accardi, Y.G. Lu, I. Volovich: *Quantum theory and its stochastic limit.* Springer-Verlag Berlin (2002).

[AL87]   R. Alicki, K. Lendi: *Quantum dynamical semigroups and applications.* Lecture Notes in physics, **286**. Springer-Verlag Berlin (1987).

[AJ07]   S. Attal, A. Joye: The Langevin Equation for a Quantum Heat Bath. *J. Func. Analysis, 247, p. 253-288 (2007).*

[AP06]   S. Attal, Y. Pautrat: From Repeated to Continuous Quantum Interactions. *Annales Institut Henri Poincaré, (Physique Théorique) 7, p. 59-104 (2006).*

[B06]   A. Barchielli: Continual Measurements in Quantum Mechanics. *Quantum Open systems. Vol III: Recent developments. Springer Verlag, Lecture Notes in Mathematics, 1882 (2006).*

[BR96]   O. Bratteli, D.W. Robinson: *Operator algebras and Quantum Statistical Mechanics II,* Volume 2. Springer-Verlag New York Berlin Heidelberg London Paris Tokyo, second edition (1996).

[C04]   R. Carbone: Optimal Log-Sobolev Inequality and Hypercontractivity for positive semigroups on $M_2(\mathbb{C})$, *Infinite Dimensional Analysis, Quantum Probability and Related Topics, Vol. 7, No. 3 317-335 (2004).*

[D74]   E.B. Davies: Markovian Master equations. *Comm. Math. Phys.* **39**, *91-110 (1974).*

[D76a]   E.B. Davies: Markovian Master Equations II. *Math. Ann.* **219**, *147-158 (1976).*

[D80]   E.B. Davies: *One-Parameter Semigroups.* Academic Press London New York Toronto Sydney San Francisco (1980).

[D76b]   E.B. Davies: *Quantum Theory of Open Systems.* Academic Press, New York and London (1976).

[DJ03]   J. Derezinski, V. Jaksic: Return to Equilibrium for Pauli-Fierz Systems. *Annales Institut Henri Poincaré 4, 739-793 (2003).*

[DJP03] J. Derezinski, V. Jaksic, C.A. Pillet: Perturbation theory of W*-dynamics, KMS-states and Liouvillean, *Rev. Math. Phys.* **15**, *447-489 (2003).*

[DF06]   J. Derezinski, R. Fruboes: Fermi Golden Rule and Open Quantum Systems, *Quantum Open systems. Vol III: Recent developments. Springer Verlag, Lecture Notes in Mathematics, 1882 (2006).*

[F06]   F. Fagnola: Quantum Stochastic Differential Equations and Dilation of Completely Positive Semigroups. *Quantum Open systems. Vol II: The Markovian approach. Springer Verlag, Lecture Notes in Mathematics, 1881 (2006).*

[F99]   F. Fagnola: *Quantum Markovian Semigroups and Quantum Flows.* Proyecciones, Journal of Math. **18**, n.3 1-144 (1999).

[F93]   F. Fagnola: Characterization of Isometric and Unitary Weakly Differentiable Cocycles in Fock space. *Quantum Probability and Related Topics VIII 143 (1993).*

[FR06]   F. Fagnola, R. Rebolledo: Nets of the Qualitative behaviour of Quantum Markov Semigroups. *Quantum Open systems. Vol III: Recent developments. Springer Verlag, Lecture Notes in Mathematics, 1882 (2006).*

[FR98]   F. Fagnola, R. Rebolledo: The Approach to equilibrium of a class of quantum dynamical semigroups. *Inf. Q. Prob. and Rel. Topics, 1(4), 1-12 (1998).*

[HP84]   R.L Hudson, K.R. Parthasarathy: Quantum Ito's formula and stochastic evolutions, *Comm. Math. Phys. 93, no 3, pp.301-323 (1984).*

[G01]   M. Gregoratti: The Hamiltonian Operator Associated with Some quantum Stochastic Evolutions *Com. Math. Phys. 222, 181-200 (2001)*

[JP96a]   V. Jaksic, C.A. Pillet: On a model for quantum friction II : Fermi's golden rule and dynamics at positive temperature. *Comm. Math. Phys. 178, 627 (1996).*

[JP96b]   V. Jaksic, C.A. Pillet: On a model for quantum friction III: Ergodic properties of the spin-boson system. *Comm. Math. Phys. 178, 627 (1996).*

[M95]   P. A. Meyer: *Quantum Probability for Probabilists.* Second edition. Lect Not. Math. **1538**, Berlin: Springer-Verlag (1995).

[P92]   K. R. Parthasarathy: *An Introduction to Quantum Stochastic Calculus.* Birkhäuser Verlag: Basel. Boston. Berlin (1992).

[R06]   R. Rebolledo: Complete Positivity and Open Quantum Systems. *Quantum Open systems. Vol II: The Markovian approach. Springer Verlag, Lecture Notes in Mathematics, 1881 (2006).*

Statistical Properties of Pauli Matrices Going Through Noisy Channels

Stéphane Attal and Nadine Guillotin-Plantard

Université Lyon 1, Institut Camille Jordan,
43 bld du 11 novembre 1918, 69622 Villeurbanne Cedex, France
e-mail: nadine.guillotin@univ-lyon1.fr; attal@math.univ-lyon1.fr

Summary. We study the statistical properties of the triple $(\sigma_x, \sigma_y, \sigma_z)$ of Pauli matrices going through a sequence of noisy channels, modeled by the repetition of a general, trace-preserving, completely positive map. We show a non-commutative central limit theorem for the distribution of this triple, which features in the limit a 3-dimensional Brownian motion with a non-trivial covariance matrix. We also prove a large deviation principle associated to this convergence, with an explicit rate function depending on the stationary state of the noisy channel.

1 Introduction

In quantum information theory one of the most important question is to understand and to control the way a quantum bit is modified when transmitted through a quantum channel. It is well-known that realistic transmission channels are not perfect and distort the quantum bit they transmit. This transformation of the quantum state is represented by the action of a completely positive map. These are the so-called noisy channels.

The purpose of this article is to study the action of the repetition of a general completely positive map on basic observables. Physically, this model can be thought of as the sequence of transformations of small identical pieces of noisy channels on a qubit. It can also be thought of as a discrete approximation of the more realistic model of a quantum bit going through a semigroup of completely positive maps (a Lindblad semigroup).

As basic observables, we consider the triple $(\sigma_x, \sigma_y, \sigma_z)$ of Pauli matrices. Under the repeated action of the completely positive map, they behave as a 3-dimensional quantum random walk. The aim of this article is to study the statistical properties of this quantum random walk.

Indeed, for any initial density matrix ρ_{in}, we study the statistical properties of the empirical average of the Pauli matrices in the successive states $\Phi^n(\rho_{in}), n \geq 0$ where Φ is some completely positive and trace-preserving map describing our quantum channel. Quantum Bernoulli random walks studied

C. Donati-Martin et al. (eds.), *Séminaire de Probabilités XLII*,
Lecture Notes in Mathematics 1979, DOI 10.1007/978-3-642-01763-6_16,
© Springer-Verlag Berlin Heidelberg 2009

by Biane in [1] corresponds to the case where Φ is the identity map. Biane [1] proved an invariance principle for this quantum random walk when $\rho_{in} = \frac{1}{2}I$.

This article is organized as follows. In section two we describe the physical and mathematical setup. In section three we establish a functional central limit theorem for the empirical average of the quantum random walk associated to the Pauli matrices generalizing Biane's result [1]. This central limit theorem involves a 3-dimensional Brownian motion in the limit, whose covariance matrix is non-trivial and depends explicitly on the stationary state of the noisy channel. In section four, we apply our central limit theorem to some explicit cases, in particular to the King-Ruskai-Szarek-Werner representation of completely positive and trace-preserving maps in $M_2(\mathbb{C})$. This allows us to compute the limit Brownian motion for the best known quantum channels: the depolarizing channel, the phase-damping channel, the amplitude-damping channel. Finally, in the last section, a large deviation principle for the empirical average is proved.

2 Model and Notations

Let $M_2(\mathbb{C})$ be the set of 2×2 matrices with complex coefficients. The set of 2×2 self-adjoint matrices forms a four dimensional real vector subspace of $M_2(\mathbb{C})$. A convenient basis $\mathcal{B}$ is given by the following matrices

$$I = \begin{pmatrix} 1 & 0 \\ 0 & 1 \end{pmatrix}, \quad \sigma_x = \begin{pmatrix} 0 & 1 \\ 1 & 0 \end{pmatrix}, \quad \sigma_y = \begin{pmatrix} 0 & -i \\ i & 0 \end{pmatrix}, \quad \sigma_z = \begin{pmatrix} 1 & 0 \\ 0 & -1 \end{pmatrix}$$

where $\sigma_x, \sigma_y, \sigma_z$ are the traditional Pauli matrices, they satisfy the commutation relations: $[\sigma_x, \sigma_y] = 2i\sigma_z$, and those obtained by cyclic permutations of σ_x, σ_y, σ_z. A state on $M_2(\mathbb{C})$ is given by a density matrix (i.e. a positive semi-definite matrix with trace one) which we will suppose to be of the form

$$\rho = \begin{pmatrix} \alpha & \beta \\ \bar{\beta} & 1 - \alpha \end{pmatrix}$$

where $0 \leq \alpha \leq 1$ and $|\beta|^2 \leq \alpha(1 - \alpha)$. The noise coming from interactions between the qubit states and the environment is represented by the action of a completely positive and trace-preserving map $\Phi : M_2(\mathbb{C}) \to M_2(\mathbb{C})$.

Let $M_1, M_2, \ldots, M_k, \ldots$ be infinitely many copies of $M_2(\mathbb{C})$. For each given state ρ, we consider the algebra

$$\mathcal{M}_\rho = M_1 \otimes M_2 \otimes \ldots \otimes M_k \otimes \ldots$$

where the product is taken in the sense of W^*-algebra with respect to the product state

$$\omega = \rho \otimes \Phi(\rho) \otimes \Phi^2(\rho) \otimes \ldots \otimes \Phi^k(\rho) \otimes \ldots .$$

Our main hypothesis is the following. We assume that for any state ρ, the sequence $\Phi^n(\rho)$ converges to a stationary state ρ_∞, which we write as

$$\rho_\infty = \begin{pmatrix} \alpha_\infty & \beta_\infty \\ \overline{\beta}_\infty & 1 - \alpha_\infty \end{pmatrix}$$

where $0 \le \alpha_\infty \le 1$ and $|\beta_\infty|^2 \le \alpha_\infty(1 - \alpha_\infty)$.

Put

$$v_1 = 2\operatorname{Re}(\beta_\infty),\, v_2 = -2\operatorname{Im}(\beta_\infty),\, v_3 = 2\alpha_\infty - 1.$$

For every $k \ge 1$, we define

$$x_k = I \otimes \ldots \otimes I \otimes (\sigma_x - v_1\, I) \otimes I \otimes \ldots$$
$$y_k = I \otimes \ldots \otimes I \otimes (\sigma_y - v_2\, I) \otimes I \otimes \ldots$$
$$z_k = I \otimes \ldots \otimes I \otimes (\sigma_z - v_3\, I) \otimes I \otimes \ldots$$

where each $(\sigma_. - v.\, I)$ appears on the kth place. For every $n \ge 1$, put

$$X_n = \sum_{k=1}^{n} x_k,\ Y_n = \sum_{k=1}^{n} y_k,\ Z_n = \sum_{k=1}^{n} z_k$$

with initial conditions

$$X_0 = Y_0 = Z_0 = 0.$$

The integer part of a real t is denoted by $[t]$. To each process we associate a continuous time normalized process denoted by

$$X_t^{(n)} = n^{-1/2}X_{[nt]},\ Y_t^{(n)} = n^{-1/2}Y_{[nt]},\ Z_t^{(n)} = n^{-1/2}Z_{[nt]}.$$

3 A Central Limit Theorem

The aim of our article is to study the asymptotical properties of the quantum process $(X_t^{(n)}, Y_t^{(n)}, Z_t^{(n)})$ when n goes to infinity. This process being truly non-commutative, there is no hope to obtain an asymptotic behaviour in the classical sense.

For any polynomial $P = P(X_1, X_2, \ldots, X_m)$ of m variables, we denote by $\widehat{P}$ the *totally symmetrized polynomial* of P obtained by symmetrizing each monomial in the following way:

$$X_{i_1} X_{i_2} \ldots X_{i_k} \longrightarrow \frac{1}{k!} \sum_{\sigma \in S_k} X_{i_{\sigma(1)}} \ldots X_{i_{\sigma(k)}}$$

where S_k is the group of permutations of $\{1, \ldots, k\}$.

Theorem 1. *Assume that*

$$(\mathbf{A}) \qquad \Phi^n(\rho) = \rho_\infty + o\left(\frac{1}{\sqrt{n}}\right). \qquad (1)$$

Then, for any polynomial P of $3m$ variables, for any $(t_1,\ldots,t_m)$ such that $0 \le t_1 < t_2 < \ldots < t_m$, the following convergence holds:

$$\lim_{n \to +\infty} w\left[\widehat{P}(X_{t_1}^{(n)}, Y_{t_1}^{(n)}, Z_{t_1}^{(n)}, \ldots, X_{t_m}^{(n)}, Y_{t_m}^{(n)}, Z_{t_m}^{(n)})\right]$$
$$= \mathbb{E}\left[P(B_{t_1}^{(1)}, B_{t_1}^{(2)}, B_{t_1}^{(3)}, \ldots, B_{t_m}^{(1)}, B_{t_m}^{(2)}, B_{t_m}^{(3)})\right]$$

where $(B_t^{(1)}, B_t^{(2)}, B_t^{(3)})_{t \ge 0}$ is a three-dimensional centered Brownian motion with covariance matrix Ct, with

$$C = \begin{pmatrix} 1 - v_1^2 & -v_1 v_2 & -v_1 v_3 \\ -v_1 v_2 & 1 - v_2^2 & -v_2 v_3 \\ -v_1 v_3 & -v_2 v_3 & 1 - v_3^2 \end{pmatrix}.$$

Remark : Theorem 1 has to be compared with the quantum central limit theorem obtained in [5] and [9]. In our case, the state under which the convergence holds does not need to be an infinite tensor product of states. We also give here a functional version of the central limit theorem. Finally, in [5] (see Remark 3 p.131), the limit is described as a so-called quasi-free state in quantum mechanics. We prove in Theorem 1 that the limit is real Gaussian for the class of totally symmetrized polynomials.

Proof. Let $m \ge 1$ and $(t_0, t_1, \ldots, t_m)$ such that $t_0 = 0 < t_1 < t_2 < \ldots < t_m$. The polynomial $P(X_{t_1}^{(n)}, Y_{t_1}^{(n)}, Z_{t_1}^{(n)}, \ldots, X_{t_m}^{(n)}, Y_{t_m}^{(n)}, Z_{t_m}^{(n)})$ can be rewritten as a polynomial function Q of the increments: $X_{t_1}^{(n)}$, $Y_{t_1}^{(n)}$, $Z_{t_1}^{(n)}$, $X_{t_2}^{(n)} - X_{t_1}^{(n)}$, $Y_{t_2}^{(n)} - Y_{t_1}^{(n)}$, $Z_{t_2}^{(n)} - Z_{t_1}^{(n)}, \ldots, X_{t_m}^{(n)} - X_{t_{m-1}}^{(n)}$, $Y_{t_m}^{(n)} - Y_{t_{m-1}}^{(n)}$, $Z_{t_m}^{(n)} - Z_{t_{m-1}}^{(n)}$. A monomial of Q is a product of the form $Q_{i_1} \ldots Q_{i_k}$ for some distinct $i_1, \ldots, i_k$ in $\{1, \ldots, m\}$ where Q_i is a product depending only on the increments $X_{t_i}^{(n)} - X_{t_{i-1}}^{(n)}$, $Y_{t_i}^{(n)} - Y_{t_{i-1}}^{(n)}$, $Z_{t_i}^{(n)} - Z_{t_{i-1}}^{(n)}$'s. Since the Q_i's are commuting variables, the totally symmetrized polynomial of the monomial $Q_{i_1} \ldots Q_{i_k}$ is equal to the product $\widehat{Q_{i_1}} \ldots \widehat{Q_{i_k}}$. Remark that since one considers product states, the increments are independent, thus the expectations factorize, which allows to reduce to prove the theorem for any polynomial Q_i.

Let $i \ge 1$ fixed, for every $v_1, v_2, v_3 \in \mathbb{R}$, we begin by determining the asymptotic distribution of the linear combination

$$(v_1^2 + v_2^2 + v_3^2)^{-1/2}\left(v_1(X_{t_i}^{(n)} - X_{t_{i-1}}^{(n)}) + v_2(Y_{t_i}^{(n)} - Y_{t_{i-1}}^{(n)}) + v_3(Z_{t_i}^{(n)} - Z_{t_{i-1}}^{(n)})\right) \quad (2)$$

which can be rewritten as

$$\frac{1}{\sqrt{n}} \sum_{k=[nt_{i-1}]+1}^{[nt_i]} \left(\frac{v_1 x_k + v_2 y_k + v_3 z_k}{\sqrt{v_1^2 + v_2^2 + v_3^2}}\right).$$

Consider the matrix

$$A = \frac{1}{\sqrt{\nu_1^2 + \nu_2^2 + \nu_3^2}} \Big(\nu_1(\sigma_x - v_1 I) + \nu_2(\sigma_y - v_2 I) + \nu_3(\sigma_z - v_3 I) \Big)$$

$$= \frac{1}{\sqrt{\nu_1^2 + \nu_2^2 + \nu_3^2}} \begin{pmatrix} -\nu_1 v_1 - \nu_2 v_2 + \nu_3(1 - v_3) & \nu_1 - i\nu_2 \\ \nu_1 + i\,\nu_2 & -\nu_1 v_1 - \nu_2 v_2 - \nu_3(1 + v_3) \end{pmatrix}$$

which we denote by

$$\begin{pmatrix} a_1 & a_3 \\ \bar{a}_3 & a_2 \end{pmatrix},$$

with $a_1, a_2 \in \mathbb{R}, a_3 \in \mathbb{C}$.

From assumption (**A**) we can write, for every $n \geq 0$

$$\Phi^n(\rho) = \begin{pmatrix} \alpha_\infty + \phi_n(1) & \beta_\infty + \phi_n(2) \\ \bar{\beta}_\infty + \phi_n(3) & 1 - \alpha_\infty + \phi_n(4) \end{pmatrix}$$

where each sequence $(\phi_n(i))_n$ satisfies: $\phi_n(i) = o(1/\sqrt{n})$.

Let $k \geq 1$, the expectation and the variance of A in the state $\Phi^k(\rho)$ are respectively equal to

$$\text{Trace}(A\Phi^k(\rho))$$

and

$$\text{Trace}(A^2\Phi^k(\rho)) - \text{Trace}(A\Phi^k(\rho))^2.$$

If both following conditions are satisfied:

$$\sum_{k=[nt_{i-1}]+1}^{[nt_i]} \text{Trace}(A\Phi^k(\rho)) = o(\sqrt{n}) \tag{3}$$

and

$$\lim_{n \to +\infty} \frac{1}{n} \sum_{k=[nt_{i-1}]+1}^{[nt_i]} [\text{Trace}(A^2\Phi^k(\rho)) - \text{Trace}(A\Phi^k(\rho))^2] = a(t_i - t_{i-1}), \tag{4}$$

then (see Theorem 2.8.42 in [3]) the asymptotic distribution of (2) is the Normal distribution $\mathcal{N}(0, a(t_i - t_{i-1})), a > 0$.

Let us first prove (3). For every $k \geq 1$, a simple computation gives

$$\text{Trace}(\Lambda\Phi^k(\rho)) - [u_1 u_\infty + a_3 \bar{\beta}_\infty + \bar{a}_3 \beta_\infty + a_2(1 - \alpha_\infty) + o(1/\sqrt{n})] = o(1/\sqrt{n}),$$

hence

$$\sum_{k=[nt_{i-1}]+1}^{[nt_i]} \text{Trace}(A\Phi^k(\rho)) = \sum_{k=[nt_{i-1}]+1}^{[nt_i]} o(1/\sqrt{n}) = o(\sqrt{n}).$$

This gives (3).

Let us prove (4). Note that the sequence $(\text{Trace}(A\Phi^n(\rho)))_n$ converges to 0 as n tends to infinity. As a consequence, it is enough to prove that

$$\frac{1}{n} \sum_{k=[nt_{i-1}]+1}^{[nt_i]} \text{Trace}(A^2\Phi^k(\rho))$$

converges to a strictly positive constant. A straightforward computation gives

$$\lim_{n \to +\infty} \frac{1}{n} \sum_{k=[nt_{i-1}]+1}^{[nt_i]} \text{Trace}(A^2\Phi^k(\rho))$$

$$= a_1^2 \alpha_\infty + a_2^2 (1 - \alpha_\infty) + |a_3|^2 + (a_1 + a_2)(a_3 \bar{\beta}_\infty + \bar{a}_3 \beta_\infty)$$

$$= \frac{(t_i - t_{i-1})}{\nu_1^2 + \nu_2^2 + \nu_3^2} \Big[\nu_1^2(1 - \nu_1^2) + \nu_2^2(1 - \nu_2^2) + \nu_3^2(1 - \nu_3^2)$$

$$- 2\nu_1\nu_2 v_1 v_2 - 2\nu_1\nu_3 v_1 v_3 - 2\nu_2\nu_3 v_2 v_3 \Big].$$

This means that, for every $\nu_1, \nu_2, \nu_3 \in \mathbb{R}$, for any $p \geq 1$, the expectation

$$w\Big[\Big(\nu_1(X_{t_i}^{(n)} - X_{t_{i-1}}^{(n)}) + \nu_2(Y_{t_i}^{(n)} - Y_{t_{i-1}}^{(n)}) + \nu_3(Z_{t_i}^{(n)} - Z_{t_{i-1}}^{(n)})\Big)^p\Big]$$

converges to

$$\mathbb{E}\Big[\Big(\nu_1(B_{t_i}^{(1)} - B_{t_{i-1}}^{(1)}) + \nu_2(B_{t_i}^{(2)} - B_{t_{i-1}}^{(2)}) + \nu_3(B_{t_i}^{(3)} - B_{t_{i-1}}^{(3)})\Big)^p\Big],$$

where $(B_t^{(1)}, B_t^{(2)}, B_t^{(3)})$ is a 3-dimensional Brownian motion with the announced covariance matrix.

The polynomial

$$\Big(\nu_1(X_{t_i}^{(n)} - X_{t_{i-1}}^{(n)}) + \nu_2(Y_{t_i}^{(n)} - Y_{t_{i-1}}^{(n)}) + \nu_3(Z_{t_i}^{(n)} - Z_{t_{i-1}}^{(n)})\Big)^p$$

can be expanded as the sum

$$\sum_{0 \leq p_1 + p_2 \leq p} \nu_1^{p_1} \nu_2^{p_2} \nu_3^{p-p_1-p_2} \sum_{\mathcal{P}} S_1 S_2 \ldots S_p$$

where the summation in the last sum runs over all partitions $\mathcal{P} = \{A, B, C\}$ of $\{1, \ldots, p\}$ such that $|A| = p_1, |B| = p_2, |C| = p - p_1 - p_2$, with the convention:

$$S_j = \begin{cases} X_{t_i}^{(n)} - X_{t_{i-1}}^{(n)} & \text{if } j \in A \\ Y_{t_i}^{(n)} - Y_{t_{i-1}}^{(n)} & \text{if } j \in B \\ Z_{t_i}^{(n)} - Z_{t_{i-1}}^{(n)} & \text{if } j \in C. \end{cases}$$

The expectation under w of the above expression converges to the corresponding expression involving the expectation $(\mathbb{E}[\cdot])$ of the Brownian motion $(B_t^{(1)}, B_t^{(2)}, B_t^{(3)})$. As this holds for any $\nu_1, \nu_2, \nu_3 \in \mathbb{R}$, we deduce that $w[\sum_{\mathcal{P}} S_1 S_2 \ldots S_p]$ converges to the corresponding expectation for the Brownian motion.

We can end the proof by noticing that $\widehat{A_i}$ can be written, modulo multiplication by a constant, as $\sum_{\mathcal{P}} S_1 S_2 \ldots S_p$ for some p. $\square$

Let us discuss the class of polynomials for which Theorem 3.1 holds. In the particular case when the map Φ is the identity map and $\rho = 1/2I$ (in that case $v_i = 0$ for $i = 1, 2, 3$ and $C = I$), Biane [1] proved the convergence of the expectations in Theorem 1 for any polynomial in $3m$ non-commuting variables. It is a natural question to ask whether our result holds for any polynomial P instead of $\widehat{P}$, or at least for a larger class.

Let us give an example of a polynomial for which the convergence in our setting does not hold. Take $P(X, Y) = XY$. From Theorem 1, the expectation under the state ω of

$$X_t^{(n)} Y_t^{(n)} + Y_t^{(n)} X_t^{(n)}$$

converges as $n \to +\infty$ to $2\, \mathbb{E}[B_t^{(1)} B_t^{(2)}]$.

Since we have the following commutation relations

$$[(\sigma_x - v_1\, I), (\sigma_y - v_2\, I)] = 2i\sigma_z, \quad [(\sigma_y - v_2\, I), (\sigma_z - v_3\, I)] = 2i\sigma_x \quad (5)$$

and

$$[(\sigma_z - v_3\, I), (\sigma_x - v_1\, I)] = 2i\sigma_y,$$

we deduce that

$$[X_t^{(n)}, Y_t^{(n)}] = 2in^{-1/2}\, Z_t^{(n)} + 2itv_3\, I, \quad [Y_t^{(n)}, Z_t^{(n)}] = 2in^{-1/2}\, X_t^{(n)} + 2itv_1\, I$$

and

$$[Z_t^{(n)}, X_t^{(n)}] = 2in^{-1/2}\, Y_t^{(n)} + 2itv_2\, I. \quad (6)$$

Then the expectation under the state ω of

$$X_t^{(n)} Y_t^{(n)} = \frac{1}{2}\Big[\widehat{P}(X, Y) + [X_t^{(n)}, Y_t^{(n)}]\Big]$$

converges to $\mathbb{E}[B_t^{(1)} B_t^{(2)}] + itv_3 \neq \mathbb{E}[B_t^{(1)} B_t^{(2)}]$, if v_3 is non zero.

Furthermore, by considering the polynomial $P(X, Y) = XY^3 + Y^3 X$, it is possible to show that the convergence in Theorem 1 can not be enlarged to the class of symmetric polynomials. A straightforward computation shows that $P(X, Y)$ can be rewritten as

$$\widehat{XY^3} + \widehat{YX^3} + \frac{3}{4}[X, Y](Y^2 - X^2) + \frac{1}{2}(Y[X, Y]Y - X[X, Y]X) + \frac{1}{4}(Y^2 - X^2)[X, Y]$$

so the expectation $w[P(X_t^{(n)}, Y_t^{(n)})]$ converges as n tends to $+\infty$ to

$$\mathbb{E}[P(B_t^{(1)}, B_t^{(2)})] + 3iv_3 t(v_1^2 - v_2^2)$$

which is not equal to $\mathbb{E}[P(B_t^{(1)}, B_t^{(2)})]$ if $v_3 \neq 0$ and $|v_1| \neq |v_2|$.

In the following corollary we give a condition under which the convergence in Theorem 1 holds for any polynomial in $3m$ non-commuting variables.

Corollary 1. *In the case when ρ_∞ is equal to $\frac{1}{2}I$, the convergence holds for any polynomial P in $3m$ non-commuting variables, i.e. for every $t_1 < t_2 < \ldots < t_m$, the following convergence holds:*

$$\lim_{n \to +\infty} w\left[P(X_{t_1}^{(n)}, Y_{t_1}^{(n)}, Z_{t_1}^{(n)}, \ldots, X_{t_m}^{(n)}, Y_{t_m}^{(n)}, Z_{t_m}^{(n)})\right]$$
$$= \mathbb{E}\left[P(B_{t_1}^{(1)}, B_{t_1}^{(2)}, B_{t_1}^{(3)}, \ldots, B_{t_m}^{(1)}, B_{t_m}^{(2)}, B_{t_m}^{(3)})\right]$$

where $(B_t^{(1)}, B_t^{(2)}, B_t^{(3)})_{t \geq 0}$ is a three-dimensional centered Brownian motion with covariance matrix tI_3.

Proof. We consider the polynomials of the form $S = \frac{1}{N} \sum_{\mathcal{P}} S_1 S_2 \ldots S_{p_1+p_2+p_3}$ where the summation is done over all partitions $\mathcal{P} = \{A, B, C\}$ of the set $\{1, \ldots, p_1+p_2+p_3\}$ such that $|A| = p_1, |B| = p_2, |C| = p_3$, with the convention:

$$S_j = \begin{cases} X_{t_i}^{(n)} - X_{t_{i-1}}^{(n)} & \text{if } j \in A \\ Y_{t_i}^{(n)} - Y_{t_{i-1}}^{(n)} & \text{if } j \in B \\ Z_{t_i}^{(n)} - Z_{t_{i-1}}^{(n)} & \text{if } j \in C \end{cases}$$

and N is the number of terms in the sum.

From Theorem 1 the expectation under the state w of S converges to

$$\mathbb{E}\left[\prod_{j=1}^{3}(B_{t_i}^{(j)} - B_{t_{i-1}}^{(j)})^{p_j}\right].$$

Using the commutation relations (6) with all the v_i's being equal to zero, monomials of S differ from each other by $n^{-1/2}$ times a polynomial of total degree less than or equal to $(p_1 + p_2 + p_3) - 1$. It is easy to conclude by induction. $\square$

4 Examples

4.1 King-Ruskai-Szarek-Werner's Representation

The set of 2×2 self-adjoint matrices forms a four dimensional real vector subspace of $M_2(\mathbb{C})$. A convenient basis of this space is given by $\mathcal{B} = \{I, \sigma_x, \sigma_y, \sigma_z\}$. Each state ρ on $M_2(\mathbb{C})$ can then be written as

$$\rho = \frac{1}{2}\begin{pmatrix} 1+z & x-iy \\ x+iy & 1-z \end{pmatrix}$$

where x, y, z are reals such that $x^2 + y^2 + z^2 \leq 1$. Equivalently, in the basis $\mathcal{B}$,

$$\rho = \frac{1}{2}(I + x\,\sigma_x + y\,\sigma_y + z\,\sigma_z)$$

with x, y, z defined above. Thus, the set of density matrices can be identified with the unit ball in $\mathbb{R}^3$. The pure states, that is, the ones for which $x^2 + y^2 + z^2 = 1$, constitute the Bloch sphere.

The noise coming from interactions between the qubit states and the environment is represented by the action of a completely positive and trace-preserving map $\Phi : M_2(\mathbb{C}) \to M_2(\mathbb{C})$. Kraus and Choi [2, 7, 8] gave an abstract representation of these particular maps in terms of Kraus operators: There exists at most four matrices L_i such that for any density matrix ρ,

$$\Phi(\rho) = \sum_{1 \leq i \leq 4} L_i^* \rho L_i$$

with $\sum_i L_i L_i^* = I$. The matrices L_i are usually called the *Kraus operators* of Φ. This representation is unique up to a unitary transformation. Recently, King, Ruskai et al. [10, 6] obtained a precise characterization of completely positive and trace-preserving maps from $M_2(\mathbb{C})$ as follows. The map $\Phi : M_2(\mathbb{C}) \to M_2(\mathbb{C})$ being linear and preserving the trace, it can be represented as a unique 4×4-matrix in the basis $\mathcal{B}$ given by

$$\begin{pmatrix} 1 & \mathbf{0} \\ \mathbf{t} & \mathbf{T} \end{pmatrix}$$

with $\mathbf{0} = (0,0,0)$, $\mathbf{t} \in \mathbb{R}^3$ and $\mathbf{T}$ a real 3×3-matrix. King, Ruskai et al [10, 6] proved that via changes of basis, this matrix can be reduced to

$$T = \begin{pmatrix} 1 & 0 & 0 & 0 \\ t_1 & \lambda_1 & 0 & 0 \\ t_2 & 0 & \lambda_2 & 0 \\ t_3 & 0 & 0 & \lambda_3 \end{pmatrix} \tag{7}$$

Necessary and sufficient conditions under which the map Φ with reduced matrix T for which $|t_3| + |\lambda_3| \leq 1$ is completely positive are (see [6])

$$(\lambda_1 + \lambda_2)^2 \leq (1 + \lambda_3)^2 - t_3^2 - (t_1^2 + t_2^2) \left(\frac{1 + \lambda_3 \pm t_3}{1 - \lambda_3 \pm t_3} \right) \leq (1 + \lambda_3)^2 - t_3^2 \tag{8}$$

$$(\lambda_1 - \lambda_2)^2 \leq (1 - \lambda_3)^2 - t_3^2 - (t_1^2 + t_2^2) \left(\frac{1 - \lambda_3 \pm t_3}{1 + \lambda_3 \pm t_3} \right) \leq (1 - \lambda_3)^2 - t_3^2 \tag{9}$$

$$\left[1 - (\lambda_1^2 + \lambda_2^2 + \lambda_3^2) - (t_1^2 + t_2^2 + t_3^2) \right]^2$$
$$\geq 4 \left[\lambda_1^2(t_1^2 + \lambda_2^2) + \lambda_2^2(t_2^2 + \lambda_3^2) + \lambda_3^2(t_3^2 + \lambda_1^2) - 2\lambda_1\lambda_2\lambda_3 \right] . \tag{10}$$

We now apply Theorem 1 in this setting. Let Φ be a completely positive and trace preserving map with matrix T given in (7), with coefficients $t_i, \lambda_i, i = 1, 2, 3$ satisfying conditions (8), (9) and (10). Moreover, we assume that $|\lambda_i| < 1, i = 1, 2, 3$. For every $n \geq 0$,

$$\Phi^n(\rho) = \frac{1}{2}\begin{pmatrix} 1+\phi_n(3) & \phi_n(1)-i\,\phi_n(2) \\ \phi_n(1)+i\,\phi_n(2) & 1-\phi_n(3) \end{pmatrix}$$

where the sequences $(\phi_n(i))_{n\geq 0}$, $i=1,2,3$ satisfy the induction relations:

$$\phi_n(i) = \lambda_i\phi_{n-1}(i) + t_i.$$

with initial conditions $\phi_0(1)=x$, $\phi_0(2)=y$ and $\phi_0(3)=z$. Explicit formulae can easily be obtained. We get, for every $n \geq 0$,

$$\phi_n(1) = \left(x-\frac{t_1}{1-\lambda_1}\right)\lambda_1^n + \frac{t_1}{1-\lambda_1}$$

$$\phi_n(2) = \left(y-\frac{t_2}{1-\lambda_2}\right)\lambda_2^n + \frac{t_2}{1-\lambda_2}$$

$$\phi_n(3) = \left(z-\frac{t_3}{1-\lambda_3}\right)\lambda_3^n + \frac{t_3}{1-\lambda_3}.$$

Hence, for any state ρ, for any $n \geq 1$,

$$\Phi_n(\rho) = \rho_\infty + o(|\lambda|_{max}^n)$$

where $|\lambda|_{max} = \max_{i=1,2,3} |\lambda_i|$ and

$$\rho_\infty = \begin{pmatrix} \alpha_\infty & \beta_\infty \\ \bar{\beta}_\infty & 1-\alpha_\infty \end{pmatrix}$$

with $\alpha_\infty = \frac{1}{2}\left(1+\frac{t_3}{1-\lambda_3}\right)$ and $\beta_\infty = \frac{1}{2}\left(\frac{t_1}{1-\lambda_1}-i\,\frac{t_2}{1-\lambda_2}\right)$. Theorem 1 applies with $v_i = \frac{t_i}{1-\lambda_i}$, $i=1,2,3$.

We now give some examples of well-known quantum channels. For each of them we give their Kraus operators, their corresponding matrix T in the King-Ruskai-Szarek-Werner's representation, as well as the vector $v = (v_1, v_2, v_3)$ and the covariance matrix C obtained in Theorem 1. It is worth noticing that if Φ is a *unital* map, i.e. such that $\Phi(I) = I$, then the covariance matrix C is equal to the identity matrix I_3.

1. *The depolarizing channel:*
   Kraus operators: for some $0 \leq p \leq 1$,

$$L_1 = \sqrt{1-p}I, L_2 = \sqrt{\frac{p}{3}}\sigma_x, L_3 = \sqrt{\frac{p}{3}}\sigma_y, L_4 = \sqrt{\frac{p}{3}}\sigma_z.$$

King-Ruskai-Szarek-Werner's representation:

$$T = \begin{pmatrix} 1 & 0 & 0 & 0 \\ 0 & 1-\frac{4p}{3} & 0 & 0 \\ 0 & 0 & 1-\frac{4p}{3} & 0 \\ 0 & 0 & 0 & 1-\frac{4p}{3} \end{pmatrix}$$

The vector v is the null vector and the covariance matrix C in this case is given by the identity matrix I_3.

2. *Phase-damping channel:*
Kraus operators: for some $0 \leq p \leq 1$,

$$L_1 = \sqrt{1-p}\, I, \quad L_2 = \sqrt{p} \begin{pmatrix} 1 & 0 \\ 0 & 0 \end{pmatrix}, L_3 = \sqrt{p} \begin{pmatrix} 0 & 0 \\ 0 & 1 \end{pmatrix}$$

King-Ruskai-Szarek-Werner's representation:

$$T = \begin{pmatrix} 1 & 0 & 0 & 0 \\ 0 & 1-p & 0 & 0 \\ 0 & 0 & 1-p & 0 \\ 0 & 0 & 0 & 1 \end{pmatrix}$$

The vector v is the null vector and the covariance matrix C in this case is given by I_3.

3. *Amplitude-damping channel:*
Kraus operators: for some $0 \leq p \leq 1$,

$$L_1 = \begin{pmatrix} 1 & 0 \\ 0 & \sqrt{1-p} \end{pmatrix}, L_2 = \begin{pmatrix} 0 & \sqrt{p} \\ 0 & 0 \end{pmatrix}$$

King-Ruskai-Szarek-Werner's representation:

$$T = \begin{pmatrix} 1 & 0 & 0 & 0 \\ 0 & \sqrt{1-p} & 0 & 0 \\ 0 & 0 & \sqrt{1-p} & 0 \\ t & 0 & 0 & 1-p \end{pmatrix}$$

The vector v is equal to $(0,0,1)$. The covariance matrix in this case is given by

$$C = \begin{pmatrix} 1 & 0 & 0 \\ 0 & 1 & 0 \\ 0 & 0 & 0 \end{pmatrix}$$

4. *Trigonometric parameterization:*
Consider the particular Kraus operators

$$L_1 = \left[\cos(\frac{v}{2}) \cos(\frac{u}{2}) \right] I + \left[\sin(\frac{v}{2}) \sin(\frac{u}{2}) \right] \sigma_z$$

and

$$L_2 = \left[\sin(\frac{v}{2}) \cos(\frac{u}{2}) \right] \sigma_x - i \left[\cos(\frac{v}{2}) \sin(\frac{u}{2}) \right] \sigma_y.$$

King-Ruskai-Szarek-Werner's representation:

$$T = \begin{pmatrix} 1 & 0 & 0 & 0 \\ 0 & \cos u & 0 & 0 \\ 0 & 0 & \cos v & 0 \\ \sin u \sin v & 0 & 0 & \cos u \cos v \end{pmatrix}$$

The vector v is equal to $(0, 0, \dfrac{\sin u \sin v}{1 - \cos u \cos v})$. The covariance matrix in this case is given by

$$C = \begin{pmatrix} 1 & 0 & 0 \\ 0 & 1 & 0 \\ 0 & 0 & 1 - v_3^2 \end{pmatrix}$$

with $v_3 = \dfrac{\sin u \sin v}{1 - \cos u \cos v}$.

4.2 CP Map Associated to a Markov Chain

With every Markov chain with two states and transition matrix given by

$$P = \begin{pmatrix} p & 1-p \\ q & 1-q \end{pmatrix}, \quad p, q \in (0, 1)$$

is associated a completely positive and trace preserving map, denoted by Φ, with the Kraus operators:

$$L_1 = \begin{pmatrix} \sqrt{p} & \sqrt{1-p} \\ 0 & 0 \end{pmatrix} = \frac{\sqrt{p}}{2}(I + \sigma_z) + \frac{\sqrt{1-p}}{2}(\sigma_x + i\sigma_y)$$

and

$$L_2 = \begin{pmatrix} 0 & 0 \\ \sqrt{q} & \sqrt{1-q} \end{pmatrix} = \frac{\sqrt{1-q}}{2}(I - \sigma_z) + \frac{\sqrt{q}}{2}(\sigma_x - i\sigma_y).$$

Let ρ be the density matrix

$$\frac{1}{2}\begin{pmatrix} 1+z & x - iy \\ x + iy & 1 - z \end{pmatrix}$$

where x, y, z are real numbers such that $x^2 + y^2 + z^2 \le 1$. The map Φ transforms the density matrix ρ into a new one given by

$$\Phi(\rho) = L_1^* \rho L_1 + L_2^* \rho L_2.$$

By induction, for every $n \ge 0$,

$$\Phi^n(\rho) = \begin{pmatrix} p_n & r_n \\ r_n & 1 - p_n \end{pmatrix}$$

where the sequences $(p_n)_{n \ge 0}$, and $(r_n)_{n \ge 0}$ satisfy the recurrence relations: for every $n \ge 1$,

$$p_n = p_{n-1}(p - q) + q$$

and

$$r_n = \sqrt{q(1-q)} + p_{n-1}(\sqrt{p(1-p)} - \sqrt{q(1-q)})$$

with initial condition $p_0 = (1 + z)/2$. Assumption (A) is then clearly satisfied with

$$\rho_\infty = \frac{1}{1+q-p}\begin{pmatrix} q & \beta \\ \beta & 1-p \end{pmatrix}$$

where $\beta = \left[q\sqrt{p(1-p)} + (1-p)\sqrt{q(1-q)}\right]$. Then, applying Theorem 1, if P is a polynomial of $3m$ non-commuting variables, for every $0 < t_1 < t_2 < \ldots < t_m$, the following convergence holds

$$\lim_{n\to+\infty} w\left[\widehat{P}(X_{t_1}^{(n)}, Y_{t_1}^{(n)}, Z_{t_1}^{(n)}, \ldots, X_{t_m}^{(n)}, Y_{t_m}^{(n)}, Z_{t_m}^{(n)})\right]$$
$$= \mathbb{E}\left[P(B_{t_1}^{(1)}, B_{t_1}^{(2)}, B_{t_1}^{(3)}, \ldots, B_{t_m}^{(1)}, B_{t_m}^{(2)}, B_{t_m}^{(3)})\right]$$

where $(B_t^{(1)}, B_t^{(2)}, B_t^{(3)})_{t\geq 0}$ is a three-dimensional centered Brownian motion with Covariance matrix Ct where

$$C = \begin{pmatrix} 1-v_1^2 & 0 & -v_1 v_2 \\ 0 & 1 & 0 \\ -v_1 v_2 & 0 & 1-v_2^2 \end{pmatrix}$$

with

$$v_1 = \frac{2}{1+q-p}[q\sqrt{p(1-p)} + (1-p)\sqrt{q(1-q)}]$$

and

$$v_2 = \frac{p+q-1}{1+q-p}.$$

5 Large Deviation Principle

Let Γ be a Polish space endowed with the Borel σ-field $\mathcal{B}(\Gamma)$. A good *rate function* is a lower semi-continuous function $\Lambda^* : \Gamma \to [0,\infty]$ with compact level sets $\{x; \Lambda^*(x) \leq a\}, a \in [0,\infty[$. Let $v = (v_n)_n \uparrow \infty$ be an increasing sequence of positive reals. A sequence of random variables $(Y_n)_n$ with values in Γ defined on a probability space $(\Omega, \mathcal{F}, \mathbb{P})$ is said to satisfy *a Large Deviation Principle* (LDP) with speed $v = (v_n)_n$ and good rate function Λ^* if for every Borel set $B \in \mathcal{B}(\Gamma)$,

$$- \inf_{x\in B^o} \Lambda^*(x) \leq \liminf_n \frac{1}{v_n} \log \mathbb{P}(Y_n \in B)$$

$$\leq \limsup_n \frac{1}{v_n} \log \mathbb{P}(Y_n \in B) \leq - \inf_{x\in \bar{B}} \Lambda^*(x).$$

For every $k \geq 1$, we define

$$\bar{x}_k = I \otimes \ldots \otimes I \otimes \sigma_x \otimes I \otimes \ldots$$
$$\bar{y}_k = I \otimes \ldots \otimes I \otimes \sigma_y \otimes I \otimes \ldots$$
$$\bar{z}_k = I \otimes \ldots \otimes I \otimes \sigma_z \otimes I \otimes \ldots$$

where each σ. appears on the kth place.
For every $n \geq 1$, we consider the processes

$$\bar{X}_n = \sum_{k=1}^n \bar{x}_k, \quad \bar{Y}_n = \sum_{k=1}^n \bar{y}_k, \quad \bar{Z}_n = \sum_{k=1}^n \bar{z}_k$$

with initial conditions
$$\bar{X}_0 = \bar{Y}_0 = \bar{Z}_0 = 0.$$

To each vector $\nu = (\nu_1, \nu_2, \nu_3) \in \mathbb{R}^3$, we associate the Euclidean norm $\|\nu\| = \sqrt{\nu_1^2 + \nu_2^2 + \nu_3^2}$ and $\langle ., . \rangle$ the corresponding inner product.

Theorem 2. *Let Φ be a completely positive and trace-preserving map for which there exists a state*

$$\rho_\infty = \begin{pmatrix} \alpha_\infty & \beta_\infty \\ \bar{\beta}_\infty & 1 - \alpha_\infty \end{pmatrix}$$

such that for any given state ρ,

$$\Phi^n(\rho) = \rho_\infty + o(1).$$

For every $\nu = (\nu_1, \nu_2, \nu_3) \in \mathbb{R}^{3,*}$, the sequence

$$\left(\frac{\nu_1 \bar{X}_n + \nu_2 \bar{Y}_n + \nu_3 \bar{Z}_n}{n} \right)_{n \geq 1}$$

satisfies a LDP with speed n and good rate function

$$I(x) = \begin{cases} \frac{1}{2}\left[\left(1 + \frac{x}{\|\nu\|}\right) \log\left(\frac{\|\nu\| + x}{\|\nu\| + \langle \nu, \nu \rangle}\right) \\ \quad + \left(1 - \frac{x}{\|\nu\|}\right) \log\left(\frac{\|\nu\| - x}{\|\nu\| - \langle \nu, \nu \rangle}\right)\right] & \text{if } |x| < \|\nu\|, \\ \\ +\infty & \text{otherwise.} \end{cases}$$

where $v_1 = 2\operatorname{Re}(\beta_\infty), v_2 = -2\operatorname{Im}(\beta_\infty), v_3 = 2\alpha_\infty - 1$.

Proof. The matrix

$$B := \nu_1 \sigma_x + \nu_2 \sigma_y + \nu_3 \sigma_z = \begin{pmatrix} \nu_3 & \nu_1 - i\nu_2 \\ \nu_1 + i\nu_2 & -\nu_3 \end{pmatrix}$$

has two distinct eigenvalues $\pm\|\nu\|$.
For every $n \geq 0$, we can write

$$\Phi^n(\rho) = \begin{pmatrix} \alpha_\infty + \phi_n(1) & \beta_\infty + \phi_n(2) \\ \bar{\beta}_\infty + \phi_n(3) & 1 - \alpha_\infty + \phi_n(4) \end{pmatrix}$$

where the four sequences $(\phi_n(i))_{n \geq 0}$ satisfy $\phi_n(i) = o(1)$.

For any $k \geq 1$, the expectation of B in the state $\Phi^k(\rho)$ is equal to

$$\mathrm{Trace}(B\ \Phi^k(\rho)) = \langle \nu, v \rangle + \varepsilon_k,$$

with $\varepsilon_n = o(1)$. As a consequence, the distribution of B is

$$p_k(\|\nu\|) = \frac{1}{2}\left[1 + \frac{1}{|\nu|}(\langle \nu, v \rangle + \varepsilon_k)\right] = 1 - p_k(-\|\nu\|).$$

Using the fact that $\nu_1 \bar{X}_n + \nu_2 \bar{Y}_n + \nu_3 \bar{Z}_n$ is the sum of n commuting matrices, we get that

$$\frac{1}{n} \log w \ (\exp t(\nu_1 X_n + \nu_2 Y_n + \nu_3 Z_n))$$

$$= \frac{1}{n} \sum_{k=1}^{n} \log \left(e^{\|\nu\|t} p_k(\|\nu\|) + e^{-\|\nu\|t}(1 - p_k(\|\nu\|)) \right)$$

Since $\varepsilon_n = o(1)$, we obtain that

$$\lim_{n \to +\infty} \frac{1}{n} \log w \ (\exp t(\nu_1 X_n + \nu_2 Y_n + \nu_3 Z_n))$$

$$= \log \left(\cosh\left(\|\nu\|t\right) + \frac{\langle \nu, v \rangle}{\|\nu\|} \sinh\left(\|\nu\|t\right) \right)$$

$$= \log \left(\cosh\left(\|\nu\|t\right) \right) + \log \left(1 + \frac{\langle \nu, v \rangle}{\|\nu\|} \tanh\left(\|\nu\|t\right) \right).$$

We denote by $\Lambda(t)$ this function of t.

For every $t \in \mathbb{R}$, the function Λ is finite and differentiable on $\mathbb{R}$, then, by Gärtner-Ellis' Theorem (see [4]), the LDP holds with the good rate function

$$I(x) = \sup_{t \in \mathbb{R}}\{tx - \Lambda(t)\}.$$

A simple computation leads to the rate function given in the theorem.   □

References

1. BIANE, P. *Some properties of quantum Bernoulli random walks.* Quantum probability & related topics, 193–203, QP-PQ, VI, World Sci. Publ., River Edge, NJ, 1991.
2. CHOI, M. D. *Completely positive linear maps on complex matrices.* Linear Algebra and Appl. 10, 285–290 (1975).
3. DACUNHA-CASTELLE, D. and DUFLO, M. *Probabilités et statistiques 2. Problèmes à temps mobile.*, Masson, Paris (1983).
4. DEMBO, A. and ZEITOUNI, O. *Large Deviations Techniques and Applications.* Springer, (1998).

5. GIRI, N. and VON WALDENFELS, W. *An algebraic version of the central limit theorem*. Z. Wahrscheinlichkeitstheorie Verw. Gebiete 42, 129–134 (1978).

6. KING, C. and RUSKAI, M.B. *Minimal entropy of states emerging from noisy quantum channels*. IEEE Trans. Inform. Theory 47, No 1, 192–209 (2001).

7. KRAUS, K. *General state changes in quantum theory*. Ann. Physics, 64, 311–335 (1971).

8. KRAUS, K. *States, effects and operations. Fundamental notions of quantum theory*. Lecture Notes in Physics, 190. Springer-Verlag, Berlin (1983).

9. PETZ, D. *An invitation to the algebra of canonical commutation relations*. Leuven Notes in Mathematical and Theoretical Physics, Vol. 2 (1990).

10. RUSKAI, M.B., SZAREK, S. and WERNER, E. *An analysis of completely positive trace-preserving maps on* $\mathcal{M}_2$. Linear Algebra Appl. 347, 159–187 (2002).

Erratum to: "New Methods in the Arbitrage Theory of Financial Markets with Transaction Costs", in Séminaire XLI

Miklós Rásonyi*

Computer and Automation Institute of the Hungarian Academy of Sciences
email: rasonyi@sztaki.hu

Unfortunately, the proof of Lemma 4.6 in [1] needs an additional assumption.

For a closed cone $C \subset \mathbb{R}^d$ let C^* denote its positive dual cone (see [1]). It is erroneously claimed in the last line of page 460 that $(G^*_{T-l} \cap X)^* = G_{T-l} + X^*$ where $G_{T-l} = G_{T-l}(\omega)$ is a random closed cone in $\mathbb{R}^d$ and $X^*(\omega) = \{\alpha\xi(\omega) : \alpha \leq 0\}$ with some $\mathbb{R}^d$-valued random variable ξ (i.e. X^* is a random ray in $\mathbb{R}^d$).

The claimed identity holds if and only if $G_{T-l} + X^*$ is a *closed* cone in $\mathbb{R}^d$ a.s., see Corollary 16.4.2 of [2]. Hence the following hypothesis must be added to the statements of Lemma 4.6 and the main Theorem 3.1 in [1]:

Assumption. For all $0 \leq t \leq T$ and for almost all ω the cone $G_t(\omega)$ is such that $G_t(\omega) + \{\alpha x : \alpha \geq 0\}$ is closed in $\mathbb{R}^d$ for each $x \in \mathbb{R}^d$.

The above Assumption is trivially satisfied when G_t is a (random) *polyhedral* cone: a ray is, in particular, a polyhedral cone and the sum of two polyhedral cones is polyhedral and hence closed.

Although restricted in generality by the Assumption given above, Theorem 3.1 of [1] still covers the cases which are relevant to financial markets with proportional transaction costs. In those models G_t are assumed to be polyhedral, see the references of [1].

References

1. Rásonyi, M. (2008) New methods in the arbitrage theory of financial markets with transaction costs. *Séminaire de Probabilités XLI*, Lecture Notes in Mathematics **1934**, 455–462, Springer, Berlin.
2. Rockafellar, R. T. (1970) *Convex analysis.* Princeton University Press, Princeton, N. J.

---

* I would like to thank Yuri M. Kabanov and Christophe Stricker for discussions.

C. Donati-Martin et al. (eds.), *Séminaire de Probabilités XLII*,   449
Lecture Notes in Mathematics 1979, DOI 10.1007/978-3-642-01763-6_17,
© Springer-Verlag Berlin Heidelberg 2009

Lecture Notes in Mathematics

For information about earlier volumes
please contact your bookseller or Springer
LNM Online archive: springerlink.com

of Quadratic Forms. Summer School, Lens, 2000. Editor: J.-P. Tignol (2004)

Vol. 1836: C. Năstăsescu, F. Van Oystaeyen, Methods of Graded Rings. XIII, 304 p, 2004.

Vol. 1837: S. Tavaré, O. Zeitouni, Lectures on Probability Theory and Statistics. Ecole d'Eté de Probabilités de Saint-Flour XXXI-2001. Editor: J. Picard (2004)

Vol. 1838: A.J. Ganesh, N.W. O'Connell, D.J. Wischik, Big Queues. XII, 254 p, 2004.

Vol. 1839: R. Gohm, Noncommutative Stationary Processes. VIII, 170 p, 2004.

Vol. 1840: B. Tsirelson, W. Werner, Lectures on Probability Theory and Statistics. Ecole d'Eté de Probabilités de Saint-Flour XXXII-2002. Editor: J. Picard (2004)

Vol. 1841: W. Reichel, Uniqueness Theorems for Variational Problems by the Method of Transformation Groups (2004)

Vol. 1842: T. Johnsen, A. L. Knutsen, K_3 Projective Models in Scrolls (2004)

Vol. 1843: B. Jefferies, Spectral Properties of Noncommuting Operators (2004)

Vol. 1844: K.F. Siburg, The Principle of Least Action in Geometry and Dynamics (2004)

Vol. 1845: Min Ho Lee, Mixed Automorphic Forms, Torus Bundles, and Jacobi Forms (2004)

Vol. 1846: H. Ammari, H. Kang, Reconstruction of Small Inhomogeneities from Boundary Measurements (2004)

Vol. 1847: T.R. Bielecki, T. Björk, M. Jeanblanc, M. Rutkowski, J.A. Scheinkman, W. Xiong, Paris-Princeton Lectures on Mathematical Finance 2003 (2004)

Vol. 1848: M. Abate, J. E. Fornaess, X. Huang, J. P. Rosay, A. Tumanov, Real Methods in Complex and CR Geometry, Martina Franca, Italy 2002. Editors: D. Zaitsev, G. Zampieri (2004)

Vol. 1849: Martin L. Brown, Heegner Modules and Elliptic Curves (2004)

Vol. 1850: V. D. Milman, G. Schechtman (Eds.), Geometric Aspects of Functional Analysis. Israel Seminar 2002-2003 (2004)

Vol. 1851: O. Catoni, Statistical Learning Theory and Stochastic Optimization (2004)

Vol. 1852: A.S. Kechris, B.D. Miller, Topics in Orbit Equivalence (2004)

Vol. 1853: Ch. Favre, M. Jonsson, The Valuative Tree (2004)

Vol. 1854: O. Saeki, Topology of Singular Fibers of Differential Maps (2004)

Vol. 1855: G. Da Prato, P.C. Kunstmann, I. Lasiecka, A. Lunardi, R. Schnaubelt, L. Weis, Functional Analytic Methods for Evolution Equations. Editors: M. Iannelli, R. Nagel, S. Piazzera (2004)

Vol. 1856: K. Back, T.R. Bielecki, C. Hipp, S. Peng, W. Schachermayer, Stochastic Methods in Finance, Bressanone/Brixen, Italy, 2003. Editors: M. Fritelli, W. Runggaldier (2004)

Vol. 1857: M. Émery, M. Ledoux, M. Yor (Eds.), Séminaire de Probabilités XXXVIII (2005)

Vol. 1858: A.S. Cherny, H.-J. Engelbert, Singular Stochastic Differential Equations (2005)

Vol. 1859: E. Letellier, Fourier Transforms of Invariant Functions on Finite Reductive Lie Algebras (2005)

Vol. 1860: A. Borisyuk, G.B. Ermentrout, A. Friedman, D. Terman, Tutorials in Mathematical Biosciences I. Mathematical Neurosciences (2005)

Vol. 1861: G. Benettin, J. Henrard, S. Kuksin, Hamiltonian Dynamics – Theory and Applications, Cetraro, Italy, 1999. Editor: A. Giorgilli (2005)

Vol. 1862: B. Helffer, F. Nier, Hypoelliptic Estimates and Spectral Theory for Fokker-Planck Operators and Witten Laplacians (2005)

Vol. 1863: H. Führ, Abstract Harmonic Analysis of Continuous Wavelet Transforms (2005)

Vol. 1864: K. Efstathiou, Metamorphoses of Hamiltonian Systems with Symmetries (2005)

Vol. 1865: D. Applebaum, B.V. R. Bhat, J. Kustermans, J. M. Lindsay, Quantum Independent Increment Processes I. From Classical Probability to Quantum Stochastic Calculus. Editors: M. Schürmann, U. Franz (2005)

Vol. 1866: O.E. Barndorff-Nielsen, U. Franz, R. Gohm, B. Kümmerer, S. Thorbjønsen, Quantum Independent Increment Processes II. Structure of Quantum Lévy Processes, Classical Probability, and Physics. Editors: M. Schürmann, U. Franz, (2005)

Vol. 1867: J. Sneyd (Ed.), Tutorials in Mathematical Biosciences II. Mathematical Modeling of Calcium Dynamics and Signal Transduction. (2005)

Vol. 1868: J. Jorgenson, S. Lang, $Pos_n(R)$ and Eisenstein Series. (2005)

Vol. 1869: A. Dembo, T. Funaki, Lectures on Probability Theory and Statistics. Ecole d'Eté de Probabilités de Saint-Flour XXXIII-2003. Editor: J. Picard (2005)

Vol. 1870: V.I. Gurariy, W. Lusky, Geometry of Müntz Spaces and Related Questions. (2005)

Vol. 1871: P. Constantin, G. Gallavotti, A.V. Kazhikhov, Y. Meyer, S. Ukai, Mathematical Foundation of Turbulent Viscous Flows, Martina Franca, Italy, 2003. Editors: M. Cannone, T. Miyakawa (2006)

Vol. 1872: A. Friedman (Ed.), Tutorials in Mathematical Biosciences III. Cell Cycle, Proliferation, and Cancer (2006)

Vol. 1873: R. Mansuy, M. Yor, Random Times and Enlargements of Filtrations in a Brownian Setting (2006)

Vol. 1874: M. Yor, M. Émery (Eds.), In Memoriam Paul-André Meyer - Séminaire de Probabilités XXXIX (2006)

Vol. 1875: J. Pitman, Combinatorial Stochastic Processes. Ecole d'Eté de Probabilités de Saint-Flour XXXII-2002. Editor: J. Picard (2006)

Vol. 1876: H. Herrlich, Axiom of Choice (2006)

Vol. 1877: J. Steuding, Value Distributions of L-Functions (2007)

Vol. 1878: R. Cerf, The Wulff Crystal in Ising and Percolation Models, Ecole d'Eté de Probabilités de Saint-Flour XXXIV-2004. Editor: Jean Picard (2006)

Vol. 1879: G. Slade, The Lace Expansion and its Applications, Ecole d'Eté de Probabilités de Saint-Flour XXXIV-2004. Editor: Jean Picard (2006)

Vol. 1880: S. Attal, A. Joye, C.-A. Pillet, Open Quantum Systems I, The Hamiltonian Approach (2006)

Vol. 1881: S. Attal, A. Joye, C.-A. Pillet, Open Quantum Systems II, The Markovian Approach (2006)

Vol. 1882: S. Attal, A. Joye, C.-A. Pillet, Open Quantum Systems III, Recent Developments (2006)

Vol. 1883: W. Van Assche, F. Marcellàn (Eds.), Orthogonal Polynomials and Special Functions, Computation and Application (2006)

Vol. 1884: N. Hayashi, E.I. Kaikina, P.I. Naumkin, I.A. Shishmarev, Asymptotics for Dissipative Nonlinear Equations (2006)

Vol. 1885: A. Telcs, The Art of Random Walks (2006)

Vol. 1886: S. Takamura, Splitting Deformations of Degenerations of Complex Curves (2006)

Vol. 1887: K. Habermann, L. Habermann, Introduction to Symplectic Dirac Operators (2006)

Vol. 1939: D. Boffi, F. Brezzi, L. Demkowicz, R.G. Durán, R.S. Falk, M. Fortin, Mixed Finite Elements, Compatibility Conditions, and Applications. Cetraro, Italy 2006. Editors: D. Boffi, L. Gastaldi (2008)

Vol. 1940: J. Banasiak, V. Capasso, M.A.J. Chaplain, M. Lachowicz, J. Miękisz, Multiscale Problems in the Life Sciences. From Microscopic to Macroscopic. Będlewo, Poland 2006. Editors: V. Capasso, M. Lachowicz (2008)

Vol. 1941: S.M.J. Haran, Arithmetical Investigations. Representation Theory, Orthogonal Polynomials, and Quantum Interpolations (2008)

Vol. 1942: S. Albeverio, F. Flandoli, Y.G. Sinai, SPDE in Hydrodynamic. Recent Progress and Prospects. Cetraro, Italy 2005. Editors: G. Da Prato, M. Röckner (2008)

Vol. 1943: L.L. Bonilla (Ed.), Inverse Problems and Imaging. Martina Franca, Italy 2002 (2008)

Vol. 1944: A. Di Bartolo, G. Falcone, P. Plaumann, K. Strambach, Algebraic Groups and Lie Groups with Few Factors (2008)

Vol. 1945: F. Brauer, P. van den Driessche, J. Wu (Eds.), Mathematical Epidemiology (2008)

Vol. 1946: G. Allaire, A. Arnold, P. Degond, T.Y. Hou, Quantum Transport. Modelling, Analysis and Asymptotics. Cetraro, Italy 2006. Editors: N.B. Abdallah, G. Frosali (2008)

Vol. 1947: D. Abramovich, M. Mariño, M. Thaddeus, R. Vakil, Enumerative Invariants in Algebraic Geometry and String Theory. Cetraro, Italy 2005. Editors: K. Behrend, M. Manetti (2008)

Vol. 1948: F. Cao, J-L. Lisani, J-M. Morel, P. Musé, F. Sur, A Theory of Shape Identification (2008)

Vol. 1949: H.G. Feichtinger, B. Helffer, M.P. Lamoureux, N. Lerner, J. Toft, Pseudo-Differential Operators. Quantization and Signals. Cetraro, Italy 2006. Editors: L. Rodino, M.W. Wong (2008)

Vol. 1950: M. Bramson, Stability of Queueing Networks, Ecole d'Eté de Probabilités de Saint-Flour XXXVI-2006 (2008)

Vol. 1951: A. Moltó, J. Orihuela, S. Troyanski, M. Valdivia, A Non Linear Transfer Technique for Renorming (2009)

Vol. 1952: R. Mikhailov, I.B.S. Passi, Lower Central and Dimension Series of Groups (2009)

Vol. 1953: K. Arwini, C.T.J. Dodson, Information Geometry (2008)

Vol. 1954: P. Biane, L. Bouten, F. Cipriani, N. Konno, N. Privault, Q. Xu, Quantum Potential Theory. Editors: U. Franz, M. Schuermann (2008)

Vol. 1955: M. Bernot, V. Caselles, J.-M. Morel, Optimal Transportation Networks (2008)

Vol. 1956: C.H. Chu, Matrix Convolution Operators on Groups (2008)

Vol. 1957: A. Guionnet, On Random Matrices: Macroscopic Asymptotics, Ecole d'Eté de Probabilités de Saint-Flour XXXVI-2006 (2009)

Vol. 1958: M.C. Olsson, Compactifying Moduli Spaces for Abelian Varieties (2008)

Vol. 1959: Y. Nakkajima, A. Shiho, Weight Filtrations on Log Crystalline Cohomologies of Families of Open Smooth Varieties (2008)

Vol. 1960: J. Lipman, M. Hashimoto, Foundations of Grothendieck Duality for Diagrams of Schemes (2009)

Vol. 1961: G. Buttazzo, A. Pratelli, S. Solimini, E. Stepanov, Optimal Urban Networks via Mass Transportation (2009)

Vol. 1962: R. Dalang, D. Khoshnevisan, C. Mueller, D. Nualart, Y. Xiao, A Minicourse on Stochastic Partial Differential Equations (2009)

Vol. 1963: W. Siegert, Local Lyapunov Exponents (2009)

Vol. 1964: W. Roth, Operator-valued Measures and Integrals for Cone-valued Functions and Integrals for Cone-valued Functions (2009)

Vol. 1965: C. Chidume, Geometric Properties of Banach Spaces and Nonlinear Iterations (2009)

Vol. 1966: D. Deng, Y. Han, Harmonic Analysis on Spaces of Homogeneous Type (2009)

Vol. 1967: B. Fresse, Modules over Operads and Functors (2009)

Vol. 1968: R. Weissauer, Endoscopy for GSP(4) and the Cohomology of Siegel Modular Threefolds (2009)

Vol. 1969: B. Roynette, M. Yor, Penalising Brownian Paths (2009)

Vol. 1970: M. Biskup, A. Bovier, F. den Hollander, D. Ioffe, F. Martinelli, K. Netočný, F. Toninelli, Methods of Contemporary Mathematical Statistical Physics. Editor: R. Kotecký (2009)

Vol. 1971: L. Saint-Raymond, Hydrodynamic Limits of the Boltzmann Equation (2009)

Vol. 1972: T. Mochizuki, Donaldson Type Invariants for Algebraic Surfaces (2009)

Vol. 1973: M.A. Berger, L.H. Kauffmann, B. Khesin, H.K. Moffatt, R.L. Ricca, De W. Sumners, Lectures on Topological Fluid Mechanics. Cetraro, Italy 2001. Editor: R.L. Ricca (2009)

Vol. 1974: F. den Hollander, Random Polymers: École d'Été de Probabilités de Saint-Flour XXXVII – 2007 (2009)

Vol. 1975: J.C. Rohde, Cyclic Coverings, Calabi-Yau Manifolds and Complex Multiplication (2009)

Vol. 1976: N. Ginoux, The Dirac Spectrum (2009)

Vol. 1977: M.J. Gursky, E. Lanconelli, A. Malchiodi, G. Tarantello, X.-J. Wang, P.C. Yang, Geometric Analysis and PDEs. Cetraro, Italy 2001. Editors: A. Ambrosetti, S.-Y.A. Chang, A. Malchiodi (2009)

Vol. 1978: M. Qian, J.-S. Xie, S. Zhu, Smooth Ergodic Theory for Endomorphisms (2009)

Vol. 1979: C. Donati-Martin, M. Émery, A. Rouault, C. Stricker (Eds.), Séminaire de Probabilités XLII (2009)

Recent Reprints and New Editions

Vol. 1702: J. Ma, J. Yong, Forward-Backward Stochastic Differential Equations and their Applications. 1999 – Corr. 3rd printing (2007)

Vol. 830: J.A. Green, Polynomial Representations of GL_n, with an Appendix on Schensted Correspondence and Littelmann Paths by K. Erdmann, J.A. Green and M. Schoker 1980 – 2nd corr. and augmented edition (2007)

Vol. 1693: S. Simons, From Hahn-Banach to Monotonicity (Minimax and Monotonicity 1998) – 2nd exp. edition (2008)

Vol. 470: R.E. Bowen, Equilibrium States and the Ergodic Theory of Anosov Diffeomorphisms. With a preface by D. Ruelle. Edited by J.-R. Chazottes. 1975 – 2nd rev. edition (2008)

Vol. 523: S.A. Albeverio, R.J. Høegh-Krohn, S. Mazzucchi, Mathematical Theory of Feynman Path Integral. 1976 – 2nd corr. and enlarged edition (2008)

Vol. 1764: A. Cannas da Silva, Lectures on Symplectic Geometry 2001 – Corr. 2nd printing (2008)

LECTURE NOTES IN MATHEMATICS

Springer

Edited by J.-M. Morel, F. Takens, B. Teissier, P.K. Maini

Editorial Policy (for Multi-Author Publications: Summer Schools/Intensive Courses)

1. Lecture Notes aim to report new developments in all areas of mathematics and their applications - quickly, informally and at a high level. Mathematical texts analysing new developments in modelling and numerical simulation are welcome. Manuscripts should be reasonably self-contained and rounded off. Thus they may, and often will, present not only results of the author but also related work by other people. They should provide sufficient motivation, examples and applications. There should also be an introduction making the text comprehensible to a wider audience. This clearly distinguishes Lecture Notes from journal articles or technical reports which normally are very concise. Articles intended for a journal but too long to be accepted by most journals, usually do not have this "lecture notes" character.

2. In general SUMMER SCHOOLS and other similar INTENSIVE COURSES are held to present mathematical topics that are close to the frontiers of recent research to an audience at the beginning or intermediate graduate level, who may want to continue with this area of work, for a thesis or later. This makes demands on the didactic aspects of the presentation. Because the subjects of such schools are advanced, there often exists no textbook, and so ideally, the publication resulting from such a school could be a first approximation to such a textbook. Usually several authors are involved in the writing, so it is not always simple to obtain a unified approach to the presentation.

   For prospective publication in LNM, the resulting manuscript should not be just a collection of course notes, each of which has been developed by an individual author with little or no co-ordination with the others, and with little or no common concept. The subject matter should dictate the structure of the book, and the authorship of each part or chapter should take secondary importance. Of course the choice of authors is crucial to the quality of the material at the school and in the book, and the intention here is not to belittle their impact, but simply to say that the book should be planned to be written by these authors jointly, and not just assembled as a result of what these authors happen to submit.

   This represents considerable preparatory work (as it is imperative to ensure that the authors know these criteria before they invest work on a manuscript), and also considerable editing work afterwards, to get the book into final shape. Still it is the form that holds the most promise of a successful book that will be used by its intended audience, rather than yet another volume of proceedings for the library shelf.

3. Manuscripts should be submitted either online at www.editorialmanager.com/lnm/ to Springer's mathematics editorial, or to one of the series editors. Volume editors are expected to arrange for the refereeing, to the usual scientific standards, of the individual contributions. If the resulting reports can be forwarded to us (series editors or Springer) this is very helpful. If no reports are forwarded or if other questions remain unclear in respect of homogeneity etc, the series editors may wish to consult external referees for an overall evaluation of the volume. A final decision to publish can be made only on the basis of the complete manuscript; however a preliminary decision can be based on a pre-final or incomplete manuscript. The strict minimum amount of material that will be considered should include a detailed outline describing the planned contents of each chapter.

   Volume editors and authors should be aware that incomplete or insufficiently close to final manuscripts almost always result in longer evaluation times. They should also be aware that parallel submission of their manuscript to another publisher while under consideration for LNM will in general lead to immediate rejection.

4. Manuscripts should in general be submitted in English. Final manuscripts should contain at least 100 pages of mathematical text and should always include
   - a general table of contents;
   - an informative introduction, with adequate motivation and perhaps some historical remarks: it should be accessible to a reader not intimately familiar with the topic treated;
   - a global subject index: as a rule this is genuinely helpful for the reader.

   Lecture Notes volumes are, as a rule, printed digitally from the authors' files. We strongly recommend that all contributions in a volume be written in the same LaTeX version, preferably LaTeX2e. To ensure best results, authors are asked to use the LaTeX2e style files available from Springer's web-server at

   ftp://ftp.springer.de/pub/tex/latex/svmonot1/ (for monographs) and
   ftp://ftp.springer.de/pub/tex/latex/svmultt1/ (for summer schools/tutorials).

   Additional technical instructions are available on request from: lnm@springer.com.

5. Careful preparation of the manuscripts will help keep production time short besides ensuring satisfactory appearance of the finished book in print and online. After acceptance of the manuscript authors will be asked to prepare the final LaTeX source files and also the corresponding dvi-, pdf- or zipped ps-file. The LaTeX source files are essential for producing the full-text online version of the book. For the existing online volumes of LNM see: http://www.springerlink.com/openurl.asp?genre=journal&issn=0075-8434.

   The actual production of a Lecture Notes volume takes approximately 12 weeks.

6. Volume editors receive a total of 50 free copies of their volume to be shared with the authors, but no royalties. They and the authors are entitled to a discount of 33.3% on the price of Springer books purchased for their personal use, if ordering directly from Springer.

7. Commitment to publish is made by letter of intent rather than by signing a formal contract. Springer-Verlag secures the copyright for each volume. Authors are free to reuse material contained in their LNM volumes in later publications: a brief written (or e-mail) request for formal permission is sufficient.

Addresses:

Professor J.-M. Morel, CMLA,
École Normale Supérieure de Cachan,
61 Avenue du Président Wilson,
94235 Cachan Cedex, France
E-mail: Jean-Michel.Morel@cmla.ens-cachan.fr

Professor F. Takens, Mathematisch Instituut,
Rijksuniversiteit Groningen, Postbus 800,
9700 AV Groningen, The Netherlands
E-mail: F.Takens@rug.nl

Professor B. Teissier,
Institut Mathématique de Jussieu,
UMR 7586 du CNRS,
Équipe "Géométrie et Dynamique",
175 rue du Chevaleret,
75013 Paris, France
E-mail: teissier@math.jussieu.fr

For the "Mathematical Biosciences Subseries" of LNM:

Professor P.K. Maini, Center for Mathematical Biology,
Mathematical Institute, 24-29 St Giles,
Oxford OX1 3LP, UK
E-mail: maini@maths.ox.ac.uk

Springer, Mathematics Editorial I, Tiergartenstr. 17,
69121 Heidelberg, Germany,
Tel.: +49 (6221) 487-8259
Fax: +49 (6221) 4876-8259
E-mail: lnm@springer.com